T0390530

NICKEL BASE SINGLE CRYSTALS ACROSS LENGTH SCALES

NICKEL BASE SINGLE CRYSTALS ACROSS LENGTH SCALES

Edited by

GEORGES CAILLETAUD
Mines ParisTech
Paris, France

JONATHAN CORMIER
ISAE-ENSMA
Poitiers, France

GUNTHER EGGELER
Ruhr-Universitat Bochum
Bochum, Germany

VINCENT MAUREL
Centre des Materiaux
Mines ParisTech
Paris, France

LOEÏZ NAZÉ
Mines ParisTech
Paris, France

Elsevier
Radarweg 29, PO Box 211, 1000 AE Amsterdam, Netherlands
The Boulevard, Langford Lane, Kidlington, Oxford OX5 1GB, United Kingdom
50 Hampshire Street, 5th Floor, Cambridge, MA 02139, United States

Notices

Knowledge and best practice in this field are constantly changing. As new research and experience broaden our understanding, changes in research methods, professional practices, or medical treatment may become necessary.

Practitioners and researchers must always rely on their own experience and knowledge in evaluating and using any information, methods, compounds, or experiments described herein. In using such information or methods they should be mindful of their own safety and the safety of others, including parties for whom they have a professional responsibility.

To the fullest extent of the law, neither the Publisher nor the authors, contributors, or editors, assume any liability for any injury and/or damage to persons or property as a matter of products liability, negligence or otherwise, or from any use or operation of any methods, products, instructions, or ideas contained in the material herein.

Library of Congress Cataloging-in-Publication Data
A catalog record for this book is available from the Library of Congress

British Library Cataloguing-in-Publication Data
A catalogue record for this book is available from the British Library

ISBN: 978-0-12-819357-0

For information on all Elsevier publications
visit our website at https://www.elsevier.com/books-and-journals

Publisher: Matthew Deans
Acquisitions Editor: Christina Gifford
Editorial Project Manager: Gabriela D. Capille
Production Project Manager: Prasanna Kalyanaraman
Designer: Miles Hitchen

Typeset by VTeX

Contents

4. Microstructure and chemical characterization

Florence Pettinari-Sturmel and Loïc Nazé

5. Mechanical characterization at high temperature

Vincent Bonnand, Jean-Briac le Graverend, and Marion Bartsch

6. Elementary deformation processes in high temperature plasticity of Ni- and Co-base single-crystal superalloys with γ/γ' microstructures

Cathie M.F. Rae, Gunther Eggeler, and Jean-Loup Strudel

II

Building SX parts

7. Processing of directionally cast nickel-base superalloys: solidification and heat treatments

Jonathan Cormier and Charles-André Gandin

16. Crystal plasticity and damage at cracks and notches in nickel-base single-crystal superalloys

Samuel Forest

IV

Application to engineering cases

17. Implementation of constitutive equations for single crystals in finite element codes

Jean-Michel Scherer and Jacques Besson

List of contributors

Katrin Abrahams Ruhr-University Bochum, ICAMS, Bochum, Germany

Benoît Appolaire Université de Lorraine, CNRS, IJL, Nancy, France

Marion Bartsch Institute of Materials Research, German Aerospace Center (DLR), Cologne, Germany

Jacques Besson Centre des Matériaux, Mines ParisTech — PSL Research University, UMR CNRS 7633, Evry, France

Erik Bitzek Department of Materials Science and Engineering, Institute I, Friedrich-Alexander-Universität Erlangen-Nürnberg, Erlangen, Germany

Francesca Boioli LEM, CNRS-ONERA, Chatillon, France

Vincent Bonnand Department of Materials And Structures (DMAS), Onera, Chatillon, France

Georges Cailletaud Mines ParisTech, Centre des Matériaux, Evry, France

Jonathan Cormier ISAE-ENSMA, Institut Pprime, UPR CNRS 3346, Futuroscope-Chasseneuil, France

Maeva Cottura Université de Lorraine, CNRS, IJL, Nancy, France

Benoit Devincre LEM, CNRS-ONERA, Chatillon, France

Ralf Drautz ICAMS, Ruhr-Universität Bochum, Bochum, Germany

Gunther Eggeler Ruhr-University Bochum, ICAMS, Bochum, Germany

Bernard Fedelich Bundesanstalt für Materialforschung und -prüfung (BAM), Division 5.2: Experimental and Model Based Mechanical Behaviour of Materials, Berlin, Germany

Alphonse Finel Université Paris-Saclay, ONERA, CNRS, LEM, Châtillon, France

Marc Fivel SIMaP, Univ. Grenoble Alpes, CNRS, Grenoble, France

Samuel Forest MINES ParisTech, PSL University, Centre des matériaux, CNRS UMR 7633, Evry, France

Charles-André Gandin MINES ParisTech, PSL Research University, CEMEF, UMR CNRS 7635, Sophia Antipolis, France

Jean-Yves Guédou Safran Aircraft Engines, Site de Villaroche, Moissy Cramayel, France

Vincent Guipont MINES ParisTech, PSL University, MAT - Centre des Materiaux, UMR CNRS 7633, Evry Cedex, France

Thomas Hammerschmidt ICAMS, Ruhr-Universität Bochum, Bochum, Germany

Yann Le Bouar Université Paris-Saclay, ONERA, CNRS, LEM, Châtillon, France

Jean-Briac le Graverend Department of Aerospace Engineering, Texas A&M University, College Station, TX, United States
Department of Materials Science Engineering, Texas A&M University, College Station, TX, United States

Vincent Maurel MINES ParisTech, PSL University, MAT - Centre des Materiaux, UMR CNRS 7633, Evry Cedex, France

Loïc Nazé Ecole des Mines de Paris, Paris, France

Fernando Pedraza La Rochelle University, LaSIE UMR 7356-CNRS, Faculty of Sciences and Technology, La Rochelle, France

Florence Pettinari-Sturmel CEMES CNRS University of Toulouse, INSA, Toulouse, Cedex 4 France

Cathie M.F. Rae Department of Materials Science and Metallurgy, University of Cambridge, Cambridge, United Kingdom

Luc Rémy MINES ParisTech, PSL University, MAT - Centre des Materiaux, UMR CNRS 7633, Evry Cedex, France

Jutta Rogal Department of Chemistry, New York University, New York, NY, United States
Fachbereich Physik, Freie Universität Berlin, Berlin, Germany

Irina Roslyakova Ruhr-University Bochum, ICAMS, Bochum, Germany

Jean-Michel Scherer Université Paris-Saclay, CEA, Service d'Etude des Matériaux Irradiés, Gif-sur-Yvette, France
Centre des Matériaux, Mines ParisTech — PSL Research University, UMR CNRS 7633, Evry, France

Ingo Steinbach Ruhr-University Bochum, ICAMS, Bochum, Germany

Jean-Loup Strudel MINES ParisTech, PSL University, Centre des Matériaux, Evry, France

Damien Texier Institut Clement Ader (ICA) - UMR CNRS 5312, Universite de Toulouse, CNRS, INSA, UPS, IMT Mines Albi, ISAE-SUPAERO, Albi Cedex 09, France

Satoshi Utada ISAE-ENSMA, Institut P', UPR 3346-CNRS, Chasseneuil-Futuroscope, France

Robert Vaßen Institute of Energy and Climate Research: Materials Synthesis and Processing (IEK-1), Jülich, Germany

Marie-Helene Vidal-Sétif Department of Materials and Structures (DMAS), ONERA Université Paris Saclay, Châtillon, France

Foreword: Ni-base superalloy single crystals, a fascinating class of high temperature engineering materials

Ni-base superalloy single crystals (SXs) are a fascinating class of structural engineering materials, needed to fabricate first stage blades in gas turbines for aeroengines and power plants, as well as other components which need to withstand high mechanical loads at temperatures in the 1000 °C range and above. Thus they represent key materials in the quest to answer two of the great challenges of mankind, namely those of global transport and safe energy supply. Even if the future of aeronautical industry is now uncertain, in the dawn following the pandemic of the early 2020, mankind will again require mobility to connect our continents in a global player world. Moreover, governments who intend to strongly lean towards renewable energies realize that gas turbines are needed as backup systems in cases where sunlight and wind force are temporarily not available. This is why SXs are important and have been in the focus of interest of scientists from the materials and engineering communities for several decades. What makes them so special is that unlike many other metallic alloys they maintain good mechanical properties up to temperatures which are close to their melting point. Unlike ceramic systems, they combine high strength with acceptable ductility. From a materials science and mechanical engineering point of view, it is fascinating that expertise from very different fields and from different length and time scales needs to be combined to understand these materials, to be able to manufacture them, and to successfully design safe components. Modern alloy design strives for better performance in terms of the most efficient use of fossil resources. There are also permanent driving forces towards better thermal efficiency in terms of higher operating temperatures (according to the second law of thermodynamics), in terms of less fuel consumption (maintaining performance) and regarding long exploitable service lives of critical components.

- From a materials research point view, the following areas are important:

First, one has to establish a basic understanding of all aspects of alloy design and performance. It is important to understand how d-shell elements like Re and W affect superalloy thermodynamics and kinetics, and the evolution of the cast microstructure (dendritic and interdentritic regions – large scale heterogeneity) and the γ/γ'-microstructure, where small ordered γ'-cubes (ordered $L1_2$-phase; cube edge length, 0.5 μm; volume fraction, 75%) are separated by thin γ-channels (fcc crystal structure; channel width, 100 nm; volume fraction, 25%). It is important to understand how these parameters affect thermodynamic equilibria, the evolution of microstructure during processing, and high temperature exposure and strength. As modern alloy design strives for better performance in terms of the most efficient use of fossil resources, it must also respect the limited availability of strategic elements (e.g., less Re, more W, Mo and others).

Second, processing technologies are continuously improving and there is a need to accompany technological efforts by fundamental research. This applies to the microstructural evolution during directional solidification and heat treatment. The potential of new processing methods like coating technologies, additive manufacturing, and hot isostatic pressing needs to be explored and validated on a microstructural basis.

Third, there is a need for advanced scale bridging microstructural characterization. Materials must be characterized starting from the atomic scale (with high resolution TEM and atom probe tomography) up to the size scale of components (optical and electron microscopy techniques and tomography). The mastering of advanced characterization techniques represents a challenge in itself and requires full dedication.

Fourth, there is a need to move towards scale bridging materials modeling where one can use information from the atomistic scale (first principle calculations) as input for calculations on the mesoscale (phase field and CALPHAD methods). There also is a need to establish modeling tools which allow us to design new materials.

- Material mechanics has to ensure that one can safely design components and make safe predictions of their life, in view of complications like crystallographic anisotropy, complex deformation mechanisms, and unavoidable softening and damage processes. There is a need for materials mechanics research in four areas:

First, one has to bridge the gap between observations of mechanisms on the nano- and microscales and micromechanical and mechanical modeling. This provides opportunities for a challenging scientific dialogue between communities. This type of approach leads to multiscale models, where physical reality can be accounted for in two ways: either explicitly, by using a structural calculation (finite elements or finite differences) to solve a problem posed on a cell of a representative material element, or, alternatively, by using the homogenization theory which considers the material as periodic or disordered.

Second, the numerical procedures have to take into account coupled phenomena, like discrete dislocation dynamics (where dislocation segments move because Peach–Köhler forces are acting) coupled with finite element solutions of mechanical problems (which can, for example, precisely identify contributions to the overall stress state, like the misfit stress). Equally important are coupled thermal-diffusion-mechanics problems, and even phenomena where the coupling between fluid and solid mechanics needs to be considered. Last but not least, structural mechanics can provide valuable contributions to process simulation, which is important when one aims for get-it-right first time approaches.

Third, in addition to approaches which are based on microstructural understanding (see the above two points), the need to provide design engineers with improved phenomenological models has not lost importance, these models are needed for the safe calculation of industrial components. These models must be robust, and operate with a reduced number of model parameters, which can be identified from experiments with reasonable effort. But they should nevertheless capture the essence of the underlying fundamental phenomena, such as the evolution of the γ/γ' microstructure during ageing, or the evolution of microstructural damage.

Fourth, scientists working for industrial R&D groups must have the possibility to carry out parametric studies at the design stage, not only on individual components, but also on component assemblies and complex systems. Such calculations, which rely on coupled nonlinear constitutive equations models, typically are associated with a heavy use of computer time. Calculations which take more than one night are still not unusual, despite the considerable progress in computer hardware and efficient calculation algorithms. Therefore, the need to develop simplified calculation procedures which involve model reduction strategies is also of utmost importance.

As a PhD student in the late 1970s, one of us (GC) had the task to perform pioneering testing of a nickel-base alloy specimen (IN100) on an advanced high temperature mechanical test rig. His mission was to show that the formalism of thermodynamics taught by Paul Germain and Jean Lemaitre was capable to account for the materials response during cyclic anisothermal mechanical testing. This took place at the Office National d'Etudes et de Recherches Aérospatiales (ONERA), where the mechanics and metallurgy departments were separate, and an interaction between mechanical engineers and material physicists was not common. It was therefore a question of conducting tests involving temporary overheating in order to estimate the effect on the width of the stress–strain hysteresis loop governing creep fatigue life. As a first result, it was found that the cycle after overheating was five times smaller than predicted [189]. At the same time, the young mechanical research engineer learned about the existence of the γ/γ' microstructure and its fundamental importance for the mechanical behavior of Ni-base superalloys. It was therefore necessary to take micromechanical aspects into account and the thesis represented an excellent opportunity to brigde the gap between the fields of mechanics and metallurgy [306]. The obvious need for these exchanges has been confirmed throughout his career, and today culminates in the present volume.

The coauthor of this introduction (GE), was trained as a material scientist in the creep field. While lacking mechanical background, he was aware of the early review paper by Decker, who already addressed many of the relevant microstructural aspects which govern superalloy behavior [324]. At that time, our teachers and predecessors had already identified the superalloy field as a challenging area for materials research (Gleiter/Hornbogen, pairwise cutting [528]) and for material mechanics/micro mechanics (Pineau, rafting [1108]). In the late 1990s, we personally met and came in working contact through European research projects (BRITE/EURAM scheme) on single crystal Ni-base superalloys.

We have kept contact ever since, exchanging ideas, helping each other with supervising students, and even writing common research papers [198]. In the last decade, we both were in charge of running collaborative programmes on high temperature materials, through a research chair at Mines Paristech,[1] and a large collaborative research center located at the Ruhr-University Bochum,[2] which enabled us to reach out and keep close contact with superalloy single crystal researchers from all over the world. And because the field moves fast and there are so many new developments which characterize this research field, we never lost our interest in superalloy single crystals.

The initiative to compile this book resulted from the wish to present an up-to-date review, which introduces all important aspects from different fields in superalloy single crystal technology. We are well aware of all previous efforts in this respect, which we could build on. We are also well aware of the fact that there is good work out there which we have not been able, for reasons of space, to give fair credit. The hope is that the information collected can be useful for materials engineers who work in the field and for bachelor, master, and graduate students who do project work in the superalloy field. The book is subdivided into four parts, which differ in volume:

- After a short history describing the development of superalloys (Chapter 1), the first part covers the basics of thermodynamics (Chapter 2), anisotropic elasticity (Chapter 3), elements of microstructure (Chap-

[1] Research Chair SAFRAN-MINES ParisTech "CRISTAL" on Materials at High Temperature (2015-2019).

[2] SFB/TRR 103, Collaborative Research Centre funded by the German Research Association (DFG, https://www.sfb-transregio103.de/).

ter 4), mechanical characterization (Chapter 5), and the fundamental deformation mechanisms (Chapter 6).
- The second part deals with processing and manufacturing of components, with load spectra during operation and with life cycle engineering processing (Chapter 7). It, moreover, covers ageing phenomena (Chapter 8), discusses possibilities for refurbishment (Chapter 9), and introduces the application of protective coatings (Chapter 10).
- The state-of-the-art scale bridging modeling techniques are presented in the third part of the book, which presents modeling strategies on all scales step by step, starting at the atomic scale (Chapter 11) and then proceeding to discrete dislocation dynamics (Chapter 12) and phase field modeling (Chapter 13). The presentation of macroscopic models differentiates between models based on dislocations (Chapter 14) and purely phenomenological models (Chapter 15). A chapter on damage evolution and crack propagation (Chapter 16) concludes this part.
- The fourth and shortest part of the book addresses structural applications. It contains one important chapter on how to implement models in finite element code (Chapter 17). The reader must keep in mind that not all details describing design and calculation codes which are used by actual engine manufacturers can be published.

The outline of the book was developed collectively during two workshops that were held at MINES ParisTech in Paris and were sponsored by the CRISTAL Chair. The authors and editors would like to thank SAFRAN for the confidence the company has shown in the project. The individual chapters have been written by experts, who work in their specific fields as indicated. We thank all of them very much for their efforts. We sincerely hope that readers who are interested in the fascinating field of Ni-base superalloy single crystals will benefit from this book.

Georges Cailletaud
Paris, France
Gunther Eggeler
Bochum, Germany
July 2021

PART I

Introduction and basics

CHAPTER

1

Past, present, and future of SX superalloys

Jean-Yves Guédou[a] *and Luc Rémy*[b]

[a]Safran Aircraft Engines, Site de Villaroche, Moissy Cramayel, France,
[b]MINES ParisTech, PSL University, MAT - Centre des Materiaux, UMR CNRS 7633, Evry Cedex, France

1.1 Introduction

The need for alloys able to sustain high temperature strongly increased by the end of the 19th century with the development of turbomachinery. The purpose of this chapter is to give a historical survey of superalloys and in particular of single crystal developments for the enhancement of basic properties. This chapter will consider successively the implementation of Ni-base alloys in the industry (1880–1940), their development for aeronautical applications (1940–1960), microstructure monitoring in cast superalloys (1960–1980), the "Golden Age" of Ni-base single crystal (SX) superalloys (1980–2005), considering alloy development and the studies oriented by component design. The introduction of single crystal components into aero-engines and the design practices vary among companies. Therefore this part is only detailed for the work carried out at Safran Aircraft Engines (formerly SNECMA) in collaboration with Centre des Matériaux MINES Paris and ONERA. The final part gives a few insights beyond the present industrial Ni-base SX (after 2005).

1.2 Implementation of Ni-base alloys in the industry (1880–1940)

The first applications in the turbomachinery industry were steam turbines for ships (C. Parsons patent in the UK, 1884), then gas turbines for power generation (G. Westinghouse in the USA, by 1900), and eventually

Nickel Base Single Crystals Across Length Scales
https://doi.org/10.1016/B978-0-12-819357-0.00008-1

aero-turboengines for planes: the first engine in operation in the UK by F. Whittle (1937) and the first engine in flight in Germany by H. von Ohain (1939) on a Heinkel 178 plane (the first jet plane) [505].

The initial alloys implemented in turbines were iron-base materials, in which metallurgists began to add chromium and copper. Nickel was available as a metallic element through the Mond reduction process (1890). This allowed designing the first nickel alloys: two basic compositions were patented in 1906 by A. Monell in the USA (70Ni–30Cu), an alloy named "Monel 400" [1330], and by A. Marsh in the UK (80Ni–20Cr) [1063,758], both exhibiting good corrosion resistance. In 1923, P. Merica in the USA added 5% Al into the Monel 400 alloy and observed a dramatic increase after aging of the tensile resistance up to 500 °C: he named this alloy "Monel K500" (US patent 1,572,744 on June 26th, 1923) which can be considered as the first real "superalloy" [1330,758], i.e., a material with high strength at elevated temperature, due to γ' (Ni_3Al, Ti) precipitates (which were identified through TEM only in 1952 by A. Taylor and R. Floyd [1402]).

By mid-1920s and up to 1940, numerous works were performed to improve Ni alloys' resistance mainly with addition of Al (USA, UK, France) and Ti (USA, Germany) [1330,1063,323].

1.3 Superalloy development for aeronautical applications (1940–1960)

During World War II, the Wiggins Company (UK) developed a basic alloy, Nimonic™ 80 (Fig. 1.1), Ni–20Cr–2Ti–1Al – up to 3Fe,1Si,1Mn – with an UTS of 1000 MPa at 500 °C [1330,1063]. It was used in Goblin De Haviland engine (1942). Moreover, there was a high requirement in the UK for manufacturing parts for military applications: this led to the fast development of industrial new processes, among which the investment casting techniques for various alloys. The real breakthrough in superalloys is due to the implementation of Vacuum Metallurgy (VIM, VAR, etc.) [1330,758] at an industrial scale (Darmara in the USA, 1952) enabling to add refractory elements into superalloys (W, Nb, Ta, later Re, etc.) and consequently to increase high-temperature resistance while avoiding detrimental precipitates (oxides and nitrides).

During the 1950s numerous Ni-base superalloys have been developed, mainly in the USA and the UK, and trademark names of grades from famous companies are still used today (Fig. 1.2): INCO™ for "International Nickel Company", MAR-M™ for "Martin Marietta", or Hastelloy™ for "Haynes Alloys" [758]. In parallel with the metallurgical progress, the design of turbine blades with internal cooling systems constituted a technological progress, which allowed higher ITT (Inlet Turbine Temperature) in turboengines. The first studies were carried out by Rolls Royce in

FIGURE 1.1 Presentation of Nimonic 80; www.gracesguide.co.uk/Henry_Wiggin_and_Co.

the 1950s which lead to the first flight (1962) of a VC10 powered Conway turboengines with cooled turbine blades [834]: the turbine blades were manufactured in forged Nimonic 115 plates in which cooling channels were machined. However, forged Ni-base superalloys exhibit limited creep resistance at high temperature (900 °C maximum allowable) due to low/medium γ' content ($< 40\%$), opposite to cast superalloys in which the high γ' content ($> 60\%$) induces a good creep resistance at 950 °C and above [834,506]. Investment casting techniques have therefore been developed at the end of the 1950s, including internal cooling systems for turbine hollow blades.

1.4 Microstructure monitoring in cast superalloys (1960–1980)

Turbine blades which are manufactured through the investment casting process (liquid alloy pouring in molds followed by solidification dur-

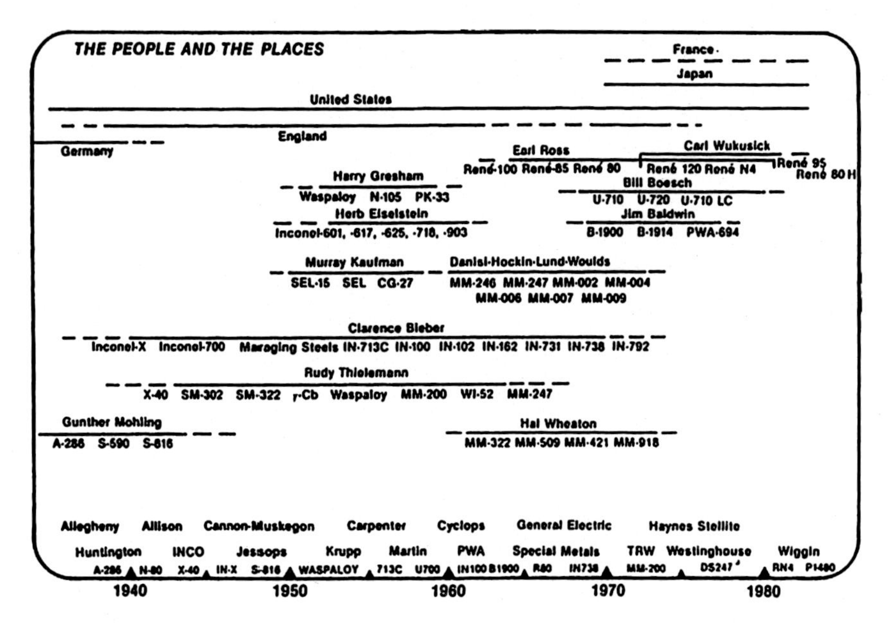

FIGURE 1.2 Superalloy companies and inventors [1330, Fig. 1.3].

ing natural cooling) exhibit multigrained microstructures which are the "basic" equiaxed microstructures. Grain boundaries are "weak points" regarding high-temperature resistance, particularly as far as creep damage is concerned. Therefore, at the beginning of the 1960s many looked at the possibilities to control grain growth and even to remove grain boundaries, i.e., to produce single crystals. The basic principle is to generate a thermal gradient in the liquid metal and to monitor the metal bath (temperature evolution at the surface) during solidification to control the microstructure. Pratt & Whitney mainly pioneered directionally solidified (DS) and single crystal casting in the USA, even though some studies were carried out in the UK by Rolls Royce [506,517]. The first step consisted in removing grain boundaries that were perpendicular to the radial centrifugal loading which generates columnar structures. On an equiaxed (EX) grade, the first DS microstructure was obtained by Versnyder and Guard on NiAlCr in 1960 [1489]: DS alloys were found better than EX alloys for creep resistance and thermal fatigue, specifically due the beneficial effect of the low elastic modulus in the axial direction. The first DS turbine blade (Fig. 1.3) was patented in 1964 by Pratt & Whitney (Versnyder, US patent 3,260,050) and devices for casting DS parts were subsequently developed [517]. In the second step, the radial grain boundaries in a DS structure alloy were eliminated, leading to a single grain microstructure, i.e., single crystal at meso–macro scale (SX). The first industrial single crystal superalloy SX was elaborated in MAR-M200 grade by Versnyder and M. Shank in 1970

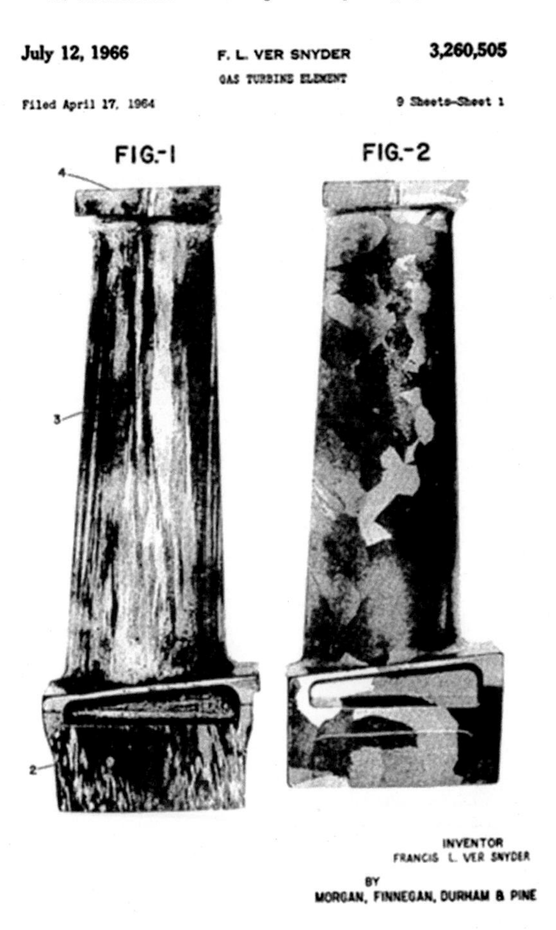

FIGURE 1.3 First DS blades (US patent 3,260,050).

[506,517]. The process is based on the Bridgman method known since 1925 [153] when American physicist Bridgman produced single crystals of various metals (W, Sb, Bi, Te, Cd, Zn, and Sn). The method involves heating polycrystalline material above its melting point and slowly cooling it from one end of its container, where a seed crystal is located. A single crystal of the same crystallographic orientation as the seed material is grown on the seed and is progressively formed along the length of the container. The process was upgraded in 1936 by D. Stockbarger with a better control over the temperature gradient at the melt/crystal interface. While developing the DS casting process, Piearcey discovered how to avoid using a seed for single grain growth and perfected the first grain selector [792] enabling him get a unique grain growth from liquid metal (US patent 3,494,709,

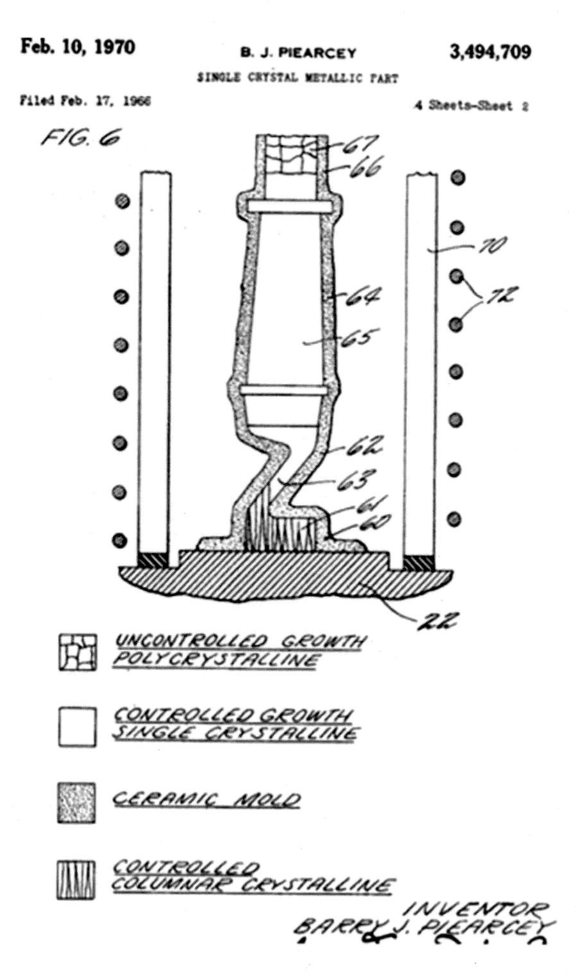

FIGURE 1.4 First industrial single crystal casting for blade device (US patent 3,494,709).

1966). The first single crystal cast blade (Fig. 1.4) was produced in 1969 (US patent 3,625,275) by S. Copley and coworkers [517].

By the end of 1970s, there was simultaneous development of processes (grain selection devices, thermal gradient monitoring, etc.) and of specific new grade compositions for SX with tailored compositions: (*i*) reduction of Zr, Hf, B, and C which are GB strengthening but induce T_m decrease, (*ii*) addition of refractory elements for γ reinforcement at high temperature. Complementary mechanical studies were carried out to take into account more complex mechanical behaviors: cyclic loading (fatigue), creep–fatigue interaction, environmental effects, etc. The DS alloy 444 (derived from MAR-M200) was introduced in Pratt & Whitney military (1969) and commercial (1974) engines [792]. The composition of alloy 444 evolved

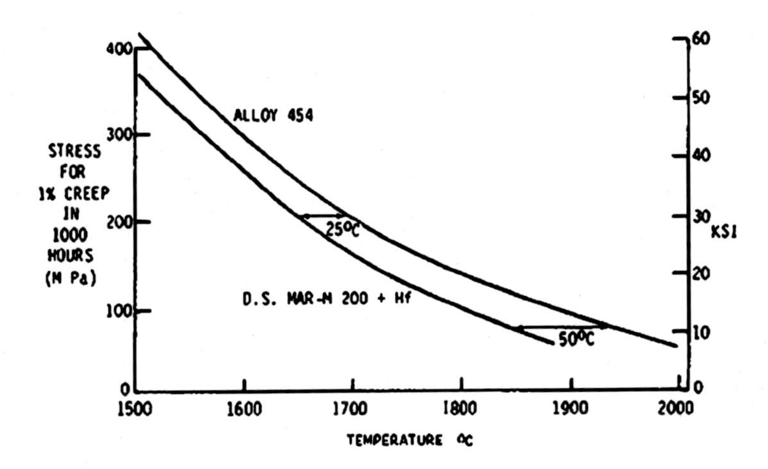

FIGURE 1.5 Advantage of SX over DS for creep [506, Fig. 1.3].

towards alloy 454 [517] that was specifically designed for SX application (Fig. 1.5): it was introduced in a military engine (1980) with the commercial name PWA 1480 (US patent 4,209,348, 1978). At the same period, derived from the experimental NASA NASAIR 100 single crystal, Cannon–Muskegon developed the CMSX-2 (US patent 4,643,982, 1984) that will be subsequently used worldwide [1489,792,718].

1.5 The "Golden Age" of SX Ni-base superalloys (1980–2005)

1.5.1 Alloy development

The chemical compositions of SX Ni-base superalloys for turbine blades continuously evolved for two decades with the development of successive generations (Fig. 1.6). A synthetic table sums up these compositions (Fig. 1.7).

The first generation. In the first generation of SX Ni-base superalloys (1975–1985), the objective was to improve creep resistance versus equiaxed and DS grades, through increasing the γ' ratio up to 68–70% [517]. Then the criteria were to avoid TCP, to secure a good resistance against corrosion (which means higher Cr content, higher than 7.5 wt%), and finally, to keep the balance between γ and γ' strength. All the turboengine manufacturers patented their own grades [506,1232,544,209]: PWA1480 for Pratt & Whitney, René N4 for General Electric (US patent 5,399,313), SRR 99 (US patent 8,223,353) for Rolls Royce, and AM1 (and derived MC2) for SNECMA (US patent 4,639,280). In the USA, Cannon–Muskegon Company launched the CMSX™ family with the CMSX-2. All these grades exhibit moderate densities (typically in the range 8.6–8.7).

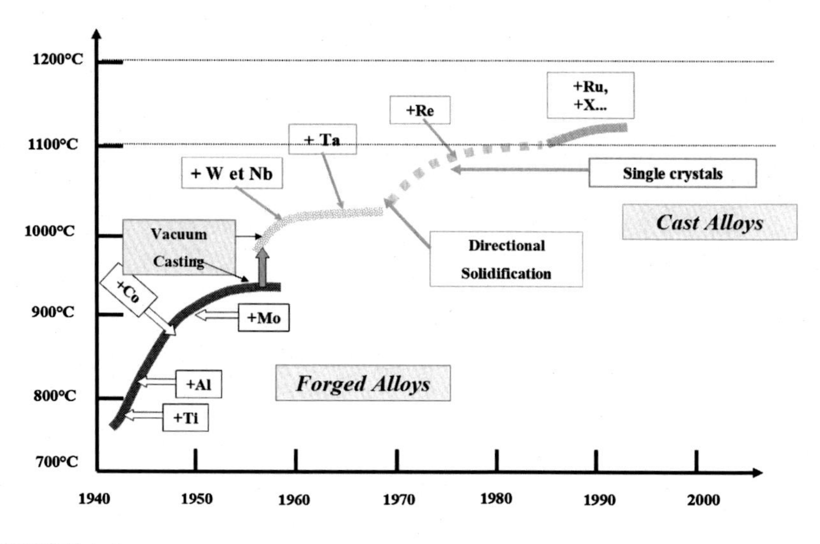

FIGURE 1.6 Evolution of the composition of cast superalloys.

%wt	Cr	Co	Mo	W	Ta	Re	Ru	Nb	Al	Ti	Zr	Hf	Pt/Ir/Rh/Pd	Ni	Density
1st generation															
PWA1480	10	6		4	12				5	1,5				Bal	8,7
RenéN4	12,8	9	1,9	3,8	4			0,5	3,7	4,2				Bal	8,55
SRR99	8	5	0	10	3				5,5	2,2				Bal	8,56
AM1	8	6	2	6	9				5,2	1,2				Bal	8,59
AM3	8	6	2	5	4				6	2				Bal	8,25
CMSX-2	8	5	0,6	8	6				5,6	1				Bal	8,56
CMSX-6	10	5	3	0	2				4,8	4,7		0,1		Bal	7,98
MC2	8	5	2	8	6				5	1,5				Bal	8,63
2nd generation															
CMSX-4	6,5	9	0,6	6	6,5	3			5,6	1		0,1		Bal	8,7
PWA1484	6	10	2	6	9	3			5,6	0		0,1		Bal	8,95
SC180	5	10	2	5	8,5	3			5,2	1		0,1		Bal	8,94
LEK 94	5,8-6,4	7,2-7,8	1,7-2,3	3 to 7	2-2,6	2,3-2,6			6,2-6,8	0,9-1,1		0,05-0,15		Bal	8,3
RenéN5	7	7,5	1,5	5	6,5	3			6,2	0		0,15		Bal	8,63
3rd generation															
CMSX10	2	2	0,4	5	8	6		0,1	5,7	0,2		0,03		Bal	9,05
RenéN6	4,2	12,5	1,4	6	7,2	5,4			5,75	0		0,15		Bal	8,98
RR2100	2,5	12	0	9	5,5	6,4			6	0		0		Bal	
TMS75	3	12	2	6	6	5			6	0		0,1		Bal	8,89
4th generation															
MCNG	4	0	1	5	5	4	4		5,2	1,1		0		Bal	8,75
MX4:PWA1497	2	16,5	2	6	8,25	5,95	3		5,55	0		0,15		Bal	9,2
RR2101	2,5	12	0	9	5,5	6,4	2		6	0		0		Bal	
TMS173	2,8	5,6	2,8	5,6	5,6	6,9	5		5,6	0		0,1		Bal	9,11
TMS138	2,9	5,8	2,9	5,8	5,5	4,9			5,8			0,1	3Ir	Bal	8,95
5th generation															
TMS162	2,9	5,8	3,9	5,8	5,6	4,9	6		5,8	0		0,1		Bal	9,04
TMS196	4,6	5,6	2,4	5	5,6	6,4	5		5,8			0,1		Bal	9,01
6th generation															
TMS238	4,6	6,5	1,1	4	7,6	6,4	5		5,9	0		0,1		Bal	
GE alloy	1,25-10	4,25-17	1 à 4	3-7,5	5,8-10,7	3 to 8	0,4-6,5	0,5-2	5-6,6	0-2	0,05-0,5	0,1-2	0,1-6	Bal	

FIGURE 1.7 Ni-base SX superalloy compositions.

Some "low density" alloys (density between 7.9 and 8.25) were developed during the same time period [718]: RR2000 in the UK, CMSX-6 in the USA, AM3 in France. The contents of heavy refractory metallic elements such as W and Ta was lowered or even removed, and Ti and Al contents was increased: due to a decrease of γ strength at high temperature, the creep resistance of those alloys was significantly inferior to that of the "basic" family, even though the specific resistance (stress over density) was rather close. Therefore, those grades found limited applications: for instance, AM3 was used in some helicopter turboshaft engines by Turbomeca (now Safran Helicopter Engines). In parallel, numerous studies were carried out on environment effects, pointing out the critical role of sulphur, and on coatings, mainly based on Ni aluminide deposition at the surface. Solidification microstructures were also improved by tailored heat treatments including total solutioning steps and by optimized aging treatments.

The second generation. The second Ni-base SX generation (1985–1990) was developed [225] to fulfill the challenge of an improved temperature capability (up to 50 °C) with respect to the first generation. This was accomplished mainly by adding 3 wt% Re into the alloy, which revealed a noticeable reinforcement of the strength of the γ matrix. The Re addition necessitated the decrease of the Cr content (lower than 7 wt%) which induced a poorer environmental resistance of the alloys. The amount of Ti was also reduced (< 1.5 wt%), which had the effect of lowering the γ' strength and consequently the alloy mechanical resistance at intermediate temperature (650–750 °C). The chemical compositions were determined to increase the γ/γ' mismatch and to decrease the γ' coalescence during creep, so as to improve the alloy stability and consequently mechanical resistance at high temperature. The main grades were developed in the USA by Cannon–Muskegon (CMSX-4), Pratt & Whitney (PWA1484) and General Electric (René N5). NIMS in Japan has started to develop its own alloy for research purposes (TMS-82). These alloys exhibited more metallurgical instabilities (they are more prone to TCP precipitation) and more casting defects (pores, freckles, etc.) than the first generation alloys. Moreover, they present higher densities (8.7 to 8.9) due to Re content. This is a critical drawback for rotating parts such as turbine blades. All the alloys have limited sulphur contents to improve oxidation resistance. In the same period, ceramic thermal barrier coatings were developed following the preliminary works of Allied Signal company which patented the columnar ceramic deposit (US patent 5 154 844) and the nickel aluminide bond coats (US patent 5 514 482).

The third generation. The third generation (1990–1995) aimed to still increase the very high temperature resistance [1527], which can be achieved by adding more Re (up to 6 wt%) into the alloy. Consequently, Cr had to be decreased (lower than 4 wt%) which had the consequence of a very poor resistance against oxidation and corrosion. Ti contents was also reduced, and eventually removed. The densities significantly increased (higher than 9). Two commercial grades were developed: René N6 by General Electric and CMSX-10 by Cannon–Muskegon [398], which was used by Rolls Royce with the RR3000 (CMSX-10 according to Rolls Royce own specifications). These alloys, which are, moreover, prone to metallurgical instabilities, exhibit more disadvantages than benefits, and thus found very few industrial applications (CMSX-10 in IP turbine blades for Rolls Royce engines). In Japan, NIMS proposed its own alloys, TMS-75 and TMS-80, that will be implemented in the basic studies.

The fourth generation. The fourth generation [1526] appeared before the end of the 20th century (1998–2005) with the objective to improve the stability of the third generation Ni-base SX alloys while keeping their high temperature resistance. Re content was reduced to about 3 wt% and Ru was added (2 to 4 wt%) which allowed keeping a good creep resistance at very high temperature while getting more stability. Grades were patented in the USA (EPM102 developed as MX4 by General Electric and PWA1497 by Pratt & Whitney) and also in France [37] by ONERA (MCNG). The most important studies were performed in Japan [739] by NIMS (TMS-138 and TMS-173) which carried on research on the subsequent generation. In those alloys, similar to the third generation, Cr content was low (2 to 4 wt%) and consequently they exhibited a very poor oxidation resistance. However, they had another advantage over the third generation alloys which is their lower densities (between 8.75 and 9.2) due to lower density of Ru as compared to Re (12.2 versus 20.8). But the economical aspect is really an issue for those alloys. Besides Re which is a rare metal with a limited production worldwide (50 t in 2018) and is so costly, Ru belongs to the platinum family and is a precious metal with an even more limited production (12 t in 2018), and consequently a very high price (about 10000 $ per kg in 2018). Therefore, these alloys present a high strategic risk for industrial production (scarcity and consequent difficulties for supply, elevated and fluctuating cost) and no industrial application has been identified in aero turboengines.

1.5.2 Mechanical and environmental characterization studies for design

The Ni-base SX alloys are intended to be implemented in structural parts in aero turboengines, i.e., blades and vanes, which are submitted to important loadings (mechanical and thermal) in severe environments (oxidation, corrosion). Therefore it is mandatory to characterize exhaustively those materials as regards to their behavior under realistic thermal and mechanical loadings, including environmental effects, and the associated damaging modes to deliver design models for durability prediction of parts. This has been achieved through studies along the development of all the Ni-base SX alloys from 1980 to 2005. As explained in the introduction, this part will be developed for the case of Safran Aircraft Engines. In 1980 most of the blades were conventionally cast in IN100 Ni-base superalloy coated with NiAlCr against oxidation and corrosion. Directionally solidified MAR-M200-Hf was introduced in M53 engine with improved cooling

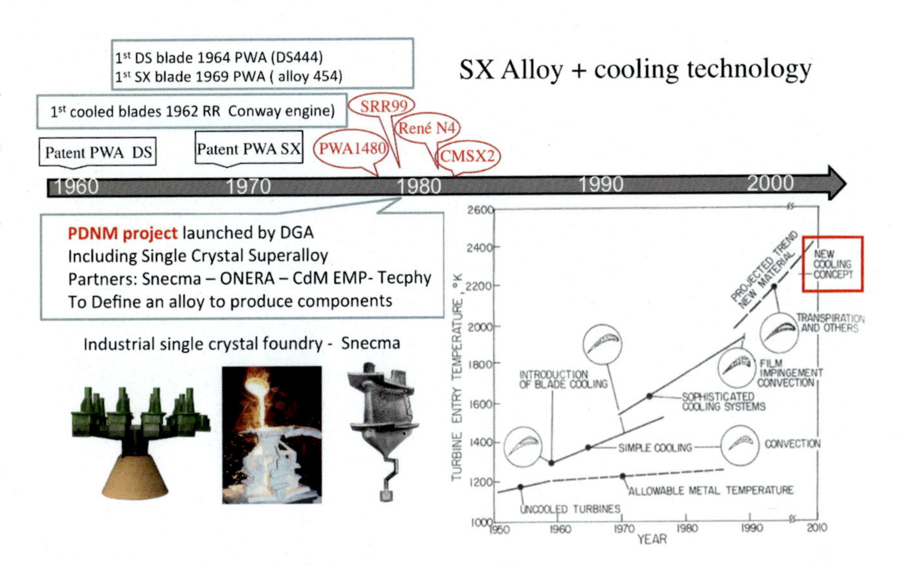

FIGURE 1.8 Summary of single crystal alloy development and evolution of cooling technology for advanced blades (insert courtesy of [251]) Research and development starts in the 1980s at SNECMA along both lines, with partners.

systems. A program for New Materials Development was launched between Safran Aircraft Engines (SNECMA at that time), ONERA, Centre des Matériaux of MINES ParisTech and Aubert & Duval (Techphy at that time). This program was supported by the French Ministry of Defense. It was aimed at defining a single crystal composition and to produce it (Fig. 1.8). An industrial foundry was built at SNECMA to cast single crystals. A composition domain was defined with a density of $8.6\,\mathrm{kg\,m^{-3}}$, a complete solution treatment of γ' phase and a creep resistance better than MAR-M200 and CMSX-2 (35 to 45 °C better). This gave alloy AM1 used by Safran Aircraft Engines and alloy MC2 used by Safran Helicopter Engines (Turbomeca). Simultaneously, the improvement of the cooling technology was the object of numerous investigations at SNECMA (Fig. 1.8, [783]).

A large number of theses were written at Centre des Matériaux and ONERA to generate databases for constitutive behavior and lifetime assessment, as environmental effects (Fig. 1.9). The new blades were more severely cooled, combining internal and film cooling. Thus thermal gradients and thermal–mechanical fatigue became prominent while the alloy composition was mostly chosen using tensile and creep resistance. Fatigue [507] and mainly low-cycle fatigue (LCF) tests [479,630,952,1573,251,448], thermal fatigue and thermal–mechanical fatigue (TMF) tests were mostly used. While the US pioneers in TMF testing [625] were mostly concerned with TMF lifing of smooth or multiperforation hollow specimens [508], the first French facilities aimed at evaluating the prediction capabilities of con-

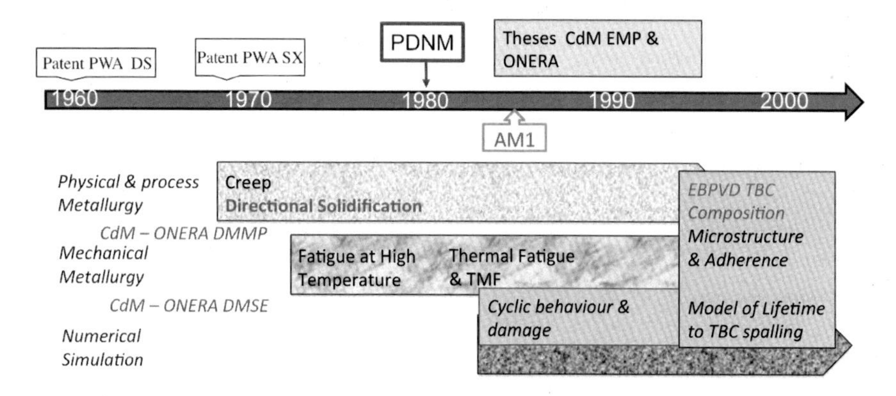

FIGURE 1.9 Summary of design oriented research for SNECMA single crystal blades at Centre des Matériaux and ONERA. The two major steps deal with (1) cyclic behavior and creep–fatigue damage, (2) introduction of TBC.

stitutive models [896,449,752,1200] and life prediction models. Such models were first developed at ONERA for polycrystals using a viscoplastic unified theory [226]. Accounting for elastic and (visco-)plastic anisotropy in single crystals raised a competition between models based on invariants and those based on crystal plasticity. Crystal plasticity models proved to be better than others [937]. The description of plasticity was made at the slip system level at the mesoscopic scale. For Ni-base superalloys with such a high volume fraction (0.6 to 0.7) of precipitates, it was necessary to use phenomenological models for octahedral slip (111) <011> (12 systems) and cube slip (100) <011> (6 systems). After identification on <001> and <111> LCF isothermal tests with some tests including hold time to describe relaxation behavior, the model was able to describe the effects of loading crystallographic orientation on the cyclic stress–strain loops as the slip systems observed in specimens [577,578] (see Chapter 15 on crystal plasticity models). Fairly good results were obtained comparing TMF stress–strain loops obtained by experiment or by modeling [1200,577], as shown in Fig. 1.10, and predicting slip band locations [1200,578] as under tension–torsion loading [1025]. Some attention was given to ageing and rafting on constitutive modeling at the research level [404] but it was not considered in the current design practice [190].

Regarding damage mechanisms under LCF and TMF, the oxidation–fatigue interaction was emphasized at Centre des Matériaux [1195,1204], while the creep–fatigue interaction was the main basis of ONERA damage models [227]. A later version of ONERA model for coated specimens distinguished between initiation in coating and damage growth in the substrate [227]. Fairly good predictions were achieved for TMF tests on coated specimens, poiting out at the major role of the database used than that of actual equations. The influence of coating against oxidation was found to

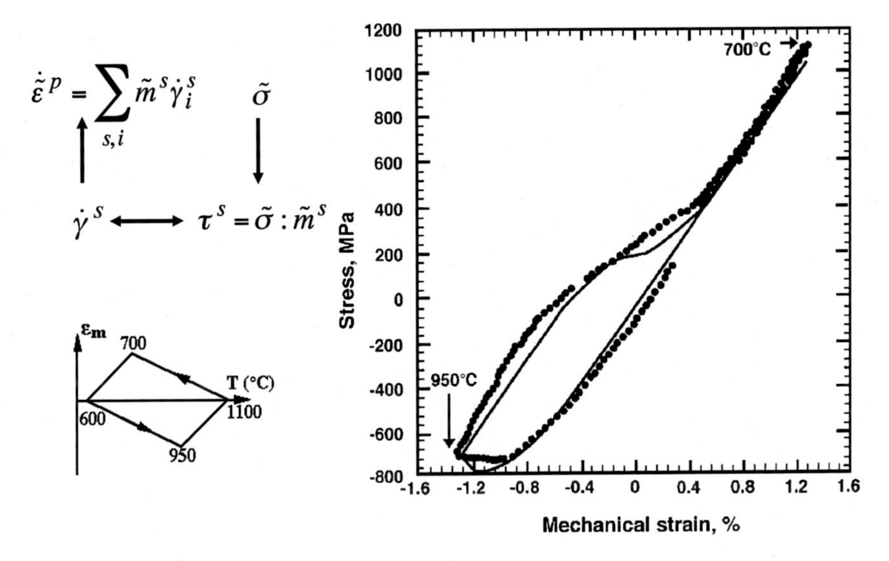

FIGURE 1.10 Basic principles of crystal plasticity model: constitutive equations are written on the slip systems. Application to TMF using a counterclockwise strain temperature cycle with a comparison between model (solid curve) and experiment (points).

be extremely sensitive to coating deposition conditions, and overall coating formulation (NiAl, NiAlCr, NiAlPt) [234,940].

The next step achieved an improvement of SX blade performance (Fig. 1.9) and was to apply a thermal barrier coating (TBC) in order to reduce thermal gradient in the substrate during severe thermal transients [1289]. This started quite lately at SNECMA since the development mostly focused on military applications at that time (Fig. 1.9). Two key points were investigated in the mid-1990s: how to improve the adherence between ceramic topcoat and the alloy substrate, and how to model the lifetime of the substrate–bond coat–ceramic topcoat system. A major difficulty was the absence of any industrial electron beam physical vapor deposition (EBPVD) industrial facility in France. The Ceramic Coating Centre was built in France thanks to a joint venture between MTU and SNECMA in 1999. A problem was not anticipated, namely the adherence of the TBC on AM1 was strongly sensitive to the sulphur content of the alloy. Tests using a thermal shock or engine bench at SNECMA and TMF tests at Centre des Matériaux gave similar damage forms [1201]. SNECMA with Aubert & Duval developed a low sulphur casting procedure to limit AM1 sulphur content to 0.1 ppm. The effect of sulphur content and TBC deposition conditions was investigated in a series of theses [1201,560,255,1203,253]. The location of delaminating interfaces was investigated using industrial cyclic oxidation tests at SNECMA and compression tests of preoxidized specimens for different conditions: isother-

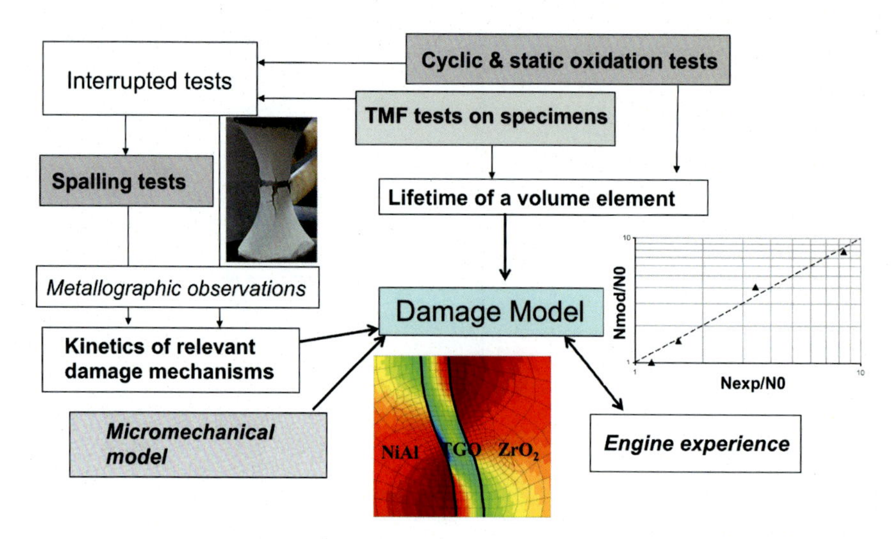

FIGURE 1.11 Lifetime prediction approach to TBC spalling: combining standard cyclic oxidation tests, compression tests on different preoxidation conditions, micromechanical modeling and building a damage model applicable to components.

mal oxidation, cyclic oxidation, and TMF oxidation [560]. The coupling between oxidation and mechanical loading or thermal transients was especially investigated [1203].The reduction of sulphur content and careful surface conditioning was shown to favor delamination at thermally grown oxide–zirconia interface instead of NiAlPt–substrate interface that was detrimental for the lifetime to TBC spallation [560,1203]. Information from service experience came only in the decade from 2000 to 2010 and confirmed observations at the laboratory scale.

The modeling of the lifetime to top-coat spallation followed in the early stages the literature progress coming from the USA. Industrial models focused on the growth kinetics of thermally grown oxide using purely empirical correlations. A significant multiscale effort was made between SNECMA, ONERA, and Centre des Matériaux (see Figs. 1.9 and 1.11). ONERA modeled the local interface roughness and local stress variation to TGO growth as being the critical step in the delamination and spalling process [406,69], following the current literature but introducing viscoplasticity in the analysis [195]. An extensive study of roughness at Centre des Matériaux suggested that this was not clearly the case. This is still a fairly open area at the academic stage. Cohesive zone models were combined with experimental data [195]. Fairly good predictions were obtained with a quite simple macroscopic engineering model at Centre des Matériaux mostly based on compression tests on preoxidized specimens [1203] to investigate the remaining life to spallation of TBC [295]. Both lifetime and

TBC spalling locations can be predicted ([295], see Chapter 10 on coated single crystals).

Some effort was devoted to the role of cooling holes, first investigated by Pratt & Whitney [508], and casting process defects in more recent years. In the mid-1990s a homogenization procedure was introduced to use an equivalent homogeneous medium in cooling holes rows [199] instead of describing all the blade fine details. The nucleation of fatigue cracks at cooling holes was reinvestigated after 2010 under LCF or TMF conditions (see Chapter 5 on mechanical characterization). This was considered using nonlocal approaches: the ONERA approach uses a stress-based approach [693], while Centre des Matériaux uses a microcrack growth model based on dissipated energy and dilatation energy rate contributions combined with an oxidation–fatigue interaction damage model [1202,141], that was validated by microcrack initiation and growth tests at notches. The account of creep–fatigue interaction needed further improvement of the constitutive model either introducing static recovery [903] or two viscoplasticity rate equations [693]. Today the design of blades is still driven by the estimated lifetime before crack initiation in the engineering viewpoint. However, the resistance to TMF crack growth may be an issue either for severe bench test conditions or for the future introduction of damage tolerance approaches, especially when microcracks nucleate at cooling holes or from casting defects. Long and short crack growth studies have been carried out mostly under linear elastic fracture mechanics conditions [630,1573,508,299,325]. The influence of hold times has been considered, and the effects of loading and crack plane/front orientations have been investigated [141,903,451]. Simple and complex criteria were proposed [937,147] but today application is still at an early stage (see Chapter 16). The development of conformal mapping of 3D crack geometry and continuous crack growth is implemented in Z-crack software [633]. The technique seems quite promising [829].

1.6 Beyond present industrial Ni-base SX (2005–)

1.6.1 Ni-base SX families

The fifth and sixth generations. Studies have been carried out to develop new generations of SX Ni alloys (Fig. 1.12), mainly by NIMS in Japan [1268,706]. The principle is to keep the basic strengthening mode in Ni-base alloys (γ/γ') with more efficiency at high temperature, if possible: it may be achieved by increasing the γ/γ' mismatch which is observed by increasing Re and Ru contents (up to 6 wt%) for the fifth generation (TMS-162 and TMS-196) and possibly replace Ru by another Pt family metal, i.e., Pt, Pd, or Ir. The sixth generation (TMS-238 and a General Electric alloy) aimed at getting a better oxidation resistance. All the densities are 9.2.

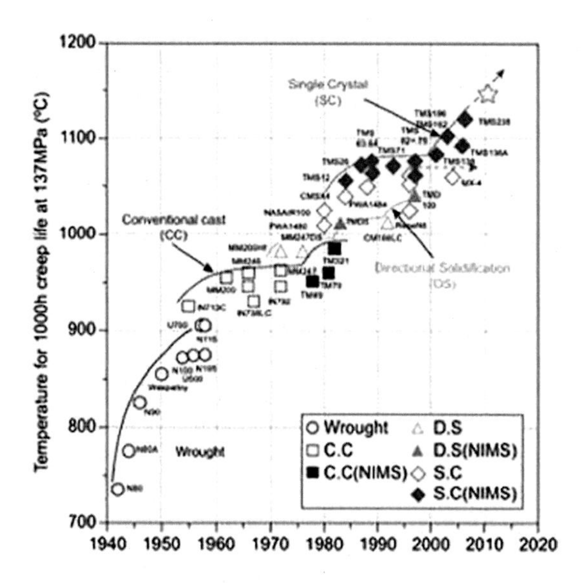

FIGURE 1.12 Fifth and sixth generation Ni-base SX (NIMS).

The creep resistance is unmatched: in TMS-196, creep rupture after 1500 h under 167 MPa loading at 1100 °C, twice the value of TMS 138 (the fourth generation). However, there are still some questions about the stability in anisothermal conditions and for cycling, which are more realistic conditions for parts. The implementation of those alloys induces high strategic considerations before introducing them in aero turboengines: very high cost, supply capacity, affordability, and the need of an efficient recycling of rare metals.

Cost reduction through Re contents decrease. The second generation Ni-base SX alloys (with 3 wt% Re) are the most commonly used in aero turboengines, and around 2010 the cost and scarcity of Re started to be viewed as an important issue [442]. Therefore alloys derived from the second generation with lower Re were included in the investigations. From René N5, General Electric developed René N500 (0%Re) and René N515 (1.5 wt% Re) that exhibited lower creep resistance as compared to René N5, yet sufficient for static parts in aero turboengines (vanes) and large rotating blades in land-based turbines. In the same approach, Cannon–Muskegon proposed two grades derived from CMSX-4, namely CMSX-7 (1.5 wt% Re) and CMSX-8 (0% Re) [1519]. Furthermore, Cannon–Muskegon launched another approach to develop an alloy intermediate between the second and third generation, with intermediate Re content, namely 4–5 wt%; the CMSX-4 PLUS [1520] exhibited a higher creep resistance as compared to CMSX-4 but also a higher density (8.93 versus 8.7). However, today it seems to be quite affordable for industrial applications.

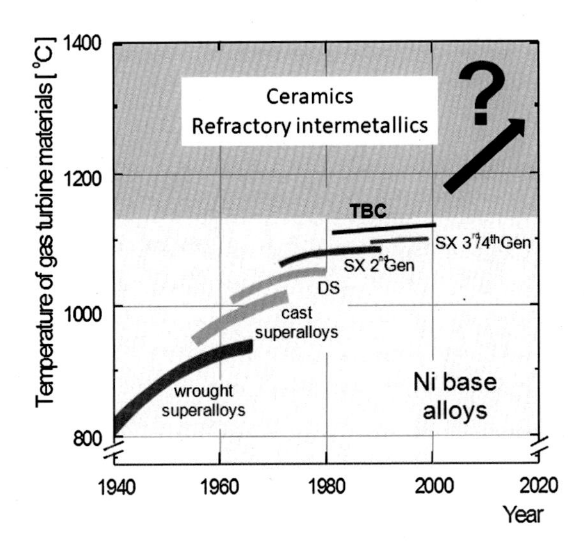

FIGURE 1.13 Beyond Ni-base SX superalloys.

1.6.2 Other material families

Ni-base SX alloys seem to have reached their maximum temperature resistance (around 1200–1250 °C, depending on duration) due to the intrinsic strengthening mechanisms (γ/γ') that have a temperature limit for efficiency. To go further, it appears that there are no solutions with metallic alloys, even refractory metals, to get a high resistance at elevated temperature. Other possibilities (Fig. 1.13) are considered as a potential route to aim for the most extreme performance in aero-engines [559,1042], and they involve intermetallics and ceramic-based materials.

Refractory intermetallics based on NbSi or MoSi systems. They could be used up to 1400 °C [361]. A lot of key points need to be investigated, e.g., processing route (cast or powder), medium temperature behavior, etc. Moreover, they present a very poor oxidation resistance that must be greatly improved.

Ceramic-based materials. Presented in [582], they include composite matrix ceramics [569] (SiC, oxide) with long fibers (C, SiC, oxide). They are lighter than metallic alloys but their use is up to now limited to 1200 °C. Investigations are carried out on structural ceramics such as eutectic solidified ceramics [1056] that are a mixture of alumina and one or two complex oxides of rare earth elements. They exhibit stable properties up to 1700 °C but the TRL (Technology Readiness Level) is still very low.

CHAPTER

2

Fundamentals: alloy thermodynamics and kinetics of diffusion

Ingo Steinbach, Irina Roslyakova, and Katrin Abrahams

Ruhr-University Bochum, ICAMS, Bochum, Germany

2.1 General considerations

2.1.1 Phase equilibria, microstructures, and interfaces

Ni-base superalloys are designed to work at elevated temperatures, i.e., temperatures above 50% of the melting temperature of the alloy, which is typically in the range of 1500 °C or 1800 K. At these temperatures, solute diffusion is "relatively fast" even in solid state, with diffusion coefficients ranging from $5.6 \times 10^{-13}\,\mathrm{m^2\,s^{-1}}$ for the fast diffusing element Al to $7.4 \times 10^{-16}\,\mathrm{m^2\,s^{-1}}$ for the slow diffuser Re (both at solvus temperature 1267 °C for the alloy CMSX4, evaluated with the databases TCNI9 and MOBNI5). So "equilibrium" is attained "relatively fast" and microstructures evolve "quickly." We will elaborate on the meaning of "relatively fast" and "quickly" in the next Section 2.1.2. Also we have to discuss "equilibrium", since this is not as obvious as it seems at first sight. For the moment let us make note of the fact that the material is close to its thermodynamic equilibrium most of its service life in the sense that the equilibrium is characterized by the condition that the diffusion potential is uniform within macroscopic regions of the material (see [786]). It is also well known that in the state of operation the material is two-phase (if we neglect for the moment precipitation third phases, carbides, or unwanted "topologically close packed" (TCP) phases, see Section 6.2.2. The ordered γ' phase precipitates from the disordered γ matrix and reaches a volume fraction above 50% in technical alloys at service temperatures. This high

Nickel Base Single Crystals Across Length Scales
https://doi.org/10.1016/B978-0-12-819357-0.00009-3

volume fraction of precipitates, which hinders dislocation glide and therefore creep, is the major strengthening mechanism, as detailed in Chapter 6. The two phases with different crystal structure and different composition are close to a common equilibrium. The dominating contribution to determine the equilibrium fraction of the phases, as well as the equilibrium composition, is the chemical part of the Gibbs free energy. The equilibrium is defined by the condition of equal chemical potential between the phases, the so-called "common tangent construction." In a technical multicomponent alloy with n of the order of 10 components, this tangent is an n-dimensional space of composition on the Gibbs energy surface of the γ and γ' phases (for constant temperature and pressure) and equilibrium can only be calculated numerically using the so-called CALPHAD (CALculation of PHAse Diagrams) approach, see Section 2.2.1. Besides the pure chemistry, i.e., alloy composition, three more effects contribute to the total free energy of the system. The first is capillarity, accounting for the interface energy between matrix and precipitates. The second is elastic energy if matrix and precipitates are coherent. The final one, which is hardly considered in thermodynamic models, is the energy of stored dislocations and other crystalline defects. These energies are divided into two parts, those which scale with the surface and those which scale with the volume. Thereby the surface part increases the energy of the precipitates relatively to the matrix, while the second part must be determined for each phase individually. It is important to note that all these energetic contributions evolve during the service life of the turbine blade, i.e., although we consider that we are close to equilibrium, the state of the material changes significantly during its life time. While it is obvious that defect energies rise during creep and elastic energy decreases if coherency between precipitate and matrix is lost, it is less obvious that the contribution interface energy changes its sign as the microstructure evolves! This relates, of course, to the so-called "topological inversion" of the microstructure: Microstructures with a volume fraction of precipitates higher than 50% must inevitably invert to reduce the total interface energy, i.e., the original precipitate, γ', becomes the matrix and the original matrix phase, γ, becomes the precipitate [541]. The processes corresponding to these microstructural processes will be detailed in later chapters. For now let us picture in Fig. 2.1 the general process of microstructure evolution during high temperature creep (see also Fig. 6.25, [58]). We distinguish three different stages of creep. In the first stage, precipitates are coherent and γ' is the precipitate phase. Interface energy increases the energy of the γ' phase with respect to the γ phase. The material is elastically stressed, which favors again the γ phase (see below). Defect energy is low. During the second stage coherency is lost. This means that interface energy increases dramatically by the formation of interface dislocation networks, while elastic energy decreases. Also γ' rafts still are disconnected in a percolating γ matrix. In the final stage, the microstructure is inverted, which means that interface energy now works against the (minority) γ phase.

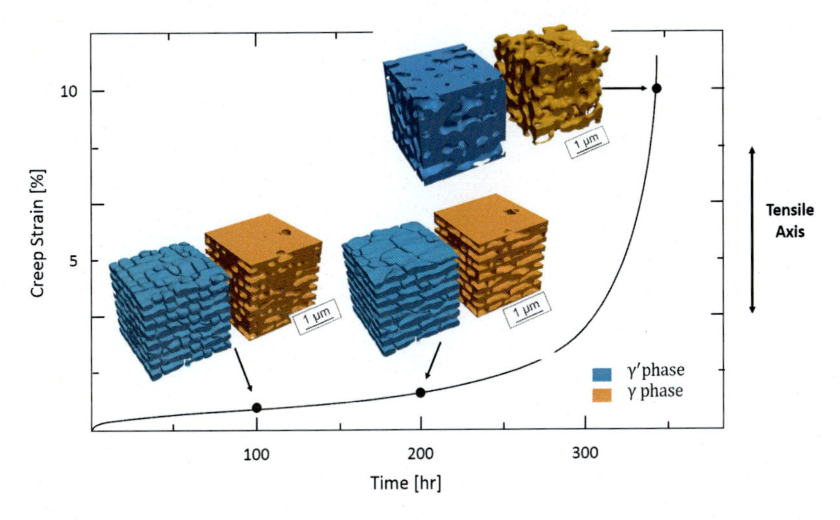

FIGURE 2.1 Creep curve at high temperature and low stress producing the rafted microstructure during low creep of 0.6% and 1.0% creep strain and inverted at higher creep strain. The microstructure images, simulated using the phase-field method (for details see [1244,17,16]), correspond to the three stages of creep as indicated by circles in the creep curve.

In the following we shall roughly estimate the different energetic contributions to be distinguished within the γ/γ' microstructure of Ni-base superalloys. The largest contribution is coherency stress for fully coherent precipitates, which reaches up to 350 MPa, or 350 J cm^{-3} as energy per volume, for alloys with misfit in the range of 0.3%. But it acts only locally at the interface with a rapid decay into the bulk. The second largest (at least in the high temperature, low stress regime) is the external load in the range of several 100 MPa. Both contributions lead to plastic activity, mainly in terms of dislocation movement, glide, climb, or cutting events of precipitates. The third largest, and the one which sets the scale, is the Gibbs energy difference between γ and γ' coming from the energy gain by ordering, see Fig. 2.2. It is estimated here to a maximum of $\approx$56 MPa. Clearly, the two phases will not have the same composition but partition according to the demand of equal chemical potential. Nevertheless, a deviation from equilibrium of 1 MPa will be already 2% of this energy difference.

The first systematic contribution to deviation from chemical equilibrium is capillarity. We neglect here interface energy anisotropy and take the interface stiffness $\sigma* = \sigma + \sigma''$, where the $''$ denotes the second derivative with respect to orientation, equal to the interface energy σ with typical value of 0.05 J m^{-2}. Since precipitates are small ($\approx$ 100 nm in diameter), the total interface area in 1 cm^3 of material is about 30 m^2 (for 50% volume fraction and precipitate size of 100 nm). As we see, already the interface energy

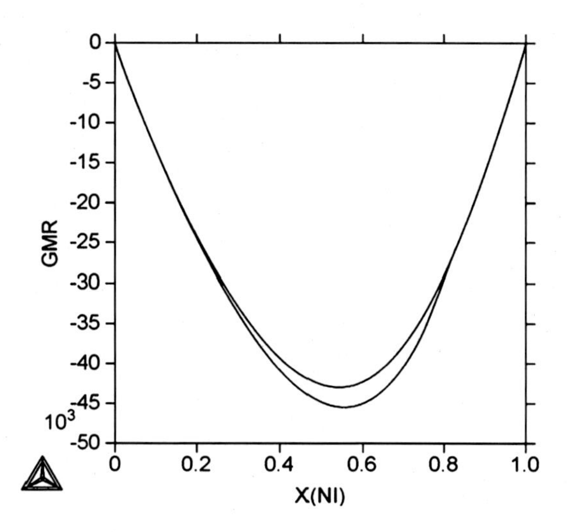

FIGURE 2.2 Plot of the Gibbs energy of γ and γ' for binary NiAl.

accounts for a deviation from equilibrium of about $1.5\,\mathrm{J\,cm^{-3}}$, or 1.5 MPa in units of stress. This is negligible compared to external loads of several 100 MPa, which act on turbine blades in service, but well in the range of driving forces for phase-transformations in metallurgical processes, or in the percent range of our scale, the Gibbs energy difference between the phases.

The average stored elastic energy in CMSX4 with coherent precipitates can be estimated to be of the order of 0.4 MPa in the more compliant γ phase and 0.1 MPa in γ', respectively. Capillarity and elastic energy therefore act with opposite sign on the equilibrium fraction of the phases. The third contribution considers the dislocation network between the γ and γ' phases if coherency is lost. The density of the dislocation network is estimated from geometrical arguments (how many dislocations are needed to remove coherency) [533]. This interface dislocation network lowers the energy density in the γ channels to approximately 0.17 MPa. Because of the discrete nature of the dislocation network, accommodation for misfit stresses is not perfect and thus the elastic energy density is not reduced to zero. Pure edge dislocations are assumed, consistently with the observations of dislocation networks found in later stages of elastic creep [439].

The quantitative consideration of these effects for technical alloys and service conditions is still a hot topic of research at the time of writing this chapter. It is obvious, however, since phase fractions and compositions in both phases are not independent, that the composition of both phases must depend on the state of coherency of the interfaces. Experimentally the dependence of the matrix composition on external load in vertical and horizontal channels is well established, reaching up to 20% variation in the Cr-content in the superalloy SRR99 in the high stress low temperature regime [1278]. The effect of coherency stress is less well investigated, but must lead to similar variations in composition between the coherent and incoherent state of the precipitates.

In consequence we have to consider that microstructures in the high-temperature material Ni-base superalloys, in particular the famous γ/γ' structure, have to be "close" to the thermodynamic equilibrium. This equilibrium, however, deviates from the chemical equilibrium and the deviation from the chemical equilibrium even may change its sign during the evolution of the microstructures. One has to accept the evolution of the microstructure as a service life limiting factor. We will try to specify as clear as possible which definitions and approaches are needed to understand the fundamentals of "thermodynamics and kinetics" related to the specific properties of Ni-base superalloys. This includes some established textbook knowledge, as well as current directions of research and unresolved questions.

Let us continue with a small exercise about diffusion, length and time scales, phase equilibria, and some aspects of processing. This will help link the reader to the respective subjects within this and related different chapters. The exercise is called "Fourier number."

2.1.2 Fourier number

The Fourier number F is a dimensionless measure of the distance a general nonequilibrium state at a given initial time t_0 has traveled towards equilibrium till time t_1. It is named after the famous French scientist Joseph Fourier (1768–1830). Here we consider only solid state systems, in particular multicomponent multiphase alloys, precisely Ni-base superalloys. The relevant transport mechanism is the diffusion of elements in a composition gradient. The initial state at time t_0 is characterized by some nonuniform distribution of elements, called microsegregation (see Section 2.2.2). Also we allow phase-separation, i.e., the system is partitioned into more than one crystallographic phase, precisely γ and γ'. The initial structure may be determined by solidification, or by heat treatment. It is characterized by a typical length scale, the dendrite spacing λ or the precipitate size $\bar{r}$. To attain equilibrium, i.e., the state where the diffusion potential is constant, we have to distinguish two different cases depending on the nominal, or average, composition c_0 of the system, as displayed in Fig. 2.3. At $t = t_0$, we have some given microsegregation profile. Letting the system evolve for given time, it may develop such that its composition simply homogenizes by diffusion. This corresponds to a homogeneous system, i.e., after a given time t_1 the average deviation Δc of the composition from the nominal composition will decrease to a factor F, which is inversely proportional to the Fourier number, defined as

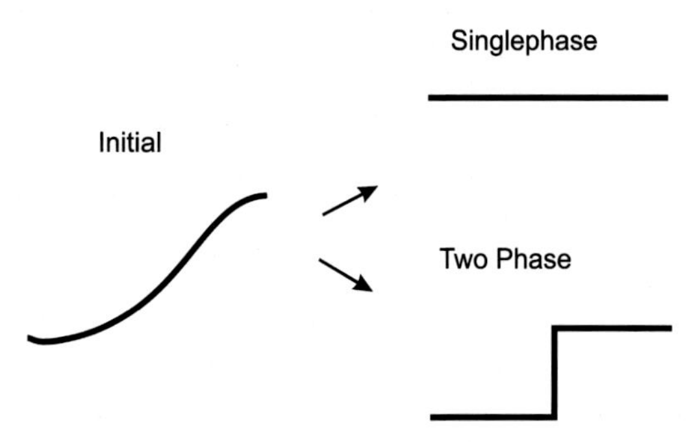

FIGURE 2.3 Sketch of the evolution of segregation from an initially inhomogeneous segregation profile towards (a) uniform distribution in a homogeneous system or (b) piecewise homogeneous distribution with a discontinuity in a two-phase system.

$$F = \frac{D\Delta t}{L^2}\,[-], \tag{2.1}$$

where $\Delta t = t_1 - t_0\,[s]$ is the time for diffusion, $D\left[\mathrm{m^2\,s^{-1}}\right]$ the diffusion coefficient, and $L\,[\mathrm{m}]$ the sample size (assuming adiabatic boundary conditions). The Fourier number simply reflects the effect that in a homogeneous system a gradient in chemical potential, connected to chemical inhomogeneity, drives the system towards the homogeneous state more given longer time and a higher diffusion coefficient, but less given a bigger system. A Fourier number of 1 characterizes the switch from an inhomogeneous system ($F < 1$) to a homogeneous system ($F > 1$). For all "real" systems, the correct value of the "distance towards equilibrium" will deviate from (2.1), but it is a very good "rule of thumb" for our purpose.

The second path (lower path in Fig. 2.3) is totally different. It is, of course, related to an inhomogeneous system, which separates into two thermodynamic phases which tend towards a common thermodynamic equilibrium. The composition seems to demix into different composition sets, which are attributed to the different phases. In the case of our Ni-base superalloys, this separation, however, is attributed to ordering if we consider γ/γ', or to a classical first-order phase transformation if we consider the precipitation of carbides or TCP phases. We see that the system behaves very differently from a homogeneous system. But can we still apply Fourier law? In fact, we can, at least with some care. Let us consider the phenomenon of "rafting" in high-temperature low-stress regime of creep deformation: The γ' precipitates, which are cubic in the initial state of precipitation, due to lattice misfit and elastic anisotropy, tend towards plate-like (n-type) or rod-like (p-type) structures under uniaxial tensile of compressive load, respectively. Although this transformation is induced by external load, it is a so-called "diffusion controlled" transformation where γ' strengthening elements have to diffuse from those faces of the initially cubic microstructure, which dissolve during rafting, towards those faces which grow in order to elongate the plates or rods. The time for individual elements needed to diffuse from one face to the other must be related to the size of the precipitates, i.e., in formula (2.1) the sample size L has to be related to the precipitate size $\bar{r}$.

To end this exercise, let us give a rough estimate of the time needed to reduce microsegregation from solidification to a decent homogeneity of the initial segregation ($F = 1$), e.g., during a heat treatment process. The temperature shall be set to the solvus temperature of CMSX4, as given above. The characteristic length l shall be set to 1/2 of the dendrite spacing λ during a directional solidification process, set to 250 μm. Using the diffusion coefficient of Al at the given temperature and requesting that the segregation is reduced to a factor which corresponds to $\alpha = 0.1$, we have $\Delta t = 2.8 \times 10^4$ s $\approx$8 h, i.e., we might expect that Al is nicely homogenized after this time. A long way from the truth! Because Al has a significant cross-relation with other slow diffusing elements, Al has to wait for the slowest diffuser. Taking Re as the slowest diffuser, we arrive at a time Δt $=2.0 \times 10^7$ s $\approx$5000 h. This means that it is literally impossible to homogenize a technical superalloy to a decent degree in a technical heat treatment process of the order of 10 h! The situation changes slightly if we reduce the spacing, e.g., to 100 μm, in a process with liquid metal cooling or significantly in additive manufacturing with spacings below 10 μm.

2.2 Thermodynamics of multicomponent multiphase systems and the CALPHAD method of CALculating PHase Diagrams

Ni-base superalloys are multicomponent and multiphase. As stated above, at their service temperature they are close to the thermodynamic equilibrium. Therefore a reliable description of the equilibrium properties of the alloys as a function of temperature, pressure, and chemical composition is the basis of every investigation and an indispensable requirement for property-targeted alloy development. Phase diagrams, which are the classical tools for alloy development in general, in this case do not work since the superalloys consist of more than five principal elements: visualization of single phase-fields and phase-equilibria in a five-dimensional and higher composition space is simply impossible. Numerical tools are needed. The idea goes back to the 1970s when computer started to be available to the scientific community. For historical remarks, see [1348]. Its success also rests on the collaboration of many researchers worldwide, as well as on the database standard which has been developed and is widely accepted by all commercial and open software tools till today. There are several good textbooks available for detailed description of the method

[705,1269,872,865,1074], so we will only shortly summarize the general philosophy of the CALPHAD method and comment on actual developments and some issues particularly important for Ni-base superalloys.

2.2.1 General philosophy of CALPHAD thermodynamics

In equilibrium, and the CALPHAD (CALculation of PHAse Diagrams) method makes statements mainly about equilibria, a material can exist in a single- or multiphase state (see Fig. 2.3). According to Gibbs phase rule, a maximum of $n+2$ phases can exist in equilibrium, where n is the number of components of the alloy. In the case of a multiphase equilibrium, the nominal concentration $\vec{c}_0$ partitions uniquely (for given temperature and pressure) into the phase concentrations $\vec{c}_\alpha$ with phase fractions f_α, and we have the concentration conservation from the sum over all phases α, $\vec{c}_0 = \sum_\alpha f_\alpha \vec{c}_\alpha$. It is important to notice that in the case a material shall be stable in a multiphase state, the nominal composition will lie in a two- or multiphase region in the phase diagram for the designated operation temperature, i.e., it cannot be stable in the nominal composition. Ni- or Co-base superalloys attain their superior high-temperature mechanical stability from the famous γ/γ' microstructure with a high fraction of γ' phase, a high fraction γ' phase is one of the major criteria for alloy development, i.e., the nominal composition must lie in the two-phase region between these phases, closer to the γ' composition on the so-called tie-line between the γ/γ' equilibrium compositions in a multicomponent phase diagram, as depicted in Fig. 2.4 for the isothermal section of the ternary model system NiAlCr at 1100 °C. The length of the sections between the nominal composition and the equilibrium compositions then determines the fractions of the two phases according to lever rule. In higher than ternary alloys, the graphical representation of phase diagrams is nearly impossible. Therefore the CALPHAD approach provides a number of options to extract information about phase equilibria, as well as the deviation of a system from equilibrium or metastable equilibria, in multicomponent alloys. In order to derive this information, one basically needs two ingredients:

(i) a suited representation of a thermodynamic potential, e.g., the Gibbs free energy as a function of composition, pressure, and temperature for all possible crystallographic phases, liquid and gas;

(ii) a so-called minimizer which searches for the global minimum of this potential for arbitrary mixtures of possible phases.

The second part is more or less technical, though not at all trivial. The different commercial software packages use different strategies to determine the global minimum of the free energy of a multicomponent multiphase system. Also good open tools, like OpenCalphad [1378,1379], pyCalphad [1039], and ESPEI (Extensible, Self-optimizing Phase Equilibrium Infrastructure) [121], are available online.

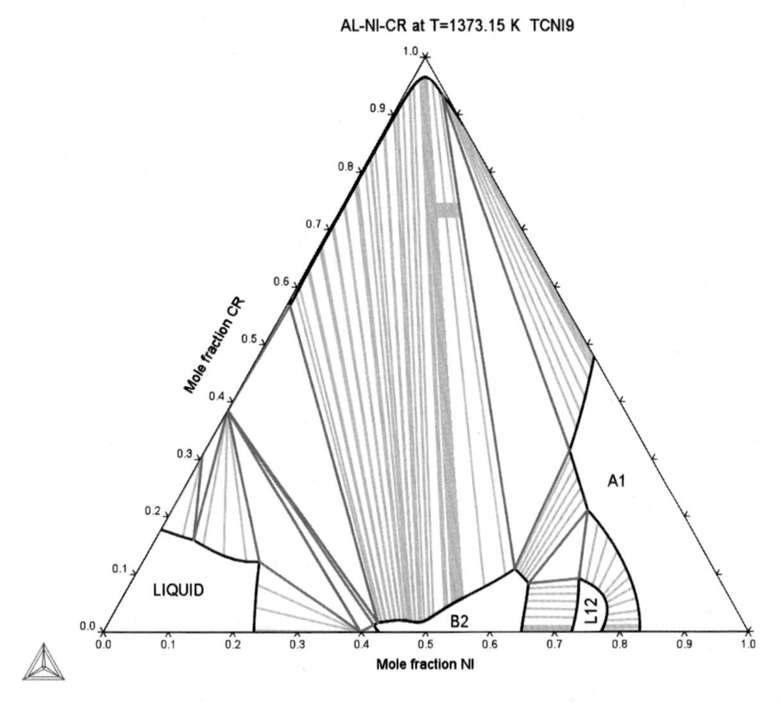

FIGURE 2.4 Isothermal section of the ternary phase diagram Ni–Cr–Al at 1100 °C calculated with Thermocalc database TCNi9. The ordered and disordered phases are labeled as $L1_2$ and A1, respectively, to be found in the right bottom corner of the diagram. Note the strong composition dependence of the tie-lines between these phases. Even liquid can be found for this temperature in the Al-rich bottom left corner.

The first part is the heart of the method. From the thermodynamic potential, all necessary information about phase equilibria can be derived by minimizing the total free energy of the system, part (ii). Since there is no analytical description of such a potential, it has to be expanded in a suitable set of basis functions and parametrized with the expansion coefficients. The set of functions and the corresponding parameters are stored in the so-called databases which are available from public or commercial sources. The databases are organized in a hierarchical manner, i.e., starting from pure substances, the so-called unaries, which then are combined with appropriate models to binary, ternary, and higher order systems. The temperature dependence of the heat capacity of pure substances (unaries) forms the basis of all available CALPHAD databases. It is relatively easy accessible experimentally by calorimetric measurements *if* the corresponding phase is stable in the given temperature range. Here it is already important to notice that the thermodynamic functional has to contain much more information than needed to describe the equilibrium state of a system. An obvious example is the determination of the melting temperature of a pure substance. This temperature is characterized by the condition that the free energy of liquid and solid are equal. Above the melting temperature, liquid has the lower free energy, below that of the solid. Now we might say that it is sufficient to store the solid free energy up to the melting temperature, and the liquid free energy above. Since, however, the melting temperature is also a function of pressure, and in case we alloy the substance it will become a function of composition, we need the free energy as a function of temperature, pressure, and concentration in the whole range of possible phase transformation temperatures for a given pressure and concentration region under consideration. In general, one would like to have the functional in the full composition range from $c_i = 0$ to $c_i = 1$ for all compositions c_i of element i, for all crystallographic solid phases, glass and liquid state, as well as gas and plasma. Finally, we would like to span a temperature range from 0 K to, say, the temperature of the Sun, and pressure from full vacuum to the pressure in the center of the Earth if, e.g., we want to investigate phase equilibria in magmatic systems. Most of this information relates to metastable or unstable phases and has to be fixed by appropriate models or extrapolations. An actual trend is to derive such information from first principles, where a crystal structure can be stabilized by appropriate means in the unstable region. This reflects in a recent development for compound phases, which are stable only in a narrow composition window, but their energy shall be parametrized in the full composition range [369]. This so-called "effective bond order formalism" uses first principle calculations to determine the energetics of un- or meta-stable endmembers which are impossible to determine experimentally since this material simply does not exist. First-principle methods are originally designed for calculating energies at 0 K, while the extrapolation to higher temperature is much more challenging [608,1652].

Traditionally, CALPHAD databases had been only developed from room temperature upward with the main argument that the experimental evaluation of phase equilibria at low temperatures is difficult due to the slow diffusion of species [346]. Another well-known problem of traditional databases is the polynomial expansion of the temperature dependence of the heat capacity and thus all related thermo-physical properties, such as enthalpy, entropy, and Gibbs energy. This expansion had been chosen in the early days of the method since it is simple and numerically cheap. This numerical costs had been a serious issue with computational resources of these days. It leads to the obvious problem that the energy diverges to $\pm\infty$ at high temperatures dependent on the odd or even character of the highest power. To overcome this problem, more physical modeling is required which is able to consider several contributions to the heat capacity (e.g., electronic, vibrational, phonon, etc.) and takes into account the recently available theoretical data. Since phonon contributions represent the main part of the heat capacity that can be described by the

Debye or Einstein model, two alternative physics-based modeling strategies based on these functions have been proposed to replace the standard polynomial description. The first modeling strategy applies a combination of the Einstein function with polynomials [233,243] and the second uses a combination of the Debye model with segmented function [1230]. Both these modeling approaches have been developed to be valid from 0 K, but considered physical effects are modeled using different sets of assumptions and thus different mathematical functions. Thermodynamic databases which build on such physically-based modeling strategies are called the "third generation" CALPHAD databases. They should deliver a robust and accurate description of the thermodynamic quantities from 0 K up to the melting point and far above it.

First, such models were proposed by the CALPHAD community during a Ringberg Workshop in 1995 [233] and were extended in [243,112,837, 111]

In contrast to these formulations, the segmented regression model [1230] has been developed under the assumption that several physical effects of the heat capacity can appear at different temperatures. Moreover, the phonon vibrations were modeled by the Debye function, and thus it provides a more accurate description of low temperatures. The segmented regression model was first applied to describe the thermodynamic properties of pure Cr, Al, and Fe [1230], where the study results were successfully validated using existing experimental and DFT data on correlated thermo-physical properties such as relative enthalpy and standard entropy. Here, a very good agreement between experimental and modeled heat capacities was observed, especially at low temperatures. Therefore, in order to identify the most appropriate model for the temperature dependency of thermodynamic properties from 0 K, all these alternative models [233,243,1230] have been applied to pure Al, Cr, Fe, and compared with each other. Up to now, this model has been successfully applied for more than 18 pure elements [1231].

Despite of this significant improvement for the pure elements' description, a recent implementation of the CALPHAD method inside of almost all existing software, however, is lacking strategies and tools for straightforward implementation of the new models and new experimental and DFT data into existing databases. This means that if we would like to update existing multicomponent thermodynamic databases with new unary models, all relevant binary, ternary, and other related systems have to be reassessed, which draws back from the desirable improvement of CALPHAD databases. One promising solution of this problem is open-source Python-based ESPEI software for automated thermodynamic database development within the CALPHAD method [1307,121]. It uses the pyCalphad software package [1039] for calculating Gibbs free energies of thermodynamic models to rapidly develop and modify databases using a combination of first-principles and experimental thermochemical and phase equilibria data. ESPEI evaluates Gibbs energy model parameters of individual phases in two steps: parameter generation and Markov Chain Monte Carlo (MCMC) Bayesian optimization. In comparison with traditional database development, the usage of MCMC allows us to optimize and quantify the uncertainty for all model parameters simultaneously. An example of its application for the automatic assessment of the Cu–Mg binary system can be found in [121]. The current version of ESPEI software is able to generate only binary and ternary parameters in the parameter generation step. However, implementations of MCMC error functions are designed in a general way to fit arbitrary multicomponent, multiphase systems. The extension of ESPEI functionality to automatize the assessment of ternary and higher order systems is currently under development.

2.2.2 Solidification

Solidification is not in particular a topic one would think of talking about equilibrium thermodynamics and the CALPHAD method. Solidification is a pronounced nonequilibrium transformation from which a microstructure will result that contains phases which would never be expected for the nominal alloy composition in equilibrium. The temperature during solidification, however, is high, which means that local equilibrium at the solid liquid interface can be assumed to establish instantaneously, i.e., the solidification front can well be approximated in a local equilibrium. There the Gulliver–Scheil model of solidification is well established [562,1272], which treats the melt as homogeneously mixed, but neglects diffusion in the solid. The microsegregation, i.e., the solute distribution frozen in solid, can be predicted by stepping down the temperature from the melting temperature of the primary phase down the nonequilibrium solidus temperature, i.e., the temperature where liquid vanishes. Since in the original Gulliver–Scheil model the primary phase never completely solidifies, the CALPHAD treatment includes precipitation of secondary or tertiary phases. In the case of Ni-base superalloys, the secondary phase is γ' precipitating from the melt (note that the solvus temperature of γ' lies significantly below (equilibrium) solidus temperature). This precipitation leads to the so-called "eutectic islands" between the primary dendrites of disordered γ phase. Thereby it depends on the alloy whether these islands are formed by a eutectic or a peritectic reaction. Stepping further down in temperature, also other phases, like carbides, μ, or laves phases, can be predicted. Fig. 2.5(a) depicts the evolution of the fraction of phases during solidification as a function of temperature for CMSX4, calculated by using the database. This evolution, of course, considers no influence of the dendritic growth morphology, kinetics of solidification, and back diffusion. To consider these effects, the phase-field method is the method of choice (see Chapter 13). Fig. 2.5(b) depicts the microsegregation of Al and Re, respectively, as calculated in a 2D cross-section through a dendritic array perpendicular to the solidification direction. Fig. 2.5 shows the corresponding segregation profiles along the scan, as indicated in (b).

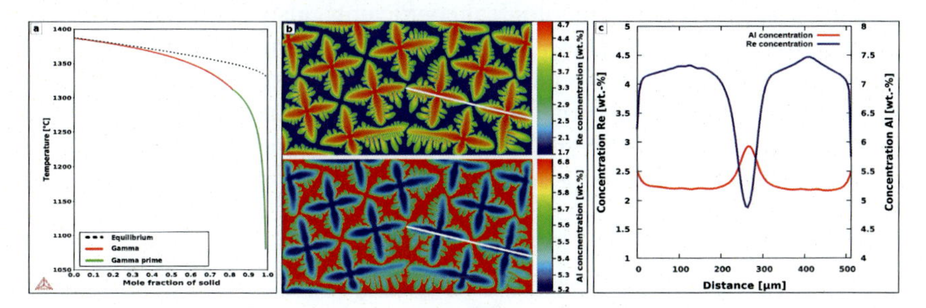

FIGURE 2.5 Simulation of solidification: (left) Scheil calculation of the solid fractions, γ and γ', with respect to temperature; (middle) microsegregation simulated by the phase-field method, Re and Al profile; (right) line scan through the segregation profile.

2.3 Multicomponent diffusion

Diffusion is in general known as the transport of matter/particles that occurs in all solid, liquid, and gas phases. Ni/Co-base superalloys consist of roughly ten different elements and in the solid state of a γ/γ' microstructure (depending on the conditions also TCP phases), and diffusion occurs in all phases. Diffusion in ordered phases, as in γ' (ordered $L1_2$), differs in its mechanism (compare Section 2.3.2.2) and has a special treatment in the modeling of the so-called atomic mobilities (compare Section 2.3.3.4).

First diffusion measurements reach back more than 150 years, when Thomas Graham conducted experiments with high-concentration salt water in contact with low-concentration water and measured diffusion from the higher concentration to lower [929]. Shortly afterwards, the first diffusion measurements in solids in the form of a diffusion couple experiment (see Section 2.3.1.2) were performed. Adolf Fick used a relation between temperature and mass flow (diffusion) and proposed that the change of concentration is proportional to its gradient [438]:

$$\frac{\partial c}{\partial t} = -\nabla J = \nabla D \, \nabla c, \tag{2.2}$$

where D is the proportionality constant (diffusion coefficient) and J is the flux of diffusing species. In 1931 Onsager proposed a phenomenological linear relation for diffusion also valid for nonideal behavior [1034,1035]:

$$J_i = -\sum_{k=1}^{n} L_{ik} \nabla \mu_k, \tag{2.3}$$

where L_{ik} are the phenomenological coefficients, also known as Onsager coefficients, and μ_k is the chemical potential of element k. Furthermore, L_{ik} is a matrix and the off-diagonal terms are the coefficients to relate the gradient in the chemical potential of element k to the flux of element i. Since chemical potentials are difficult to measure in experiments, compared to concentrations, the following general diffusion flux equation is used commonly:

$$J_i = -\sum_{m=1}^{n} D_{im} \nabla c_m, \tag{2.4}$$

which is basically the same equation as (2.2) including off-diagonal terms; D_{im} can be related to the Onsager coefficients by $D_{im} = \sum_{k=1}^{n} L_{ik} \partial \mu_k / \partial c_m$. The key characteristic to all diffusion coefficient definitions is the composition out of a purely kinetic part (L_{ik}) and a thermodynamic contribution ($\partial \mu_k / \partial c_m$).

2.3.1 Experimental methods

From an experimental side, there are mainly two different setups to measure and determine diffusion properties, namely radiotracer and interdiffusion experiments.

2.3.1.1 Radiotracer method

To conduct radiotracer experiments, a sample without a concentration gradient is chosen and radioactive isotopes of one or more species are placed onto one side of the sample [1067]. During annealing the isotopes diffuse into the alloy. Afterwards the sample is sectioned parallel to the surface on which the radioactive isotopes were placed and the radioactivity of each section is measured. The radioactivity is proportional to the concentration. Since the amount of radioactive isotopes is in the range of ppm, they do not change the overall concentration during diffusion, the diffusion coefficient is constant, and the concentration profile can be described with the analytical solution of Fick's law (Eq. (2.2)):

$$c(x,t) = \frac{M}{\sqrt{\pi D t}} \exp\left(-\frac{x^2}{4Dt}\right), \tag{2.5}$$

with M being the initially placed amount of tracer atoms on the surface, x the distance, and D the self- or impurity-diffusion coefficient. As the annealing time t and temperature T are known, the concentration profiles can be measured and then the self-diffusion coefficient (if the tracer atom is also in the overall composition of the sample) or rather the impurity-diffusion coefficient (if the tracer atom is not in the overall concentration of the sample) can be determined.

2.3.1.2 *Diffusion-couple method*

For the second experimental method, the so-called diffusion-couple method, two samples with different compositions are brought into contact and annealed [1067,929]. Due to initial concentration gradients, diffusion takes place and changes the overall concentration. This method has the advantage that also different phases can be brought into contact, new phases can form in the interdiffusion zone, uphill diffusion can be investigated, and in general a broad composition range can be covered. The disadvantage is the difficulty of evaluation of these diffusion couples. In binary couples the interdiffusion coefficient can be determined using, for example, the Boltzmann–Matano or Sauer–Freise method [1067]. In a higher-order system, it is difficult, or not possible anymore, to obtain interdiffusion coefficients from one couple. If marker atoms are used to mark the initial contact plane, also the intrinsic diffusion coefficients can be determined at this plane. This plane will move due to the different diffusion speeds of the atoms.

2.3.1.3 *Relation between diffusion coefficients and atomic mobility*

When dealing with diffusion, four different terms came up until now: D_i^*, the self-diffusion coefficient of element i; $\tilde{D}_{ij}$, the interdiffusion coefficient between element i and j; D_i^{in}, the intrinsic diffusion coefficient; and M_i, the atomic mobility. In 1948 Darken published relations between these coefficients for a binary system [311], which was later extended to multicomponent systems [728].
The self-diffusion coefficient D_i^* is a purely kinetic term and can be measured using the radiotracer method (compare Section 2.3.1.1). It is directly related to the atomic mobility (velocity of a species under unit force) via the Einstein equation [1067]:

$$D_i^* = M_i RT, \tag{2.6}$$

with R being the ideal gas constant and T the temperature. Both M and D^* are composition and temperature dependent. The intrinsic diffusion coefficient is related to the self-diffusion coefficient by the thermodynamic factor Φ in the following way:

$$D_i^{in} = D_i^* \Phi. \tag{2.7}$$

In the binary case there is only one interdiffusion coefficient which can also be related to the self-diffusion coefficients (or rather atomic mobilities):

$$\tilde{D}_{ij} = x_i D_j^{in} + x_j D_i^{in} = (N_i D_j^* + N_j D_i^*)\Phi, \tag{2.8}$$

with $\Phi = (x_i x_j / RT)(dG^2/dx_i^2)$ being the thermodynamic factor and N_i the number of atoms of element i. In multicomponent alloys, the relation between the interdiffusion coefficient and the intrinsic diffusion coefficient is given as [1067]:

$$\tilde{D}_{ij}^{n} = D_{ij}^{in,n} - N_i \sum_{k=1}^{n} D_{kj}^{in,n}, \tag{2.9}$$

with n being the reference element.

2.3.2 Diffusion mechanisms

2.3.2.1 *Disordered phases*

Until now we only looked at phenomenological models which describe diffusion without further knowledge about the underlying atomic mechanism. Over the years many different atomic mechanisms were suggested, but here only the most important ones for solids will be described. Diffusion in solids mainly happens via an atom–vacancy-exchange, which means that an atom is jumping into an adjacent vacancy. The amount of equilibrium vacancies is temperature dependent, and close to the melting point the site fraction of vacancies in most metals is around 10^{-3} to 10^{-4} [929]. Besides this monovacancy diffusion mechanism, a divacancy mechanism exists. A divacancy is an aggregate of two vacancies, and an atom can then jump into one of the vacancies. The amount of divacancies is also temperature dependent but increases faster with increasing temperature than the amount of monovacancies and therefore often becomes the dominating mechanism around 2/3 of the melting temperature.

2.3.2.2 *Ordered phases*

In ordered phases, e.g., the γ' phase, diffusion also happens via vacancy–atom exchange but is more complicated due to ordering. If we assume a perfectly ordered Ni_3Al $L1_2$ phase, we would have one sublattice occupied only by Ni and the second one occupied by Al with a relation of 3 : 1 [1098]. Diffusion mechanisms in ordered phases depend on different contributing factors. First of all, since it is still a vacancy-exchange mechanism, one has to keep in mind that the formation energy of vacancies on different sublattices can differ. For example, in this case vacancies are mainly formed on the Ni-sublattice [929]. Furthermore, atom–vacancy exchanges are favorable for nearest neighbor exchanges. In this case Ni has 8 Ni atoms and 4 Al atoms as neighbors and therefore can diffuse to 8 positions without creating an antisite defect (Ni atom on an Al sublattice or the other way around). Al has accordingly less Al neighbors and therefore fewer possibilities for diffusion without defect creation. If the formation energy of these defects is too high, other mechanisms are possible, e.g., jumps to the second-nearest neighbors or the six-jump-cycle mechanism (six consecutive jumps where in the end the order is maintained) [1067].

2.3.3 Diffusion simulations

Diffusion simulations can be conducted either on an atomic scale, where all atoms are resolved, or on a continuum scale, where only a concentration field is given and no single atoms or vacancies are resolved and therefore also no details about the diffusion mechanism are given. This section will focus on continuum simulations since they have the advantage to simulate the dimensions of an experiment in time, temperature, and size and also deal with multicomponent and even with multiphase (e.g., using the phasefield) alloys. Two different diffusion models will be introduced in the following: the DICTRA model, which was developed in 1992, and a new model, the pair-diffusion model, developed in 2019.

2.3.3.1 *Frames of reference*

The diffusion flux can be simulated and measured with respect to different references, and to make this clear, two different frames of references will be introduced shortly, namely the lattice- and volume-fixed frames of reference [311,728].
The lattice-fixed frame, as the name says, takes the underlying atomic lattice as a reference. This frame is especially used in atomic simulations and the intrinsic flux (compare Section 2.3.1.3) is measured with respect to this frame. The following condition describes the lattice-fixed frame:

$$J_{Va} = -\sum_{i=1}^{n} J_i, \tag{2.10}$$

which means that there is a vacancy flux J_{Va} balancing the flux of all other elements.

In both diffusion simulation models, the DICTRA model and the pair-diffusion model, diffusion is simulated with respect to the volume-fixed frame of reference. In this frame there is no net flow of volume,

$$\sum_{i=1}^{n} J_i V_i = 0, \tag{2.11}$$

with V_i being the partial molar volume of i. If the molar volume of all elements is assumed to be equal, this frame of reference is equal to the number-fixed frame of reference,

$$\sum_{i=1}^{n} J_i = 0. \tag{2.12}$$

In this frame of reference, there is no net flow of vacancies. Different frames of reference can be related to each other by determining the velocity between them [728].

2.3.3.2 DICTRA model

The DICTRA (Diffusion Controlled Phase Transformations) model was developed in 1992 by Ågren and Andersson [24,196]. In this model it is assumed that all substitutional elements have the same molar volume, and therefore diffusion is described with respect to the number-fixed frame of reference. The concentration change of element i is given as

$$\frac{\partial x_i}{\partial t} = \sum_{j=1}^{n-1} \tilde{D}_{ij}^{n} \nabla x_j, \tag{2.13}$$

where $\tilde{D}_{ij}^{n} = D_{ij} - D_{in}$ with n as the reference element:

$$\tilde{D}_{ij}^{n} = \sum_{k} (\delta_{ki} - x_i) x_k M_k \left(\frac{\partial \mu_k}{\partial x_j} - \frac{\partial \mu_k}{\partial x_n} \right). \tag{2.14}$$

The atomic mobility M_k and the chemical potential μ_k are composition and temperature dependent, and therefore mostly taken from CALPHAD type databases.

2.3.3.3 Pair-diffusion model

The pair-diffusion model is a generalization of the DICTRA model with the advantage of its consistency with self-diffusion models [480]. To make diffusion models usable for highly alloyed systems also with highly concentrated alloying elements, the use of a reference element is not always advantageous. Therefore diffusion is described in a pairwise manner,

$$\frac{\partial x_i}{\partial t} = \nabla \sum_{\substack{j=1 \\ j \neq i}}^{n} x_i x_j M_{ij} \nabla (\mu_i - \mu_j), \tag{2.15}$$

with M_{ij} being the chemical mobility. This term can be derived by transformation from a lattice-fixed to a volume-fixed frame of reference under consideration of mass conservation:

$$M_{ij} = x_i M_j + x_j + \sum_{\substack{k=1 \\ k \neq i \\ k \neq j}}^{n} x_k (M_i + M_j - M_k). \tag{2.16}$$

Although M_{ij} is the chemical mobility between i and j, all other elements also contribute since they all influence the movement of the volume-fixed frame with respect to the lattice-fixed frame.

This model is also consistent with self-diffusion. Assuming a system i with tracer elements i^*, the Gibbs energy is given as

$$G = G_i^0 + RT(x_i \ln x_i + x_{i^*} \ln x_{i^*}). \tag{2.17}$$

Implementing this into Eq. (2.15) and assuming that the mobilities of the tracer and nontracer atoms are equal, the pair-diffusion model reduces to Fick's law (compare Eq. (2.2)). Therefore this model can be also used to simulate radiotracer experiments, and atomic mobilities can be consistently used.

2.3.3.4 Atomic mobility databases

Atomic mobility databases store parameters to describe the composition and temperature dependence of atomic mobilities in a broad range of alloys with different compositions. They were developed following the example of thermodynamic databases in the CALPHAD style [24]. The temperature dependence is given in an Arrhenius law [196]:

$$M_i = M_i^0 \exp\left(-\frac{Q_i}{RT}\right)^{mag}\Omega, \tag{2.18}$$

with M_i^0 being the frequency factor, Q_i the activation energy, and ${}^{mag}\Omega$ containing magnetic contributions. The composition dependence is given as the Redlich–Kister expansion

$$\Phi_i = \sum_j x_j \Phi_j^i + \sum_k \sum_{k>j} x_j x_k \sum_{r=0}^{m} \Phi_i^{j,k;r} (x_j - x_k)^r, \tag{2.19}$$

where $\Phi_i = Q_i$ or $\ln(RT M_i^0)$.

To include the ordering phenomena into atomic mobility databases, as it is necessary for an atomic mobility database for γ', Helander and Ågren [598] divided the activation energy into the activation energy of the disordered phase and an ordering contribution,

$$Q_i = Q_i^{dis} + \Delta Q_i^{ord}. \tag{2.20}$$

The ordering contribution is given as

$$\Delta Q_i^{ord} = \sum_j \sum_{k \neq j} \Delta Q_{ijk}^{ord} [y_i^\alpha y_j^\beta - x_i x_j], \tag{2.21}$$

with ΔQ_{ijk}^{ord} as the ordering contribution to i due to j–k ordering.

2.4 Conclusion

In this chapter we have discussed (i) the general mechanisms needed to understand in order to control thermodynamic and kinetic stability of the microstructure of superalloys and (ii) the theoretical tools to predict thermodynamic and kinetic properties of superalloys. It must be highlighted again that microstructural transformations during processing and service must be considered as "diffusion controlled." Thereby microstructural stability rests on slowly diffusing elements and on a large partitioning of these elements between the phases, i.e., that a large amount of these elements has to diffuse in order to change the microstructure. The theoretical tools for thermodynamic stability and diffusion are implemented in CALPHAD-type databases and software. The most advanced tools for simulation solidification, heat treatment, and rafting under creep conditions is the phase-field method coupled to CALPHAD databases. The phase-field method will be explained in Chapter 13.

CHAPTER

3

Crystal orientation and elastic properties

Bernard Fedelich

Bundesanstalt für Materialforschung und -prüfung (BAM), Division 5.2: Experimental and Model Based Mechanical Behaviour of Materials, Berlin, Germany

3.1 Introduction

The elastic constants are the most basic mechanical properties of a material and are needed for any structural analysis of a component. For example, they have a major influence on the eigenfrequencies of vibrating parts. Single crystals of Ni-base superalloys are strongly anisotropic, which means that the observed properties are orientation dependent. Tensor algebra is then required to mathematically formulate the elastic properties and their relations to the crystal orientation. Hence, this chapter first summarizes some basic definitions and calculation rules for rotation matrices, including the definition of the Euler angles, which are most commonly used to define the relative orientations of the crystal and the component. Parts of this chapter closely follow the lines of the excellent exposition of the topic by Olschewski [1033].

3.2 Euler angles and rotation matrices

3.2.1 Axes and rotation matrices

The orientation of a specimen or component in space is usually defined by an orthonormal basis (or global axes)

$$\mathbf{e}_i = \{\mathbf{e}_1, \mathbf{e}_2, \mathbf{e}_3\}, \tag{3.1}$$

Nickel Base Single Crystals Across Length Scales
https://doi.org/10.1016/B978-0-12-819357-0.00010-X

with the property $\mathbf{e}_i \cdot \mathbf{e}_j = \delta_{ij}$ (Kronecker symbol). The crystal axes

$$\mathbf{m}_i = \{\mathbf{m}_1, \mathbf{m}_2, \mathbf{m}_3\} \tag{3.2}$$

form another orthonormal basis and are usually attached to the crystallographic directions $\mathbf{m}_1 = [100]$, $\mathbf{m}_2 = [010]$, and $\mathbf{m}_3 = [001]$. The rotation tensor $\mathbf{Q}$ provides the connection

$$\mathbf{m}_i = \mathbf{Q} \cdot \mathbf{e}_i \tag{3.3}$$

between both axis systems. It admits the representation

$$\mathbf{Q} = \sum_{j=1}^{3} \mathbf{m}_j \otimes \mathbf{e}_j = \mathbf{m}_1 \otimes \mathbf{e}_1 + \mathbf{m}_2 \otimes \mathbf{e}_2 + \mathbf{m}_3 \otimes \mathbf{e}_3. \tag{3.4}$$

The representation (3.4) makes use of the dyadic operator $\otimes$, with which new tensors can be defined from vectors according to

$$(\mathbf{a} \otimes \mathbf{b}) \cdot \mathbf{c} = \mathbf{a} (\mathbf{b} \cdot \mathbf{c}) \tag{3.5}$$

for any vector $\mathbf{c}$. Use is often made of the Einstein summation convention for repeated indices to abbreviate the notations. For example, Eq. (3.4) can be rewritten as $\mathbf{Q} = \mathbf{m}_j \otimes \mathbf{e}_j$. With this convention, one checks that $\mathbf{Q} \cdot \mathbf{e}_i = (\mathbf{m}_j \otimes \mathbf{e}_j) \cdot \mathbf{e}_i = \mathbf{m}_j (\mathbf{e}_j \cdot \mathbf{e}_i) = \mathbf{m}_j \delta_{ij} = \mathbf{m}_i$. Introducing the transpose $\mathbf{Q}^T = \mathbf{e}_j \otimes \mathbf{m}_j$ of $\mathbf{Q}$, the rotation $\mathbf{Q}$ has the additional property

$$\mathbf{Q} \cdot \mathbf{Q}^T = \sum_{i=1}^{3} \mathbf{m}_i \otimes \mathbf{m}_i = \mathbf{Q}^T \cdot \mathbf{Q} = \sum_{i=1}^{3} \mathbf{e}_i \otimes \mathbf{e}_i = \mathbf{1}, \tag{3.6}$$

which is the second-order identity tensor. The inverse rotation $\mathbf{Q}^{-1}$ is thus identical with its transpose $\mathbf{Q}^T$ and provides the reverse mapping

$$\mathbf{e}_i = \mathbf{Q}^{-1} \cdot \mathbf{m}_i. \tag{3.7}$$

Due to the identities $\mathbf{m}_i = \sum_{j=1}^{3} (\mathbf{m}_i \cdot \mathbf{e}_j) \mathbf{e}_j$ and $\mathbf{e}_i = \sum_{i=1}^{3} (\mathbf{e}_i \cdot \mathbf{m}_j) \mathbf{m}_j$, the components Q_{ij} of $\mathbf{Q}$ in either the basis $\{\mathbf{m}_i\}$ or $\{\mathbf{e}_i\}$ can be expressed as a function of the direction cosines

$$Q_{ij} = \cos \alpha_{ij} = \mathbf{e}_i \cdot \mathbf{m}_j, \tag{3.8}$$

that is,

$$\begin{aligned} \mathbf{Q} &= \sum_{i,j=1}^{3} (\mathbf{e}_i \cdot \mathbf{m}_j) \mathbf{m}_i \otimes \mathbf{m}_j = \sum_{i,j=1}^{3} (\mathbf{e}_i \cdot \mathbf{m}_j) \mathbf{e}_i \otimes \mathbf{e}_j \\ &= \sum_{i,j=1}^{3} Q_{ij} \mathbf{m}_i \otimes \mathbf{m}_j = \sum_{i=1}^{3} Q_{ij} \mathbf{e}_i \otimes \mathbf{e}_j. \end{aligned} \tag{3.9}$$

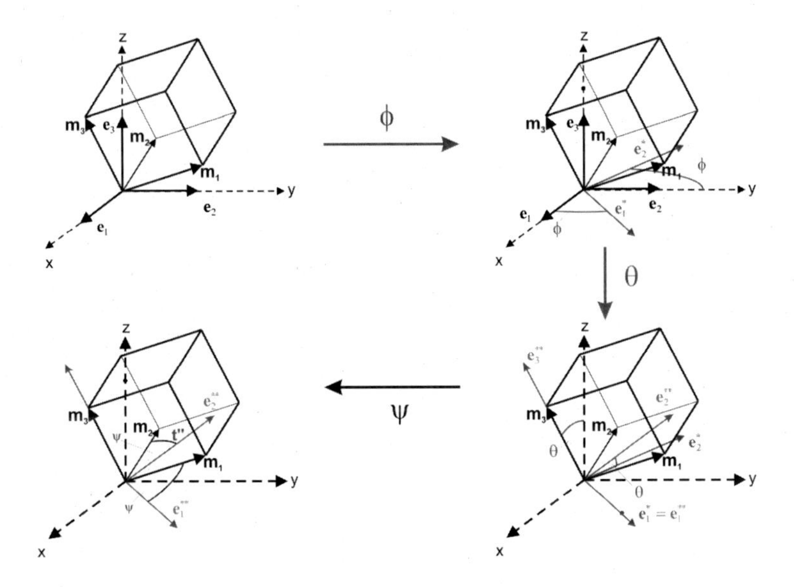

FIGURE 3.1 Euler angles defined by a succession of 3 rotations.

3.2.2 Euler angles

The use of Euler angles is a common way to characterize the relative orientations of two systems of axes. They can be related to a decomposition of the rotation $\mathbf{Q}$ into three elementary rotations, i.e., rotations around the axes of a coordinate system or its transformed (see Fig. 3.1). Note that different choices of the rotation axes are possible, which lead to different definitions of the angles. We describe hereafter the so-called ZXZ-convention, which after Bunge [172] is most commonly used in texture analysis. The first transformation $\mathbf{Q}_1$ is a rotation around the $\mathbf{e}_3 = Z$ axis of the angle ϕ such that $\{\mathbf{e}_i\} \rightarrow \{\mathbf{e}_i^*\}$, i.e., $\mathbf{e}_i^* = \mathbf{Q}_1 \cdot \mathbf{e}_i$, where

$$
\begin{aligned}
\mathbf{e}_1 &\rightarrow \mathbf{e}_1^* = \cos\phi\ \mathbf{e}_1 + \sin\phi\ \mathbf{e}_2, \\
\mathbf{e}_2 &\rightarrow \mathbf{e}_2^* = -\sin\phi\ \mathbf{e}_1 + \cos\phi\ \mathbf{e}_2, \\
\mathbf{e}_3 &\rightarrow \mathbf{e}_3^* = \mathbf{e}_3.
\end{aligned} \tag{3.10}
$$

The second transformation $\mathbf{Q}_2$ is the rotation of the angle θ around the new axis $\mathbf{e}_1^*$ such that $\{\mathbf{e}_i^*\} \rightarrow \{\mathbf{e}_i^{**}\}$, i.e., $\mathbf{e}_i^{**} = \mathbf{Q}_2 \cdot \mathbf{e}_i^*$, where

$$
\begin{aligned}
\mathbf{e}_1^* &\rightarrow \mathbf{e}_1^{**} = \mathbf{e}_1^*, \\
\mathbf{e}_2^* &\rightarrow \mathbf{e}_2^{**} = \cos\theta\ \mathbf{e}_2^* + \sin\theta\ \mathbf{e}_3^*, \\
\mathbf{e}_3^* &\rightarrow \mathbf{e}_3^{**} = -\sin\theta\ \mathbf{e}_2^* + \cos\theta\ \mathbf{e}_3^*.
\end{aligned} \tag{3.11}
$$

The third transformation $\mathbf{Q}_3$ is the rotation of angle ψ around the axis $\mathbf{e}_3^{**} = \mathbf{m}_3$ such that $\{\mathbf{e}_i^{**}\} \rightarrow \{\mathbf{m}_i\}$, i.e., $\mathbf{m}_i = \mathbf{Q}_3 \cdot \mathbf{e}_i^{**}$, where

$$\begin{aligned}
\mathbf{e}_1^{**} &\rightarrow \mathbf{m}_1 = \cos\psi\ \mathbf{e}_1^{**} + \sin\psi\ \mathbf{e}_2^{**}, \\
\mathbf{e}_2^{**} &\rightarrow \mathbf{m}_2 = -\sin\psi\ \mathbf{e}_1^{**} + \cos\psi\ \mathbf{e}_2^{**}, \\
\mathbf{e}_3^{**} &\rightarrow \mathbf{m}_3.
\end{aligned} \tag{3.12}$$

Introducing (3.10)–(3.12) into Eq. (3.8) yields the matrix form of the complete rotation $\mathbf{Q} = \mathbf{Q}_3 \cdot \mathbf{Q}_2 \cdot \mathbf{Q}_1$ in either the axes $\{\mathbf{e}_i\}$ or $\{\mathbf{m}_i\}$, namely

$$Q_{ij} = \begin{bmatrix} \cos\phi\ \cos\psi - \sin\phi\ \cos\theta\ \sin\psi & -\cos\phi\ \sin\psi - \sin\phi\ \cos\theta\ \cos\psi & \sin\phi\ \sin\theta \\ \sin\phi\ \cos\psi + \cos\phi\ \cos\theta\ \sin\psi & -\sin\phi\ \sin\psi + \cos\phi\ \cos\theta\ \cos\psi & -\cos\phi\ \sin\theta \\ \sin\theta\ \sin\psi & \sin\theta\ \cos\psi & \cos\theta \end{bmatrix}. \tag{3.13}$$

Note that the inverse rotation $\mathbf{Q}^{-1}$ corresponds to the reversed sequence of elementary rotations with the angles $-\psi \rightarrow -\theta \rightarrow -\phi$.

3.2.3 Transformation of tensor components

The practical evaluation of Hooke's law for an anisotropic material requires frequent transformations of the components of vectors or tensors between the crystal and global axes. Let us start with a vector $\mathbf{v}$ and its components in each basis

$$\mathbf{v} = \sum_{i=1}^{3} v_i\ \mathbf{e}_i = \sum_{k=1}^{3} v_k^*\ \mathbf{m}_k. \tag{3.14}$$

Applying relation (3.7), we obtain

$$\begin{aligned}
\mathbf{v} &= \sum_{i=1}^{3} v_i\ \mathbf{e}_i = \sum_{i=1}^{3} v_i\ \left(\mathbf{Q}^T \cdot \mathbf{m}_i\right) \\
&= \sum_{i=1}^{3} v_i \left(\sum_{l,k=1}^{3} Q_{lk}\ \mathbf{m}_k \otimes \mathbf{m}_l \right) \cdot \mathbf{m}_i = \sum_{i,k=1}^{3} v_i\ Q_{ik}\ \mathbf{m}_k,
\end{aligned} \tag{3.15}$$

from which the transformation rules for the vector components immediately follow:

$$v_k^* = \sum_{i=1}^{3} v_i\ Q_{ik}, \quad v_k = \sum_{i=1}^{3} v_i^*\ Q_{ki}. \tag{3.16}$$

The transformation rules for the components of second-order tensors like stresses or strains with their components

$$\boldsymbol{\sigma} = \sum_{i,j=1}^{3} \sigma_{ij}\ \mathbf{e}_i \otimes \mathbf{e}_j = \sum_{i,j=1}^{3} \sigma^*_{ij}\ \mathbf{m}_i \otimes \mathbf{m}_j \tag{3.17}$$

can be obtained in the same way. It can be checked that

$$\sigma^*_{kl} = \sum_{i,j=1}^{3} \sigma_{ij}\ Q_{ik}\ Q_{jl}\ , \quad \sigma_{kl} = \sum_{i,j=1}^{3} \sigma^*_{ij}\ Q_{ki}\ Q_{lj}. \tag{3.18}$$

The case of fourth-order tensors like the elastic stiffness or the compliance is also of practical importance. The respective components of the fourth-order tensor **C**, namely

$$\mathbf{C} = \sum_{i,j,k,l=1}^{3} C_{ijkl}\ \mathbf{e}_i \otimes \mathbf{e}_j \otimes \mathbf{e}_k \otimes \mathbf{e}_l = \sum_{i,j,k,l=1}^{3} C^*_{ijkl}\ \mathbf{m}_i \otimes \mathbf{m}_j \otimes \mathbf{m}_k \otimes \mathbf{m}_l, \tag{3.19}$$

can be transformed according to

$$\begin{aligned} C^*_{mnpq} &= \sum_{i,j,k,l=1}^{3} C_{ijkl}\ Q_{im}\ Q_{jn}\ Q_{kp}\ Q_{lq}\ , \\ C_{mnpq} &= \sum_{i,j,k,l=1}^{3} C^*_{ijkl}\ Q_{mi}\ Q_{nj}\ Q_{pk}\ Q_{ql}\ . \end{aligned} \tag{3.20}$$

3.3 Representations of the elastic stiffness for cubic symmetry

3.3.1 Elastic stiffness

Hooke's law relates the stress $\boldsymbol{\sigma}$ to the strain tensor $\boldsymbol{\varepsilon}$ in linear elasticity (the case of infinitesimal transformations). It is primarily formulated in its tensorial form with the fourth-order stiffness tensor **C** or its component form,

$$\boldsymbol{\sigma} = \mathbf{C} : \boldsymbol{\varepsilon} \quad \text{or} \quad \sigma_{ij} = C_{ijkl}\ \varepsilon_{kl}. \tag{3.21}$$

The later form makes use of the summation convention and is valid in any orthonormal basis. Voigt's notation was introduced to rewrite this relationship in the matrix form. The components of second-order tensors are thereby rewritten as vector components, i.e., $\sigma_{ij} \to \sigma_I$, and fourth-order tensors are denoted by matrices, i.e., $C_{ijkl} \to c_{IJ}$. It allows us to take advantage of the symmetries $\sigma_{ij} = \sigma_{ji}$ and $\varepsilon_{kl} = \varepsilon_{lk}$, thereby reducing the number of variables. Accordingly, a connection between the tensor and vector components is introduced. Clearly, the order of shear components is arbitrary. The following order is used by Abaqus [3] (see Table 3.1):

TABLE 3.1 Relation between tensor indices and Voigt's notation (Abaqus convention).

(i, j)	(1, 1)	(2, 2)	(3, 3)	(1, 2)	(1, 3)	(2, 3)
I	1	2	3	4	5	6

Due to the symmetries $C_{ijkl} = C_{klij} = C_{jikl} = C_{ijlk}$, the stiffness tensor depends at most on 21 independent constants. In the case of cubic symmetry, this number is reduced to three independent constants [461,583].

It is most convenient to formulate Hooke's law in the crystal axes since the three axes are equivalent and the components C_{iikl} $(k \neq l)$ (without summation) corresponding to c_{IJ} $(I \leq 3, J > 3)$ and C_{ijkl} $(i \neq j, k \neq l, \{i, j\} \neq \{k, l\})$ corresponding to c_{IJ} $(I \neq J, I > 3, J > 3)$ vanish. The only remaining constants are

$$\begin{aligned} c_{11} &= C_{1111} = C_{2222} = C_{3333}, \\ c_{12} &= C_{1122} = C_{1133} = C_{2233}, \\ c_{44} &= C_{1212} = C_{1313} = C_{2323}. \end{aligned} \tag{3.22}$$

It is worth mentioning here that the notation c_{44} is also in use for $c_{44} = 2C_{1212}$ in the engineering literature. In accordance with the notations (3.22), Hooke's law in the matrix form reads

$$\begin{bmatrix} \sigma_{11} \\ \sigma_{22} \\ \sigma_{33} \\ \sigma_{12} \\ \sigma_{13} \\ \sigma_{23} \end{bmatrix} = \begin{bmatrix} c_{11} & c_{12} & c_{12} & 0 & 0 & 0 \\ c_{12} & c_{11} & c_{12} & 0 & 0 & 0 \\ c_{12} & c_{12} & c_{11} & 0 & 0 & 0 \\ 0 & 0 & 0 & c_{44} & 0 & 0 \\ 0 & 0 & 0 & 0 & c_{44} & 0 \\ 0 & 0 & 0 & 0 & 0 & c_{44} \end{bmatrix} \begin{bmatrix} \varepsilon_{11} \\ \varepsilon_{22} \\ \varepsilon_{33} \\ 2\varepsilon_{12} \\ 2\varepsilon_{13} \\ 2\varepsilon_{23} \end{bmatrix} \tag{3.23}$$

in the crystal axes.

3.3.2 Elastic compliance

The fourth-order tensor elastic compliance $\mathbf{S}$ with its components S_{ijkl} provides the inverse relationship to (3.21), namely

$$\boldsymbol{\varepsilon} = \mathbf{S} : \boldsymbol{\sigma}, \quad \varepsilon_{ij} = S_{ijkl}\sigma_{kl}. \tag{3.24}$$

Following the previous section, the components S_{ijkl} can be rewritten in the matrix form s_{IJ} with Voigt's notation. In the crystal axes, the only non-vanishing components are $S_{iiii} = s_{11}$, $S_{iijj} = s_{12}$ $(i \neq j)$, and $S_{ijij} = s_{44}$ $(i \neq j)$

(in all cases without summation). The following relationships between stiffness and compliance components can be derived after some algebra:

$$s_{11} = \frac{c_{11}+c_{12}}{(c_{11}+2c_{12})(c_{11}-c_{12})}, \quad s_{12} = \frac{-c_{12}}{(c_{11}+2c_{12})(c_{11}-c_{12})}, \quad s_{44} = \frac{1}{4c_{44}}, \tag{3.25}$$

$$c_{11} = \frac{s_{11}+s_{12}}{(s_{11}+2s_{12})(s_{11}-s_{12})}, \quad c_{12} = \frac{-s_{12}}{(s_{11}+2s_{12})(s_{11}-s_{12})}, \quad c_{44} = \frac{1}{4s_{44}}. \tag{3.26}$$

Here again, some care needs to be exercised when referring to published data since the notation $s_{44} = 1/c_{44}$ is also in use in the literature.

3.3.3 Spectral representation

The decomposition of the elasticity stiffness tensor in orthonormal modes can be traced back to the historic work of Kelvin [1427]. It first received little attention and was only rediscovered in the late 1980s apparently independently by Walpole [1522,1523] and Mehrabadi and Cowin [928] (see also [1380]). The use of this decomposition leads to a partition of the elastic energy in a hydrostatic mode and several deviatoric modes and it allows for an elegant derivation of various theoretical results as well as the formulation of elastic and inelastic models of a given symmetry class. The following compact presentation closely follows [1287] (see also [892] for an application to anisotropic creep modeling).

It starts from the recognition that the eigenvalue problem $\mathbf{C} : \mathbf{N} = \lambda \mathbf{N}$ has at most 6 distinct real and positive eigenvalues λ_i and 6 orthogonal normalized eigentensors (Kelvin modes) $\{\mathbf{N}_i\}, i = 1, 2, \ldots, 6$ satisfying $\mathbf{N}_i : \mathbf{N}_j = \delta_{ij}$. Let K be the number of distinct eigenvalues. The eigenvalues are renumbered in such a way that the first K eigenvalues are distinct. Let now K_m be the multiplicity of the eigenvalue λ_m, $m = 1, 2, \ldots, K$, so that $\sum_{m=1}^{K} K_m = 6$. For each distinct eigenvalue $m = 1, 2, \ldots, K$, one can construct the operator

$$\mathbf{P}_m = \sum_{j \in \mathcal{K}_m} \mathbf{N}_j \otimes \mathbf{N}_j , \tag{3.27}$$

where $\mathcal{K}_m$ is the set of indices associated to the eigenvalue λ_m. It can be shown that $\sum_{m=1}^{K} \mathbf{P}_m = \mathbf{I}$, where $\mathbf{I}$ denotes the symmetric fourth-order identity tensor, and furthermore that $\mathbf{P}_m : \mathbf{P}_n = \delta_{mn}\mathbf{P}_m$ and $\mathbf{P}_m :: \mathbf{P}_n = \delta_{mn} K_m$ (in both cases without summation). From these definitions and properties, it follows that each elastic stiffness tensor $\mathbf{C}$ has a spectral decomposition of the form

$$\mathbf{C} = \sum_{m=1}^{K} \lambda_m \mathbf{P}_m , \tag{3.28}$$

from which the representation of the compliance, namely

$$\mathbf{S} = \sum_{m=1}^{K} \lambda_m^{-1} \mathbf{P}_m, \tag{3.29}$$

can be inferred since $\left(\sum_{m=1}^{K} \lambda_m \mathbf{P}_m\right) : \left(\sum_{n=1}^{K} \lambda_n^{-1} \mathbf{P}_n\right) = \mathbf{I}$. In general, the eigentensors $\mathbf{N}_i$ depend on $\mathbf{C}$. However, the eigenvalue problem of the cubic symmetry case has a particularly simple solution. Three distinct eigenvalues $\lambda_m, m = 1, 2, 3$ exist with respective multiplicity 1, 2, and 3. The corresponding eigenmodes $\mathbf{N}_i$ and hence the associated operators $\mathbf{P}_m$ are universal, i.e., they do not depend on the elasticity constants c_{11}, c_{12}, c_{44}. These modes respectively correspond to dilatation, deviatoric tetragonal distortion in the crystal axes, and pure shear (see Fig. 3.2). The general decomposition (3.28) in particular leads to

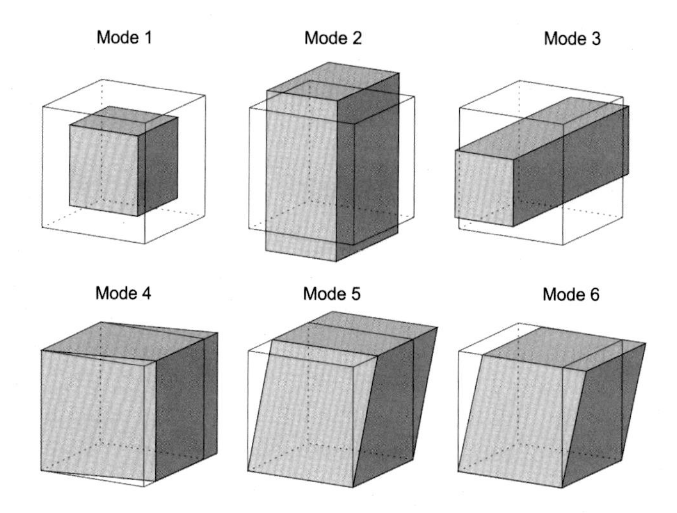

FIGURE 3.2 The six Kelvin deformation modes. Taken from Fig. 1 in [892].

$$\mathbf{C} = \lambda_1 \mathbf{P}_1 + \lambda_2 \mathbf{P}_2 + \lambda_3 \mathbf{P}_3 = 3\kappa \mathbf{P}_1 + 2\mu_2 \mathbf{P}_2 + 2\mu_3 \mathbf{P}_3, \tag{3.30}$$

while for the compliance one has

$$\mathbf{S} = \lambda_1^{-1} \mathbf{P}_1 + \lambda_2^{-1} \mathbf{P}_2 + \lambda_3^{-1} \mathbf{P}_3 = \frac{1}{3\kappa} \mathbf{P}_1 + \frac{1}{2\mu_2} \mathbf{P}_2 + \frac{1}{2\mu_3} \mathbf{P}_3. \tag{3.31}$$

The 6 eigentensors of the cubic symmetry case can be found, for example, in [1287]. The Voigt's representation of the resulting operators $\mathbf{P}_m$ is

$$\mathbf{P}_1^V = \begin{pmatrix} \frac{1}{3} & \frac{1}{3} & \frac{1}{3} & 0 & 0 & 0 \\ \frac{1}{3} & \frac{1}{3} & \frac{1}{3} & 0 & 0 & 0 \\ \frac{1}{3} & \frac{1}{3} & \frac{1}{3} & 0 & 0 & 0 \\ 0 & 0 & 0 & 0 & 0 & 0 \\ 0 & 0 & 0 & 0 & 0 & 0 \\ 0 & 0 & 0 & 0 & 0 & 0 \end{pmatrix}, \quad \mathbf{P}_2^V = \begin{pmatrix} \frac{2}{3} & -\frac{1}{3} & -\frac{1}{3} & 0 & 0 & 0 \\ -\frac{1}{3} & \frac{2}{3} & -\frac{1}{3} & 0 & 0 & 0 \\ -\frac{1}{3} & -\frac{1}{3} & \frac{2}{3} & 0 & 0 & 0 \\ 0 & 0 & 0 & 0 & 0 & 0 \\ 0 & 0 & 0 & 0 & 0 & 0 \\ 0 & 0 & 0 & 0 & 0 & 0 \end{pmatrix},$$

$$\mathbf{P}_3^V = \begin{pmatrix} 0 & 0 & 0 & 0 & 0 & 0 \\ 0 & 0 & 0 & 0 & 0 & 0 \\ 0 & 0 & 0 & 0 & 0 & 0 \\ 0 & 0 & 0 & \frac{1}{2} & 0 & 0 \\ 0 & 0 & 0 & 0 & \frac{1}{2} & 0 \\ 0 & 0 & 0 & 0 & 0 & \frac{1}{2} \end{pmatrix}. \tag{3.32}$$

The component κ is the bulk modulus, while the components μ_2 and μ_3 with multiplicity 2 and 3, respectively, correspond to two different shear moduli. The stiffness components and the components of the spectral decomposition are related by

$$c_{11} = \kappa + \frac{4\mu_2}{3}, \quad c_{12} = \kappa - \frac{2\mu_2}{3}, \quad c_{44} = \mu_3, \tag{3.33}$$

and

$$\kappa = \frac{c_{11} + 2c_{12}}{3}, \quad \mu_2 = \frac{c_{11} - c_{12}}{2}, \quad \mu_3 = c_{44}. \tag{3.34}$$

The material is isotropic if the Zener ratio defined by $A = \mu_3/\mu_2 = 2c_{44}/(c_{11} - c_{12})$ is equal to 1. Finally, the stored elastic energy $\mathcal{E}(\boldsymbol{\varepsilon}) = 1/2\, \boldsymbol{\varepsilon} : \mathbf{C} : \boldsymbol{\varepsilon}$ is also additively decomposed in the contributions of three deformation eigenmodes, namely

$$\begin{aligned} \mathcal{E}(\varepsilon_{ij}) = {} & \frac{\kappa}{2}(\varepsilon_{11} + \varepsilon_{22} + \varepsilon_{33})^2 \\ & + \frac{\mu_2}{3}\left((\varepsilon_{11} - \varepsilon_{22})^2 + (\varepsilon_{11} - \varepsilon_{33})^2 + (\varepsilon_{33} - \varepsilon_{22})^2\right) \\ & + 2\mu_3\left(\varepsilon_{21}^2 + \varepsilon_{31}^2 + \varepsilon_{32}^2\right). \end{aligned} \tag{3.35}$$

Since the deformation eigenmodes are linearly independent, the condition of stability $\mathcal{E}(\varepsilon_{ij}) > 0$, $\forall \varepsilon_{ij} \neq 0$ is satisfied if $\kappa > 0$, $\mu_2 > 0$, and $\mu_3 > 0$.

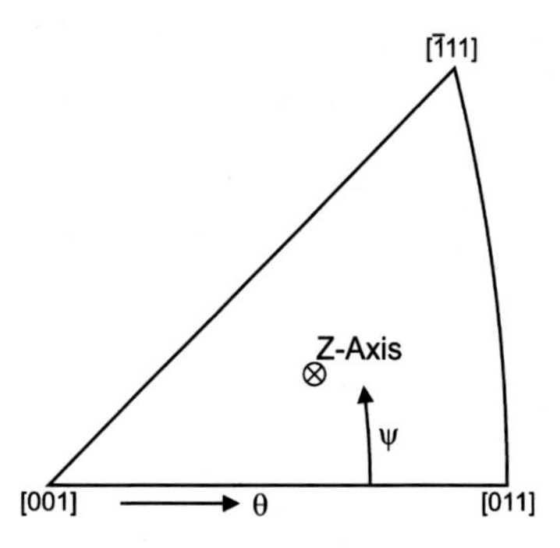

FIGURE 3.3 Tension axis in the standard stereographic triangle.

3.4 Application to the tension test

3.4.1 General relations

In arbitrary axes, the relations (3.21) imply that extension and shearing are coupled since components of the type C_{iikl} for which $k \neq l$ (without summation) do not necessarily vanish. These components relate a shear strain to an extensional stress in the direction of one of the axes. A pure tension stress applied in a direction $Z \equiv \mathbf{e}_3$ (see Fig. 3.3) corresponds to the stress tensor $\boldsymbol{\sigma} = \sigma \mathbf{e}_3 \otimes \mathbf{e}_3$. If Z is not identical with a crystallographic direction with privileged symmetries, the resulting strain tensor contains nonvanishing shear components in addition to the extensional strain ε_{33}. The ratio $E = \sigma/\varepsilon_{33}$ is then termed the free Young's modulus [594]. If a stress system is applied to suppress the shear strain components, the resulting ratio σ/ε'_{33} corresponds to the pure Young's modulus. Applying Eq. (3.18), we obtain the stress components in the crystal axes

$$\left\{\sigma^*_{ij}\right\} = \sigma \begin{pmatrix} \sin^2(\theta)\sin^2(\psi) & \sin^2(\theta)\sin(\psi)\cos(\psi) & \sin(\theta)\cos(\theta)\sin(\psi) \\ \sin^2(\theta)\sin(\psi)\cos(\psi) & \sin^2(\theta)\cos^2(\psi) & \sin(\theta)\cos(\theta)\cos(\psi) \\ \sin(\theta)\cos(\theta)\sin(\psi) & \sin(\theta)\cos(\theta)\cos(\psi) & \cos^2(\theta) \end{pmatrix}, \tag{3.36}$$

for which relationship (3.24) written in the crystal axes can be applied to obtain the strain components. Rotating it back to the global axes, we eventually obtain the axial strain $\varepsilon_{33}(\theta, \psi)$ and the free Young's modulus σ/ε_{33}. Alternatively, the free Young's modulus is also given by the inverse $1/S_{3333}$ of the compliance component S_{3333} in the global axes, which can be obtained by transformation of the compliance tensor in the crystal axes following the rule (3.20). After some algebra it is found that

$$E(\theta,\psi) = \frac{1}{S_{3333}}$$
$$= \frac{1}{s_{11} - 2(s_{11} - s_{12} - 2s_{44})\left(\sin^4(\theta)\sin^2(\psi)\cos^2(\psi) + \sin^2(\theta)\cos^2(\theta)\right)}. \tag{3.37}$$

Two different representations of the variations of Young's modulus with the orientation of the tensile axis are plotted in Figs. 3.4 and 3.5 for the alloy CMSX-4 at room temperature. The constants are given in Table 3.2. The $\langle 111 \rangle$ orientation is almost three times stiffer than the $\langle 001 \rangle$ orientation, in which $E(\theta, \psi)$ reaches its minimum value.

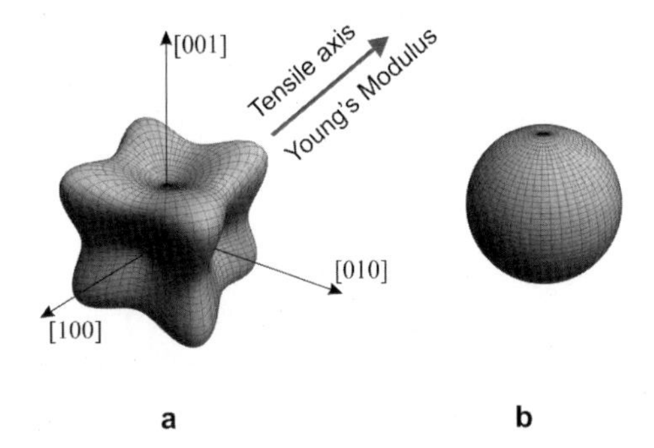

FIGURE 3.4 (a) 3D-representation of Young's modulus for a typical single crystal of superalloy (CMSX-4) as a function of the orientation of the tension axis. (b) Case of an isotropic material.

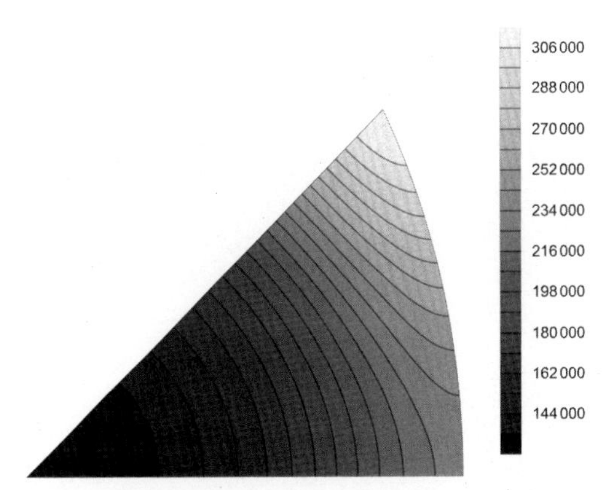

FIGURE 3.5 Young's modulus of the alloy CMSX-4 at room temperature (in MPa) in the standard triangle of the stereographic projection.

The lateral contraction is in general not constant around the circumference of the specimen. Hence, the extensometer location must be specified by the angle γ made by the direction $\mathbf{e}_r = \cos\gamma\ \mathbf{e}_1 + \sin\mathbf{e}_2$ with the X-axis (see Fig. 3.6). The measured contraction $\varepsilon_{rr} = \mathbf{e}_r \cdot \boldsymbol{\varepsilon} \cdot \mathbf{e}_r$ and Poisson's ratio $\nu(\gamma, \theta, \psi) = -\varepsilon_{rr}/\varepsilon_{33}$ depend on three angles.

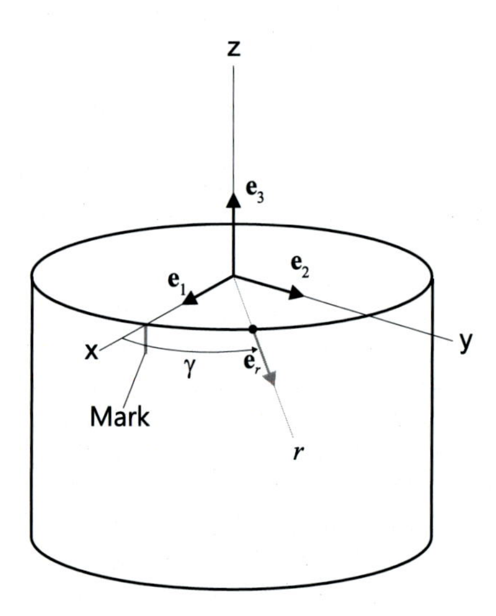

FIGURE 3.6 Definition of the angle γ for the lateral extensometer location.

3.4.2 Case of a ⟨001⟩ specimen

For definiteness, we consider a [001] specimen, for which the tension direction along the $\mathbf{e}_3$ axis coincides with the crystal axis $\mathbf{m}_3 = [001]$ and the second Euler angle θ is zero. Usually, the other angles φ and ψ are also taken to be zero, i.e., $\mathbf{e}_1 = \mathbf{m}_1$ (see Fig. 3.7).

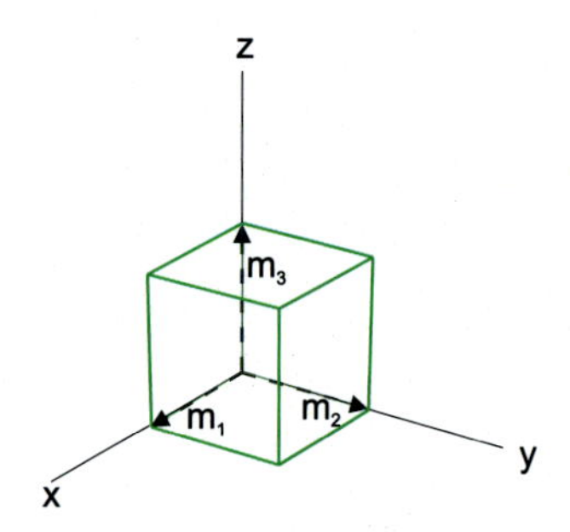

FIGURE 3.7 Tension test along the [001] direction.

The Young's modulus in the ⟨001⟩ direction immediately follows from the general result (3.37) as

$$E_{001} = \frac{1}{s_{11}} = \frac{(c_{11} - c_{12})(c_{11} + 2c_{12})}{(c_{11} + c_{12})}, \tag{3.38}$$

while the lateral contraction is constant around the circumference so that Poisson's ratio becomes

$$\nu_{001} = -\frac{s_{12}}{s_{11}} = \frac{c_{12}}{c_{11} + c_{12}}. \tag{3.39}$$

3.4.3 Case of a ⟨111⟩ specimen

In the particular case of a [111] specimen, the extension takes place along $\mathbf{e}_3 = 1/\sqrt{3}\,(\mathbf{m}_1 + \mathbf{m}_2 + \mathbf{m}_3)$, while $\theta = \arccos\left(1/\sqrt{3}\right)$ and $\psi = \pi/4$. The first angle φ can be arbitrarily set to zero, so that $\mathbf{e}_1 = 1/\sqrt{2}\,(\mathbf{m}_1 - \mathbf{m}_2)$ (see Fig. 3.8).

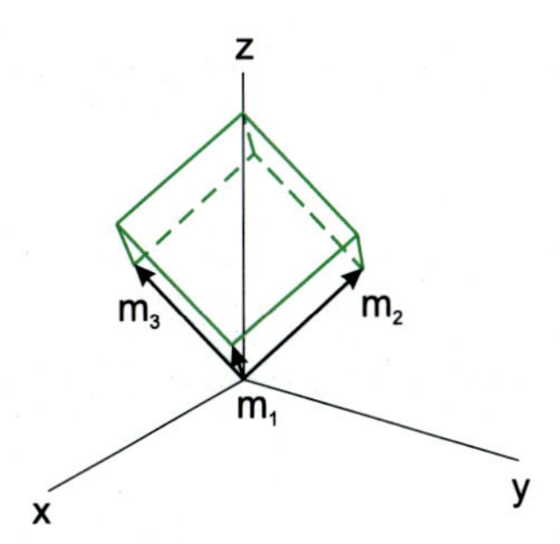

FIGURE 3.8 Tension test along the [111] direction.

The Young's modulus is then

$$E_{111} = \frac{3}{s_{11} + 2s_{12} + 4s_{44}} = \frac{3c_{44}(c_{11} + 2c_{12})}{c_{11} + 2c_{12} + c_{44}}, \tag{3.40}$$

and Poisson's ratio is also constant around the circumference of the specimen, i.e.,

$$\nu_{111} = -\frac{s_{11} + 2s_{12} - 2s_{44}}{s_{11} + 2s_{12} + 4s_{44}} = \frac{c_{11} + 2c_{12} - 2c_{44}}{2(c_{11} + 2c_{12} + c_{44})}. \tag{3.41}$$

3.4.4 Case of a ⟨011⟩ specimen

We consider, for example, a [011] specimen for which the elongation occurs along $\mathbf{e}_3 = 1/\sqrt{2}\,(\mathbf{m}_2 + \mathbf{m}_3)$, while $\theta = \pi/4$ and $\psi = 0$. The angle φ can be arbitrarily set to zero, so that $\mathbf{e}_1 = \mathbf{m}_1$ (see Fig. 3.9).

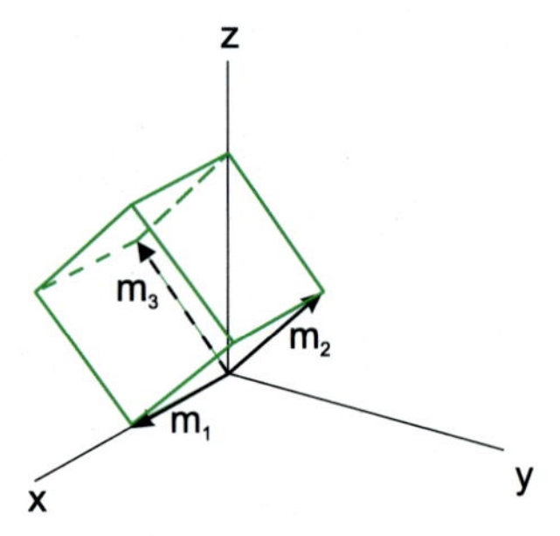

FIGURE 3.9 Tension test along the [011] direction.

The Young's modulus is then

$$E_{011} = \frac{2}{s_{11} + s_{12} + 2s_{44}} = \frac{4c_{44}(c_{11} - c_{12})(c_{11} + 2c_{12})}{c_{11}^2 + c_{11}(c_{12} + 2c_{44}) - 2c_{12}^2}, \tag{3.42}$$

while the Poisson's ratio is no longer constant around the circumference. It depends on the angle γ between $\mathbf{e}_1$ and $\mathbf{e}_r$ (see Fig. 3.6) as

$$\begin{aligned}\nu_{011}(\gamma) &= -\frac{\sin^2(\gamma)(s_{11} - 2s_{44}) + s_{12}\left(\cos^2(\gamma) + 1\right)}{s_{11} + s_{12} + 2s_{44}} \\ &= \frac{\sin^2(\gamma)\left(c_{11}^2 + c_{11}(c_{12} - 2c_{44}) - 2c_{12}(c_{12} + c_{44})\right) + 2c_{12}c_{44}\cos^2(\gamma) + 2c_{12}c_{44}}{c_{11}^2 + c_{11}(c_{12} + 2c_{44}) - 2c_{12}^2}.\end{aligned} \tag{3.43}$$

The variations of ν_{011} around the circumference, together with the values ν_{001} and ν_{111}, are plotted in Fig. 3.10 for the alloy CMSX-4 at room temperature. In particular, it is seen that directions exist with a negative Poisson's ratio, which correspond to local lateral dilatation under tension, a phenomenon that has been often reported for cubic metals [84].

3.5 The shear coefficients

The (free) shear modulus provides the shear strain $\gamma = 2\,\mathbf{n}\cdot\boldsymbol{\varepsilon}\cdot\mathbf{t}$ response to a tangential force of direction $\mathbf{t}$ acting on the plane of normal $\mathbf{n} \perp \mathbf{t}$, provided the other resulting strain components are not constrained. This state is described by the stress tensor $\boldsymbol{\sigma} = \tau(\mathbf{t}\otimes\mathbf{n} + \mathbf{n}\otimes\mathbf{t})$. For example, $\mu = 1/\left(4S_{ijij}\right)$ $(i \neq j)$ corresponds to a shear stress acting on the plane normal to $\mathbf{e}_j$ in the direction of $\mathbf{e}_i$ or vice versa. For an anisotropic material, the shear modulus $\mu = \tau/\gamma$ depends in general on both directions $\mathbf{n}$ and $\mathbf{t}$ and is therefore not easily experimentally accessible. In the case of cubic symmetry, the shear modulus is independent of the force direction $\mathbf{t}$ if the

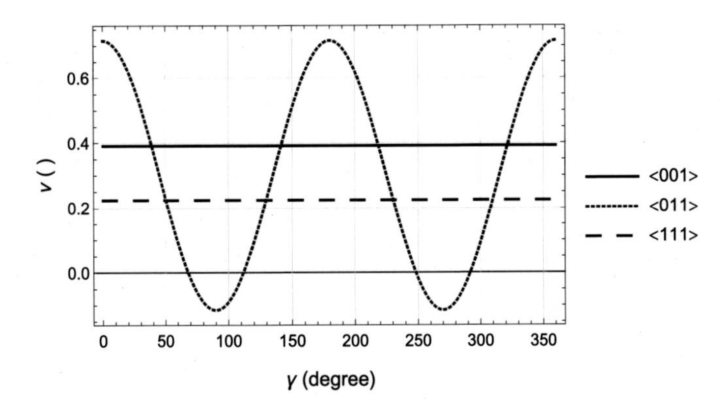

FIGURE 3.10 Poisson's ratios along the circumference.

shear plane is either {001} or {111} and

$$\mu_{001} = \frac{1}{4s_{44}} = c_{44} = \mu_3, \tag{3.44}$$

while

$$\mu_{111} = \frac{3}{4s_{11} - 4s_{12} + 4s_{44}} = \frac{3c_{44}(c_{11} - c_{12})}{c_{11} - c_{12} + 4c_{44}} = \frac{3\mu_2\mu_3}{\mu_2 + 2\mu_3}. \tag{3.45}$$

As a consequence, the interpretation of static torsion tests is not straightforward if the specimen axis is arbitrary [594]. Olschewski [1032] proposed an approximate solution for a circular hollow and thin cylinder of section $\mathcal{A}$ (inner radius a and outer radius b) and axis along an arbitrary direction Z. The applied torque is denoted by $M_T = \int_{\mathcal{A}} (x\sigma_{23} - y\sigma_{13})\, dA$. An evaluation is possible under the assumption of a pure torsion stress system: Then the relative displacements of two points on the outer surface such that $\mathbf{x}_A = b\,\mathbf{e}_r$ (see Fig. 3.6) and $\mathbf{x}_B = b\,\mathbf{e}_r + h\,\mathbf{e}_z$ are given by [1032,1033]

$$\begin{aligned}
\Delta u_1 &= u_{B1} - u_{A1} = -\frac{M_T}{I_P} 2b\,(S_{1313} + S_{2323})\, h \sin\gamma, \\
\Delta u_2 &= u_{B2} - u_{A2} = \frac{M_T}{I_P} 2b\,(S_{1313} + S_{2323})\, h \cos\gamma,
\end{aligned} \tag{3.46}$$

where $I_P = \int_{\mathcal{A}} r^2 dA = \pi/2\,(b^4 - a^4)$. The relative tangential displacement $\Delta u_t = \left(\Delta u_1^2 + \Delta u_2^2\right)^{1/2}$ is related to the relative rotation $\Delta\theta$ of the same two points by $\Delta\theta = \Delta u_t / b$, which is proportional to the rate of twist $D = \Delta\theta / h$. The apparent shear modulus defined by $G = M_T/(D I_P)$ is finally obtained as

$$G = \frac{1}{4\,(s_{11} - s_{12} - 2s_{44})\left(\sin^4(\theta)\sin^2(\psi)\cos^2(\psi) + \sin^2(\theta)\cos^2(\theta)\right) + 4s_{44}} \tag{3.47}$$

and only depends on the direction of the specimen axis. In the case of $\langle 001 \rangle$ or $\langle 111 \rangle$ specimens, the solution is exact and one retrieves relations (3.44) and (3.45).

3.6 Determination of the elastic stiffness components

3.6.1 Static tests

An inspection of Eq. (3.37) readily shows that measurements of the Young's moduli of three differently oriented specimens are not sufficient to determine all constants since at most s_{11} and $s_{11} - s_{12} - s_{44}/2$ can be unambiguously derived from such measurements. At least an additional independent measurement is required. As an example, we assume that tension test data are available for $\langle 001 \rangle$ and $\langle 111 \rangle$ specimen, as well as measurement of lateral contraction of the $\langle 001 \rangle$ specimen. From Eqs. (3.38)–(3.40), the compliance coefficients result as

$$\begin{aligned} s_{11} &= \frac{1}{E_{001}}, \\ s_{12} &= -\frac{\nu_{001}}{E_{001}}, \\ s_{44} &= -\frac{-3E_{001} - 2E_{111}\nu_{001} + E_{111}}{E_{001}}. \end{aligned} \tag{3.48}$$

Due to the dependence of Poisson's ratio on the exact placement of the lateral extensometer when a crystal deviates from the perfect orientation $\langle 001 \rangle$ and lower precision of lateral contraction measurements, it is rather recommended to perform torsion tests to complete the data. Fig. 3.11 shows an experimental device for torsion testing at elevated temperature. As an example, we assume that test results from tension of $\langle 001 \rangle$ and $\langle 111 \rangle$ specimen and torsion of $\langle 001 \rangle$ specimen are available. From Eqs. (3.38), (3.40), and (3.47), the compliance coefficients result as

$$\begin{aligned} s_{11} &= \frac{1}{E_{001}}, \\ s_{12} &= -\frac{E_{001}E_{111} - 3E_{001}\mu_{001} + E_{111}\mu_{001}}{2E_{001}E_{111}\mu_{001}}, \\ s_{44} &= \frac{1}{4\mu_{001}}. \end{aligned} \tag{3.49}$$

FIGURE 3.11 Torsion specimen heated by induction with extensometer.

3.6.2 Sonic resonance method (SR)

3.6.2.1 Testing procedure

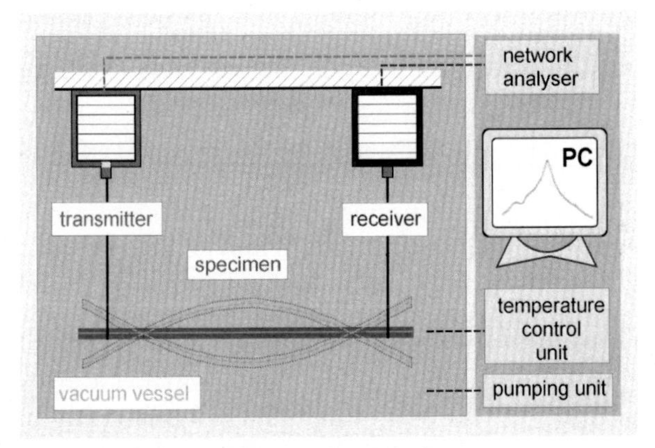

FIGURE 3.12 Principle of the resonance measurements.

The principle of the sonic resonance (SR) method, which was developed by Förster [459], is also described with more details in the ASTM E1875 standard [657]. The method has been widely applied to single crystals, directionally solidified and conventionally casted (CC) Ni-base superalloys [419,575,1324]. As shown schematically in Fig. 3.12, a sample is suspended with specially prepared carbon fiber threads between a piezoelectric transmitter and a receiver. The suspension system should ensure free vibrations and provide a reliable mechanical coupling between signal transmission and reception. A sinusoidal voltage signal is applied to the transmitter. The amplitude of the mechanical vibration is detected by the receiver, which converts the mechanical signal into an electrical one. The amplitude spectrum of the specimen, which reveals several sharp peaks, is obtained by continuously varying the exiting frequency. Due to oxidation, high-temperature measurements are usually performed in vacuum. An experimental device suitable for high temperature measurements is shown in Fig. 3.13.

FIGURE 3.13 View of the Elastroton 2000 (company HTM Reetz, Berlin), which allows for measurements between 20°C and 1900°C at frequencies between 0.1 and 100 kHz.

An example of measured amplitude spectrum is shown in Fig. 3.14. The specimen length is typically about 80–100 mm and the resonance frequencies usually lie between 1 and 50 kHz.

3.6.2.2 *Evaluation of the elastic constants*

In the case of isotropic materials, the eigenfrequencies can be estimated with good accuracy by Timoshenko extended beam theory, and closed-form solutions are available to estimate the elastic constants from the resonance peaks [657]. Since shear and bending modes are generally coupled for anisotropic materials, sufficiently accurate analytical estimates of the eigenfrequencies are not known for arbitrary oriented crystals. The evaluation can be performed by an inverse finite element analysis. The eigenfrequencies ω_k are solution of the problem

$$(\omega_k \mathbf{M} + \mathbf{K}) \cdot \mathbf{u}_k = \mathbf{0}, \quad k = 1, 2, \ldots, \tag{3.50}$$

where $\mathbf{M}$ and $\mathbf{K}$ are the inertia and stiffness matrices of the FE model and $\mathbf{u}_k$ are the kth eigenmodes. The eigenfrequencies depend on the elastic constants of the alloy, the mass density, and on the geometry of the specimen. The unknown elastic constants c_{11}, c_{12}, and c_{44} are determined by

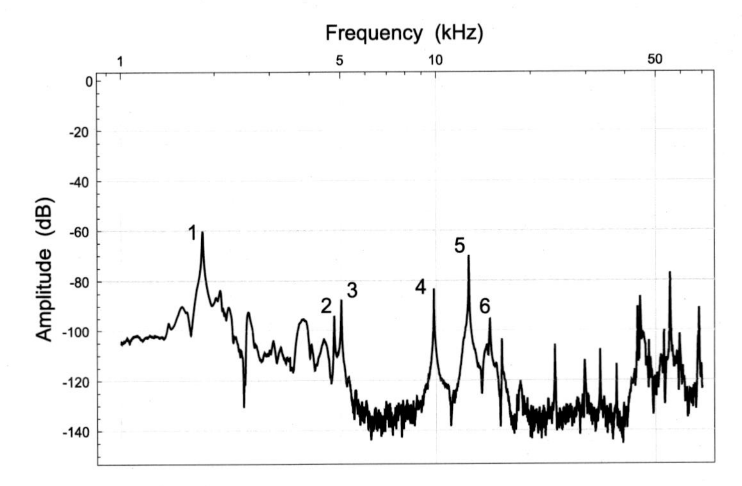

FIGURE 3.14 Example of measured spectrum for a ⟨001⟩ rectangular specimen of the alloy CMSX-4 at RT. The six first resonance peaks are marked and the corresponding modes are shown in Fig. 3.15.

minimizing the square of the deviation between measured and calculated frequencies, i.e.,

$$R\left(c_{11}, c_{12}, c_{44}\right) = \sum_{i=1}^{N_s} \sum_{k=1}^{M_i} \left(\frac{\omega_{k,i}^{FE} - \omega_{k,i}^{exp}}{\omega_{k,i}^{exp}} \right)^2, \tag{3.51}$$

where N_s is the number of tested specimens, M_i the number of resonance modes for specimen i, and $\omega_{k,i}$ the kth resonance frequency of the ith specimen. The computed resonance modes of a rectangular ⟨001⟩ specimen of the alloy CMSX-4 are shown in Fig. 3.15. They include flexural and torsional deformation modes of increasing order. Note that it is essential to verify that torsion modes are included in the testing data.

3.6.3 Resonant ultrasound spectroscopy (RUS)

The principle of the resonant ultrasound spectroscopy (RUS) is similar to the SR method previously exposed. A presentation of the method can be found in [945]. However, the specimens have much smaller dimensions (typically rectangular with edge lengths <10 mm) and the resonance frequencies are higher (usually > 50 kHz). The piezoelectric elements are in physical contact with the specimen. The advantage of the method is that only one specimen is needed. There is also no closed-form solution for the eigenfrequencies, and approximate numerical methods are required to estimate them for given elastic constants. As a consequence, the identification of the elastic constants from the measured resonance frequencies is

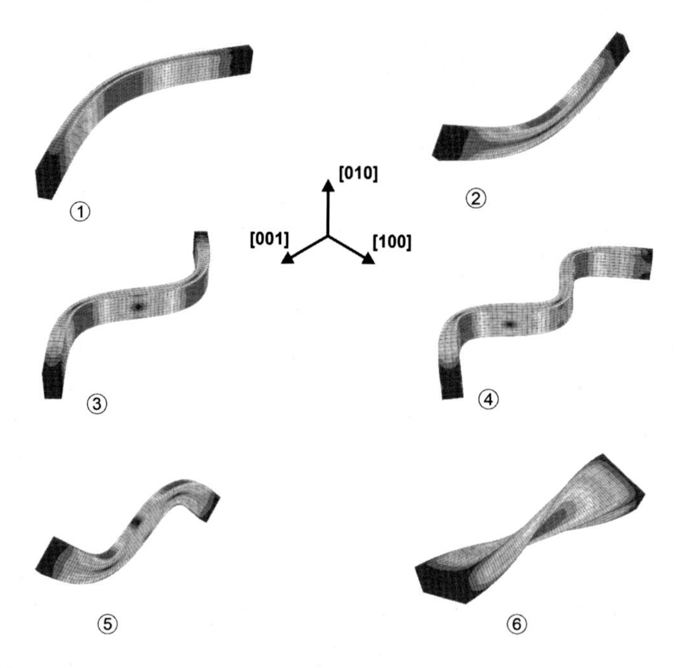

FIGURE 3.15 Resonance modes of a ⟨001⟩ specimen identified by FE analysis.

usually carried out by solving a least squares problem of the type (3.51). Applications of the method to single crystals of superalloys are reported in [330,68].

3.6.4 Examples of results

The constants of the alloy CMSX-4 obtained by sonic resonance are presented in Table 3.2. The testing and the FE evaluation were performed with 3 rectangular specimens of dimensions about $(80 \times 8 \times 3)\,\mathrm{mm}^3$ with the main axis respectively along the ⟨001⟩, ⟨011⟩, and ⟨111⟩ direction [429].

TABLE 3.2 Elastic constants (in MPa) of the alloy CMSX-4 estimated by SR measurements [429].

T (°C)	c_{11}	c_{12}	c_{44}	E_{001}	ν_{001}	κ	μ_2
24.	250175.	160707.	128721.	124461.	0.391	190530.	44734.
200.	243197.	157880.	122907.	118901.	0.393	186319.	42658.
400.	235848.	155497.	116582.	112277.	0.397	182281.	40175.
600.	226833.	151833.	109994.	105073.	0.400	176833.	37500.
800.	218106.	150109.	102292.	95717.	0.407	172774.	33998.
1000.	198900.	141077.	93419.	81818.	0.414	160351.	28912.

In Figs. 3.16–3.18, the constants of the alloy CMSX-4 (second generation alloy) presented in Table 3.2 are represented graphically with the values obtained in [1324] by SR with cylindrical specimens and in [1655] by surface Brillouin scattering (SBS). For comparison, the values obtained for the alloy SRR-99 (first generation alloy) by SR in [574] are also plotted. The differences for the Young's moduli and the shear moduli of the alloy CMSX-4 between the different measurements are limited. It can also be observed that the differences between the constants of the two alloys are not significantly higher than the dispersion of the measurements for the alloy CMSX-4. A higher scatter is, however, observed in Fig. 3.18 for the bulk modulus κ and the Poisson's ratio ν_{001}. This increased dispersion could be due to a lower sensitivity of the resonance frequencies with respect to the bulk modulus, since pure dilatation modes are not directly triggered by SR.

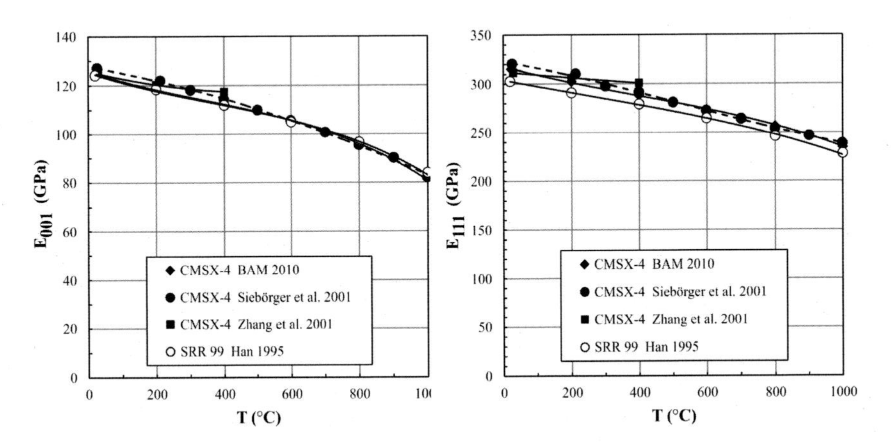

FIGURE 3.16 Young's moduli in the ⟨001⟩ and ⟨111⟩ directions of the alloys CMSX4 and SRR99 between RT and 1000°C (BAM 2010 [429], Siebörger et al. 2001 [1324], Zhang et al. 2001 [1655], Han 1995 [575]).

More results are available in the open literature for the constants at room temperature (RT). In addition to the results obtained by the SR method, measurements with the RUS method [68,330], with the pulse-echo-technique (PE) [309], with surface Brillouin scattering (SBS) [1655], and with the RUS method combined with electromagnetic acoustic resonance (EMAR) to identify the resonance modes [654] were published. Note that the alloy ERBO/1 has a similar composition to CMSX-4. The alloy ERBO/1A is in the as cast state while the alloy ERBO/1C is fully heat treated, which results in a more homogeneous chemical composition [1059]. The constants obtained for the CMSX-4 and its variants are compiled in Table 3.3, while the constants of other alloys are presented in Table 3.4.

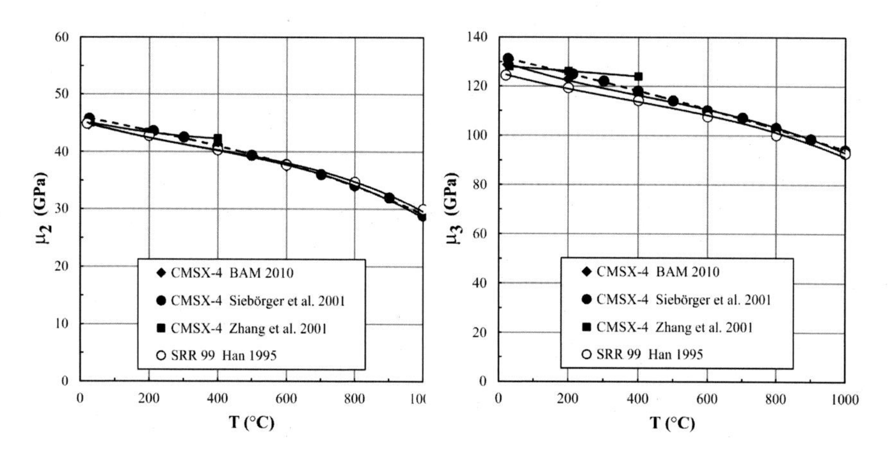

FIGURE 3.17 Shear moduli μ_2 and $\mu_3 = \mu_{001}$ of the alloys CMSX4 and SRR99 between RT and 1000°C (BAM 2010 [429], Siebörger et al. 2001 [1324], Zhang et al. 2001 [1655], Han 1995 [575]).

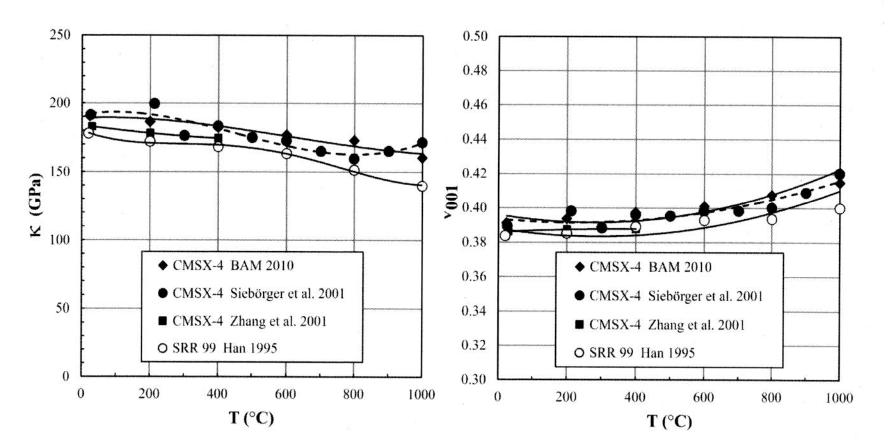

FIGURE 3.18 Compression modulus and Poisson's ratio for the ⟨001⟩ direction of the alloys CMSX4 and SRR99 between RT and 1000°C (BAM 2010 [429], Siebörger et al. 2001 [1324], Zhang et al. 2001 [1655], Han 1995 [575]).

The relative dispersion $\Delta p/p := (p - p^{average}) / p^{average}$ of the data presented in Tables 3.3 and 3.4 is shown in Fig. 3.19. The average of the CMSX-4 constants (columns 1 to 4 in Table 3.3) is taken herein as the reference $p^{average}$ for the parameter p. Aside from the SBS measurements, all constants for the alloy CMSX-4 and its variants are within a scatterband of 2%. For the other alloys, deviations up to 5% for individual constants can be observed. However, the methodologies behind these identifications differ, and it remains questionable whether these deviations are significant for the considered alloy. For example, the constants of the alloy SRR-99 re-

TABLE 3.3 Elastic constants (in MPa) of the CMSX-4 and its variants at RT.

Nr.	1	2	3	4	5	6
Alloy	**CMSX-4**	**CMSX-4**	**CMSX-4**	**CMSX-4**	**ERBO/1A**	**ERBO/1C**
Method	**RUS**	**SR**	**SR**	**SBS**	**RUS**	**RUS**
Reference	**[68]**	**[1324]**	**[429]**	**[1655]**	**[330]**	**[330]**
c_{11}	249200	252408	250175	242885	252100	247600
c_{12}	157444	160946	160707	152833	160900	156200
c_{44}	131450	131213	128721	128000	131500	130900
κ	188029	191433	190530	182850	191300	186700
μ_2	45878	45730	44734	45026	45600	45700
E_{001}	127282	127074	124461	124832	126800	126700
E_{111}	319822	320429	315185	311349	321000	316200

TABLE 3.4 Elastic constants (in MPa) of other alloys than CMSX-4 at RT.

Nr.	7	8	9	10	11
Alloy	**LEK94**	**TMS-26**	**MAR-M200**	**RenéN4**	**SRR99**
Method	**RUS**	**RUS/EMAR**	**PE**	**PE**	**SR**
Reference	**[330]**	**[654]**	**[309]**	**[309]**	**[574]**
c_{11}	242000	251800	248700	246900	237600
c_{12}	150900	158000	158500	152300	147900
c_{44}	130500	132000	131400	128100	124400
κ	181300	189300	188567	183800	177800
μ_2	45550	46900	45100	47300	44850
E_{001}	126200	129965	125310	130691	124114
E_{111}	315800	321304	319895	311862	302622

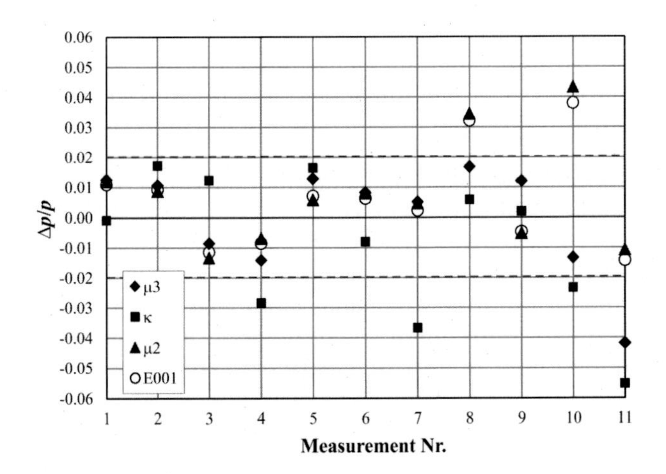

FIGURE 3.19 Relative dispersion of the elastic constants from Tables 3.3 and 3.4. The measurement number refer to the number in the tables.

ported in [574] were obtained with a coarser mesh than in [429]. It is thus expected that the alloy stiffness was underestimated to compensate the effect of a stiff FE model.

3.7 The elastic properties of the isolated γ and γ' phases

3.7.1 Relationships between properties of the isolated phases and the compound

The methods presented in Section 3.6 allow us to determine the effective or macroscopic elastic stiffness $\mathbf{C}^{eff}$ and compliance $\mathbf{S}^{eff}$ of the γ/γ' compound. These tensors relate the average stresses and strains, i.e.,

$$\langle \boldsymbol{\sigma} \rangle = \mathbf{C}^{eff} : \langle \boldsymbol{\varepsilon} \rangle \,, \tag{3.52}$$

$$\langle \boldsymbol{\varepsilon} \rangle = \mathbf{S}^{eff} : \langle \boldsymbol{\sigma} \rangle \,, \tag{3.53}$$

where $\langle \boldsymbol{\sigma} \rangle$ and $\langle \boldsymbol{\varepsilon} \rangle$ represent the average stresses and strains over a representative volume element (RVE) of the γ/γ' microstructure. The derivation of relationships between the elastic stiffnesses (or compliances) of the isolated γ and γ' phases $\mathbf{C}^{\gamma}$ or $\mathbf{S}^{\gamma}$, resp. $\mathbf{C}^{\gamma'}$ or $\mathbf{S}^{\gamma'}$, and their macroscopic counterparts is a special case of the main task of the homogenization theory [1004]. Roughly, one can distinguish:

- The full field methods. The fields at the microscopic scale $\{\boldsymbol{\sigma}(\mathbf{x}), \boldsymbol{\varepsilon}(\mathbf{x})\}$ and their spatial average are evaluated either by Fourier transforms [974] or finite elements in a periodic microstructure. In general, these methods are numerically demanding and not readily amenable to analytic solutions.
- The mean field methods. Here, relationships between the average values of the stresses and strains in each phase are searched. The self-consistent method makes use of the Eshelby solution for ellipsoidal inclusions and its validity is therefore questionable for cuboidal precipitates. The use of the bounds of Voigt and Reuss and their average after a proposal by Hill [609] provides a way to derive very simple estimates of the effective properties. As a matter of fact, the difference between the properties of both phases is small in Ni-base superalloys, and it was shown [773] that the estimate by Hill is sufficiently accurate for practical purposes. This procedure will be exposed in more details in the following.

The bounds of Voigt and Reuss are based on the following inequalities initially derived by Hill [610]:

$$\mathbf{E} : \mathbf{C}^{eff} : \mathbf{E} \leq \mathbf{E} : \langle \mathbf{C} \rangle : \mathbf{E}, \tag{3.54}$$

for all second-order tensors **E** and

$$\boldsymbol{\Sigma} : \mathbf{S}^{eff} : \boldsymbol{\Sigma} \leq \boldsymbol{\Sigma} : \langle \mathbf{S} \rangle : \boldsymbol{\Sigma}, \tag{3.55}$$

for all second-order tensors $\boldsymbol{\Sigma}$. These two inequalities are most conveniently evaluated by using a spectral representation of the stiffness tensors

$$\mathbf{C}^{eff} = \sum_{i=1}^{3} \lambda_i^{eff} \mathbf{P}_i, \quad \mathbf{C}^{\gamma} = \sum_{i=1}^{3} \lambda_i^{\gamma} \mathbf{P}_i, \quad \mathbf{C}^{\gamma'} = \sum_{i=1}^{3} \lambda_i^{\gamma'} \mathbf{P}_i. \tag{3.56}$$

Due to the orthogonality properties of the tensors $\mathbf{P}_j$, the following bounds are obtained for the coefficients λ_i:

$$\lambda_i^R = \left\langle \lambda_i^{-1} \right\rangle^{-1} = \left((1-f)\lambda_i^{\gamma-1} + f\lambda_i^{\gamma'-1} \right)^{-1} \leq \lambda_i^{eff} \leq \lambda_i^V = \langle \lambda_i \rangle = (1-f)\lambda_i^{\gamma} + f\lambda_i^{\gamma'}. \tag{3.57}$$

An estimate of the coefficients λ_i^{eff} is then obtained by averaging λ_i^R and λ_i^V. The geometric average

$$\lambda_i^{GH} = \sqrt{\lambda_i^R \lambda_i^V} \tag{3.58}$$

has the advantage of satisfying the consistency condition

$$\left(\lambda_i^{GH} \right)^{-1} = \sqrt{(\lambda_i^R)^{-1} (\lambda_i^V)^{-1}}, \tag{3.59}$$

which is not the case of an arithmetic average of the two bounds.

3.7.2 Results for the alloy CMSX-4

It is a difficult task to test the γ and γ' phases separately due to the small dimensions of the precipitates and matrix channels. By using the results of in-situ neutron diffraction tension experiments and solving an inverse problem, an estimate of the constants c_{11} and c_{12} in each phase of the alloy CMSX-4 was obtained in [373]. The shear modulus $c_{44} = \mu_3$ is not accessible through this method since only the first 3 Kelvin modes are activated in tension along the $\langle 001 \rangle$ direction (see Fig. 3.2).

In another approach, alloys of similar composition to either the γ or γ' phase have been produced to estimate the constants of the isolated phases in the compound alloy. Fahrmann et al. [419] have produced two allows with a low γ' solvus temperature and their corresponding compound alloys. The elastic constants were determined between RT and 1200°C by the SR method (see Section 3.6). Around their γ' solvus temperature, the low γ'-solvus temperature alloys contain little or no γ' phase and can be regarded as a pure matrix alloy. Eventually, the elastic constants of the

γ' phase were determined from a mixture rule of the Hill type. Siebörger et al. [1324] produced pure γ and γ' phase alloys and determined the elastic constants of these alloys and of the CMSX-4 by the SR method. The results of these estimates are plotted in Figs. 3.20 and 3.21. It can be observed that the constants κ and μ_2 of the alloy CMSX-4 are not bounded by their counterparts of the single phase alloys, in contradiction to the inequalities (3.57) since $\min\left(\lambda_i^{\gamma}, \lambda_i^{\gamma'}\right) \leq \lambda_i^R \leq \lambda_i^{eff} \leq \lambda_i^V \leq \max\left(\lambda_i^{\gamma}, \lambda_i^{\gamma'}\right)$. Siebörger et al. [1324] attribute these deviations to the fact that the composition of the single phase alloys necessarily deviate from that of the phases in the com-

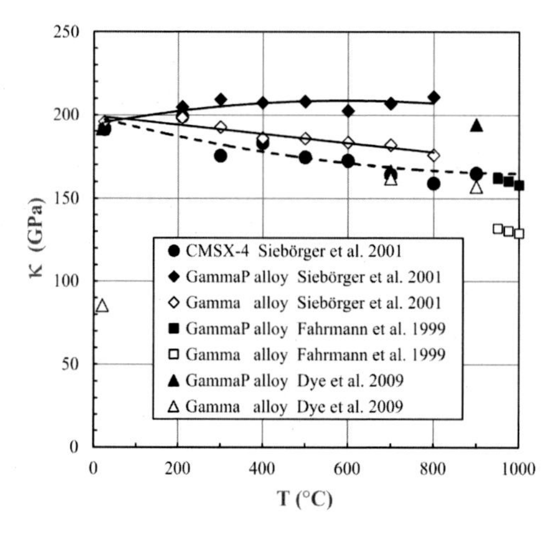

FIGURE 3.20 Estimates of κ for the alloy CMSX-4 [1324] and the isolated phases (Siebörger et al. [1324], Fahrmann et al. [419], Dye et al. [373]).

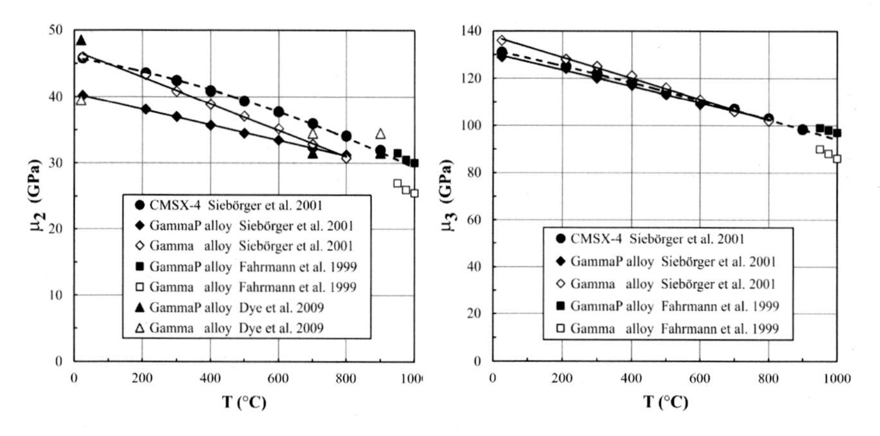

FIGURE 3.21 Estimates of μ_2 and μ_3 for the alloy CMSX-4 [1324] and the isolated phases (Siebörger et al. [1324], Fahrmann et al. [419], Dye et al. [373]).

pound, which would not be stable taken as isolated. Indeed, the γ phase alloy tested in [1324] is almost pure Ni_3Al, in contrast to the γ' phase in the compound, which is more heavily alloyed. In addition, the [111] specimen used for the SR measurements departed somewhat from the ideal orientation, a deviation that could not be accounted for in the analytical procedure followed in the paper.

In summary, reliable determinations of the elastic properties of matrix and precipitate phases are still missing. Indirect indications, like the amount of deviation from the cubic symmetry induced by a plate microstructure after rafting, suggest that the difference between both phases is minor, which, of course, largely hampers an unambiguous identification of these constants. Indeed, in [654] a tetragonal symmetry was inferred from RUS measurements for the alloy TMS-26 after rafting, which with the values (c_{11}=252.5 GPa, c_{33}=250.4 GPa, c_{12}=158.5 GPa, c_{13}=157.2 GPa, and c_{44}=131.8 GPa, c_{66}=132.4 GPa) hardly departs from the cubic symmetry.

CHAPTER

4

Microstructure and chemical characterization

Florence Pettinari-Sturmel[a] *and Loïc Nazé*[b]

[a]CEMES CNRS University of Toulouse, INSA, Toulouse, Cedex 4 France,
[b]Ecole des Mines de Paris, Paris, France

Glossary

Scanning Electron Microscopy SEM
Transmission Electron Microscopy TEM
Atom Probe Tomography ATP
(TEM) HAADF (TEM) High Angle Annular Dark Field
ECCI Electron Channeling Contrast Imaging
(SEM or TEM) EDX or EDS (SEM or SEM) Energy Dispersive X-Ray Spectroscopy
(TEM) EELS (TEM) Energy Electron Loss Spectroscopy
CBED Convergent Beam Electron Diffraction
SF Stacking Fault
SFE Stacking Fault Energy
APB Antiphase Boundary
APE Antiphase Boundary Energy
SISF / SESF Super-Intrinsic/Extrinsic Stacking Faults
CSF Complex Stacking Fault

4.1 General description of the microstructure

Ni-base superalloys are essentially two-phase alloys: they consist in a γ-phase matrix strengthened by γ'-phase precipitates. The γ-phase matrix is a nickel-based solid solution. It has the face-centered cubic structure (fcc) of pure nickel (see Fig. 4.1(a)) but solute elements such as Cr, Co, Mo, W, Re, and Ru are substituted for Ni atoms. The γ'-phase which precipitates in the matrix is based on the $L1_2$ structure of the Ni_3Al ordered intermetallic compound (see Fig. 4.1(b)). That structure uses the same atom positions as the fcc structure but the atoms at the corners of

Nickel Base Single Crystals Across Length Scales
https://doi.org/10.1016/B978-0-12-819357-0.00011-1

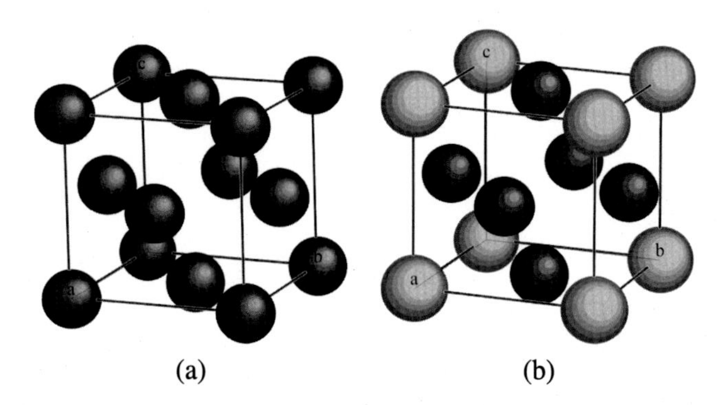

FIGURE 4.1 (a) Face-centered cubic structure, (b) $L1_2$-ordered structure.

the cubic cell are Al and those at the center of the faces are Ni. The γ' phase is an ordered solid solution, and solute elements such as Ti or Ta substitute Al, Co substitutes Ni, and Cr substitutes both Ni and Al. The γ matrix cubic lattice parameter, a_γ, and the γ' precipitated phase one, $a_{\gamma'}$, are close. The small mismatch between the lattice parameters of the two phases is defined as $\delta = 2(a_{\gamma'} - a_\gamma)/(a_{\gamma'} + a_\gamma)$. Then the γ' phase precipitates nucleate and grow respecting the cube–cube cristallographic relationship which insures a high crystal coherency between both phases. This is actually a necessary requirement for a two-phase alloys to be considered as a single-crystal. In its ready-to-be-used condition, a single-crystal nickel-based superalloy employed, e.g., as a turbine blade, consists of a continuous single-crystal matrix with regularly distributed cuboidal shape precipitates. The parameters required to quantitatively describe the microstructure are:

- Chemical parameters – the chemical composition of each phase;
- Crystallographic parameters – cubic lattice parameter of each phase;
- Morphological parameters – size of the precipitates (i.e., edge length of the cuboidal precipitate), width of the γ matrix channels. To be noted is that two parameters which are the size of the precipitates, the width of the γ matrix channels and the γ' phase volume fraction, are interdependent.

Some comments have to be formulated regarding this list:

- At thermodynamic equilibrium, all those parameters depend on the nominal composition of the alloy and the temperature. Chemical composition analysis, lattice parameter determinations, microstructure observations, and morphological measurements are often made on a material which has been quenched from the temperature of the specimen at the end of the heat treatment sequence or at the end of a mechanical test. The results of the measurements are more or less close to the value at service temperature or even at the mechanical test temperature. (Moreover, during service or during testing, thermodynamic equilibrium is rarely reached.)

- The microstructure is not morphologically as perfect as the former description suggests: variations in the γ' precipitate edge length and the γ matrix channel width result in distributions of those parameters from which mean, modal, or even median values are deduced. However, these local irregularities of the microstructure are uniformly distributed in the microstructure and are seldom taken into account when those microstructural parameters are used, for instance, in the description of the mechanical behavior.
- Heterogeneities, relevant to a larger scale of observation than the local irregularities of the microstructure, might be necessary to be described. Those heterogeneities are inherited from the solidification of the single-crystal and are directly correlated to the dendritic microstructure.
- Phases other than γ and γ' may be present in the microstructure according to the heat treatment condition, mechanical testing parameters (temperature, stress, deformation, cycling frequency, and testing time), or in-service aging history. Those, exceptionally, are carbides in René N6, but, more often, topologically close packed (TCP) phases after exposition at high temperature and stress for long times.
- Metallurgical singularities may also be present in the microstructure, such as eutectic microconstituents nucleated at the end of solidification, and not always solutionized during heat treatments. Also inherited from solidification are shrinkage porosities. Other porosities are the result of condensation of vacancies during heat treatments or during deformation at high temperature. Another important microstructural feature resulting from high temperature deformation is rafting.

All these "foreign" phases, metallurgical singularities, and microstructural features need to be documented by chemical and physical analysis, and morphological measurements. Finally, a rather different aspect of microstructural analysis in single-crystal superalloys concerns dislocation microstructures. Tools used to observe and analyze all the microstructural features listed above are obviously all the (more or less) standard tools used for microstructural analysis in metallurgy. However, their use on single-crystal nickel-based superalloys involves specific goals, experimental methods, and test results' analysis that are described in this chapter. More recent experimental techniques involving new TEM, SEM, or synchrotron X-Ray technology are also considered.

4.2 Microstructure characterization studied from the macroscopic to microscopic scale

4.2.1 Introduction

In metal alloys, macroscopic properties can result from a singularity, heterogeneity, or defects that need to be carefully identified. Thus accurate and quantitative investigation is of great importance in the characterization of the structure and composition of Ni-base superalloys. Metallographic analysis is used in order to identify if an alloy was correctly processed, to examine multiple phases within the material, to locate and characterize imperfections such as porosities or impurities, or simply to characterize the microstructural features. Metallographic analysis can also be used either to observe damaged or degraded areas in failure analysis investigations in order to establish the reason why a material failed, or to interpret the origin of a particular mechanical property.

Metallographic characterization relies on both the technique itself and on specimen preparation. The appropriate specimen preparation is a critical part of material characterization. Sample preparation should satisfy some criteria as follows: all the structural elements must be entirely retained, the surface must be without scratches or deformation, no foreign matter should be introduced on the specimen's surface. Sample preparation requires a certain degree of skill and experience, due to the strong chemical resistance of most superalloys. A detailed presentation of the existing knowledge about metallographic techniques and their application to the study of metals can be found in the book by Vander Voort [1467].

Metallographic preparation involves standard mechanical grinding followed by polishing. Selected chemical or electrochemical etchants are then applied for revealing the microstructural features of superalloys. In the case of electron microscopy, since the differences in composition between the γ matrix and the γ' precipitates do not result in a significant difference of the mean atomic number of each phase, the contrast between both phases in the backscattered electron imaging mode is slight. It is usually necessary to etch the specimens and perform microstructure observations using a secondary electron imaging mode [1293]. Various techniques were developed for characterizing the different phases that can be observed in Ni-base superalloys. A review of the different techniques that can be used is proposed by Wusatowska-Sarnek et al. [1589]. More recently, Szczotok and Reichel [1389] have made a detailed analysis of different chemical and electrochemical etching methods used for revealing the microstructure. Their paper is focused on those etchants' effect on the microstructure characterization in the case of CMSX-4 superalloy.

As the characterization of the γ–γ' microstructure has to be performed with accuracy and with high statistical significance, developments have been carried out combining a new imaging technique (biased for backscattered electron detection, which is useful for image processing and enables the collection of a relatively large sample) and image processing for rapid quantitative characterization of γ/γ' superalloy microstructures. An example of a combination of sample preparation, imaging technique, and segmentation algorithm that can be used to characterize the various phases in Ni-base superalloys was proposed by Payton et al. [1069].

4.2.2 Crystallographic orientation of single-crystals

In the case of single-crystal Ni-base superalloys, one important property is their single-crystalline characteristic. Crystalline materials are characterized in the reciprocal space by their diffraction peaks related to their average crystallographic structure according to the Bragg's law which is given by

$$2d_{hkl} \sin \theta_{hkl} = n\lambda, \tag{4.1}$$

where n is an integer, λ is the wavelength of the incident radiation, d_{hkl} is the interplanar spacing of the $\{hkl\}$ planes and θ_{hkl} is the angle between the incident probe and the $\{hkl\}$ planes. This law is based on the fact that constructive interference occurs when the path differences between scattering waves by adjacent planes are integer multiples of the wavelength of the radiation. When such constructive interference occurs, it is said that the Bragg condition for the given diffracting planes is satisfied. The diffraction patterns (using X-rays, neutrons, or electrons) allow determining the atomic coordinates (the center of the scattering objects) and thus the crystal structure, crystallography, and lattice parameters of the constituent phases in the investigated crystal. The basic theory, standard methods, and many of the more classical applications are well described in various books; see, for example, the papers from Lipson and Cochran, Warren, Schwartz and Cohen, Cullity [1077,1541,1293,307]. A recent review on X-ray and neutron scattering applied to physical metallurgy is documented in details by Kostorz in [753].

The Laue diffraction method is the most reliable way and the most commonly used experimental tool to determine or even just to check the crystal orientation of metal alloys.

Recrystallization was reported in single-crystal Ni-base superalloys after specific mechanical tests by Moverare et al. [975] (during thermal-mechanical fatigue in CMSX-4) and by Le Graverend et al. [811] (after creep test at 1050°C on a notched specimen generating large local strains, which, at high temperature, triggers dynamic recrystallization). The recrystallization concerns rather small areas (< 1 mm wide). The local crystal orientation determination, illustrated by Fig. 4.2, shows a large change of orientation of the new grains grown by recrystallization, from the [001] orientation of the load axis of the test specimen. The most appropriate

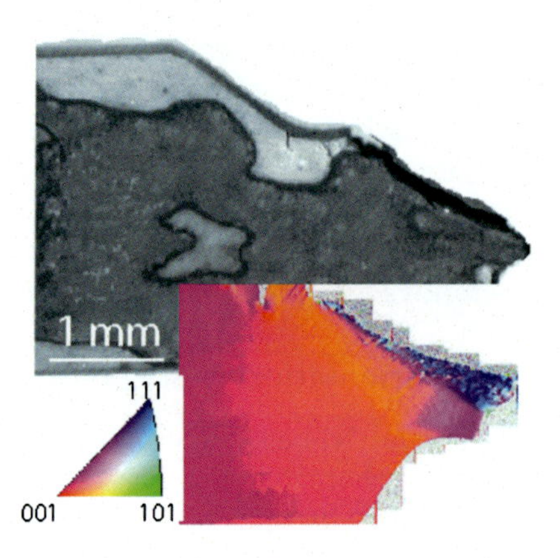

FIGURE 4.2 SEM image of the notched area of a creep specimen (upper part of the figure) and corresponding EBSD inverse pole figure (IPF) of the crystal direction of the test specimen loading axis (lower part of the figure), adapted from [811]. The single-crystal Ni-base superalloy creep test specimen has its axis along [001] (red indexed area in the IPF). Recrystallization of the strongly strained area of the notched specimen causes growth of grains with <111> direction along the specimen loading axis (blue indexed grains in the IPF).

tool in the case of such a crystallographic change of crystal orientation is electron backscatter diffraction (EBSD), which is a diffraction technique performed with a scanning electron microscope (SEM). The principles of scanning electron microscopy were given in the 1930s by Max Knoll [731] who first obtained scanned electron images from the surface of a solid, and by Manfred von Ardenne who established the principles underlying the SEM [1505,1506]. The accelerating voltage used is typically in the 15–40 kV range. As an example, with a conventional SEM, the resolution ranges from 0.5 to 1 nm with an accelerating voltage of 30 kV. Nowadays, more and more SEM are fitted with a field emission gun (FEG SEM). The various signals used to generate images or diffractograms or X-ray spectra by means of different detectors in a scanning electron microscope result from the elastic or inelastic interaction between the incident electrons and the specimen. Among all the emitted radiations resulting from these interactions, secondary electrons (SE) and backscattered electrons (BSE) are used the most for imaging microstructures. The SE mode mainly creates a topographic contrast resulting from inelastic interactions, between incident electrons and atoms of the specimen, whereas the BSE mode produces contrasts due to chemical composition or crystal structures and orientation variations in the specimen. Backscattered electrons are also used for the diffraction mode (EBSD) which gives access to crystal orientations of the specimen. Electron backscattered diffraction (EBSP) has been applied for many years when determining crystal orientations with good spatial resolution by SEM (see, for example, the papers of Venables and Harland [1483] or Dingley [344]). In the past decades, several research groups have worked to advance the EBSD technique and make it an important analytical tool. Some of the most important steps in the development of the technique were the introduction of online analysis of EBSPs [345], the development of user-friendly computer-aided indexing software [1277], and the automation of the analysis with both precision and reliability [1577,779,795,766].

4.2.3 Solidification microstructure

Monocrystalline solidification, which is the first step of the manufacturing process of a single crystal, proceeds as a dendritic solidification in a temperature directional gradient. Various microstructural features, distinctive of dendritic solidification, are clearly identifiable in a cross-section perpendicular to the temperature gradient direction (see Fig. 4.3):

- Dendrites which nucleate and, as temperature decreases, grow first (white contrast);
- Interdendritic regions which solidify at lower temperature than dendrites (grey contrast);
- Eutectic aggregates, which are the last feature to solidify.

For more details regarding dendritic segregation with respect to the chemical composition of alloys, see Chapter 7.

In single-crystal Ni-base superalloys, the largest microstructural feature is actually this dendritic structure. It can be observed by optical microscopy, including confocal microscopy, and scanning electron microscopy as illustrated by Figs. 4.3 and 4.4 that a rough estimate of their characteristic dimension is in the range of one to several hundred micrometers.

During solidification, the chemical composition of the high-temperature solid single phase evolves between the dendrite core and the interdendritic regions because the diffusion is too slow to continuously and instantly homogenize the composition of that solid solution as it solidifies (as it would be predicted by a hypothetical equilibrium diagram). Very different chemical compositions can be found between the dendrite cores and the interdendrite spaces.

In Fig. 4.5, Al, Ta, and Ti are observed to segregate into the interdendritic region, unlike W which segregates into dendrite cores and secondary arms. This distribution of chemical elements can be determined by EPMA (electron probe microanalysis), which is an analytical tool used to determine chemical compositions attached to a scanning electron microscope: an electron beam is fired at the specimen and the X-ray photons produced by the electron beam interaction with the sample are characteristic (energy/wavelength) of the atoms excited by the incident electrons. This technique was initiated by Baker and Hillier in the 1940s in the RCA lab

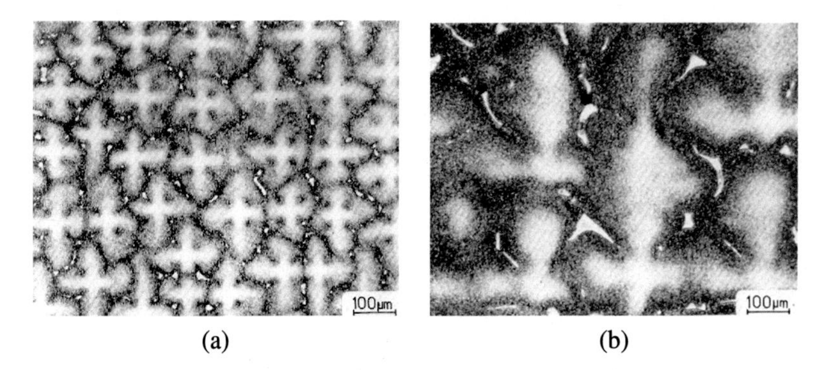

FIGURE 4.3 Dendritic structures of as-cast CMSX-2 single-crystals (optical microscopy; sections normal to the [001] growth direction): (a) high thermal gradient process (25 K mm^{-1}); (b) low thermal gradient process (5 K mm^{-1}) [717]. Courtesy of P. Caron.

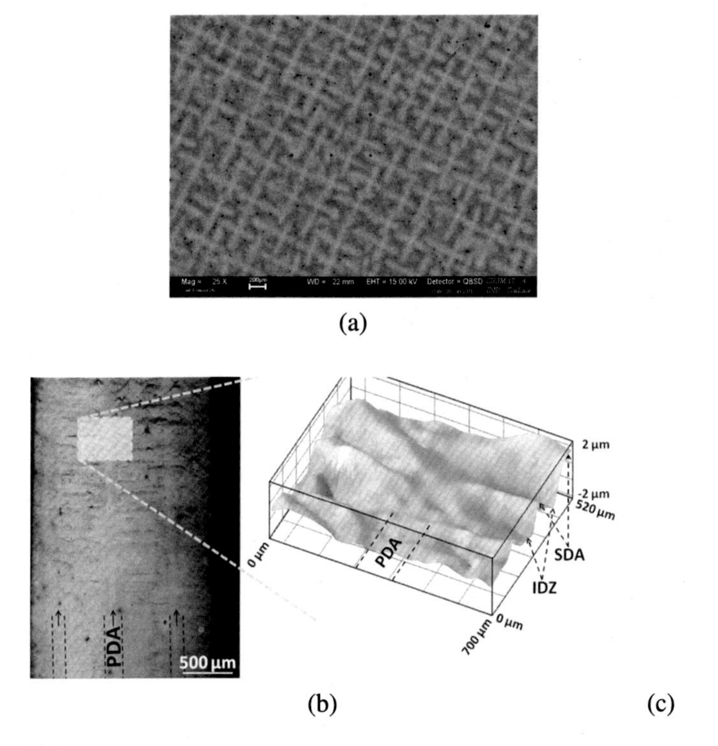

FIGURE 4.4 (a) SEM observations of dendrites in MC2 superalloy; (b) Optical microscopy observations of the surface. The dotted lines highlight different stripes with significantly lower roughness, corresponding to the primary dendrite axis; (c) Local topography of the specimen observed by confocal microscopy. Courtesy of D. Texier [1417].

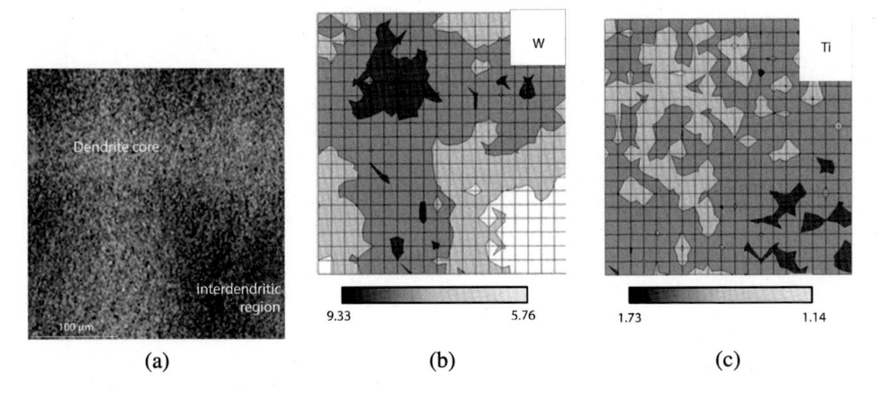

FIGURE 4.5 Alloying element distribution at the dendrite scale using EPMA: (a) SEM image of a dendrite core and surrounding interdendritic regions; (b)–(c) EPMA analysis showing the heterogeneous repartition of W and Ti. Courtesy of D. Texier. Adapted from [1417].

at Princeton and further developed by Castaing and Guinier in 1951 at ONERA French lab [615,616,273,1318].

4.2.4 γ/γ' cuboidal microstructure observed at the mesoscopic scale

A suitable technique to characterize the cuboidal γ/γ' microstructure is the scanning electron microscopy (SEM), as the scale of the γ' precipitation is typically lower than a micrometer, more precisely between 400 and 500 nm. Nevertheless, transmission electron microscopy (TEM) can also be used as illustrated by Fig. 4.6.

In order to have an accurate characterization of the γ'-precipitates, the microstructure can be highlighted by selective dissolution of the γ-phase

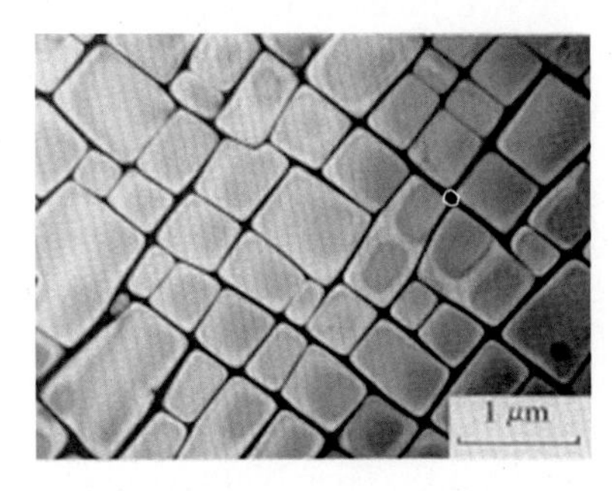

FIGURE 4.6 TEM dark field image of γ/γ' microstructure in fully heat-treated CMSX-2 Ni-base superalloy [211]. Courtesy of P. Caron.

(a) (b)

FIGURE 4.7 SEM image and 3D image of the γ/γ' microstructure in MC2 superalloy after selective dissolution. Courtesy of (a) D. Texier and (b) J. Cormier.

after electrochemical polishing as described, for example, by Le Graverend et al. [804] and illustrated Fig. 4.7.

4.2.5 Characterization of the under creep γ/γ' microstructure at the mesoscopic scale

During creep at high temperature (above 900°C for low stress condition) the γ/γ' cuboidal microstructure of single-crystal Ni-base superalloys evolves from uniformly distributed cubes to stacks of rafts as illustrated by Fig. 4.8. This microstructural evolution and the physical origin of the γ' phase coarsening during isothermal creep have been widely documented [1432,1108,207,467,1266,602,1322]. Some key elements have been summarized in the following.

The early observations of rafting were interpreted in a quantitative way by Pineau [1108]. Using the equivalent inclusion method of Eshelby [401], Pineau suggests that the major factors which affect the stress-coarsening behavior are the direction and value of the applied load, the ratio between the elastic moduli of the two phases, as well as the γ/γ' misfit. Considering only two parameters (misfit and stress), it has been shown that when the superalloy has a negative misfit ($\delta < 0$) and the stress is tensile, the rafts tend to expand perpendicularly to the stress axis. On the contrary, if the stress is compressive, the rafts tend to expand parallel to the stress axis. This microstructural evolution is reverse when the misfit is positive. Then, Fredholm and Strudel [467] showed that many simplifying assumptions have been made in this analysis (e.g., the elastic interaction between particles remains unaccounted for) so that the experimental observations are in agreement with Pineau model only in the case of γ' particles which are elastically stiffer than the matrix. Then Nabarro et al. [995] have developed an adaptation of the initial model of Pineau. Later, various models (purely elastic, i.e., without the presence of slip dislocations, or taking channel dis-

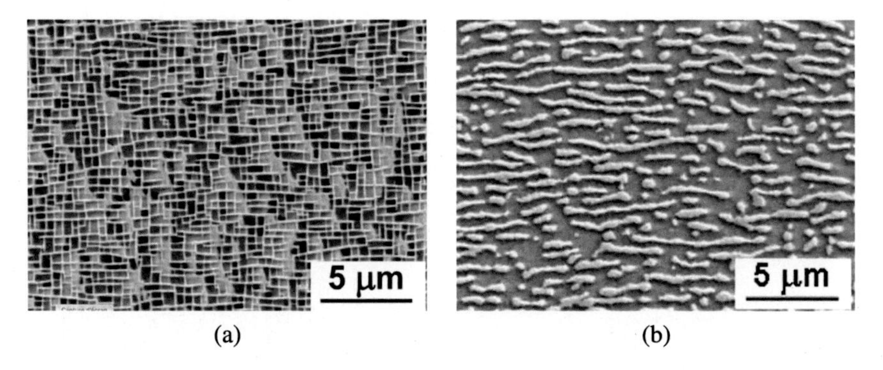

FIGURE 4.8 Microstructures observed in the MC-NG single-crystal alloy after creep at 1050°C and 150 MPa: (a) $t = 20$ h; (b) $t = 217$ h (SEM; longitudinal sections) [213]. Courtesy of P. Caron.

location slip into account) have been proposed to identify the driving force for the morphological evolution. These investigations rely on different experimental procedures combining TEM observations and small angle X-ray or neutron scattering (SAXS or SANS) [711,1487,1488,1055,1486,89,88]. The advantage of SANS experiments is that they generate bulk averaged information in a nondestructive way through relatively thick samples, while artifacts such as elastic stress relaxation in thin foils extracted from bulk specimens may result in wrong conclusions deduced from TEM observations. As an example, Bellet et al. [88] were able to analyze about 10^{11} precipitates using SANS instead of about 10^3 precipitates by SEM. From the analysis of the position and shape of the correlation peaks obtained by SAXS or SANS experiments, the coarsening and ordering of the precipitates during heat treatments are determined, which allows for a precise description of the γ' phase evolution. Indeed, the characteristics of the SANS patterns change from fourfold to twofold symmetry when the precipitates evolve from cuboids to rafts and the position of the correlation peaks corresponding to the coalescence of the precipitates during annealing is modified.

On the basis of these papers, it is obvious that the determination of the creep controlling deformation mechanisms requires a precise characterization of the rafted microstructure and the precise determination of the γ/γ' misfit during creep.

First, a proper quantitative stereological characterization of the rafted structures is needed to clarify the rafting kinetics [1124]. It is hoped that, once the rafting kinetics has been properly quantified, the effect of rafting on creep strain can be properly evaluated. Different procedures have been described by Ignat et al. [655], and then adapted by Matan et al. [909].

Recent methods to properly image the rafted microstructures are based on FEG-SEM experiments associated with image analysis software.

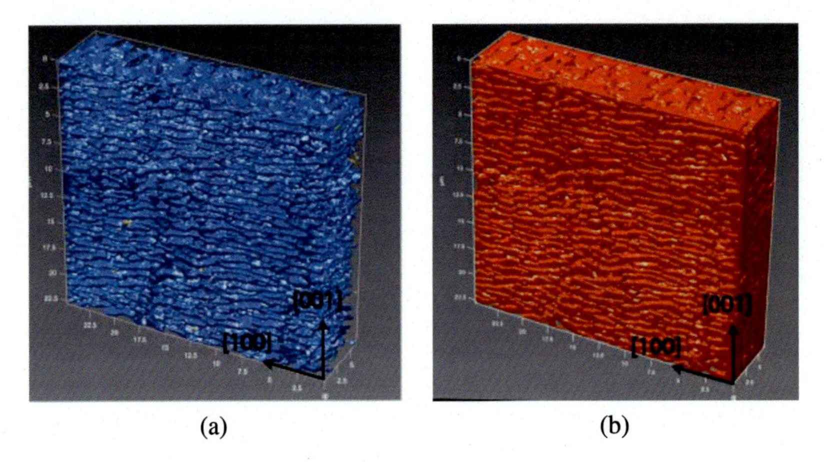

FIGURE 4.9 3D reconstruction of the crept microstructure at 850°C, 400 MPa split into (a) the γ phase and (b) the γ' phase. The load is along [001] and the scale is in µm. Imaging of a 6×28 µm^2 area using a 50 nm step size. Adapted from Antonov et al. [27].

Nowadays, in order to provide full microstructure imaging, several 3D reconstruction techniques are combined. Different experimental procedures have been reported in the literature. The characterization of the γ/γ' cuboidal microstructure of a $Ni_{70}Cr_{20}Al_{10}$ superalloy by focused ion beam (FIB) slicing and imaging by SEM was reported by Uchic et al. in 2009 [1453]. Later a reconstruction method by X-ray tomography and 3D reconstruction by slicing and imaging in a high resolution SEM of MC2 superalloy have been reported by Jouiad et al. [684]. Furthermore, similar attempts were carried out by Link et al. [848] who used X-ray synchrotron tomography to study porosity in nickel-based superalloys. More recently, new developments in this imaging technique and reconstructed image processing software have allowed to image the large-scale plasticity assisted topological inversion in a Ni-base superalloy after creep at 850°C. An illustration is proposed Fig. 4.9. FIB was used for serial sectioning and imaging. The image set was analyzed and reconstructed with the image processing software MIPARTM, using a custom-development as described by Sosa et al. in 2014 [1344].

Next, one of the important microstructural parameters involved in the creep controlling deformation mechanisms is the γ/γ' misfit. The experimental method suitable for its determination is developed in Section 4.2.7.3.

4.2.6 Short overview of numerical approaches for describing the γ/γ' microstructure evolution

See associated chapters of this book for more details.

A further investigation of the γ/γ' microstructure has been developed using different numerical approaches. First, the origin of the cuboidal microstructure has been well documented as resulting from competition between the isotropic interface energy and the anisotropic elastic energy during the growth of the γ' phase [716,876,877,1426], which induces a morphological evolution from spheroidal to cuboidal morphology of the precipitates. The elastic interactions between the precipitates favor their progressive alignment according to the cubic directions of the matrix. This elasticity induced reconfiguration leads to the formation of almost periodic arrangements of cuboidal precipitates. However, regular mistakes in the alignment of γ' precipitates have been reported. At a high volume fraction, the periodicity of the arrangement is broken [202,347,500,1412]. At low volume fraction, rows of precipitates aligned along [100] and [010] tend to be adjacent, thus forming defects in the alignments of the precipitates [34,836]. Despite numerous studies on the formation and the evolution of the γ/γ' microstructure, the origin and the dynamics associated with these defects still have to be understood.

In this context, there was a need to develop numerical approaches to calculate simulations of the γ/γ' microstructure evolution and to correlate this evolution with the mechanical behavior. Both the cuboidal and rafted microstructures have been studied. Different scales have been investigated.

For the macroscopic scale, mechanical modeling including a description of the microstructure has been carried out (see Chapters 15 and 17). At the intermediate scale, named the mesoscopic scale, different approaches are used to model the coupling between elasticity and microstructure. The main purpose is to model the evolution of a microstructure consisting of dispersed ordered intermetallic precipitates [291,634,1671,1458,1533,327]. The phase field method is the reference approach to model this coupling at the mesoscopic scale. In this approach, the evolution of the microstructure is simulated using continuous fields. Recently a new formulation of phase field models known as the sharp phase field method (S-PFM) has been developed by Finel et al. [441]. The advantage is that the S-PFM allows for the modeling of extended microstructures, with large numbers of precipitates both in two and three dimensions, and provides access to statistical information on the microstructure.

At the microscopic or atomic scale, several approaches make it possible to study the mechanisms involved in the evolution of the microstructure, for example, the molecular dynamics [832,1615] or Monte Carlo type approach [1472,1473].

4.2.7 Characterization of the γ and γ' phases using diffraction techniques

4.2.7.1 *Crystallographic identification of the γ and γ' phases*

As already mentioned, the γ' phase has an ordered $L1_2$ structure whereas the γ phase has an fcc structure. These two phases can be easily distinguished and identified using one of the various diffraction techniques because the diffraction patterns associated with each phase is different.

Both structures are built on the same atom positions but ordering of the atoms in the $L1_2$ results in the fact that the crystal lattice used to describe the 3D periodicity of this structure is a primitive cubic lattice. The fcc structure whose all atomic positions are homologous, as there is no distinction between the atoms of various elements occupying these positions, is periodicity described by a face-centered cubic lattice. The crystallographic conventional cubic cells of those two crystal lattices have both the same cubic parameter. However, the difference of the crystal lattices of these two structures results in discrepancies between crystal lattice plane families characteristic of each structure. As a matter of fact, a triplet of integers (hkl) would correspond to the Miller indices of a crystal lattice plane of the fcc structure if and only if h, k, and l have the same parity. This means that, for instance, the triplet (100), which corresponds to Miller indices of the lattice plane family normal to one of the edges of the primitive cubic cell of the $L1_2$ structure of the γ' phase, does not correspond to the Miller indices of the lattice plane family normal to one of the edges of the face-centered cubic cell of the fcc structure of the γ phase, as the numbers 1 and 0 do not have the same parity. For this fcc structure, the Miller indices must be written (200). The crystal lattice interplanar spacing of the lattice plane family (hkl) in cubic lattices is expressed as $d_{hkl} = a(h_2 + k_2 + l_2)^{-1/2}$, where a is the cubic lattice parameter. By using this equation in both phases, it can be checked that the crystal lattice interplanar spacing of (200) crystal plane family in the γ phase is $a/2$ and that of (100) is a, both values corresponding to what can be easily determined by just looking at the respective unit cells. This difference of interplanar latttice spacings of, again for instance, $(100)_{\gamma'}$ and $(200)_{\gamma}$ translates into two different Bragg first-order diffraction angles for γ' and γ:

- for γ', $\sin\theta_{\gamma'} = \lambda/2d_{(100)\gamma'} = \lambda/2a$,
- for γ, $\sin\theta_{\gamma} = \lambda/2d_{(200)\gamma} = \lambda/a = 2\sin\theta_{\gamma'}$.

In conventional electron diffraction in TEM of both phases in "cube–cube" relationship, this description results in diffractions patterns in a row of diffraction spots whose first spot (the closest to the spot corresponding to the transmitted beam) corresponds to the first-order diffracted beam on (100) in γ', as illustrated by Fig. 4.10. The second spot in the row corresponds to the superimposition of the first-order diffracted beam on (200)

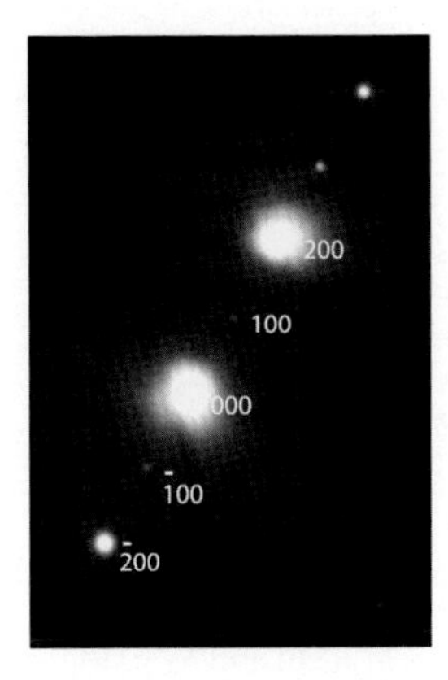

FIGURE 4.10 Experimental diffraction pattern in MC2 Ni-base γ/γ' superalloy showing the 100-type superstructure diffraction spot. Note that their intensity is much lower than the Bragg diffraction spot.

in γ and of the second-order diffracted beam on (100) in γ'. The third spot corresponds to the third-order diffracted beam on (100) in γ' (and is indexed (300) or [300]* in the reciprocal lattice). The fourth spot in the row corresponds to the superimposition of the second-order diffracted beam on (200) in γ and of the fourth order diffracted beam on (100), indexed (400), in γ', and so on.

By analogy to supplementary diffraction spots appearing when a superstructure develops from an initial crystal structure (this superstructure having lattice parameters larger than those of the original structure), and though the $L1_2$ structure of γ' phase is not a crystal superstructure of the fcc structure of the γ phase, those spots, specific to the γ' phase, are called "superstructure spots" and are a signature of the γ' phase. Their intensity is lower than the 200 Bragg diffraction spot. An example of the diffraction pattern carried out in the MC2 γ/γ' Ni-base superalloy is given Fig. 4.10(b). The superstructure spots are distinguishable, although their intensity is lower than that relative to both the γ and γ' phases.

4.2.7.2 *Determination of the γ/γ' misfit using TEM conventional and CBED modes*

The γ/γ' misfit, denoted by δ, can be estimated by measuring the average dislocation spacing in γ/γ' interface dislocation networks imaged by conventional TEM experiment as illustrated by Fig. 4.11. It can be shown that the average dislocation spacing d is inversely proportional to the magnitude of the lattice mismatch following the equation, $\delta = b/d$, with b being the Burgers vector of the dislocations.

Another experimental method which allows us to determine the γ/γ' misfit uses convergent beam electron diffraction (CBED) technique,

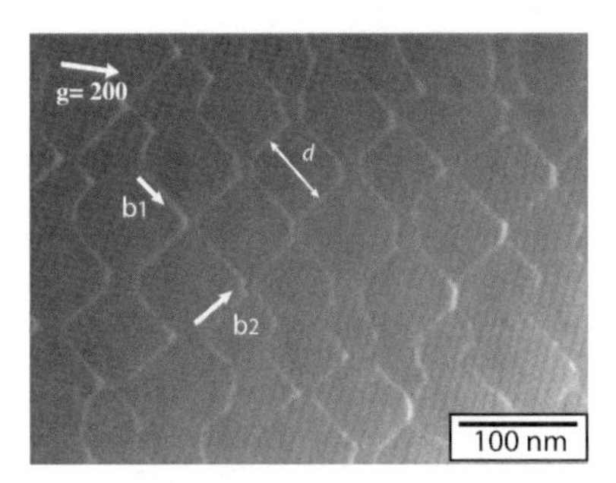

FIGURE 4.11 TEM image of interfacial dislocation network in MC2 superalloy after creep test at 1150°C / 80 MPa using a diffraction vector $g = 200$. Two types of dislocations are mainly observed, with the Burgers vectors $b_1 = \bar{1}10$ and $b_2 = 110$. The distance d corresponds to the spacing between the dislocations with b_2 Burgers vector. Adapted from [579].

which has been developed by the Bristol Group in 1984 and by Tanaka and Terauchi in 1985. CBED is performed with a convergent incident beam (with a beam convergence in the range from 0.1° to 1°) focused on the specimen plane. The patterns are observed in the back focal plane of the objective lens and consist of disks where excess and deficiency lines are in contrast. This technique is described in details by Tanaka et al. [1399,1400] or Morniroli [967]. CBED is known for the measurement of foil thickness, lattice parameters, and strains, and can also be applied to the study of crystal defects, especially of stacking faults and antiphase boundaries. Due to the high lateral resolution of the method, lattice parameters of the γ matrix and γ' phase can be directly measured. At coherent γ/γ' interfaces, the determination of the misfit in the alloys can be carried out without any microstructural model, which is needed when X-ray or neutron diffraction is used. This method has been used, for example, by Völkl et al. [1503,1504] in the case of single crystal Ni-base superalloys. They were able to distinguish between the misfit values in dendrites and in interdendritic regions. The advantage of this method is that it allows a local misfit determination.

4.2.7.3 *Determination of the misfit evolution during creep using synchrotron*

The determination of the misfit evolution during a creep test requires specific experimental tools: a very high angular resolution of the diffraction angles is necessary to distinguish the γ and γ' phases. As the misfit evolves during creep test, it is essential to develop in situ experiments, that is to say, creep experiments performed in an equipment allowing crystal lattice parameter determination with a very high data recording rate. The high intensity of synchrotron radiation, combined with increasingly fast high-resolution detectors, allows a diffraction peak to be recorded in a few hundredths of a second. The development of high-energy lines in

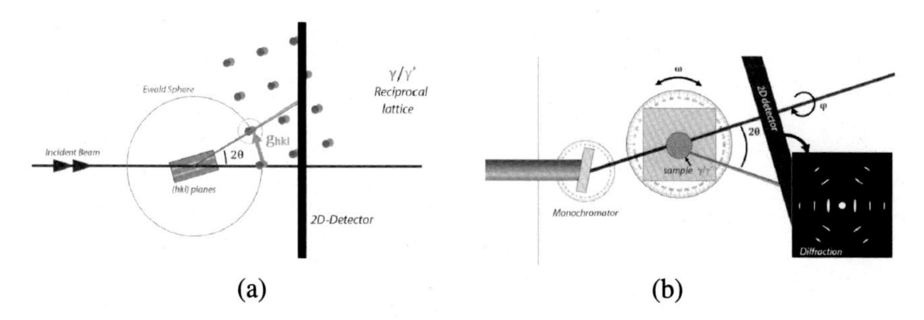

FIGURE 4.12 (a) Schematic representation of the principle of the Bragg diffraction from a single-crystal by a monochromatic X-ray radiation. (b) Experimental setup of an X-ray diffraction experiment. Courtesy of A. Jacques, T. Schenk, and R. Trehorel. Adapted from the PhD thesis of R. Trehorel [1447].

synchrotrons such as the ESRF (Grenoble, France), DESY (Hamburg, Germany), APS (Chicago, USA), or even sPring-8 (Hyōgo, Japan), makes it possible to provide an X-ray beam with a short wavelength which insures a low X-ray absorption rate by the material and allows transmission experiments on relatively thick specimens. Thanks to the development of suitable test devices, it is now possible to study the behavior of a material during in situ experiments [431,849]. Some experimental details about the up-to-date synchrotron diffraction experiments are given in the following.

The interaction between X-ray synchrotron radiation and a crystal produces a diffraction pattern or a diffractogram, depending on the type of X-ray detector used in the experiment. As reported in Section 4.2.2, the interaction between a monochromatic X-ray (with a wavelength λ) and a perfect single-crystal gives rise to a significant diffracted intensity in the direction defined by the angle θ_{hkl} when the Bragg's law, $2d_{hkl}\sin\theta_{hkl} = n\lambda$, where d_{hkl} corresponds to the spacing between the diffracted planes and θ_{hkl} is the incidence Bragg angle on the (hkl) lattice family plane, is fulfilled. If this condition is fulfilled, the diffraction vector g_{hkl} (defined as the difference between the incident wave vector $\vec{k}$ and the diffracted wave vector $\vec{k}'$) is a reciprocal lattice vector as illustrated Fig. 4.12(a) [683]. The lattice plane family (hkl) is then said to be under Bragg condition and to diffract. The corresponding diffraction spots can be recorded using a 2D-detector as illustrated in Fig. 4.9(b).

When the microstructure of the two-phase single-crystal superalloy γ/γ' presents rafts of γ' precipitates as the lattice parameters of both phases are close, a single diffraction spot is recorded when the X-ray detector is too close to the specimen (as illustrated in Fig. 4.12(b)). It is not possible then to discriminate the contributions of each phase.

The accurate measurement of the Bragg angle θ is necessary to precisely determine the lattice parameter of each phase, by derivation from the interreticular distances d_{hkl}. The use of a high-resolution diffraction tech-

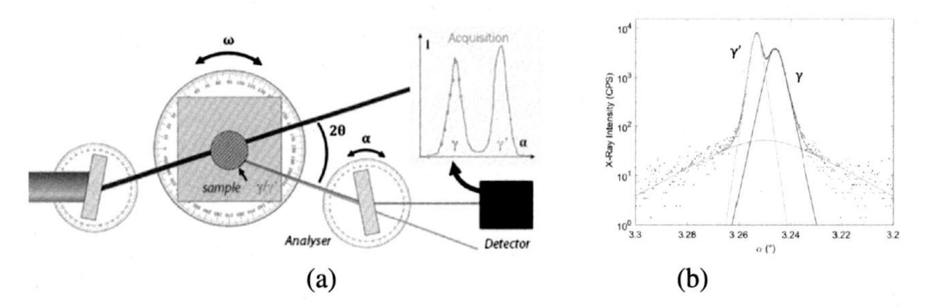

FIGURE 4.13 (a) Three crystal diffraction experimental setup combined with in situ creep testing experiment. (b) Experimental TCD scan of the 020 diffraction peak. Courtesy of A. Jacques, T. Schenk, and R. Trehorel. Adapted from the PhD theses of R. Trehorel and L. Dirand [1447,352].

nique allows sufficient angular resolution to discriminate both phases and have a good delineation of the diffraction peaks. In the research group of A. Jacques and T. Schenk, high resolution diffraction techniques combined with in situ creep tests have been developed in order to determine the γ and γ' lattice parameters and their evolution during creep [1448,1273]. In the last years, two diffraction techniques have been used, namely the triple crystal X-ray diffraction (TCD) and a far-field double-crystal X-ray diffraction (DCD) techniques.

TCD is a high-resolution reciprocal lattice mapping technique with a triple-crystal diffractometer for high energy X-rays [850,1299]. A drawing of the experimental setup is given Fig. 4.13(a). The specificity of this technique is that the angle measurement is limited to one node of the reciprocal lattice. The TCD profiles are fitted with mathematical adjustment, so that the lattice parameters can be determined with a relative precision $\Delta 2\theta/2\theta$ of 10^{-5} (see Fig. 4.13(b)). The acquisition of one angular scan is relatively short ($\sim$ 5 min) allowing us to measure misfit changes within the microstructure.

Double crystal X-ray diffraction (DCD) is usually performed with a detector (a CCD flat panel, for instance) relatively close to the specimen (as illustrated by Fig. 4.12(b)) to record the full Laue diffraction pattern [849]. Diffraction spots are the sum of the X-ray intensities diffracted by both phases, $I_{hkl}(\gamma) + I_{hkl}(\gamma')$. When a similar detector is placed farther away from the specimen (see Fig. 4.14(a)), the diffraction spots relative to γ and γ' can be differentiated. Actually, the longer the sample–detector distance, the larger the diffraction spot and the better the resolution [694,122].

Both techniques provide access to misfit measurements during an in situ creep test. Angular accuracy large detector distance DCD is equivalent to that of TCD, but the acquisition of a diffraction profile with this last technique is much faster (an acquisition only takes 7 seconds) [351]. TCD provides a 1D diffractogram of the measured relative intensity function of

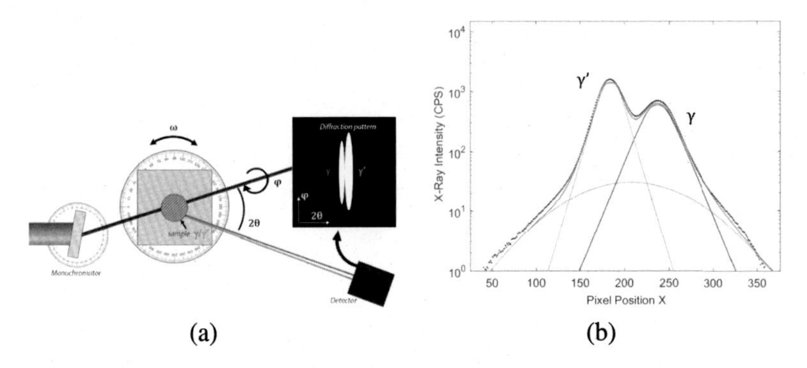

FIGURE 4.14 (a) Double crystal X-ray diffraction experimental setup. (b) Experimental DCD scan for the 020 reflection. The dots correspond to the experimental data, the lines correspond to the mathematical adjustments. Courtesy of A. Jacques, T. Schenk, and R. Trehorel. Adapted from the PhD thesis of R. Trehorel [1447].

the angle $\alpha(\approx \theta)$ during an ω scan, while by DCD, the diffraction patterns are recorded in 2D or 3D. The volume irradiated by the X-ray incident beam is typically in the range of 1 mm^3.

4.2.7.4 Characterization of short range order (SRO) in the γ phase

The γ phase is generally described as a "disordered" phase (random substitutional solid solution) although it has been reported to be subject to short-range order at temperature lower than 700–800°C [1089]. The identification of SRO is important for at least two reasons:

(i) Short-range order in a solid solution is associated with a strong localization of the deformation, leading to a heterogeneous deformation. This local atomic arrangement may also influence the macroscopic mechanical properties. The effect of local ordering on the deformation micromechanisms has been widely documented in binary alloys, such as Ni–Cr, Cu–Mn, or Ni–Mo alloys, as well as in the γ phase of Ni-base superalloys [1293,1031,1296,220,774,955,511,1115,1292,294, 1091,1286,1197,1228].

(ii) SRO is strongly influenced by the alloy chemical composition, so that the occurrence of SRO may be controlled by the choice of solute contents. For instance, W or Re have been pointed out as SRO-promoting elements [522,1092].

The experimental tools to investigate SRO in a binary solid solution are neutron diffraction and X-ray diffraction using synchrotron X-ray radiation. These diffraction techniques are very well documented both from theoretical and experimental points of view in the recent textbook by Kostorz [753]. Whereas the Bragg diffraction peaks intensity is simply correlated to the average structure factor of the unit cell, diffuse scattering will reflect deviations from a random arrangement to short-range order or to the formation of clusters. The diffuse scattered intensity consists of two components: (i) the inelastic component that can be experimentally discriminated by the analyzer and (ii) the elastic component which is the intensity of interest. The elastic coherent intensity results from different contributions. One comes from the local atomic arrangement (ordering) named ISRO. This intensity is written as follows:

$$I_{SRO} = \sum_n NC_A C_B |f_A - f_B|^2 \sum \alpha_{lmn} e^{ikr_{lmn}}, \tag{4.2}$$

where C_i stands for the atomic concentrations of the different elements in each sample and f_i denotes the respective coherent scattering lengths (which correspond to the atomic scattering factors in the case of X-ray). The use of neutron scattering instead of X-ray or electron diffraction in the case of the γ phase is justified by a better contrast in the diffuse intensity resulting from a maximization of the difference between the scattering lengths f_{Ni} and f_{Cr} using neutron rather than X-ray or electron. One of the challenging objectives in the investigation of SRO is to be able to separate within the diffracted intensity the component due to SRO. As the determination of I_{SRO} requires a large development based on theoretical background in diffuse scattering using X-rays and neutrons, this will be not developed in this chapter. A complete description of the separation techniques used to isolate each term within the scattering intensity and identify the SRO component can be found in the review proposed by Schönfeld [1285].

The γ phase is described as a complex NiCr solid solution. Describing this phase as a short-range ordered one means that, considering the atomic arrangement of the Ni and Cr atoms, there is a probability to form Ni–Cr atom pairs higher than in a solid solution where atoms are randomly distributed. The signature of SRO in the γ phase is thus the presence of {1 1/2 0}-type diffuse diffraction maxima, as observed in Ni–Cr alloy. An example of diffuse neutron scattering experiment performed in the γ phase containing Re is illustrated in Fig. 4.15. The investigated single phase is named γ_{MCRe} as its chemical composition is close to the γ matrix of the MC2 superalloy with an addition of 4 at%. Re. This single phase has been developed and casted by ONERA [211,202,210].

It is worth mentioning that, because of the presence of many solute atoms in the γ-phase (more than five alloying elements), there is no theoretical model which would be available to deduce quantitative data on SRO from diffuse neutron scattering experiments. A method based on dislocation pile-up analysis performed by TEM has been developed in the case of the γ-phase of Ni-base superalloys [1091]. As the occurrence of SRO induces the formation of dislocation pile-ups, these dislocation configurations can be used as a signature of SRO. Then, the quantitative

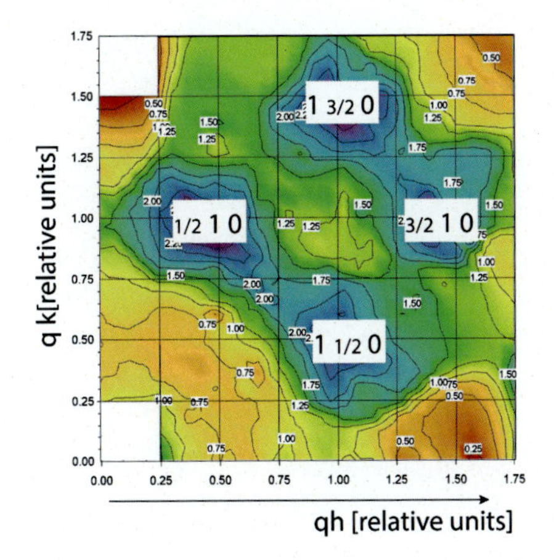

FIGURE 4.15 2D map of the diffuse scattering observed in γ_{MCRe} as an example for 25°C. Four peaks are clearly visible at the positions of the (1 1/2 0) family in this quadrant.

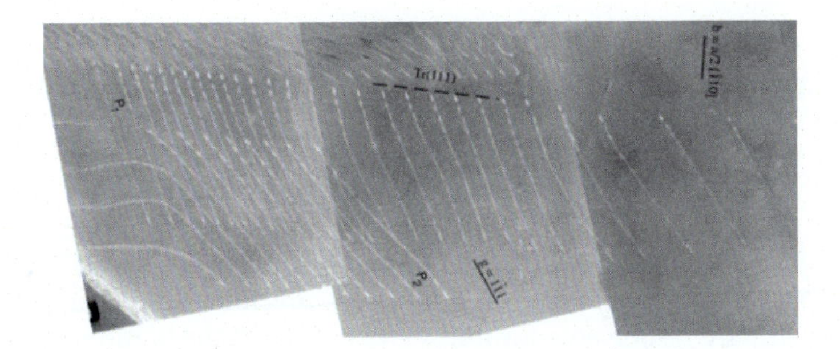

FIGURE 4.16 TEM observation of dislocation pile-ups observed in γ_{MCRe} after tensile test at 750°C. The glide plane is (111) and their Burgers vector is $\frac{1}{2}[\bar{1}10]$.

analysis of dislocation pile-ups (the determination of the experimental positions of the dislocations) gives access to the quantification of the degree of SRO and eventually its evolution as a function of the temperature. An example of dislocation pile-ups resulting from SRO observed in γ_{MCRe} after tensile test is illustrated by Fig. 4.16.

4.2.8 Characterization of the TCP phases

Some refractory solute elements, when they are present in relatively high concentration, may promote the precipitation of undesirable intermetallic phases, named as "topologically close-packed," or TCP, phases. The general chemical formula of these phases is AxBy, where A and B are transition metals. Various kinds of TCP phases can precipitate in Ni-base superalloys, depending on their chemistry and crystallographic structure (e.g., μ, P, σ, or R phases) [1163]. The μ and P phases have structures whose

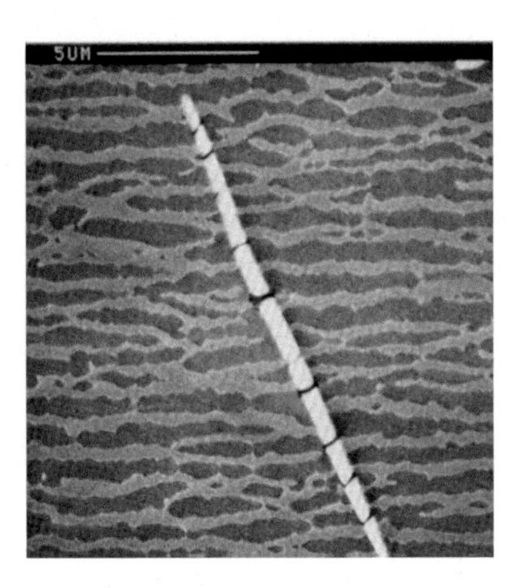

FIGURE 4.17 Fractured μ-phase particle in a preaged MC2 tensile specimen tested at room temperature (SEM on a longitudinal section) [1081]. Courtesy of P. Caron.

lattice is rhomboedral, while the R phase is orthorhombic, and the σ phase is tetragonal. Therefore these phases can be quite easily identified by electron diffraction in TEM.

These phases must be identified as they are known to be brittle and favor crack initiation in Ni-base single-crystals [1329]. In addition, they are known to induce a solid-solution depletion in refractory elements and to locally disturb the regularity of the γ' precipitate distribution [1329,801]. In the Ni-base single-crystal MC2 alloy, the precipitation of the μ-phase was potentially responsible for the decrease of both the high-temperature ductility and creep life if its volume fraction was over 1% [1329,801]. An example of the μ-phase observed in the MC2 superalloy by Simonetti and Caron is illustrated by Fig. 4.17.

4.3 Crystal structure defects' observation and characterization

4.3.1 Introduction

In crystalline materials, various defects can be observed:

- 0D-defects or point defects such as vacancies,
- 1D-defects or linear defects such as dislocations,
- 2D-defects or planar defects such as grain boundaries,
- interfaces, stacking faults or antiphase boundaries.

Depending on the scale of these defects, different experimental tools are used. Even if all these defects – except grain boundaries which are not present in a single-crystal – may have an effect on the mechanical behavior of single-crystal Ni-base superalloys, this section will be limited to the description of the dislocations observed in these alloys. The investigation of the dislocations allows us to identify the relevant parameter controlling the deformation. In such crystalline materials, the thorough understanding of the mechanical behavior requires a full comprehension of the interactions between the dislocations and the microstructural features and identification of the deformation micromechanisms associated with the motion of the dislocations.

This part is also aimed at giving some basic knowledge about the dislocations in the γ and γ' phases and some examples of experimental results, without going into the details about the different fundamental deformation mechanisms as these are detailed in Chapter 6. Besides, the theory of dislocations is exposed in the textbook written by Friedel [474].

Dislocation characteristics and dislocation microstructures are commonly studied by transmission electron microscopy. Many textbooks have been written which describe the fundamental aspects of the interaction between an electron beam and a crystal structure, and of the various imaging or chemical analysis techniques attached to transmission electron microscopy [217,378,1561,91].

The principle of dislocation imaging by TEM can be simplified as follows: a dislocation generates a strain field around its line (as illustrated Fig. 4.18(a)). As crystal lattice planes are bent in a neighborhood of the dislocation line, the Laue diffraction condition (or the Bragg law, which is its equivalent) is not fulfilled in the same way all around the dislocation line for a given lattice plane family: this diffraction condition might be fulfilled in a very limited volume along the dislocation line, where a small area of the bent crystal lattice planes of a lattice plane family has a right orientation in relation to the incident electron beam. It then produces a diffracted electron beam with a significant intensity, though the unstrained part of this lattice plane family is far from any right orientation to fulfill the diffraction condition. This difference of diffracted intensity results in a contrast between a limited zone close to and along the diffraction line and the surrounding perfect crystal in an image formed either by the transmitted electron beam or by the diffracted electron beam. The right condition to produce a contrast of dislocations in diffraction contrast TEM images is to orientate the crystal in relation to the incident electron beam so as to have only one crystal lattice plane family close to fulfilling the diffraction Laue condition (or, equivalently, the Bragg condition). Then, electrons exiting the specimen are either found in the transmitted beam or in the only diffracted beam produced for this crystal orientation. The images formed either by the transmitted or diffracted beam are called "two-beam diffraction contrast images." When the orientation of the crystal in relation to the incident electron beam is adjusted in order to correctly fulfill the diffraction condition in the sole limited zone where the crystal lattice planes of the plane family approaching the diffraction condition are bent due to their closeness to the dislocation line, the image formed by the transmitted beam would be a bright-field (BF) image. Actually, the orientation of the perfect crystal would not be close enough to the diffraction condition to allow for a strong diffracted intensity in the perfect crystal and a significant loss of electrons in the transmitted beam. The perfect lattice would then appear almost as bright as if the electron beam would not have had to cross the thin foil specimen. However, that bright field image would show a dark contrast where (or rather very close to where) the dislocation line crosses the crystal because diffraction would have occurred in this part or the strain crystal and, consequently, a loss of electrons would have occurred in the transmitted beam. The dislocation line then appears as a dark line in a bright field.

In the same way, the image formed by the diffracted beam would be a dark-field (DF) image because the diffracted intensity in crystal would be low except for the zone close to the dislocation line where a part of the crystal lattice bent planes close to the dislocation line would perfectly fulfill the diffraction condition. This image would show a bright contrast close and along the actual location of the dislocation line. By adjusting further the orientation of the crystal in relation to the incident beam to increase the deviation from the Bragg angle for the considered lattice plane family and then to lower the diffracting portion of the bent lattice plane down to its most bent part, the dislocation line contrast narrows and appears as a fine bright line in a dark field. Such an image is called a "weak-beam dark-field image" (WB-DF), with "weak-beam" referring to the low intensity diffracted beam produced by the small most-bent part of the lattice planes close to dislocation line core. Weak beam images allow for a precise identification of dislocation characteristics (Burgers vector, line direction) of partial dislocations of a slightly dissociated dislocation, or of dislocations at the interface with precipitates, or of individual dislocations in high dislocation density microstructures.

This imaging technique has been first reported by Cockayne et al. and later by de Ridder and Amelinckx, and Stobbs and Sworn [268,319, 1187,1363,267]. An example of WB-DF image of a dislocation is shown in Fig. 4.18(b).

4.3.2 Defects in Ni-base superalloys (dislocations and planar defects)

4.3.2.1 Dislocations in the γ-phase

The γ phase is an fcc structure and can be described as an ABCABC stacking of {111} compact planes where the dislocations preferentially slip. Therefore a perfect matrix dislocation has a Burgers vector of the type

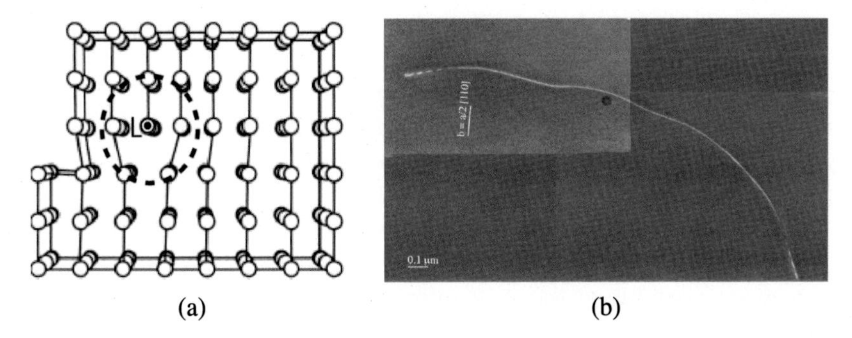

FIGURE 4.18 (a) Schematic representation of an edge dislocation in a crystal, L points out the dislocation line; (b) TEM image of a singled out dislocation in the γ-phase of an Ni-base superalloy, using weak-beam dark-field condition. The perfect crystal appears here in dark and the dislocation line in bright contrast. Adapted from [1088].

$\vec{b} = 1/2 < 110 >$. The $1/2 < 110 >$ dislocations can dissociate in their $\{111\}$ slip plane into two partial dislocations whose Burgers vector has the type $1/6 < 112 >$, according, for instance, to the dissociation reaction

$$1/2 < 110 > \rightarrow 1/6[211] + SF + 1/6[12\bar{1}]. \tag{4.3}$$

Those partial dislocations are "Shockley" partial dislocations. As a $1/6 < 112 >$ vector is not a crystal lattice vector, the shearing of the crystal propagated by the glide of a partial dislocation generates the formation of a stacking fault (SF) in the fcc structure. The glide of the second partial in the wake of the first one achieves the shearing of the crystal and restores the perfect stacking of the fcc structure. Such a reaction with the creation of stacking fault was observed in the γ matrix of AM1 superalloy by Décamps et al. [321]. The stacking fault energy (SFE) can be determined from the measurement of the distance between the two Shockley partial dislocations [1088,321,322,272]. In single-crystal Ni-base superalloy, the SFE ranges from 20 to 30 mJ/mm^2.

4.3.2.2 Dislocations and planar defects in the γ'-phase

As the γ' phase is an ordered phase, type vectors are not crystal lattice vectors. The shortest crystal lattice vectors in these directions are <110> type vectors. A dislocation with a <110> type Burgers vector in the γ' phase can dissociate into two partial dislocations with 1/2<110> Burgers vectors, generating an antiphase boundary (APB), according to the equation

$$< 110 > \rightarrow 1/2 < 110 > + APB + 1/2 < 110 > . \tag{4.4}$$

This shear mechanism is commonly observed in the γ' phase after tensile test at temperatures close to room temperature and after creep test at high temperature [950,1275,769]. Two perfect matrix 1/2 <110 > dislocations shear the γ' precipitate with the creation of an APB. The antiphase boundary energy (APBE) can also be determined from the separation width between these two 1/2 <110> dislocations. The APBE has been estimated by different authors either in Ni_3Al model alloy, or in the γ' phase of the CMSX2 or CMSX6. It ranges from 100 to 160 mJ/m^2 [525,72,357].

Another shear mechanism of the γ' phase was also observed in the intermediate temperature range (around 760°C) [709]. Two perfect matrix dislocations combine at the γ/γ' interface and then dissociate into two Shockley superpartials. A dislocation having the Burgers vector 1/3<112> is called super-Shockley. This dissociation reaction is associated with the creation of super-intrinsic/extrinsic stacking faults (SISF/SESF) according to the reaction

$$1/2[10\bar{1}] + 1/2[10\bar{1}] \rightarrow 1/3[2\overline{11}] + SISF + 1/3[2\overline{11}] + SESF + 1/3[\bar{1}2\bar{1}]. \tag{4.5}$$

The defect is intrinsic (SISF) if it occurs on a single {111} plane and extrinsic (SESF) if it occurs on two planes. SESF were first observed by Kear, Oblak, and Giamey [709]. Lours et al. observed, during a TEM in situ deformation experiment carried out in the γ' phase of CMSX2, the formation and propagation of SISF [869]. These energies have been estimated in different studies. As an example, the value of the energy of an SISF in CMSX-2 has been found to range from 15 to 28 mJ/m^2 and in the AM1, the SISF energy has been estimated to be 15 mJ/m^2 and the SESF energy as 25 mJ/m^2 by Véron et al. [1492,129,1485]. In general, the energy of an SESF is shown to be slightly higher than that of an SISF.

Finally, a complex stacking fault (CSF) can be observed. This defect results from a shear by a dislocation with Burgers vector $\vec{b} = 1/6 < 112 >$ which disturbs not only the order but also the stacking sequence of the {111} planes (Fig. 4.19). Thanks to weak-beam imaging and using the framework of isotropic elastic theory, Condat and Décamps estimated this energy to be 80 mJ m^{-2} in AM1 [272] and Décamps et al. found an energy of 70 mJ m^{-2} in AM3 [322].

To sum up, the perfect stacking of planes in the $L1_2$ structure is illustrated in Fig. 4.19. As in the matrix, a dislocation can dissociate to minimize energy and produce several types of defects:

- Antiphase boundary (APB),
- Superlattice stacking fault (intrinsic or extrinsic, SISF or SESF),
- Complex stacking fault (CSF).

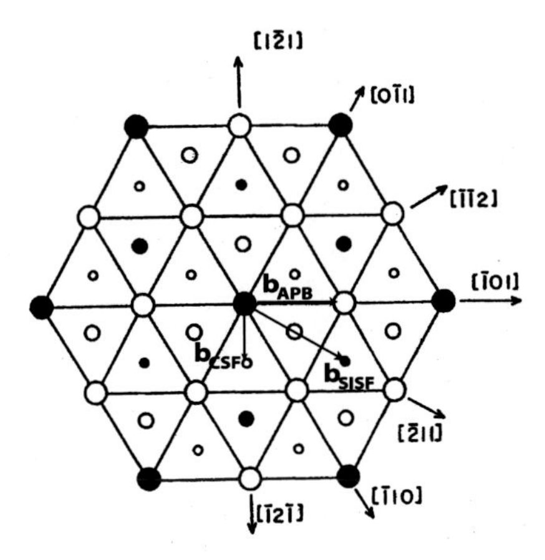

FIGURE 4.19 Schematic representation of the plane stacking in the $L1_2$ structure. The projection contains three successive {111} planes of Ni_3Al, depicted by the different sizes of the atom sites. The white points represent the Ni sites and the black points represent the Al sites. Different planar defects resulting from the slip of dislocations are pointed out: the slip of a 1/2<110> dislocation leading to the generation of an antiphase boundary (APB), the generation of a superlattice intrinsic stacking fault (SISF), and the generation of a complex stacking fault (CSF).

4.3.3 Observation of planar defects and dislocations

4.3.3.1 Conventional TEM experiments

The direct observation of crystal lattice defects is usually thought to fall within the domain of transmission electron microscopy (TEM). TEM experiments are commonly performed to observe the dislocations and the planar defects but also to identify the Burgers vectors of the dislocations. As the electron beam in TEM goes through the specimen, it must be a thin foil less than 100 nm thick. This technical constraint can be a disadvantage for transmission electron microscopy compared to the new imaging techniques (such as ECCI, described in the next paragraph, which is performed on bulk samples). Nevertheless, the advantage of TEM is the high image resolution (usually between 0.04 and 0.08 nm) it can provide and the easy way to identify the Burgers vector of a dislocation. The Burgers vector of a dislocation can be determined using the familiar criterion for dislocation invisibility [378,709], which is $\vec{g} \cdot \vec{b} = 0$ where $\vec{g}$ is the diffraction vector (see Figs. 4.20–4.25).

Some example of dislocations observed in single-crystal Ni-base superalloys are shown in the following images. No details are given about the elementary deformation mechanisms at the origin of these dislocation configurations. They will be developed in detail in Chapter 6. All these TEM images have been made by P. Caron and his coworkers.

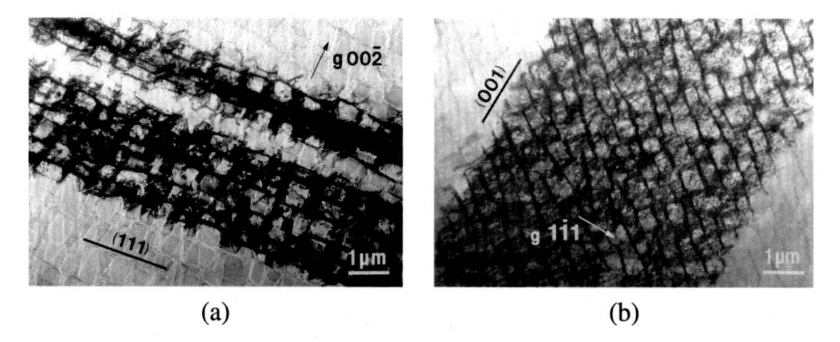

FIGURE 4.20 Deformation bands with high densities of dislocations in CMSX-2 tensile specimens: (a) [011] oriented single-crystal specimen strained to 2% at room temperature; octahedral slip on a {111} plane; (b) [111] oriented single-crystal specimen strained to 2% at 650°C; cube slip on the primary (001) [110] slip system [208]. Courtesy of P. Caron.

FIGURE 4.21 TEM observations in CMSX-2 creep specimens tested at 760°C (γ' precipitate size, 450 nm): (a) dark-field TEM image in the γ-channels where the deformation operates by 1/2<110> dislocation glide and climb mechanisms; observation of pairs of matrix dislocations bounded by an antiphase boundary as they enter the precipitates as <110> superdislocations; (b) observation of stacking faults resulting from the shearing of the γ' precipitates by {111}<112> slip [207]. Courtesy of P. Caron.

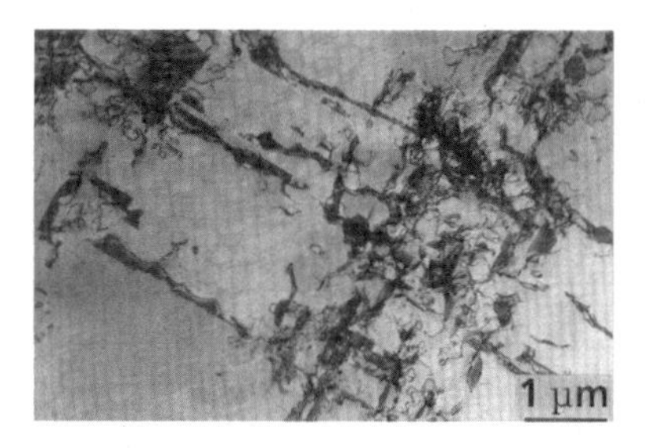

FIGURE 4.22 TEM observation in CMSX-2 after creep at 760°C (γ' precipitate size, 230 nm): heterogeneous deformation due to cooperative shearing of the γ/γ' structure by {111}<112> slip generating extended superlattice intrinsic and extrinsic stacking faults [207].

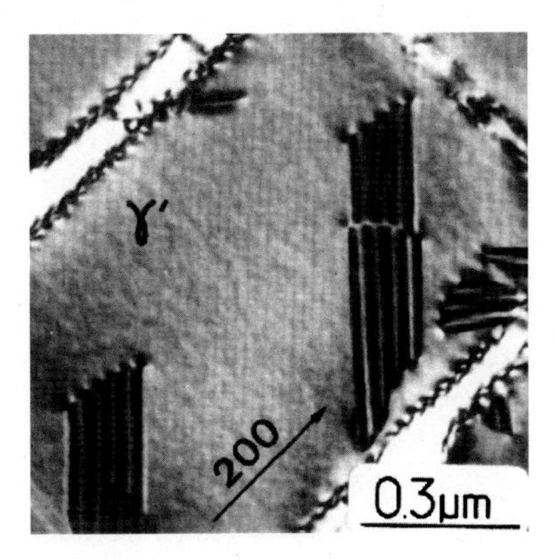

FIGURE 4.23 SESF–SISF observed within a γ' precipitate of the CMSX-2 Ni-base superalloy. Courtesy of P. Caron.

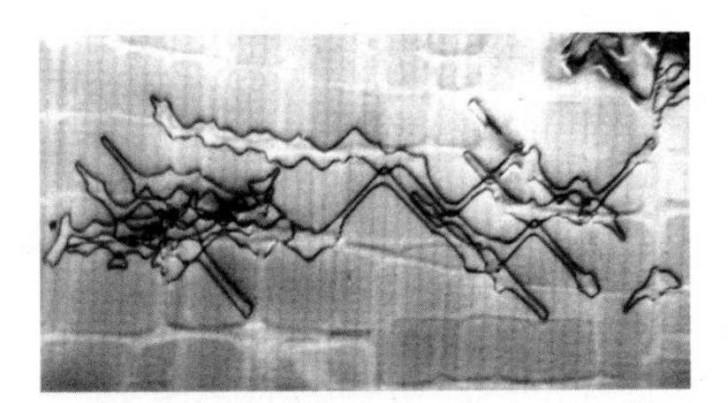

FIGURE 4.24 By-passing of γ' precipitates by cross-slip of perfect matrix dislocations in the CMSX-2 superalloy. Courtesy of P. Caron.

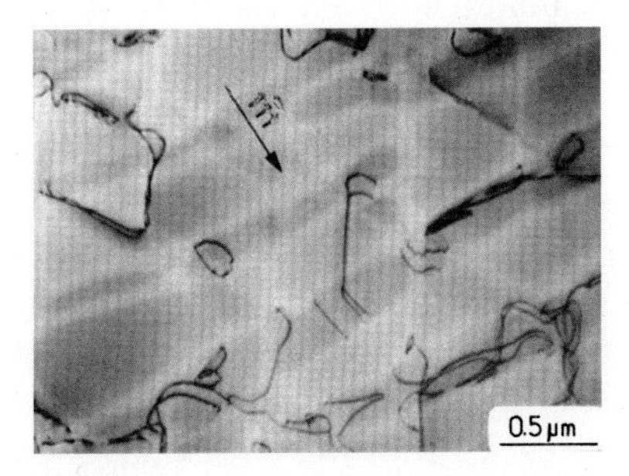

FIGURE 4.25 Pairs of 1/2<110> dislocations observed in the CMSX-2 superalloy after low cycle fatigue testing at 760°C. Courtesy of P. Caron.

4.3.3.2 *Defect 2D and 3D imaging using STEM and HAADF detector*

Defect analysis in crystalline materials has been and still is essentially performed by conventional transmission electron microscopy (CTEM) imaging methods. However, as early as during the 1970s, it was shown that scanning transmission electron microscopy (STEM) could also be successfully used for defect analysis [641,640]. These early publications highlight the many benefits of STEM versus CTEM.

Scanning transmission electron microscopy (STEM), as CTEM, is performed on thin foil specimens. The incident electron beam generally forms a fine probe (down to 0.05 nm diameter) on the specimen, much less spread than the "parallel" incident beam usually illuminating the imaged area in CTEM. The area considered for imaging or analysis is scanned by this probe according to a raster, as in SEM, but in STEM, the incident beam is kept parallel to the microscope axis, as in TEM (and unlike in SEM). In that way, a "stationary" diffraction pattern forms in the back focal plane of the objective lens of the TEM. The STEM detectors are located under the specimen, after the transmitted and diffracted beams have traveled in the magnifying lenses of the TEM which are worked in diffraction mode. The magnifying lenses are then only be used to modify the camera length which results in increasing the distance between the transmitted beam and the diffracted beams (or other scattered electrons) when they reach the level of the microscope column where the detectors are located. A small-diameter disk detector receives the transmitted beam which intensity is transformed in a signal sent to a display (computer display or, at one time, a CRT screen) submitted to the same scan as the specimen. A bright-field image is then formed on the display. An annular detector is used to receive the diffracted beams and other scattered electrons and to form a dark-field image in the same way as for the bright-field detector. In combination with various spectroscopies (EDS, EELS), elemental maps can be recorded simultaneously with the image with a nanometric resolution.

The development of this technique is due to the limitation in the resolution of conventional TEM because of the aberrations in the electromagnetic lenses, which increases with the image magnification [1561]. In STEM, as the magnifying lenses are not used to form the image, the resolution is limited by the quality and size of the electron probe, while the magnification is simply controlled by varying the area of the thin foil specimen scanned by that probe. Thus, the ability to obtain a fine and high-intensity coherent probe is paramount in HR STEM. Recent improvements in spherical aberration correction of the probe-forming lenses have made the subnanometer high-resolution scanning transmission electron microscopy (HR STEM) a useful technique for imaging the atomic columns.

STEM also allows atomic mass contrast (Z-contrast) imaging, which is achieved through the use of a high-angle annular dark-field (HAADF) detector. This type of detector collects electrons scattered by the sample foil at high angles, greater than 50 milliradians. At such great angles, Bragg scattering becomes negligible, and most of the signal intensity arises from incoherent elastic scattering, which is only sensitive to the elemental atomic mass. Given a sufficiently small electron probe and correct specimen–beam alignment, the selection of the lone Z-contrast enables the formation

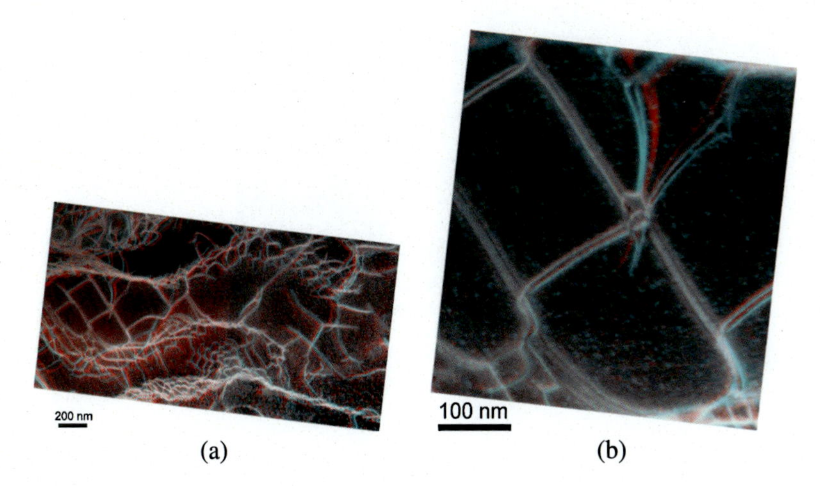

FIGURE 4.26 3D images of dislocations observed in a Ni-base superalloy. A γ' raft with strong interactions between superdislocations is observed. The technique allows us to easily distinguish between dislocations that interact in the volume and those that overlap in each of the two-dimensional images, even when high dislocation densities are present. Courtesy of L. Agudo Jacome.

of images with atomic resolution. Spaces between atoms nuclei are regions of very low Z and thus appear with a dark contrast in the image. HAADF STEM is quite ideal for high-resolution imaging, as resulting images should not suffer from drawbacks inherent to phase-contrast TEM, such as contrast reversals due to strong effects of specimen thickness and image defocus [1003]. Thus, STEM imaging can be performed on thicker samples than CTEM, and bend contour and auxiliary contrast effects can be suppressed while retaining defect contrast.

As a consequence, the high resolution provided by STEM imaging enables one to study the detailed structure of defects that are usually difficult to image in conventional weak-beam TEM [1100,1101]. As an example, the detailed structure of the 1/3 <112> superpartials at the endings of 1/2 <112> dislocations, which are the main source of strain accumulation in superalloys subjected to conditions favoring the primary creep deformation mode, has been first observed by Vorontosov et al. in CMSX-4 by high-resolution scanning transmission electron microscopy [1512].

Furthermore, during the recent years, efforts have been made to develop 3D imaging of dislocations. This has been made possible by the enhancement of TEM equipments (analytical 200 kV TEM fitted with a field emission gun (FEG TEM) and a high-angle annular dark-field (HAADF) detector). The 3D images are built from stereo pair images combined into one anaglyph. A more detailed description of the method developed by L. Agudo Jácome et al. can be found in [5]. One example is given in Fig. 4.26. Such experiments are very useful to visualize the location of the

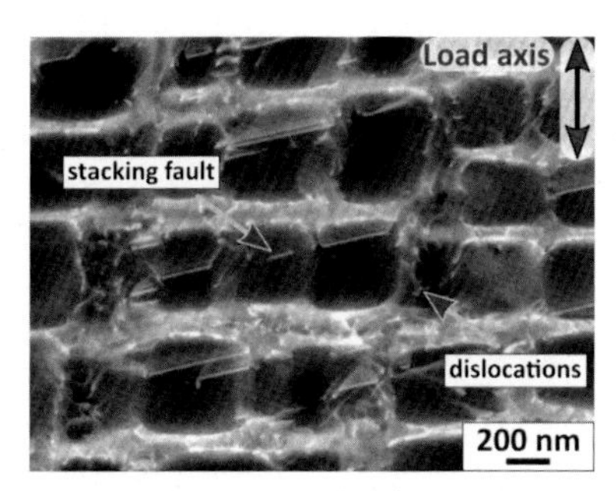

FIGURE 4.27 ECCI-imaging of deformed Mar-M200. A dense array of dislocations is visible. Courtesy of P. Kontis.

dislocations in relation to the whole microstructure and to identify the deformation micromechanisms.

4.3.3.3 *Electron channeling contrast imaging (ECCI)*

Electron channeling contrast imaging (ECCI) is a powerful technique for observing crystal defects, such as dislocations, stacking faults, twins, and grain boundaries in the scanning electron microscope. Scanning electron microscopy provides detailed information on surface topography via secondary electrons (SE) and surface chemistry via backscattered electrons (BSE) and energy dispersive X-ray spectroscopy (EDS) [535]. In addition, SEM can provide crystallographic information via electron backscattered diffraction. With a new generation of SEM instruments, the direct observation of dislocations using scanning electron microscopy by electron channel contrast has become an alternative to the TEM. ECCI is more and more used as it offers a direct observation of dislocations in bulk samples. ECCI provides contrast features comparable to dark-field TEM, with the advantage that the images are obtained on bulk samples rather than on thin foils. However, resolution and contrast are not as good as in TEM. An overview of the literature on channeling contrast of crystal defects and on theoretical backgrounds is given in the paper of Zaefferer et al. [1634]. An example of an ECCI image of dislocations observed in the MC2 superalloy by P. Kontis is illustrated Fig. 4.27.

4.3.3.4 *TEM in situ straining experiments*

TEM in situ straining experiments allow deforming a small tensile specimen inside a transmission electron microscope at various temperatures using a specific sample holder. An example of such an experimental tool is illustrated Fig. 4.28.

In situ straining experiments allow observing in real time the dynamics of the dislocations and recording the nucleation and the slip of dislocations during straining. At the present time, in situ straining experiments are still

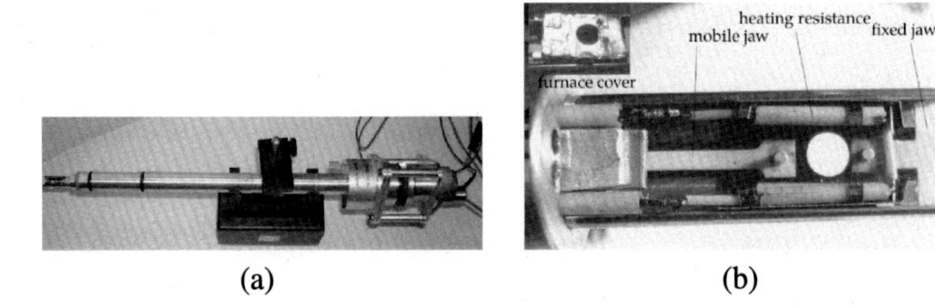

FIGURE 4.28 "Home-made" straining holder developed at CEMES-CNRS, Toulouse, France: (a) general overview of the straining holder allowing tensile test at high temperature (>1000°C); (b) enlargement of the tip of the straining holder where the specimen is located between the heating resistances. Courtesy of F. Mompiou.

developed by the group in Toulouse (at CEMES-CNRS) and were also performed by the group at Halle with a high voltage microscope [939,938]. Experimental details concerning in situ straining experiments have been presented in different papers [422,1087].

From a scientific point of view, an in situ straining experiment has been used with a Ni-base single-crystal superalloy to identify and quantify dynamic processes associated with the controlling deformation mechanisms [1090,1093]. For example, in-situ TEM tensile tests performed by Legros et al. at 850°C in the MC2 rafted structure have shown microtwinning as well as unexpected superdislocations splitting according to a dissociation mode that involves stacking fault ribbon creation [815]. Moreover, the {100} slip of superdislocations in the γ' phase of the CMSX-2 superalloy has been studied from 250°C to 350°C in order to understand the increase of the critical resolved shear stress when the temperature increases [261]. The controlling deformation micromechanisms have been also investigated from 35°C to 500 °C in an fcc γ phase alone whose composition was close to that of the matrix of the MC2 superalloy. In this case, the short-range order acting as the prevalent strengthening mechanism has been confirmed. A quantitative approach for characterizing SRO has been developed using the experimental positions of dislocations within a pile-up (as illustrated in Fig. 4.29) and taking into account the effects resulting of the small thickness of the TEM specimen [1091,1094].

4.3.4 Chemical characterization

4.3.4.1 Chemical analysis using SEM

Scanning electron microscopy (SEM) allows observing the surface of a sample using various forms of interaction between electron and matter. In SEM, the beam is focused on the specimen and scans the sample according to a raster. An image is formed using the secondary electrons produced by the interaction between the beam and the specimen. This

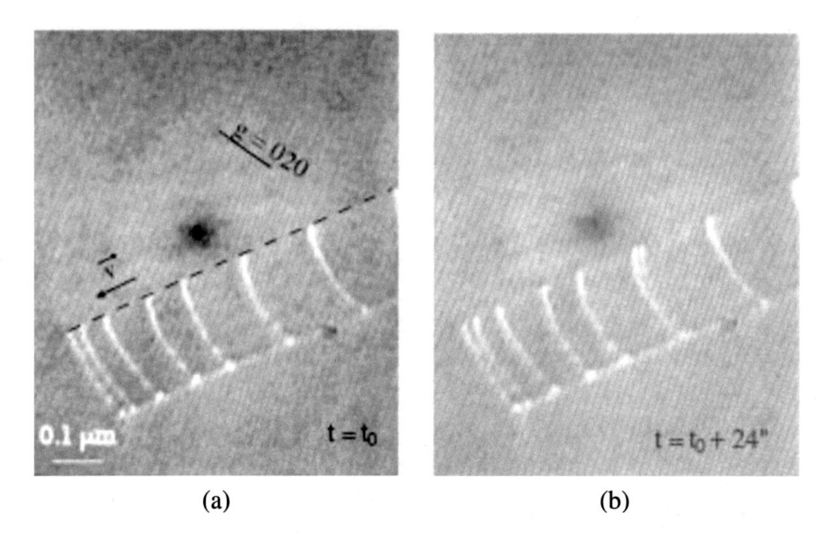

(a) (b)

FIGURE 4.29 Example of dislocation pile-up observed at $t = 0$ s and $t = 1'47''$ during TEM in situ experiment in a γ-phase of Ni-base superalloy. The slip plane is (111) and the Burgers vector is 1/2 [011]. The diffraction vector is $g = 020$. The dislocations are imaged using weak-beam dark-field conditions.

method provides images of areas of the specimen which are much larger than what is usually done by TEM. It is then possible to couple this imaging technique with a technique allowing chemical analysis. X-ray energy dispersive spectroscopy EDX (or EDS, energy dispersive X-ray spectrometry) provides qualitative or quantitative elemental analysis, depending on the experimental conditions, of the chemical elements in the specimen with the resolution corresponding approximately to the electron interaction volume (a sphere of a few hundred nanometers to a few micrometers). The X-ray photons resulting from the interaction between the incident electrons and the material are collected by an energy dispersive spectrometer which provides a spectrum of X-ray intensity versus energy. The peaks of this spectrum are characteristic of the elements present in the sample.

4.3.4.2 TEM spectroscopies: EELS and EDX

Energy dispersive X-ray spectroscopy (EDX) uses the energy distribution of the X-ray photons emitted by a specimen submitted to an electron beam. These X-ray photons have an energy characteristic of the element the electrons have encountered, which will then be identified. The studied X-ray energy domain is generally quite large (0–20 keV) and it allows global chemical information to be obtained rather quickly. However, the resolution in EDX is much lower than that reached by EELS. Different information can be obtained: either a 2D map showing the distribution of the chemical elements within the investigated specimen, or a profile line which makes it possible to quantitatively identify the chemical composition variations along a chosen line within the sample. This method enables one to obtain a better precision in the analysis because it focuses on a smaller area by sweeping the line several times in order to improve the statistics. However, the light elements cannot generally be accurately detected by EDX. Electron energy loss spectroscopy (EELS) is an analysis technique allowing a measurement of the local chemical composition in a specimen at a subnanometer scale. This analysis results from the measurement of the energy lost by the electrons by inelastic interactions with the atoms of the specimen. An EELS analysis leads to a spectrum composed of several peaks. The first peak at 0 eV is the zero loss peak from electrons which have not lost any energy during their travel through the thin foil specimen. Then there are the plasmon peaks resulting from the collective oscillation of the free electrons of the specimen following the passage of the incident electrons. These peaks lie at the low energy level on the spectrum. Finally, around several hundred eV, other peaks are observed that correspond to core loss, reflecting interactions with core electrons of the atoms. Each peak is characteristic of a chemical element. EELS analysis is not efficient for heavy elements, which is why it is necessary to often combine EDX and EELS experiments. EELS experiments can be done using localized probe or filtered image mode and also using STEM mode.

4.3.4.3 *Atom probe tomography*

Field ion microscopy (FIM) and atom probe tomography (APT) are microscopy and microanalysis techniques based on field emission. These instruments allow us to produce 2D and 3D maps at the atomic scale. The FIM was invented in 1951 by Erwin Wilhelm Müller [984]. This microscope was the first which allowed visualizing the organization of atoms on the surface of a metal alloy. The principle of this imaging technique rests upon the electric field induced ionization of inert gas atoms in the vicinity of a very sharp metallic needle. An imaging-gas is introduced into a vacuum chamber containing a specimen shaped as a sharp needle. A high voltage of a few kilovolts is then applied that generates an intense electric field at the apex of the specimen of the order of 10^{10} V m^{-1}. The gas atoms are then polarized and attracted by the sample surface where they are ionized. These positive ions are then repelled from the tip surface on a trajectory close to normal to the tangent plane of the specimen surface. A detector, consisting of a phosphor screen and a charge amplifier, is placed in front of the sample in order to collect the ions which form an image of the specimen surface. The ionized gas acts as the "imaging gas." However, the images do not allow us to identify the chemical nature of the atoms at the surface of the specimen.

The initial atom probe was devised by Muller in 1969. It combined an FIM in which a strong electrical field allowed for evaporation and field

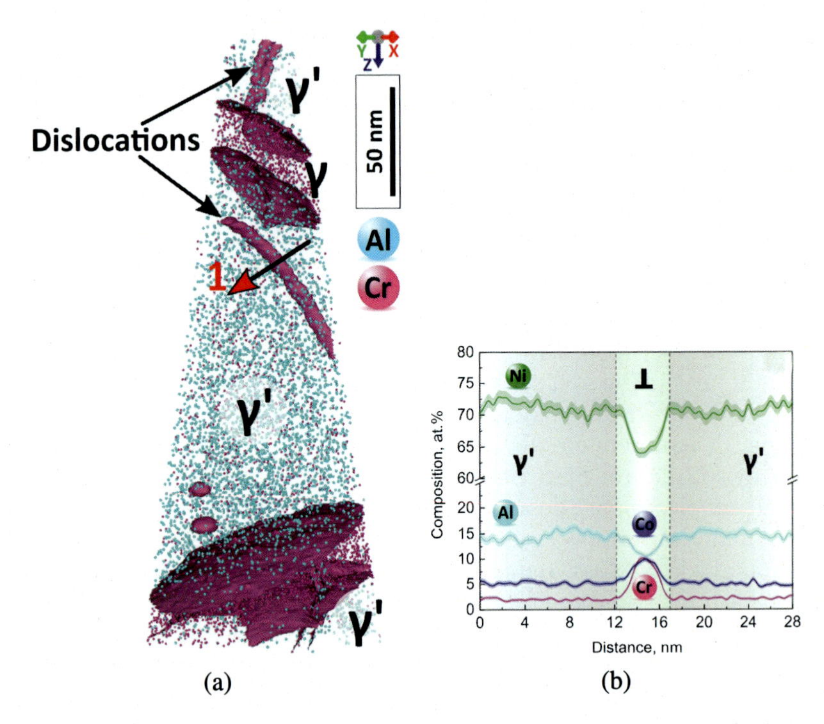

FIGURE 4.30 Example of 3D APT reconstruction illustrating the chemical element distribution across a dislocation (arrow 1) observed in a Ni-base superalloy. Courtesy of P. Kontis.

emission of the atoms from the surface of the needle specimen tip with a time of flight mass (TOF-MS), and produced concentration profiles of a fixed element. It was developed in France since 1974, first by Gallot, Sarrau, and Bostel and then by Blavette, Deconihout, and Menand [483,137]. Atom probe tomography (APT) uses a combination of a position-sensitive detector and a time of flight mass spectrometer which identify the chemical nature of each evaporated atom. The spatial resolution is less than a nanometer. The evaporated volume can be reconstructed atom by atom. It is also possible to measure a concentration profile across the reconstructed volume. Atom probe tomography analyses must be performed on very sharp needle specimens with a curvature radius of their tip smaller than a few nanometers. Nowadays this technique is widely used for the chemical characterization of Ni-base single-crystal superalloys [117,119,118,504,746].

Thanks to its spatial resolution, the chemical segregation at lattice defects (e.g., dislocation) can be showed as illustrated by Fig. 4.30. It is worth mentioning that over the years, the various types of lattice or structure defects observed in Ni-base superalloys have been investigated in detail by TEM, and the first evidences of solutes segregating to stacking faults based on TEM observations were reported in 2012 [1512]. Direct observation and quantitative near-atomic scale segregation of solute elements at dislocations in superalloys was performed by wide-field-of-view APT much later in 2018 [746] and has become the pertinent experimental tool for such quantitative nanoscale analysis.

4.4 Conclusion

Depending on the microstructural feature that has to be characterized, different experimental tool can be used for Ni-base superalloy investigation. During the last decades, technological breakthroughs in high-resolution characterization have enabled structural and compositional imaging at near-atomic scale. They have revealed the interactions of solute atoms with crystal defects. It is then obvious that current superalloy design needs precise characterization and prediction of the microstructure and of the local chemical composition at near-atomic scale in order to give a thorough understanding of the high-temperature deformation micromechanisms, where the solute atoms diffusion is active. This topic has been recently discussed in the viewpoint article by P. Kontis [745]. In addition, the dynamic approach (using in-situ synchrotron or TEM) becomes more and more required in order to access the in-service life conditions.

Acknowledgments

The authors fully acknowledge all the colleagues who have nicely agreed to provide some of their remarkable scientific results to illustrate this chapter: P. Caron, J. Cormier, P. Kontis, L. Agudo Jácome, A. Jacques, T. Schenk, D. Texier, and R. Trehorel.

CHAPTER

5

Mechanical characterization at high temperature

Vincent Bonnand[a], *Jean-Briac le Graverend*[b,c], *and Marion Bartsch*[d]

[a]Department of Materials And Structures (DMAS), Onera, Chatillon, France, [b]Department of Aerospace Engineering, Texas A&M University, College Station, TX, United States, [c]Department of Materials Science Engineering, Texas A&M University, College Station, TX, United States, [d]Institute of Materials Research, German Aerospace Center (DLR), Cologne, Germany

5.1 Introduction

Single-crystal superalloys are preferred materials for gas turbine blades and play a crucial role in their performance due to their high strength and creep resistance at high temperature. The requirements on materials for turbine blades are highly challenging. In many gas turbines, the turbine inlet temperatures exceed the material capability, so that the blades have to be internally cooled with pressurized air from the compressor and protected with thermal barrier coatings. The resulting thermal gradient between the outer and inner surface of a cooled blade causes multiaxial stresses, mainly in-plane biaxial compressive at the hot surface and tensile at the cooled surface. Further, tensile mechanical loads are superposed in the case of rotor blades due to rotation. Thermal and mechanical loads vary during an operating cycle, which can last from several hours in the case of gas turbines in aero-engines up to several hundred hours in the case of stationary gas turbines. Materials are, therefore, subjected to thermo-mechanical loadings. In addition, aerodynamic forces generate further high cycle fatigue. Besides the mechanical loading, the environment of hot combustion gas plays a significant role in mechanical variations and damage evolutions. In most cases the surfaces of turbine blades

Nickel Base Single Crystals Across Length Scales
https://doi.org/10.1016/B978-0-12-819357-0.00012-3

are protected by metallic oxidation protective layers, and additional ceramic thermal barrier coatings are deposited in the case of first turbine stages. The role of such coating systems on damage and failure behavior is considered in Chapter 10.

Before materials get into service, laboratory testing is mandatory and serves different purposes. In materials development, it is important to obtain information on the effect of microstructural features and chemical composition on relevant mechanical properties. In this context, experiments are designed to elucidate deformation and damage mechanisms under defined conditions. For numerical modeling, mechanical data at different length scales are necessary, and for safe component design, reliable information on the materials' performance under realistic conditions is required. This chapter aims at providing an overview of advanced testing and characterization methods developed for evaluating the mechanical properties of superalloys in relevant testing conditions. At first, macroscopic testing with increasing complexity is discussed, starting with uniaxial monotonic mechanical loading at constant temperature and further presenting techniques for applying multiaxial mechanical loads. Concerning fatigue loading, one section is dedicated to high and very high cycle fatigue. A separate section deals with tests capturing the material behavior as in-service conditions, viz. nonisothermal loading conditions. Further, examples for small scale testing down to distinct phases are presented, and the final section focuses on observation and characterization of damage and crack evolution. Concerning elastic properties and advanced test methods for their characterization, the interested reader can refer to Chapter 3.

5.2 Environment effects in testing at high temperatures

Most laboratory high-temperature experiments are performed in air and, therefore, subjected to oxidation. As a result, oxide scales grow and affect the mechanical behavior and lifetime. Several studies demonstrated that the oxide scale growth comes along with vacancy injection [25]. This process creates sub-scale pores and new phases [388]. Further, experiments performed by Dryepondt et al. [366] revealed that oxidation modifies the plastic strain rate in creep during the secondary creep stage.

It can be summarized that oxidation induced changes in the microstructure, phase composition, and mechanical behavior, which impair the representative volume approach used to characterize material properties from a continuum mechanical standpoint. Therefore, some efforts are necessary to overcome oxidation effects in creep and in fatigue by performing mechanical experiments at high temperature in vacuum or inert atmosphere, as in [94,93]. The lack of experimental data is probably because of the experimental challenge of sealing the test chamber: the load train in the test frame consists of different materials with differing thermal expansion coefficients. Further, hutches and fittings for extensometers or quartz glass windows for optical strain measurements are difficult to seal when heating the specimen in the test chamber.

It is hereby essential to point out that laboratory atmospheres are different from combustion gases affecting components in service. Therefore, studies of degradation and failure behavior during mechanical testing at high temperature in air have limited informational value if oxidation- or corrosion-assisted-damage mechanisms have a significant influence on the mechanical behavior and lifetime. Experiments under realistic atmospheres at high temperature are similarly challenging to experiments in vacuum or inert atmosphere. The interested reader can refer to [1366].

5.3 Uniaxial testing under isothermal conditions

5.3.1 Creep and monotonic mechanical loading

Creep appears when a constant load is applied for a long period of time at a temperature that is above one-third of the melting point. Creep is always applied using a dead weight. In fact, hydraulic test machines are not suitable to apply a constant load for a long time. The applied force is controlled by servo-pumps resulting in fluctuations of the applied load around the controlled value and, consequently, to a superposed fatigue load. It is important to not constrain the specimen in transversal displacements during creep tests. Indeed, it is known that lattice rotation, i.e., a large deformation phenomenon can occur during creep at high temperature [516,31]. Thus, constraining the specimens would lead to extra stresses modifying the mechanical behavior and damage evolution of the material. Depending on the generation of the alloy investigated, the shape of the creep curve can be totally different: from three creep stages well delimited to what may appear to be a continuous increase in the plastic strain rate without any noticeable secondary creep stage. Constitutive models need to have well-defined experimental secondary creep stages to calibrate the exponent of the viscous flow. The uncertainty in the viscous flow exponent forces the experimentalists to couple creep tests with other characterization methods, such as monotonic tensile tests at different strain rates. Regarding the strain measurement, two options are available: the use of extensometers with ceramic rods which contact the specimen either at the grips or in the strain gage [1568] and a laser extensometer that measures the displacement between two flags that are positioned next to the strain gage [912].

Strain-controlled monotonic tests are used to obtain either stress–strain or relaxation curves. Contrary to creep tests, an electromechanical test rig or a hydraulic frame is necessary to perform these tests along with an ex-

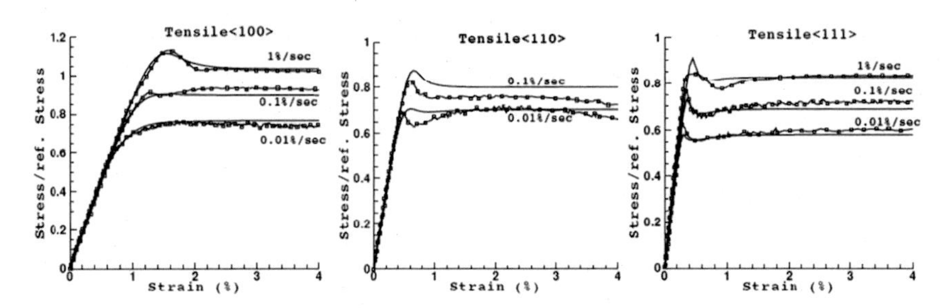

FIGURE 5.1 Monotonic tensile tests performed on CMSX-4 at 950°C for different crystallographic orientations and different strain rates. Adapted from [826].

tensometer that is in contact with the specimen. In fact, the use of a laser extensometer is too risky since a loss of signal can lead to a dramatic end of the test. In case of monotonic tests that are performed by controlling the total strain rate and measuring the force necessary to ensure it (see Fig. 5.1), the higher the plastic strain rate is, the higher the yield stress. It is a direct effect of time-dependent mechanisms that modify the apparent yield stress. For relaxation tests, specimens are deformed up to a specific total strain value. As soon as this value is reached, the total strain is maintained and the force variation is measured. There is a large stress drop at the beginning, which will ultimately reach an asymptote. The asymptotic value corresponds to the value of the internal stresses. If a strain decomposition is used, viz. $\varepsilon^t = \varepsilon^e + \varepsilon^p$, then $\dot{\varepsilon}^p = -\dot{\sigma}/E$ since $\dot{\varepsilon}^t = 0\,\mathrm{s}^{-1}$. Thus, the relaxation curve gives, at all time, plastic strain rates that can range from 10^{-3} to $10^{-10}\mathrm{s}^{-1}$ within 24 h. This test is very sensitive to temperature variations that can operate in a laboratory and may appear very noisy, which may require the use of a fitting curve (see Fig. 5.2). It is here important to point out that the mechanical response during monotonic loading may be affected by the microstructural state depending on the temperature investigated. For instance, rafting affects the mechanical response at 950°C, but does not affect it at 650°C. The interested readers are invited to take a look at the chapter dedicated to aging.

5.3.2 Cyclic testing

Cyclic tests are commonly performed with hydraulic frames. Cyclic tests can be stress or strain controlled, which gives access to different information on the nature of the mechanical behavior. A stress-controlled test will lead to ratcheting (Fig. 5.3(a)), i.e., a plastic strain accumulation at each cycle [1028,803]. A strain-controlled test, for which the experimentalist has to define a strain rate and a total strain amplitude $\Delta\varepsilon^t$, will allow determining any Baushinger effect, i.e., the development of a tension/compression asymmetry in the yield curve [604,676,1606]. This effect is commonly modeled via kinematic hardening. Furthermore, cyclic hard-

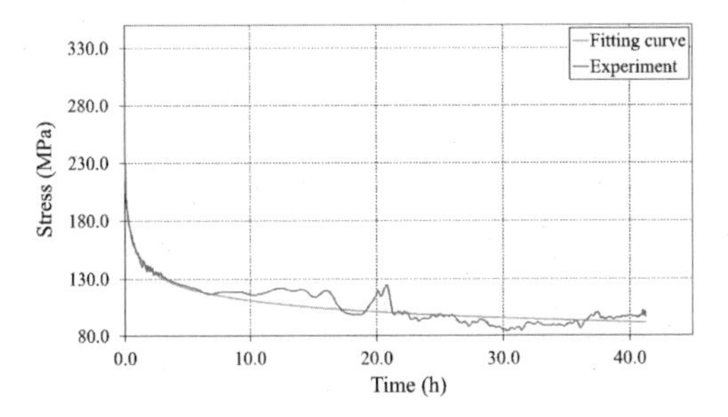

FIGURE 5.2 Relaxation test at 1050°C for a total strain of 2% on the MC2 alloy [799].

ening (Fig. 5.3(b)) corresponds to a total stress amplitude that increases at each cycle and reaches an asymptote [479,937]. This phenomenon is commonly modeled by isotropic hardening.

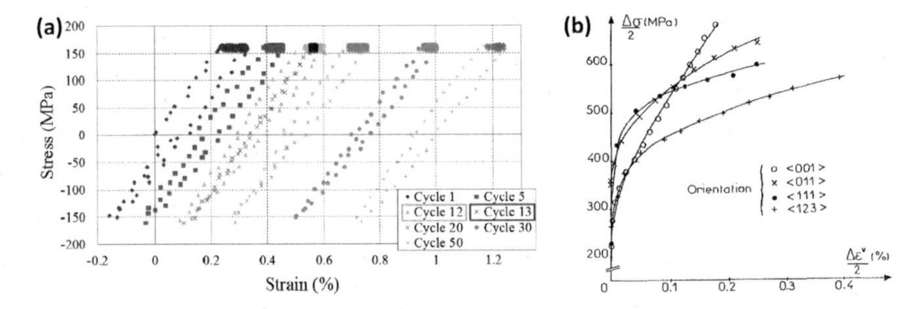

FIGURE 5.3 (a) Ratcheting of MC2 during a dwell-fatigue test at 1050°C (adapted from [803]) and (b) cyclic hardening curves of AM1 for different orientations (adapted from [937]).

Cyclic testing allows us to simultaneously investigate mechanical behavior and fatigue degradation. Nevertheless, for turbine blade applications, the introduction of a hold time during cycling, called dwell-fatigue, is considered more representative of in-service loading conditions. It is remarkable that such cyclic loading leads to lifetimes that are neither bounded by creep nor fatigue lives, in a Wöhler representation creep life being the lower bound, and are associated with a ratcheting effect, as observed in [1078,1354,803]. Indeed, dwell-fatigue loading appears much more damaging than creep.

5.3.3 High cycle fatigue (HCF)

The most obvious motivation to perform high cycle fatigue (HCF) tests is to save testing time and to obtain a large number of cycles in a reasonable amount of time. For instance, a 10^9-cycle test takes over 1 year in a traditional 20 Hz servo-hydraulic testing machine, but only 11 days at 1 kHz and less than one day at 20 kHz. In the context of superalloy turbine blades, high frequencies have to be considered since the blades are subjected to vibration loading due to a wide range of stimuli, such as structural excitations or turbulent air flow. The frequency regime of such vibration typically ranges from 50 Hz to 20 kHz. Contrary to LCF, HCF represents a mechanical loading domain where macroscopic plastic strain is not measurable or strongly limited. While servo-hydraulic test machines are typically limited to frequencies of about 20 Hz, it is possible to achieve test frequencies up to 400 Hz by means of rotary bending machines or by resonant electromagnetic machines, also called vibrophores.

Rotary bending is based on the principle of beam technique where a bending is superimposed to a continuous rotation. Such a device, similar to Wöhler's fatigue testing machine, allows reaching 3000 rotations per minute, corresponding to 50 Hz. With such devices, Shi et al. [1314] and Yu et al. [1627], respectively, have explored fatigue between 10^5 and 10^7 cycles at 700°C of DD6 alloy and the anisotropy effect on the SRR99 alloy. According to the stress gradient induced by bending fatigue, cracks always initiate on the surface of the specimen.

At higher frequency (around 100 Hz), resonant fatigue machines can also be employed to explore HCF [887,873,785]. Vibrophore excites the specimen with an electromagnet and, as a servo-hydraulic device, without stress gradient or intentionally with it. MacLachlan and Knowles [887] have performed a test combination with creep, low cycle fatigue and high cycle fatigue (loading ratio of 0) at three temperatures, 750, 850, and 950°C. At 750°C, cracks initiated at pore or subsurface pore. At highest temperature, small cracks initiated from oxide spikes and were joined by shear of the remaining ligament. Furthermore, effects of casting conditions on the HCF properties were investigated by the authors of [785]. They show that smaller dendrite arm spacing results in considerably increased high-cycle fatigue life. Fatigue tests were conducted using a resonance-testing machine in the study of MacLachlan and Knowles. Fatigue cracks are found to originate from shrinkage porosity in the high cycle fatigue regime at an elevated temperature (800°C). Indeed, the fatigue life increased with the pore size and decreased with the dendritic spacing. Consequently, the authors underline the pore-to-surface distance as being a critical parameter for scatter in fatigue results.

To extend the frequency limit of conventional servo-hydraulic fatigue machine, an specific improvement of servovalve was done. A 1 kHz servo-hydraulic fatigue testing system was exploited for single crystal superalloy at 1038°C by Wright et al. [1575] up to 900 Hz. It is noticeable that it constrains specimens to have small dimensions (4.06 mm diameter by 8.1 mm gage length). The authors showed the potentials of such device to explore the effects of stress amplitude at different loading ratio. For $R = -1$, no effect of loading frequency was observed. In contrast, it was significant for a positive load ratio. Their analysis indicates a transition frequency between time dependent and time independent fatigue behavior, strongly dependent on mean stress. For high mean stress, authors concluded that the failure process appears dominated by creep rupture with a significant elongation and necking.

5.3.4 Very high cycle fatigue – ultrasonic loading

To overcome the limitation of cycling frequency, ultrasonic fatigue testing machines with 20 kHz are considered. Such a facility was first presented in the scientific community by Mason in 1950 and allows exploring the very high cycle fatigue regime, where the number of cycles can reach 10^{10} cycles. The principle of the testing system is based on the stimulation of the specimen at the resonance. This technique is largely employed in the fatigue community [82], and all devices share the same basic mechanical principle for the load train. A piezoelectric actuator is connected to the booster which is connected to an amplifying horn. At one of its ends, the specimen is clamped to the horn, the other end being free. A particular issue of ultrasonic fatigue testing is the determination of stress state in the resonating specimen. In gigacyclic fatigue, the classical approach used to determinate the stress cycles consists in considering the loading in the elastic regime, and therefore in employing a modal analysis.

In the context of single crystal superalloys, an ultrasonic fatigue machine was firstly used by Tien [1431] at room temperature and more recently, extended for high temperature by Yi et al. [1619]. The specimen was then heated by induction to 1000°C, as presented in Fig. 5.4. A rod was introduced to protect the transducer and the horn from the heat of the specimen. To prevent the specimen from self-heating, the loading could be applied in an intermittent manner. The optimal pulse parameters were chosen as pulse duration of 250 ms and pause duration of 2 s. Ultrasonic fatigue tests were conducted under fully reversed loading, which corresponds to the natural response when the end of the specimen is free. The main conclusions of these works are the consistence of fatigue results generated from the ultrasonic testing system and those generated from servo-hydraulic fatigue testing. This point was confirmed by Furuya et al. [477] at the same temperature on another superalloy. Despite this comparability of fatigue for alternated loading, fracture surfaces obtained by an ultrasonic loading on ⟨001⟩-oriented specimen exhibit octahedral slip planes, rather than fracture perpendicular to the loading, as expected at high temperature with conventional fatigue device.

In order to explore the effect of mean stresses on ultrasonic fatigue, a constant load have to be applied at a point with no displacement, referred to as node. The setup developed by Cervellon et al. [222] enables exploring the impact of microstructural degradation on VHCF. It is observed at 1000°C that the microstructural state of CMSX-4 (coarsened γ' and γ'-rafted) has an impact only on the VHCF life with a positive stress ratio by the multiplication of mode I cracks and no impact for loading with $R = -1$.

In this study, ultrasonic failure is affected by larger casting pores, contrary to [1619] where crack initiation occurred at Ta-rich carbides in hot isostatically pressed (Hipped) PWA1484 specimens.

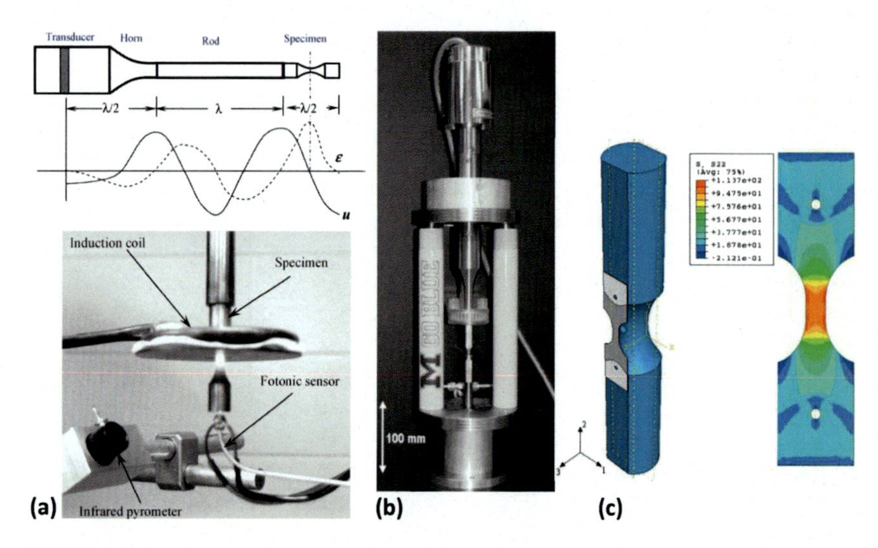

FIGURE 5.4 Ultrasonic fatigue facilities: (a) standard version [1619], (b)–(c) portable version for in situ crack observation by synchrotron radiation for different stress ratio [861].

Liu et al. [861] have developed an original portable version of ultrasonic device to image in situ fatigue crack by high-brilliance synchrotron X-radiation. The modifications are based on using two hydraulic cylinders, with a combined load capacity of 22 kN for positive loading ratio and the design of a specimen configuration to observe thin single crystal specimens (200 μm). For that, a carrier specimen was designed for resonance at 20 kHz, and a microspecimen was attached rigidly at the carrier specimen shoulders, as presented in Fig. 5.4(b). This modified device allows studying crystallographic crack growth at ambient temperature.

5.4 Multiaxial loading under isothermal conditions

5.4.1 Multiaxiality through multiaxial mechanical loads

Multiaxial tests are conducted by applying a torsional load on tubes [228,1025,1023,695,619,1536,127], a shear stress on samples [476,922], or by a combination of uniaxial loads in multiple directions [1256,1536]. Only a few papers are dedicated to these types of test, maybe because it often requires large and, therefore, very expensive specimens. Multiaxial characterizations using nonuniaxial methods are almost all on fatigue, except [922] on creep and [793] on monotonic shear testing, with the latter being suitable to study deformation mechanisms, i.e., dislocation structures. The complexity of multiaxial testing lies in the testing machines that have to be used. For instance, torsional load will have to be applied using a servo-hydraulic frame with torsional capabilities (Fig. 5.5), shear testing requires in-house setups, biaxial or triaxial loading necessitates very expensive frames. In addition, the complex geometry of the specimens to perform such tests leads to difficulties in applying homogeneous temperature fields. Nevertheless, multiaxial tests are necessary to validate or invalidate constitutive models: multiaxial loading conditions result in deformation constraints that are even more complex if the material properties are anisotropic. Indeed, models are commonly calibrated using uniaxial data at multiple temperatures and are often extended to multiaxial loading without further considerations.

FIGURE 5.5 Hydraulic frame used to perform torsion or tension/torsion at ONERA [799].

A well-known phenomenon is, for instance, the observed hard and soft zones (every 90° along the ⟨110⟩ orientations for ⟨001⟩-oriented specimens) that are revealed during torsion and tension/torsion tests [1023]. These experiments reveal that the deformation is heterogeneous (see Fig. 5.6), according the Schmid law, and that cube slip is predominant for pure torsion at room and high temperatures.

5.4.2 Multiaxiality through a designed geometry

If it is not possible to perform torsional or biaxial loading, multiaxiality can be obtained through a designed geometry, commonly by notches that are either symmetric [875,1529,30,853,197], as in Fig. 5.7(a), or, more rarely, asymmetric [805,181,811], as in Fig. 5.7(b).

Stress triaxiality ratio is defined as $T_\sigma = \sigma_H/\sigma_{eq}$ (where σ_H is the hydrostatic stress and σ_{eq} is the von Mises stress). This ratio can also be

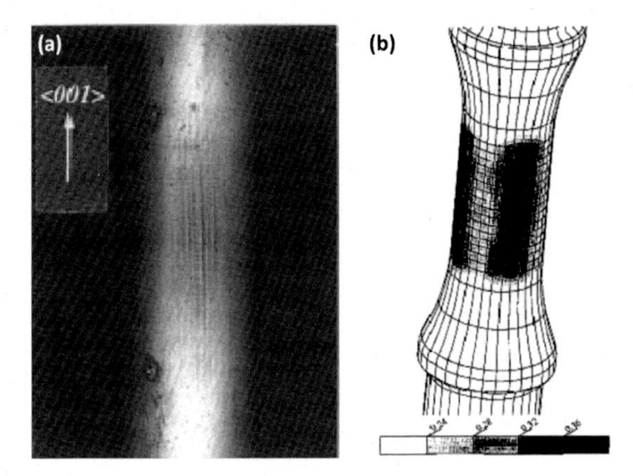

FIGURE 5.6 Torsion test: (a) slip traces observed on the specimen and (b) plastic shear strain contours predicted by a micromechanical model (%). Adapted from [1025].

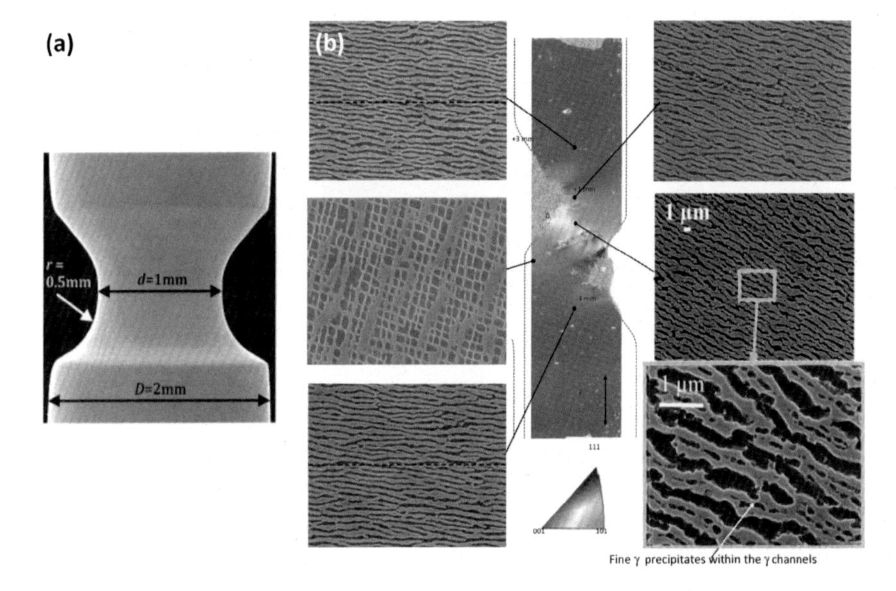

FIGURE 5.7 (a) Symmetric notched specimen for creep test from [197] and (b) asymmetric notched specimen with the rafting states along the specimen.

controlled by the geometry of the notch following the Bridgman's equation [154], $T_\sigma = 1/3 + \ln(1 + r/(2R))$, where r is the radius of the minimum cross-section and R is the radius of the circumferential notch. When it comes to symmetric notch, it was found that microstructural evolution in the circular notched specimen is coupled to the kinetics of stress redistribution during creep. Rafting and the increase of dislocation density start in the notch root before the center of the specimen is affected. Furthermore, the magnitude and the direction of the maximum principal stress are more important than stress triaxiality in affecting the nature of rafting, which was already obtained in [286] when studying rafting around voids. Lukas [874] also found that the lifetime of ⟨001⟩-, ⟨011⟩-, and ⟨111⟩-oriented notched specimens can be longer than that of smooth specimens at 850°C, as already expressed in the section dedicated to HCF.

The asymmetric notched specimen as in Fig. 5.7(b) revealed lattice rotations where the multiaxiality and the shearing is maximum. The microstructural observations showed tilted rafts, consistent with the local crystallographic orientation after lattice rotation, a phenomenon already observed during uniaxial creep tests [516,31,514]. The massive shearing between the notches can even lead to mechanical twinning at high temperature, namely 1050°C [811]. This is unusual since mechanical twinning is commonly observed at lower temperatures for which the shearing of the γ' phase is the principal deformation mechanism. In addition to the necessary multiaxial condition to activate mechanical twinning, a high plastic strain rate, which makes the accommodation of the deformation difficult, i.e., not enough available slip systems, is also essential to observe mechanical twins at such very high temperature.

5.5 Nonisothermal loading

5.5.1 Thermo-mechanical fatigue (TMF)

5.5.1.1 Uniaxial strain-controlled TMF

In the first decade of 2000s, lots of works have been published on the consideration of simultaneous cyclic thermal and mechanical loadings, the so-called thermo-mechanical fatigue (TMF) testing ([1205,567] and references therein). Nonisothermal loading is a major issue for turbine blade applications. Standard isothermal tests (creep or fatigue) are used to give input data for mechanical response assessments and lifetime predictions but can appear unrealistic compared to in-service loading. Nonisothermal tests are able to validate fatigue life prediction including constitutive behavior and even to reveal peculiar damage being developed during thermal cycling. The resulting damage to these nonisothermal repeated loadings is essential information to better grasp damage in gas turbine blades that are subjected to simultaneously high temperature gas and centrifugal force.

TMF aims to idealize thermal and mechanical conditions of a critical location of a component on a uniaxial laboratory test specimen. In principle, the TMF conditions consist of the overlay of a mechanical strain cycle and a thermal cycle. In such tests, as for all kinds of nonisothermal testing methodologies, a key point is to ensure the accurate control of temperature cycling for the duration of the tests.

TMF tests are commonly conducted under strain control. The difficulty is then to impose the mechanical strain while only the total strain can be measured by an extensometer. This means that the thermal strain cycle must be subtracted from the total strain at each instant of the cycle, allowing us to impose a definite mechanical strain cycle. To do that, the recommended approach [566] is based on three steps:

1. Thermal cycling which aims to record the thermal strain;
2. Thermal compensation intended to validate the thermal strain control at a zero mechanical stress;
3. Real TMF test where the desired mechanical strain is added to the thermal strain.

While the basic TMF cycle consists of temperature and strain loads of triangular wave form, thermo-mechanical cycling offers several degrees of freedom. A dwell-time can be introduced at the maximal temperature [975]. Moreover, an arbitrary phase shift can be defined between temperature and mechanical loading. The cases most frequently encountered are in-phase cycling (IP), where the maximum mechanical strain coincides with the maximal temperature, and, conversely, out-of-phase cycling (OP), where the mechanical strain maximum occurs at the temperature minimum. Basically, the phase shift leads to a change of the hysteresis loop and also to a change of the prevailing damage mechanisms. In OP testing, oxidation and fatigue are the dominating damage mechanisms, while creep and fatigue dominate in IP test. This last configuration appears the most damaging, especially for long life.

Among TMF investigations on single crystal superalloys, one can cite, for example, Lautridou et al. for AM1, AM3, MC2 and CMSX-4 [797], Okazaki et al. for CMSX-4 [1030], Amaro et al. for PWA 1484 [19], Liu et al. for DD8 [858], and Moverare [975,977] for CMSX-4 and STAL-15, respectively. Whatever the studied material, the main difficulty in the understanding of TMF damage processes comes from their multiplicity, and their interactions in ways that are not observed in isothermal tests. Indeed, TMF damage and failure are affected by the difficulty to accommodate plastic strain at low temperature in combination with creep and oxidation at high temperature. Moverare et al. [976] have finely described the TMF damage process according to the testing conditions. They described the influence of localized deformation occurring during TMF loading and in a more pronounced manner for OP loading. Deformation bands by mechanical twinning on the {111} planes act as a preferable path for crack propagation. Moreover, they appeared as preferential locations where TCP formation can occur, as recrystallization phenomena. This last microstructural aspect is described as particularly detrimental for TMF resistance.

Furthermore, IP TMF loading can exhibit rafting phenomena when the maximal temperature is high enough. It results in an increase of the inelastic strain that reduces the fatigue life.

5.5.1.2 Complex TMF

More representative TMF testing is still challenging and entails more complex loading and/or more realistic environmental conditions. A first practicable degree of complexity is multiaxiality. At the moment, multiaxial TMF testing is poorly explored and limited to axial and torsional loadings [1678,128]. Torsional loading has two main specificities according to the investigated SX superalloy. First, no thermal compensation is required because of cubic symmetry of the thermal expansion coefficient, and second, a strong dependence of the secondary direction of the hollow specimen is observed (see Fig. 5.6). Another alternative beyond multiaxial issue is to apply the TMF methodology on other types of classical experiments. It was done under load control on cyclic four-point bending experiments by [1284,1600] favoring crack initiation on the side with the positive bending moment.

If TMF acts as a simplified view of turbine blade operating conditions, such as start-up and shut-down cycle, resulting damage processes cannot be considered fully representative of in-service operations, and can interact with other damage processes observed in real component. The effect of high frequency loading previously evoked can be suspected to affect fatigue life under nonisothermal conditions. The introduction of HCF loading on TMF was performed by Hirsch et al. [617] during high temperature strain holds. By changing the baseline waveform, the authors showed that adding HCF during a high temperature hold produces a decrease in the environmental effects, as characterized by γ' depletion measurements. Nevertheless, the high frequency loading may have either deleterious or beneficial effects on specimen life depending on the conditions. The understanding of coupled TMF-HCF tests may require more analyses to evaluate the effects of HCF loading on TMF life and the effects of the nonmonotonous stress–strain behavior due to the Portevin–Le Chatellier effect, which was shown to intensify damage.

Integrating more realistic environmental conditions can make experiments more representative of in-service conditions. For this purpose, burner rig facilities have been developed, but most of the testing do not include mechanical loading. As mentioned by Mauget et al. [911], only few TMF facilities were developed due to the high costs and intricate implementations to allow testing in realistic gas turbine environments (France, Japan, USA). Fig. 5.8 shows an example of such facility where specimens may be mechanically loaded under creep as well as monotonic and cyclic conditions, respectively, with a creep frame or an electromechanical testing device [911]. Specimens can also be designed as standardized general

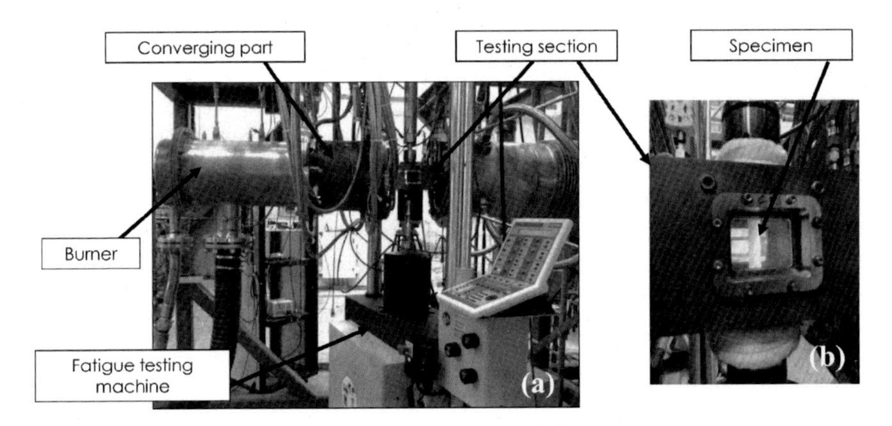

FIGURE 5.8 (a) Global view of Maatre test bench and (b) view of heated specimen. Adapted from [911].

purpose specimens or to represent a feature encountered in an industrial component with specific cooling hole arrangements or applied thermal barrier coatings. For their high degree of representativeness, such facilities may be exploited to simulate accelerated mission under combustion environment. Special care has been paid on the metrology under hot gas environments to be able to accurately measure the temperature and the deformation.

5.5.2 Thermal gradient mechanical fatigue (TGMF)

In industrial gas turbines and jet engines, the turbine blades in the first rows behind the combustion chamber are internally cooled to allow an increase of the turbine entry temperature, and consequently to improve turbine ability and efficiency. Few studies have investigated the impact of air cooling and the resulting thermal gradient in laboratory. An early concept was published by Marci et al. [904] and a description of the first test rig by Bartsch et al. [79] in the context of coated turbine blades. The main requirements for the new test methodology were (i) generating a thermal gradient such that the temperature and stress distributions over the wall of a coated specimen are close to that at the leading edge part of a turbine blade, (ii) fast heating and cooling rates in order to reproduce heating conditions in the jet engine and to ensure a reasonable short cycle duration, and (iii) mechanical loads representing the centrifugal forces acting on a rotor blade should be applied simultaneously and independently to the thermal loading.

Three main facilities of thermal gradient mechanical fatigue, the so-called TGMF, can be mentioned: they are in China [628], France [326,127], and Germany [83,80]. The common characteristic is the use of thin-walled tubular specimens as a geometrical simplification of cooled component sections allowing the injection of cooling air flow. Nevertheless, in these works, three different heating systems are used, namely resistive furnace [628], induction [127], and lamp furnace [83], respectively. At ONERA, special attention was paid to evaluate the internal wall temperature because it is not possible to measure it by using common thermocouples. In the course of developing the TGMF technique, the thermal gradient over the wall of the tubular specimen was increased by raising the internal cooling air flow and further introducing an alumina sleeve. By means of a so-called thermochromic paint on the inner surface, only high temperature differences could be experimentally assessed. The surface roughness relative to paint was modified, improving the heat exchange between cooling air and specimen wall. An alternative for a smooth configuration without paint is to estimate the temperature distribution in the specimen by a numerical approach. For that, the thermal problem is solved by a weak coupling between fluid and solid solvers, based on two software applications developed at ONERA, Cedre and Z-set, respectively (Fig. 5.9). The induction heating system offers a very high heating power required by the thermal losses due to the internal cooling. Nevertheless, this heating method has the disadvantage of heating the material itself by the Joule effect which affects wall temperature distribution. This volumetric heating effect was taken into account by the introduction of penetration depth in previous thermal calculations. The evaluated thermal gradient was then computed by a numerical method, showing the underestimation of analytical models. Finally, with the second configuration with smooth internal surfaces, an estimated thermal gradient of 40°C/mm was achieved, whereas with the first configuration, using thermochromic paint, nearly 120°C/mm was reached.

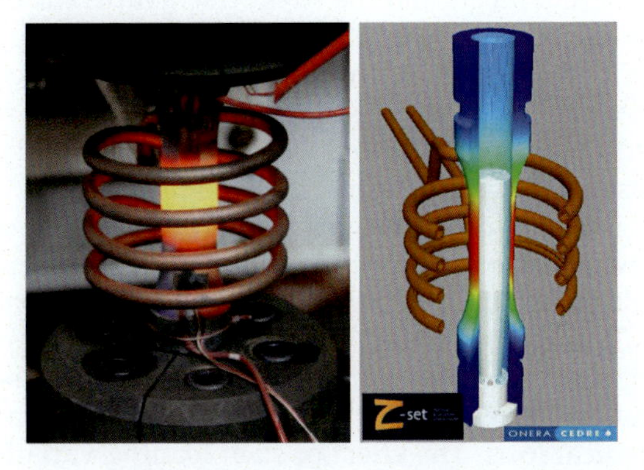

FIGURE 5.9 TGMF at ONERA: (left) a view of the heated specimen and (right) thermal calculations by coupling Z-set software for the solid part and Cedre software for the fluid part (source: ONERA).

To control this numerical approach for internal temperature assessment, the air temperature at the output of the specimen is a key issue. Indeed it is the only value which can be compared with an experimental measurement as thermal paints increase the heat exchange rate. The global thermal exchange can be characterized by two temperature probes set upstream and downstream from the specimen. A correct correlation was observed by the authors. Finally, force-controlled thermo-mechanical tests are performed on smooth and perforated specimens. Additionally to simple sinusoidal waveform, complex thermal gradient-mechanical cycles are also applied to reproduce a close-to-real engine mission with high heating and cooling rates, up to 50°C/s [326,127].

At the German Aerospace Center (DLR), the outer surface of the tubular gauge length is heated by means of a radiation furnace, as presented on Fig. 5.10. The radiation ensures that the temperature is maximal at the surface also in the case of ceramic coatings which are electric insulators. In contrast, when using inductive heating, the highest temperature would occur underneath the coating in the metallic substrate [127]. The radiation of multiple cylindrical quartz lamps was focused onto the specimen with mirrors shaped as segments of an ellipse. Each a quartz lamp is located in one focus line of the ellipse and the specimen in the other focus line. Test rigs have been built with 4 and 16 lamps, respectively. The specimen is cooled by pressurized air, which can be optionally preheated up to 500°C to reproduce the state of cooling air in a jet engine which is diverged from the high pressure compressor. The thermal gradient generated over the wall of the tubular specimen gauge length results in multiaxial stresses. Considering the metallic substrate the in-plane stresses in the axial and circumferential direction are compressive at the outer surface while they are tensile at the inner surface. For the complete coating system, the stresses are more complex and depend on the mismatch of thermal expansion coefficient of the different layers and the stress-free temperature of the system which depends on the processing temperature of the coating system. More information on TGMF-testing of coatings is given in Chapter 10. The radiation heating with focusing mirrors allows for high heating rates such that a specimen can be heated from ambient temperature of 25°C to 1000°C within 30 s. Fast cooling with similar rates is provided by an additional cooling system using a shutter which during the cooling sequence encloses the specimen while blowing air onto its surface through multiple vents. For applying mechanical loads, the specimen is inserted in the load train of a servo-hydraulic testing machine. Due to a flexible control system, a wide range of thermal and mechanical loading can be adjusted, including some thermo-mechanical load sets, as well as thermal gradient fatigue without external mechanical load. The effect of such loading variation, as well as the microstructural evolution as a function of thermal gradient and local stress field over the specimen wall, has been reported for CMSX-4 single crystalline specimens in [83].

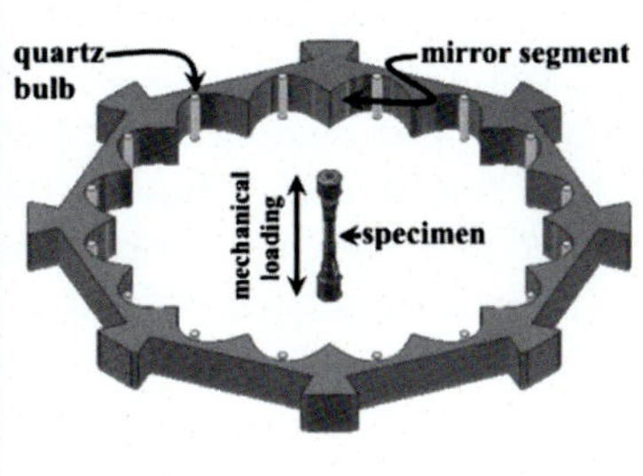

FIGURE 5.10 TGMF testing device at DLR (adapted from [83] and [80]).

5.5.3 Fast and very high thermal jumps

Robust life-prediction methods for the design of high temperature aero-engine components are highly desired by engine makers, especially when considering certification procedures. Indeed, the thermo-mechanical paths encountered by components during certification are far more complex and severe than during their service life. As an example, the procedures for the certification of turboshaft engines for helicopters consist in mixing different engine regimes and repeated short very high temperature jumps [285,802,912]. During these temperature peaks, high pressure turbine blades are exposed to extreme conditions (temperatures close to their melting point). However, the short durations of these thermal jumps (typically, from 5 to 150 s) leave the microstructure of the material out of equilibrium, leading to a transient mechanical response of the alloy [806,803]. These tests are performed with a burner rig, as described in Section 5.5.1.2. Thus, the temperature rate and the thermal gradient because of the heat flow coming from each side are representative.

Most of the studies on the subject were performed with a simple thermal jump introduced somewhere in the lifetime of the material during creep [283,284,514,802] or fatigue [806,803]. These authors found that longer thermal jumps lead to longer lifetimes due to temporarily-reduced subsequent plastic strain rates. This observation was attributed to an enhanced dynamic recovery either by dislocation climb along the γ/γ' interfaces or by annihilation induced by γ' cutting. In addition, it was found that the plastic strain rate is increased after a thermal jump at 140 MPa while the plastic strain rate is the same as before at 160 MPa. It was credited to different deformation mechanisms activated depending on the applied stress level. Finally, the authors observed a counterintuitive result, namely that a thermal jump can extend the lifetime of SX superalloys. It happens when the thermal jump is introduced at a specific time corresponding to a maximum constrained lattice misfit in absolute value, i.e., to a microstructure not yet rafted but not cuboidal anymore.

A natural extension of tests with one thermal jump is thermal cycling tests with fast heating and cooling rates [1165,281,1496,520,579,1361,279]. It was found that a higher thermal cycling frequency leads to an increased average creep strain rate and to a reduced creep life. P-type rafted microstructures have proved to have the highest resistance against non-isothermal creep because dislocation climb is more strongly impeded due to larger areas of vertical γ/γ' interfaces. Similarly to what was found with one thermal jump, γ' precipitation in the γ channels has a large effect on the mechanical properties and lifetime: the finer and the more regular the γ' precipitation, the better the creep resistance. Also, the higher the frequency, the more the creep properties decay. Thermal cycling at temperatures above the temperature leading to a drop in the γ' volume fraction, namely 975°C, modifies the γ/γ' dislocation network. It seems that this effect depends on whether the jumps are heating or cooling jumps [283,284,579].

5.6 Small-scale testing

5.6.1 Miniaturization of testing representative volumes

Several motivations are driving the development of miniaturized test techniques. Processing of large test specimens is expensive and appears inadequate in the development of new alloys. Furthermore, when using small specimens, they can be extracted out of thin walled components either in as-processed state to account for effects of manufacturing processes or after service to assess damage evolution and the respective effect on residual life. In miniaturization, the size of microstructural features of the material investigated has to be considered. Material heterogeneities exist on different length scales depending on processing routes and parameters. In single crystalline superalloys with dendritic solidification, the size of defects is related to dendrite spacing which is in investment cast alloys typically about 500 µm [1059]. The γ/γ' microstructure within the dendrites and in the interdendritic space differs, and solidification porosity occurs in the interdendritic space. If local properties are of interest, micromechanical methods have to be applied, and some examples are discussed in the following section. If properties, representative of the material on large scale, have to be determined, miniaturization is limited. For capturing the effect of heterogeneities associated to dendrite spacing in cast material, miniature specimens have sizes in the millimeter range, which allows using geometries directly inspired by standard specimens. However, the reduced size of the specimens entails challenges such as specimen clamping, alignment, and strain measurement, especially for experiments at high temperatures.

For creep testing, Mälzer et al. [897] present a testing methodology for miniature specimens at high temperatures up to 1150°C. The specimen length is 20 mm (Fig. 5.11(a)). An extension of this test equipment for experiments in inert atmosphere is presented by Peter et al. [1083] where surface investigation is facilitated due to the absence of oxidation. Details are given on grips made of oxide dispersion strengthened (ODS) Ni-base alloy, and the strain was indirectly measured by two alumina rod-in-tube extensometers [897].

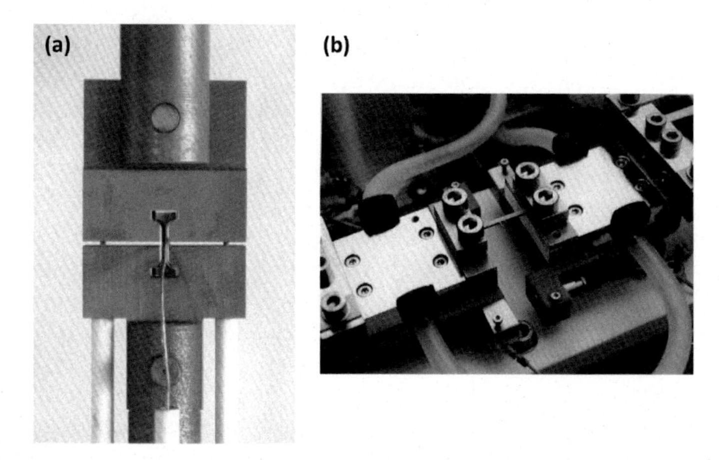

FIGURE 5.11 Experimental devices for miniature specimens: (a) adapted creep test facility [897], (b) electro-thermo-mechanical test facility [1226].

An alternative to miniaturization of conventional testing methods was chosen by Roebuck et al. [1225] by the development of an electro-thermo-mechanical test system for miniature specimens with cross-sections as small as 1 mm^2 (Fig. 5.11(b)). The specimen can be heated up to 1000°C by means of DC electrical current utilizing the Joule heating effect. The test device is encapsulated and can be used under inert atmosphere. Examples of physical and mechanical characterization on CMSX-4 are the solidus temperature [1226] or γ' volume fraction [1225], both deduced by electrical resistivity measurements. Such device also allows exploring recrystallization phenomena of indented or uniaxially deformed specimens and getting a strain threshold for recristallization [296].

An approach for low cycle fatigue (LCF) testing of millimeter-size specimens at temperatures up to 1000°C is published by Meid et al. [932]. The specimens have a dog-bone shape with a total length of 20 mm which is the largest size allowing the extraction of specimens in all crystallographic orientations from commonly available raw material samples. The material was provided as cast alloy blocks with dimensions of about

$(20 \times 100 \times 115)\,mm^3$ for laboratory experiments. The restriction in design space as well as specific requirements made it necessary to deviate from general recommendations of available standards for low cycle fatigue testing. Therefore, it was not possible to keep proportions of standard specimens and just downsizing them. Optimization challenges in specimen dimensioning were (i) to make the specimen head for gripping as small as possible while allowing maximum diameter of the gauge length for providing a statistically relevant test volume with respect to dendrite spacing, (ii) to make the length and diameter of the gauge length as large as possible while ensuring buckling stability under compressive load, and (iii) to design the filet between gauge length and gripping end such that minimal stress concentrations occur. For the filet shape, a cubic spline with three nodes was selected, and shape optimization was performed for the three main crystallographic orientations by means of a parametric finite element model by variation of the spline node positions. The optimized filet shape has proved to be suitable for testing specimens with different crystallographic orientation [930].

For miniature LCF testing, not only the specimens have to be designed but also an adapted clamping system. Some general requirements are (i) a stable and homogeneous specimen temperature, (ii) alignment of the specimen length axis with respect to the load, and (iii) reliable strain measurement, especially for strain controlled experiments. In [932], temperature stability is provided by inserting the specimen with the grips into an electric resistive furnace. The gripping system is modular, allowing to mount and align the specimen in fixtures outside the furnace and further clamp the fixture into quite massive grips. The massive grips and load train further ensured that a high temperature extensometer could be applied without bending the load train. Since the specimen is very small, the extensometer contacts are not applied directly on the specimen but on the fixture. Therefore, strain measurement needs to be calibrated. Fig. 5.12(a) displays the miniature LCF specimen, and Fig. 5.12(b) provides a view of the respective fixture with an applied extensometer.

5.6.2 Micromechanical testing considering microstructural heterogeneity

The assessment of properties at a local scale by considering the material heterogeneity, as the dendritic structure and the interdiffusion layers at the interface between substrate and coating, requires further miniaturization. In a single-crystal superalloy, the dendritic structure is widely suspected to play a role in plastic strain and on fatigue properties. The typical dendritic arm spacing is around 400 μm, which means that a micrometer-scale test needs to be employed. One solution was proposed by Texier et al. [1414,1417] who developed a creep test on ultrathin ribbon specimens obtained by a lapping machine followed by polishing. The sam-

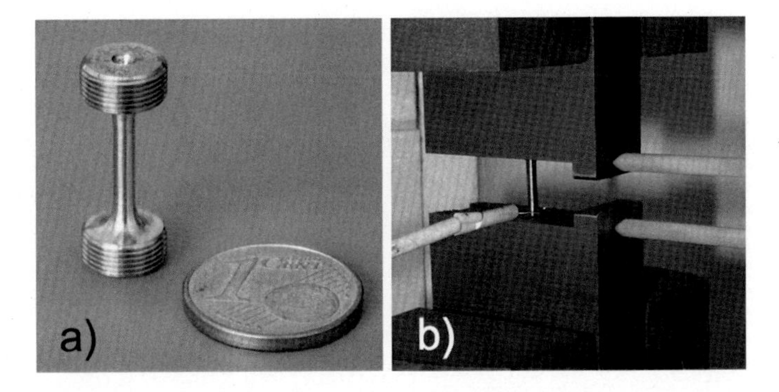

FIGURE 5.12 Miniature specimen and its fixture with extensometer (from [931]).

ples illustrated in Fig. 5.13 have a maximal thickness of 100 μm. To reach this thickness, an intricate preparation is necessary. However, it is possible to extract specimens with different volume fractions of dendrite core and interdendritic material with this method, allowing to investigate the effect of local variations in microstructure and chemical composition.

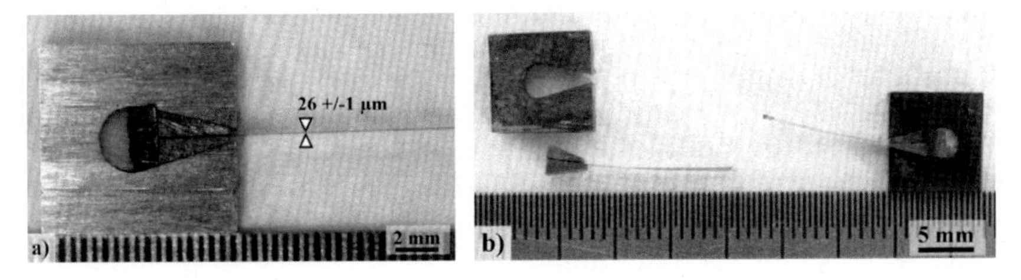

FIGURE 5.13 Ultrathin ribbon test at elevated temperature with grips based on clamping by friction (from [1418]).

The assessment of intrinsic properties of ultrathin samples at high temperature required performing creep experiments in ultra-high vacuum. Results show differences in high temperature creep behavior at the dendrite length scale: specimens with high volume fractions of dendrite cores have lower creep resistance than those with higher volume fractions of interdendritic material. The authors attributed the variations in the creep resistance to microstructural features such as pore density or size of γ' precipitates, to name but a few. The ultrathin ribbon specimens have been also utilized to investigate the properties of interdiffusion zones at the interface between substrate and metallic coatings [1418]. Other solutions for micromechanical testing considering microstructural heterogeneity have been designed to be integrated in a scanning electron microscope (SEM). Different in-situ testing devices have been developed, such as microdrawing mill [852] and hydraulic system [1656,881]. The in-situ approach also allows obtaining precious information on plastic deformation mechanisms (see Fig. 5.14) or growth of short cracks ([881]).

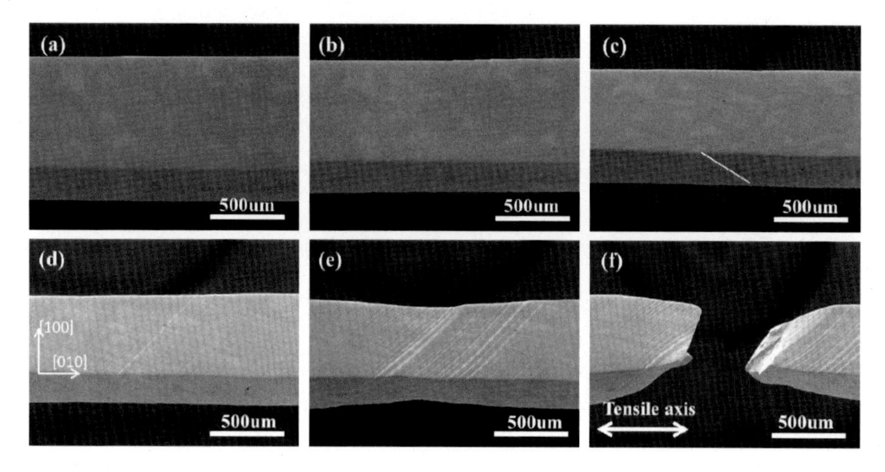

FIGURE 5.14 SEM images of a series of in-situ tensile experiments on a Ni–Al–Re model superalloy: (a) initial state before loading, (b) elastically deformed, (c) occurrence of the first slip line appearing as a thin white line, (d) activation of slip lines of the second slip system, (e) necking of the specimen due to stress concentrations, and (f) fractured specimen. Adapted from [852].

For very small specimens, a recent solution for micromechanical testing is micropillars, usually fabricated via focused ion beam (FIB) micromilling. The seminal work in this domain is from Uchic et al. [1454]. Few works have yet been done with micropillars on single crystal superalloys, compared to those on polycrystalline superalloys (see [303], for instance). Arora [39,40] did microcompression experiments on single crystal superalloys, and the first microtension experiments were performed by Shade [1303]. The micropillar technique is particularly effective to investigate plasticity at the microscale (see Fig. 5.15) since the micropillars can have a diameter as small as only 1 μm. In fact, the basis for mechanical analysis that spans different length scales remains heuristic because the effects of microstructure at the dislocation and larger scales are not yet well represented within plasticity or deformation theories and simulations. To better understand and advance plasticity models so that they may include the fundamental materials science of deformation at small scales, new experimental methods that permit better coupling to all aspects of theory and simulation are needed. These methods can reveal new intrinsic mechanisms operating when the physical dimensions of a sample approach those of the dislocation processes.

The samples are not freestanding but remain integrally attached to the bulk substrate to eliminate the need for micromanipulations that may lead to undesired stresses. As a result, the substrate acts as the lower com-

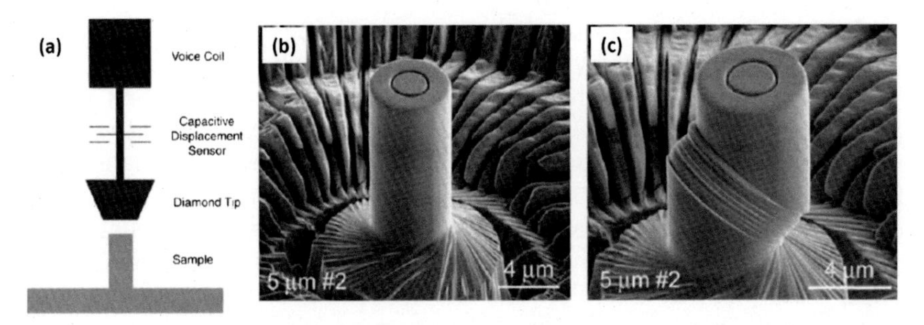

FIGURE 5.15 (a) Schematic of the microcompression test that highlights the spatial relationship of the sample relative to the primary components of a commercially available nanoindentation system, which are colored in black; (b) Scanning electron microscope (SEM) image of a 5 µm-diameter microcrystal sample of pure Ni oriented for single slip; (c) SEM image of (b) after testing. Adapted from [1455].

pression platen during the test. Commercial nanoindentation systems are commonly used as the mechanical test frame, in which the sharp indentation tip is replaced with a flat-punch tip. The micromachining by FIB allows one to cut samples near the surface of a bulk crystal with extreme control over the location and the overall size of the sample. However, the samples always have some degree of taper, such that the diameter at the top is smaller than that at the base. This complex geometry makes the interpretation of the microcompression experiments difficult. Indeed, studies showed that the tapered samples result in inhomogeneous deformation, which can lead to inaccurate flow stresses and artificial increase in strain-hardening rates. Another practical concern with micropillar testing is to ensure that there is minimal misalignment between the flat-punch tip and the top surface of the sample. Misalignments greater than 1° lead to a number of test artefacts: underestimation of the yield point and elastic modulus, changes the strain-hardening response, and buckling. To avoid the latter, systems in tension have also been developed. To date, typical microcompression experiments span quasistatic strain rates ranging from 10^{-6} to $10^{-3}s^{-1}$. All reported tests have been performed at room temperature. The effect of thermal activation on small-scale deformation remains an open question. However, an effort to do micropillar testing at high temperature has to go through thermal drift issues, i.e., maintaining a constant temperature for a long time, which is not perfectly controlled as of now.

5.6.2.1 Micromechanical tests in TEM

At a lower scale, for observing the elementary deformation mechanisms, such as dislocation gliding and climbing, micromechanical tests with simultaneous observations of a specimen in a transmission electron microscope (TEM) have been performed, but are very seldom. One of the original works on observing dislocations evolving at high temperature while applying a load is by Benyoucef et al. [96] who determined the localization of deformation in channels perpendicular to the applied stress. In addition, the authors noticed an important work hardening of the channels originating from extensive multiplication processes. The expansion of tridimensional loops was also observed while both cross-slip and climb events appear to occur, giving rise to an intricate tridimensional array of dislocations where all the possible Burgers vectors are present. Benyoucef et al. [97] also revealed that the creation and propagation modes of moving dislocations evidence the important role of the misfit dislocation network as well as the strength of the γ/γ' interfaces. Legros et al. [815] observed that the γ/γ' interfacial dislocations play a different role depending on the direction of propagation of the dislocations on rafted microstructures. In addition, the authors also noticed super stacking faults and microtwinning in the γ' phase. Other authors [381] also performed high-temperature in-situ TEM investigations to better understand the micromechanisms associated to creep, but without applying a mechanical load [1647].

5.6.2.2 Indentation

Indentation testing is an instrumented characterization often employed to assess hardness and Young's modulus. This testing method consists of penetrating an indenter until an impression is formed. The simplicity of the indentation process entails significant interest to explore nonlinear phenomena. As such, several tests exist according to the geometry of the indenter and its scale. At the macroscale, indentation appeared as an interesting means to investigate elastic limit and plasticity. For instance, experimental investigations were presented by Eidel on CMSX-4 [383]. At the macroscopic scale, finite element simulations can be used to determine crystallographic creep parameters from experimental results of indentation tests [1598] or also for the investigation of the differences between slip activity at the free surface and in the bulk of the specimen [1250]. Room temperature indentation may be considered to provide limited information, except to explore peculiar cases. For instance, indentation was considered by Zambaldi et al. [1635] to relate the plastic strain field induced by indentation at room temperature and recrystallization phenomena occurring at higher temperatures. The high temperature indentation was developed and used by several groups to characterize the creep behavior of various materials (for a review on impression creep, see, for example, [828]). More recently, the literature revealed the development of micro- and nanoindentation to understand the effect of microstructure or property gradients. Takagi et al. [1392] present an experimental study of microindentation of CMSX-4 up to 1073 K and determine elastic properties. A similar facility was developed by Passily et al. [1062] to characterize thin coatings of turbine blades, see Fig. 5.16(a). At a smaller scale,

Neumeier et al. [1007], Gan et al. [484], and Zietara et al. [1679] investigated superalloys by the nanoindentation technique in an atomic-force microscope (NI-AFM), see, e.g., Fig. 5.16(b). They have measured the hardness of the γ and γ' phase in specimens in as-received condition and after creep deformation at 982°C without finding a clear difference. In parallel, the nanoindentation technique was also developed for high temperature by Sawant et al. [1270]. Beyond elastic properties, the assessment of constitutive behavior laws at a local approach seems practical, despite the complexity of necessary simulations [1062].

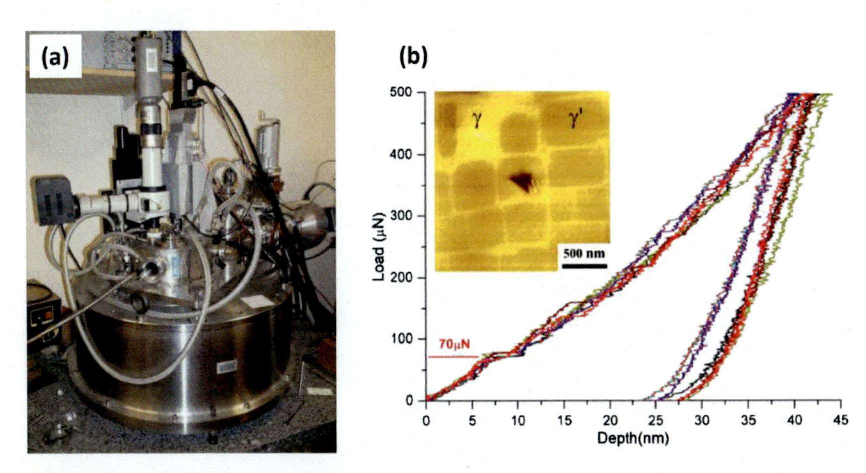

FIGURE 5.16 Small-scale indentation: (a) instrumented microindentation equipment for experiments up to 1000°C [1062], (b) load–displacement curves obtained by indentations made in γ' precipitates of CMSX-4, with an AFM image of a nanoindent (from [484]).

5.7 In-situ testing by means of high energy radiation

5.7.1 X-ray diffraction

X-ray diffraction is performed in a synchrotron facility (Fig. 5.17). Diffraction is when an incident X-ray beam interferes with the crystalline structure causing it to diffract into many specific directions. This powerful light source allows obtaining information on the structure and chemical properties of materials at the atomic and molecular length scales. More precisely, it is an electron accelerator intended to produce radiation in the electromagnetic spectrum, namely from infrared to X-rays, that can be used in all scientific fields. The synchrotron consists of a set of components used to produce, channel, and maintain a high-energy electron beam. X-ray diffraction has been already extensively used to study single crystal superalloys and is based on the Bragg equation, $\lambda = 2d \sin\theta$, where

n is a positive integer, λ is the wavelength of the incident wave, d is the interplanar distance, and θ is the scattering angle. X-ray diffraction has been employed to determine the evolution of the lattice parameters with temperature of cuboidal and rafted microstructures [999,1237,108,1239]. The diffraction generates two Gaussian-shaped profiles, one per phase. Unfortunately, it is only possible to observe two distincts profiles when the microstructure is rafted [351]. A third peak can be generated during nonisothermal loading. The latter corresponds to the fine precipitates that are formed in the γ channels. They possess a different chemical composition compared to the large γ' precipitates leading to a different lattice parameter. In addition, the areas under the diffraction peaks correspond to the volume fraction of the phases. Therefore, X-ray diffraction can also be employed to determine the evolution of the phase volume fractions and validate dissolution/precipitation models [808].

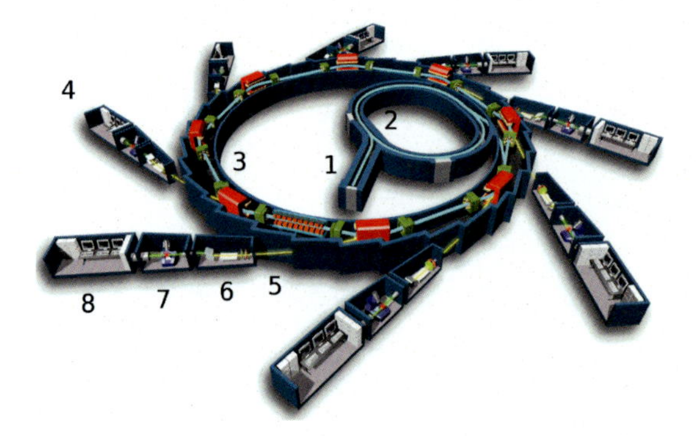

FIGURE 5.17 Synchrotron Soleil (France) consisting of (1) production of electrons in an electron gun and brought to several thousand keV in a linear accelerator (Linac), (2) circular accelerator (booster): the energy of particles coming out of the Linac is of the order of 0.1 GeV to reach a few GeV out of the booster, (3) storage ring: the particle injected into the storage ring has a limited lifetime of a few hours under a high vacuum, (4) beamline, including (5) front part of the beamline, consisting of a lead enclosure to ensure security of people working, (6) optical booth: a polychromatic beam is transformed into a monochromatic one, (7) experimental booth, and (8) workstation.

Other studies have been dedicated to understand the effect of stress (constant or variable) on the evolution of the lattice misfit [349,665,666, 667,351]. The lattice misfit under stress is called a constrained lattice misfit, contrary to the natural lattice misfit, that is, the misfit between the two phases before deformation. As it can be noticed in Fig. 5.18, the constrained lattice misfit decreases in absolute value to reach a minimum during the primary creep stage. The subsequent evolution of the constrained lattice misfit is different for the two alloys investigated that are not from the same generation of single-crystal superalloys: AM1 is first generation whereas MCNG is fourth generation. The misfit evolution rates of the two alloys are different, which can be due to different rafting kinetics (much slower in MCNG that has also a shorter secondary creep stage). In addition, the larger peak width observed for MCNG is probably due to a more disordered microstructure and, therefore, to its chemical composition.

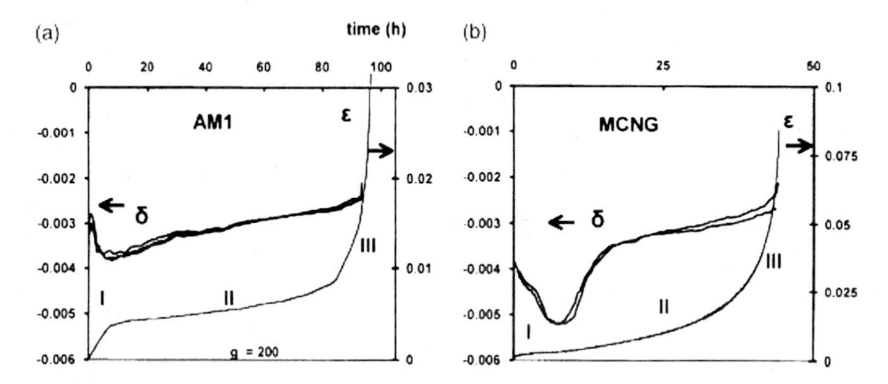

FIGURE 5.18 Evolution of the constrained lattice misfit and the plastic strain for (a) the AM1 alloy and (b) the MCNG alloy during creep tests at 1100°C/150 MPa obtained by synchrotron X-ray diffraction. Adapted from [349].

Jacques et al. [666,667] did varying-stress experiments and found that stress variations above 150 MPa on ⟨001⟩-oriented specimens at 1070°C lead to larger stress changes in the γ' phase than in the γ phase. This was attributed to a partial transfer of the stress from the γ to the γ' phase. Thus, larger variations of the constrained lattice misfit are measured for stresses above 150 MPa. le Graverend et al. [808,809] did thermal jump tests with different ΔT on rafted microstructures and for temperature regimes that trigger dissolution/precipitation of the γ' phase, namely above 975°C, on the AMI alloy. They found that if ΔT is sufficiently large (above 130°C), viz the drop in the γ' volume fraction is important, the constrained lattice misfit is decreased in absolute value compared to its value prior to the thermal jump. This means that a thermal jump equal or above 130°C leads to a decrease in the internal stresses, which was attributed to shearing, annihilation, and climbing of dislocations at the γ/γ' interfaces.

5.7.2 Neutron diffraction

The evolution of the natural lattice misfit with temperature [987,523], the stress distribution between the phases during stress varying tests [271, 373], defects such as lattice distortion, misorientation, and subgrains [1578, 1579], constrained lattice misfit evolution [265], and rafting [1681] can also be obtained with neutron diffraction. Because neutron radiation can go through much larger specimen, it is even possible to analyze blades [1445].

5.8 Damage and crack characterization

5.8.1 Ultrasonic waves

In the broad scope of nondestructive evaluation, ultrasonic waves are used for a wide range of applications. They can firstly be used to assess the elastic properties of an alloy (see Chapter 3) but give also interesting opportunities for damage characterization. Lane has recently developed a 2D ultrasonic array to inspect single crystal turbine blades [790]. It was specifically enhanced for the inspection in the root section of high-pressure turbine blades. This technique has been significantly improved by taking into account the propagation waves in an anisotropic solid. Analytical models were developed to correctly simulate the variation of wave velocity with direction and also the variation in signal amplitude form a point-force acting on the surface of the solid. The writing of these models and the accounting of propagation direction is essential in these works to have a correct use of ultrasonic array imaging algorithms. Finally, the developed 2D ultrasonic array and deployment fixtures have been tested successfully in situ on turbines blades in a real engine. Beyond in-service inspection on turbine blades, ultrasonic methods can also be employed to investigate defects coming from manufacturing process. To achieve that, a resonant ultrasonic spectroscopy system [1212] is elaborated to investigate recrystallization induced by shot peening. Once again, a coupled ultrasonic and FE calculation appeared necessary. They both have proven promising to detect recrystallization by resonance frequency shifts on real component, and consequently to perform valuable manufacturing process.

5.8.2 Digital image correlation (DIC)

In the far field characterization technique in mechanics, digital image correlation (DIC) is well known as a recent tool for full-field displacement measurement. This powerful methodology consists of comparing pairs of images from a reference state to a strained state using DIC techniques to measure the surface displacement field. It needs to be extended to high-temperature applications. In this sense, many difficulties, as optical contrast and its stability, convection, optical measurement access, etc., limit the use in a lab, despite of promising outlook. For that, Rabbolini et al. [1157] limit their investigation of fatigue crack growth to room temperature. However, a recent work by Rossmann et al. [1233] presents strain measurements on tensile test specimens at about 1000°C. The developed technique is based on special blue gel filters and utilizes the illumination of the specimen by a quartz lamp heating system. One can notice punctual works on single crystal superalloys, such as [820] at the macroscopic scale, and in [558] at the microscale in an SEM. This approach can even be expanded in the presence of thermal barrier coating. Indeed, Maurel et al.

[914,917,919] have demonstrated with the DIC technique the correlation between the oxide spallation pattern and the local strain level dependent on the substrate single crystal orientation. Pictures of an oxide surface taken during mechanical compressive testing show that the experimental local strain can reach 2% for an imposed average level of 0.2% and that locations of maximal strain influence the spallation propagation.

5.8.3 X-ray tomography

X-ray tomography is a recent tool to obtain detailed information on the 3D microstructure and the defects in a nondestructive way. An X-ray computed tomograph (Fig. 5.19) is made up of three parts:

X-ray source. X-ray tomography can be performed either in a synchrotron with a monochromatic beam or more frequently with X-ray tubes. In the latter case, the emitted beam is polychromatic, which leads to a "hardening phenomenon" of the spectrum. Indeed, the attenuation coefficient, for an energy considered, is less sensitive to the nature of the materials for higher energy ranges (greater than 200 keV). Thus, using a polychromatic source, attenuation is observed for low-energy spectral components due to the modification of the interaction between the beamline and the material.

Acquisition system. The detectors are placed in the axis of the material, behind the object to be analyzed. They deliver analog signals resulting from the creation of a current proportional to the number of received photons. The type of detector most often used for very-high energy installations are semiconductors.

Rotational platform. It allows a precise rotation of the sample between 0° and 360°.

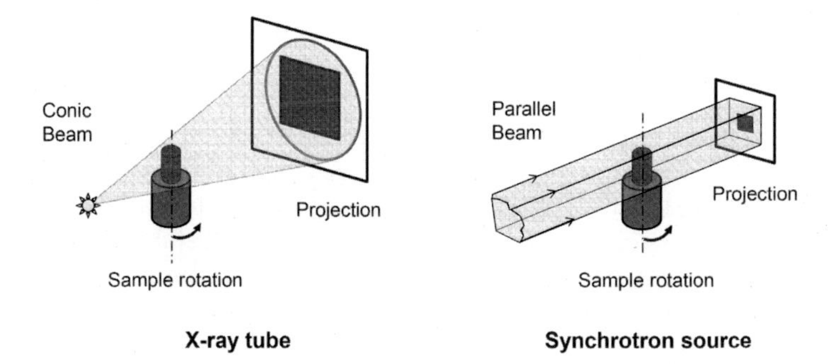

FIGURE 5.19 Schematic diagrams of X-ray tube based and synchrotron radiation-based tomography.

Computed tomography systems allow the synthesis of 3D images by acquiring a series of 2D X-ray images during a progressive rotation of the sample up to 360° with an increment less than 1°. These projections contain information describing the respective position and density of the different details in the object. This information is necessary for the numerical reconstruction of the volumes.

In order to obtain an accurate reconstruction of the volume, the depth or the diameter of the sample must be completely contained in the field of view and in the beam during the full 360° rotation. The magnification is limited by the diameter of the sample d as well as by the width D of the detector following $M = D/d$. For a pixel size of the detector P, the resolution of the voxel is denoted by $V = Pd/D$. The size of each point, viz. the voxel (volume element, extension of the pixel in 3D), is, therefore, conditioned by the resolution of the imaging system. The reconstruction step is based on a filtered back-projection algorithm. Three types of errors can occur during the test, (i) the variations of the beam hardening, which influence the incident flux at the detector and, therefore, the resolution in density, because of the simultaneous variations of the linear absorption coefficient of the material, (ii) the parallax errors, and (iii) the processing artefacts. All can induce errors in the reconstructed densities.

Tomography has been essentially used in absorption to get the volume fraction of pores [848,800,835] or to get the morphology of the microstructure, such as dendrite [671,684,646]. When used to follow the volume fraction of voids and determine which type of pores, i.e., solidification, homogenization, and deformation, as defined by [389], it was found that the growth of porosity is based on the enlargement of preexisting pores in the interdendritic areas and on the generation of new small pores. Also, growth processes during creep at high temperature result in polyhedral pores. This level of description for the shape of the pores (Fig. 5.20) was made possible by Link et al. [848] who did synchrotron tomography allowing for high spatial resolution, which is not the case with computed tomography using an X-ray tube [800]. The authors of [848] followed the volume fraction of pores during creep tests by doing ex-situ tomographic characterizations and found that the plastic strain rate increase, when transitioning from secondary to tertiary creep stage, was not directly related to an accelerated void growth process. Thus, it would appear that the increase in the plastic strain rate from the secondary to tertiary creep stage is not only due to voids, but could be also due to a destabilization of the microstructure, such as topological inversion, and of the γ/γ' interfacial dislocation network allowing a massive shearing of the γ' phase.

5.8.4 Short and long cracks

During the 20th century, crack growth in single-crystal superalloys was investigated, for instance, by [299,325,448,26,601]. The crack length was often measured from direct current potential drop (DCPD), well adapted for high-temperature measurements. The potential drop method works in a furnace without optical or extensometer access and can also be used with induction heating, when filtering the electrical interferences from the high

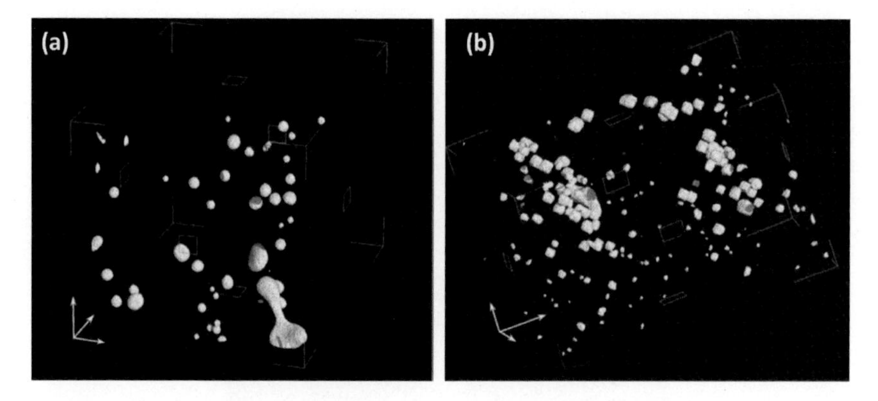

FIGURE 5.20 3D rendering of the shape of the pores in a CMSX-4 specimen showing the change from (a) spherical pores after heat treatment to (b) polyhedral pores after creep deformation at 1100°C/117 MPa for 392 h (adapted from [848]).

frequency current. More recently, a few studies used imaging approaches, the optical imaging [1255], SEM imaging [881], and IR imaging [714], respectively. The specimens used are typically notched, such as the compact tension (CT) [26,601,1255] or single edge notched (SEN) [325] geometry.

The crack growth in single crystalline superalloys depends on the crystal orientation with respect to the load direction [949,881,686,1660,1386]. Two types of crack paths can occur depending on the testing conditions: either the crack propagates perpendicular to the applied load in mode I along noncrystallographic planes or along crystallographic, octahedral $\{111\}$, slip planes. In the latter case, the macroscopic crack planes and propagation directions can be defined in analogy to the Miller indices of crystallographic slip systems in terms of $\{hkl\}$ or $\langle uvw \rangle$, respectively. As described by Neu [1006], the observed differences in crack path and growth rate are due to the experimental parameters such as temperature, environment, and characteristics of the mechanical load (frequency, load ratio, average load, and dwell time). A change in the prevailing crack paths is typically observed with increasing temperature due to enhanced oxidation at the crack tip.

At room temperature, crack propagation exhibits a combination of octahedral crystallographic planes, leading to mixed mode propagation even if the specimen is nominally in mode I loaded [1386]. Depending on the crack growth rate, the crack path can be more or less marked by crack deflection and branching. The orientation of the crack planes depends on the combination of crystal orientation with respect to the load direction and the stress state. It is notable that alternations of the crack system at local length scale occur while the macroscopic crack still propagates in mode I. As a result of the anisotropy of the single crystal, the propagation rate of the crack on local length scale depends on the propagation direction with respect to orientation. As a consequence, the crack growth rate of the macroscopic crack propagating in mode I depends on the propagation direction as well. At elevated temperature, like 950°C, macroscopic crack growth in a single crystal, oriented with the [001] direction parallel to the applied load, is observed in mode I [822] with the crack growth rates much slower along the ⟨110⟩ direction than along the ⟨100⟩ direction, which was the direction of the fastest crack propagation [11].

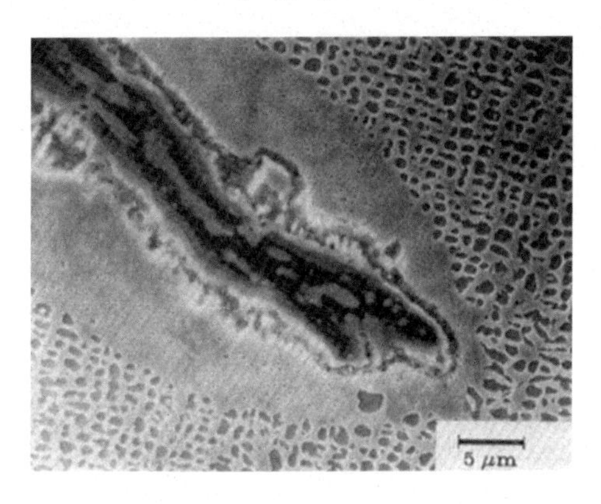

FIGURE 5.21 Oxidation assisted crack growth in RENE N4 at 1093°C [822].

At high temperature, fatigue testing in lab atmosphere can be characterized by oxidation-assisted crack growth. To understand the impact of oxidation on propagation mechanisms, fatigue crack growth test can be performed under vacuum. The comparison between oxidizing and nonoxidizing environment shows an increase of fatigue crack growth rate with oxidation especially at low growth rate [325,1029,1288,1189]. Since in oxidizing atmosphere the material in front of the crack tip oxidizes (Fig. 5.21), it is more brittle than the initial superalloy so that at the local length scale the crack does not propagate along the slip planes but in mode I.

Fatigue crack growth in single crystal superalloy is commonly studied under uniaxial and isothermal conditions with or without hold time, far different from in-service conditions. Under nonidealized loading conditions, quantitative assessment of the crack path and growth rates is complicated because of the interaction between mechanical, thermal, and chemical loads. For industrial needs, it is required to understand well the mechanisms of fatigue crack growth under multiaxial and/or nonisothermal loading. Studies barely begun to address these topics in the context of crack propagation. Crystallographic crack growth implies understandably local mixed mode under mode I loading. Crack growth under macroscopic mixed mode axial–torsional loading was studied by [228] using a notched specimen at room temperature. In this case, crack growth rate and path were not altered when the ratio between tension and torsion was changed.

Thermo-mechanical fatigue crack growth starting at laser drilled holes in hollow specimens was investigated by [714]. In OP-TMF tests was observed that cracks, which initiated at the holes, propagated initially along crystallographic planes, but during crack growth the propagation mode changed to noncrystallographic mixed mode. Concerning crack growth rate, the authors have demonstrated OP-TMF conditions promote faster crack propagation than isothermal LCF conditions at similar temperature and loading conditions of the tensile part of the TMF cycle. Thermo-mechanical fatigue crack growth test was also investigated by Palmert et al. [1048] on SEN specimens with their axes oriented along the crystallographic ⟨100⟩ directions. The authors of [1048] performed different force-controlled TMF tests for temperatures ranging from 100°C to 750°C or 850°C in order to evaluate the effect of phase shift (in-phase or out-of-phase thermo-mechanical loading) and the impact of dwell-time ranging from 10 s to 6 h. A dedicated compliance-based method was proposed to experimentally evaluate the crack opening stress. If simple IP-TMF loading results in similar crack growth rates like isothermal loading at the maximal temperature, dwell-time at the maximal temperature leads to an increase of crack growth rate. For OP-TMF tests, the maximal temperature and the eventual hold time duration do not affect the crack growth rate. Nevertheless, the OP-TMF tests show a significantly higher crack growth rate than isothermal tests at equivalent temperature. The results show that the microstructural evolution induced by the thermal cycle and the phase shift between thermal and mechanical cycle may affect the fatigue crack growth mechanisms.

CHAPTER

6

Elementary deformation processes in high temperature plasticity of Ni- and Co-base single-crystal superalloys with γ/γ' microstructures

Cathie M.F. Rae[a], *Gunther Eggeler*[b], and *Jean-Loup Strudel*[c]

[a]**Department of Materials Science and Metallurgy, University of Cambridge, Cambridge, United Kingdom,** [b]**Ruhr-University Bochum, ICAMS, Bochum, Germany,** [c]**MINES ParisTech, PSL University, Centre des Matériaux, Evry, France**

6.1 Introduction and focus

Superalloy single crystals (SX) are principally used to make blades for gas turbines operating in aeroengines and power plants. They must withstand static and cyclic mechanical loads at varying temperatures to well above 1000°C, where oxidation, creep, and fatigue limit the service life of components. In the present chapter we focus on elementary deformation processes which determine the mechanical behavior under monotonic loading. Central to this is the movement of dislocations through the γ channels, constrained by the necessity to form high energy faults if regular fcc dislocations enter the ordered γ' phase. Yield and low-temperature, high-stress creep represent conditions where stresses and thermally activated processes are sufficient to form such faults, allowing shear in both phases. On the other hand, the exceptional creep performance at high temperatures is made possible by the directional coarsening of the cuboidal γ' precipitates, driven by γ/γ' misfit, to inhibit climb around these obstacles

Nickel Base Single Crystals Across Length Scales
https://doi.org/10.1016/B978-0-12-819357-0.00013-5

[59,58]. These mechanisms are controlled by the stress state in the γ phase, determined from the combination of intrinsic misfit stresses and the externally imposed stresses. The structure of the chapter reflects this central issue, focusing first on the microstructure of single crystals and the internal stresses controlling dislocation movement in the narrow γ channels. Dislocations experience extraordinary constraint, moving in spaces less than 100 nm in width and subject to intrinsic misfit stresses approaching 500 MPa. We go on to describe how yield and creep mechanisms develop from this basic process.

The complex relations between the mechanical properties of Ni-base superalloys and their chemical, physical, and microstructural characteristics have been reviewed by many authors [324,1330,996,1192,209,1125,370, 1190,1129,1383], and we endeavor to give fair credit to all researchers who have contributed to our understanding over the last six decades. Here we identify the key elementary processes which determine the response of single-crystal materials to monotonic uniaxial mechanical loading at elevated temperatures. We take a microscopist's point of view and address yield and creep over the full range of temperatures and stresses investigated. We review some of the seminal papers on deformation mechanisms and also highlight some recent results in both Ni- and Co-base alloys.

6.2 Fundamentals

6.2.1 Heterogeneity of cast structures

The γ/γ' microstructure of Co-and Ni-base single-crystal superalloys is key to the deformation of superalloys as it imposes and controls the manner and scale of the deformation processes and hence the mechanical properties at all temperatures. Not only does the starting microstructure evolve during processing, specifically, casting, solidification, and heat treatment [1190], it continues to develop during, and in response to, the imposed mechanical and thermal stresses and the resulting deformation mechanisms.

Single crystal superalloys are produced in a Bridgman-type solidification process [1190]. During solidification, primary dendrites grow in $\langle 001 \rangle$ directions. This is fortuitously the preferred axis orientation for the blades, as it has the lowest modulus and hence the lowest stresses resulting from thermal strains. The dendrite spacing depends on the cooling rate [1190,1562]. The as-cast microstructure (Fig. 6.1) is subjected to a post cast heat-treatment which can be separated into a homogenization and into a precipitation phase [1058]. From a mechanical point of view, it is important to appreciate that even after a full heat treatment at a temperature 20°C to 50°C below solvus temperature (typically 1300°C to 1340°C for 5 to 10 h) one finds significant differences between the microstructure of a prior den-

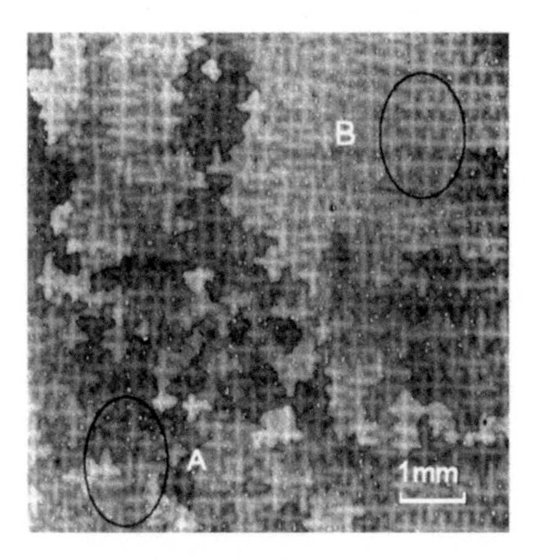

FIGURE 6.1 Dendritic (in B) and mosaic (around A) microstructures in CMSX-4 ⟨001⟩ oriented single crystal [163].

FIGURE 6.2 Initial γ/γ' microstructure of AM1 fully heat treated.

dritic (D) and interdendritic regions (ID). Thus, the alloy element Re, for example, tends to segregate to dendritic regions [1058,1613], where Re-levels are slightly higher than in interdendritic regions. This represents a dendrite-scale heterogeneity which characterizes Ni-base SX superalloys. In addition, misorientations between the dendrites or groups of dendrites lead to a mosaic structure where the small misorientations between the dendrites are accommodated by low angle boundaries. These are a ready sources of dislocations during the early stages of plastic deformation.

After solution and precipitation heat treatments, the well-known γ/γ' microstructure forms, typically with a 75% volume fraction of the ordered simple cubic $L1_2$ phase γ'. It consists of small γ'-cuboids some 0.5 µm in length separated by thin coherent γ-channels typically 100 nm in width, Fig. 6.2. The volume fraction can differ between dendritic and interdendritic regions by nearly 10%, thus, for example, the γ'-volume fraction can

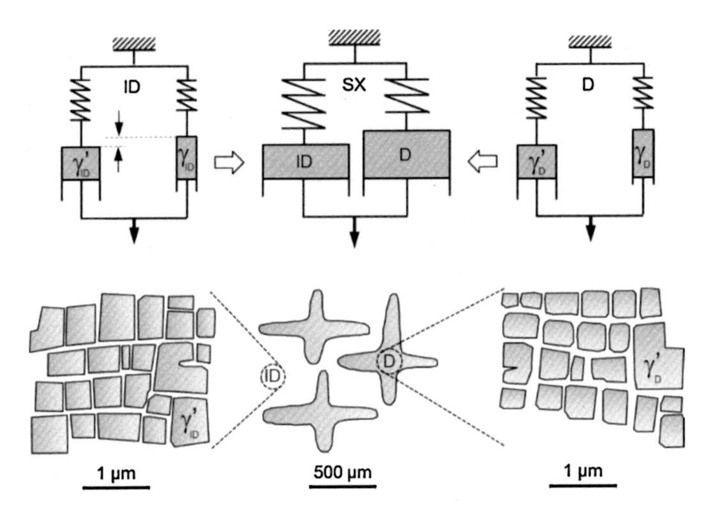

FIGURE 6.3 Cast Ni-base single crystal superalloys represent materials with composite character on the mm- and µm-scales [1587].

be 72% of the volume in the dendritic regions and 77% in the interdendritic regions.

As a result of these differences in composition and hence properties, cast superalloy single crystals represent mm-scale composite materials, where stresses redistribute in the early stages of creep between dendritic and interdendritic regions [1587]. This leads to significant effects such as deformation occurring first in the dendritic cores where the volume fraction of γ' is lower [296]. Observations at high resolution should therefore identify their location within this structure. The assembly of these regions in turn, represents a microcomposite composed from the γ/γ' microstructure. This is schematically illustrated in Fig. 6.3.

6.2.2 Micromechanical implications of coherent γ/γ' microstructures

When the ordered γ' phase forms, it precipitates quasicoherently. Its atoms occupy the same lattice sites as the atoms of the fcc matrix phase (Chapter 4). The fact that the lattice constants of the γ and γ' phases differ results in a lattice misfit δ defined as

$$\delta = \frac{2(a_{\gamma'} - a_{\gamma})}{a_{\gamma'} + a_{\gamma}}, \tag{6.1}$$

where $a_{\gamma'}$ and a_{γ} are the lattice constants of the γ and γ' phase, respectively. Ni-base superalloys typically have a smaller lattice parameter in the γ' phase, leading to a negative misfit of about -0.1% to -0.3%, while Co-base SX with very similar microstructures exhibit positive misfit [1383]. For both cases the misfit is generally not sufficient to lead to semicoherency at the typical γ' sizes produced. The equilibrium volume fraction of the γ' phase decreases with increasing temperature and, for most commercial alloys, there is a critical temperature above which the γ' phase is no longer stable (see recent thermodynamic calculations [1613]). One drawback of most Co-base SX is that the dissolution temperature of its γ' phase is much lower ($\ll$1000°C) than in Ni-base SX (> 1000°C) [1190,1383]. Recently, new compositions containing 6–8 at.% Ti+W have been developed, which exhibit a γ' solvus temperature above 1100°C with a yield stress larger than 1000 MPa [1016].

The shape of the precipitates is determined by a competition between the tendency to minimize surface energy (which dominates for small particles) and the tendency to minimize the overall strain energy, which increases with particle size [1134]. The anisotropy of the elastic modulus, with a minima in the $\langle 100 \rangle$ directions [1108,41], drives the cubic shape of the precipitates (see recent overview and new elastic data published by Demtröder et al. [330]). Moreover, local misfit stresses enter into the thermodynamic potential of atoms [1134], and can result in driving forces for diffusion of atomic species which plays an important role in the evolution of the γ/γ' microstructure under creep load or "rafting" (see Section 6.4.4). In the discussion of microstructural evolution in Ni- and Co-base superalloys the term "constrained misfit" is commonly used [981,264,266]. This term must be interpreted with care. The unconstrained lattices of the two phases have different lattice constants: this defines the misfit δ (Eq. (6.1)). When the two phases occupy one common crystal lattice, elastic distortions constrain the two phases. It is this constrained lattice misfit which governs the results obtained in diffraction experiments of the two-phase alloy. It gives rise to multiple values of the γ lattice parameter depending on the location and orientation where the measurements are made and on the microstructural state of the material, as well as on the local and external stress fields [352,353].

The early stages of both tensile yield and creep in $\langle 001 \rangle$-oriented nickel-base superalloys are characterized by the penetration of the matrix channels, perpendicular to the applied stress, by one or several slip systems. In this period, interfacial dislocation networks evolve rapidly. They help decrease the overall strain energy, associated with the crystallographic misfit between the two phases. It is an essential part of creep under all conditions and also determines the course of rafting processes at higher temperatures. At the higher stresses characteristic for low temperature creep (T < 800°C), dislocations go on to cut through the precipitates, and at high temperatures and correspondingly low stresses the microstructure evolves through rafting. As a result, we look at the movement of the dislocations confined to the γ phase in some detail [232].

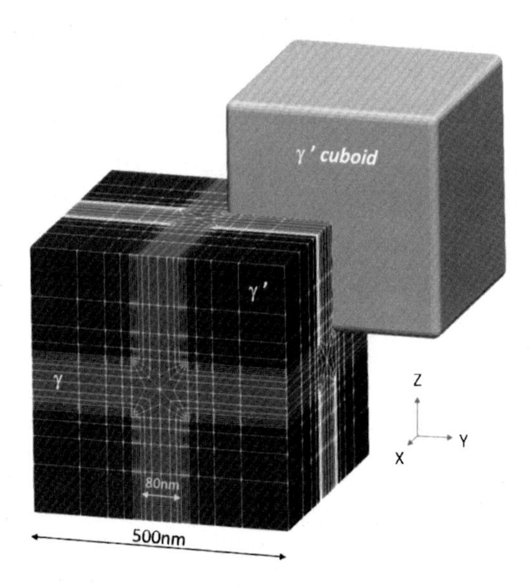

FIGURE 6.4 Elementary cell of an idealized 3D periodic distribution of γ' cuboids with γ matrix channels in red (mid grey in print version) color [232].

In order to estimate the stress and strain fields produced by coherent misfit and the subsequent addition of dislocations to relax the interface, we use an idealized 3D distribution of equal size cuboids (Fig. 6.4) periodically aligned along the cube directions. The details of these fields can be obtained using the FEM approach (finite element method) to compute the local nonuniform stress and strain fields, with appropriate meshing of all the constituents of the elementary cell, particular attention being paid to the rounded edges and of the cuboids. The first such results were presented by Pollock and Argon [1126] in a 2D scheme. Here the γ' cuboids edges are 420 nm long and the matrix channels 80 nm wide; the γ/γ' misfit was chosen as $\delta = -3.0 \times 10^{-3}$, this being typical of commercial nickel-base alloys.

Two of the components of the misfit stress field appear in Figs. 6.5(a) and 6.5(b), the third σ_{22} is identical to σ_{11}, turned 90° around the z-axis. A striking feature is that the matrix channels are subjected to a strong compressive biaxial plane stress (dark blue) parallel to the γ/γ' interface, since the matrix, with a larger lattice parameter, is constrained in two orthogonal directions by the smaller lattice of the surrounding γ' precipitates. The resulting strain field of the matrix channels is sketched in Fig. 6.5(c): it is elastically elongated along a direction normal to the γ/γ' interfaces. All channels are elastically thickened by this internal stress field distribution. According to Hooke's law, the plane stress exerted by the misfit across the

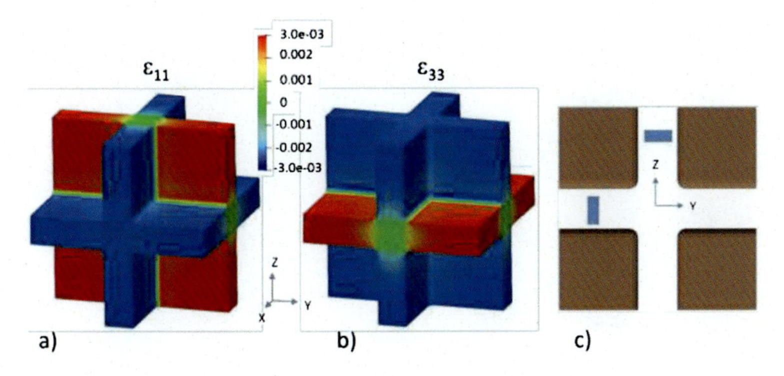

FIGURE 6.5 (a)–(b) γ/γ' misfit stress field in the matrix channels; (c) the resulting prestrain [232].

horizontal channel, for instance, reads as

$$\varepsilon_{33}^{\text{misfit}} = -\frac{\nu}{E}\left(\sigma_{11} + \sigma_{22}\right) = -2\nu\delta > 0, \tag{6.2}$$

since $\sigma_{11} = \sigma_{22} = E\delta$ in that channel. Note that all misfit quantities are negative, hence the strain increases the width of the channel. The strain in the γ' precipitates is uniform and tensile.

The elastic stress fields localized in the matrix channels (and γ' cuboids), being of the same order of magnitude as the stresses applied during the service life of the material, have important consequences for the mechanical behavior at high temperature [846]. Since we are examining this material at the nanometer scale, the von Mises stress concept, useful when considering that straining of the material is uniform and homogenizing techniques are appropriate, is no longer relevant since heterogeneous stress fields will develop inhomogeneous plasticity as individual slip systems respond to the local stress state. A better approach is to use the resolved shear stress induced by the eight potentially active slip systems. In applying the tensile stress along the z axis, we shall consider the Burgers vector normal to the y axis, i.e., $(1/2)[011]$ gliding in the slip plane $(1\bar{1}1)$ as shown on Fig. 6.6. The intensity of the shear stress in the P-channel lying perpendicular to the tensile axis indicates that the γ/γ' interfacial stress field is able to attract dislocations to the horizontal channel even in the absence of an imposed stress Fig. 6.6(a). The effect is even stronger when a stress of 500 MPa is applied, Fig. 6.6(b). Notice that in the C channel, where the dislocations lie in a screw orientation, there is no effect of the misfit stress field (khaki color). Similar considerations apply to all the eight potentially active systems with Burgers vectors having a component parallel to the tensile axis along [001].

Having established that the misfit imposes an asymmetry on the resolved shear stress of each dislocation with respect to the orientation of

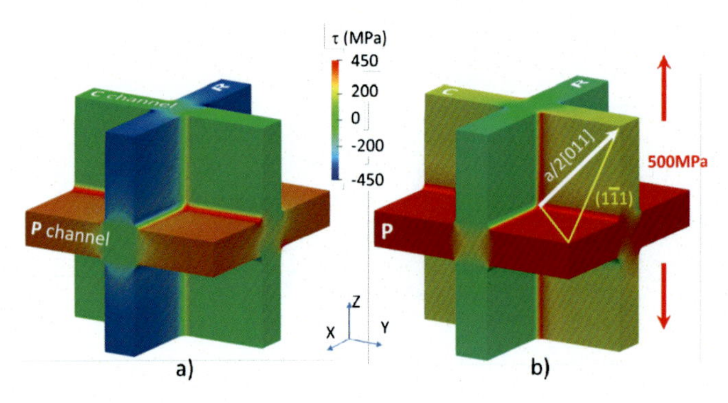

FIGURE 6.6 Resolved shear stress induced on slip system [011] (1$\bar{1}$1) or (11$\bar{1}$): (a) by misfit alone; (b) by misfit and tensile stress of 500 MPa along the vertical axis [232].

the γ channel, we can consider how dislocations move in the early stages of creep. There is no need for the nucleation of dislocations, since the subgrains of the mosaic structure characteristic of annealed single crystals contain, in their boundaries, dislocation networks with the full range of Burgers vectors, see Fig. 6.1. For simplicity we ignore diffusional processes, such as climb and pipe diffusion, and assume that the γ' cuboids are nonshearable, nonpenetrable by dislocations, and have elastic properties identical to that of the γ matrix.

The behavior of individual dislocations in the misfit-controlled stress fields (when misfit stress field are ignored, see Chapter 14) can be computed by FEM and coupled with the discrete dislocation dynamics (DDD) technique by introducing Frank–Read sources in the activated P-channel and applying an increasing uniaxial stress along the ⟨001⟩ direction, step-by-step. Observed in its glide plane, the [011] dislocation first explores the horizontal or P-channel (Fig. 6.7(a)), where it leaves a 60° segments in the γ/γ' interface, then spreads into the C channel where it leaves segments in pure screw orientation (Fig. 6.7(b)), but it is repelled from channel R where it would aggravate the misfit. Only under very high stresses, could it bow out into the R channel, Fig. 6.7(c). The situation clearly differs from a simple Orowan-type stress analysis, namely $\tau_{Or} = A\mu b/w$ where w is the width of the channel, μ the shear modulus, and A is a constant approximately equal to 1, where all channels would have the same opportunity of being penetrated. As a consequence, shear loops developing during the first stages of creep will have the shape of parallelograms as shown in Fig. 6.7(c).

For a given dislocation such as that illustrated in Fig. 6.7, the resolved shear stress will be the same whichever of the two possible slip planes it moves on, (1$\bar{1}$1) or (11$\bar{1}$). Hence, as it moves in the horizontal P channel, the leading screw segment of the loop can cross slip between the two slip planes. TEM observations in several alloys made using thin foils taken

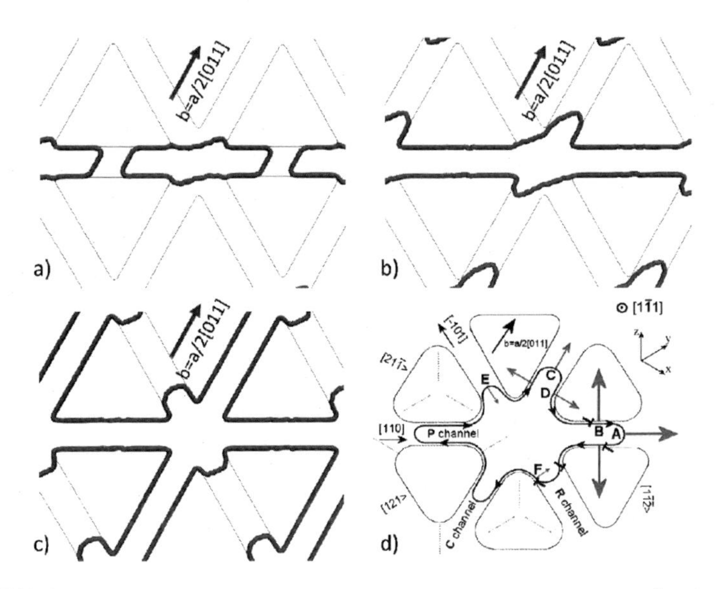

FIGURE 6.7 Shear loop [011] spreading through various channels of its $(1\bar{1}1)$ slip plane. Snapshots of DDD simulations: (a) under 300 MPa; (b) under 400 MPa; (c) under 600 MPa; (d) sketch of glide forces [232].

from high temperature creep tests interrupted at strains less than 1% have shown that the leading screw segments of dislocations bowing out into horizontal matrix channels are moving in the cube directions [100] and [010], normal to the stress axis, by repeated cross-slip as sketched and shown in Fig. 6.8(a)–(b) [846,215,1368]. This has the advantage of allowing the expanding loop to grow overall in the [100] direction along the center of the γ channel where the stress advantage from the misfit is greatest. The micrographs show that the cross-slip occurs as the dislocations approach the corners of the precipitates. In the case of very low creep rates, or higher temperatures, when climb mechanisms become effective, the dislocation segments, initially oriented at ±45° from the cube directions, as at B in Fig. 6.8(b), have been observed to straighten out (C), shortening the length of these interfacial dislocations and thus improving their efficiency to relax misfit stresses [215].

Due to frequent cross-slip, the dislocations propagating on either side of a precipitate will generally be moving on different, but parallel slip planes. This process aids the propagation and a more uniform distribution of slip throughout the γ phase resulting in several dislocation loops wrapping the precipitates. Activation of further slip systems creates loops on different planes and the γ' precipitates will become surrounded on all sides. The glide forces acting on the various segments of this spreading loop are sketched on Fig. 6.7(d). The climb forces are of opposite sign at B in channel P and at D in channel C [232], this is significant during rafting

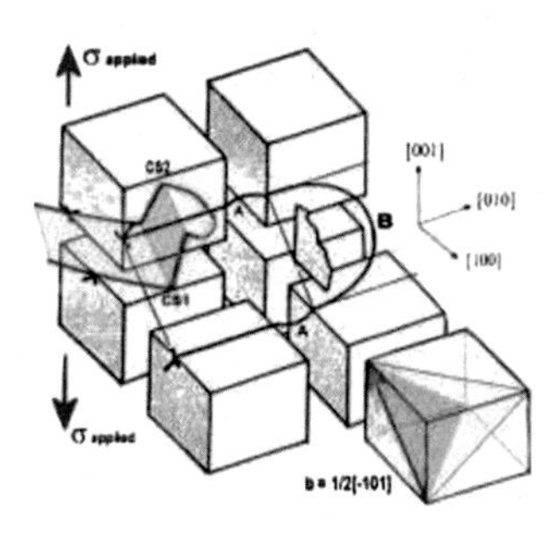
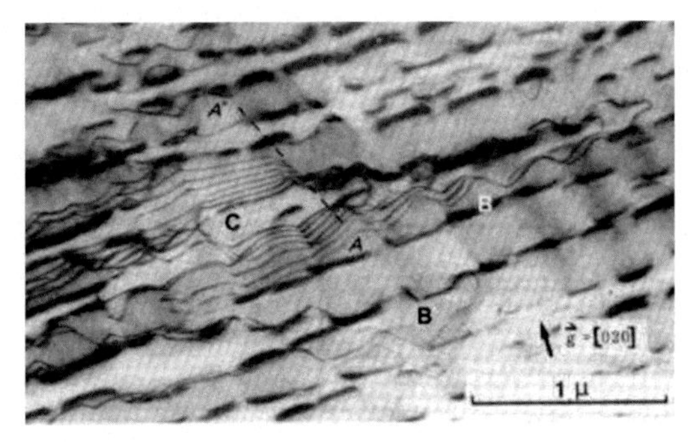

FIGURE 6.8 Matrix dislocations repeatedly cross slipping in the P-channel during initial creep stage: (a) sketch; (b) TEM observation of cross slip alone; (c) straightening of the 60° segments into pure edge interfacial dislocations (in C) by climb [215].

section 6.4.4. Chang et al. [232] illustrate this using dislocation dynamics. As the plastic strain increases, dislocations from two different slip systems zigzag by cross-slip, multiply, and interact in channel P, leaving screw segments in vertical C channels, but are repelled from the R channels (Fig. 6.6). Many lie in the γ/γ' interfaces, others are immobilized by attractive junctions, but all contribute to a biaxial ($\sigma_{11} \approx \sigma_{22}$) contraction of the sample during the very early stages of creep. This strain rate is of the order of $10^{-4}\,s^{-1}$ and was indeed observed with very careful experimentation by Reppich et al. who termed the phrase "negative creep" to describe this phenomenon [1208].

6.2.3 Formation of misfit networks at high temperatures

The different dislocation densities in the horizontal and vertical channels compared to vertical ones also lead to anisotropic coarsening of the γ' precipitates as they coalesce in continuous rafts perpendicular to the tensile stress. This process of rafting is critical to the high temperature performance of Ni-base superalloys with a negative misfit, as it limits the extent to which dislocations can climb in the γ/γ' interface and it thus greatly improves creep performance. The processes of glide, cross-slip, and climb, driven by the modified stresses in the individual channels populate the γ channels with dislocations which generally reduces elastic strain energy associated with misfit.

Ideally, in order to best compensate for the misfit, dislocations lying in the interface should be oriented pure edge [794]. This is what happens when a homogenized nickel-base superalloy single-crystal blade material is submitted to a high temperature heat treatment (here 1050°C for 100 h) in the absence of any externally applied stress (Fig. 6.9).

Its cuboidal shape has been truncated along the eight apexes of the cube by $\langle 111 \rangle$ type facets, the precipitate takes on the shape of a tetrakaïdecahe-

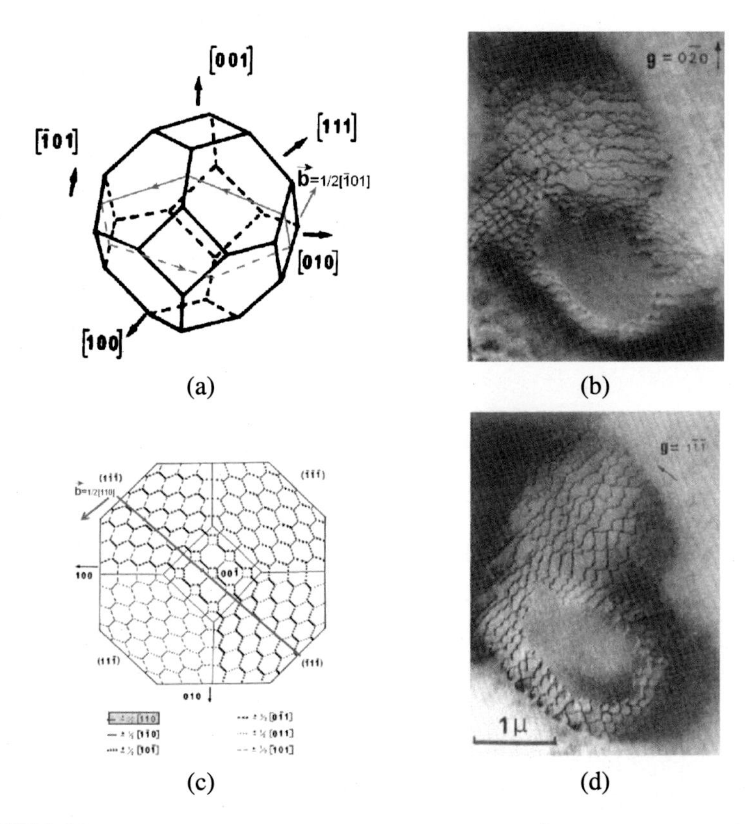

FIGURE 6.9 Minimum free energy configuration of a large γ' precipitate: (a) tetrakaïdecahedron shape surrounded your pure edge loop; (b) and (d) TEM micrographs of the same γ/γ' interface observed under different diffracting conditions; (c) notice that the interfacial dislocation networks made of interwoven pure edge loops produces lines perpendicular to the diffraction vector on the various facets of the rounded γ' precipitate (alloy AM1) [794].

dron (6 + 8 = 14 facets), thus minimizing the surface area of a polyhedron made of low index crystalline planes. In this manner, the two-phase material spontaneously minimizes its surface energy together with the elastic strain energy caused by its misfit, by attracting edge dislocations and placing them in the position of a pure edge segment.

Notice that when these interfacial dislocations are optimally arranged to compensate the γ/γ' misfit, their average perpendicular distance, d, is given by Brook's formula

$$d = \frac{b \cos\phi}{|\delta|}, \tag{6.3}$$

where b is the Burgers vector of the edge dislocation, ϕ the angle with the interface, and δ the misfit parameter between the phases. Here, for instance, with the parameters $b = 0.254\,\text{nm}$, $\phi = 45°$ and $\delta = -3.0 \times 10^{-3}$, the spacing should be $\approx$ 60–70 nm. In this configuration, they are a perfect example of geometrically necessary dislocations (GND) since they minimize the free energy of the two-phase system in the absence of any applied stress. After hundreds of hours of creep under an applied stress, they evolve into denser networks, in general less regular than those presented in Fig. 6.9. This is well illustrated by the Japanese-developed alloys TMS 75, TMS 138, and TMS 162, with respectively 2%, 3% and 4% Mo [739], exhibit increasingly negative misfit and correspondingly smaller interdislocation spacings after creep rupture as given by Brook's formula.

6.3 Yield phenomena

6.3.1 Yield anomaly

Nickel- and cobalt-base superalloys exhibit an exceptional yield stress and, more generally, a flow stress which stays almost constant in the temperature range from 20°C to 750°C followed by an anomalous peak stress at around 900°C [1304,604] Fig. 6.11. This unusual mechanical resistance at elevated temperature is related to the presence and behavior of the γ' hardening phase itself, and, crucially, to its specific microstructural architecture in the form of quasiperiodic assembly of coherent cuboids densely packed in a ductile γ matrix (Fig. 6.2). Fig. 6.11(a) shows that the increase in strength of the alloy is due to the "anomalous" rise in the yield strength of the ordered intermetallic phase and rises approximately with the increase in the proportion of that phase.

The yield anomaly has been extensively studied in single phase $L1_2$ intermetallic alloys and is attributed to the cross-slip of the dislocation pairs such that the antiphase boundary (APB) between them moves from the high energy {111} plane to the lower energy {001} plane, where the ordering does not violate nearest neighbor bonds [450,710,1620]. This process, known as Kear–Wilsdorf locking (Fig. 6.10) [341], is assisted by the anisotropy of the elastic properties of Ni [1622] which can contribute a torque equivalent to as much as 60% of the repulsive force between the two dislocations. This torque, proposed by Yoo [1621], has the effect of rotating the dislocation pair from the {111} plane (which is not a mirror plane) to the {001} plane, thus adding to the APB energy advantage. Once cross-slip has occurred, the pair can no longer glide on the {111} plane and must undergo a stress-induced activation to return to glide, the flow stress in intermetallics is derived from the stress required to repeatedly activate this process [187]. As the temperature rises, movement on the cube plane through thermal activation becomes possible and the yield stress drops rapidly above this temperature. These processes take place over many microns, in two phase materials, i.e., on a scale rather larger than the 500 nm

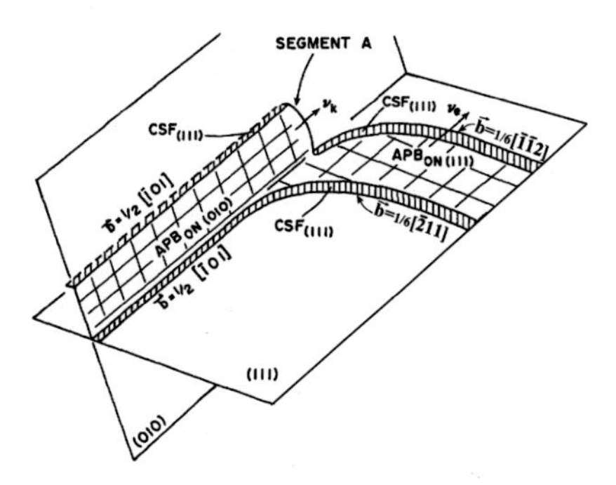

FIGURE 6.10 Schematic of Kear–Wilsdorf lock: cross-slip of dislocations pair from (111) slip plane to (010), after [341].

of the γ' precipitates. The short segments of dislocations in each precipitate, combined with the necessity for continuous movement through both phases, means that the factors determining yield in two-phase superalloys are different.

For polycrystalline nickel alloys used in aerospace applications, yield is the most critical property, and hence it is in these alloys that the strength is best understood. Yield has had less attention in single-crystal alloys where creep performance is usually the limiting property. Superalloys have a unique property of a steadily rising yield stress with temperature, in the temperature range from RT to 900°C attributable to the high-volume fraction of the ordered γ' phase (Fig. 6.11). Yield and creep are inextricably linked and, as we will show later in the section on creep, the mechanisms form a continuum, depending on temperature and strain rate. Many factors contribute to high yield strength and include order strengthening, solid solution strengthening, misfit of precipitates, Kear–Wilsdorf locking and grain boundary (Hall–Petch) strengthening in polycrystalline materials [529,161]. Of these, order strengthening is thought to have a major effect.

6.3.2 Weak and strong coupling in low volume fraction materials

The first quantitative approach to modeling the strength of superalloys was described by Gleiter and Hornbogen [529,528] who recognized that the interaction of dislocation pairs with ordered precipitates was the key hardening mechanism, a single dislocation creating an APB fault in the ordered precipitate where the Burgers vector $a/2\langle 110\rangle$ of the matrix dislocation is not a lattice vector. Hence the impediment to flow comes from the need to force the leading dislocation partially into the precipitate, creating an energetically expensive APB fault, before the second is able to enter, reordering the lattice. They distinguished between single dislocations cut-

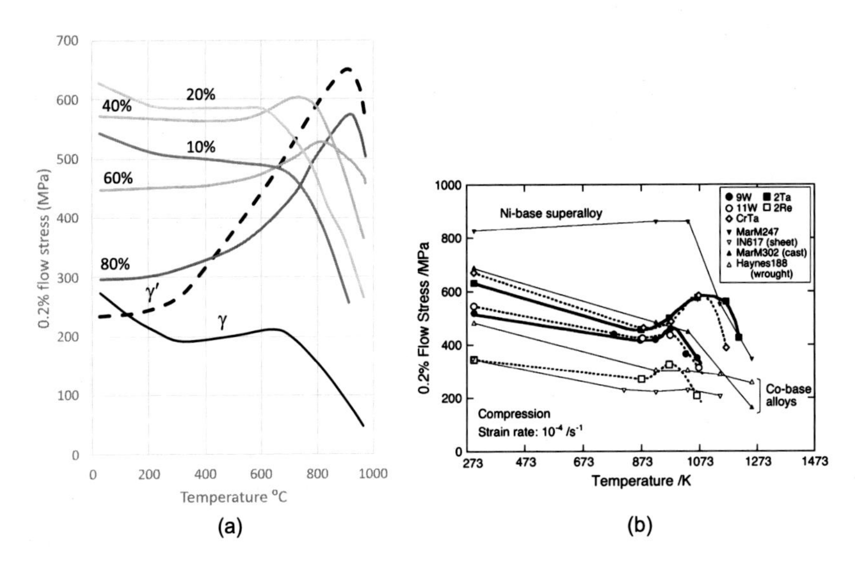

FIGURE 6.11 Flow stress variations with temperature of: (a) Ni-base tie-line alloys with γ and γ' phases based on the composition of MarM200 with various proportions of γ' precipitates (after [85]); (b) several Co-base alloys [1383].

ting the γ' and pairwise cutting by closely-spaced pairs minimizing the effect of the APB energy. Single dislocations creating an APB fault across the precipitate section give a shear stress increasing parabolically with precipitate size as the dislocations bow to interact with increasing numbers of precipitates (see curve ① in Fig. 6.13). Alternatively, when precipitates are large enough, hence distant enough, matrix dislocations can bypass precipitates by bowing out between them rather than, shearing them and the Orowan-type flow stress of this mechanism decreases as $1/r$, where r is the distance between the precipitates (see curve ③ in Fig. 6.13). The maximum strength occurs at the precipitate size where these two cutting processes have equal flow stress, i.e., at about $d = 2r = 120$ nm under a predicted shear stress larger than 500 MPa. Yet, these predicted values were in excess of the experimentally observed values, i.e., $\simeq 380$ MPa.

This approach has been refined and modified by a number of authors [161,1207,1210,1209,757,481,1188], but the most comprehensive treatment of the effect has been made by Reppich and coauthors in a number of landmark papers. Reppich [1005,1207,1210] identified two distinct configurations of the dislocation pairs, depending on the APB energy and the precipitate size. Weak coupling (curve ①, Fig. 6.13) is characterized by the dislocations lying in different precipitates, and strong coupling (curve ②, Fig. 6.13) by the dislocations lying in the same precipitate. Fig. 6.12 shows both configurations with the precipitates strongly pinning the dislocation pairs highlighted by arrows, but the dislocations clearly separated in other

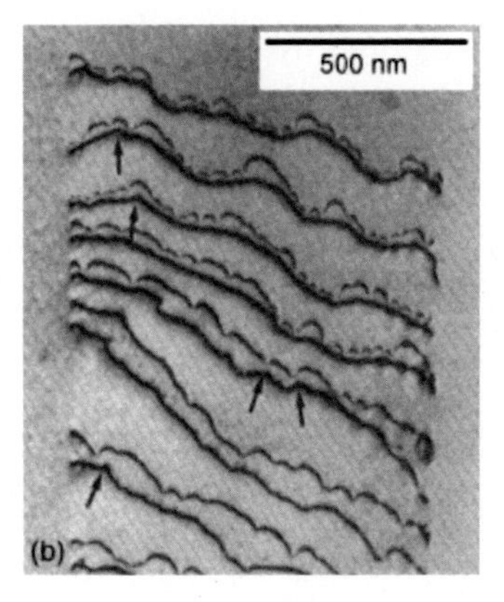

FIGURE 6.12 Pairs of dislocations moving through a fine dispersion of ordered precipitates, dislocation pairs strongly coupled are indicated by arrows [1005].

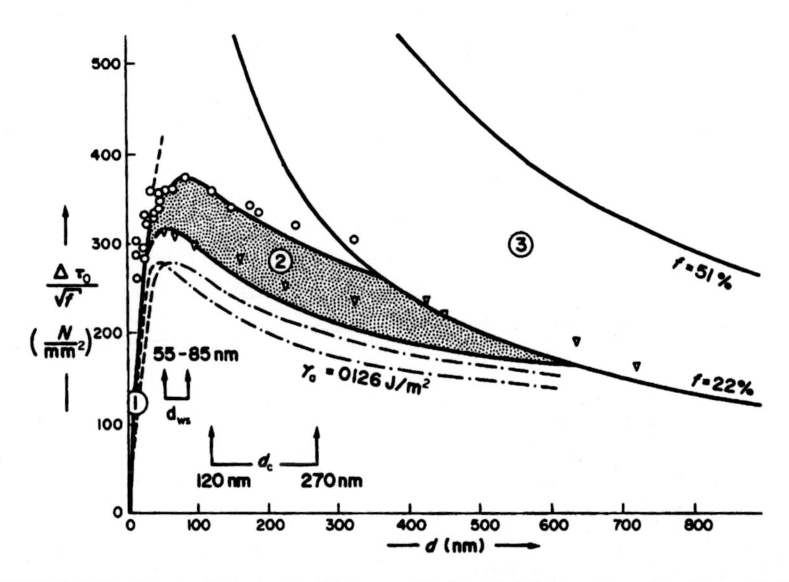

FIGURE 6.13 (Curve 1) Cutting stress for weakly coupled dislocation pairs; (Curve 2) Cutting stress for strongly coupled dislocation pairs; (Thick full lines) Curves of Labusch and Schwarz model [781,1295]; (Curve 3) Orowan stress for different f. Experimental data from Nimonic 105 [1207,1210].

locations. In both cases, the maximum strengthening effect (and hence yield strength) comes at the optimal precipitate radius r_m where the leading dislocation is passing through the greatest possible combined length of precipitate compared to the trailing dislocation. For weak coupling, this occurs because the leading dislocation is interacting with a greater number of precipitates, calculated from the angle of bowing as it is pinned by passing through the diameter of the precipitate; in the case of strong coupling, the maximum stress is observed at the point where the second dislocation just touches the edge of the precipitate already penetrated by the leading dislocation. Hence the weak coupling increases approximately as the square root of the precipitate size at constant volume fraction as shown in Fig. 6.13. A number of modifications are made, for example to take account of precipitate shape [651] and finite precipitate size as in the Kozar model [757], Fig. 6.14(a).

The strong coupling diminishes with precipitate size, r, as approximately $1/r$, Fig. 6.13. Hence the maximum strength is produced at the intersection of the two [1210]. Reppich recognized that Orowan looping of the precipitates occurs not only at very large precipitate sizes (curve ③, Fig. 6.13), but also at stresses in the range d_c (120–270 nm) marked on Fig. 6.13, where the curvature of the precipitates is insufficient to assist entry into the precipitates by the leading dislocation and the volume fraction is low enough to provide space for the dislocation to bow around the precipitates. The deviation of the experimental results in this size range suggests that this is occurring in the Nimonic 105 and PE16 alloys used in this work. This only becomes significant as the γ' size becomes very large, or the volume fraction very low, because it depends on the spacing between the precipitates; the effect is generally not seen in single crystal alloys as will be discussed in Section 6.3.3. Hence, in superalloys, the regime of strongly coupled dislocations effectively replaces the looping mechanisms in defining the strength maximum conventional precipitation hardened alloys.

This approach works well for low volume fractions of γ', typically in early polycrystalline superalloys for which the theory was developed, Fig. 6.14 (b), but overestimates yield stress for alloys with higher volume fractions, Fig. 6.14 (c). It extrapolates weak and strong coupling to the point r_m where they intersect, but at this point is not a valid description of the dislocation–precipitate configuration. This issue is addressed by Galindo-Nava [481]. Weak coupling is limited by the maximum force the line tension can exert, Fig. 6.15(a), i.e., it is maximized at the point at which the bowing angle ϕ is zero and the two dislocations are parallel to each other: the maximum pinning force (the APB energy, γ_{APB}, times the diameter of the precipitate, r_m) is equal to twice the line tension,

$$2r_m\gamma_{\mathrm{APB}} = \mu b^2, \tag{6.4}$$

where μ is the shear modulus, and b the Burgers vector. The radius penetrated, r_m, also represents the equilibrium spacing of two parallel screw dislocations in γ' separated by APB, and thus the precipitate radius of maximum strength for the strong coupling equations. Hence it forms a natural transition between weak and strong pair coupling. Beyond r_m the

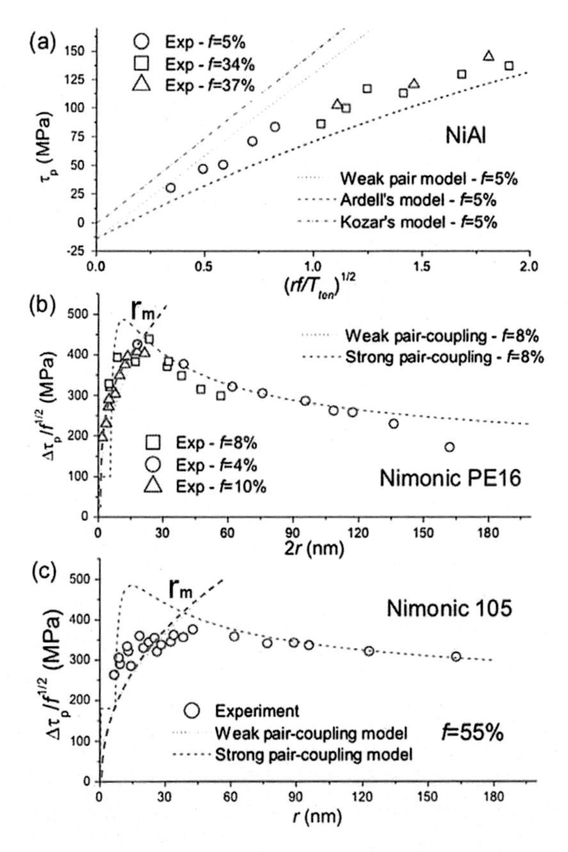

FIGURE 6.14 Predictions of the room temperature critical resolved shear stress due to particle shear: (a) Ni–Al alloys employing the weak pair-coupling; (b) Nimonic PE16; (c) Nimonic 105 employing weak and strong coupling [481].

conventional strong coupling stress equation reduces with increasing precipitate size, but it is not fully representative of the configuration of the precipitates in this transition region because the bowing of the dislocations in the γ channels adds to the applied stress and allows a lower cutting stress, i.e., $\phi < 180°$, Fig. 6.15(b)–(c).

The common force equations relate to the critical configuration where the leading dislocation is in the first half of the precipitate and the trailing on the point of entering, i.e., at maximum pinning force. The transitions are managed by the definition of the dislocation line length λ_1 in the γ and l_1 in the γ'. This leads to a continuous curve and a lower cutting stress better fitting high-volume fraction data [481].

To accurately replicate the yield strength, it is necessary to add other strengthening mechanisms such as solid solution hardening, and grain boundary stabilization. Solution strengthening is estimated using Labusch

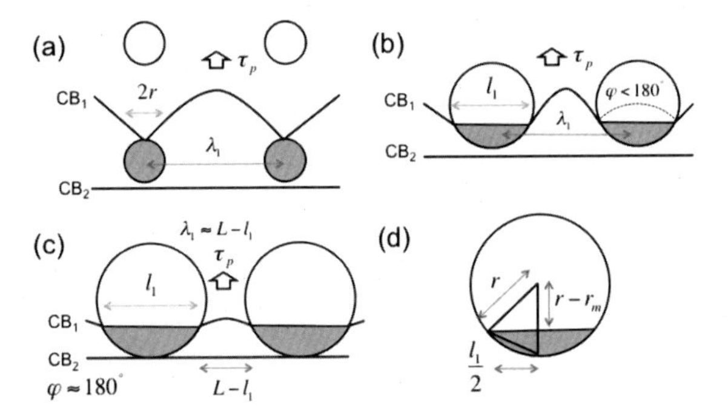

FIGURE 6.15 Schematic illustrations of the dislocation configurations in the case of: (a) weak pair-coupling; (b) intermediate pair coupling; and (c) strong pair coupling. (d) Geometric configuration between l_1, r and r_m in the case of partial shear.

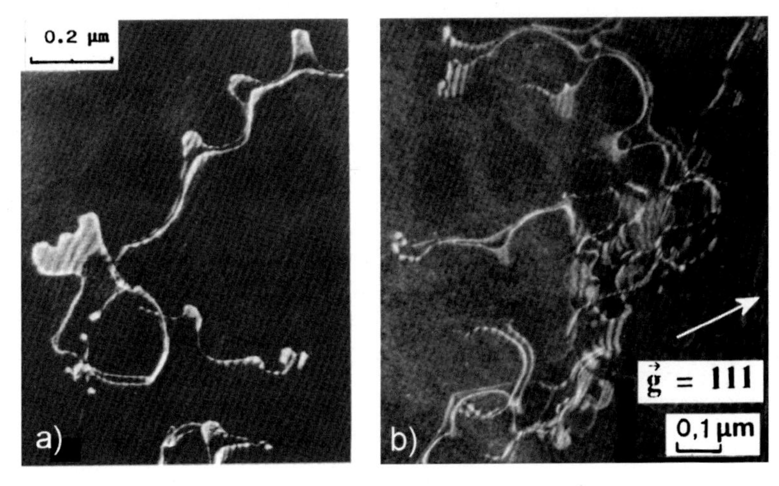

FIGURE 6.16 Astroloy at 650°C: Decoupling of partial dislocations (a) of perfect matrix dislocations after 0.2 % plastic relaxation, (b) of pairs of dislocations after 1% creep, after [1254].

theory [781] and the Hall–Petch equation is used for grain boundary effects [1209]. Elastic strains due to lattice mismatch between the γ and γ' phases can also make a small contribution [537,538]. No element of Kear–Wilsdorf locking has been included. Combining these different mechanisms was achieved by Kozar et al. [757] and Galindo-Nava [481] by a simple linear addition, and it can accurately replicate measured yield strength for evolving microstructure. Comparing the effect of order strengthening, calculated and experimental yield stresses in the alloy RR1000 suggests the 55% volume fraction of γ' precipitates contribute approximately 75% of the yield stress excluding Hall–Petch strengthening effects.

Shearing of the precipitates to give an APB is not the only option. For precipitates larger than r_m, it is easier for the leading dislocation to loop around the precipitate leaving perfect crystal and a dislocation loop which will be pushed through the precipitate when the second dislocation arrives. This further reduces the yield stress at higher precipitate spacings, and hence becomes the preferred bypass mode for coarsened precipitates where the γ' volume fraction is very low.

An alternative mechanism observed in nickel-base superalloys is the bowing out of partial dislocations of the matrix, with a/6 $\langle 112 \rangle$ Burgers vector, into these channels, leaving behind an intrinsic stacking fault (ISF) bounded by the trailing partial, as shown on Fig. 6.16(a) [1254]. Shockley partials, with their smaller Burgers vectors, exhibit a lower line tension that increases their ability to penetrate between neighboring obstacles (distance d), according to Orowan rule,

$$\tau = \alpha \frac{\mu b}{d}, \tag{6.5}$$

where μ is the shear modulus and τ the shear stress necessary for bypassing. Similar decoupling of partials may take place when dislocation pairs are formed during creep at $10^{-6}s^{-1}$, at a later stage of deformation (Fig. 6.16(b)). These mechanisms occur more easily in matrices with lower SF energy (Astroloy, CMSX-4), see Section 6.4.3. Repeated on several consecutive atomic planes, they lead to the often observed microtwinning taking place during creep of most superalloys ([1498,740,870]). They have also been observed in NR3 single crystals, crept at 700°C under 500 MPa, in connection with the formation of superlattice intrinsic and extrinsic stacking faults (S-ISF and S-ESF) [1093].

6.3.3 Yield in two-phase single crystals

For single-crystal superalloys, the precipitates are up to 0.5 μm in diameter, and the volume fraction can be as high as 75%, giving a γ channel width of the order of 50–100 nm. The separation of two dislocations bracketing an APB fault is about 10 nm, or 2% of the precipitate diameter, neither weak nor strong coupling is appropriate for this morphology, and hence a modified approach to yield is required.

Yield normally marks the start of plastic deformation by the movement and multiplication of dislocations, but in superalloys tested at elevated temperatures, dislocation activity can be detected before the sudden increase in strain marking the onset of gross plasticity. In constant strain rate tests, interrupted before this apparent yield point, dislocations are moving in the narrow γ channels weaving and cross-slipping as shown in Fig. 6.8 and discussed in Section 6.2.2. This leads to a slope less steep than would be expected from the elastic modulus, and a strain at yield of about 1%. This mechanism can be regarded as Orowan looping, as discussed in Section 6.3.2, but the expressions developed for lower volume fractions of γ' particles are inappropriate where the precipitates are approaching 75% of the volume fraction and the gaps between the cuboid precipitates are about one-eighth of the precipitate width. Here the shear stress τ for a single dislocation loop to penetrate a channel can be approximated by Eq. (6.5), replacing d by channel width w, $\tau = \mu b/w$. The shear stress is inversely proportional to the channel width and is of the order of 190 MPa, equivalent to 460 MPa tensile stress for a generous channel width of 100 nm. This assumes a $\langle 001 \rangle$ shear modulus, μ, at 700°C of 105 MPa [1324], a Burgers vector b of 0.253 nm, and takes account of the inclination of the slip plane to the channel. The dislocation population of the channels is not equal for two reasons: firstly, the uneven distribution of the width of the channels, and secondly, the effect of misfit stress combined with the imposed macroscopic stress. It has been argued that the horizontal and vertical channels "yield" at different stresses [910], giving rise to multiple slight changes in the initial rise of the stress–strain curve before abrupt yield at 1% strain.

Fig. 6.17(b) shows single dislocation loops in the horizontal γ channels of CMSX-4 interrupted before yield at 750°C. The characteristic zigzag shape comes about as the dislocation loops cross-slip between {111} planes within the channel. The definition of the corners produced by cross-slip decreases with temperature as climb becomes easier and increases in more creep resistant alloys due to reduced climb. Note the similarity between this microstructure and which evolved during creep in Fig. 6.8. This deformation mechanism leads to relatively low values of strain as the area sheared is restricted to the matrix and the dislocation supply limited. In a tensile test, the strain achieved is insufficient to match the strain rate imposed in the tensile test and the stress continues to rise until the dislocations cut into the γ'.

The large increase in strain at the yield point is associated with the stress-driven entry of the dislocations into the γ' precipitates as dislocation pairs of similar Burgers vector separated by an APB shear the ordered particle. Localized slip is seen on octahedral planes and both phases are cut by dislocation pairs which can be seen in interrupted test-pieces just after yield. Fig. 6.18 shows the localized nature of the slip in the early stages of yield in one slip plane and the nature of the paired loops in the octahedral plane parallel to the specimen surface. In this test, interrupted just after yield, the dislocation pairs are largely located in the γ phase but pushed up against the γ'. There is little ingress into the narrow channels either by single or pairs of dislocations; the stress would be doubled by the paired dislocations, and the channel width effectively reduced. The yield stress is hence principally associated with pushing the leading dislocation into the precipitate aided by the reaction from the trailing dislocation.

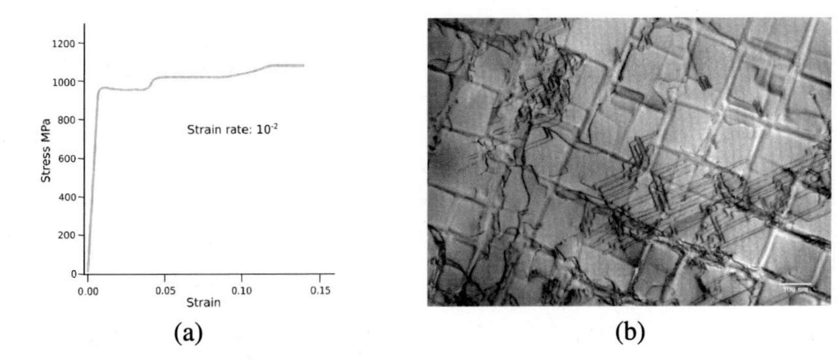

FIGURE 6.17 (a) Stress–strain curve for CMSX-4 deformed at 750°C and the strain rate of $1 \times 10^{-2}\,s^{-1}$. (b) Dislocations cross-slipping in the horizontal γ channels prior to yield in the alloy CMSX-4 as in (a).

FIGURE 6.18 Dislocation structure immediately after yield showing slip localized to slip bands [1539]. Arrows 1 mark active pairs of dislocations in the plane of the specimen, 2 shows an isolated loop. CMSX-4 strained 1.8 % at 750°C and $\dot{\varepsilon} = 10^{-4}\,s^{-1}$.

Hence, ignoring the effects of dislocation curvature at the corners of the precipitates, the force per unit length, τb, is equal to half the APB energy, $\gamma_{APB}/2$, times the volume fraction of γ' in the slip plane, as this will also control the proportion of a dislocation pair that needs to create an APB. The second dislocation effectively doubles the stress on the leading. For a plausible value of the γ_{APB} of $200\,\mathrm{mJ\,m^{-2}}$, and a γ' volume fraction of 75%, this gives an approximate shear yield stress of 300 MPa, or, using a Schmid factor of 0.408, a tensile stress of 726 MPa in the $\langle 001 \rangle$ direction. This is somewhat below the observed yield maximum of approximately 1000 MPa.

Thus, the main contribution to yield for two-phase superalloys comes from the difficulty dislocations have overcoming the APB energy, and rightly, alloy designers seek to increase this value by appropriate additions of elements such as Ti, Ta, and Nb [302]. As the γ' volume fraction increases, the APB energy, rather than the size and distribution of the γ' precipitates, becomes the most important factor. Crudden et al. [538] have looked at the effect of the various ternary alloying additions on the APB energy comparing their DFT results with Calphad calculations and empirical data and conclude that the most important alloying additions raising the APB energies are Ti, Ta, and Nb. Complex alloys are modeled by linear addition of the ternary elements, ignoring any synergistic effects. Comparison with yield stress data from several disc alloys gives correct trends but a wide scatter of some 200 MPa, in large part because an optimal distribution of precipitates sizes was assumed rather than the real microstructure.

Misfit stresses must also play a role, increasing the stress in the horizontal channels and decreasing it in the vertical channels. (But as penetration from both channels is required for the dislocation to propagate beyond a single precipitate, the effect may be averaged out.) Solid solution strengthening from refractory elements prevalent in the γ, such as tungsten, molybdenum, and rhenium, will also raise the yield stress. The various contributions to the mechanism of dislocations bowing into the γ channels have an effect on yield; by wrapping around the precipitate, the component of the line tension of the leading dislocation assists entry into the γ'. This effect was highlighted in the models of Parthasarathy et al. [1061]. Solid solution strengthening from refractory elements prevalent in the γ, such as tungsten, molybdenum, and rhenium, will also contribute to raising the yield stress.

That two-phase alloys show an anomalous yield effect (Fig. 6.11) demonstrates that the same Kear–Wilsdorf locking must be occurring in the γ' of two-phase alloys. This holds not only for single crystals, but also in lower volume fraction alloys such as IN100, as was neatly demonstrated by Nitz et at. [1017]. They showed that the two-phase alloys exhibited the same Schmidt factor anomalies with respect to the orientation dependence and anisotropy of the yield stress as a single-phase alloy of the same γ' composition. Thus the γ' precipitates, which were less than 100 nm in diameter, are large enough to allow cross-slip and Kear–Wilsdorf locking. This calls into question the assumption by Kozar [757] that only γ' precipitates larger than 200 nm can accommodate these locks.

Evidence that the dislocation pairs do indeed sometimes cross-slip to the (001) plane was given by the observation of dipole pairs in tensile deformed single crystal alloys by Feller-Kniepmeier and coworkers [1275]. Working with the first-generation alloy SRR99, they observed a yield maximum at 850°C for a strain rate of $10^{-3}s^{-1}$, at 760°C for $10^{-4}s^{-1}$, and 550°C for $10^{-5}s^{-1}$. Paired dislocation loops and paired screw dislocation dipoles were seen in the γ' and were assumed to be expanding to deform the γ' and add to the extensive deformation seen in the γ phase.

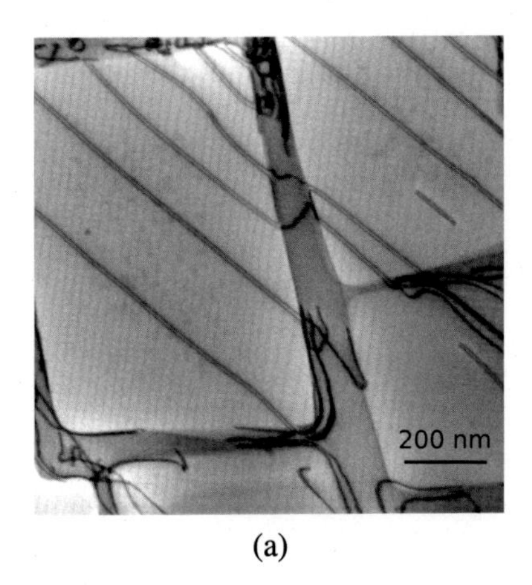

(a)

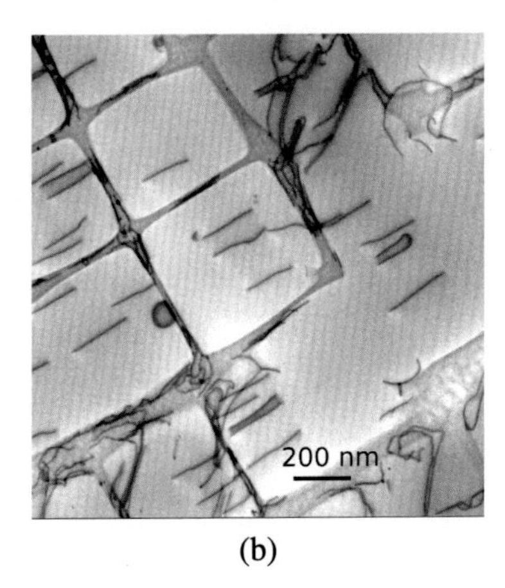

(b)

FIGURE 6.19 (a) Dipole loops of paired cross slipped dislocations in γ' viewed from above, parallel to the tensile axis [001]. (b) Dislocation pairs (part of dipole) cross slipped onto the vertical [010] plane viewed perpendicular to the tensile axis along [010] showing that they are closely spaced pairs, the loops are formed by the annihilation of the dipole dislocations in the γ phase, the inner pair can be seen to have disappeared in the center γ channel [1539]. CMSX-4 strained 1.8% at 750°C and $\dot{\varepsilon} = 10^{-4}\ s^{-1}$.

Subsequent work on the second-generation alloy CMSX-4 [1539] has established that these elongated loops are shrinking rather than expanding. The loops form by the cross-slip of the screw components onto the lower energy (001) plane rendering these parts sessile, and are left behind after the dislocation pair cutting through the γ', rejoins after cutting through the adjacent precipitates, Fig. 6.19. The sessile loop will impede further slip on this plane. It is suggested that the observation of numerous loops, some spanning several precipitates and others only a few tens of nm in size, can be explained by the loops gradually shrinking through diffusion-assisted reordering the APB between the paired dipoles.

The frequency of these sessile dipole loops suggests that only a fraction of the dislocation pairs passing through the γ' becomes Kear–Wilsdorf locked, Wang-Koh [1539] suggests that those dislocations constrained by the γ/γ' interface into a screw orientation as they enter, are more liable to cross-slip. This is consistent with the mobile loops seen in Fig. 6.20(f) which are elongated in the screw orientation suggesting that movement of the edge and mixed components is faster, as in single-phase metallics. For a loop of paired dislocations expanding through the γ/γ' microstructure, one-third of the γ' interfaces encountered constrain the dislocation pair into the screw orientation, Fig. 6.19(a). The presence of the cross-slipped loops in the slip plane will have an effect on the flow stress, forcing following dislocation pairs to move to unimpeded octahedral planes widening the slip traces but not introducing significant hardening whilst undeformed material remains available. This leads to a degree of easy glide

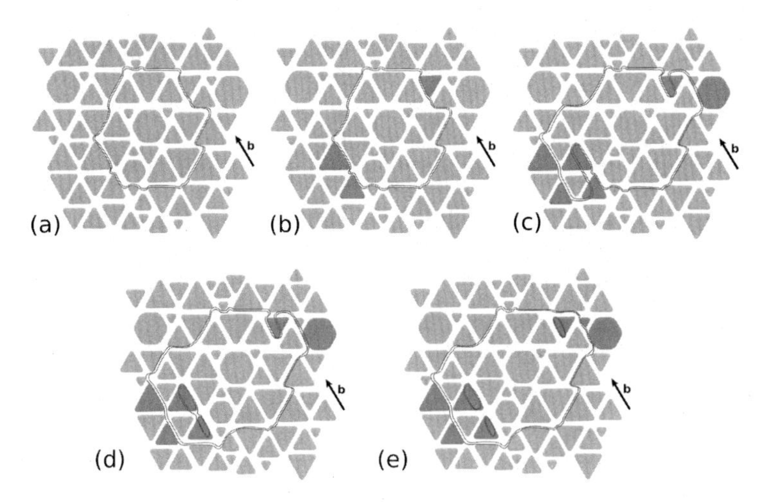

FIGURE 6.20 (a)–(e) Schematic of dislocation pairs moving on the slip plane with a fraction of the pairs cross slipping to form dipoles, read from left to right. The green (dark grey in print version) precipitates are those where the adjacent dislocation pair is screw-oriented.

seen in Fig. 6.17(a) followed by an increase in flow stress as the slip traces merge.

It is also observed that the larger the precipitates, the more likely cross-slip is to occur. The evidence for this is that the dipoles are not located adjacent to the γ/γ' interface and hence do not cross-slip immediately after entering the γ' phase. The longer the passage through the γ' precipitate the higher the chance of becoming locked. This is consistent with the observations of Sengupta et al. who observed a distinct and increasing effect of the yield anomaly around 700–900 °C for increasing γ' size in the alloy CMSX-4, Fig. 6.21 [1301].

The effect of Kear–Wilsdorf locking in the γ' on the flow stress has not yet been included in theoretical treatments of the yield stress, but the insights into the dislocation processes give a strong basis for this to be done. One approach that is able to include the effects of the APB, the curvature of the dislocations into the γ channels, the effect of misfit stresses, and the statistical chance of cross-slip is afforded by phase field modeling [1668,1665,1513].

At the very lowest strain rates both Scheunemann-Frerker [1275] and Wang-Koh [1539] observe the formation of stacking faults similar to those seen during creep. This in not entirely surprising as the slower strain rates used for testing ($\approx 10^{-6}\text{s}^{-1}$) approach those observed in creep and the formation and movement of stacking faults require diffusion. Wang-Koh observes the appearance of the stacking faults is associated with a drop in the flow stress to values more typical of creep stresses where stacking fault shear would be expected. The alloy CMSX-4 deformed at a strain rate of

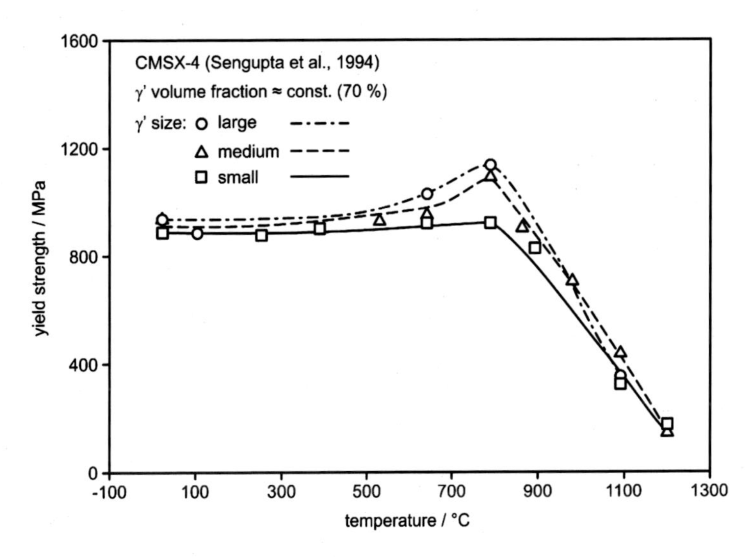

FIGURE 6.21 Yield stress anomaly in CMSX-4 reported for ⟨100⟩-tensile tests in [1301]. For larger γ'-sizes at constant γ'-volume fractions, the effect is more pronounced.

$10^{-6}s^{-1}$ initially forms narrow slip bands of paired a/2⟨110⟩ dislocations but when sufficient density and diversity of Burgers vectors is achieved, stacking faults can form and flow at lower stress, across both phases, along ⟨112⟩ directions. However, as the stacking faults require thermal activation to move in the form of shuffles in the ordered γ' phase [709], these effects are only seen at lower strain rates and higher temperatures. These mechanisms, which clearly form a continuum, are basic to the creep behavior. They will be dealt with in the next section on creep deformation of single crystal superalloys.

6.4 Creep

6.4.1 Introduction

In this section, we will concentrate on the response of ⟨001⟩, ⟨011⟩, and ⟨111⟩ oriented single crystals to uniaxial loading under various stress levels and temperatures, but all below the stress necessary to produce the rapid plastic deformation produced by cutting of dislocation pairs described in the previous section on yield. Over the temperature range of creep, typically between 650°C to 1200°C, a wide range of specific deformation mechanisms lead to radically different plastic and visco-plastic regimes. Creep rupture life increases with volume fraction of the γ' up to a maximum of about 70% [85]. The strengthening effect of the high volume-fraction of

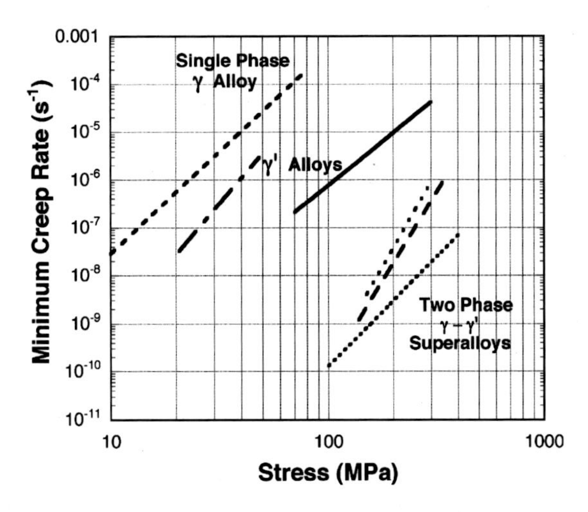

FIGURE 6.22 Best creep resistance of single phase γ and γ' vs. two-phase γ/γ' alloys, at 850°C [1129].

γ' precipitates in the γ/γ' superalloys is clearly shown in Fig. 6.22 where the creep rate of these two-phase microstructures at 850°C appears to be 10^7 to 10^8 times lower than that of a corresponding material hardened only by solid solution elements.

Experimentally, it is difficult to collect creep data in the range from 10^{-9} to $10^{-10}\,\mathrm{s}^{-1}$, so close to the background noise of the sensing and recording devices, but this regime is very important since it encompasses most service strain rates and may even be considered as an upper bound for some components. Indeed, taking a minimum creep rate of $1 \times 10^{-9}\,\mathrm{s}^{-1}$ under 185 MPa on the CMSX-4 test (Fig. 6.23) and assuming this minimum creep rate is maintained for 100 h, the material would reach a plastic strain of 0.4%, an underestimate as a steady state creep rate is assumed. In service, even for military engines, strain rates at least two orders of magnitude lower are usually required to give reasonable engine life. Typical strain/time $\varepsilon(\mathrm{t})$ creep curves at intermediate temperature and stress (Fig. 6.23) show, at first glance, a creep rate gradually increasing in an approximately logarithmic manner. However, on closer examination the presence of a creep rate minimum just after loading is confirmed by looking at the creep rate as a function of the creep strain. It is often more revealing to use creep strain, as the parameter on the horizontal axis, rather than time, since the duration of creep tests can vary by orders of magnitude when varying the applied stress. This is illustrated dramatically in this dataset [1569]. Figs. 6.23(a) and 6.23(b) show CMSX-4 creep curves plotted as strain vs. time and as logarithmic strain rate vs. strain, respectively.

This form of creep curve is dominated by the propagation of dislocations through the γ phase determined by the imposed stress and the mis-

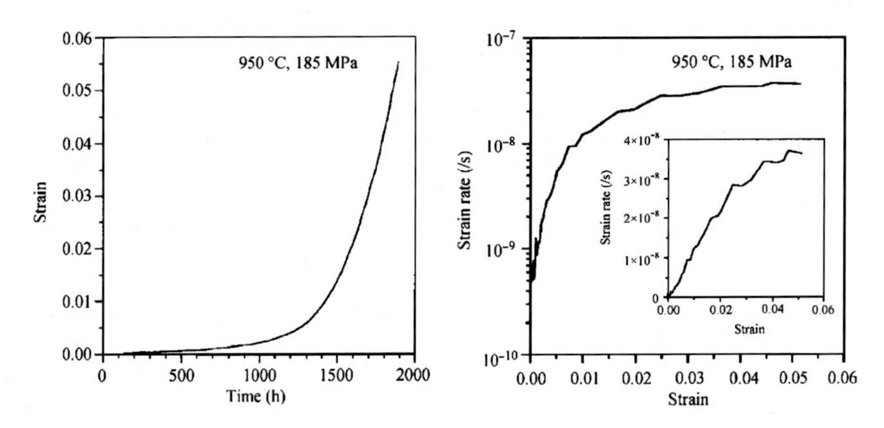

FIGURE 6.23 CMSX-4 under low stress: (a) strain vs time creep curve; (b) CMSX-4 derivative creep curve [908,907].

fit. At these intermediate temperature conditions there is little dislocation activity in the γ'. At lower temperatures, associated with high stresses, the dislocations are able to cut the precipitates, leading to sigmoidal creep with a strong primary creep contribution in the early stages. At high temperatures, creep is determined by the formation of rafts as the precipitates coalesce and eventually coarsen. These conditions lead to very different forms of the creep curve illustrated in Fig. 6.24.

Tensile data, reported in [1301], were used as guidelines for designing a miniature creep test programme for CMSX-4. The results of creep tests performed at 750°C and 1050°C for a full set of ⟨100⟩, ⟨011⟩, and ⟨111⟩ orientations were recently published [1569], and are shown in Fig. 6.24. For a wide range of single crystal alloys, it is observed that high temperature plasticity of single crystals is strongly dependent on the crystallographic loading direction and on the γ' volume fraction [1190,709,823,1160,212,1305,362]. It can be seen that at 750°C and 800 MPa (low temperature/high stress creep regime), the ⟨011⟩ specimen oriented for shear in the twinning direction, deforms orders of magnitude faster than the ⟨001⟩ specimen, oriented in the antitwinning direction, as observed earlier on macrosamples [769].

The ⟨001⟩ orientation shows a distinctive double minimum in the strain rate and a very high primary strain in the first 1% of deformation. At these lower temperatures under higher applied stress levels (or higher imposed strain rates), different mechanisms of plasticity develop as the dislocations deform both phases. This requires the formation of complex dislocation combinations and rearrangements, resulting in additional strengthening mechanisms, which are associated with the second minimum creep rate. Small changes in the composition of an alloy can have a profound effect on the ease of formation of those complex configurations involving stacking faults in the ordered structure of the γ' phase. Here, for instance, alloy

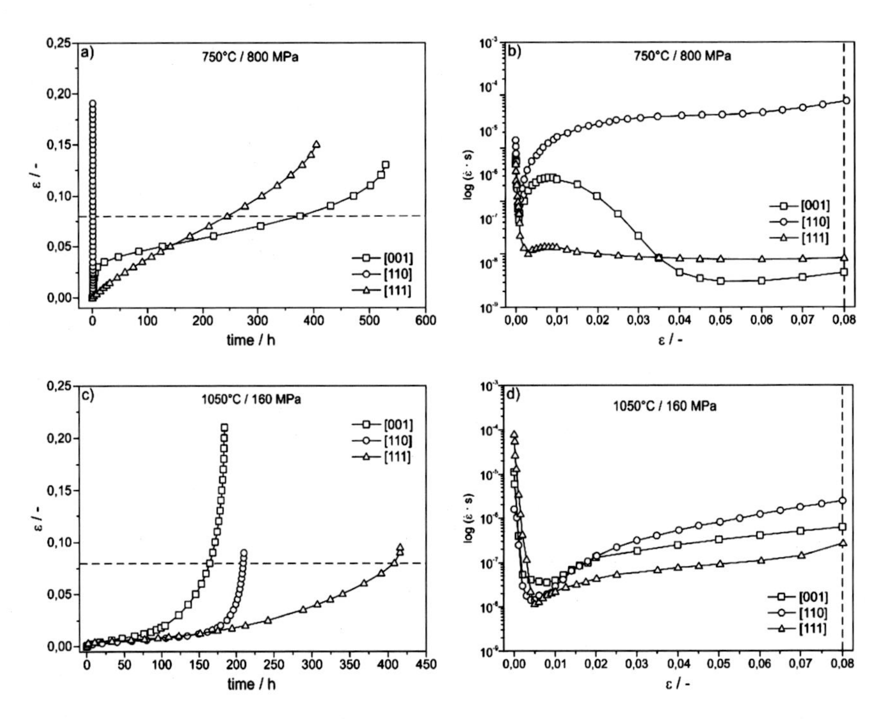

FIGURE 6.24 CMSX-4 creep curves from miniature tensile creep tests performed in ⟨001⟩, ⟨110⟩, and ⟨111⟩ crystallographic directions: (a)–(b) low temperature and high stress creep regime; (c)–(d) high temperature and low stress creep regime; (a) and (c) strain plotted as a function of time; (b) and (d) strain rate plotted as a function of strain. As reported in [1569].

AM1 does not exhibit a second creep minimum (see Fig. 6.36) [57,468], whereas the alloy CMSX-4 does. The principal differences in composition for the alloys are: CMSX-4 with 9 wt.% Co, 6.5 wt% Ta, 3 wt% Re, and AM1 with only 6 wt.% Co, 9% wt. Ta, and 0% Re; but it is not clear which of these minor elements has a greater effect. The high stress regimes producing this characteristic viscoplastic flow may also appear in service, either during an emergency power excursion or at a point of local stress concentration: in the vicinity of a fillet or a crack tip, for instance.

In contrast, at high temperatures and low stresses (1050°C and 160 MPa, Fig. 6.23(c)–(d)), one observes similar minimum creep rates for all three major crystallographic loading directions. In the high temperature low stress creep regime, crystallographic differences are much less pronounced. However, at the lower temperature, the minimum creep rates are again established at very small strain levels less than 1%.

The defining feature of the creep behavior of these highly strengthened γ/γ' superalloys at the higher temperatures, is the rapid change of their microstructure as the test progresses, Figs. 6.25 and 6.36. The initially quasiperiodic distribution of cuboidal γ/γ' precipitates rapidly forms

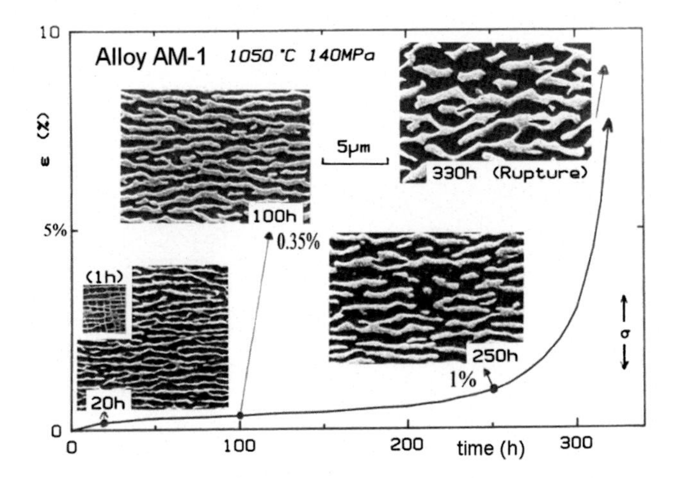

FIGURE 6.25 Strain versus time creep curve at high temperature low stress exhibiting the rapid rafting of the γ/γ' microstructure and finally its inversion of connectivity (γ' surrounding γ) after 1% strain [58].

rafts normal to the applied tensile stress. First, the horizontal γ channels thicken, at the same time the vertical channels thin down and disappear, finally, after less than 1% strain, the γ' phase, which still occupies more than 50% of the volume fraction at that temperature, becomes continuous and the γ phase is surrounded by γ', a process known as topological inversion of the connectivity of the phases (Fig. 6.25).

In conclusion, this section has shown that the behavior of superalloy single crystals subjected to high-temperature creep is complex and cannot be fully understood merely on the basis of creep textbook knowledge and Norton-law creep behavior.

6.4.2 Dislocation configurations during creep

To understand and interpret the plastic and viscoplastic behavior of these complex materials over such a large temperature range and under the various applied stress levels and strain rates, we must consider the interaction of individual, or groups of, dislocations with the stress field of precipitates generated by their misfit, or cutting through both phases creating planar defects with various surface energies. These mechanisms operate simultaneously, interact and depend critically on alloy composition. But they also have a common core process, namely that of the propagation of dislocation loops through the narrow γ channels. From this core process, the temperature and stress prevailing during creep lead to very different mechanisms and outcomes. Hence, we look at the movement of dislocations through the γ channels in some detail in this section, and then

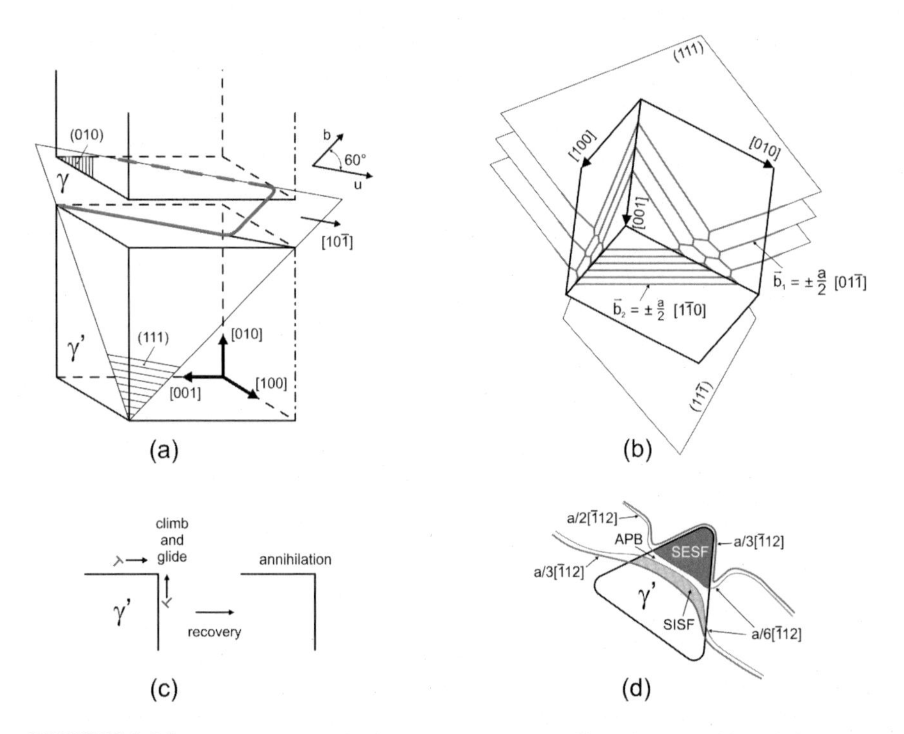

FIGURE 6.26 Key scenarios in high temperature superalloy plasticity. (a) A dislocation squeezes into a narrow γ channel, as reported in [1142,433]. (b) Formation of dislocation networks around γ' particles [741]. (c) Diffusion controlled climb and dislocation annihilation at corners of γ' cuboids (high-temperature low-stress creep regime) [1350,380]. (d) Movement of an a⟨112⟩ ribbon through the γ/γ' microstructure and, after rafting has closed the vertical channels, by diffusion assisted cutting of the ordered phase (low-temperature high-stress creep regime) [1142].

go on to look at how this behavior develops at low temperature and high stress, and finally at high temperature and low stress.

This can be summed up in four elementary dislocation scenarios, Fig. 6.26(a)–(d). First in the early stages of deformation, dislocations avoid the γ' phase [215,216,1126,1142,1190]. Fig. 6.4(a) shows how an ordinary fcc 1/2 ⟨110⟩ {111} dislocation squeezes into a narrow γ-channel normal to the tensile direction. It glides on a {111} plane and has Burgers vector b of type a/2⟨110⟩. A leading screw segment glides on its {111} planes in a ⟨100⟩ direction by repeated cross-slip [215,1126] and deposits 60° dislocations in the γ/γ'-interfaces. As was shown earlier, the tendency for dislocations to enter γ-channels depends on the type of channel [1143,232], this needs to be considered in a quantitative treatment of the process. At a higher temperature, there is generally more than one microscopic slip system activated and climb is rapid. As a result, one observes the formation of dislocation networks which sur-round the γ/γ' interfaces [362,1142,433,741,214,439,711,432,921], as illustrated schematically in Fig. 6.26(b). Their formation requires multiple slip in the γ-channels as well as dislocation climb and reactions between interface dislocations to establish regular networks. Fig. 6.26(c) illustrates schematically how the dislocation climb/glide process and the annihilation of dislocations at corners of γ' cuboids can represent possible recovery processes, which allow creep to proceed [1142,1350,1643]. These processes are correlated with rafting at high temperatures. Fig. 6.26(d) illustrates how the combination of dislocations allows cutting into the γ' by lower energy planar stacking faults at stresses well below the yield stress [709,908,1160,708,206,1266,1158,1159,823]. In Ni-base superalloys at low temperature and high stresses, groups of two matrix dislocations can combine twice to form three $a/3\langle 112\rangle$ partial dislocations of the γ' phase, bounding superlattice stacking faults:

$$a/2[011]+a/2[\bar{1}01]=a/2[\bar{1}12]\longrightarrow a/3[\bar{1}12]+a/6[\bar{1}12], \qquad (6.6)$$

repeated twice between consecutive $(1\bar{1}1)$ planes.

They build up, in the $(1\bar{1}1)$ plane of the γ' precipitates, a ribbon which moves in a plastically compatible manner through the γ/γ' microstructure but requires thermal activation to allow the superlattice intrinsic stacking fault (S-ISF) and superlattice extrinsic stacking fault (S-ESF) to move through the γ' phase. The overall displacement vector of this ribbon is $a\langle 112\rangle$, Fig. 6.26(d). This schematic drawing applies to the low temperature high stress creep regime of Ni-base superalloy single crystals.

Finally, cutting of the γ' phase has also been observed under conditions of high temperature and low stress creep. It is well-established that γ-channel dislocations with different Burgers vector can combine and form an $a\langle 100\rangle$ superdislocation in the γ' phase (Fig. 6.26(c)) which moves by a combination of glide and climb processes [1350,868,129,380,1264,6,7]. These studies provide strong evidence that dislocation activity in the γ' phase controls activity in the γ phase and hence the overall creep rate.

6.4.3 Low-temperature ($< 800°C$), high-stress creep ($> 600\,MPa$)

Under conditions of low temperature and high stress creep (here, 750°C, 800 MPa), Ni-base single crystal superalloys can show creep curves with a distinctive shape [362,1158,1282,1267,1266]. An example is given in Fig. 6.27, which shows creep data from five interrupted creep tests from ERBO1/C [1587,1059] (CMSX-4 type of SX). When plotted as strain vs. time, one finds the high primary creep strains which are characteristic for this creep regime, Fig. 6.27(a).

But one cannot appreciate a second feature, which becomes apparent when the logarithmic creep rate is plotted as a function of strain or as

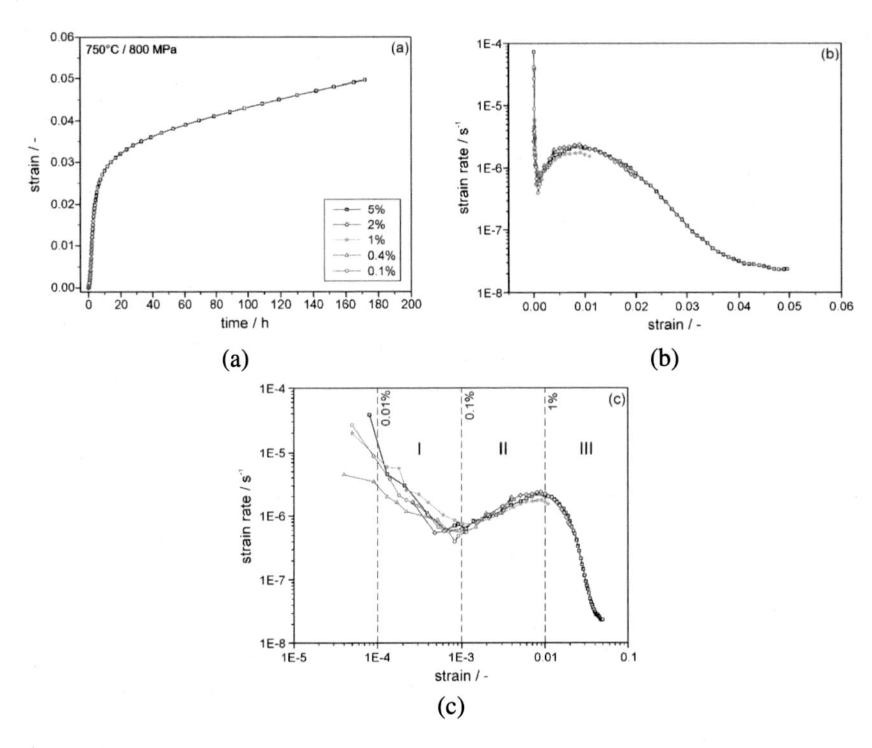

(a) (b) (c)

FIGURE 6.27 Creep data from five CMSX-4 tests performed in the low temperature (750°C) high stress (800 MPa) creep regime: (a) strain vs. time. Note the double creep minimum; (b) logarithmic creep rate vs. time; (c) logarithmic creep rate vs. strain [1587,1059].

a function of logarithmic strain, Fig. 6.27(b)–(c). The curves reveal that there are two minima, separated by an intermediate maximum. The occurrence of two minima was first reported by Schneider et al. [1282]. It did not receive much attention. The results shown in Fig. 6.27 raise two questions. The first is related to the term "primary creep." Single-crystal researchers often paid little attention to the very early stages of [001] low-temperature high-stress tensile creep. They took an engineering view and assessed the dependence of primary creep on stress, crystallographic orientation, and on microstructural details such as γ' volume fraction [709,908,823,1160,212,1305,362] based on strain time curves as shown in Fig. 6.27(a). Without further analysis, the results presented in Fig. 6.27(a) may well lead to the conclusion that there is an extended period of classical primary creep, which accounts for a strain interval of around 3% (methods to determine this strain interval have been proposed in the literature [823]). There is no doubt that this strain interval is important from a technological point of view. However, this type of analysis neglects other subtle features, such as the decrease of creep rate towards the first local

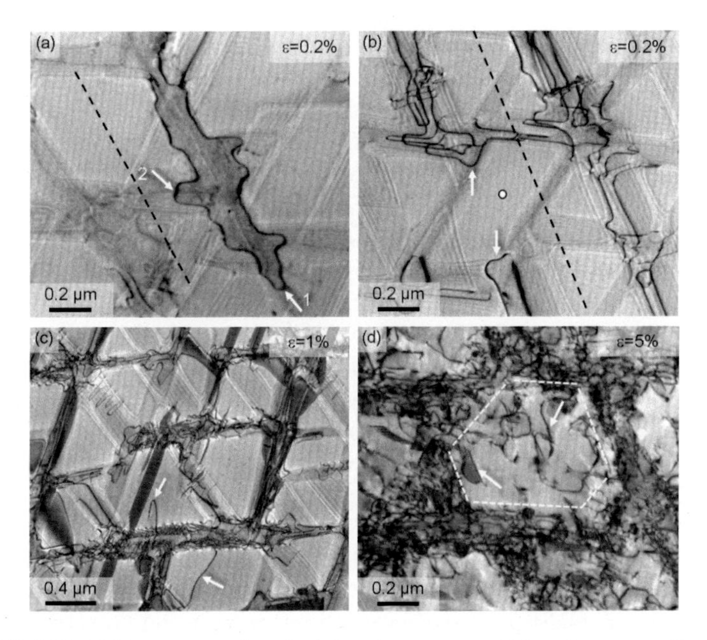

FIGURE 6.28 TEM micrographs of dislocation events: dislocation expanding along γ channel in (111) plane of TEM-foil for a strain of 0.2%. CMSX-4 strained 1.8 % at 750°C and $\dot{\varepsilon} = 10^{-4}\ s^{-1}$.

minimum and the presence of an intermediate maximum, which can be clearly seen in Fig. 6.27(b)–(c). Fig. 6.27(c) subdivides the creep curve into three regimes: I (decreasing creep rates towards, first minimum), II (increasing creep rates towards intermediate maximum), and III (decreasing creep rates towards global minimum). It is clear that classical definitions of primary, secondary and tertiary creep [499,656,1123,414,182] are not sufficient in describing the creep curves shown in Fig. 6.27(b)–(c).

Previous researchers have reported that there is an incubation period of low temperature and high stress creep (e.g., for AM1 tested at 300 MPa and 850°C [232], for Mar-M200 tested at 690 MPa and 760°C [433], and for CMSX-3 tested at 552 MPa at 800°C [1126]). During these incubation periods no strain accumulated during the few hours before creep started. It is difficult to comment on these results [1126,709], without a precise description of the creep test procedures (loading, heating). In none of the five experiments shown in Fig. 6.27 were such incubation periods observed. Creep always started directly after the specimens were rapidly loaded to 800 MPa. Indeed, it is difficult to understand why no microscopic strain would be measured when dislocations spread into previously dislocation-free areas [1126]. The duration of the incubation periods reported in [1126,709] are of the same of order of magnitude as the time required to reach the first minimum shown in Fig. 6.27. Three microstructural processes are responsible for the decrease of dislocation activity towards the first early minimum. First, any misfit dislocations present in unfavorable channels are pulled away from the γ/γ' interfaces. Second, ingrown dislocations in the favorable γ-channels start to move until they hit an irregularly placed γ'-precipitate. Fig. 6.28 shows a TEM micrograph taken from a {111} foil. The dashed vertical line indicates the direction of the favored γ-channels, called P-channels. Two dislocations (leading segments marked with white arrows) hit an irregularly placed γ' particle where they come to an abrupt stop, but attempt and succeed to escape into the C channel oriented horizontally in the figure; notice that the alternative escape through the R channel is inhibited by the local misfit stress field and this unsuccessful (compare with model described in Fig. 6.7). Regularly-arranged γ' microstructures represent an idealized configuration and the fact that there is irregularity can have effects on the mechanical behavior. And finally, in the early stages there is a redistribution of stresses from harder to softer regions, for the case schown schematically in Fig. 6.3 from dendritic (lower γ' volume fraction) to interdendritic regions (higher γ' volume fraction). Wu et al. [1587] suggest that it is the combination of these three processes (exhaustion mechanism, run-and-stop mechanism, and stress transfer mechanism), which accounts for the decrease of the creep rate towards the first early minimum.

An alternative interpretation of this first creep rate minimum encountered at a very small total strain (0.1%) consists in looking at the evolution of the two components of the total strain measured during stage I in stage II. The strain measured during creep is the total strain taking place in the specimen, hence a value integrated and averaged over the entire gauge length of the test piece,

$$\varepsilon = \varepsilon_e + \varepsilon_p. \tag{6.7}$$

As soon as dislocations start moving in the specimen, leaving the subgrain boundaries and entering the most favorable P channels, plasticity is taking place and ε_p is locally increasing; but simultaneously, the relaxation of elastic strain fields, generated by the misfit, is taking place in the opposite direction as shown in Section 6.2.2 and Eq. (6.1). Chang et al [232], combining FEM and DDD in a numerical model, demonstrated that the amplitude of the elastic relaxation produced is at least equivalent or even larger than the plastic forward strain. Hence, the total measured strain ε will be null, as well as the strain rate, during this early creep stage, explaining the often reported "incubation period" [216,1126] or even "negative creep" [1208] may be recorded if this simple, elementary, initial process is operating alone a for long period of time on several active slip systems (a low enough applied stress for instance).

The question then arises: why after this first early minimum there is an increase of creep rate towards the intermediate creep rate maximum? As

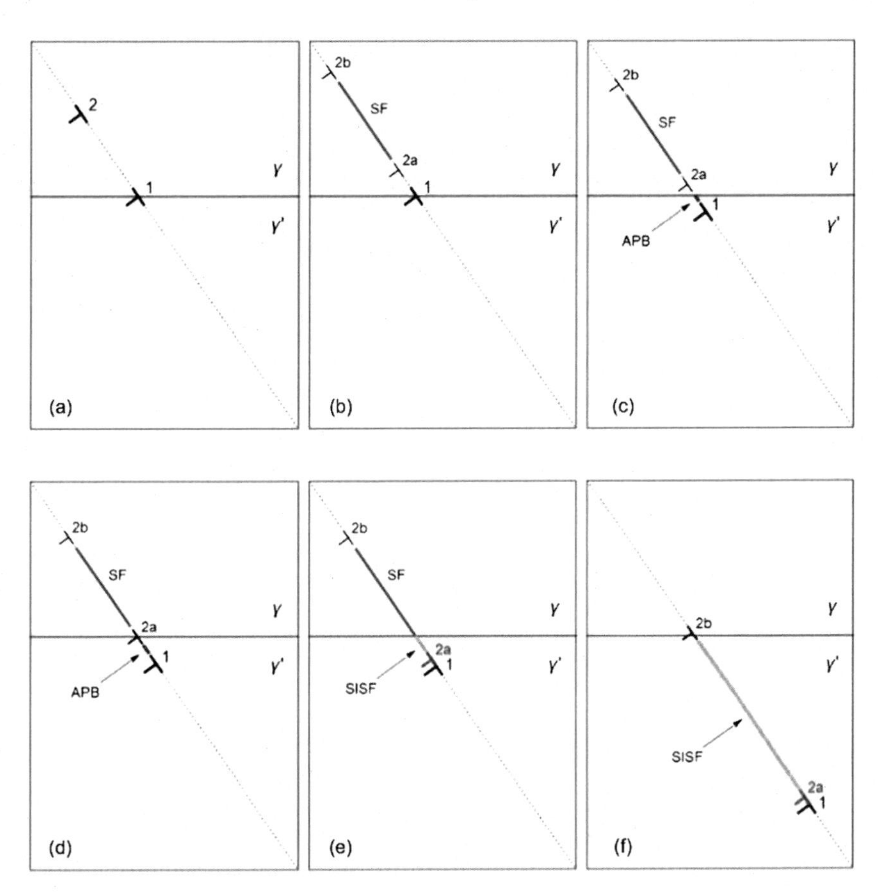

FIGURE 6.29 Nucleation of a ⟨112⟩ dislocation ribbon, as proposed by [1585]. (a) A γ-channel dislocation (2) and an interface dislocation (1) are on one common (111) glide plane; (b) the channel dislocation splits up; (c) the interface dislocation enters the γ' particle and creates a small APB; (d) the first partial 2a (from dislocation 2) immediately follows and reduces APB energy; (e) dislocations (1) and (2a) move on and creates a S-ISF; (f) the intrinsic stacking fault grows.

has been elegantly described by Rae et al. [1160,1158,1159], a window of opportunity opens at the beginning of stage II. As γ channel dislocation densities increase, dislocations belonging to different active slip systems recombine and the number of dislocation reactions which facilitate the formation of ⟨112⟩ ribbons for γ' particle cutting increases. The dislocation density at the beginning of stage II is high enough such that there are partners which can react but also not so high such that dislocation reactions can occur to halt the progress of the complex dislocation ribbons. The γ channel dislocation reactions which precede γ' cutting have recently been analyzed by Wu et al. [1585] using diffraction contrast transmission electron microscopy and 2D discrete dislocation modeling. They suggest a scenario schematically illustrated in Fig. 6.29. This scenario is new and is in line with what can be observed in the TEM and what can be rationalized with a simplified micromechanical assessment in a 2D discrete dislocation model [1585], while it differs from the two conflicting classical views, which can be attributed to either the Kear-scenario [709,823,708,824] (three $1/6\langle 112\rangle$-type partial dislocations in the γ phase approach the γ/γ' interface) or the Decamps scenario [321,272,322] (full $1/2\langle 110\rangle$ γ dislocations cut directly into the γ' phase). In line with previous qualitative interpretations, the work of Wu et al [1585] provides a self-consistent dislocation mechanism-based view on the increase of the creep rate between the early minimum and the intermediate maximum. At a strain of 0.4%, the Rae window [1160,1158,1159] is wide open and creep rates increase.

The window of opportunity for rapid creep closes as channel dislocation densities become so high that the movement of ribbons is suppressed [1159,1585]. The closing coincides with the beginning of stage III. In this context it is interesting to compare the TEM micrographs presented in the montage shown in Fig. 6.30 (0.4% strain, before the intermediate maximum) and Fig. 6.31 (5% strain, at the global minimum). In the (111) TEM montage shown in Fig. 6.30 (0.4% strain), a dashed line marks the orientation of the γ channels, which are referred to as horizontal channels or P channels during [001] tensile creep loading Fig. 6.7 forming V's or diamond patterns with the C channels.

When trying to interpret macroscopic mechanical curves and results, in relation with microscopic observations (TEM or SEM), it is important to keep in mind that plasticity is taking place in a largely heterogeneous manner across the volume of the specimen, starting preferentially in the dendrite cores, for instance, where the γ channels are wider or in regions where the γ' precipitates are more perfectly aligned. Some regions of the test piece may be in stage I when, on the contrary, important volume fractions of the specimen are already evolving in stage II or even stage III. Similarly, on a finer scale, shearing may appear much sooner in regions of the microstructure where small tertiary γ' precipitates are initiating the process which can spread more easily into larger size γ' precipitates. The straining mechanism that occupies the largest volume fraction and provides the higher strain rate is the one that determines which stage is observed experimentally; nevertheless, several mechanisms of plasticity are acting simultaneously at any time: a situation that causes identification of plasticity mechanisms by measurement of their thermal activation parameters (energy, volume) during tensile creep tests to be erratic and highly questionable.

Fig. 6.30, a TEM macrograph taken on a CMSX-4 single crystal deformed to 0.4% strain, shows that the dislocation density in the P channels is high, while other channels show significantly lower densities. The

FIGURE 6.30 Montage of TEM micrographs obtained from a material state which was [001] tensile creep deformed to 0.4% strain (CMSX-4, 750°C, 800 MPa). Bright field image, two beam contrast: $g = (\bar{1}\bar{1}1)$. For details, see [1587].

local stress states in these other channels keep dislocations out. In contrast, the montage of TEM micrographs in Fig. 6.31 (5% strain) shows that mainly two out of three γ-channels are filled with dislocations, thus forming diamond shape patterns, as confirmed by the DDD model presented earlier (Fig. 6.7). Dislocations cannot simply enter all channels by glide. The movement of ⟨112⟩ dislocation ribbons (stacking fault shear) and especially the process when the dislocations of the ribbon leave the γ' phase represents an alternative mechanism which accounts for the increase of dislocation densities in channels with unfavorable local stress states. As empty channels (receiving channels) fill up with dislocations associated with stacking fault shear (initiated by dislocations from sources in P channels), back stresses build up and it becomes more and more difficult to inject dislocations into the receiving channels (either by dislocation glide or by stacking fault shear). Therefore, the decrease of creep rate in stage III can be qualitatively rationalized by a combination of three elementary processes, the closing of the Rae window, the build-up of back stresses in the receiving channels and an increase in the inherent resistance of the γ' phase to stacking fault shear related to chemical changes on the nanoscale, e.g., [1512,755,969], and also to the increase of the density of dislocations in the γ' phase, Fig. 6.31.

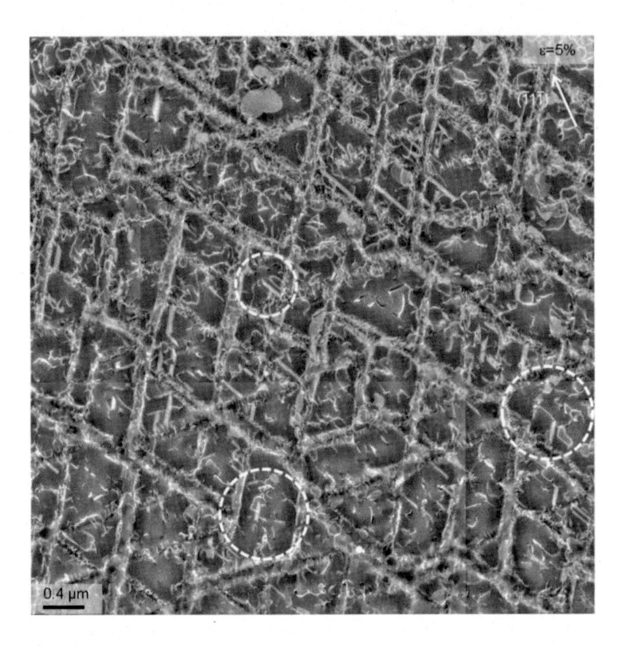

FIGURE 6.31 Montage of STEM HAADF images obtained for a material state which was deformed in a [001] tensile creep test to 5% strain (CMSX-4, 750°C, 800 MPa); $g = (\bar{1}1\bar{1})$. For details, see [1587].

While it is well accepted that at lower temperatures and higher stresses planar faults form, which types of planar faults form depends on stress, temperature, and crystallographic loading direction. For example, the $\langle 110 \rangle$ orientation in tension and the $\langle 100 \rangle$ orientation in compression are particularly prone to the formation of microtwins. This was investigated as a function of stress and temperature by Barba et al. [74]. They observed twinning of as much as 40% volume fraction of the sample at temperatures of 800°C and stress of 675 MPa. Chen and Knowles [242] showed that, for specific SX orientations, stacking faults in larger γ' particles coexist with microtwins, and that SESFs are more frequent in compression and associated with the formation of twins. Whichever planar fault forms, it is important to consider how, and under what conditions, dislocations from the γ-channel enter the γ' phase. This, and the type of planar faults formed, depends on the appropriate dislocation reactions occurring in the γ-channels prior to cutting [1585]. While these early events have received less attention, more work has been done on faults which have already formed in superalloys. Thus Kovarik et al. [755,756] confirmed an earlier suggestion by Kolbe [740]: they showed that the operative twinning dislocations are identical Shockley partials of type $1/6\langle 112 \rangle$, which propagate through the γ' phase in closely-separated pairs on consecutive $\{111\}$ planes. The rate-limiting elementary process is diffusion-controlled

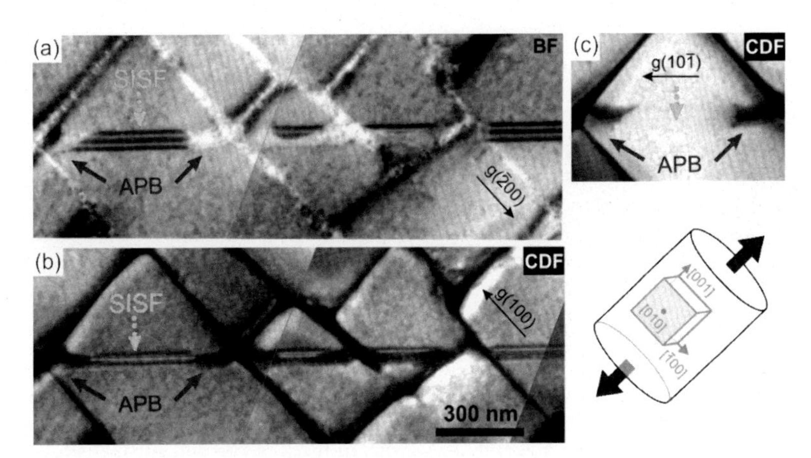

FIGURE 6.32 Diffraction contrast TEM micrographs of planar faults which extend over multiple γ'-precipitates after creep of a Co–Ni-SX at 900°C and 300 MPa to 0.5%: (a) stacking faults in contrast, APB effectively invisible; (b) stacking faults and APB in contrast; and (c) APB in contrast, stacking fault effectively invisible [381].

reordering in the γ' phase behind the leading partials. This view is also a basic element of later publications where the segregation of the elements Cr, Co, Ta, and Mo to planar faults is demonstrated [870,1499,1340]. One should expect that the segregation of atoms to planar faults and the reordering mechanism discussed above would decrease all planar fault energies, and this does, indeed, appear to be the case. DFT calculation suggest a strong link between the chemical composition at the fault and the fault energy [405,1462]. The diffusion of alloy elements to planar faults adds another dimension of diffusion control to the temperature dependence of creep, in addition to the reordering processes and the short-range climb processes in the γ-channels postulated in [1585].

In the case of Co-based single crystal superalloys with γ/γ' microstructures, much lower fault energies have been reported [1439,1440,381, 699]. Therefore, it is not surprising that other planar fault structures are observed [1439,381,382]. Specifically, a much lower APB energy [381] in Co-base superalloys rationalizes the appearance of extended APBs in Co-based SX. Figs. 6.32 and 6.33 document the importance of planar faults during uniaxial creep of a CoNi-based superalloys at 900°C and 310 MPa (accumulated strain, 0.5%) [381]. Fig. 6.32 shows a typical planar defect configuration. The [010] foil normal is perpendicular to the tensile axis, as shown in the schematic on the lower right of the image. Three TEM micrographs are shown taken at different tilt positions (diffraction contrast conditions, g-vectors). Each γ' precipitate along the extended planar defect (dark horizontal contrast in the middle of the images) exhibits the same characteristic APB/S-ISF/APB sequence, Fig. 6.32(a)–(c). The TEM

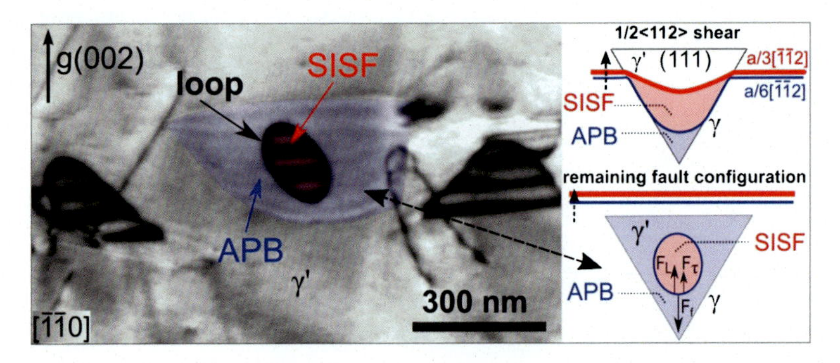

FIGURE 6.33 Diffraction contrast TEM micrographs of planar faults ((1̄1̄0) foil normal), corresponding to the planar faults shown in the previous figure ({010} foil normal). The micrograph shows an S-ISF, which is fully embedded in the γ' phase and surrounded by an APB. The schematic figures on the right show how the movement of dislocations in a {111} glide plane leads to this type of configuration [381].

image, shown in Fig. 6.33, was taken from a thin foil with a [1̄1̄0] normal. The image shows, a central S-ISF which is surrounded by an APB. The mechanism which leads to the formation of this configuration is illustrated on the right of Fig. 6.33. This mechanism was analyzed using an advanced diffraction contrast method (LACBED [968]). A leading partial dislocation with a Burgers vector of 1/3 [1̄1̄2] moves on a {111} plane (upwards in the schematic illustration on the right). It is followed by a trailing 1/6 [1̄1̄2] partial. When entering the γ' particle, the leading dislocation forms a S-ISF on the (111) glide plane. This is transformed into an APB, as the trailing partial follows. As the leading partial leaves the γ' particle, the trailing partial creates a loop, which is fully contained in the γ' phase, as can be seen on the TEM-image. The authors of [1439] could further show that the APB has tendency to move from its original {111} plane to an energetically favorable {100} plane.

As reported for Ni- [1512] and Co-base superalloys [1440], segregation to planar faults was also observed for this NiCo-base superalloy [381]. In Fig. 6.34 the planar faults shown in the previous two figures are viewed edge on. Line scan EDX results are shown below the TEM micrographs. It can be seen that the elements Co, W, and Cr segregate to the S-ISF. In contrast, only Co and Cr are enriched at the APB, which loses all other alloy elements in the process. It appears that the APB attempts to reach the composition of the γ phase (right column of Fig. 6.34). One can look at this scenario as segregation to a planar fault in the γ' phase. But equivalently, one can also take the view that a local phase transformation from γ' to γ (APB) and from γ' to D019 (S-ISF) takes place. In Ni-base superalloys, planar faults form during low temperature and high stress creep. In Co-base alloys, they are also observed at higher temperatures. Similar phenomena can be observed in both types of alloys [894,1341,472]. The for-

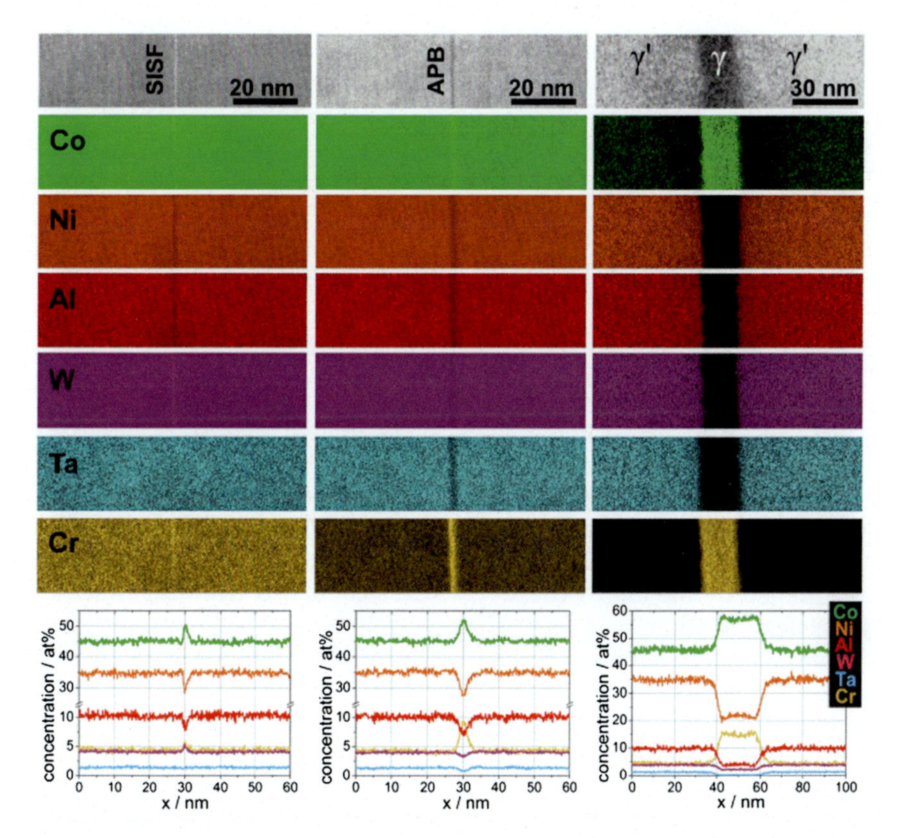

FIGURE 6.34 High-resolution scanning transmission images (top row) and corresponding EDXS element maps (below) of an S-ISF, an APB which forms during creep and a narrow γ-channel obtained in edge on orientation [381]. For details, see text.

mation of planar faults in Ni- and Co-superalloys is presently a very active research topic.

6.4.4 High-temperature (> 1000°C), low-stress (< 300 MPa) creep

Uniaxial high temperature and low-stress creep of single-crystal superalloys is not as strongly dependent on the crystallographic loading direction as low-temperature high-stress creep (Fig. 6.24 [897,692]). Also, it does not react so strongly to differences in γ' volume fractions because, at high temperature, equilibrium volume fractions are more quickly established. But it is by no means less complex. One reason is that we take the alloy into a temperature range where the γ' volume fraction decreases towards a new thermodynamic equilibrium while the material deforms [1613,1302]. Most importantly, at this temperature, one observes rafting, the directional coarsening of the γ' phase. The micromechanics, thermodynamics, and kinetics of rafting have been studied theoretically and experimentally over many years [1108,41,981,264,266,6,897,1343,994,1487,995,565,503,1668,428, 427,1127,1488,1486,909,393,690,1191,946,691,1396,1570,854,853,1630,181,589, 1020].

It has been shown that the evolution of a raft microstructure during creep depends on the lattice misfit, e.g., [981,264] (which can cause misfit stresses as high as 500 MPa [1127]), the crystallographic loading direction during uniaxial testing, e.g., [6,7], and the nature of a stress state in multiaxial experiments [41,6,7,1302,1191,946,691,1396,1570] as discussed in Section 6.2. In the early stages of creep, the dislocation density increases to relieve the misfit stresses in those γ channels with the highest resolved shear stress. Differences in the chemical potential of the species in the different channels drive the diffusion, leading to closure of the unrelieved channels. It is significant that rafting continues in the absence of that stress once a minimum critical strain is achieved [909]. This is consistent with the establishment of dislocation networks relieving misfit in specific channels determined by the imposed external stress combined with local internal stress fields [232].

Rafting also depends on local microstructural conditions such as, for example, the features of the cast microstructure (dendritic or interdendritic regions, Fig. 6.1), e.g., [589,1020], or the presence of elevated dislocation densities next to a hardness indent [1488]. As can be seen in Fig. 6.35, during [001] tensile testing of a negative misfit Ni-base superalloy, γ' rafts form perpendicular to the direction of the applied stress [6]. However, it has been shown that in positive misfit Co-base superalloys, γ' rafts form parallel to a [001] tensile direction (P-type rafts) [981,264,266,1441,1385] or normal to the direction of an [001] compression test [1604] (N-type rafts).

At high temperature and low stress, plasticity mechanisms are identical to those described previously, at low temperature during the early stage of creep leading to the first creep rate minimum (see Section 6.2.2) but their kinetics is significantly increased by temperature since they are thermally activated. It appears clearly (Fig. 6.36) that the morphology of the matrix channels is changing faster (filling up of the vertical channels and widening of the horizontal P channel) than that of the γ' precipitates. Soon after the creep rate minimum is reached, for a creep strain of hardly 0.3%, the connectivity of the two phases is inverted, hence the γ' phase starts surrounding the former matrix γ: the so-called connectivity inversion also implies that the percolation of plasticity, from then on, becomes easier in the γ' phase than in the γ islets.

As plastic strain increases, stresses build up [439], and it becomes increasingly difficult to squeeze new dislocations into the channel [1142]. This results in a decrease of creep rate. When the channels are full, such that the back stress equals the imposed stress, recovery processes are

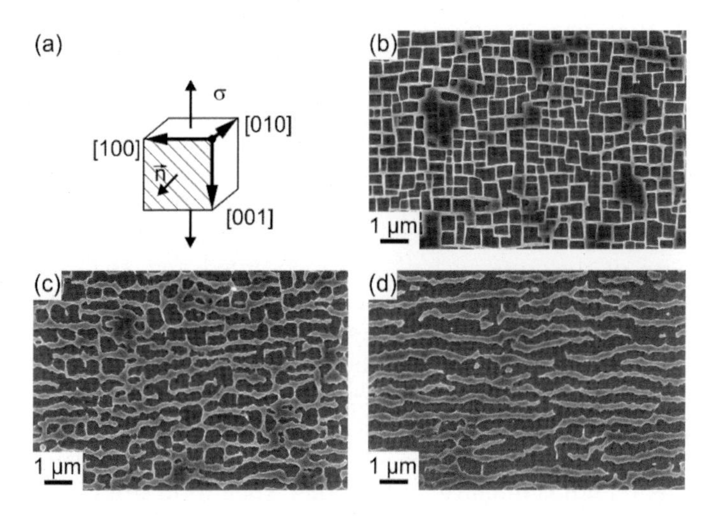

FIGURE 6.35 Rafting in the Ni-base superalloy LEK 94 tested at 1020°C and 160 MPa. (a) Loading geometry, hashed area represents position of SEM cross-section. SEM micrographs: (b) microstructure prior to creep; (c) 19 h creep (strain, 0.1%); (d) 48 h of creep (strain, 0.4%).

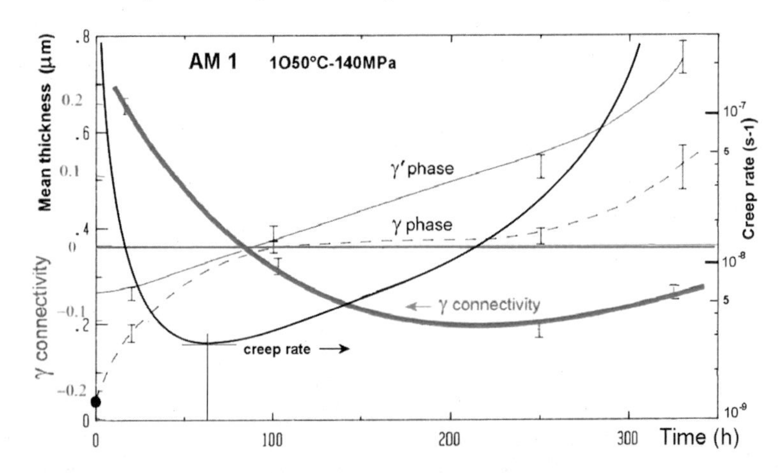

FIGURE 6.36 Creep rate of AM1 at 1050°C (see Fig. 6.25) in relation with rafting kinetics of phases and change in connectivity of the γ matrix (after [468]).

needed for creep to continue. In the early stages of creep, recovery can occur at the corners of γ' cuboids. Climb and glide processes first contribute to the straightening of the interfacial dislocations into the pure edge ⟨010⟩ orientations, then participate in the formation of loops surrounding individual or groups of γ' precipitates. By climb/glide along γ/γ' interfaces, they either shrink or annihilate one another [1643]. Pipe diffusion along dislocation lines joining different faces of γ' cuboids are expected to amplify and accelerate the diffusion-controlled oriented coalescence of the rafting phases driven by gradients of stress and chemical potential [1367,600]. Considering the overwhelming role of various diffusional processes in its implementation, rafting can be thought of as a time-hardening process. Another type of recovery process has been identified [1305]. Two dislocations jointly shear a γ' particle and annihilate with two dislocations of opposite sign in the channel on the other side of the γ' particle.

A variety of dislocations have been observed and analyzed in the γ' following creep at high temperatures. These include superdislocation pairs as seen during yield, but also dislocations having a Burgers vector of the type a[100]. Dislocations of this Burgers vector were first reported by Louchet and Ignat [868] and confirmed by Bonnet and Ati [129]. These distinctive segmented dislocations were studied in greater detail by Eggeler and Dloughy [380] in a specimen subjected to a pure shear strain in accurately aligned high symmetry orientations. They form from two dislocations in the γ channels with Burgers vectors at 90°, for example, $a/2[011] + a/2[01\bar{1}] \rightarrow a[010]$. Each dislocation is trapped at the γ/γ' interface, necessarily from different γ slip planes, and often with different line directions. Eggeler and Dloughy demonstrate that the combination of the imposed shear stress and the misfit stresses, estimated in the mid-γ channels, allow the dislocations to approach to within a calculated 20 nm of each other, by a combination of glide and climb.

The resulting dislocation pair has a perfect Burgers vector of the γ' phase, $a\langle 100\rangle$. It is observed to enter the precipitate as a distinctive dislocation of straight segments with a strong preference for a $\langle 011\rangle$ type line direction having either an edge or mixed character. The core structure was examined in more detail by Srinivasan et al. [1350] and shown to be two distinct cores consistent with the $a/2\langle 110\rangle$ type Burgers vectors of the component dislocations, separated by 2.5 nm. Both edge and mixed configurations were observed, each with line vectors along $a/2\langle 110\rangle$ directions. Conservative movement of the dissociated dislocation core, $b = a/2[010]$, on the (101) noncompact glide plane is proposed. This would require a self-contained flux of vacancies over the short distance between the two cores, each moving by a combination of glide and climb. Estimates of the expected creep rate based on this mechanism being rate-controlling for the steady state creep rate give values within an order of magnitude of the observed creep rate.

Looking at more conventional (but more complex to interpret [001] tensile creep configurations) also shows similar segmented dislocations in the γ' with Burgers vectors verified as $a\langle 010\rangle$, Fig. 6.37 [1264,847]. No Burgers vector of the type $a\langle 010\rangle$ can make a direct impact on the strain for a tensile strain applied along the [001] direction, but, by bridging the γ', these dislocations are able to eliminate dislocation density in the γ channels as they leave one interface, and annihilate equal and opposite dislocation density

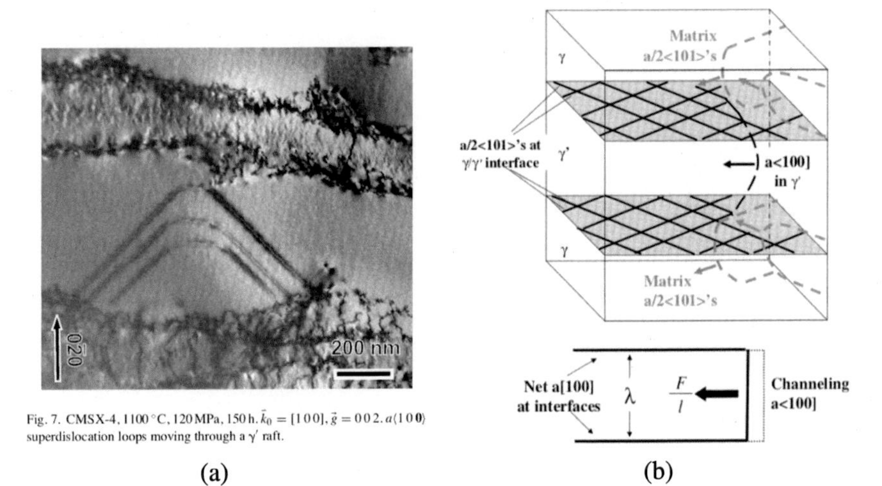

FIGURE 6.37 [100] dislocations climbing through the rafted microstructure of CMSX-4 after creep at 1100°C, 120 MPa for 150 h [1264].

in the other interface as they reach the far side. In this way the dislocation density in the γ/γ' interface is reduced allowing further ingress of dislocations into the γ channels as the backstress is reduced. This is illustrated in Fig. 6.7 from Sarosi et al. [1264] where the [100] dislocation acts like a wire cutting a cheese. An analysis of a longitudinal section of a [001] tensile specimen of Nasair 1000 shows numerous dislocations bridging the γ' host with Burgers vectors [010] or [100]. Movement of these dislocations during [001] tensile creep hence appears to be largely by pure climb, although some instances of segmented [100] edge and mixed dislocations, as seen in the high-resolution work done on sheared specimens were also reported [1350].

A dramatic example of the action of one of these dislocations is shown in the anaglyph of Fig. 6.38. This can be viewed with the use of red–cyan glasses to give the full 3D effect. The effect of the tensile creep axis is also important as deviations as much as 20° from the exact [001] orientation can be permitted for blades and also creep stresses can be very complex in components. A comparison was made of the dislocations present in the alloy LEK94 stressed in the [001] and [110] tensile directions. For the [001] direction, almost all the dislocations in the γ' were observed to be of the type a[100] where the Burgers vector was perpendicular to the tensile axis. In contrast, for the [110] oriented specimen, all the dislocations in the γ' (and some 45 were analyzed) were superdislocation pairs of the type a[101] as seen during deformation above the yield point. The authors rationalize this by looking at the dislocations most likely to be active in the γ channels in the two orientations. The Burgers vectors in the [001]

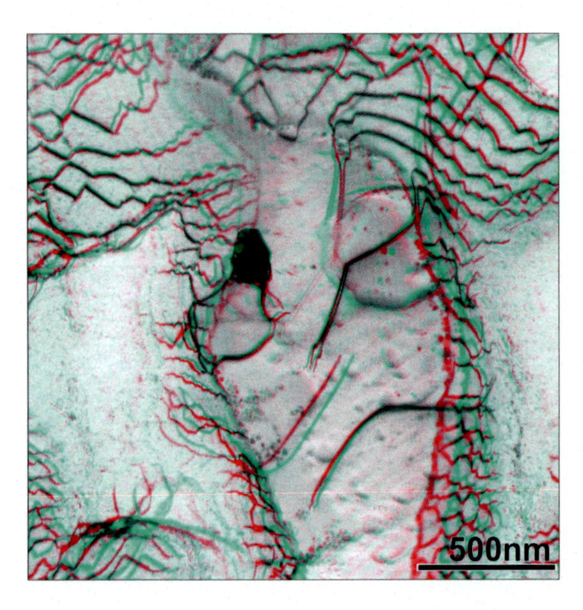

FIGURE 6.38 3D anaglyphs of the microstructure of a single crystal Ni-base superalloy after creep. $g = (111)$. One can appreciate the microstructure in 3D when viewing the micrograph with cyan glasses (for details, see [5]). Alloy LEK 94 strained 2% at 1020°C under 160 MPa.

oriented test are more diverse and more likely to lead to the formation of the [100] type dislocation. Hence, in less precisely oriented tests, or in real components, a mixture of dislocations cutting the γ' is likely to be observed. This process is important, not only in allowing and controlling continuing steady state creep after rafting is effectively complete, but also in shearing the γ rafts and leading to the gradual breakdown of the linear rafted structure as strain progresses. As can be seen in Fig. 6.25, this results in a fractured raft structure with the two phases more vulnerable to extensive dislocation shear on octahedral planes. This structure becomes particularly pronounced close to the fracture surface.

6.5 Summary and strategies to improve mechanical properties

Nickel-base superalloys are exceptionally strong and creep resistant at high temperatures because they contain high volume fractions of the ordered γ' phase. The necessity to form an APB fault makes cutting the γ' difficult and is the principal contributor to the yield strength in terms of the APBE. Creep occurs, by definition, at stresses lower than the yield strength, and the principal role of the γ' is to keep the dislocation activity confined to the γ phase and to modify the stress state in the γ by the application of the misfit stresses to the best advantage of the creep performance. Misfit also has a crucial role at high temperatures as the interaction between the misfit and the deformation pattern in the γ phase results in the formation of rafts during high temperature creep. For the specific case of tensile stress and negative misfit, this leads to planar rafts normal to the tensile axis and hence reduced mobility by climb of the dislocations in the γ and a long plateau of exceptionally low creep strain. But, as we have seen, there are lower-stress alternative dislocation configurations that allow entry into the γ' with the contribution of thermal activation. At high temperatures, dislocations combine to form $\langle 010 \rangle$-type Burgers vectors mobile by climb, which although they do not contribute directly to strain in a [001] tensile creep test, by reducing dislocation density in the interface allow further movement of dislocations in the γ phase and hence further creep strain. At low temperatures, where creep stresses operating are much higher, a separate process whereby the γ' can be sheared by the movement of superlattice partial dislocations formed from those in the γ, which are able to enter trailing lower energy stacking faults. The energy penalty is lower, and these dislocations move by glide, but the ordered nature of the γ' requires localized reordering of the atoms at the dislocation cores to form the low energy faults. In both cases there is good evidence that the cutting processes become rate-controlling.

Understanding the mechanisms of yield and creep guides our approach to the systematic design of alloys. Increasingly this is approached through modeling, both of the physical processes involved [757,302], but also at a fundamental level through the use of atomistic modeling of the interactions between atoms in the alloy. This allows access to vital input parameters, such as diffusion coefficient, activation energies, and fault energies, which are otherwise difficult to measure and also to predict as a function of composition.

For strength, the main approach is to maintain sufficient γ' and ensure that the APB energy is as high as possible [481,302]. As the γ' volume fraction increases, the APB energy, rather than the size and distribution of the γ' precipitates, becomes increasingly important. Crudden et al. [302] conclude that the most important alloying additions raising the APB energies are Ti, Ta, and Nb, and, indeed, strength is strongly correlated to Ti/Al ratios [677]. In addition to the segregation noted earlier to APB faults in Co alloys, segregation is also found in Ni-base superalloys. Barba et al. [75] observe that APB faults on $\{111\}$ planes are heavily segregated by Co and Cr, locally mimicking the γ composition. There is clearly much more to understand, one issue is that if the APB energy of Co-base alloys can be so low, why do these compositions retain good strength? The role of the composition in controlling APB energy both at equilibrium and dynamically is of great interest and also lends itself to atomistic modeling. Increasing yield stress expands the envelope in which creep operates, but the strategies that increase yield stress do not necessarily also increase creep resistance; they are different.

Improving creep performance is a more complex question. All creep processes are diffusion controlled and the addition of elements which slow down diffusion is of primary importance. This is particularly important in the intermediate temperature range, around 900°C. As an example, we consider the role of Re in Ni-base single crystal superalloys. It is well known that Re has a beneficial effect on creep strength of Ni-base single crystal superalloys [1190]. Suggestions that this was due to Re clustering have been largely dismissed [971,972], evidence pointing instead to the effect it has in lowering the diffusion rate in the γ phase and in the γ/γ' interface. Measurements in binary Ni/Ni–X diffusion couples show that it has a lower diffusion coefficient than other refractory 5d elements, especially at temperatures above 1000°C [702,703].

So, it is clear that the slow diffusion of Re is largely responsible for this beneficial effect. Re is particularly effective because it partitions more strongly than any other component to the γ phase, where lowering diffusion has the most effect on creep [1058,1613,1542]. Therefore, it seems reasonable to assume that it delays the climb processes of dislocations in the γ-channels and atomistic scale modeling provides useful information regarding these diffusion processes [1567]. DFT calculations for a binary Ni–Re solid solution system in the solute limit yield an apparent activation energy for diffusion of Re-atoms of 278 kJ mol^{-1}. This value was not too far off from the apparent activation energy of creep in the low-temperature/high-stress creep regime which was measured to be 382 kJ mol^{-1}. It remains to be clarified how Re affects dislocation climb [1058]. However, the alloy complexity (number of alloy elements in alloys) represents a challenge for all types of atomistic approache. Repartitions to prior dendritic regions during solidification result in a distribution leaving the dendritic regions prone to TCP precipitation and the interdendritic regions lower in Re than would be optimal for creep resistance [701]. The introduction of 2–3 wt% Ru, in the 4th generation of alloys for single crystals, homogenizes the distribution of Re between the γ and γ' phase [1007], thus avoiding the formation of TCP phases and the same time increasing the amplitude of the γ/γ' misfit and the creep resistance at high temperature (up to 1100°C) [1630].

Controlling misfit also has a strong effect in producing rafts which impede high temperature creep [1007,1153] and also in managing the stresses that the dislocations in the γ channels experience. For example, consider the positive misfit between the γ' phase and the γ phase in Co-base single-crystal superalloys [1383,981,1439,1440,472,1382,1638,1151]. Atomistic calculations have shown that the energies of planar faults (stacking faults and APB) are lower in Co- than in Ni-base superalloys. As a result, processes of the type shown in Fig. 6.26(d) occur more easily in Co- than in

Ni-base systems. Apparently, as the γ' particles form in Co-base systems, the atoms of the $L1_2$ phase move further apart (positive misfit), while they approach each other in Ni-base systems (negative misfit). The chemical bonds between the atoms in Co-base SX seem to be weaker than in Ni-base SX. This is consistent with the lower planar fault energies and lower dissolution limits in the $L1_2$ phase of the Co-base alloy.

Finally, advances in atomic-scale characterization are leading to a much deeper fundamental understanding of mechanical properties. Recent atom probe work has been able to identify segregation of Cr and Co to dislocations in first generation Ni-base alloys [746] and the addition of Re and Mo in higher generation alloys [1586]. High resolution chemical analysis has also found segregation and ordering on stacking faults where individual atom species replicate the ordered hexagonal phases D024 and D019 created at the core of extrinsic and intrinsic stacking faults [1339,1340]. It is becoming clear that defect energies and structures are not fixed but can be influenced by the time-dependent supply of appropriate atomic species. An understanding of how these factors are likely to affect mechanical properties and be incorporated into models is beginning to emerge [894]. This is an exciting and fast developing field where there remain numerous challenges and opportunities posed by these complex alloys.

PART II

Building SX parts

CHAPTER

7

Processing of directionally cast nickel-base superalloys: solidification and heat treatments

Jonathan Cormier[a] *and Charles-André Gandin*[b]

[a]ISAE-ENSMA, Institut Pprime, UPR CNRS 3346, Futuroscope-Chasseneuil, France, [b]MINES ParisTech, PSL Research University, CEMEF, UMR CNRS 7635, Sophia Antipolis, France

7.1 Introduction

The thermodynamic cycle of gas turbine engines reveals that better efficiency can be achieved when increasing the operating temperature. This simple rule remains the main driver for designing and engineering parts, which includes the development of materials as well as the processing technologies, both for aeroengines and land-based power generation gas turbines. More precisely, the cold compressed gas entering the combustion chamber must leave it with a maximum temperature when entering the subsequent turbine. The first blading row of the gas turbine, just behind the combustion chamber, thus experiences the highest temperature and pressure. It is certainly the most significant constraint when designing the turbomachinery. Thus, the solutions for these turbine components must withstand very high loads at elevated temperatures, not to mention the severe gas environments and the long periods of time, all repetitively.

Before the Second World War, at the beginning of the development of gas turbine for aeroengine applications, iron-based cold wrought materials were first used [370]. The superiority of *nickel-base superalloys* was yet very soon recognized, notably for their exceptional creep and fatigue properties at high operating temperature, as well as resistance to oxidation and corrosion [1190]. Nickel-base superalloys were first shaped by extrusion

Nickel Base Single Crystals Across Length Scales
https://doi.org/10.1016/B978-0-12-819357-0.00015-9

and forging. But with always higher strength and the introduction of hollow blade designs, *investment casting* based on the lost-wax process was developed. It is the current processing route for all turbine blades made of nickel-base superalloys, although concurrent methodologies are being investigated based on additive manufacturing [749].

The hollow *design* aims at defining cooling passages for fresh air, as well as weight reduction. This major breakthrough in turbine blade engineering started in the 1960s and led to a dramatic increase of the turbine entry temperature. It is still a topic of optimization by engine manufacturers. But the other main innovation over the second half of the 20th century was due to alloy *elaboration* by metallurgists. With uninterrupted demand for higher mechanical properties at more elevated temperatures, alloy cleanliness was a crucial requirement. This was achieved by introducing multiple successive melting of alloys under vacuum, thus considerably reducing nonmetallic inclusions and chemical inhomogeneity and achieving a very high control of composition [370]. Another main step was the constant search for better *compositions* of the nickel-base superalloys. The basic trend consisted in replacing Cr by γ'-forming elements such as Al, Ti, and Ta, increasing the γ'-solvus temperature by additions of Ta and W, and finally introducing creep-strengthening elements Re, W, Ta, Mo, and Ru [1190,202]. The design of the alloys could then become more systematic by defining automatic procedures that scan the chemical species and its content, access computed properties using thermodynamic and physical databases, and make use of criteria to select compositions [1193]. Nowadays, the *microstructure* leading to the highest mechanical properties for turbine blades are directionally solidified *single crystal* (SX) with chosen crystallographic orientation so as to suppress the weakening effects of grain boundaries. The SX components are heat treated to reach around 70% of coherent γ'-precipitates in a γ-matrix phase while maintaining a small lattice misfit, high creep strengthening, and oxidation resistant properties. Finally, multilayer coatings are applied to enhance the corrosion and oxidation properties, increasing the durability of these expensive parts when operating at a temperature close to the γ'-solvus temperature in severe gas environments, for long periods of time and thousands of cycles. With typically 10 alloying elements and use of noble and rare chemical species, the SX nickel-base materials are complicated metallic alloys. They challenge our metallurgical and mechanical understanding as most of theoretical studies are developed for simple model alloys, often dilute with only few chemical species.

Historical development of directional solidified SX components made of nickel-base superalloys is a paramount example of how materials properties and processing of products are intimately linked. The present contribution aims at giving the status of the interplays between processing and material properties for SX components. It will start by introducing the general concepts of the casting technologies and the related solidification defects. Modeling efforts will be cited. Heat treatments of as-cast components contributing to the definition of the required (service) mechanical properties are important steps in the processing route. They will thus be introduced and discussed, also focusing on related defects. Conclusions will be drawn on perspectives for further improvements and innovations.

7.2 Directional solidification and related defects

7.2.1 Generalities on the casting processes

With the advent of nickel-base superalloys for gas turbines, shaping from the liquid has become essential. The reasons are multiple. They include the limitations of machining techniques for complicated hollow geometries due to the design of internal cooling passages for in-service temperatures higher than the compressor-chamber gas temperature. The high-strength properties of the superalloys also infer dedicated and expensive tooling for machining. So direct casting of a liquid melt into a mold has quickly become the usual route.

Due to the demand for highly precise geometries, including thin walls with complex shape, the lost-wax casting process was adopted. It starts by making a replica of the desired nickel-base superalloy component by injecting wax around ceramic cores in to a metallic die. Several such wax plus core models are assembled as a cluster by adding gates and runners forming a tree like structure later defining the filling path of the molten metal. Ceramic filters are integrated as part of the gates/runners so as to block large inclusions that may be present in the melt and to reduce melt turbulence. The next step consists of making a ceramic mold. The wax assembly is immersed into slurries of colloidal silica binder and sand-like stuccos. The operation is repeated several times so as to form a thick and rigid coating. The mold is dried and the wax is removed, thus forming the ceramic shell with its embedded ceramic cores. Casting is operated in a vacuum induction melting furnace. This permits thorough control of superheat and composition through elimination of dissolved gases, elements' traces, and oxides [29]. After casting, the ceramic shell and cores are removed, the gates and runners are cut off so only the metallic component plus cores remain. Cores are eliminated prior to heat treatment (see Section 7.3 of this chapter).

In this section we present the main features and issues related to casting of nickel-base superalloy to produce single crystalline turbine blades by the lost-wax process. The latter is of particular importance in the processing route of the material as it defines the initial metallurgical microstructures, i.e., the thermodynamic phases, its amount, crystallography, size, morphology, composition, and stresses, as well as defects, all taking part in dictating the subsequent forming steps and final properties of the components.

7.2.1.1 The Bridgman–Stockbarger casting process

In a conventional lost-wax casting, the liquid metal is poured in a static mold and cooling takes place by heat exchange with the surrounding environment in all directions. As a consequence, heat flow is not directional and the structure is mainly isotropic or consist of columnar and equiaxed grains. Components produced with this static mold technology are named "equiaxed." In fact, this denomination is somewhat abusive: while copious nucleation of grains takes place at the mold surface, growth proceeds toward the central part of the casting geometry. Seen from its skin after chemical etching, the casting skin indeed looks like an equiaxed structure while most of the grains grew columnar in a direction opposite to the heat flow and meet at the center part of the geometry [487,488]. In fact, for such equiaxed component, nucleation at mold surface is induced by application of a special coating at the internal surface of the mold containing an inoculant compound (e.g., cobalt aluminate). This is achieved by adding the compound as part of the first slurry of colloidal silica binder forming the ceramic mold. In order to produce directionally solidified components, both nucleation and growth of the solidifying structure need to be well controlled. Directional solidification of nickel-base superalloys to produce single crystal components was first introduced in the 1960s [1490]. The schematics of a typical setup based on the so-called Bridgman–Stockbarger directional solidification process is shown in Fig. 7.1(a). It consists of (top) a hot zone and (bottom) a cold zone separated by a baffle, as well as a chill plate on which is located the mold. Unlike conventional lost-wax casting, the mold is no longer static. Initially located in the hot zone, it is dynamically moved downward into the cold zone by withdrawing the chill plate at a rigorously controlled velocity. With the hot/cold zones respectively above/below the melting temperature of the alloy, the solidification front is located in a narrow window close to the baffle and growth of the solid proceeds upward at a velocity roughly equal to the withdrawing rate of the chill plate, v. Note that, in principle, the temperature gradient at the position of the growth front, G, can also be adjusted by controlling the temperature of the hot zone. Thus both the velocity and the temperature gradient could be controlled, which are the main processing parameters for casting (the cooling rate $\dot{T}$ is simply the product of the last-two parameters when a steady regime is established as $\dot{T} = -Gv$).

As a result of directional solidification, a columnar dendritic microstructure is growing into the melt. For single-crystal Ni-base superalloys, they consist of a solid disordered fcc-phase. Its characteristic length scales after complete solidification are the primary dendrite arm spacing, λ_1, and the secondary dendrite arm spacing, λ_2. According to simple liter-

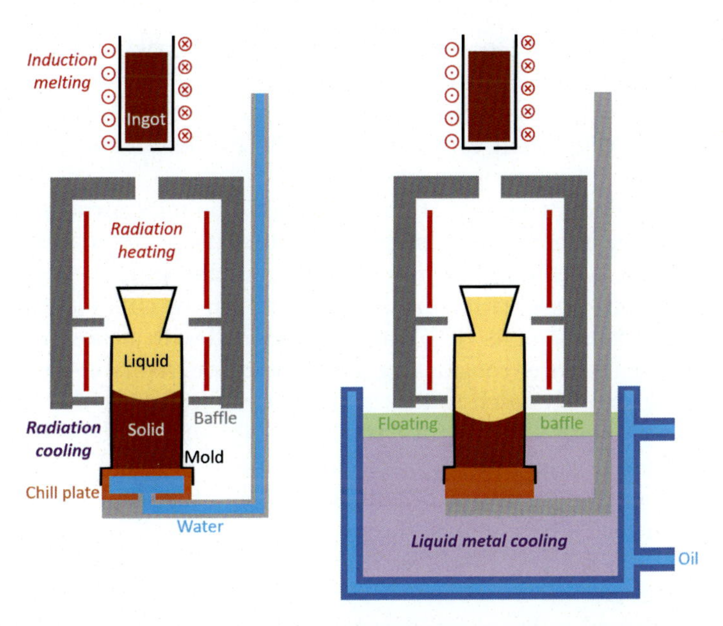

FIGURE 7.1 Schematics of (a) the Bridgman–Stockbarger process and (b) the liquid metal cooling process for directional solidification of nickel-base superalloy to produce single-crystal turbine blades [385,165].

ature models [780], they depend on the processing parameters as:

$$\lambda_1 \propto G^{-1/2} v^{-1/4}, \tag{7.1}$$

$$\lambda_2 \propto G^{-1/3} v^{-1/3}, \tag{7.2}$$

while derivation of the first relation involves a dendrite tip kinetics theory, the second relation is linked to the physics of coarsening taking place in the mixture region of liquid and solid phases called mushy zone. The latter region is approximately defined by the solidification interval, i.e., the difference between liquidus temperature, T_L, at which the first solid phase forms from the melt and the solidus temperature, T_S, at which the last liquid solidifies. Such relations are approximate and dependencies with alloying elements are not given. More detailed experimental studies and relationship coming from direct measurements for nickel-base superalloys are available [142]. Although compositions of single crystal Ni-base superalloys are defined to avoid the formation of secondary interdendritic phases formed in the residual liquid located at the end of the mushy zone, formation of peritectic and/or eutectic microstructures could happen, involving an ordered fcc-phase. The solidification structure thus generally consists of a primary dendritic disordered fcc-phase and a mixture of secondary disordered and ordered fcc-phases, as will be further described in the present contribution. Another feature of the as-cast microstructure includes microporosity. It is naturally presents as its origin is related to the density variations between phases in the mushy zone. Directional solidification is well suited to reduce it but its full elimination could be challenging.

Another important feature of a single crystalline microstructure is the crystallographic orientation. Its selection requires using either a solid seed or/and a grain selector. In the first configuration, the seed is a single crystal with controlled orientation, positioned between the chill plate and the component, with one of its ⟨100⟩ directions aligned with a chosen component axis. For instance, considering a turbine blade, its main airfoil axis (along its height) is chosen to host a ⟨100⟩ crystallographic direction by positioning the seed accordingly. Partial remelting of the seed takes place after the liquid is poured in the ceramic mold and epitaxial growth proceeds, propagating the crystal orientation through the entire component. Note that in nickel-base alloys, the preferred growth directions defined by the trunks and arms of the dendritic microstructure correspond to the ⟨100⟩ crystallographic directions. Thus, a ⟨100⟩ crystal orientation is chosen to correspond to the airfoil direction, usually the normal to the chill plate and thus the direction of the temperature gradient defined by the hot and cold zones of the furnace. It is also worth noting that the secondary ⟨100⟩ directions, the dendrite arms perpendicular to the trunk and main airfoil axis, can also be chosen to correspond to selected component directions. The second configuration makes use of a grain selector. It consists of a constriction designed as part of the ceramic mold from the wax assembly located between the chill plate and the component. Its role is to select a given single crystal orientation from a polycrystalline columnar grain structure arising from randomly nucleated grains in contact with the chill plate. The grain selector geometry can be a simple helix (the so-called "pig tail"), as schematized in Fig. 7.2, or it can take more sophisticated shapes. In any case, its development is based on the fact that the preferred growth direction of the columnar dendritic grains follows a ⟨100⟩ crystallographic direction more or less aligned with the temperature gradient.

7.2.1.2 Alternate casting processes

While the Bridgman–Stockbarger directional solidification process is by far the most commonly used for the production of nickel-base superalloy single-crystal components, alternate methods have been proposed. Considering that a nonstatic mold and the presence of hot and cold zones separated by a baffle was a somewhat complex casting process relying on heavy investments, Walser et al. proposed to rely on existing conventional vacuum casting equipment for static molds, still using a chill plate to ensure directional heat flow [1524,754]. This main idea was to optimize cooling from the chill plate and insulation of the ceramic shell using layers of blanket materials. Despite its interest, the so-called SMCT-process was not widely adopted. In fact, upon directional growth of a mushy zone from a chill, the temperature gradient at the dendritic growth front progressively decreases while the velocity increases [485]. In such conditions, the microstructure length scales (Eqs. (7.1) and (7.2)) can hardly be controlled and, upon low values of the ratio (G/v), nucleation of equiaxed grains takes place, obviously breaking the condition for single crystal growth [310].

The need for higher temperature gradient and velocity was very soon identified. Not only it produces finer microstructures, but we also shall see in Section 7.2.2 its roles on various defects that could happen during solidification. This is why other cooling media were proposed. Fig. 7.1(b) gives an illustration of the liquid metal cooling process [385,165,518]. The cold zone is made of a low melting and low vapor pressure liquid metal (e.g., Sn), maintained at a temperature always much lower than the solidus temperature of the nickel-base superalloy. The ceramic mold filled with the superalloy is thus progressively immersed into the liquid metal. Both the combination of the heat flow through the liquid metal and a baffle floating above it ensure a temperature gradient at the solidification front 2 to 3 times higher compared to the standard Bridgman–Stockbarger casting process in the 1500°C to 900°C temperature range. This also offers the possibility to increase the withdrawal rate while maintaining a constant (G/v) ratio, thus increasing productivity. It is worth noticing that a seed is preferred to define the single crystal orientation as using a grain selector implies direct contact of the superalloy melt with the chill plate and risk of mixture with the liquid metal bath. Interaction of the nickel-base superalloy with the liquid metal is also an issue. A variant was thus introduced where the liquid medium is replaced by solid particles (e.g., carbon balls) held in suspension and transported by a continuous flow of inert gas (e.g., argon) [385,549,621]. All these processes with higher solidification rates are also known to produce alloys with a smaller casting pore size, and hence better fatigue properties (see the last subsection of this chapter) [385,165,785,1362].

Other parameters and manufacturing processing are being considered. One could cite additive manufacturing of single crystal superalloys [749]. Using electron beam, advantages include an even finer microstructure due to always higher temperature gradients, more homogeneous chemical composition, and reduced shrinkage porosity compared to investment casting, while recrystallization is an issue if heat treatment is still to be considered. Magnetic field is another direction being investigated for improvement of the superalloy microstructure [1603]. These researches can be considered as new directions for potential breakdown technologies.

7.2.2 Casting defects

The features identified as defects in the production of single crystal components are numerous and have very diverse origins. Fig. 7.2 is a classical schematic that reports microstructure related defects observed upon inspection of a turbine blade component after casting. We shall only consider such defects that are directly related to solidification issues, meaning that inclusions, surface scales, critical dimensions, and tolerances, and other defects due to non-well-controlled foundry practices and/or quality issues with the core, shell and superalloy materials are not considered hereafter.

7.2.2.1 Crystal structure

There are several origins for the breakdown of the single crystalline structure. When nucleation proceeds on the chill plate, crystal orientation is random and copious. Dendritic growth starts from each nucleus. It requires undercooling, i.e., a driving force identified by the temperature difference between liquidus temperature of the alloy given by thermodynamic equilibrium and the dendrite tip temperature. This undercooling is due to several contributions, including solute diffusion of chemical species segregated at the solid–liquid interface [780,310]. Because the liquid is undercooled, nucleation of new grains could further happen in the melt in front of the columnar dendrites. In the absence of nucleation, a columnar grain structure grows in the starter block shown in Fig. 7.2. The growth front is then composed of several grains, each grain containing several dendrite trunks with the same crystal orientation, all dendrite tips at some undercooling, i.e., on an isotherm surface lower than the liquidus temperature of the alloy and perpendicular to the chill plate if one assumes a vertical temperature gradient opposite to the heat flow. The grain density decreases upon growth as only the grains with a ⟨100⟩ dendrite trunk direction well aligned with the temperature gradient survive, giving birth to a fiber texture [489,1109,1394]. At the exit of the starter block, i.e., at the entry of the grain selector, there still exist many grains.

7.2.2.1.1 Primary misorientation

The helix geometry aims at selecting a single crystal. It acts as a heat flow channel, where the direction of the temperature gradient basically follows the metal path during cooling. This is due to the insulating role of the ceramic shell, the difference in thermal properties between the mold and the superalloy, as well as cooling from the underneath chill plate [218]. As a consequence of growth competition between secondary branches also propagating the ⟨100⟩ directions of the grains, only a single grain and hence a single crystal orientation remains at the exit of the grain selector. Not only a single crystal must be selected at the exit of the grain selector but also its primary crystallographic direction defined by the direction of

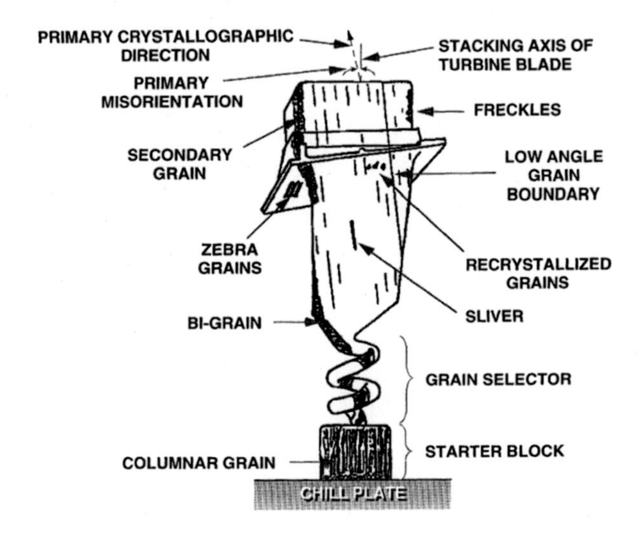

FIGURE 7.2 Schematics of casting defects in a turbine blade [978].

the primary dendrite trunks with respect to the stacking axis of the turbine blade. In Fig. 7.2, the primary misorientation is used to define the possible deviation with the desired crystal orientation. In principle, simulation of dendritic growth can be directly carried out using various methodologies. The phase field (PF) method and the cellular automaton (CA) methods are the actual two extremes in terms of objectives and scales. While PF is able to directly produce complicated dendritic patterns, CA only aims at propagating grain envelopes. Consequently, PF is much more demanding in terms of computer resources and cannot assess simulation for a whole grain selector. This was yet achieved long time ago by the CA method [486] and is still used for optimization of both the starter block and the grain selector [1215,1601]. Nowadays, PF and CA were demonstrated to converge towards satisfying results when considering growth competition of bi-grains, yet limited to binary dilute alloys [1109].

7.2.2.1.2 Bi-grain, secondary grain, and high-angle grain boundaries

As suggested in Fig. 7.2, it could also happen that the grain selector does not work properly so a bi-grain or more grains survive the growth competition in the helix geometry and propagate in the turbine blade. In this situation, a single crystal may never have been present in the helix. This situation may happen when the heat flow is not well controlled, the helix geometry is not well designed, or if heterogeneous nucleation takes place in the undercooled region ahead of the dendrite tip or in a cold part of the grain selector. Spurious or stray adjectives are used for several types of crystal defects, including undesired grains due to nucleation. It is worth noting that such grains are often seen in large section changes, for instance, when leaving the airfoil to enter the platform of a turbine foot geometry (top-left regions in Fig. 7.2). Simple explanations with observations and analytical models were provided [490,316]. They involve the time required for the dendritic growth front to propagate in the platform, thus increasing the total undercooling of the liquid together with the risk for spurious nucleation. These spurious grains are often large and with random crystallographic orientation.

7.2.2.1.3 Zebra grains

This defect is, again, a region of the component with crystal orientations that differ from the desired single crystal. Their appearance is not similar to secondary grains. Zebra grains are surrounded by the single crystal, they are of limited size and can be made of several distinct crystal orientations. In case of zebra grains located in the platform of Fig. 7.2, they form as a result of partial remelting of the dendritic structure after its propagation in the undercooled region of the platform [1068,1271]. In fact, upon growth of the single crystal, recalescence can take place when the temperature gradient could not be maintained directional and high enough to evacuate the latent heat. Partial remelting leads to detachment of dendrite arms and rotations in the liquid thus creating new crystal orientations. The occurrence of such recalescence can also be modeled when coupling the CA model with heat flow [90].

7.2.2.2 Microstructure and microsegregation

Solidification is accompanied by segregation: solute species are redistributed between the phases present in the mushy zone. In case of no phase transport by convection, the evolution of the solute composition in all phases leads to the so-called microsegregation. The path of the interfacial compositions upon cooling and solidification, even if it follows thermodynamic equilibrium, can be very complicated as full equilibrium does not necessarily take place. Fig. 7.3 presents a typical microstructure for the single crystal nickel-base superalloy AM1. Its composition can be read in Table 7.1. Full equilibrium using only thermodynamic data concludes that full γ-phase solidification takes place. This is obviously not the case as γ'-phase is present as part of an interdendritic eutectic microstructure.

In practice, the microsegregation path is a consequence of diffusion kinetics and interfacial equilibrium at the γ/l-interface. Explanations involve limited diffusion in the γ-phase while the l-phase can be assumed to be in complete mixing or of uniform composition. This is due to the ratio of the diffusion coefficients for the chemical species i in phase γ and the $D^{\gamma}_{i'}/D^{l}_{i'}$ ratio easily reaching 10^{-3}. Consideration of diffusion in the solid with cross-diffusion terms is also necessary for nickel-base superalloys. This could be handled with advanced PF models [1540]. Fig. 7.4 presents a typical distribution of four chemical species in a quaternary nickel-base

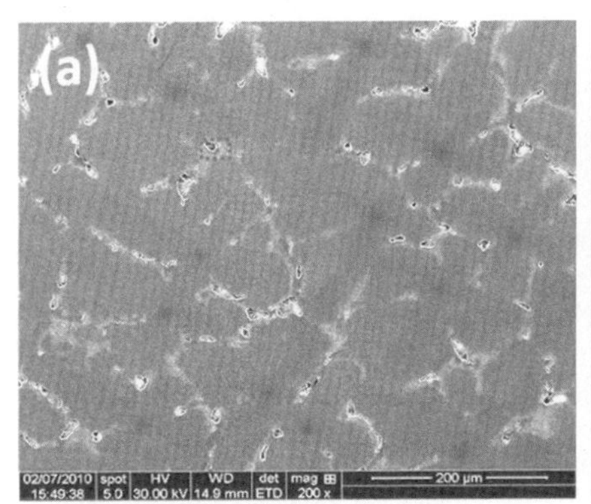

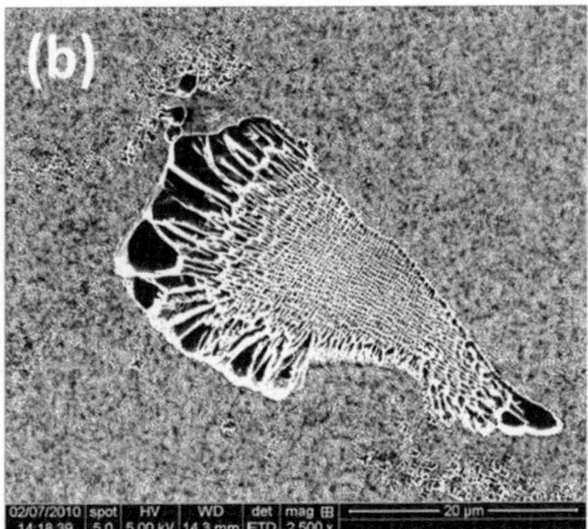

FIGURE 7.3 As-cast microstructure of directionally solidified nickel-base superalloy AM1 as observed by electron microscopy revealing (a) primary γ-phase dendritic solidification and (b) an interdendritic eutectic region made of a mixture of the γ- and γ'-phases [90].

TABLE 7.1 Chemical composition (in wt.%) of nickel-base superalloys cited in this chapter (Quaternary in Fig. 7.4).

Alloy/Elemt	Ni	Al	Ti	Ta	W	Cr	Co	Mo	Re	Nb	Hf
AM1	Bal.	5.3	1.2	8.0	5.7	7.8	6.5	2.0	/	/	0.05
AM3	Bal.	6.0	2.0	3.5	5.0	8.0	5.5	2.2	/	/	/
MC2	Bal.	5.0	1.5	6.0	8.0	8.0	5.0	2.0	/	/	/
CMSX-4	Bal.	5.6	1.1	6.5	6.4	6.5	9.7	0.6	3.0	/	0.1
CMSX-4 Plus	Bal.	5.7	0.9	8.0	6.0	3.5	10.0	0.6	4.8	/	0.1
CMSX-10K	Bal.	5.7	0.2	8.2	5.5	2.2	3.3	0.4	6.3	0.1	0.03
Quaternary	Bal.	5.8	/	7.94	8.84	8.98	/	/	/	/	/

superalloy. As can be seen, because the dendritic microstructure is due to γ-phase solidification, it inherits from the solubility of chemical species in the γ-phase. This property is given by the segregation coefficients, i.e., the ratio of the γ-phase composition over the liquid composition at the interface. As stated above, assuming thermodynamic equilibrium at the interface, it can be directly read from the phase diagram. In practice, the CALPHAD method has become a standard for determining these coefficients, although it relies on databases that are mostly fitted from experimental data. In the case reported in Fig. 7.4, measurements of composition reveal an Al-, Cr-, and Ta-depleted and W-enriched dendrite, as well as nonuniform distributions in the dendrite. Also small secondary solidified regions are seen with clear composition differences, here only representing around 1% in volume. Of course, the reported behavior for the solute species in the dendrite is retrieved with Al-, Cr-, and Ta-enriched and W-depleted secondary solidified regions. As-cast microstructure and microsegregation are essential outputs of the lost-wax casting process of nickel-base superalloys. As the eutectic is not generally desired, it could be seen as a defect,

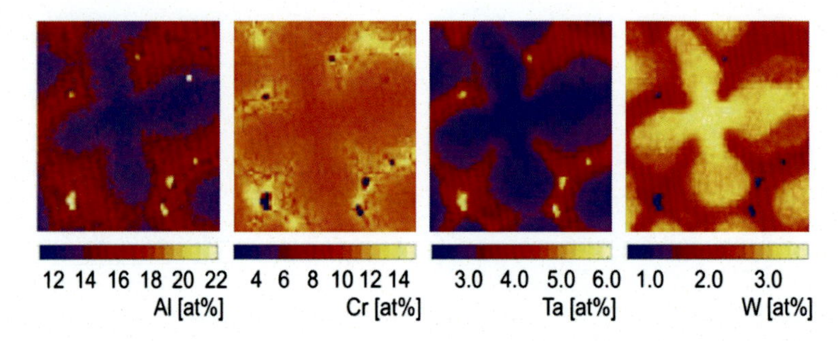

FIGURE 7.4 Distribution of Al, Cr, Ta, and W between the dendritic γ-phase and the interdendritic region after directional solidification of the nickel-base superalloy (in wt.% | at.%) Ni – 5.8 | 13.06 Al – 8.98 | 10.49 Cr – 7.94 | 2.67 Ta – 8.84 | 2.92 W as measured by wavelength dispersive X-ray spectroscopy [1540]. Processing conditions: temperature gradient $G = 20\,\mathrm{K\,mm^{-1}}$ and withdrawal rate $v = 5\,\mathrm{mm/min}$.

similarly as nonuniform chemical compositions in the solidified γ-phase. We shall see later the role of heat treatments for the control of the microstructure and microsegregation.

7.2.2.3 Freckles

Freckles defects are shown in Fig. 7.5. They appear as chains of small equiaxed grains in the turbine airfoil of Fig. 7.5(a) where variations of grey colors compared to the rest of the geometry clearly reveal different crystallographic orientations. Their origin is the redistribution of chemical species by liquid transport. During solidification, due to microsegregation, the interdendritic liquid regions are enriched or depleted as a result of solidification and interfacial equilibrium (see the explanation in Section 7.2.2.2). Variations in the liquid composition induce buoyancy forces due to inhomogeneous density. In case of density inversion close to the dendritic growth front, liquid can flow upward, leave the mushy zone, and accumulate in the liquid melt, thus creating segregation at the scale of the product, named macrosegregation. The transport of solute and its effect on dendritic growth are illustrated in Fig. 7.6 where the first in-situ real-time X-ray radiography of plumes of solute ahead of a columnar dendritic growth front observed in the 2nd generation CMSX-4 (see the composition in Table 7.1) nickel-base SX superalloy were observed (variations of the grey level in the liquid region) [1198]. These observations also reveal the strong interaction between the flow of solute and the growth front. Following the dendrite tip identified by the white arrow (Fig. 7.6(a)) with time, one can draw its position (Fig. 7.6(b)), velocity (Fig. 7.6(c)) and above liquid composition qualitatively given by the grey level (Fig. 7.6(d)). The latter information reveals a periodic oscillation due to the propagation of solute convection cells shown in Fig. 7.6(a). The correlation with the growth velocity evolution is immediate when considering the verti-

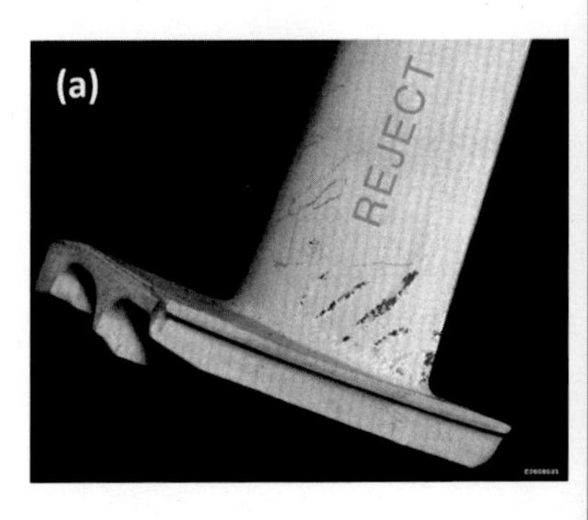

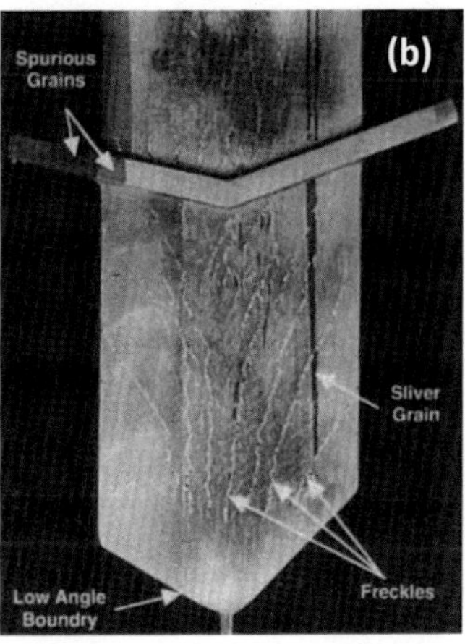

FIGURE 7.5 Examples of the freckles defect formed upon directional solidification of (a) a turbine blade geometry and (b) a dummy geometry also exhibiting spurious grains, low angle grain boundaries and sliver grains [73].

cal dashed lines: the times at maximum grey level correspond to the time at minimum velocity and oscillation of the growth front propagation. Of course, this is in link with the main contribution for dendritic growth, previously referred to as the solutal undercooling. When the intensity of the grey level increases, the average supersaturation decreases and so does the velocity. Although not reported in this figure, it could be seen that each dendrite has an oscillation regime that travel with the solute plumes, i.e., not only upward but also from right to left due to a larger convection cell in the above melt. On the right-hand-side of the mushy zone, in contact with the border of the sample, is a larger channel that is progressively forming. Closer look shows that this channel is the origin of a higher upward solute flow. Later on, it solidified with equiaxed grains formed by fragmentation of the existing dendritic microstructure and/or nucleation in the liquid channel [1198], giving rise to a freckle.

Several criteria have been proposed in the literature for the formation of the freckles. The simplest considered the effect of the processing parameters previously introduced [274]. Freckles appear when the product $G^{-m} \times v^{-n}$ becomes too low, where m and n are positive values lower than unity. Note that such relation is comparable to Eq. (7.1) and (7.2) and correspond to the observations that a critical value of the primary spacing exist above which freckles form. This observation is nothing but the manifestation of the mushy zone permeability, directly proportional to the microstructure length scale. The larger the dendrite arm spacing, the larger the permeability, thus permitting circulation of a convective flow into the mushy zone and transport of solutal species above the dendritic growth front. An experimental correlation was later derived based on a

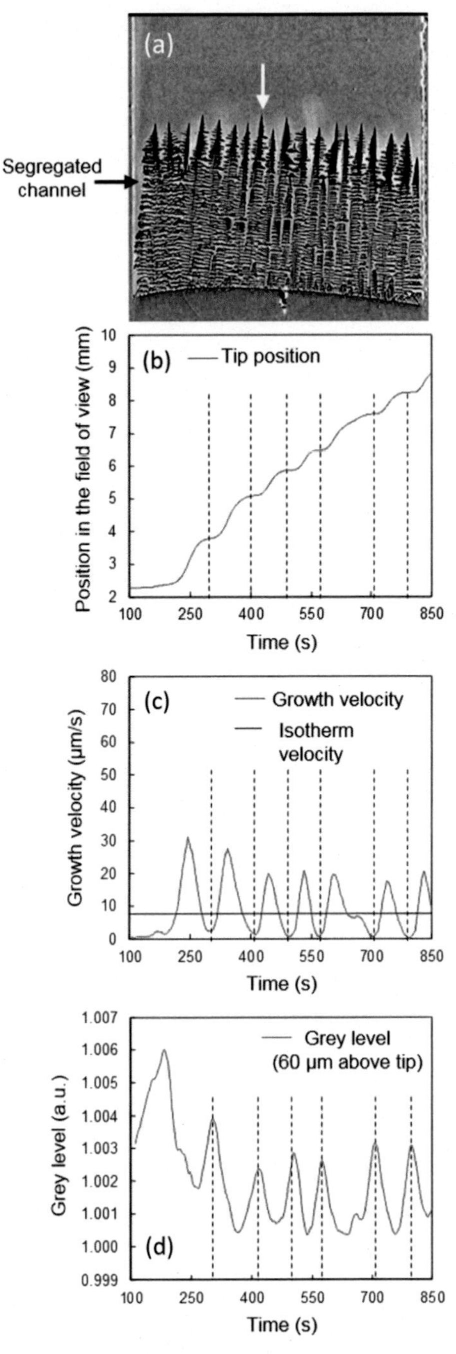

FIGURE 7.6 Experimental analyses of a directionally solidified CMSX-4 superalloy sample processed with in situ and in real-time observations by means of synchrotron X-radiography at the European Synchrotron Radiation Facility (ESRF, Grenoble, France). Temperature gradient $G = 44\,\mathrm{K\,cm^{-1}}$, growth rate $v = 7.5\,\mathrm{\mu m\,s^{-1}}$ [1198].

large experimental database with systematic characterization of the primary dendrite arm spacing, the liquid density variation due to segregation and the solidification interval [1434]. Finally, Rayleigh-type criteria were proposed, where the influence of fluid flow was estimated [86,1171]. In principle, such criteria are better suited as they compare buoyancy forces to the friction forces in the presence of a mushy zone. So properties like liquid viscosity, mushy zone permeability, and thus microstructure length scale and solidification path, liquid density variations with composition and temperature, as well as heat diffusion, could be included. In practice, however, properties are often not very well established and the role of the solidifying geometry is difficult to account for. This is why efforts continue to account for more physics in direct simulations of thermo-solutal convection, i.e., the interaction of the solidifying microstructure with liquid flow [1246].

7.2.2.4 Deformation related defects

7.2.2.4.1 Slivers and low angle grain boundaries

An example of a sliver grain is show in Fig. 7.5. While its primary ⟨100⟩ dendrite trunk direction only slightly differs from the surrounding single crystal, the other ⟨100⟩ directions could be more misoriented. This means that a rotation has taken place about the primary ⟨100⟩ direction. The literature proposes various origins for this defect. However, the most recent explanations suggest that deformation of the growing dendritic microstructure could be the best explanation [1376]. In fact, thanks to tremendous progress made by direct observations of phase transformations using in-situ real-time X-ray radiography, the deformation can now be imaged in metallic alloys [54]. Direct correlation was thus possible between deformation and rotation of sliver-type defects. In fact, the role of thermo-mechanical stresses built with solid state precipitation of the ordered γ' phase in the primary dendritic disordered γ phase below the solidus temperature was soon identified to explain cumulated deformation and low angle grain boundaries also referred to as subgrains, shown in Fig. 7.5 [1323].

7.2.2.4.2 Recrystallized grains

This defect is built during cooling of the metal trapped between the ceramic parts, i.e., the core and the shell of mold. All three materials have different dilatation coefficients. Upon cooling, the thermo-mechanical histories are thus different, leading to the build-up of stresses in the metal. These stresses are at the origin of the nucleation and growth of recrystallized grains during later heat treatments. Fig. 7.7 presents recrystallized grains as observed in the airfoil of a component designed to be a single crystal. The abrupt color changes reveal the presence of several grains, i.e., variations of the crystallographic orientations. The part not being a single crystal must simply be discarded, decreasing the yield rate of the

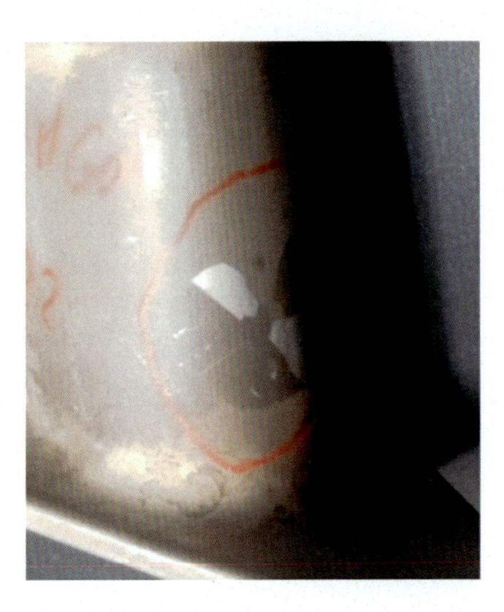

FIGURE 7.7 Recrystallized grains in the airfoil of a component designed as a single crystal.

production route. Formation of recrystallized grains following single crystal casting has only been recently studied. Deformations were first imposed by indentation at room temperature of single crystal samples. A minimum value of the plastic strain was identified above which recrystallization takes place during the solution heat treatment, e.g., typically 6 h at 1315°C in case of CMSX-4 alloy [296]. But compressive and tensile tests were also tested at several temperatures, followed by isothermal annealing heat treatments for various parameters (holding time and duration) [296,838,1054]. The presence of recrystallized grains is found to depend on (*i*) the plastic strain, (*ii*) the annealing temperature, and (*iii*) the temperature at which the specimens were deformed. Most of the experiments are conducted with plastic strains of 4% [296,1054] and of 5% [838] when the annealing and deformation temperatures are studied. But depending on the plastic strain, the annealing temperature seems to influence the formation of recrystallized grains. Similarly for CMSX-4, if the temperature at which 4% plastic strain is introduced is lower than 950°C, no recrystallized grains were formed for annealing treatments below 1260°C [1054]. Finally, the recrystallized area is a direct function of the annealing temperature. Above the solvus of the γ' precipitates in the γ matrix phase (around 1300°C), the recrystallized area increases rapidly. Below this temperature, it would be a direct function of the fraction of γ' precipitates. Despite these observations and electron microscopy investigations, the author's explanations do not converge to explain the recrystallization mechanisms [838]. A special feature of the recrystallized grains is that they are not immediately detectable in the as-cast state. They develop during the heat treatments that follow casting which will be the object of the next section.

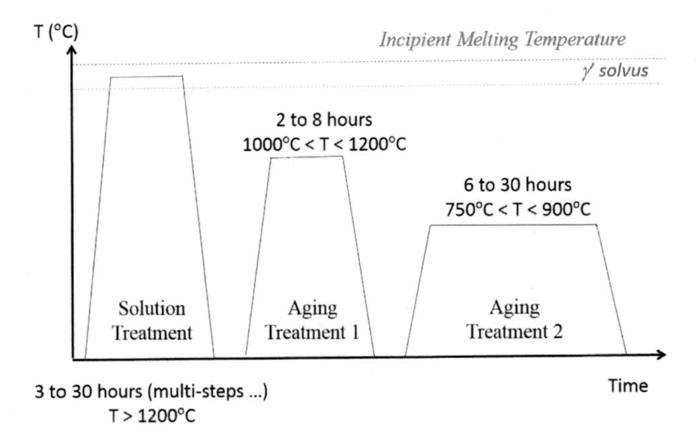

FIGURE 7.8 Typical heat treatments schedule.

7.3 Heat treatments and microstructure optimizations

7.3.1 Generalities on heat treatments

Solution and aging heat treatments (see the schematic illustration in Fig. 7.8) are usually applied to most of the single crystalline components (i.e., blades and vanes) for which a given level of mechanical properties in creep and fatigue is expected. It has to be noted that several components may be used in as-cast state (e.g., vanes or shrouds) to reduce production costs and/or to limit the development of defects during processing, such as recrystallization (Fig. 7.10(b)). During production of SX components, typical heat treatments are composed of a solution heat treatment and of two aging heat treatments at least (see Fig. 7.8). Moreover, other operations (desulfurization, deoxidation, preoxidation of the bond-coat before top-coat deposition, etc.) lead to additional temperature cycles of the components before being serviced. In the following, we will mainly focus on key heat treatments steps controlling the microstructure and mechanical properties of SX alloys, namely, the solution heat treatment and agings.

Typical metallurgical defects that can be encountered/should be avoided in SX components are presented in Figs. 7.9 and 7.10. Casting pores are shown in Fig. 7.9(a) in the case of MC2 (see composition in Table 7.1) 1st generation SX alloy whose solidification process was done with issues in withdrawal rate. These casting pores are forming in the interdendritic spacings during the dendritic growth in the last stages of solidification of the mushy zone [390,395,800,848]. These pores can reach a size of up to 100 μm to 150 μm, depending on the casting parameters and alloy's chemistry [1362,224,133]. It will be shown in the last subsection of this chapter how these pores are critical in controlling fatigue properties if no hot isostatic pressing (HIP) is applied. In Figs. 7.9(b) and 7.10(a),

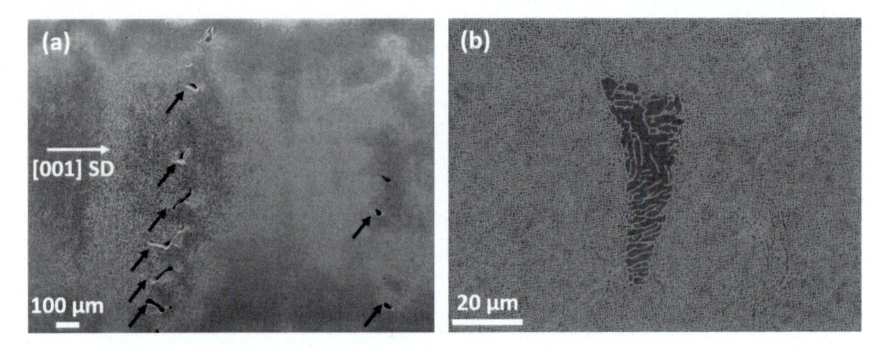

FIGURE 7.9 Typical metallurgical defects that can be encountered in SX components after investment casting: (a) casting pores (see red arrows – dark grey arrows in print version) and (b) γ/γ' eutectic pool. The nearly [001] solidification direction (SD) is indicated by a white arrow in (a) (authors' own works).

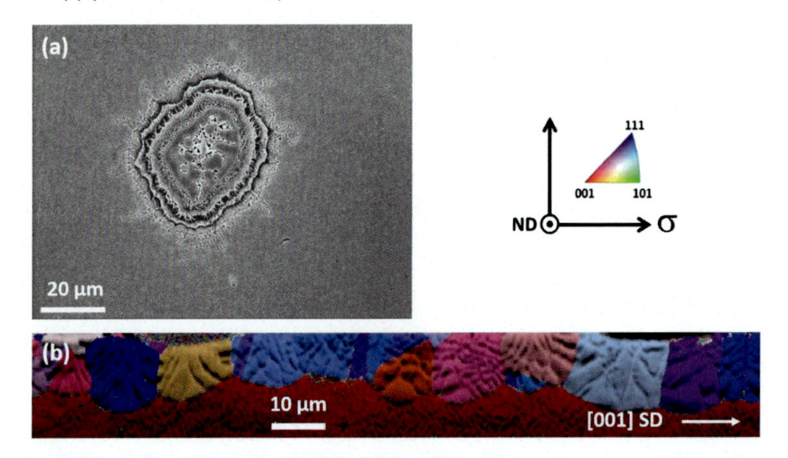

FIGURE 7.10 Typical metallurgical defects that can be encountered in SX components after solution heat treatment: (a) incipient melting in interdendritic spacings and (b) surface recrystallization due to local surface plasticization. The nearly [001] solidification direction (SD) is indicated by a white arrow in (b). (Authors' own works).

typical microstructures of γ/γ' eutectics and local incipient melting are presented, respectively. Finally, new grains nucleated at the surface of a component profile after solution heat treatment due to, e.g., a local surface plasticization during shell removal or local residual stresses introduced during the solidification are shown in Fig. 7.10(b). The authors will refer to these figures in the following when referring to such typical metallurgical defects.

7.3.1.1 Solution heat treatment

The main objective of solution heat treatments (SHT) is to break the dendritic chemical inhomogeneity inherited from the solidification and to

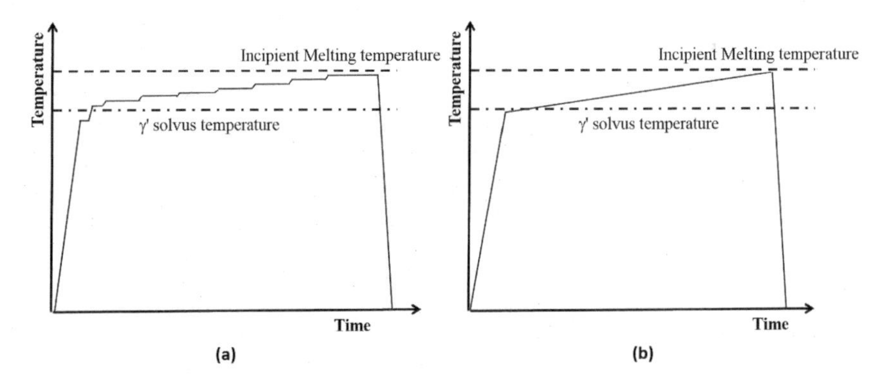

FIGURE 7.11 Traditional stepwise (a) and ramp (b) solution heat treatments. Adapted from [597].

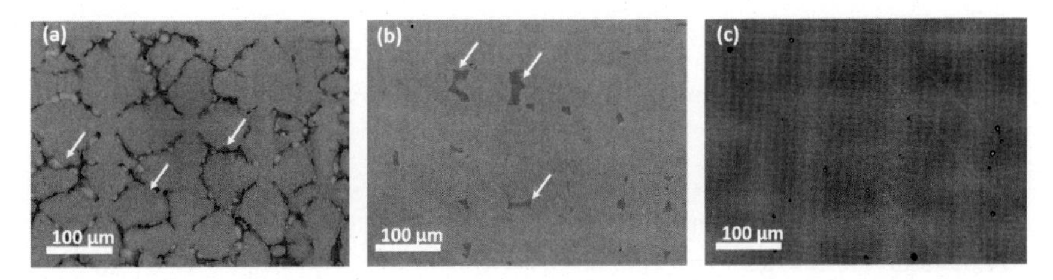

FIGURE 7.12 Evolution of the γ/γ' eutectic pools (see white arrows) as a function of the applied solution treatment in third generation CMSX-4 Plus SX alloy: as-cast structure (a), after an ST of 15 h at 1330°C (b), and after 15 h at 1340°C (c). Adapted from [133].

dissolve the γ/γ' eutectics pools [596]. For this, the alloy is usually heated above the γ' solvus temperature and the closest possible to the incipient melting temperature (or solidus) to achieve the best chemical homogeneity in a reduced time (see Fig. 7.8). Typical SHT are either performed in a stepwise manner (see Fig. 7.11(a)) or using a fixed heating ramp (see Fig. 7.11(b)) or a combination of both. These stepwise or ramped SHT procedures are especially mandatory for alloys with a high amount of refractory elements (typically, for the 2nd, 3rd, and 4th generation SX alloys with more than 3.0 wt.%. Re containing alloys) since a progressive chemical homogenization will be achieved with each temperature increment meanwhile increasing progressively the incipient melting temperature [1052,1053,1659]. These progressive (stepwise) increases in temperature during SHT as presented in Fig. 7.11 are also more convenient in industrial heat treatment furnaces to ensure almost similar temperature profiles for each component treated in a same batch.

In principal, a SHT is considered to be optimal when almost all γ/γ' eutectics pools have been dissolved, as illustrated in Fig. 7.12 in case of CMSX-4 Plus alloy (see the composition in Table 7.1), and when the

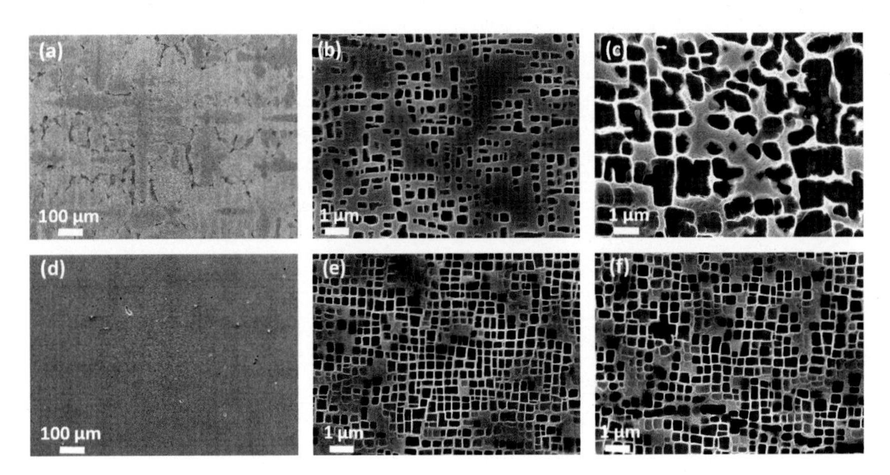

FIGURE 7.13 Cast dendritic structure (a, d) and γ/γ' microstructure in primary dendrite arms (b, e) and in interdendritic spacings (c, f) of an experimental third generation 5.4 wt.% Re-containing SX alloy. An optimized ST+ agings schedule has been applied for (d, e, f) while only agings (no ST) were applied for (a, b, c) microstructures. (Authors' own works).

γ' precipitate size distribution is homogeneous across the dendritic structure, as shown in Figs. 7.13(e) and 7.13(f). In fact, when a residual chemical microsegregation is still present, mainly in heavy elements such as Re, W, and Mo, the γ' precipitate size distribution is inhomogeneous across the dendritic structure, with coarser precipitates and a higher volume fraction of γ' phase within interdendritic spacings as compared to primary dendrite arms (compare Figs. 7.13(b) and 7.13(c)). Such a difference in precipitate size is known to degrade tensile and creep strengths of the alloy in a wide temperature range according to pioneering works of Caron and Khan [205,207,212]. Finally, limiting or ideally suppressing γ/γ' eutectics pools is highly desirable from a creep strength point of view. Indeed, the lower the γ/γ' eutectics pools fraction, the higher the fraction of efficient fine γ' precipitates as those presented in Figs. 7.13(e) and 7.13(f), overall improving creep strength [661]. It has, however, been shown in a recent study of L.M. Bortoluci Ormastroni et al. that a non-negligible fraction of γ/γ' eutectics pools can be kept without inducing a debit in creep, tensile, and fatigue properties [133]. Another very important part of the SHT is the final cooling before subsequent agings. The impact of this final cooling on microstructure and mechanical properties will be presented in a forthcoming subsection of this chapter. If the main aim of the SHT consist in optimizing bulk microstructure precipitation through elemental chemical interdiffusion across the cast dendritic structure, one should also mention that achieving a good chemical homogeneity is also required to further limit/suppress the development of secondary reaction zone (SRZ) during the bond coat deposition process in the 2nd, 3rd, and 4th generation SX alloys [1528] and to improve their oxidation/corrosion resistance [635,831]. Finally, the SHT also leads to the nucleation of the so-called "homogenization" pores in interdendritic spacings due to the imbalance of diffusing elements across the dendritic structure, in addition to the growth of solidification pores [800,848,124]. These homogenization pores being far smaller (typical size of $\approx$ 10 µm to 20 µm) than solidification pores, they do not have any noticeable impact on the mechanical properties [394].

7.3.1.2 Aging heat treatments

Differently from other precipitate strengthened materials like aluminum alloys, maraging steels, or Inconel 718, aging heat treatments applied to γ/γ' SX alloys do not trigger the γ' precipitation. Indeed, almost all the γ' precipitates have already formed during cooling from the ST temperatures, even for fast cooling rates of up to 500°C/min [554]. Industrially, typical cooling rates from the solution treatment temperatures are in the 200°C/min to 400°C/min for aeroengines components [257,1361], and slower for large blades and vanes used in industrial gas turbines due to their higher thermal mass [1103,1104]. Hence, from a bulk microstructure point of view, aging heat treatments have the main purpose to adjust the size and morphology of γ' precipitates, so as to achieve a regular array of cubical precipitates having an average edge length of nearly 400 nm to 500 nm. In fact, it is well admitted in the literature that such a microstructure is the most desirable to maximize creep, tensile and fatigue properties $\approx$ [001] oriented components in a very wide temperature range (i.e., from 700°C up to 1000–1100°C), whatever the alloy composition [1190,205,207,212,1361,1129], by favoring an homogeneous plastic activity in the γ matrix [1126].

A typical example of the to γ/γ' microstructure evolution in primary dendrite arms of the AM3 1st generation Ni-base SX alloy is given in Fig. 7.14 (see the composition in Table 7.1). Starting from a coarse and erratic precipitation state after Bridgman casting (due to slow cooling rate), a fine precipitation of nearly 100 nm to 200 nm in size is obtained after the ST followed by a 300°C/min cooling rate. During the first aging at 1080°C for 6 h, followed by air quench (AQ), a very regular array of cubical precipitates is obtained, as observed in Fig. 7.14. Even if it is recalled that the duration and temperature of the first aging heat treatment are mainly defined to ensure a good interdiffusion between the substrate and the bond coat of either β-(Pt)NiAl or MCrAlY (M = Ni, Co, Ta) type (see Chapter 10) and a sufficient preoxidation of the bond coat before ceramic top coat deposition, in addition to cost considerations, most of first aging heat treatment temperatures are within the 1050°C to 1170°C temperature range [355], ensuring this regular microstructure. In fact, in such a temperature range, the natural lattice mismatch (i.e., without any application of external stress) is close to being maximum in absolute value

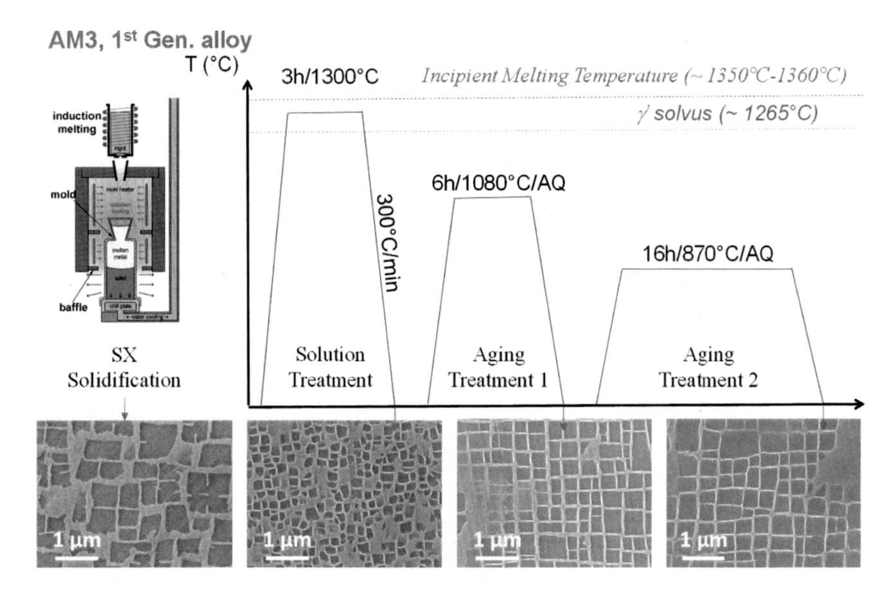

FIGURE 7.14 γ/γ' microstructure evolution in primary dendrite arms of AM3 1st generation Ni-base SX superalloy from as cast to fully heat treated state. (Authors' own works).

[200,349,353,809,665,1639,1641,1642,1643], ensuring a very huge level of long range internal stresses [587,590], hence favoring this very regular cuboidal γ/γ' microstructure. Finally, the last aging heat treatment performed at lower temperatures (typically between 750°C and 900°C) has the main purpose to relax residual stresses introduced during all processing steps (e.g., broaching of the blade root, laser cooling holes perforation, minor profile grinding to correct geometrical defects, etc.). From a microstructure point of view, this last aging leads to thinner γ matrix channels (see Fig. 7.14) thanks to the higher γ' content at temperatures below 900°C (the γ' content is of nearly 70% in most of Ni-base SX alloys from room temperature up to $\approx$ 800°C to 900°C, and then slowly decrease up to the γ' solvus temperature [213,1227,279,521,280,282,808] compared to the temperature employed during the first aging heat treatment. This decrease of the γ channel width is known to limit dislocation mobility [1126,1143] and, consequently, to improve creep and tensile properties [1361,280,222,337,805].

7.3.2 Defects and optimizations

7.3.2.1 Incipient melting

The occurrence of incipient melting (see the example in Fig. 7.10(a)) during solution treatment becomes an issue for Ni-base SX alloys with a high amount of refractory elements such Re, W, Ta, and Mo. Indeed, by increasing the content of such elements, the degree of chemical segregation is increased, overall leading to a lower temperature for incipient melting within interdendritic spacings while the γ' solvus temperature is increased. Overall, the heat treatment "window" becomes (very) narrow if one wants to dissolve as much as possible γ/γ' eutectics pools without reaching incipient melting [1052,1053,1051]. This is typically the case for alloys such as CMSX-10 alloys family (see CMSX-10K composition in Table 7.1) for which intricate ST schedules have to be designed [1052,1053,475,1562,1563]. Despite very few studies have been performed on the consequences of incipient melting on subsequent mechanical properties, it is interesting to notice that one group in China is proposing a new type of ST approach, including incipient melting and resolutioning afterwards at a close-to-solvus temperature so as to rejuvenate all melted areas [1658]. Improved creep strength have been obtained for a third generation alloy with such a modified ST, since the chemical homogeneity is better across the dendritic structure compared to "standard" stepwise ST ending just below the incipient melting temperature [1658]. Very recently, it has even been shown by L.M. Bortoluci Ormastroni that incipient melting can be tolerated in CMSX-4 Plus third generation SX alloy after ST provided that localized melting is not accompanied by a detrimental pore growth [133]. In fact, local melted areas resolidify in epitaxial relationship with the unmelted metal during th final ST cooling and subsequent aging heat treatments restore an acceptable γ/γ' microstructure within (and in the vicinity of) melted areas. It was shown that an area fraction of up to 10% of incipient melting has no detrimental impact on tensile, creep and fatigue properties [133].

7.3.2.2 Optimizations

Due to the cost added to the overall price of SX components by heat treatments, especially the solution heat treatment, reducing the duration and/or temperature of ST to the strictly necessary in terms of microstructure/mechanical properties quality is highly desirable. Usually, in such complex multi-element alloys, one of the simplest way to estimate the time necessary for ST is to assume that the completeness of this heat treatment is controlled by the diffusion of the slowest diffusion element [124,659,701], which is often rhenium for the 2nd, 3rd, and 4th generation alloys, while it is tungsten for most of the 1st generation alloys. An ST is considered completed once such an element has been able to diffuse over a characteristic length L which generally corresponds to the distance between the primary dendrite arm and interdendritic spacing centers. This distance can be calculated according to Eq. (7.3), knowing the apparent diffusion coefficient D of the least diffusive element [124,196,700], which is temperature dependent:

$$L = \sqrt{Dt}. \tag{7.3}$$

From this equation, it is then possible to obtain an estimate of the required ST duration t to achieve acceptable chemical homogeneity. It is clearly seen that the smaller the L value, the shorter/cooler the solution heat treatments. This is clearly the case for AM1 SX alloy, in which almost no more remaining eutectics has been observed after an accelerated (LMC) solidification compared to a standard one, using the same solution treatment (see Fig. 7.7 later on in this section) [1362]. More recently, even more impressive possibilities in ST time/temperature reduction have been observed considering quasi-SX solidification achieved by selective electron beam melting [749,1057,1245].

Over the last 10 years, different numerical tools have been developed to achieve much more reliable previsions of the chemical homogenization during ST, based on multicomponent diffusion databases and taking into account the initial dendritic chemical heterogeneity in a phase field approach [1214]. The limitations of these approaches are that they are much more time consuming and they require the knowledge of the initial distribution of elements in the as-cast state. The same applies regarding the optimization of the precipitate size during aging heat treatments using phase-field approaches [291,1533,1552,447] at the expanse of very simple and efficient mean-field Lifshitz–Slyozov–Wagner (LSW) models [841,1518].

7.4 Mechanical properties sensitivity to the processing parameters

7.4.1 Creep and tensile properties

As already mentioned earlier in this chapter, solution and aging heat treatments have mainly been designed so as to achieve the best degree of chemical homogeneity across the dendritic and a regular array of cuboidal γ' precipitates with an average edge length of nearly 400 nm to 500 nm. No exception to this "rule of thumb" can be noticed in the open literature for $\approx \langle 001 \rangle$ oriented components (the usual crystallographic orientation along the profile of blades) while it is known that the optimal precipitate edge length in terms of creep resistance for $\approx \langle 111 \rangle$ oriented SX specimens is around 200 nm [212]. This optimal precipitate size and regularity is also one of the targets to be reached for serviced components using rejuvenation procedures involving a ST and/or HIPping procedure after several hours of use [1244,1213,1243,1610].

However, processing of components involves many different steps in addition to the "state-of-the-art" solution and aging heat treatments and industrial practices may lead, sometimes, to variation in cooling rates at

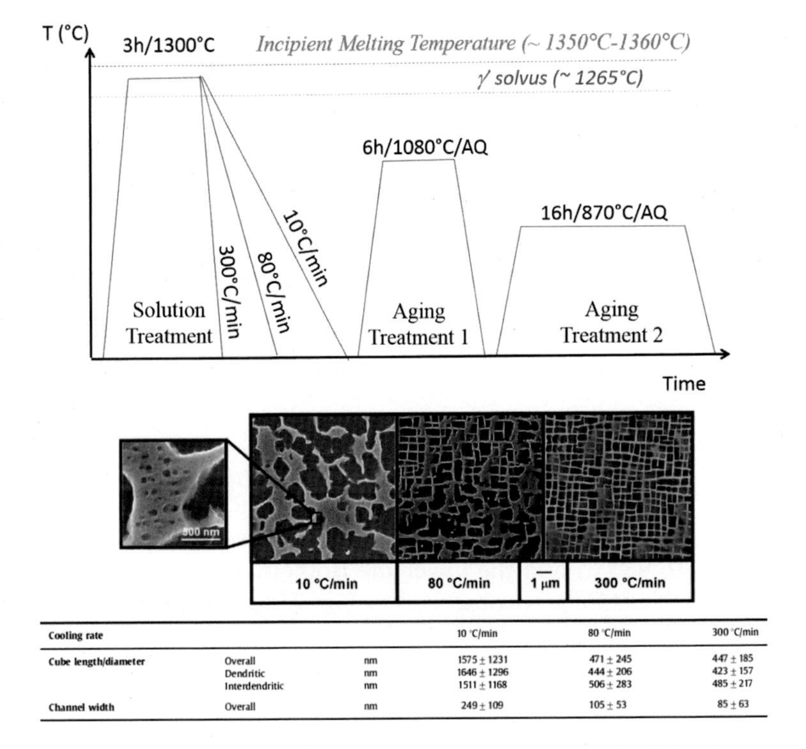

Cooling rate			10 °C/min	80 °C/min	300 °C/min
Cube length/diameter	Overall	nm	1575 ± 1231	471 ± 245	447 ± 185
	Dendritic	nm	1646 ± 1296	444 ± 206	423 ± 157
	Interdendritic	nm	1511 ± 1168	506 ± 283	485 ± 217
Channel width	Overall	nm	249 ± 109	105 ± 53	85 ± 63

FIGURE 7.15 Evolution of γ/γ' microstructure as a function of the ST cooling rate for AM3 1st generation SX alloy. Adapted from [1361].

the end of, e.g., the solution treatment. If the effect of the ST cooling rate has been widely studied for polycrystalline disk alloys [105,796,1420], very few papers have attempted to investigate this parameter on creep properties of Ni-base DS/SX alloys [257,1361]. Fig. 7.15 presents the consequences of this parameter on creep properties of AM3 alloy. The solution cooling rate was varied from 10°C/min up to 300°C/min, covering possible cooling rates encountered for high pressure turbine blades used in industrial gas turbines, large civil aeroengines, and turboshaft engines for helicopters. The slower the cooling rate, the coarser the γ' precipitation and the more irregular the morphology of precipitates. At the fastest cooling rate, the optimum cuboidal microstructure is obtained as observed in Fig. 7.15. For an intermediate cooling rate of 80°C/min, the precipitation is irregular across the dendritic structure, with cuboids within primary dendrite arms (due to a locally higher γ/γ' misfit resulting from the remaining chemical microsegregation) and a coarser precipitation within interdendritic spacings. For the slowest cooling rate, very coarse and interconnected precipitates are obtained, with a second population of very fine secondary γ' precipitates. In this study, it was shown that the slower

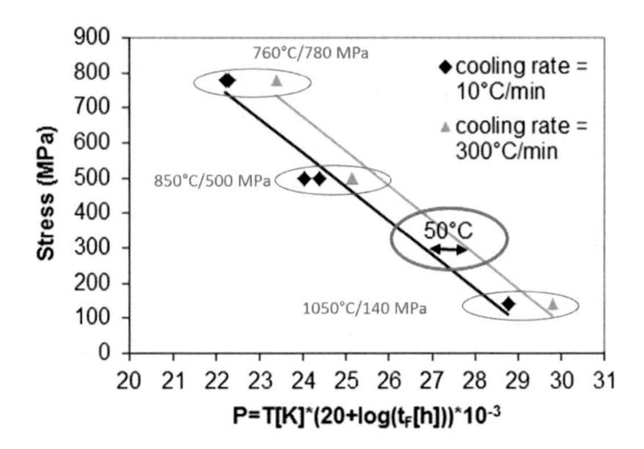

FIGURE 7.16 Creep properties dependence (Larson–Miller diagram) of AM3 1st generation SX superalloy as a function of the ST cooling rate. Adapted from [1361].

the cooling rate, the lower the yield stress in tension from room temperature up to 950°C, and the lower the isothermal and non-isothermal creep properties [1361]. As observed in Fig. 7.16, having an ST cooling rate of 10°C/min corresponds to a spectacular decrease in creep strength of nearly 50°C in all the investigated temperature range (i.e., from 750°C up to 1050°C) compared to an ST cooling rate of 300°C/min. The creep properties with an ST cooling rate of 80°C/min lie in-between these two extremes, closer to 300°C/min cooling rate [1361]. It is then of the utmost importance to have the fastest possible cooling rate after the solution heat treatment to achieve a fine distribution of precipitates that can be adjusted in morphology and size afterwards by aging heat treatments. This is the reason why most of solution heat treatment furnaces are equipped with forced convection devices (e.g., gas fan cooling systems) to achieve the desired cooling rates, especially for large industrial gas turbines components. Coupled fluid-thermic modeling of the cooling gas flow in such furnaces is highly desired to optimize location of components and their cooling histories [287,288].

To summarize, creep properties of Ni-base SX superalloys are mainly dependent on the regularity of the γ/γ' microstructure across the dendritic structure and the ability to reach a cubical γ' whose average edge length is in the 400 nm to 500 nm range, i.e., on the solution heat treatment (mainly) and agings. It is, however, weakly dependent on the size of the casting pore size, as observed by Steuer et al. [1362]. A large casting pore size and heterogeneous distribution of pores as observed in Fig. 7.9(a) may only affect creep ductility and lifetime, but with a limited effect on both primary and secondary creep stages. A typical maximum acceptable porosity level for SX components from a creep point of view is in the 0.5–1% range.

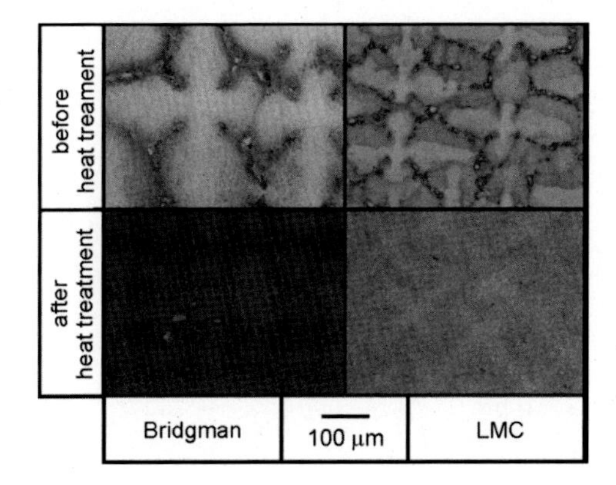

FIGURE 7.17 Dendritic structure of AM1 1st generation SX alloy after Bridgman and LMC processing. Adapted from [1362].

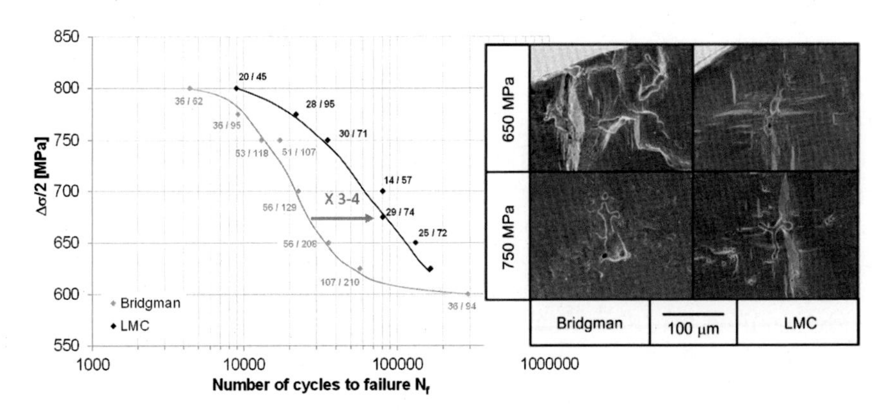

FIGURE 7.18 LCF properties of AM1 1st generation SX alloy at 750 °C/R_σ = 0.05/f = 0.5 Hz as a function of the solidification process (a) and crack initiation sites (b). Numbers in (a) for each datapoint correspond to the size of the pore serving as a crack initiation site expressed as either the diameter of the smallest circle in which the pore can be included or the projected are of this pore. Adapted from [1362].

7.4.2 Fatigue properties

Fatigue properties, in the absence of any contribution of oxidation, are mostly dependent to the solidification defects, and more particularly to the casting pore size [1362,251,887]. One of the pioneering study has been conducted by Lamm and Singer using PWA 1483 1st generation SX alloy used for land-based gas turbines for power generation, showing that the finer the dendritic structure (especially primary dendrite arm spacings), the higher the high cycle fatigue (HCF) at 800°C/R_σ = −1/f = 95 Hz [785]. Such an issue was deepened later on by C. Brundidge, using Rene N5 2nd generation SX alloy, and a modified Ta-rich version of this alloy and

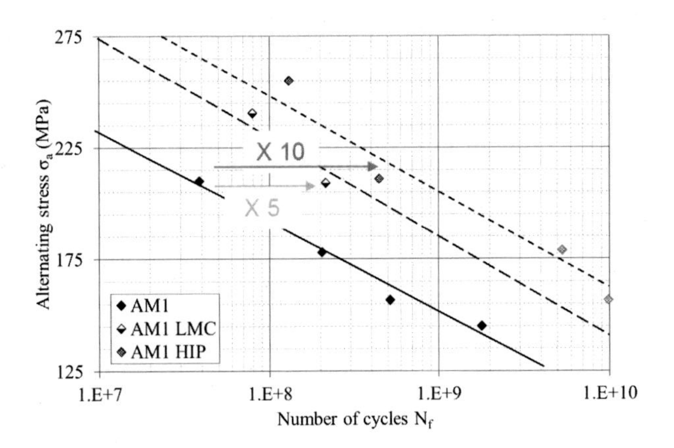

FIGURE 7.19 Effect of the pore size on VHCF properties of AM1 1st generation SX alloy at 1.000°C/$R_\varepsilon = -1/f = 20000$ Hz. Adapted from [223,221].

both "standard" Bridgman solidification and liquid metal cooling (LMC) [165,164,166]. As expected, increasing the solidification rate led to a finer dendritic structure (see Fig. 7.17) and to smaller casting pores, especially the largest ones, the volume fraction of pores being almost the same. As a consequence of this decrease in the largest pore size, an improvement in low cycle fatigue (LCF) properties was obtained, while creep properties were hardly affected by the change in solidification process. A better understanding was then obtained by S. Steuer et al. [1362] and, more recently, by A. Cervellon et al. [222,223]. Indeed, as observed Fig. 7.18, the smaller the pore size, the higher the LCF life of AM1 1st generation SX alloy (see the composition in Table 7.1) at 750°C/$R_\sigma = -1/f = 0.5$ Hz. An improvement of up to a factor 4 can be achieved in this condition using the LMC solidification process. At higher temperatures (e.g., 950°C and more), when oxidation is more active, this benefit is not observed anymore since crack initiation occurs at the surface. It was, however, shown that the casting pore size (and hence the solidification process) intrinsically controls the fatigue life at such a high temperature if tests are performed in high vacuum [1362] or at high frequencies [224,222,223], such as the ones encountered by airfoils during transient engine regimes corresponding to the excitation of one of their fundamental resonance mode. Indeed, Fig. 7.19 shows that the very high cycle fatigue (VHCF) life of AM1 1st generation SX alloy can be improved by a factor of 5 using an LMC solidification process compared to a Bridgman one. Applying a HIPping process is even more efficient according to this figure, by (almost) closing all casting pores. As a fundamental understanding, this improvement in fatigue properties by reducing the casting pore size results from a longer time spent in the very short crack propagation (the so-called micropropagation domain) which can be taken into account using newly developed fatigue life criteria [1362,224,133,222,223].

In summary, fatigue properties are mainly dependent on the solidification process when environment has a limited contribution to the crack initiation processes. Accelerating the solidification process is then a very reliable way to increase service life of Ni-base SX components, especially for internally cooled ones known to be very sensitive to fatigue cracking during their service life [1648].

7.5 Conclusions

Presentation has been given of the general features dealing with casting and heat treatments used for the production of SX components, including historical background and current trends, mainly dealing with modeling and simulation. Based upon historical perspectives, two main improvements can be highlighted considering the use of nickel-base superalloys: (i) alloys with always better creep properties at increased operating temperature and (ii) complicated air cooling systems as part of the component design. These were the results of collaborations between metallurgists and mechanical engineers, while processing had to adapt and find solutions to deliver the products. The yield rate is still low for single crystal casting, typically reaching 70% as a consequence of the demand for very high quality. And because of the evolution of the alloy compositions, further steps are required for the products, such as the deposition of a thermal barriers coating. Yet this was not part of the criteria for the design of the last generation superalloys and many such alloys could not be introduced in engines because of the absence of thermal barriers coatings. So despite the fact that metallurgist could demonstrate always better intrinsic mechanical properties of new alloys composition, other factors prevented their use. Processing capabilities and the entire processing route were not or very little considered so engineers and processing teams had to adapt forming processes of the alloys from its original formulation. As a consequence, alloys were proposed with enhanced creep properties at high temperature. However, no coating could be designed for these alloys which are not in-use for SX blades. This shows that an SX component is a very sophisticated item that, more than in the past, needs collaborative efforts to bring innovative ideas into a working product. Enhanced synergetic strategies are required that would bring engine efficiency always higher as a result of a team work gathering teams of process engineers, metallurgists, and physicists to create enhanced materials science products.

Acknowledgments

The authors would like to thank Dr. Boyd Mueller of Arconic Inc. and Dr. Virginie Jaquet of SAFRAN Tech for providing some of the illustrations. Dr. Jérémy Rame (SAFRAN Aircraft Engines), Dr. Zéline Hervier (SAFRAN Helicopter Engines), and Dr. Lorena Mataveli Suave (SAFRAN Tech) are also acknowledged for fruitful discussions and technical exchanges on heat treatments and mechanical characterizations of Ni-base SX superalloys. JC is grateful to Pr. Tresa M. Pollock (University of California – Santa Barbara/Materials Department) for sharing joint experiments on LMC processed Ni-base SX superalloys and to Dr. Alice Cervellon (formerly at ISAE-ENSMA/Institut Pprime and now at the University of California – Santa Barbara/Materials Department) for her "VHCF" contributions. Luciana Bortoluci Ormastroni (M.Sc. from ISAE-ENSMA, currently PhD student at ISAE-ENSMA Institut Pprime) is acknowledged for her contributions on solution heat treatments.

CHAPTER

8

Aging

Jean-Briac le Graverend[a,b], *Damien Texier*[c], *and Vincent Maurel*[d]

[a]Department of Aerospace Engineering, Texas A&M University, College Station, TX, United States, [b]Department of Materials Science Engineering, Texas A&M University, College Station, TX, United States, [c]Institut Clement Ader (ICA) - UMR CNRS 5312, Universite de Toulouse, CNRS, INSA, UPS, IMT Mines Albi, ISAE-SUPAERO, Albi Cedex 09, France, [d]MINES ParisTech, PSL University, MAT - Centre des Materiaux, UMR CNRS 7633, Evry Cedex, France

8.1 Microstructure evolution in the bulk and effect on the mechanical performance

SX superalloys are exposed to diverse temperature/stress conditions, as well as various environments. This section is dedicated to what happens in the bulk when the applied temperature/stress condition triggers microstructure evolutions that modify the mechanical performance, i.e., mechanical behavior and lifetime. The most well-known microstructure evolution is the directional coarsening of the γ' phase, also known as rafting, and takes an important part of this section. However, other microstructure evolutions, such as the precipitation of intermetallic phases and the dissolution/precipitation of the γ' phase, are also presented. For each microstructure evolution, the effects on the mechanical performance are discussed.

8.1.1 Topologically close-packed phases

The improvement of the mechanical properties of single-crystal superalloys, especially creep, is often obtained by an increase in the content of refractory elements, such as W, Mo, and Re. However, an excessive concentration of these elements, which are aimed at slowing deforma-

Nickel Base Single Crystals Across Length Scales
https://doi.org/10.1016/B978-0-12-819357-0.00016-0

tion rates, can lead to the formation of intermetallic phases (σ, μ, P, and R), called topologically close-packed phases (TCP phases) [202,395]. TCP phases preferentially precipitate in the dendrites due to the residual segregation of alloying elements, such as Re, W, Mo, and Co [1162,1500,1161], that cannot be completely suppressed by heat treatments [1563,701]. From the chemical composition of an alloy, it is possible to predict the formation of TCP phases via a PHACOMP [991] or newPHACOMP [966] calculation. The methods are based on atom solubilities estimated from their electronic structure, more precisely, the nature of their atomic orbital d. These unidirectional calculations are, however, erroneous in the case of new generations of superalloys. This is why 2D calculations were developed to better predict the formation of these phases [1297].

In the case of the first-generation superalloy MC2, which contains neither rhenium nor ruthenium alloying elements, the TCP phase that preferentially precipitates at high temperature (950°C $< T <$ 1150°C) is the μ phase (rhombohedral crystal structure, $4.75 \leq a \leq 4.80$ Å and $\alpha = 31°$). This phase is rich in W, Mo, Co, and Cr [372,1329], precipitates along the $\langle 110 \rangle$ orientation on $\{111\}$-type planes, and has three types of morphology (needle, plate, and sphere).

The μ phase precipitates semicoherently with the matrix, which can explain its strong influence on the surrounding microstructure. Moreover, the kinetics of precipitation of the μ phase is fast. For instance, it appears only after 20 to 50 hours at 1050°C without any mechanical load for the second-generation alloy RR2071 (composition close to the CMSX-4 [1162]). The precipitation is favored by a stress in tension (unfavored in compression), as shown for the CMSX-4 alloy and a similar alloy at 1050°C and 950°C [249,248]. It should be pointed out that the precipitation nose of the μ phase is often close to 1050°C, as shown in Fig. 8.1 for the RR2071 alloy.

The TCP phases seem to be at the origin of three damage modes:

- A loss of ductility due to their fragile nature that favors the initiation and propagation of cracks, and hence accelerates the final failure [975, 1373],
- A softening of the γ matrix because of a decrease in the concentration of refractory element contained in the matrix next to the TCP phases [1161, 885],
- A local distortion of the rafted microstructure, as observed during high-temperature creep [1329,1152,801].

In addition, some authors [297,1194,807] showed that TCP phases are often accompanied by the nucleation of pores in their neighborhood. Studies also pointed out that their effect on the mechanical behavior was noteworthy from a certain critical volume fraction [1082,1163]. TCP phases are always considered as fragile in the context of studies conducted around 900°C [1105,1501]. However, Keitz et al. [1508] showed that certain TCP

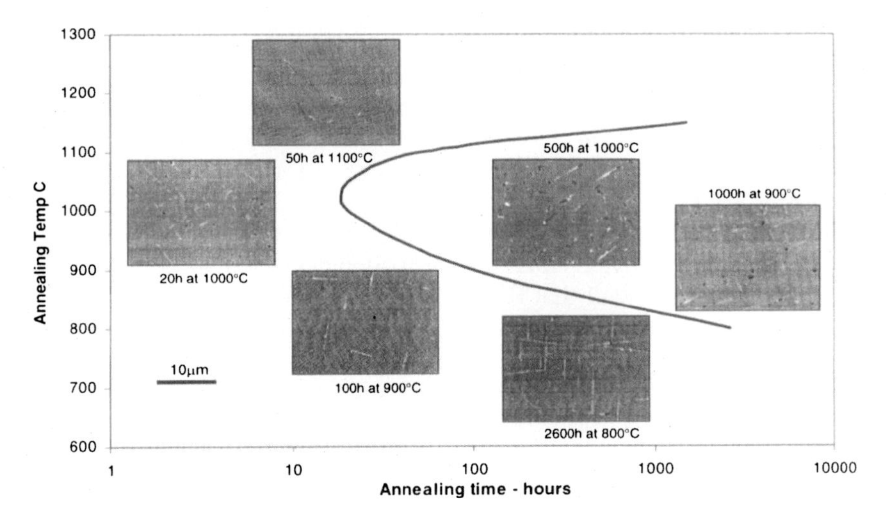

FIGURE 8.1 TTT diagram of the μ phase in the RR2071 alloy. Adapted from [1162].

phases, namely Laves phases, can plastically deform above 1050°C. As a result, it is important to determine the ductile-to-brittle-transition temperature (DBTT). The previous authors showed that the DBTT of TCP phases is about $0.65T_m$ (T_m is the melting temperature that can be obtained via a phase diagram). Despite no available data for TCP phases, it is, however, possible to use what is known for the phase Co_7W_6 that is isomorph to the μ phase. The Co_7W_6 melting point is between 1471 and 1689°C [1425]. It is, therefore, possible to deduce that the DBTT for the μ phase is about 1030°C. Nonetheless, Machon et al. [883] demonstrated that the DBTT is strain-rate dependent: the slower the plastic strain rate, the lower the DBTT. For instance, an alloy having a DBTT of 1400°C for a strain rate of 10^{-3} s^{-1} will have a DBTT of 980°C for a plastic strain rate of 10^{-7} s^{-1}. These phases introduce microstructure destabilizations, as it was observed after creep tests at 1050°C on coated [1166] and uncoated [801] $\langle 001 \rangle$-oriented specimens made of the first-generation MC2 alloy. Indeed, it is usually considered that TCP phases cannot be cut by slipping dislocations [1618], which produces inhomogeneous deformations and stress concentrations around the TCP phases. In fact, wide bands (1 μm to 4 μm thick), starting at the TCP tips and propagating parallel to the phase plane over the entire thickness of the sample, was observed. Despite several carried-out studies, no consensus could be reached on whether or not the local microstructure destabilizations induced by the TCP phases affect the mechanical creep properties [1500,1530].

8.1.2 Isotropic coarsening of γ'

The isotropic coarsening of the γ' phase is when the coarsening happens the same way in all directions, contrary to directional coarsening that will be discussed in the next section. Isotropic coarsening is due to Ostwald ripening, as shown by Chen and Immarigeon [244]. It is a time-dependent process that does not depend on the strain/stress state of the material, but depends on the temperature since it is a diffusion process. The driving force for the isotropic coarsening is the minimization of the γ/γ' interfacial energy.

If some authors accounted for isotropic coarsening in their rafting models [430,1438,805], Oswald ripening is still often neglected, although experiments [624,396] showed that the microstructure periodicity increases up to 20% during rafting. This increase in the microstructure periodicity indicates that the γ/γ'-microstructure coarsens on a more global scale: the larger precipitates increase in size at the expense of the smaller ones. The theory of volume diffusion-controlled coarsening developed by Lifshitz and Slyozov [841], as well as by Wagner [1518], known as the LSW theory, is employed to describe the isotropic coarsening. The LSW theory predicts that the average particle radius increases with time according to the following equation:

$$r^n - r_o^n = k(t - t_0), \tag{8.1}$$

where r_0 is the initial average particle radius before the coarsening starts at t_0, k is a temperature-dependent rate parameter, and n is an exponent.

In the case of nickel-base single-crystal superalloys, the γ' volume fraction varies from 70% at 950°C to 30% at 1200°C, depending on their composition. The γ' volume fraction is therefore high, whereas the LSW theory assumes a low volume fraction. Moreover, this theory was developed for a binary system, in particular Ni–Al, and considered spherical particles. This is not the case for superalloys both in the initial state (cuboidal precipitates) and in the rafted states. Nevertheless, Ardell et al. [35] showed that the hypothesis of a diffusion-controlled mechanisms for the precipitate growth remains valid in the case of Ni-base single-crystal superalloys with a high volume fraction of nonspherical γ' precipitates. The experimental results from Brailsford et al. [148] suggest that it is still possible to use an exponent equal to 3 for single-crystal superalloys. Recently, Fedelich et al. [428] modified the isotropic coalescence law to have an analytical model with an exponent of the power function equal to 0.0745. The authors justified the less-than-1 exponent by taking into account the shape of the γ' precipitates in the coalescence process. For temperatures less than or equal to 900°C, the effect of the isotropic coarsening is considered negligible. More details on Ostwald ripening for superalloys can be found in [66,67].

8.1.3 Directional coarsening of γ'

Rafting is a phenomenon that occurs for temperatures above 800°C, even if rafting was also observed at lower temperatures and sufficiently high applied stresses [1432,1127,1321]. This phenomenon is triggered once some amount of plastic deformation is reached (e.g., 0.10% ± 0.03% at 950°C) [600,909,1456]. Beyond a threshold, the rafting process can continue even in the absence of a mechanical load. During rafting, the initial cuboidal microstructure is transformed and there is a directional coalescence of the precipitates. This directional coalescence leads to precipitates forming platelets perpendicular (N-type)/parallel (P-type) to the loading axis in the case of ⟨001⟩-oriented specimen with a negative/positive lattice misfit and subjected to a uniaxial positive applied stress, and vice versa for compressive loads [1488]. For ⟨011⟩ [1430] and ⟨111⟩ [1626], the platelets form parallel to the {011}-type plane. This morphological evolution was also observed on turbine blades by Draper et al. [358] on the NASAIR 100, or more recently by Sujata et al. [1374] on a first-generation alloy. The microstructures observed on structural components are more complicated to interpret because of the complex thermo-mechanical and multiaxial loading introduced by the intricate shapes of the turbine blades, cooling channels, and thermal gradients. These complex turbine blade designs have led researchers/engineers to take interests in simpler loads responsible for microstructure evolutions, such as uniaxial creep and fatigue, to ultimately perform simulations on realistic structural components [374].

8.1.3.1 Directional coarsening during creep

A thorough description of the rafting process during creep was done by Fredholm and Ayrault [466,57]. They studied the deformation mechanisms put into play during high temperature tests ($T \geq 1050°C$) and correlated them with microstructure evolutions. Three stages can be discerned for high-temperature/low-stress creep loading, even if alloys with large amounts of refractory elements, i.e., high natural lattice misfit values, may show a limited primary creep stage and a reduced secondary creep stage [348]. In the following, the rafting process is described for uniaxial creep tests in tension on ⟨001⟩-oriented samples with negative lattice misfits:

- Microscopically, the local stress is different in the horizontal and vertical channels: compression in the vertical channels and tension in the other ones [690]. Because of the stress gradient, diffusion phenomena are taking place: the γ-forming elements diffuse towards the horizontal channels while the γ'-forming elements move towards the vertical channels. This diffusion process aims at establishing a thermodynamic equilibrium and is the reason for the directional coalescence starting at the corners of the precipitates [1387,372]. The primary stage also leads to an increase in plastic deformation, i.e., in the density of dislocations that propagate in the γ matrix until saturation [1142] and their blockage at the γ/γ' interfaces by creating a network that relaxes the coherency stresses [1372,847]. The interfacial dislocation network also locally modifies the γ chemical composition that helps the directional coarsening [170]. The end of the primary stage corresponds to a well-established rafted microstructure, as showed by image analysis in [886]. Indeed, rafts will prevent the glide of perfect dislocations in the vertical γ matrix channels. Thus, glide and climb of dislocations will be reduced, which leads to the decrease of the strain rate [1126]. The rafting kinetics depends on the chemical composition (slower for alloys with heavy alloying elements [520,1438]) and the elastic anisotropy [1324] (faster for larger values of the anisotropy) of the alloy.

- The secondary stage is characterized by a constant and minimum steady strain rate (of the order of 10^{-5}–10^{-6} h^{-1} in the case of the first generation superalloy MC2 at 1050°C for a stress between 140 and 200 MPa [277,283]). The duration of the secondary stage is as long as the rafting process is fast. From a microstructural standpoint, the secondary regime is characterized by a competition between directional coalescence and Ostwald's ripening, as demonstrated on the SRR 99 at 980°C/200 MPa [397] by revealing the mechanisms responsible for the evolution of the period λ representing the width of a γ' precipitate and a γ channel along the [001] direction during a creep test. At the deformation-mechanism level, the primary–secondary transition corresponds to the establishment of stable dislocation networks at the γ/γ' interfaces. In fact, Probst-Hein et al. [1143,354] showed that a limited number of dislocations is admissible in the γ matrix by calculating the Peach–Koehler force. They found a critical shear of 0.54%, which is significantly lower than the strain values experimentally observed for a shear creep ($\tau = 85$ MPa) at 1025°C [922]. The only way to reach this amount of plasticity is the existence of dynamic recovery processes due to either the climb of dislocations along the γ/γ' interfaces or the shearing of the γ' precipitates by dislocations that subsequently annihilate [1388,1350]. The interfacial dislocation networks are particularly important for the performance of superalloys. In fact, they create a local field of constraints that is as high as the network is fine. In fact, the more homogeneous and fine the dislocation network is, the more the network protects the γ' precipitates from being sheared by dislocations propagating in the γ channels. Zhang et al. [1642,1640,1641] also correlated the deformation rates during the secondary regime with the size of the cells constituting the interfacial dislocation networks on TMS alloys (developed at the National Institute for Materials Science). Concomitantly, the same authors also showed that the size of the cells depends on the concentration of refractory elements, i.e., on the lattice misfit value: the higher the concentration of refractory elements, the finer the dislocation cells [1642]. The transition from secondary to tertiary stage is linked to a massive shearing of the γ' precipitates. This shearing is potentially due to two factors: (1) destabilization of the γ/γ' dislocation network, which will allow the shearing of the γ' precipitates [1372] and (2) destabilization of the microstructure during which the γ' phase becomes the matrix. This last phenomenon is called topological inversion [393] and opens up new sites for γ' shearing, as shown by Clément et al. at 850°C during a comparative *in-situ* study on the effects of cuboidal and rafted microstructures on the mechanical response [260]. This process is not observed when the γ' volume fraction is smaller than 50% [213,580].

- The tertiary stage is characterized by a rapid increase in the plastic strain rate, which ultimately leads to the final failure. It results from multiple factors. One is related to the deterioration of the microstructure (topological inversion, for example) and the massive shearing of the γ' precipitates, which causes mobile dislocation multiplications. The environment (oxidation), which modifies the load bearing section of the material by creating a γ' depleted zone [366] and favores pore growth due to vacancy injections, also contributes to the damage process. Indeed, as expressed before, the γ' volume fraction is of great importance when it comes to mechanical performance at high temperatures. The effect of oxidation is even more critical when the aspect ratio of the samples (perimeter/section) is important [1236]. The transition from secondary to tertiary creep stage can also be linked to a higher effect of pores whose volume fraction continuously increases during the previous stages. When their volume fraction is sufficiently high, i.e., when the effective stress in the material is high enough based on a continuum damage mechanics (CDM) approach, microcracking occurs at pores, which ultimately leads to macrocracking. Cracking starts at the largest pores [807,169] that are usually from casting. However, there are two types of pores in the bulk [395,1353]:
  - The pores that are formed during casting and still present after solutioning and aging in the interdendritic regions, namely solidification and homogenization pores. They are formed because of the difference in the diffusion coefficients between alloying elements. They are relatively large, viz. 15 μm equivalent radius [124,1560].
  - The pores formed because of plasticity and, more specifically, the coalescence of vacancies transported by dislocations [392]. This is why they are known as deformation pores. They are present in both dendrite and interdendritic regions, are easily recognizable due to their faceted surfaces, and are relatively small, viz. below 5 μm equivalent radius [848].

8.1.3.2 Directional coarsening during fatigue

Rafting is not only observed during creep. In fact, it happens when a mechanical load is applied for a sufficiently long period of time and for a temperature high enough ($T > 850$°C). Thus, mechanical and thermal fatigue loading can lead to rafting. Rafting was observed during mechan-

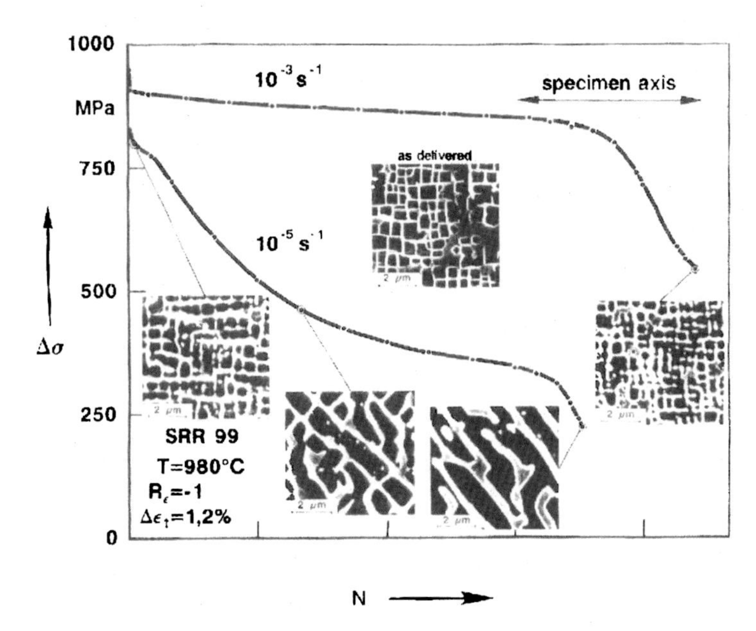

FIGURE 8.2 Microstructure evolution during strain-controlled alternate fatigue loading ($R_\varepsilon = -1$) for several stress amplitudes at 980°C on the alloy SRR 99. Adapted from [470].

ical fatigue [470,623,803] and during thermo-mechanical fatigue [975,759]. Fig. 8.2 shows the microstructure evolution that could be observed during a slow alternate cyclic loading ($\dot{\varepsilon} = 10^{-5}$ s^{-1} and $R_\varepsilon = \frac{\varepsilon_{\min}}{\varepsilon_{\max}} = -1$) at 980°C on the SRR 99 alloy. The microstructure evolves to be oriented at 45° from the loading axis. Thus, half of the time the microstructure tends to be perpendicular to the loading axis and the other half to be parallel to the loading axis leading to the 45° tilt. For faster cyclic loadings ($\dot{\varepsilon} = 10^{-3}$ s^{-1} and above), the time necessary for rafting is not met. Therefore, directional coarsening was not observed, but rounding at the corners of the precipitates was noticed. It is due to a higher density of dislocations leading to a greater amplitude of the thermo-elastic stresses because of the coherency between both phases [524,491,985]. However, Brien et al. [156,155] observed by TEM a coalescence normal to the loading axis in the AM1 alloy at 950°C and for a frequency of 0.25 Hz ($\dot{\varepsilon} \in [10^{-3}, 10^{-2}]$ s^{-1})). Furthermore, the same authors also noticed a difference between repeated and alternate cycling: rafting occurs more quickly for alternate (tension and compression components) cycling (in about 113 cycles) than for repeated (tension or compression component) cycling (in about 200 cycles). Finally, if the deformation amplitude is high, the mean stress is also high. Such severe loading conditions prevent rafting because of the massive shearing of the γ' precipitates, leading to short lifetime. In the case of in-phase and out-of-phase thermo-mechanical fatigue tests, Kraft showed that the orientation of the rafts, namely parallel or perpendicular to the loading axis, depends on the nature of the loading (traction or compression) during the hot phase of the heat cycle [759].

8.1.3.3 Directional coarsening during multiaxial loading

A few studies were carried out to understand the rafting process under multiaxial stress conditions. The effect of stress multiaxiality on γ' rafting is usually studied via notched specimens that create multiaxial mechanical fields without the need to have a torsional or a biaxial frame [1555,805,81, 198,1446]. It is, however, possible to find papers on torsion [1025,1023,888, 1256] and pure shear [379,921,742,741].

Basoalto et al. [81] performed creep tests at 850°C on CMSX-4 specimens with a double Bridgman notch [154]. The authors performed EBSD analyses and noticed crystallographic rotations of up to 20° for ⟨001⟩- and ⟨111⟩-oriented samples (see Fig. 8.3). The same result was also observed by MacLachlan et al. [888] and le Graverend et al. [805,811]. It was also pointed out that a crystallographic rotation is accompanied with a rotation of the rafts of the same magnitude. Crystallographic rotations are attributed to large deformations as mechanical twins when plasticity cannot be accommodated for temperatures up to 950°C [130,31], but also above [859,811]. In addition, MacLachlan et al. [888] noted on the CMSX-4 alloy that tension/torsion creep tests have longer lifetime than pure torsion creep tests, themselves longer than uniaxial tension creep tests for an effective stress of 360 MPa at 950°C. The authors argued that the kinetics of damage is slower in torsion than in tension which could be related to a nucleation of pores slower in torsion than in tension, as Dyson et al. already pointed out in the case of polycrystalline alloy Nimonic 80 at 750°C [376]. The effect of stress multiaxiality on the directional coalescence was also investigated by considering the microstructural state in the vicinity of pores [286] or holes [1661]. In fact, pores generate multiaxial mechanical fields that locally modify the kinetics and orientation of rafts, as showed by Cormier et al. for a specimen made of the MC2 alloy and crept at 1050°C/140 MPa [286]. It is, hereby, essential to mention that the rafts formed under multiaxial loading verify the same rule as uniaxial loading, viz. they are oriented perpendicular/parallel to the local positive/negative maximum principal stress vector.

Even if recent studies were dedicated to predict rafting in 3D [337,1438], Valles and Arrell were the first trying to predict rafting under complex loading conditions on SRR 99 [1460,41]; while Pineau paved the way for uniaxial loading [1108]. In fact, they defined a criterion based on the minimization of energy at the γ/γ' interfaces by dislocations, which allows

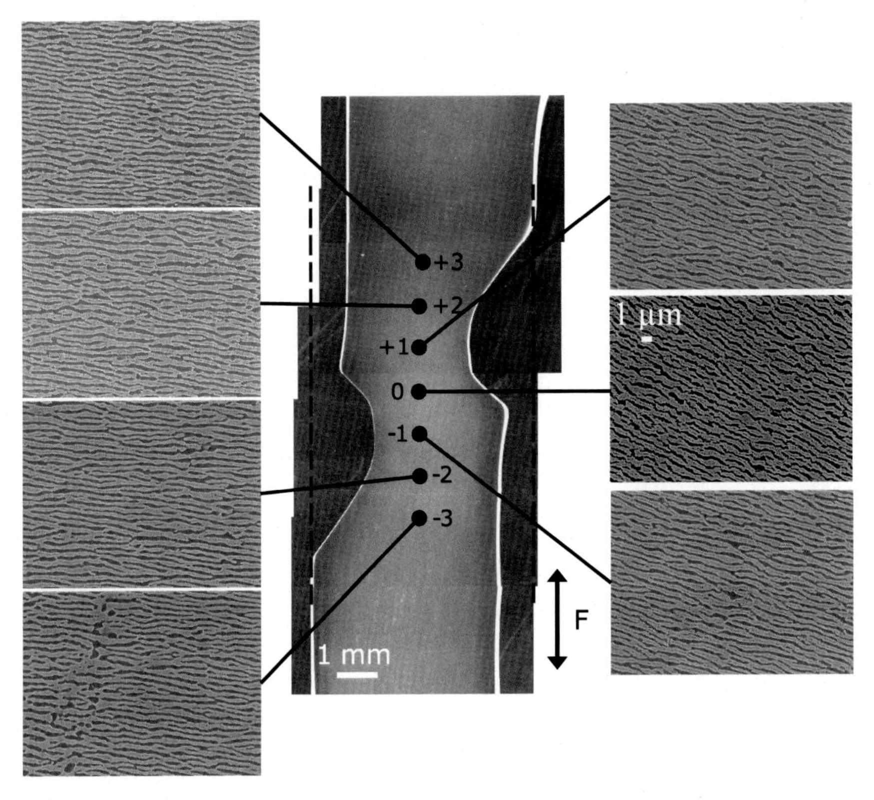

FIGURE 8.3 Longitudinal observations (1 mm apart) of the γ' rafted microstructure at the end of a nonisothermal creep test at 1050°C, $F = 410\,N$. Adapted from [805].

determining the orientation of the rafts according to the applied multiaxiality. Their study highlighted three results:

- It is possible to determine a domain in the principal stress space where the morphological evolution of the γ' is isotropic. This domain exists whichever the temperature, but its area decreases with temperature.
- The application of a third positive stress component translates, depending on the sign of this one, the domain for which the morphological evolution is isotropic.
- Stick-shape precipitates are favored at low temperatures (700°C) while platelet-shape precipitates are privileged at high temperatures (900°C). Therefore, rafting is a thermally-activated process.

Yue et al. [1633] have improved this model to take into account shear stresses to account for the microstructure evolutions and crystalline orientations during pure torsional loads.

8.1.4 Dissolution/precipitation

During nonisothermal tests at very high temperatures, the γ' volume fraction is modified. It is, therefore, essential to study the evolution of the γ' volume fraction depending on the temperature, holding time, and applied load. In fact, the γ' volume fraction has a strong effect on the mechanical response, as shown by Murakumo et al. [990] for creep on the TMS-75 alloy that optimized creep performance for 70% of γ' and for temperatures below 950°C. Numerous studies were conducted to determine the γ' volume fraction at its thermodynamics equilibrium via different techniques, such as image analysis [1302], X-ray diffraction [1238], electrical resistivity [1227], atom probe [347], and thermodynamics calculations [1345]. See Fig. 8.4.

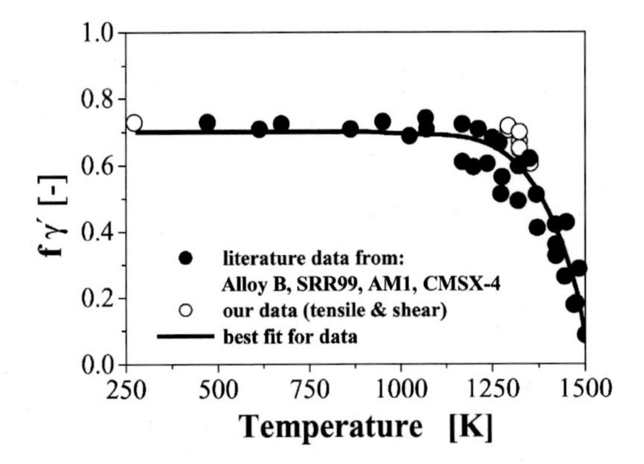

FIGURE 8.4 Evolution of the γ' volume fraction for several alloys as a function of the temperature. A best-fit curve was added considering that the γ' thermodynamic equilibrium is the same for all the alloys. Adapted from [1302].

It is important to point out that the volume fractions obtained at the thermodynamic equilibrium, i.e., for long thermal holds, are not realistic of thermo-mechanical loading involving rapid temperature variations for which equilibrium is not reached. The material is in a transient metastable regime. Cormier [277] studied the temporal evolution of the γ' volume fraction for a given temperature and showed that a massive dissolution of the precipitates is being achieved for short hold times and for temperatures above 1050°C (Fig. 8.5(a)). The author also showed that the higher the mechanical loading, the faster the dissolution kinetics. This phenomenon is due to a modification of the state energy of the γ/γ' system, which would modify the state of thermodynamic equilibrium and, consequently, the kinetics of dissolution. This observation has to be compared with the mechanism of dislocation-assisted precipitation developed by Embury et al. [386] and better known as pipeline diffusion. In fact, when dislocations move in the γ channels, they bring behind them an atomic cloud, also known as Cottrell cloud, which favors their diffusion. Giraud et al. [521] also revealed a difference of dissolution kinetics between dendrites and interdendrites because of the inhomogeneity of the microstructure. This difference, however, would not be related to differences in the diffusion coefficients between the two areas that have different chemical compositions.

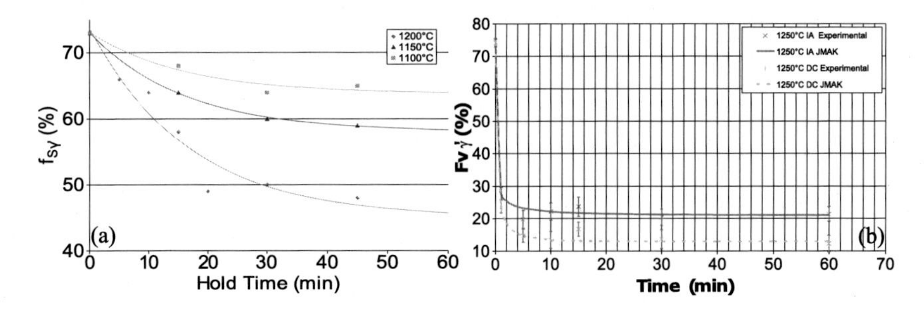

FIGURE 8.5 (a) Evolution of the γ' volume fraction for three levels of temperature, viz. 1100°C, 1150°C, and 1200°C, for the MC2 alloy – adapted from [282]. (b) Evolution the γ' volume fraction between dendrite and interdendrite at 1250°C for the CMSX-4 alloy – adapted from [521].

It is hereby important to point out that non-isothermal loadings come with the dissolution and the precipitation of fine γ' precipitates in the γ channels for temperatures above 975°C, i.e., for temperatures for which the γ' volume fraction evolves. This fine precipitation plays an important role with regards to deformation mechanisms. It decreases the γ channel width and, therefore, modifies dislocation glide on the slip systems, as shown for the $111\langle 112\rangle$ systems in the René 95 at 700°C [687,688], as well as for the MC2 alloy at 1050°C by recording a transient decrease in the plastic strain rate after a 30-second thermal jump at 1200°C [806,803]. The decrease in the plastic strain rate was attributed to a blockage of the dislocations by the fine precipitates that requires either an Orowan by-passing mechanism or a shearing of the fine precipitates depending on their size for the deformation process to continue. Indeed, the fine γ' precipitates can either coalesce with each other during thermal holds to form micro-rafts (Fig. 8.6) or be absorbed back by the larger γ' precipitates [144]. During the coalescence of the fine precipitates, their growth will follow the Lifshitz–Slyozov–Walter (LSW) law already discussed earlier in this chapter.

8.1.5 Effect of the directional coarsening on the mechanical behavior

As we saw in the previous sections, many microstructural evolutions happen in single-crystal superalloys at high temperatures. The effects of the TCP phases on the mechanical behavior were already discussed in Sec-

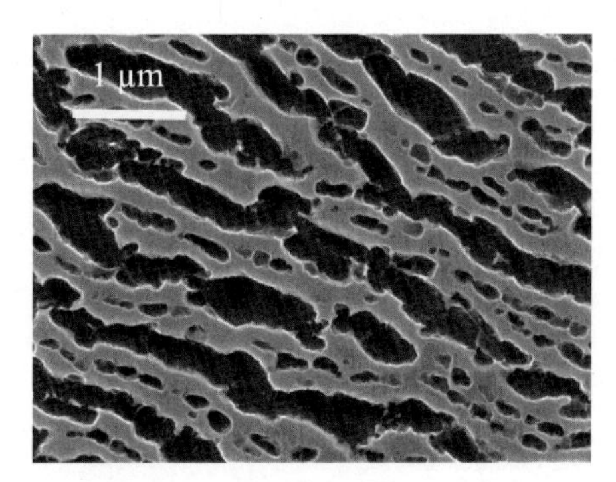

FIGURE 8.6 Fine precipitates that coalesced with each other to form micro-rafts and located at the position 0 in Fig. 8.3. Adapted from [805].

tion 8.1.1. This is why the following sections are dedicated to the effects of rafting on the mechanical behavior, more specifically on monotonic, low cycle fatigue (LCF), and creep loading.

8.1.5.1 Effect on monotonic loading

Mackay and Ebert [886] showed that rafted microstructures in the single-crystal Ni–Al–Mo–Ta alloy seem to have smaller yield (at 0.2%) and rupture stresses in tension for temperatures ranging from 930°C to 1040°C. At these levels of temperature, the shearing of the γ' phase is the primary deformation mechanism that is favored by rafted microstructures showing wider channels. Pessah-Simonetti reported the same trend in the first generation superalloy MC2 [1080]. Espié [404] studied in more details the effect of microstructure on the monotonic behavior at 650°C and 950°C on the AM1 alloy. If no effect was noticed at 650°C because of the shearing of the γ' particles occurring independently of their shape, changes were observed at 950°C. The author found that P- and N-type microstructures obtained via either creep or cyclic aging ($\Delta\varepsilon = 2\%$) had smaller ultimate strengths compared to cuboidal microstructures: the softening of the cyclic-obtained rafts being the highest (185 MPa vs. 125 MPa) for a monotonic tensile test at 950°C and a strain rate of $9 \cdot 10^{-4}$ s^{-1}. These results did not depend on the orientation of the rafts, namely P- or N-type. For the same experiment, it was observed that the yield stress for cyclic-obtained rafts was smaller than that for the creep-obtained. These results were also obtained by Gaubert [500] for the same alloy at the same temperature, as well as Fedelich et al. [428] for the CMSX-4 at 950°C. Espie [404] explained the greater softening of rafts formed by cyclic hardening by a higher density of dislocations in the γ channels. Similarly, Gaubert [500] showed that rafts formed by cyclic hardening result in greater softening, but also in stronger effects on viscosity. However, the difference between P- and N-type rafts fades away for higher levels of deformation for which the microstructural state is a second-order effect compared to deformation. See Fig. 8.7.

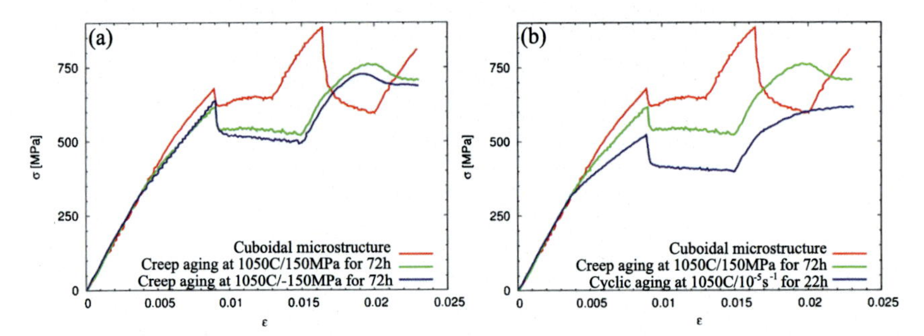

FIGURE 8.7 (a) Effect of creep aging in tension and compression at 1050°C/150 MPa on the complex mechanical behavior of the AM1 alloy at 950°C. (b) Effect of creep aging in tension (1050°C/150 MPa) and cyclic aging on the complex mechanical behavior of the AM1 alloy at 950°C. The complex monotonic tensile test consists of applying 10^{-3} s^{-1} up to 1% of deformation, 10^{-5} s^{-1} from 1 to 1.5% of deformation, and 10^{-3} s^{-1} from 1.5 to 2% of deformation. Adapted from [500].

8.1.5.2 Effect on low cycle fatigue

In a similar fashion as the effect of microstructure on the monotonic behavior, the effect of microstructure on low cycle fatigue behavior is seldom. Even if it was reported by [404] in the previous section that N- and P-type rafts give the same mechanical behavior during monotonic tensile tests, it is not the case when it comes to cyclic loading. Cyclic-obtained N-type and creep-obtained P-type rafts will have the same cyclic hardening, while creep-obtained N-type and cyclic-obtained P-type rafts will give the same softening, but softer than for the other two, at 950°C/$\dot{\varepsilon} = 9.10^{-4}$ s^{-1}). Ott et al. [1040] also investigated the influence of rafting on the LCF behavior of the CMSX-4 and CMSX-6 alloys at 950°C and 1050°C. They showed, considering that the service life is controlled by the crack micro-propagation regime, that N-type microstructures led to shorter service life compared to cuboidal microstructures, contrary to P-type microstructures (Fig. 8.8(a)–(b)). This phenomenon is due to the fact that cracks preferentially propagate along the γ channels or the γ/γ' interfaces while not shearing the γ' precipitates (Fig. 8.8(c)). Thus, their propagation is facilitated in N-type microstructures and slowed down in P-type microstructures in which the γ' rafts act as barriers to crack propagations. In addition, the mechanical behavior of the three types of microstructure, viz. cuboidal, N-, and P-type, is slightly different throughout the life of the specimens when the effect of damage can still be neglected. Indeed, the microstructural state seems to affect the average stress at the steady state cycle (Fig. 8.8(a)). Similar results have been obtained for thermo-mechanical loading in [387,726].

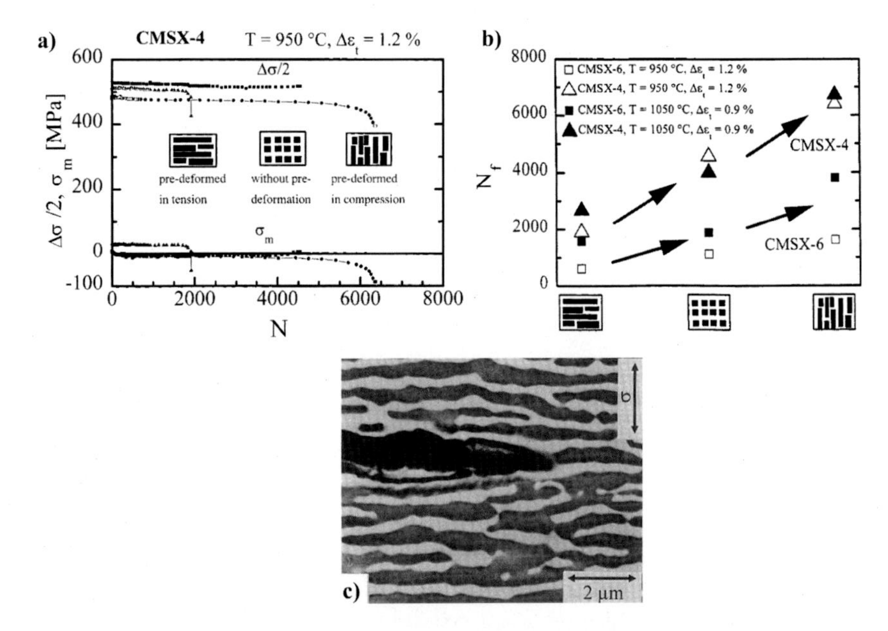

FIGURE 8.8 (a) Evolution of the stress amplitude ($\Delta\sigma/2$) and the mean stress (σ_m) for an amplitude of the imposed total strain $\Delta\varepsilon_t$ = 1.2% at 950°C on the CMSX-4 alloy and three different microstructural states, namely cube, P-, and N-type rafts. (b) Comparison of the number of cycles to failure for the CMSX-4 and CMSX-6, the two experimental conditions, and for the three microstructures. (c) Microstructure revealing a crack tip propagating along the γ/γ' interfaces during fatigue at 1050°C and $\Delta\varepsilon_t$ = 0.9%. Adapted from [1040].

8.1.5.3 Effect on creep

8.1.5.3.1 Effect on isothermal creep

The effect of a predeformation obtained by tension or compression creep on the mechanical behavior and lifetime during isothermal tension creep was respectively studied by [1000] and [1321,1411]. A predeformation in compression (0.52% and 1.46%), which leads to the formation of P-type rafts, is detrimental to the tensile properties of the AM1 alloy at 1050°C/170 MPa (Fig. 8.9). For example, the 0.52% predeformation in compression results in a lifetime reduction of a factor of 2: 100 hours without predeformation and 43 hours with predeformation under compression. This lifetime reduction is even greater than the amount of predeformation is important: 43 hours compared to 26 hours for a predeformation of 1.46%. Fig. 8.9(b) also highlights that a predeformation in compression decreases the amplitude of the primary creep stage. The author attributed this phenomenon to a reduction in the mean free path of dislocations when the microstructure is P-type. Such a microstructure would reduce the slip deformation and limit climb mechanisms. The drawback of the predeformation is the faster and shorter secondary creep stage.

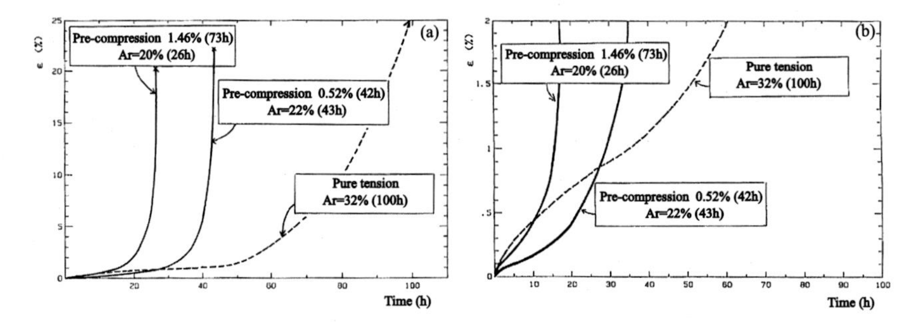

FIGURE 8.9 Effect of predeformations in compression on the mechanical behavior and lifetime of tension creep at 1050°C/170 MPa: (a) full creep curves and (b) magnification of (a) to the first 2% of deformation. Adapted from [57].

The typical microstructural evolution for creep test with a predeformed microstructure in compression was presented by Shui and Xingfu for the first generation alloys [1321,1596]. The authors showed that the precipitates, initially in the form of rafts parallel to the loading axis, are gradually fragmenting and go through a spheroidal state, before recoalescing perpendicularly to the mechanical loading (Fig. 8.10). The same results were obtained in [1321] during a creep test in tension with a predeformation in compression. However, the authors noted that at 1000°C/200 MPa, the 0.32%-precrept sample showed a longer lifetime than the undeformed one. In a similar fashion, Su et al. [1369] found that predeforming ⟨011⟩-oriented samples in compression improves the lifetime. This seems contradictory with the results from Ayrault [57]. This phenomenon would be due to a slow-down of the climb of dislocations at the γ/γ' interfaces, which is a preferential mechanism at this level of temperature but which mainly depends on the time spent at high temperature [1041], as well as the amount of predeformation reached [57]. Similarly, Nathal and MacKay investigated the effect of a predeformation in tension on the creep response in compression [1000]. The authors found the same results as in the previous case, viz. faster plastic strain rates and shorter lifetimes than for undeformed samples. See Fig. 8.11.

8.1.5.3.2 Effect on nonisothermal creep

Similarly to isothermal creep, rafting has an effect on nonisothermal creep. In fact, Cormier et al. [283] exhibited that the introduction of a 30 seconds thermal jump at 1200°C after various periods of creep time at 1050°C/140 MPa on the MC2 alloy leads to highly nonlinear residual lives. They also showed that residual lives are much longer when overheating is introduced in the primary creep stage, i.e., when the N-type microstructure is not established, yet [281]. The same result was obtained by le Graverend et al. [802] at 1050°C/160 MPa on the same alloy. In addition, the latter authors [802] also revealed that it is even possible to improve the life-

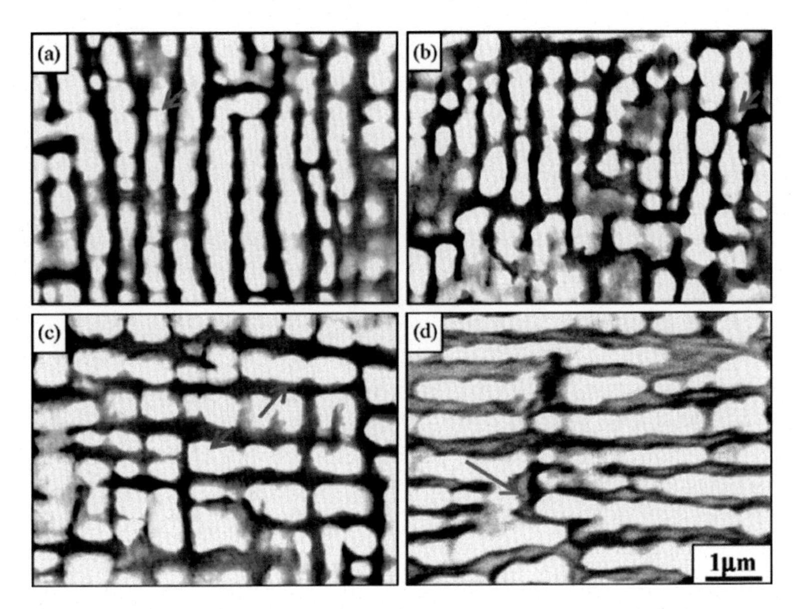

FIGURE 8.10 Microstructure evolution of a microstructure initially P-type and subjected to creep in tension at 1040°C/140 MPa after (a) 6 h, (b) 10 h, (c) 16 h, and (d) 25 h. Adapted from [1596].

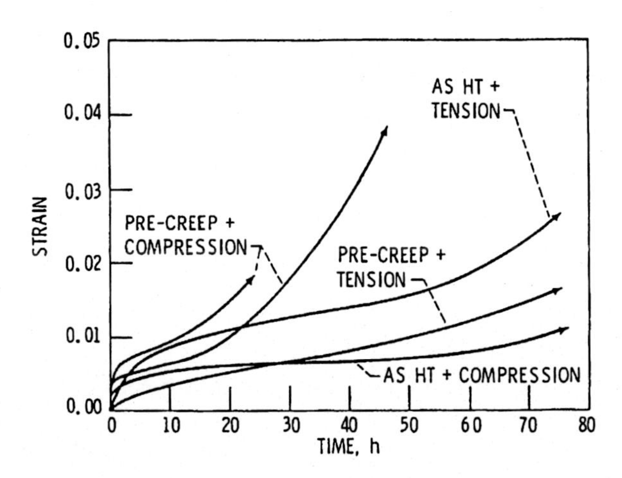

FIGURE 8.11 Effect of a predeformation in tension on the creep behavior in compression at 1000°C/207 MPa for the NASAIR 100. Adapted from [1000].

time when the thermal jump is introduced at the maximum of constrained lattice misfit[1] in absolute value. Thermal cycling tests were also performed

[1]It corresponds to the lattice misfit when a stress is applied. The definition is the same as that of the natural lattice misfit.

on the CMSX-4 alloy for a nominal condition of 1050°C/120 MPa [520]. The authors noticed that the N-type microstructure is detrimental to the mechanical response of samples that are thermally cycled (Fig. 8.12). Indeed, in the case of experiments with a N-type microstructure, it is almost impossible to define a period for which the plastic strain rate is a constant since there is a continuous acceleration of the plastic strain rate from the start. It was not the case with initial P-type microstructures. In fact, these showed plastic strain rates slower than those with an initially cuboidal microstructure: strain rate 2.5 times slower than the cuboidal microstructure and 6.5 times slower than the N-type microstructure. The authors suggested that dislocation climb is slowed down to explain these observations. This beneficial effect of P-type microstructures in terms of plastic strain rate does not correlate with a lifetime improvement, which suggests that the fragmentation of the P-type rafts controls the lifetime.

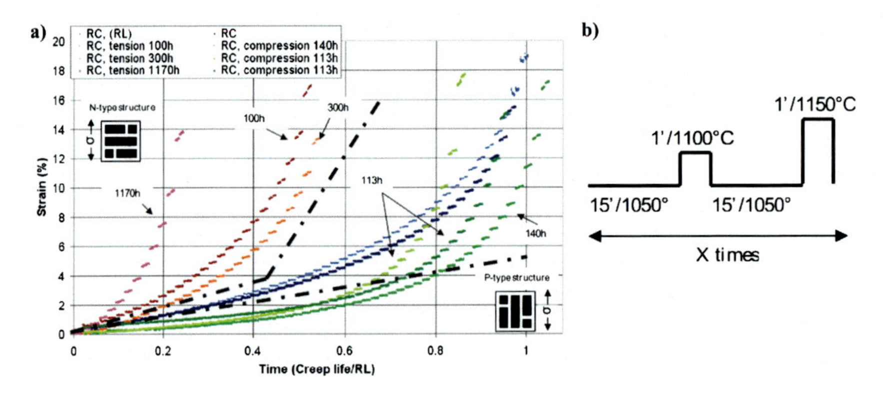

FIGURE 8.12 (a) Effect of the γ' morphology on the mechanical behavior during a thermal cycling shown in (b) and with a stress of 120 MPa. The time is normalized by the rupture time of an originally-undeformed sample. Adapted from [520].

8.1.6 Dendritic heterogeneities

Weight reduction for turbine applications imposes turbine blades to always be thinner. In specific locations of airfoil blades, thin-walled sections are as thin as 500 µm, i.e., the dimension of a dendrite pattern inherent to the solidification of monocrystalline components (Fig. 8.13(a)–(b)).

Chemical microsegregations at the dendrite scale are significant after solidification due to element partitioning leading to a significant content of γ-forming elements in the dendrite cores [1129,194,646]. High-temperature solution treatments above γ'-solvus temperature are subsequently performed to smooth chemical microsegregations, i.e., eutectics and dendritic/interdendritic segregations. Ayrault investigated the effect of the duration of the solution heat treatment on the element partitioning of the CMSX-2 and AM1 alloys [57]. Cr, Al, Ti, and Mo tend to rapidly

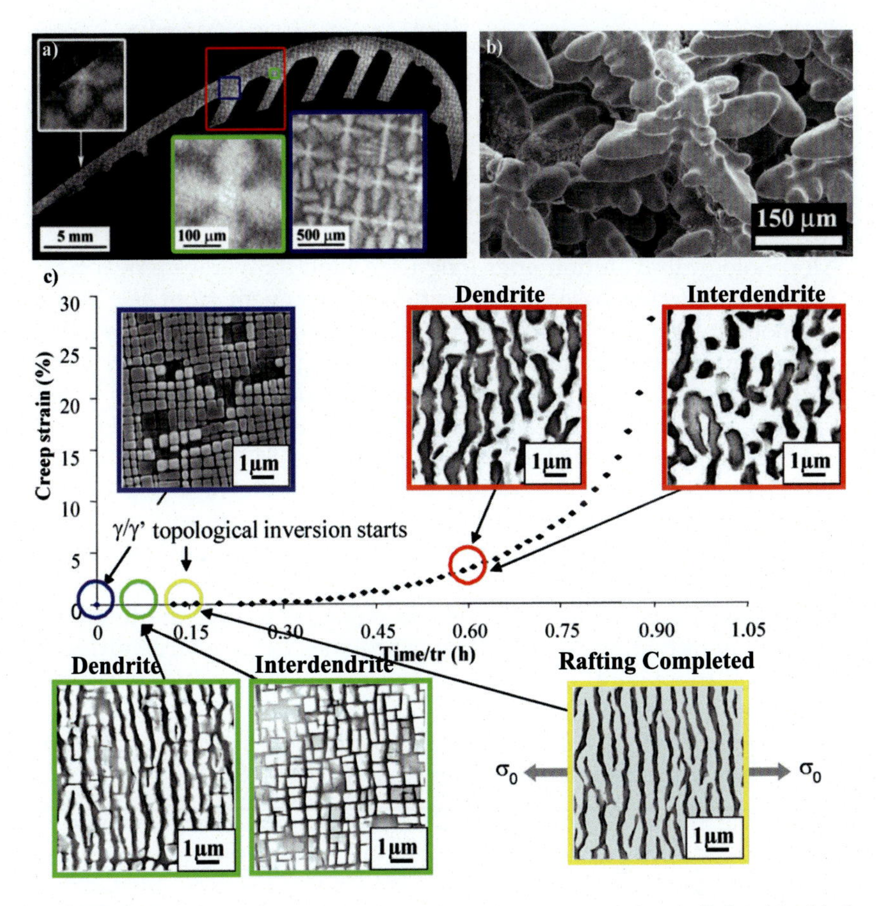

FIGURE 8.13 (a) Distribution and size of dendrite patterns in a thin-walled turbine blade (adapted from [1450]); (b) 3D representation of a dendrite after solidification (adapted from [891]); (c) Evolution of the γ/γ' microstructure at the dendrite/interdendrite scale showing delay in rafting and topological inversion in both the regions under creep stresses at high temperature. Adapted from [946].

homogenize in comparison with Ta and W for the 1st generation SX superalloys. Residual segregation of Ta (interdendritic regions) and W (dendrite regions) were still pronounced after 16 hours at 1315°C. Such a trend was confirmed by other authors [207,1080,553,701,1300,1414,946]. However, industrial solution heat treatments are not sufficiently long to fully reduce the chemical microsegregation [588], more particularly in the presence of refractory elements, such as Re and Ru. Therefore, the dendrite pattern remains after standard solution treatments and leads to differences in microstructural states, as well as chemical and physical properties between dendrite and interdendritic regions:

- Since the lattice parameter of the γ and γ' phases strongly depends on their chemistry, microsegregation at the dendritic scale affects the lattice parameters of both phases, and, consequently, the misfit between the γ and γ' phases [162,845,1144,1503]; The γ' precipitate size, morphology [553,1414,1412], and volume fraction [946,1414,1412], as well as the γ'-solvus temperature [203] after standard heat treatment, varies from dendrite to interdendritic regions;
- Morphological evolution after long-term high temperature heat treatment are different based of the difference in misfit from one region to another: Dendrite cores and the central region of secondary dendrite arms experienced directional coarsening/rafting parallel to the secondary arm direction, the interdendrite regions can develop irregular rafts, and intermediate dendrite/interdendritic regions could remain cuboidal even after long term aging [201,553,589,860]. Such an observation depends on the temperature, the degree of chemical segregation, and thus, the misfit value at the investigated temperature;
- A coarsening and rafting history during high temperature creep experiments [250,980,393,1075]. Rafting first occurs in dendrite cores while the interdendritic regions still remain cuboidal during the first stage of creep deformation. Then, rafting propagates in the interdendritic regions and topological inversion of the γ/γ' microstructure initiates in this latter region [946] (Fig. 8.13(c));
- As stated previously, solidification and homogenization pores preferentially form in the interdendritic regions [848,1500,169].

Such local chemical, physical, and microstructural inhomogeneities at the dendrite scale result in differences in mechanical properties between both regions at room temperature [1303,946,947], but also at elevated temperature [947,1414,1412,1417]. Shade et al. [1303] investigated the behavior in compression of micropillars prepared by focused ion beam (FIB) in either dendrite or interdendritic regions of a René N5 along the $\langle 123 \rangle$ crystal direction. The authors demonstrated that dendrite cores exhibit higher engineering strength than interdendritic regions (Fig. 8.14). Arnoux et al. [38,947] measured differences in the mechanical behavior at room temperature between dendrite and interdendrite regions on "as-received" and precrept MCNG (a 4th generation superalloy) specimens using nanoindentation.

In addition, Milhet et al. tracked nanoindentation markers prior and after a high temperature creep test at 1050°C/140 MPa on MCNG when loading the monocrystalline specimen perpendicular to its withdrawal direction [947]. The authors reported a strengthening effect of the dendrite primary axis at high temperature. This finding was confirmed by Texier et al. [1412,1417] testing ultrathin monocrystalline MC2 specimens along the withdrawal direction for various specimen thicknesses less than the primary dendrite axis spacing. The authors identified a power-law function

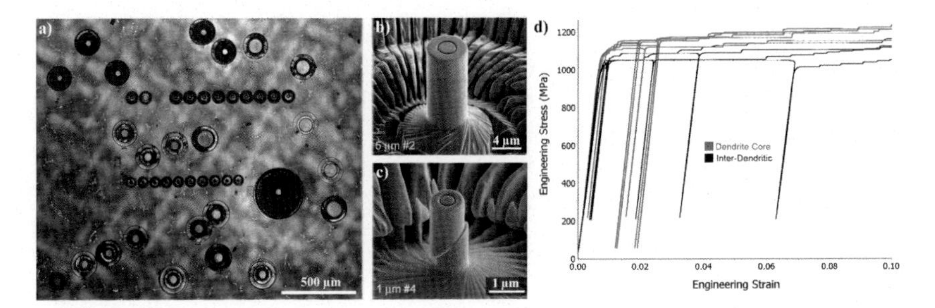

FIGURE 8.14 Compression tests of micropillar at the dendrite scale: (a) region of interest showing the preparation of focused ion beam (FIB)-machined micropillar in relation to the dendrite/interdendrite microstructure of a ⟨123⟩ oriented René N5 SX superalloy (adapted from [1303]), micropillar (b) prior and (c) after compression test (adapted from [342]), (d) compression tests of 20 µm-diameter pillar of a ⟨123⟩ oriented René N5 SX superalloy, comparing response for dendrite core and interdendritic regions. Adapted from [1303] (Courtesy to Paul Shade).

correlating the number of primary dendrite axes within the gauge section to the steady-state creep rate. This clearly indicates the strengthening role of primary dendrite axes on the high temperature creep behavior.

Topographic evolutions at the surface of the creep-tested specimens in accordance to dendrite and interdendritic locations and the creep-rate sensitive to the fraction of dendrite cores present in the specimen demonstrated the differences in high temperature mechanical behavior of both the regions at the dendrite scale [1412,1417]. All these studies converged on the strengthening effect of primary dendrites at room and high temperatures.

8.2 Oxidation, diffusion, and mechanical coupling for bare and coated SX superalloy

In service, nickel-base superalloys are subjected to high-temperature oxidation and/or corrosion in severe atmospheres. They demonstrate good resistance to surface reactivity at elevated temperatures, i.e., oxidation and corrosion, due to their aluminum and chromium contents sufficiently high to form a continuous, dense, slow-growth rate, and adherent external oxide layer of Cr_2O_3 and/or Al_2O_3 [1625,519,527,1338]. Oxidation products depend on different parameters, e.g., the temperature window, exposure time, partial pressure of oxidizing/corroding elements, but also on the aluminum and chromium activity of the superalloy, as depicted in Fig. 8.15. The oxidation behavior of multielement materials, such as monocrystalline superalloys, is complex, and several oxides could develop during the life of the components at high temperature and ruin the integrity of the protective oxide scale. Damage of the oxide scale is particularly enhanced during thermal cycling due to oxide cracking and/or spallation. The integrity and damage mechanisms of the protective oxide scale is addressed in Chapter 10.

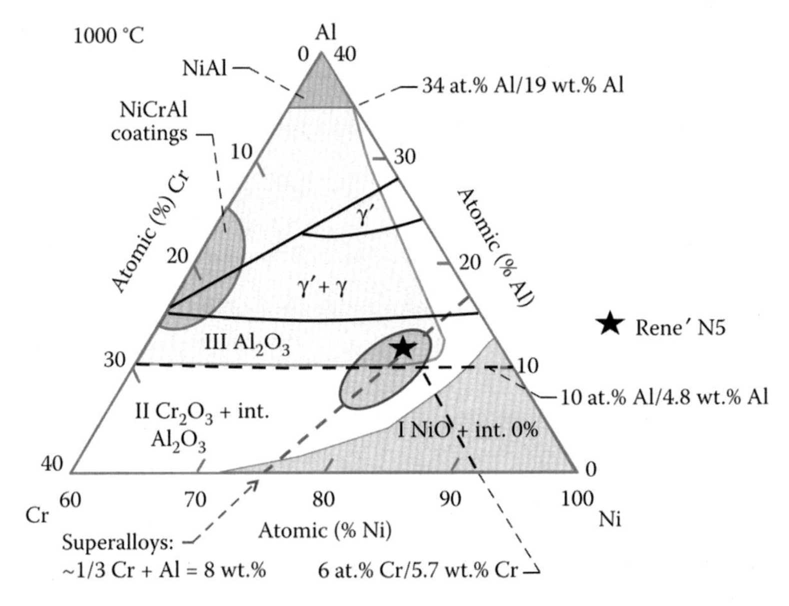

FIGURE 8.15 Map of oxidation products formed at 1000°C for NiAlCr materials (Ni-rich corner) showing three domains with different types of oxidation behavior: (i) Type I: An external NiO oxide scale + internal oxidation products, (ii) Type II: An external Cr_2O_3 oxide scale + internal Al_2O_3, and (iii) Type III: An external Al_2O_3 oxide scale. Adapted from [527, 1338]. Copyright (©) and Imprint. Reproduced by permission of Taylor & Francis Group.

Surface reactivity at high temperature can sometimes be limiting, especially with the regular increase in combustion temperatures, thermal cycling, and the size reduction of thin-walled turbine blades. Therefore, superalloys are generally coated with MCrAlY (where M = Co, Ni or Co/Ni) overlay coatings or aluminide coatings modified or not with platinum to promote the formation of a protective α-Al_2O_3 oxide [940,1217,135,545]. For MCrAlY coatings, other oxides such as fast-growing spinels or intrusive Y-rich oxides (also denoted pegs) could form at high temperature despite the formation of the external protective α-Al_2O_3 layer. These oxides are deleterious for the integrity of the protective oxide scale under thermal cycling. Controlled addition of reactive elements (e.g., Y, Hf, Zr, La, Ce) are reported to lower the growth rate of the Al_2O_3 scale and favors the adherence of the Al_2O_3 scale onto the coated system, property particularly beneficial under thermal cycling to limit oxide spallation [1110,1001]. While Al_2O_3-forming bare and coated superalloys have similar isothermal oxidation behaviors, coatings significantly improve the thermal cycling performance of superalloys components [1166]. Their oxidation behavior has been extensively investigated under isothermal and cyclic conditions and adequately reviewed in [1331,355,1338,1625,1334,413,1001]. The present section emphasizes the consequences of the oxidation on the microstructure evolutions and the subsequent changes in mechanical behavior of bare and coated superalloys at high temperature.

8.2.1 Oxidation of bare superalloys and effects on high-temperature mechanical behavior

High temperature oxidation and corrosion are selective phenomena leading to the consumption of chemical elements participating in the oxide formation, e.g., Al for Al_2O_3 and Cr for Cr_2O_3. The chemical composition of the superalloy beneath the oxide scale is, thus, modified along with the growth of the oxide scale. Al consumption results in a lower fraction of the Ni_3Al-γ' strengthening phase (γ'-reduced layer) and the formation of a γ'-free layer beneath the oxide scale, as depicted in Fig. 8.16 [95,92]. This oxidation-affected region is often considered as a non-load bearing section in comparison to the bulk region and significantly impairs the mechanical integrity of superalloys, especially for thin-walled components. Bensch et al. [95] investigated the extension of the oxidation-affected region in thin-walled components on an M247LC and a René N5 alloys confronting thermodynamic simulations with ThermoCalc, kinetic simulations with DICTRA, and experimental results. The gradient of Al concentration beneath the oxide scale and the extension of the γ'-reduced and γ'-free layers due to oxidation were successfully predicted. After 100 hours oxidation at 980°C, specimens thinner than 400 μm are fully affected by the oxidation and the Al concentration at the mid-section plane of the specimen starts to significantly decrease with the exposure time and specimen thickness [95,94]. Bensch et al. [92] also examined the creep response of thin-walled specimens at high temperature. They purposely manufactured and creep-tested model monocrystalline superalloys with nominal compositions matching at different locations within the oxidation-affected region (Fig. 8.16 [95,92,94]). This approach aimed to identify the local mechanical behavior within the gradient of chemical composition/microstructure due to oxidation. The minimum creep rate at 980°C between 50 and 230 MPa was determined for samples with γ' fraction comprised between 62 and 0%, the minimum creep rate gradually increasing with the decrease in γ' fraction. The non-load bearing behavior of both the γ'-reduced and γ'-free layers was, therefore, demonstrated and aimed to explain thickness debit effects of thin-walled specimens under oxidizing and inert atmospheres. Similar thickness debit effects due to the non-load bearing behavior of the γ'-free layers were reported by [366,1351,1631,810]. Voids formation and dynamic recrystallization resulting in the formation of subgrains in the γ'-free layer were also observed at high temperature and participate as well in the impairment of the mechanical performance of the SX superalloy [126,1351,1412].

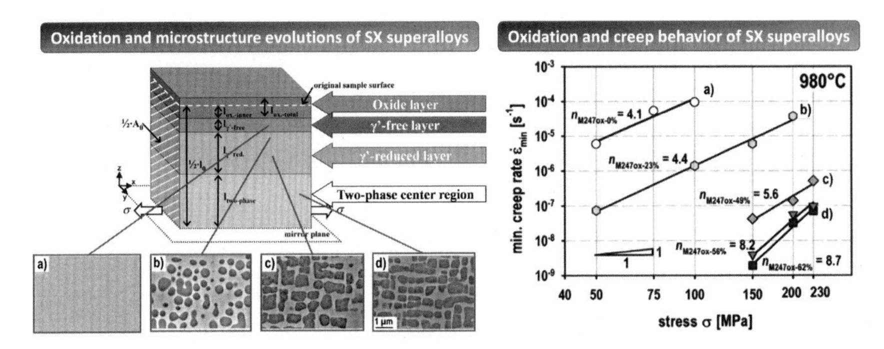

FIGURE 8.16 Effect of the high temperature oxidation on the microstructure evolution of an M247LC SX superalloy beneath the oxide scale and on the local creep behavior within the oxidation-affected region. Adapted from [95,92]. Copyright 2012 by The Minerals, Metals & Materials Society. Used with permission.

Cassenti et al. [219] attributed thickness debit effects to damage confined to the surface layer using a dislocation-based model and considering (i) dislocations interaction with the surface, and (ii) the effect of nucleated intrusions or/and extrusions on the creep strain rate. Damage confined to the surface layer could arise from either the motion of dislocations to the surface, the preferential generation of voids or microcracks near the surface, or even a direct result of surface oxidation.

Dynamic aspects on the coupling between creep behavior and oxidation were reported by Dryepondt et al. [366] by changing atmospheres from oxidizing (synthetic air) to "protective" atmospheres (hydrogenated argon) during high temperature creep tests. Despite similar extension of the γ'-free layer, they observed a significant decrease of the minimum creep rate after switching atmosphere. They attributed this change in creep behavior to the modification in oxidation mechanisms from partial cationic to anionic mechanism to form the Al_2O_3 scale, thus reducing vacancy flux. Vacancy injection under air induces long range diffusion that promotes dislocation climbing and, therefore, an increase in creep strain rate [395].

8.2.2 Oxidation and interdiffusion of coated SX superalloys

As aforementioned, coatings are used to protect SX superalloys from environmental degradation by forming a protective oxide scale, the so-called thermal grown oxide (TGO).

The formation of thin external oxide scales on the coating surface, i.e., Al_2O_3 and/or Cr_2O_3, aims to limit access of oxygen, sulfur, and corroding salts to the superalloy. The chemistry and processing routes of coatings are adapted to (i) the environmental constraints, i.e., corrosive versus oxidizing atmospheres, (ii) the geometry of the superalloy component to be coated, i.e., line-of-sight and non-line-of-sight processes, (iii) the operating conditions, i.e., temperature window, short versus long thermal exposure, lifetime expectations, etc. Despite different needs from applications to applications, common requirements have to be satisfied to ensure the successful environmental performance of metallic coatings [1010,135,1397]:

- The oxidation/corrosion resistance relies on thermodynamics, kinetics, and mechanical aspects. The oxide scale has to (i) be thermodynamically stable, (ii) uniformly grow on the coating surface with a slow growth rate to ensure uniform surface attack, (iii) be adherent to limit oxide spallation. Furthermore, the coating is required to (iv) contain high content of oxidizing/corroding elements participating in the formation of the oxide scale for reservoir and self-healing purposes, and (v) exhibit an appreciable ductility and mechanical strength to accommodate strain mismatch between the oxide and the superalloy;
- The interface stability relies on (i) a low rate of diffusion across the interface at operating temperatures, (ii) a limited interdiffusion process across the interface, i.e., a reduced compositional change, and (iii) the prevention from embrittling phase formation during the component life such as TCP phases;
- The good adhesion of the coating requires (i) compatible/similar thermomechanical properties between the coating and the superalloy to minimize thermal strain mismatch and stress generation at coating/substrate interface, (ii) surface preparation (rough and smooth, cleaned) before coating, (iii) low growth stress during the deposition process;
- The mechanical strength and ductility of the coating is highly needed to withstand creep, fatigue, and impact generally encountered in service. Matched thermal expansion properties between the coating and the substrate aim to minimize thermal stressing, especially under thermal cycling and thermal fatigue. A low ductile-to-brittle transition temperature (DBTT) is also preferable to avoid premature cracking of the coating under thermomechanical loading. However, this impairs its high-temperature mechanical strength.

Many variants of high-temperature coatings are in use today and can be divided into three categories: (i) diffusion coatings, (ii) overlay coatings, both used to protect the coated superalloy from oxidation and corrosion, and (iii) thermal barrier coatings (TBC), to protect the superalloy and the bond-coating from thermal degradation.

The intent of this section is to better understand the effect of aging on the microstructural evolution and ultimately the mechanical behavior when coated with diffusion and overlay coatings. This is particularly important because the increase need in coating solution pushes the need of a design step including both coating and substrate seen as a whole system. The case of overlay coating will be mostly addressed in the TBC chapter, since interactions are mostly limited to induce brittleness by the coating layer itself.

8.2.2.1 *Microstructure of coated single crystal superalloys*

Prior to coating, monocrystalline superalloys are solution heat-treated to homogenize chemical segregation and the dentrite/interdendritic scale inherent to the solidification process. The components are then grit blasted with alumina particles to remove the oxide scale forming both during the solidification process and the solution heat-treatment. Despite cleaning, remaining alumina grit-particles are still observed at the interface between the coating and the so-called interdiffusion zone (IDZ) and could occupy a significant surface fraction of this interface [1416]. It is worth noting that the compressive cold working associated with the grit-blasting operation generally induces recrystallization in the IDZ but also rafting of the γ/γ' microstructure of the superalloy parallel to the coating/superalloy interface after standard heat treatments [126,109,1412,1416]. The extension of the rafted microstructure in the monocrystalline substrate was found to be deeper than the extension of the chemical profile due to interdiffusion. Such surface preparations are necessary to enhance further adhesion of the coating onto the superalloy, regardless of the deposition route.

8.2.2.1.1 The case of β-(Ni,Pt)Al diffusion coatings

For Pt-modified nickel aluminide diffusion coatings, Pt is electrodeposited and Al is often applied by high-temperature CVD process [941]. Depending on the Pt and Al amount deposited, diffusion coatings are categorized in low-activity and high-activity diffusion coatings. High temperature standard heat treatments are subsequently performed to allow interdiffusion of the deposited layers and the SX superalloy but also to confer the adequate γ/γ' microstructure optimized for structural properties. In the as-received state, (Ni,Pt)Al is mostly observed as a columnar outer layer made of B2-β-NiAl grains in solid solution. IDZ is typically a B2 matrix including many (W, Cr)-rich precipitates.

8.2.2.1.2 The case of γ/γ' diffusion coatings

Pt-modified γ/γ' diffusion coating are usually manufactured by only depositing a Pt layer onto the superalloys prior to high temperature standard heat treatments. Pt-enrichment of the extreme surface of the superalloys promotes exclusive α-Al_2O_3 scale growth [586]. Such diffusion coating are highly beneficial in terms of oxidation behavior but also mechanical integrity of the coating layer, i.e., less brittle and high creep-strength properties compared to B2-β-(Ni,Pt)Al diffusion coatings. The high-temperature oxidation resistance could be substantially improved by hafnium (Hf) addition [527].

8.2.2.1.3 The case of MCrAlY overlay coatings

In comparison to diffusion coatings, overlay coatings are directly deposited onto monocrystalline superalloys with the custom-designed chemical composition. This process is limiting interaction with the substrate in terms of diffusion. Different compositions of MCrAlY-type coatings (where M = Co, Ni or Co/Ni) are in use and their chemistry is optimized depending on the environmental constraints and the substrate composition to limit both oxidation and interdiffusion. Several line-of-sight and non-line-of-sight processes are generally used depending on the complexity of the component geometry and detailed in [1217,135]. Overlay coatings generally consist in B2-β-(Ni,Co)Al and/or $L1_2$-γ'-$(Ni,Co,Cr)_3(Al,Ti)$ and/or DO_{22}-γ-Ni microstructures depending on its chemistry. Secondary phases such as α-Cr and (Ta,Ti)C phases could be formed in the coating. Furthermore, thermal spray processes are extensively used to deposit MCrAlY coatings and produce complex microstructural features such as melted and resolidified powder particles, unmelted powder particle (UMPP), pores, and dispersed alumina oxides. High-temperature standard heat treatments are also applied to MCrAlY overlay coating to improve adhesion properties due to interdiffusion. In comparison to diffusion coating, the external MCrAlY coating is much thicker (50 to 150 μm) than the interdiffusion zone (tens of micrometer). The microstructure of the interdiffusion zone is quite complex and depends on the chemistry of the overlay coating. However, γ, γ', and (W, Cr)-rich precipitates are generally found in the IDZ. Recrystallized grains and rafted γ/γ' microstructure are generally found beneath the coating/IDZ interface [1166,1416,1164].

8.2.2.2 Phase transformation under isothermal and cyclic thermal loading

When exposed to high temperature and oxidizing atmosphere, NiAl or (Ni,Pt)Al coatings are undergoing phase transformation from B2-β-NiAl to γ'-Ni_3Al and to γ-Ni for long-time exposure, according to the binary NiAl diagram [545]. For the first generation superalloys, the IDZ could be divided between the IDZ close to the IDZ/substrate interface and the IDZ close to the coating/IDZ interface. The one close to the IDZ/substrate interface is the first to transform into γ', that could be defined as a γ'-IDZ layer. The one close to the coating/IDZ interface is observed to be of B2-β-NiAl, that could be defined as a β-IDZ layer. For isothermal oxidation, it is observed that this γ'-IDZ layer grows both inward, i.e., increasing the substrate transformation, and outward, i.e., together with a decreasing of the thickness of the initial β-IDZ layer [1258]. Increasing time and/or temperature exposure, the outer BC layer thickness decreases to the benefit of the γ'-IDZ layer. In addition to microstructural evolution, vacancy flux resulting from the interdiffusion between the coating and the substrate, but also due to oxidation, is reported to promote the formation of Kirkendall voids [1493, 51].

Similarly to nickel aluminide coatings, β/γ, $\beta/\gamma/\gamma'$, and γ/γ' coatings exhibit phase transformation that occurs from β to γ' and/or γ depending on the initial composition of the overlay coating. Sloof and Nijdam developed a coupled thermodynamic-kinetic oxidation model to predict both phase transformation and chemical composition due to oxidation for multicomponent multiphase alloys, such as MCrAlY coatings [1012,1334]. After high-temperature exposure, the microstructure is affected both by the oxidation and interdiffusion. A layered microstructure develops within the coating with a progressive decrease of Al-rich phases beneath the oxide scale and adjacent to the coating/IDZ interface [1166,1412,1164,1334]. The β-free then β–γ'-free layers develop from both these surface/interface. Coarsening of the microstructure and a loss of interconnectivity between β phases are also observed with the increase of temperature and exposure time. This topological inversion improves the ductile behavior of the aged coating for temperatures below the DBTT [1413]. The extension of the interdiffusion generally progresses within the monocrystalline superalloy above 1000°C, i.e., a temperature comparable or higher to the highest standard heat treatment temperature [1412]. Increasing time and/or temperature exposure, the IDZ is mainly constituted of a polycrystalline γ layer due to recrystallization, grain boundaries being decorated with (W, Cr)-rich precipitates. Beneath this polycrystalline layer, a monocrystalline γ layer develops and progresses within the substrate, thus reducing the effective load-bearing section. It is worth noting that the γ-IDZ layer (polycrystalline + monocrystalline layer) could progress deeper than the γ'-depleted region formed due to oxidation [531,1412], as depicted in Fig. 8.17 [1412]. Such a reduction of the effective load-bearing section, due to the presence of a γ'-depleted layer (or γ'-reduced), reduces high-temperature creep performance. Furthermore, the occurrence of the non-load-bearing coating onto the blade increases the effective stress applied in the case of rotating blades. The influence of the coating on the mechanical integrity of coated superalloy components is further addressed in this chapter.

In addition, projected MCrAlY overlay coatings present a complex lamellar microstructure due to melted and resolidified powder particles also called splats. Splat boundaries can favor gas-access channels despite the growth of the external protective oxidation and lead to oxide intrusion within the coating [1013]. Such oxide intrusions severely embrittle coatings under mechanical loading [1413].

Thermal cycling promotes the spallation of the oxide layer during the cooling phase, thus making the coating surface regularly bare of oxides. Due to the parabolic growth of the α-Al_2O_3 scale at elevated temperature,

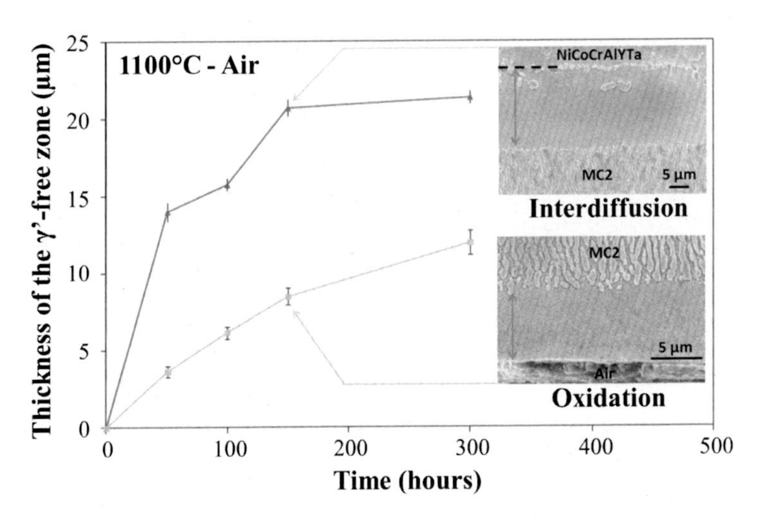

FIGURE 8.17 Extension of the γ'-free layer with the time for a bare and NiCoCrAlYTa coated MC2 SX superalloy at 1100°C. Adapted from [1412].

oxide spallation increases the effective oxidation of the material but also its progressive consumption due to the loss of Al. Therefore, microstructure evolution within coatings are much more sensitive to oxidation than interdiffusion in comparison with isothermal oxidation. For cyclic oxidation, especially considering short dwell time at high temperature ($\leq$ 10 minutes), the B2-β-NiAl to γ'-Ni_3Al transformations are mainly observed in the outer layer of the BC. The typical morphology of γ' phase is made of precipitation localized at former B2-β-NiAl grain boundaries, close to the IDZ interface and close to the TGO interface (Fig. 8.22). The thermal cycling of a (NiAl) diffusion coating, especially in the presence of Pt, induces a large undulation of the coating surface, the so-called rumpling effect [1444] (see Chapter 10), as compared to isothermal oxidation. Rumpling subsequently promotes cracking of the coating due to thermal cycling [1076,410,1128] and/or voids due to oxide/coating debonding [409]. See Fig. 8.18.

Besides this important modification of morphology in phase transformation, during cooling, it has been evidenced that martensitic transformation takes place into B2-β-NiAl phase [240,1657]. The martensitic transformation induces volume changes favoring rumpling. MCrAlY and γ/γ' coatings are less sensitive to rumpling, this latter coating being a promising solution due to its higher creep performance at elevated temperatures [527].

For the 3rd and 4th superalloy generations, using (Ni,Pt)Al yields an additional diffusion zone, the so-called secondary reaction zone (SRZ) [798,1161,313,1037]. This zone laying at IDZ/substrate interface associates

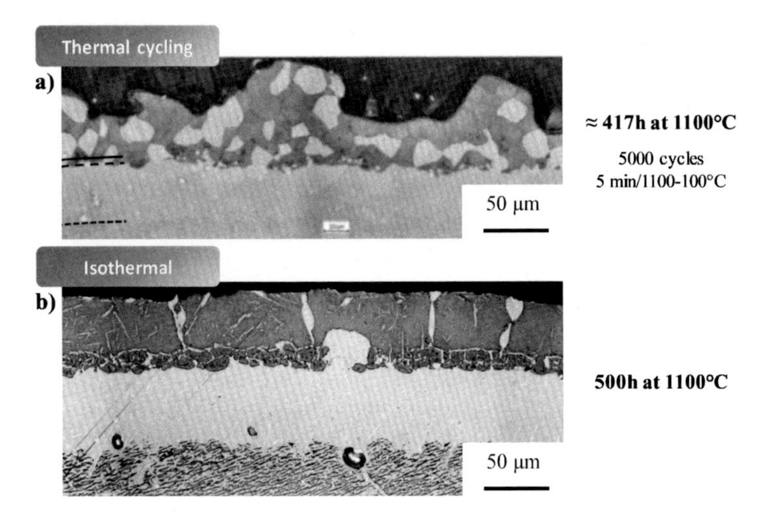

FIGURE 8.18 Comparison of microstructure evolution for thermal cycling with 5 min dwell at 1100°C and isothermal oxidation for direct (Ni,Pt)Al on AM1 single crystal superalloy. Adapted from [1258].

TCP precipitation and large recrystallization. Das et al. have shown that TCPs were P-phase for Re-rich and β-RuAl, B2 and Pm3m structures for Ru-rich TCPs [313]. Independently of SRZ, phase fraction of TCPs are observed to be higher in presence of diffusion coating than for bare alloys, for depth reaching hundreds of micrometers. With SRZ, the affected depth by TCPs could reach the whole specimen thickness (up to few hundreds of micrometers) (Fig. 8.19).

TCPs, with or without SRZ, are detrimental for the lifetime of the coated superalloy [798]. In the presence of SRZ, cracks can be induced either at grain boundaries for long TCPs with Ru- and Re-rich alloy [798] or at the interface between IDZ and SRZ associated to the presence of σ-phase, as in the CMSX-10 [1196] (Fig. 8.20).

Besides, it was observed some premature rafting in the substrate close to SRZ. For instance, for René N5 coated by NiAl, Gong et al. have shown that 15 μm in depth were affected by rafting after only 10 hours at 1100°C [536].

Some authors did suggest that the use of a direct electroplating of Pt without aluminization process could yield enhancement of the TBC lifetime [313,52]. This class of coatings is indicated as γ/γ' coatings corresponding to coarse γ/γ' microstructures developed after high-temperature annealing in the presence of electroplated Pt layer. This kind of coating is observed to drastically limit the rumpling effect observed during thermal cycling. Despite an increase of the voids volume fraction

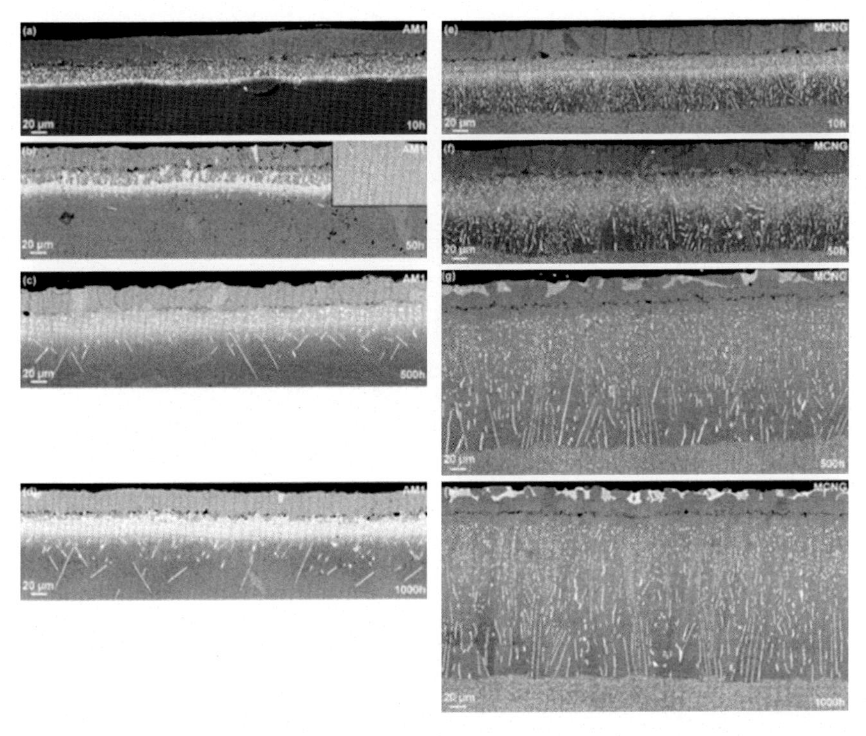

FIGURE 8.19 Microstructure evolution function of time spent at 1100°C for (Ni,Pt)Al coating AM1 (a)–(d) and MCNG (e)–(h) single crystal superalloys. Adapted from [254].

associated to Kirkendall effect, the complete TBC (including the top-coat (TC)) lifetime to spallation is clearly improved [52].

8.2.2.3 *Influence of thermo-mechanical loading on oxidation and phase transformation kinetics*

Considering the addition of mechanical loading on high-temperature oxidation yields some evidence of strong mechanical-oxidation-diffusion-phase transformation-damage coupling in coated superalloys.

Firstly, the oxidation behavior, i.e., both the kinetics and oxidation products, is a function of mechanical loading. For instance, recent results showed that the oxidation of a CoNiCrAlY coating is sensitive to creep condition, by modification of the nature of oxide formed [246]. Besides, it has been evidenced that local oxide spallation was associated to mechanical loading induced during thermal transient by the coefficient of thermal expansion (CTE) mismatch between oxide, coating and substrate, effect enhanced by the oxide strain growth due to difference in molar volume of oxide and metals [1625]. This combination of effects has been fairly described by the "*p–kp*" model proposed by Poquillon and Monceau, in-

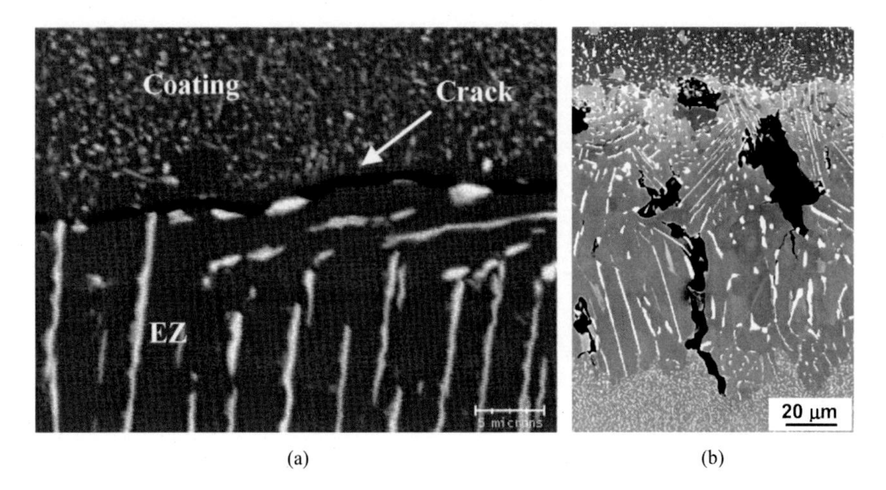

FIGURE 8.20 Crack for (a) isothermal aging after 375 h at 1100°C for (Ni,Pt)Al coating CMSX-10 (adapted from [1196]) and for (b) creep condition 199 h at 1050°C for (Ni,Pt)Al coating MC544. Adapted from [798].

cluding a spallation probability "p" to model the increase in oxide growth rate "kp" induced by loss of passivation [1131] (as detailed in Chapter 12). If an oxidized specimen bears a compressive stress/strain at low temperature, this compression will increase the spalled area fraction as a direct function of local strain (Fig. 8.21) [581,917]. Subsequently, the increase in the oxide spallation will speed up both the oxide growth rate and the outward flux of aluminum in the presence of mechanical loading.

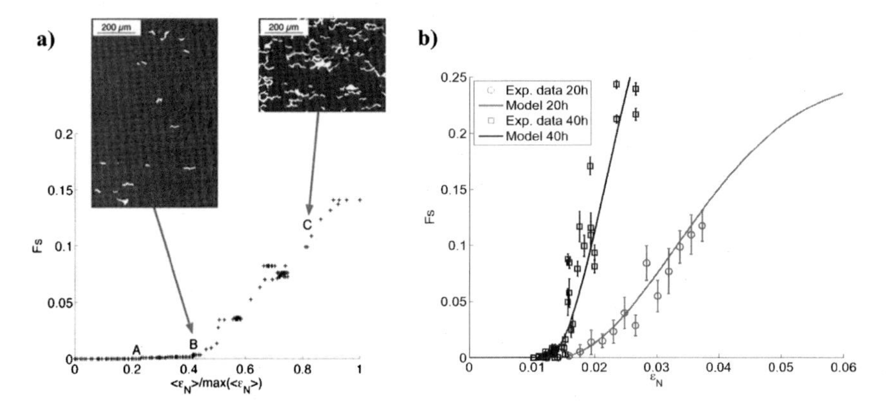

FIGURE 8.21 Oxide spallation assessment through spalled area fraction function of applied strain (left) measurement after 20 h of isothermal oxidation at 1050°C, inserts correspond to oxide spallation localized at oxide ridges (B) and ridges plus surface of grains (C) (right) comparison of spalled area fraction for 20 and 40 h of isothermal oxidation at 1050°C. Adapted from [917,919].

For thermal cycling combined to mechanical loading, typical of thermo-mechanical-fatigue (TMF) loading, it is also obvious that the rate of phase transformation is highly sensitive to the loading parameters (time, temperature, and stress/strain history). For (Ni,Pt)Al coating, the β to γ' transformation rate in the outer bond-coating is increased for both tensile and compressive stress applied at high temperature (Fig. 8.22) [1258]. Besides, the morphology of γ' precipitates depends on applied TMF loading. The increase of growth rate of γ'-IDZ layer was also observed in presence of mechanical loading as compared to pure thermal cycling (see the variations of height between dashed lines added to the microstructure in Fig. 8.22). It is worth noting that the rumpling effect (undulation of the surface without TC) is also sensitive to TMF loading parameters. Subsequently, the damage rate due to Al loss (through oxide loss and breakaway regimes) is improved without TC [1258] and TC spallation is observed sooner [1346] when the mechanical loading is added. This latter effect is also concomitant to μ and σ phases, respectively, at the β grains boundaries and at the IDZ/substrate interface, as for the CMSX-4 alloy under TMF loading [1662].

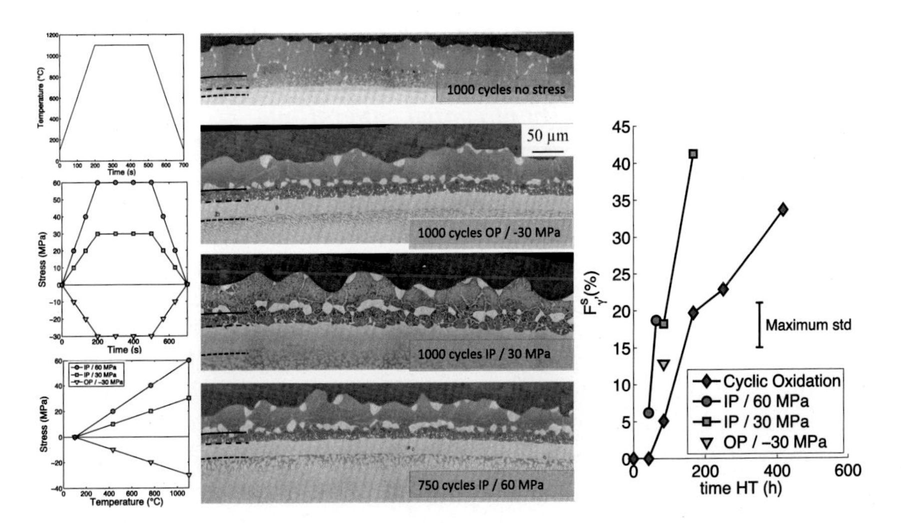

FIGURE 8.22 Sensitivity to TMF loading condition comparing thermal cycling (5 min dwell at 1100°C) w/o tensile (in-phase IP test) or compressive (out-of-phase OP test) stress applied during dwell for (Ni,Pt)Al coating AM1 (1st generation superalloy): (left) applied TMF conditions: temperature cycle, stress history as a function of time and stress as a function of temperature, (center) microstructure evolution observed by optical microscopy and (right) γ' phase fraction evolution in the outer BC as a function of time and TMF conditions. Adapted from [1258].

8.2.2.4 *Influence of superalloy composition*

The presence of Re and Ru in the substrate induces the formation of SRZ for NiAl coatings, as already evoked above. For simplified substrate composition, in the presence of NiCoCrAlY coating, it has been shown that W and Ta prevent from detrimental effect of rapid diffusion process and limit B2-β-NiAl dissolution [482].

For industrial substrate alloys, γ/γ' coatings improve the lifetime in comparison to (Ni,Pt)Al coatings for the René N5, whereas a 7 µm Pt layer deposited on the CMSX-4 yields to the shortest lifetime [1111]. This effect is mainly caused by the difference in Ti content comparing both generations of alloys (respectively second and first generation) [586].

The sensitivity of both microstructure stability and global lifetime of the TBC system to superalloy composition is also a function of the chosen coating strategy: for (Ni,Pt)Al, transition from CMSX-4 to CMSX-10 improves the lifetime of EB-PVD TC spallation, whereas for CoNiCrAlY-Si it decreases the lifetime of APS TC spallation [724]. This example proves that a TBC system, and more widely a single crystal superalloy, could not be optimized in composition without consideration of the global strategy of use: if coated, it implies the codevelopment of optimal composition and thermal treatment of the superalloy and of the process and specifications of the TBC, that definitively comprises the substrate alloy.

8.2.2.5 *Influence of the coating and interdiffusion zone on the mechanical integrity of coated superalloys*

Despite their environment-protective function, coatings affects the mechanical integrity of coated SX superalloy components due to:

- The brittle behavior of the coating below the ductile-to-brittle transition temperature (DBTT) leading to premature cracks within the coating layer [1365,1281,1415,237,660,1419,14,1050,403,13,1010]. The B2-β-NiAl phase is particularly brittle at low temperature in comparison with the γ and γ' phases. The β to γ' phase transformation leads to an increase of the local ductility despite surface roughening induced by the so-called rumpling mechanism [402,1413]. Cracking in the coating lead to premature failure of the coated component, then rapidly propagating under cyclic thermomechanical loading [1281,1491,1571,410,1128,1371, 1258,1365];
- The loss in mechanical strength of the coating but gain in ductility above the DBTT leading to non-load bearing regions and, thus, additional weight onto rotating components [599,1415,237,660,1419,15,14,1050,167, 652,1405];
- Their interdiffusion with the Ni-based substrate affecting its mechanical strength at high temperature [1416]. The loss in mechanical strength also participates in the non-load bearing behavior compared to the SX superalloy;

- Brittle phases present in the interdiffusion zone leading to potential cracking [798].

The DBTT depends on the chemical composition of the coating [1010]. Nickel aluminide coatings used to have a higher DBTT than MCrAlY coatings. DBTT of MCrAlY coatings, ranging from 500°C to 800°C varies as a function of its composition with a high effect of the Cr content [1281, 599,1415,595,1010]. Low activity nickel aluminide coatings used to have a lower DBTT (600°C) compared to high activity nickel aluminide coatings (750–840°C) [1010,13,403,1497]. Pt-addition to coatings used to increase the DBTT of low activity nickel aluminide coatings up to 760°C and the one of high activity nickel aluminide coatings above 840°C [1010,1050,13,1419]. A good trade-off has to be found to have a sufficiently low DBTT to avoid low temperature cracking of the coating but also sufficiently high to limit creep and loss of mechanical strength at high temperature.

CHAPTER

9

Refurbishment

Fernando Pedraza[a] *and Satoshi Utada*[b]

[a]La Rochelle University, LaSIE UMR 7356-CNRS, Faculty of Sciences and Technology, La Rochelle, France, [b]ISAE-ENSMA, Institut P', UPR 3346-CNRS, Chasseneuil-Futuroscope, France

9.1 Introduction

The civil aviation sector has been increasing over the past two decades, particularly in Asia. In 2012, the market share was dominated by North America (33%), Western Europe (21%), Asia-Pacific (15%), and China (8%) with about 45 000 engines operating worldwide. By 2022, about 62 500 engines running are forecasted [653] with an increasing share of the Asia-Pacific and China regions.

The engine market will be driven (in decreasing order) by the original engine manufacturers (OEM) CFM International (Safran/General Electric), General Electric, Pratt & Whitney, and Rolls Royce through their most advanced engines like, e.g., the LEAP (leading edge aviation propulsion) engine that makes extensive use of single crystal (SX) nickel-base superalloys among other evolution of materials. Indeed, SX superalloys are now also employed in the first stage of the low pressure turbine (LPT) in addition to their applications in the high pressure turbine (HPT) components (nozzle guide vanes – NGVs – and blades or buckets) of the engines currently flying. This choice derives from the increased turbine entry temperature (TET) to improve the thermodynamic yield of the engines that allows decreasing fuel consumption and emissions of greenhouse gases. For instance, an increase of 10°C allows a reduction of 1% of fuel consumption and a drop of about 150 000 t of CO_2, i.e., about 20 M€ of cost reduction per year for an airline with about 300 planes. Therefore, the current temperature of the first stage of an LPT is now about 50°C to 100°C higher (1050°C during take-off, 900°C to 950°C close to the blade tip dur-

Nickel Base Single Crystals Across Length Scales
https://doi.org/10.1016/B978-0-12-819357-0.00017-2

ing cruise) than before and that of the HPT reaches about 1300°C during take-off and 1000°C to 1100°C during cruise [1071].

The increased operating temperatures foster mechanical degradation of the components, including creep and thermomechanical fatigue (TMF). Further, the surrounding environments where the planes fly over are more and more aggressive. For instance, NASA provides updates on the evolution of human footprint on the global quality of air that show increased share of contaminants in the air [270]. Further details for the evolution of NOx emissions can be retrieved from https://svs.gsfc.nasa.gov/12094 while details for the CO_2 emissions can be found at http://www.asc-csa.gc.ca/eng/satellites/mopitt.asp. However, the most important contaminants that reduce lifespan of coated and uncoated nickel-base superalloy components are SO_2, NaCl, and particulate materials. Sulphur dioxide (SO_2) is naturally released to the air from volcanic regions, by phytoplankton, and from fires. Human activities related to fossil-fired (mostly coal-fired) power plants, oil and gas production, and from smelters also contribute to a great extent of today's SO_2 emissions (https://so2.gsfc.nasa.gov/). Over the past few years, the SO_2 emissions have decreased in Western countries, while a dramatic increase has been reported in Asia, in particular in India and China. The main issues with SO_2 are related to the formation of both sulphuric acid and of sulphates through (mainly) Eqs. (9.1) to (9.4):

$$SO_2\,(g) + 1/2O_2\,(g) \longrightarrow SO_3\,(g), \tag{9.1}$$

$$SO_3\,(g) + H_2O\,(g) \longrightarrow H_2SO_4\,(l), \tag{9.2}$$

$$NaCl(aq) + SO_2\,(g) + O_2(g) \longrightarrow Na_2SO_4\,(s)^* + Cl_2\,(g), \tag{9.3}$$

$$4NaCl(aq) + 2SO_2\,(g) + 2H_2O\,(g) + O_2\,(g) \longrightarrow 2Na_2SO_4\,(s)^* + 4HCl\,(g). \tag{9.4}$$

* *can be also molten if the temperature is over* 884°C *(melting temperature) or in the presence of other salts that form eutectic mixtures.*

It can be noticed that Eqs. (9.1), (9.3), and (9.4) result in gaseous by-products and therefore, the equilibrium is displaced to the right, i.e., to the formation of the S-containing derivatives. Eq. (9.1) often requires a catalyzer to occur, which can be an oxide at the surface and therefore most tests in the lab employ iron oxide zeolites [550] or Pt honeycombs [1401] to ensure the transformation of SO_2 into SO_3 (g). It can be also noticed that Eqs. (9.3) and (9.4) involve NaCl as the main salt contaminant in sea water [782] although mixtures with other salts can also occur. The presence of these salts in the combustion products result in significant degradation through hot corrosion.

A significant number of works have been carried out to ascertain the mechanisms of hot corrosion. These are strongly dependent on the type and amount of salt, temperature, gas flow, composition, and substrate, but they are less sensitive to the thermal regime, i.e., isothermal vs. cyclic [1096,136]. All the authors agree on Type I (above the melting point of Na_2SO_4), i.e., in the presence of a molten salt, and Type II (under deposit) corrosion. The former is often less aggressive than the latter because in Type II hot corrosion sulphation of NiO occurs at sufficiently high SO_3 partial pressure to result in $NiSO_4$. Thereafter, $NiSO_4$ is dissolved in either Na_2SO_4 solid or liquid and ends up by forming a eutectic Na_2SO_4–$NiSO_4$ at just about 671 °C that accelerates the degradation of nickel-base superalloys [956,842]. Further, acidic fluxing of protective Al_2O_3 or Cr_2O_3 scales can also occur in Type II conditions with sufficient P_{SO3} following Eqs. (9.5) and (9.6), respectively:

$$Al_2O_3 + 3Na_2SO_4 \longrightarrow Al_2(SO_4)_3 + 3Na_2O, \tag{9.5}$$

$$Cr_2O_3 + 3Na_2SO_4 \longrightarrow Cr_2(SO_4)_3 + 3Na_2O. \tag{9.6}$$

In contrast, in Type I hot corrosion, where Na_2SO_4 is molten, basic fluxing of the protective oxides Al_2O_3 or Cr_2O_3 occurs following Eqs. (9.7) to (9.10), respectively:

$$2Al_2O_3 + 2Na_2O(+O_2\,(g)) \longrightarrow 4NaAlO_2, \tag{9.7}$$

$$2Cr_2O_3 + 2Na_2O(+O_2\,(g)) \longrightarrow 4NaCrO_2, \tag{9.8}$$

$$2Cr_2O_3 + 4Na_2O + 3O_2\,(g) \longrightarrow 4Na_2CrO_4, \tag{9.9}$$

$$2Cr_2O_3 + 4Na_2O + 2O_2\,(g) \longrightarrow 2Na_2Cr_2O_7. \tag{9.10}$$

In both types of hot corrosion, incubation phenomena are followed by propagation stages that can ruin the hot section components.

At higher temperatures where the salts vaporize and are expelled, "pure" oxidation mechanisms occur by air [139] and by wet air in uncoated and aluminized superalloys [160]. At some stage of the exposure, the protective Al_2O_3 scale formed at the surface to provide environmental protection cannot be maintained. This results in degradation of the substrate by, e.g., internal oxidation, alloy depletion, appearance of cracks, etc., which lowers the overall mechanical resistance of the components. At room temperature, the engine-run components may suffer from "moisture induced spallation" in wet environments, in particular when coated with a thermal barrier coating (TBC) [1335]. In this situation, either the oxide scales or the ceramic yttria stabilized zirconia (YSZ) composing the TBC, or both spall off.

Further, the corrosion phenomenon called CMAS (calcium magnesium aluminium silicate) is of increasing occurrence because of the higher TET. The precise CMAS composition is of varying nature as it depends on the composition of the volcanic ashes, sand, sediments and dust that are ingested in the engine during take-off, approach, and landing. Such CMAS can induce wear of compressor components, clogging of internal cooling channels of the turbine components, and corrosion of the TBC [1525,1547].

In essence, the aluminium diffusion based coatings deposited by CVD (chemical vapor deposition) related techniques [1072] and the TBCs deposited by EB-PVD (electron-beam physical vapor deposition) [1070] degrade and cannot respectively provide corrosion and oxidation protection and thermal insulation anymore. This brings about overheating and internal corrosion/oxidation of the superalloy that leads to the mechanical failure of the component by loss of the γ/γ' microstructure and decrease of the load-bearing section. Therefore, the components must be refurbished before they become completely ruined in which case they will be scrapped. Figs. 9.1 and 9.2 show some typical attacks in high-pressure turbine blades and vanes after service.

9.2 Refurbishment

The MRO (maintenance, repair, overhaul) sector is big today (22 billion $) and expected to grow (32 billion $) by 2022 [653]. The MRO business will be driven by the same originally engine manufacturers (OEM) mentioned above. The inspection periods range from checks (levels):

(A) 400 h–600 h of flight (about 300 cycles)
(B) 6–8 months
(C) 20–24 months
(D) 5–6 years: full repair.

The times for inspection depend mostly on (i) the areas where the aircraft used to fly, (ii) the OEM technical specifications and (iii) the maintenance contract an airline signed with the OEM. After inspection, small repair can be accomplished on-wing. Full repair or overhaul is practiced when the time of use reaches the threshold imposed by airworthiness authorities like the FAA (Federal Aviation Administration, USA) and the EASA (European Union Aviation Safety Agency) but the OEM criteria are often more conservative. The overhaul may also be carried out in specific circumstances after inspection like, e.g., after collision of a flock of birds with the engine.

Overhauling of the components requires full disassembly of the engine, record of all components with their serial number and production of the traveller card that will follow each component (or a set of parts) upon the

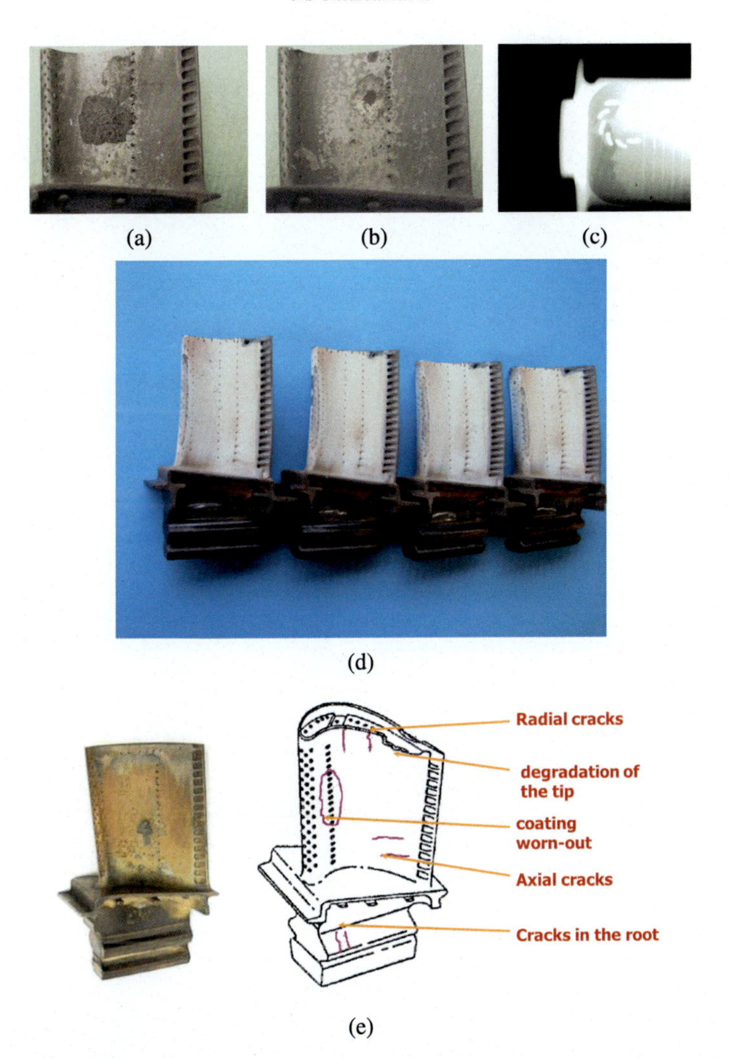

FIGURE 9.1 Engine run blades showing different degrees of attack. The brightest surfaces have been previously cleaned (stripped) to identify the extent of the attack: (a) and (b) corroded and oxidized; (c) internal clogging (radiography); (d) TBCs with CMAS deposits and detachment and the tip and leading edge; (e) schematics of major defects found in HPT blades after service.

whole repair process. When the parts to be repaired arrive in the shop, the conventional procedure follows the cycle summarized in Fig. 9.3. The very purpose of repair is schematically shown in Fig. 9.4. After service, the components often display a layer of corrosion products, a degraded coating, and very often oxidized cracks that penetrate into the material. Therefore, alloy depletion around the crack occurs.

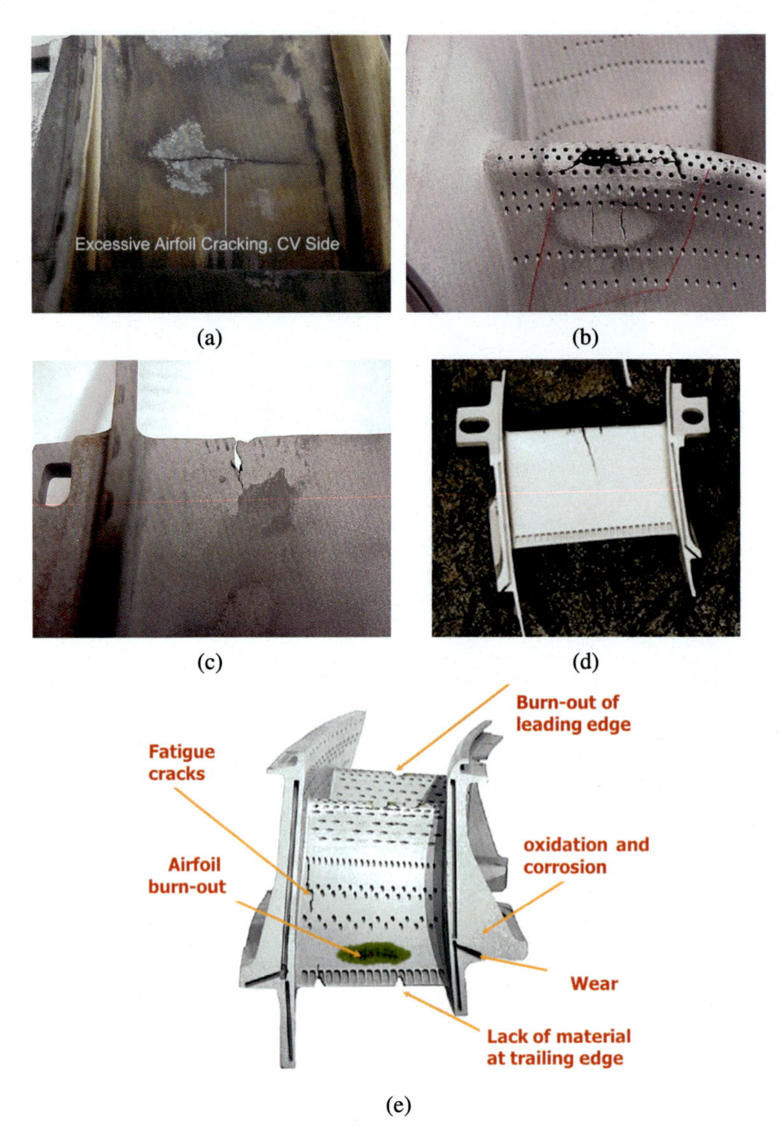

FIGURE 9.2 Engine-run nozzle guide vanes showing different degrees of attack. The brightest surfaces have been previously cleaned (stripped) to identify the extent of the attack: (a) corroded and cracked; (b) overheating and corroded; (c) and (d) creep cracks; (e) schematics of major defects found in HPT NGVs after service.

All the components are received in the repair shop, referenced, and visually inspected. In most cases, nondestructive inspection by radiography is conducted to identify the defects hidden under the worn coating, the dirty surface, and under the wall thickness (e.g., the cooling holes). The

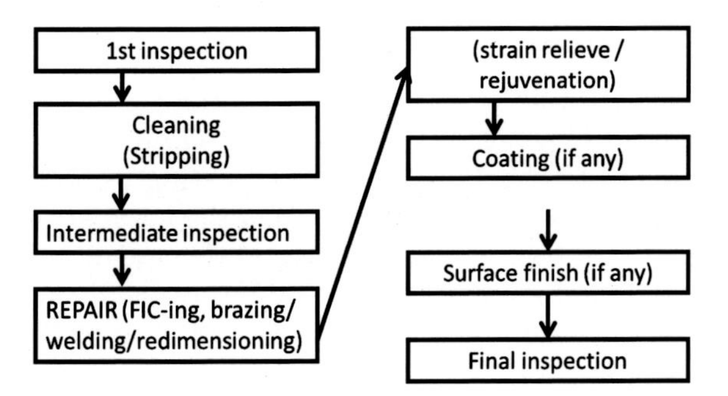

FIGURE 9.3 The repair cycle (after [1071]).

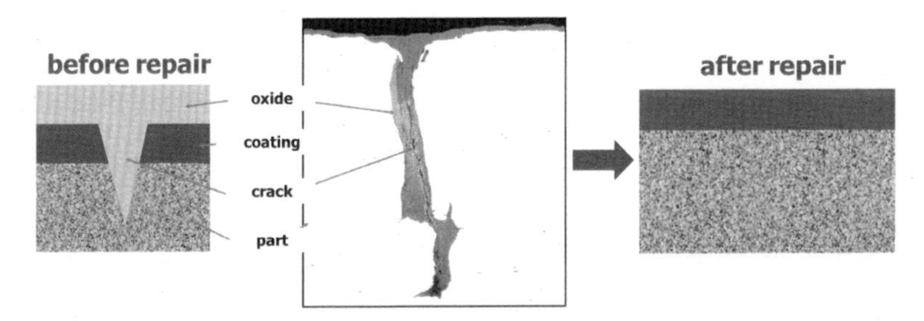

FIGURE 9.4 Schematics of the very goal of repair. The orange color (light grey color in print version) represents the corrosion and oxidation products, the blue (dark grey in print version) shows a diffusion coating, and the grey moiré is the substrate.

parts are then sorted out according to the type of damage for specific repair steps. In this chapter, the repair of fully damaged components will be explained, but the reader may well understand that some of the steps may not be applied if not needed. However, sometimes the technical specification of the OEM requires the application of all the steps regardless of the damage.

9.3 Cleaning

While all the repair steps are considered critical for the overall final performance of the component, cleaning has been often considered as the most trivial one although it is not. For instance, R. Muñoz demonstrated major failures in the repair process, hence on the quality of some repaired components when the initial cleaning was not adequately applied [993]. The cleaning processes can be classified according to their main mechanism of action, namely solubility effect, thermal effect, or physico-chemical aspects.

9.3.1 Solubility effect

The main mechanism is chemical dissolution. Therefore, REACh-compliant chemical substances shall be employed but their action is less rapid than that of old chemicals. Vapor degreasing in organic solvents is employed to remove grease and oils. Then the parts are immersed in aqueous cleaners that include surfactants and detergents, as well as soft acids (like citric acid), to remove further surface contaminants. However, the most commonly employed chemicals are hot alkaline baths (NaOH, $KMnO_4$) to dissolve oxides (including TBCs) and corrosion products followed by acid stripping to dissolve the metal coatings. There are many acid strippers depending on the specific coating type. For instance, simple NiAl coatings are dissolved in nitric-sulphamic acids [275] while methanesulfonic acids are employed to dissolve the Pt-modified NiAl counterparts [1349].These acid mixtures have the reputation of not extensively attacking the superalloy substrate, in particular the methanesulfonic acid. In practical cases, the components shall be grit-blasted intermittently to remove the smut formed at the surface that blocks dissolution of the coating. Blasting and immersion are repeated for a number of times till the coating is removed. Conventional chemical stripping has major advantages of being relatively cheap and quick, as well as requiring little investment. In contrast, the use of hazardous chemicals and the fact that the process is quite empiric are clear limitations of this process. Further, chemical stripping is not selective enough, which results in intergranular corrosion of the coating (interdiffusion zone, secondary reaction zones – SRZ) and alloy depletion can occur. Therefore, more environmentally friendly chemical stripping baths [1135] and electrochemical processes [140,812] have been proposed and patented recently [1136] to strip in a controlled and quick manner aluminium-based diffusion coatings without affecting the substrate.

9.3.2 Thermal effect

The thermal effect includes all processes where temperature is involved. Pyrolysis and thermal decomposition can be employed to burn organic matter and to decompose molecules with low sublimation points. However, the most common thermal methods are dry ice stripping and laser cleaning, in particular the latter.

Dry ice stripping requires preheating the parts at a given temperature so that when sprayed with dry ice (solid – frozen – CO_2) a thermal shock occurs between the sublimation temperature of CO_2 ($\approx$78.5°C at 1 bar) and the hot surface. The shock brings about the detachment of oxides, TBCs preferentially with respect the bond coats (overlay MCrAlYs and diffusion aluminides). The process is very efficient, but the cost of dry ice is relatively high. Therefore, laser cleaning is preferred. Indeed, there is a great variety of lasers whose power can be adjusted as a function of the cleaning requirements. Over the past few years, different French labs have developed the LASAT (laser adhesion test) as a nondestructive method to evaluate adhesion of EB-PVD TBCs [146,99], but over a threshold power the TBC can also detach. However, access of the laser to the internal surface of hollow components is not possible.

9.3.3 Physico-chemical aspects

The processes rely on joint physical and chemical effects, but the share of the physical contribution is by far larger than the chemical one.

Wet and dry blasting employ a jet at different pressure and flow of media that include alumina, sand, glass beads, nut shells, plastics, etc., made of different shapes (spherical, irregular, dendritic, etc.). They strike against the surface to be cleaned at a given pressure. The most conventional process is grit blasting (dry) using alumina and compressed air expelled through nozzles of different size. The process can be conducted manually or automatically, the latter being the preferred method to avoid deformation and distortion of the aeronautical profiles of the turbine components. In the absence of solid media, water jet stripping provokes also abrasion of the surfaces at very high pressures so that the blasting is performed in closed rooms. The process is though very cheap and allows removing TBCs (EB-PVDs and plasma-sprayed coatings) as well as overlay MCrAlYs, bond coats, the latter being difficult to dissolve using chemical means without affecting the substrate. Other abrasive processes, like brushing and vibropolishing, are often reserved to clean very small areas and to obtain smooth surfaces, respectively, which are not required at this stage.

Fig. 9.5 shows an engine-run blade before and after conventional chemical stripping. One can observe a mat surface due to some fine residues

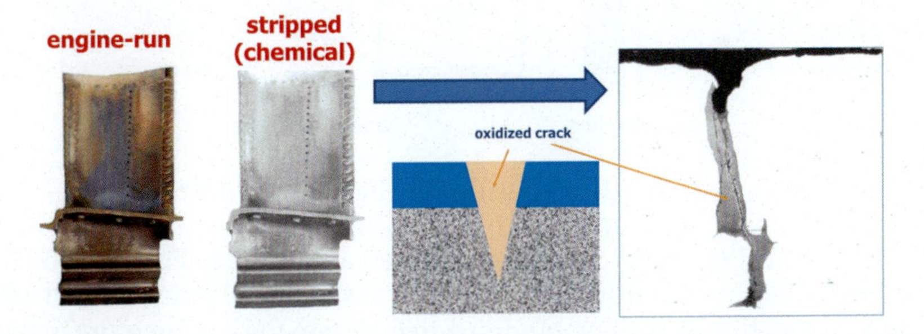

FIGURE 9.5 Engine-run blade before and after conventional chemical stripping. The stripping solution has removed the superficial corrosion products. The stripped surface shows different contrast from smut and overheating. In the cross-sections, the oxide of the cracks has not been removed and remnants of coating (blue color) are still visible.

of smut that are often removed by additional (manual) grit blasting with alumina. Overheating has also reduced the amount of the strengthening γ'-Ni_3Al of the superalloy substrate. In the cross-section, the corrosion products have been removed but the cracks remain oxidized. These cracks are often sufficiently thin so that the chemical solution cannot penetrate, or else the extended period of the immersion may damage the substrate. Also, in the example shown here, part of the diffusion coating (blue color) has been left on purpose to avoid excessive reduction of the wall thickness of the component but in most occasions the coating is fully removed. The removal of the oxides from the cracks is thus performed by fluoride ion cleaning (FIC).

9.4 Fluoride ion cleaning (FIC)

Fluoride ion cleaning is a very efficient process to remove oxides from thin and thick cracks. As shown in Fig. 9.6, HF (g) and H_2 (g) are introduced into the reaction chamber by a carrier gas (Ar). The chamber is heated at temperatures often ranging between 950°C and 1050°C [750] in a very similar manner as for the CVD processes [1072]. Stagnant and dynamic flowing conditions of the gas at low pressure (about 100 mbar to 150 mbar) can be employed depending on the substrate and types of oxides to be removed. FIC is basically employed to remove Al_2O_3 and TiO_2 oxides, but Cr_2O_3 can also be removed. The formation of these oxides requires low oxygen partial pressures like those encountered within the crack, in particular at the tip.

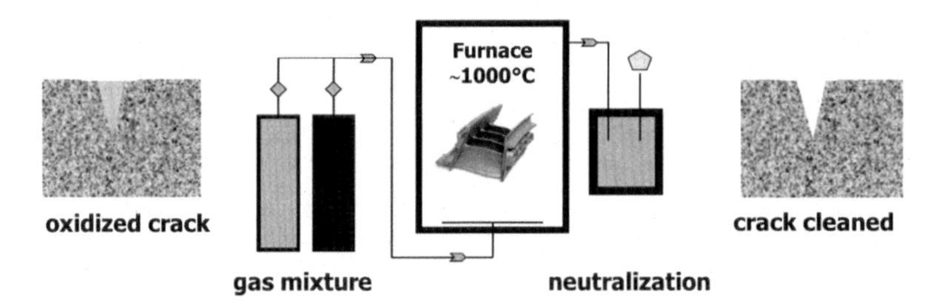

FIGURE 9.6 Schematic representation of the fluoride ion cleaning (FIC) process.

The chemical reactions can be summarized according to Eqs. (9.1)–(9.3). The oxides react with HF (g) and result in gaseous metal fluorides that are expelled from the crack upon pumping to maintain the low pressure in the chamber. Water vapor also forms, but the amount and residence time are too short to oxidize the material again. In contrast, once the oxide layers have been removed or if they are sufficiently porous, the strengthening elements (Al, Ti) of the γ'-$Ni_3(Al, Ti)$ readily react with HF (g) (Eqs. (9.4) and

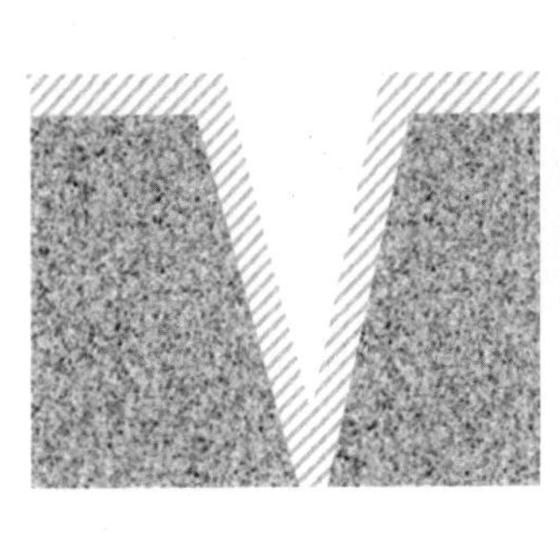

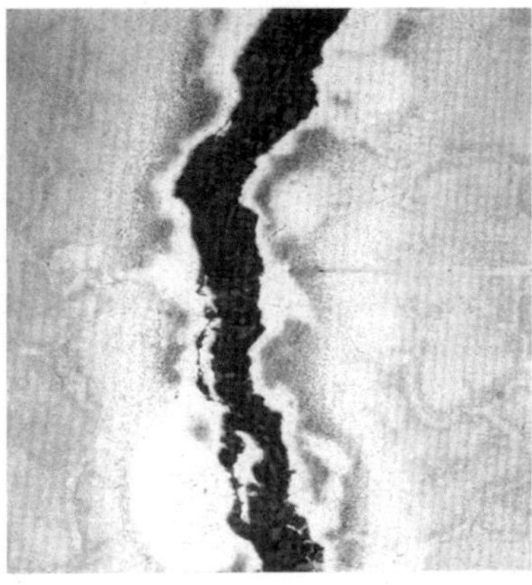

FIGURE 9.7 Schematic drawing and optical picture of alloy depletion after FIC process (here for an equiaxed superalloy).

(9.5)) which brings about alloy depletion like the one shown in Fig. 9.7. In addition to the very significant hazards of the process, alloy depletion is a major drawback but as of today it is the most practical solution to produce pristine metal surfaces as a compulsory requirement for brazing:

$$6HF\,(g) + Al_2O_3 \longrightarrow 2AlF_3\,(g) + 3H_2O\,(g), \tag{9.11}$$

$$4HF\,(g) + TiO_2 \longrightarrow TiF_4\,(g) + 2H_2O\,(g), \tag{9.12}$$

$$6HF\,(g) + Cr_2O_3 \longrightarrow 2CrF_2\,(g) + F_2\,(g) + 3H_2O(g), \tag{9.13}$$

$$6HF\,(g) + 2Al \longrightarrow 2AlF_3\,(g) + 3H_2(g), \tag{9.14}$$

$$8HF\,(g) + 2Ti \longrightarrow 2TiF_4\,(g) + 4H_2(g). \tag{9.15}$$

9.5 Patching and brazing

Once the surfaces are rendered chemically and physically cleaned, the surface defects (lack of material) and the cracks can be filled in with superalloy material by patching and brazing, respectively. In both cases, superalloy powders (fillers) with a melting temperature lower than the one of the parent material are mixed with a binder to prepare a slurry. The powders often contain melting depressant elements like B, Si, and Ge and there are cents of these slurries in the shops like "RBD (remetalling brazing diffusion)" to ensure brazing and patching depending on the superalloy substrate. Liquid slurries are injected with a syringe (Fig. 9.8(a))

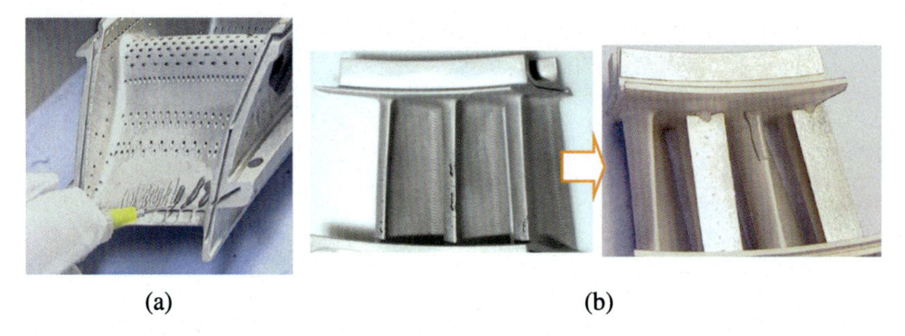

FIGURE 9.8 Application of braze on single crystal nozzle guide vanes: (a) application by syringe of liquid slurry over the cracks; (b) application by patch to cover larger areas.

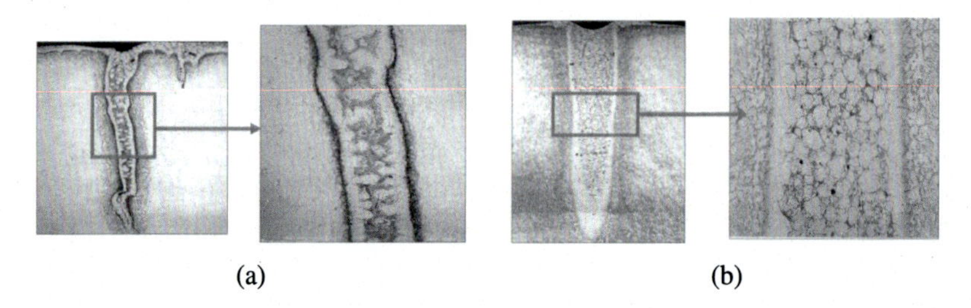

FIGURE 9.9 Microstructures of brazed and annealed: (a) thin cracks; (b) thick cracks. Note the eutectic microstructure in the equiaxed grains formed after solidification and the diffusion affected zone.

in the cracks provided they are sufficiently wide and the operator is highly skilled. Otherwise, they are placed at the top of the crack. The slurries can be also dried to obtain different shapes (patches) that are applied on larger areas like a plaster over a wound (Fig. 9.8(b)). Both in brazing and in patching, annealing is carried out under vacuum so that the molten superalloy filler wets the gap provided this wide enough. Alternatively, it enters by capillary forces if the gap is thin ($< 100\,\mu m$).

One major advantage of melting the metal is that potential remnants of oxide in the crack are dissolved. The melting temperature depressant diffuses into the parent material which allows the filler alloy to increase in temperature, hence to solidify. Finally, homogenization between the parent and filler materials occur with annealing time. One major drawback for single crystal materials is that the resulting braze is equiaxed (Fig. 9.9). Other major issues are the compounds like Ni-rich and Cr-rich borides, ternary eutectics of Ni–Si–B and diffusion affected zones all being relatively brittle [585]. It thus derives that the mechanical properties of the brazed components are lower than those of their fully single crystals counterparts.

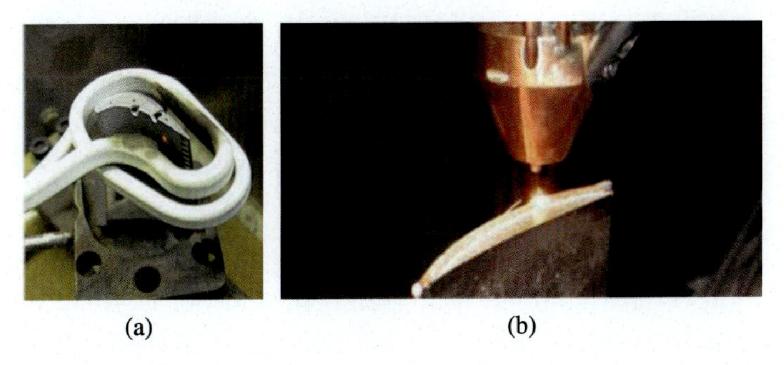

(a) (b)

FIGURE 9.10 Welding processes for tip repair of single crystal turbine blades: (a) SWET; (b) laser cladding.

9.6 Redimensioning

Redimensioning includes all the operations needed to recover the original shape of the component. In most cases, matter is supplied by brazing as discussed above and by welding. Among the welding processes, TIG (tungsten inert gas) is vastly employed. Superalloy welding at elevated temperature (SWET) is often applied for tip repair as it allows building up the worn tip of the blade by addition of matter (Fig. 9.10(a)). The tip of the blade is surrounded by an induction coil so that there is barely any affected zone off the tip of the blade and the process does not induce cracks like with more conventional welding processes. In addition, laser cladding can be employed (Fig. 9.10(b)).

The material deposited needs afterwards to be deburred usually by blending and polishing. In addition, the cooling holes of the blade shall be reopened, which is often done by electro-discharge machining (EDM). Like with any other welding process, the microstructure is equiaxed and is thus dissimilar with respect the single crystal initial nature of the blade. The same applies with current laser cladding processes although the mechanical performance, i.e., the lifetime of the component is extended with respect tip blades repaired by SWET.

Therefore, many investigations and developments have been recently conducted taking advantage of additive manufacturing processes. Processes like scanning laser epitaxy (SLE) to deposit single crystal superalloy material onto single crystal superalloy substrates are now patented [276]. Alternatively, the Laser Zentrum in Hannover (Germany) has recently demonstrated that complete reorientation of the regions previously misoriented can be achieved in tip repair of CMSX-4 nickel-base superalloy despite the difficulties of welding this superalloy [168]. The process is in fact a combination of laser cladding with the tip surrounded by induction coils like in SWET.

9.7 Rejuvenation

Rejuvenation is a process to restore microstructural characteristics of a turbine component after geometrical refurbishment processes. The process was first established for polycrystalline turbine component [1469, 332,840,669] and has been continuously developed for SX applications [1213,1235,1610,1315,1047,839,1243,1244,626]. The technique is usually a combination of near solution heat treatment and hot isostatic pressing (HIP), trying to avoid possible recrystallization during such a process. As explained previously, geometrical damages appear to turbine components after service operation. There are also internal damages to the microstructure of the components induced by long-duration high-temperature exposure and creep loading [358,110,1507,957,1311]. Overaging precipitation coarsening, creep void formation, precipitation of brittle intermetallics known as topologically close-packed (TCP) phase [312,1162,1329], and changes in carbides [862,878] are the main damages in single crystal superalloys. Additionally, changes in grain boundaries occur in polycrystalline alloys. The types of damage and the changes of microstructure differ depending on the area of the component due to the variety of temperature/stress conditions. An example of after-service microstructure in a turbine blade is shown in Fig. 9.11. Generally, overaged coarsening

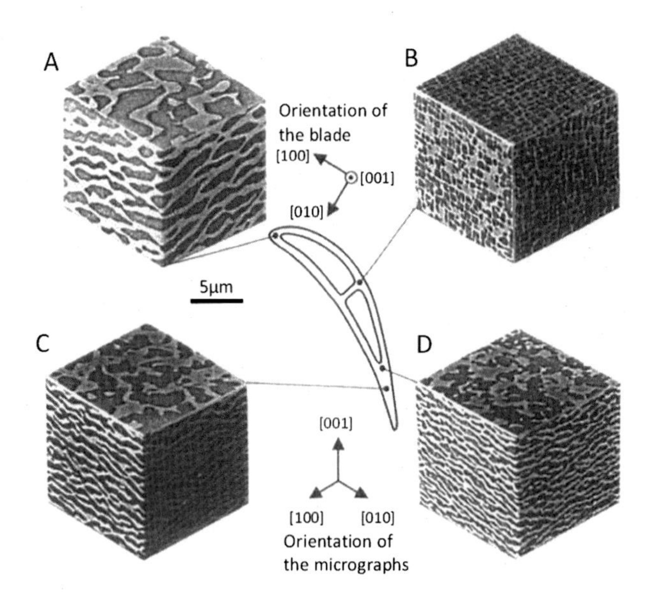

FIGURE 9.11 Schematic drawing of the cross-section with SEM images of the microstructures for different positions after an accelerated mission test. Adapted from [1507] with permission from Taylor & Francis Ltd. (www.tandfonline.com).

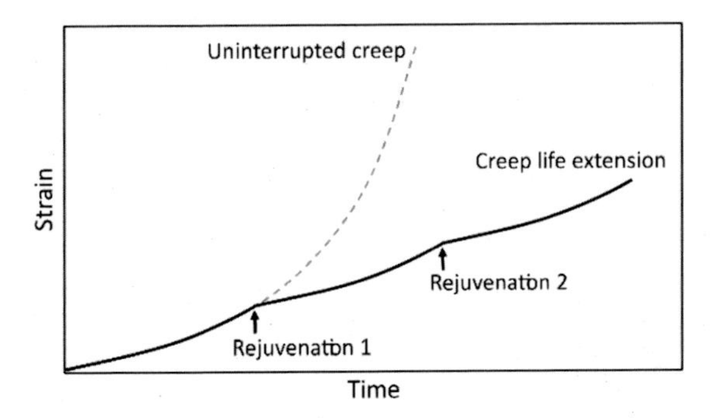

FIGURE 9.12 Schematic representation of repetitive rejuvenation cycles on creep life of a superalloy.

decreases creep resistance under lower-temperature/higher-stress condition, and creep void formation occurs and become a failure source under higher-temperature/lower-stress condition. Rejuvenation that refurbishes bulk mechanical properties of SX superalloy is an important process to extend service life of a turbine component as schematically shown in Fig. 9.12, especially large blades and vanes of industrial gas turbines or components with complex geometries with associated elevated cost.

The main idea of reheating up to solvus temperature is to dissolve γ' precipitates into γ phase and to regenerate nearly optimal γ' cuboids and γ matrix by following precipitation aging treatments just like original solution + aging treatments [207,554]. As shown in Fig. 9.13, microstructure of SX superalloy CMSX-4 can be restored by rejuvenation heat treatment after overage coarsening at 1050°C for 1000 h [1610,1612]. During re-solution stage, dislocations introduced during deformation will be annihilated or rearranged because the γ' phase that blocks movement of dislocations does not exist and the very high temperature that enhances diffusion for rearrangement. Hence, the rejuvenation heat treatment can be also applied to SX components with a room temperature plastic deformation to remove internal stress (Fig. 9.14) [1457]. Dissolution of TCP phases [1315,1047] and partial decomposition of carbides [1213] also occur during re-solution with higher solubility of γ phase at such high temperature. However, the solvus temperatures for some TCP phases and carbides are higher than that of the γ' phase and especially complete dissolution of carbide can be very close to the solidus temperature (DTA curves for solvus are shown in Fig. 9.15) [1433]. Because of possible existence of TCP phases [1161] and chemical composition gradient in the interdiffusion zone [1235], complete removal of metallic bond coating and interdiffusion zone before re-solution treatment is also important. When precipitation aging treatments are applied, γ' phase precipitates from full γ phase as normal aging treatment. This rejuvenation heat treatment is known to effectively extend

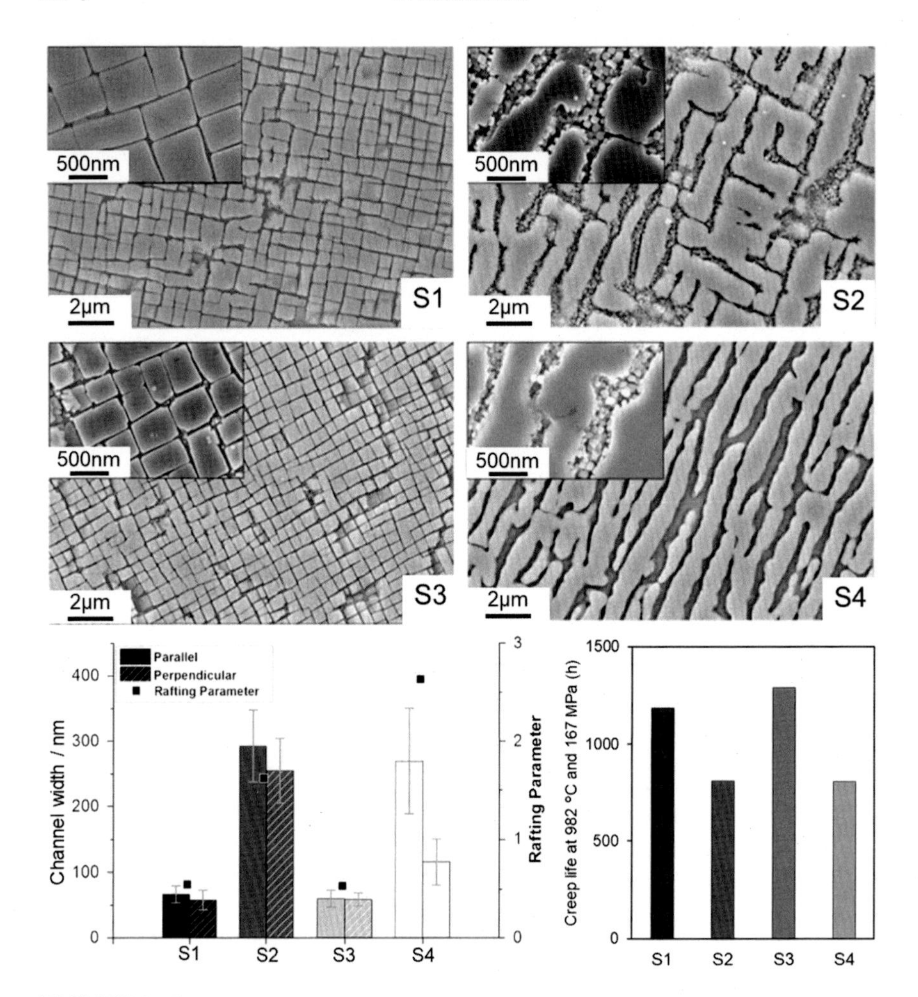

FIGURE 9.13 SEM images of the CMSX-4 samples after different heat treatment conditions: (S1) preservice, (S2) rafted by aging at 1050°C for 1000 h, (S3) S2 + solution heat treatment, (S4) S3 + rafted by aging at 1050°C for 1000 h. Adapted from [1611] (licensed under CC BY 4.0).

creep life of SX superalloys (Fig. 9.16) [1243,626] and maximum service life can be achieved if the process is applied before the accumulation of microstructural damages [1244,626], which mainly occurs during creep acceleration stage after minimum rate creep period. Fig. 9.17 shows the accumulation of creep damage that is exposed after second creep loading on the rejuvenated SX samples. Faster coarsening was observed after second creep loading in the sample with 5.0% strain during first creep loading [626].

Re-solution treatment can only restore overaged precipitates into original form, but creep voids cannot be eliminated. Such voids in damage con-

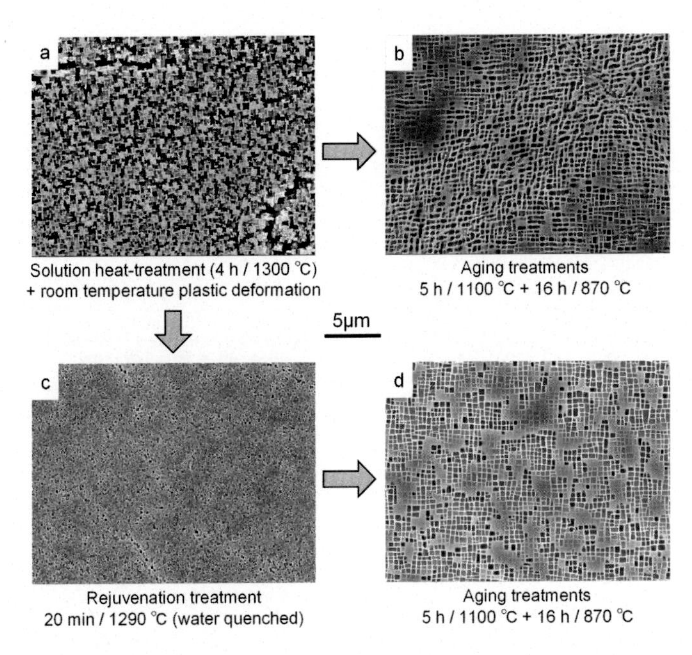

FIGURE 9.14 Microstructures of AM1 SX superalloy after (a) solution heat treatment + room temperature plastic deformation (plastic strain 0.8%), (b) aging treatments on (a), (c) rejuvenation treatment on (a), and (d) aging treatments on (c) [1457].

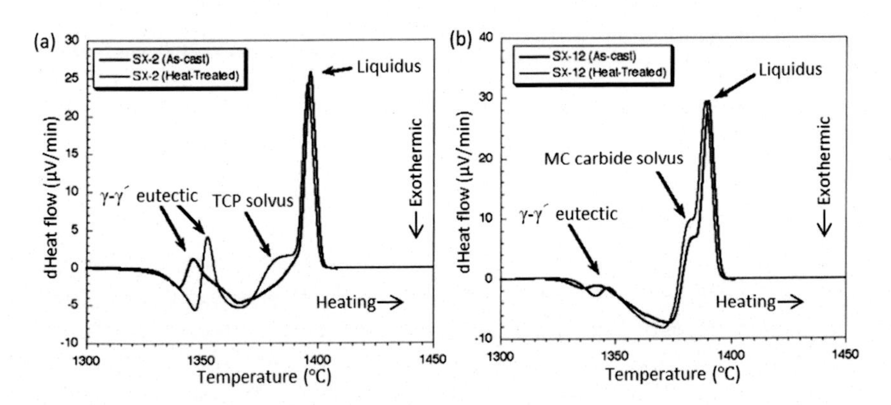

FIGURE 9.15 Comparison of DTA heating curves for experimental alloys: (a) without carbon content; (b) with carbon content. Adapted from [1433] with permission from Elsevier.

centrating area can limit high cycle fatigue life of SX component [785,223], as well as creep life [10,970,743,982]. Since HIP procedure is known for its advantage of increasing properties of superalloys by densifying and removing casting voids [223,982] this method is also useful for solving creep void issue if it is applied in the refurbishment process. Therefore,

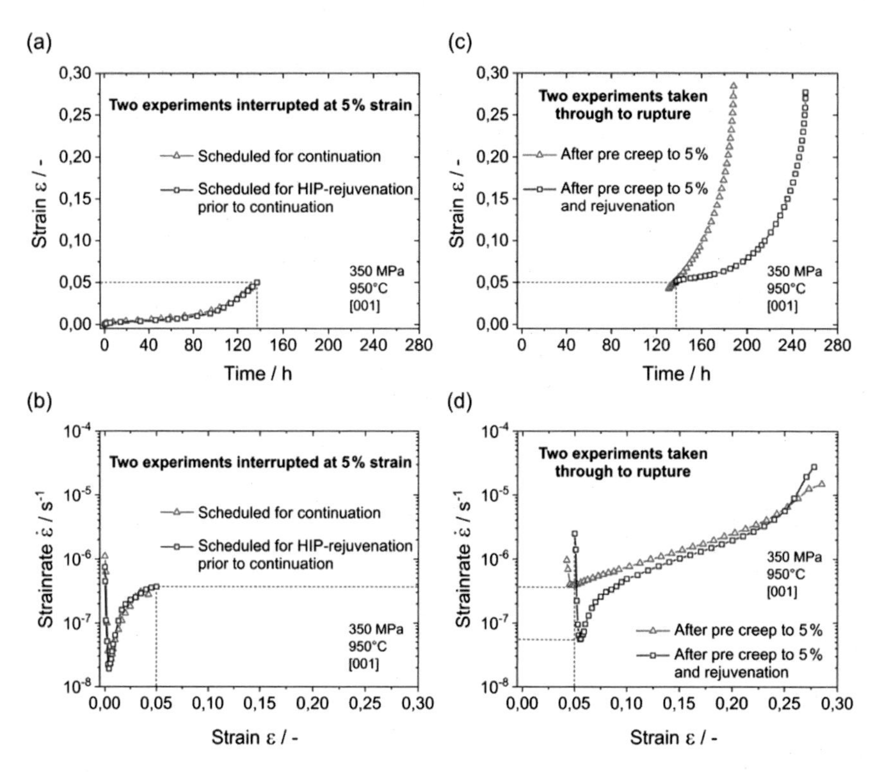

FIGURE 9.16 Creep curves of two experiments on ERBO/1 SX superalloy: (red triangle) interrupted at 5.0% strain, fast cooling to room temperature, and then continued at same condition; (blue square) interrupted at 5.0% strain, fast cooling to room temperature, HIP-rejuvenation treatment, continued at the same condition. (a, c) Creep and creep strain rate curves for first creep loading. (b, d) Creep and creep strain rate curves for second creep loading. Adapted from [626] with permission from Elsevier.

HIP treatment is processed at high-temperature (above 1200°C) which is ideally re-solution temperature so that microstructure restoration can proceed simultaneously. Example of a SX superalloy after creep deformation followed by rejuvenation treatment consisting of re-solution and HIP is shown in Fig. 9.18. However, the cooling rate after HIP treatment is often limited by the HIP facility and this must be considered to obtain optimal precipitate size and shape after all heat treatments.

The simplest way of determining re-solution temperature is to follow original heat treatment procedures. However, the temperature can be lower than original solution temperature for two reasons. First, the SX component underwent solution homogenizing treatment in the production, and therefore, the solvus temperature should be lower than the as-cast alloy. Second, plastic deformation can create internal elastic strain field with sub-grain and local lattice misorientation. With such kind of in-

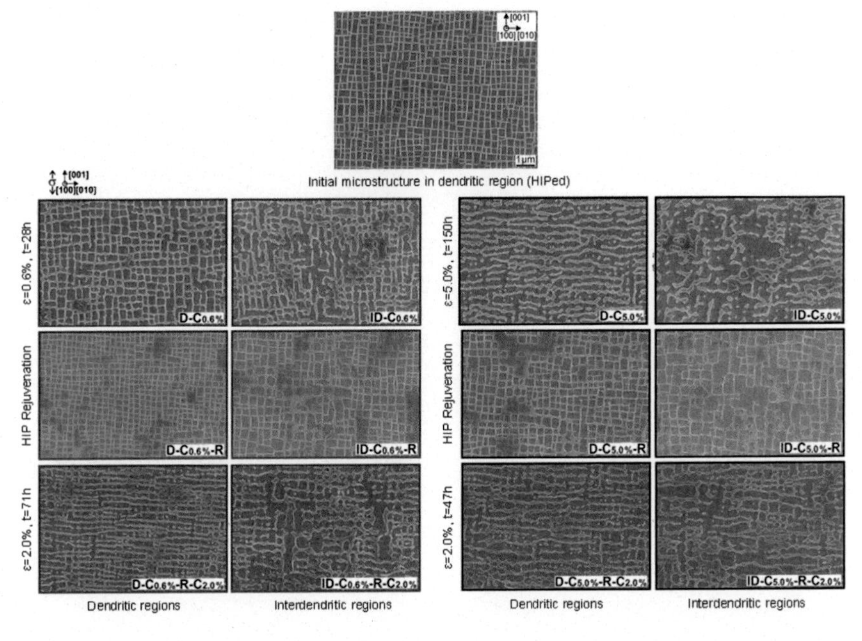

FIGURE 9.17 Microstructures of ERBO/1 SX superalloy at different stages. Initial microstructure in a dendritic region is on the top. Six images on the left and six images on the right are after creep test (950°C/350 MPa) interrupted at 0.6% and 5.0% strain, respectively. After creep interrupting test, samples are treated by HIP Rejuvenation and applied creep load again at the same condition until they reach 2.0% strain, then interrupted. Adapted from [626] with permission from Elsevier.

ternal strain, exposing a component at very high temperature can activate recrystallization in the microstructure (Figs. 9.19 and 9.20) [1235], which is detrimental to the mechanical properties of SX superalloy and must be avoided. Extra care for solution treatment is necessary if a component previously underwent geometrical repair such as machining and cladding. It is obvious that parts with additive redimensioning are locally polycrystalline and re-solution temperature must be lower than that of SX. Even if nearly SX laser surface-melting is applied, there is still a risk of recrystallization at near solution temperature (Fig. 9.21) [1014]. Residual stress which may be introduced on the component surface during machining is another cause of recrystallization when exposed to high-temperature (Fig. 9.22) [933,934,1316]. Re-solution temperature in the rejuvenation process should be determined by considering thermal history, operation history, and process history in order to estimate stored damages and to avoid recrystallization. There are not many open researches available on the threshold of creep damage and the biggest factor of activating recrystallization during re-solution treatment. In addition, efficiency of HIP treatment is affected by the process temperature because internal diffusion rate

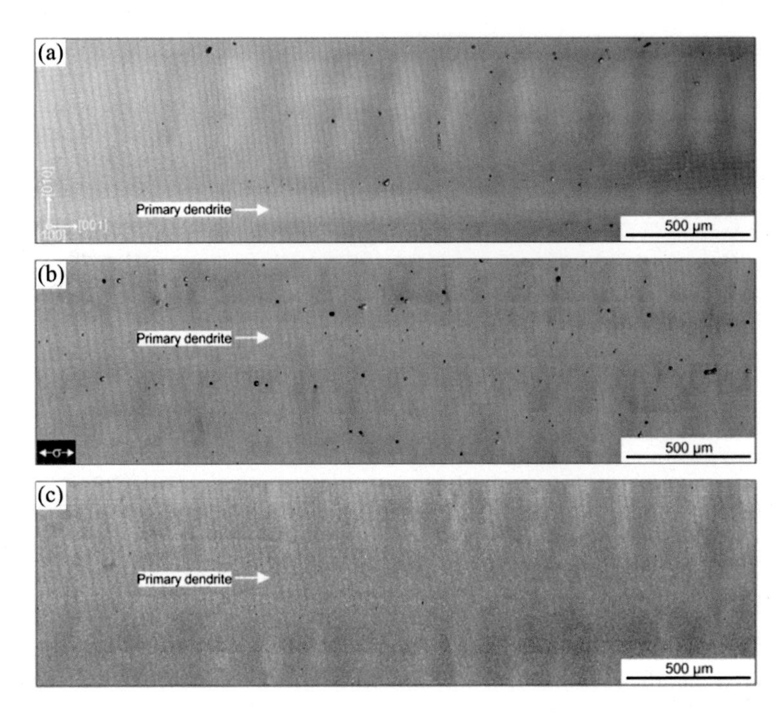

FIGURE 9.18 SEM-BSE images of the same region of the ERBO/1C SX creep testing sample: (a) the initial state prior to creep; (b) after creep test at 1050°C/160 MPa interrupted at 5% creep strain; (c) after HIP rejuvenation treatment. Adapted from [1243] with permission from Elsevier.

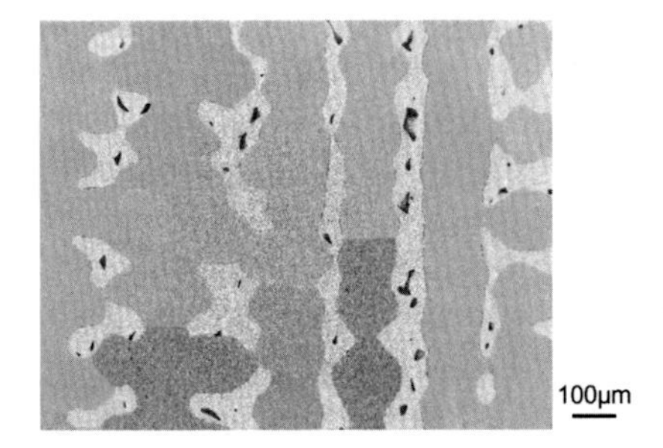

FIGURE 9.19 SEM image of AM1 SX superalloy after standard solution treatment + tensile plastic deformation (plastic strain 2.17%) at 950°C + rejuvenation treatment at 1290°C for 20 min + aging treatments 5 h/1100°C and 16 h/870°C [1457].

is the dominating factor for densification [982]. However, re-solution temperature is high enough for this purpose and recrystallization risk is the most important factor for the determination of re-solution temperature.

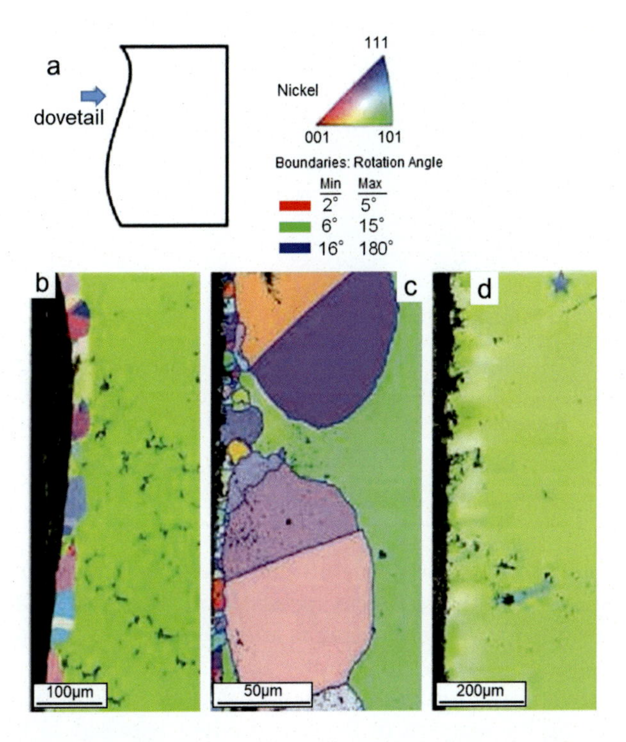

FIGURE 9.20 (a) Schematic drawing showing the dovetail location of after-service René N5 SX turbine blade (minimum 4000 engine cycles). EBSD KAM images revealing: (b) the occurrence of recrystallization along the dovetail edges after the rejuvenation heat treatment; (c) having a bimodal distribution of grains; (d) the nonmachined surface on the dovetail without recrystallization. Adapted from [1235].

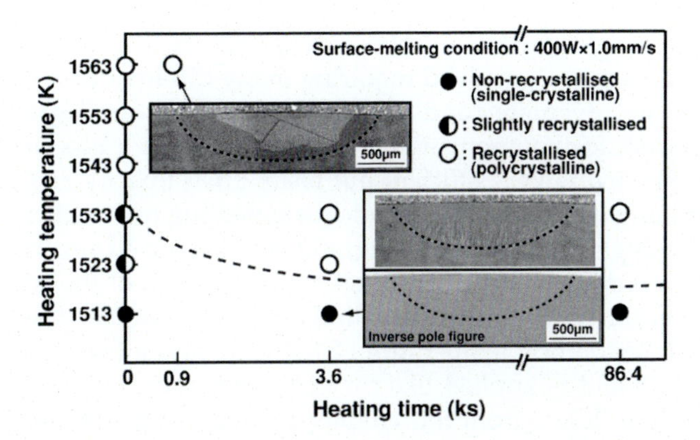

FIGURE 9.21 Effect of heat treatment temperature and time on recrystallization behavior in SX surface-melted region. Adapted by permission from RightsLink: Springer Nature, Welding in the World [1014] ©2008.

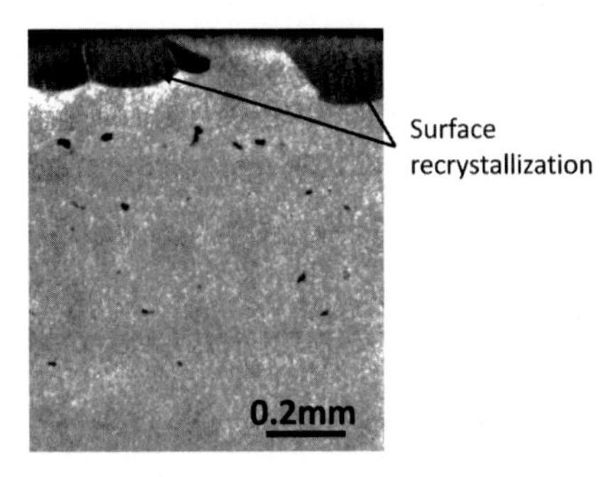

FIGURE 9.22 Optical microscope image of AM1 SX superalloy after standard solution treatment + tensile plastic deformation (plastic strain 0.8%) at 750°C + rejuvenation treatment at 1290°C for 20 min + aging treatments 5 h/1100°C and 16 h/870°C (authors' own works).

9.8 Recoating

Once the component has been fully repaired, a final coating may be applied if indicated in the technical specification. The application of coatings is the same as for new components and will not be developed on this chapter. The reader can refer to Chapter 10. However, in some situations, incorrect application of brazes may be revealed after the application of a diffusion coating by reaction of the braze when exposed to the coating atmosphere. In these cases, the coating shall be stripped, the component FIC-ed and the process shall restart upon consent of the OEM.

9.9 Conclusions

Repair of single crystal nickel-base superalloys is a clear challenge where many steps are involved requiring many different processes. Each process convey advantages and limitations. Stripping can degrade drastically the superalloy component and is critical because it blocks the other processes. FIC-ing is very efficient but bears great toxicity and alloy depletion. Brazing of cracks requires adequate wetting of the filler material. Redimensioning by brazing, patching, and welding are all very convenient treatments but do not restore the single crystal nature of the component. Rejuvenation can restore microstructural properties but may cause recrystallization at re-solution temperature. Finally, recoating does not bear any particular risk and is applied like for new components. Some final considerations that derive from this chapter can also be made. For instance, most of the repair and rejuvenation processes shall be adjusted for each single component and material they are made of. Therefore, there is a vast territory of investigation on the mechanical properties and environmental resistance of the repaired components. Further, it shall be recalled that economic factors, i.e., final price, will determine the final decision of a component being repaired. For instance, cracks developed from the cooling holes close to the platform can be technically repaired but the overall cost is far more significant than replacing the component, let alone that the repaired component is not mechanically adequate.

Acknowledgments

Dr. J. Cormier (ENSMA, France) is gratefully acknowledged for the thorough reading and critical comments of this chapter. Some of the figures presented in this chapter are adapted from their originals kindly provided by SR Technics Airfoil Services, Ltd. (Cork, Ireland) and Safran Aircraft Engines (Châtellerault, France).

CHAPTER

10

Coated single crystal superalloys: processing, characterization, and modeling of protective coatings

Vincent Maurel[a], *Marion Bartsch*[b], *Marie-Helene Vidal-Sétif*[c], *Robert Vaßen*[d], and *Vincent Guipont*[a]

[a]MINES ParisTech, PSL University, MAT - Centre des Materiaux, UMR CNRS 7633, Evry Cedex, France, [b]Institute of Materials Research, German Aerospace Center (DLR), Cologne, Germany, [c]Department of Materials and Structures (DMAS), ONERA Université Paris Saclay, Châtillon, France, [d]Institute of Energy and Climate Research: Materials Synthesis and Processing (IEK-1), Jülich, Germany

Glossary

APS air plasma spray
BC bond coat
BSE backscattered electrons
CMAS calcia-magnesia-alumino-silicate
DIC digital image correlation
EB-PVD electron beam physical vapor deposition
NMG net mass gain
OM optical microscopy
RE rare earth element
RT room temperature
SE secondary electron
SEM scanning electron microscopy
SX single crystal
TBC thermal barrier coating
TC top coat
TGO thermally grown oxide
Y(P)SZ yttria (partially) stabilized zirconia

Nickel Base Single Crystals Across Length Scales
https://doi.org/10.1016/B978-0-12-819357-0.00018-4

10.1 Introduction

Nickel-base superalloys, especially single crystalline (SX) materials, are developed for parts operating under demanding conditions, such as gas turbine vanes and rotor blades. While the superalloys are optimized to sustain high thermal and mechanical loads, they need protective coatings against oxidation and corrosive attack. Further, the request for increasing the turbine inlet temperatures beyond the melting temperature of the superalloys motivated the implementation of active internal cooling of turbine vanes and blades. Applying an additional thermal barrier coating (TBC) system improved the efficiency of internal cooling, resulting in lower substrate temperatures or allowing either higher turbine inlet temperatures or reducing cooling air mass flow. Examples of TBC coated vane and blade for an aeroengine are displayed in Fig. 10.1.

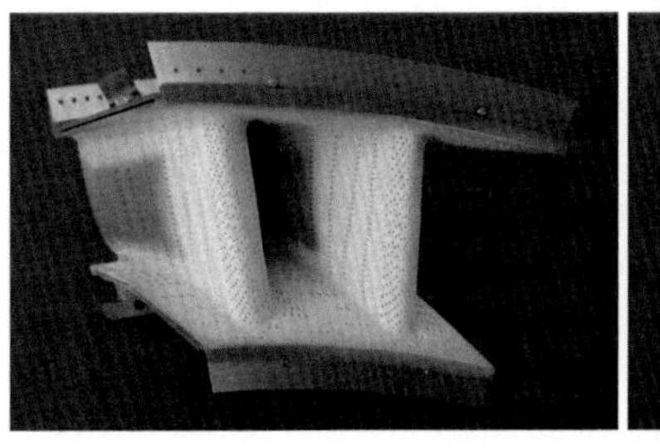

(a) TBC coated vane

(b) TBC coated blade

FIGURE 10.1 TBC coated vane (a) and blade (b) of a GP7000 aeroengine.

The most widely used material for thermal insulation is partially yttria stabilized zirconia (YSZ) with about 6–8 wt% yttria. The main coating technologies currently employed are electron beam physical vapor deposition (EB-PVD) and air plasma spraying (APS), leading to highly porous coatings with a columnar or lamellar microstructure, respectively. The porosity entails a low thermal conductivity and a low elastic modulus, the latter providing a high-strain tolerance. Since the YSZ allows fast diffusion of oxygen, as well as transport of oxygen and corrosive media via open porosity, it is mandatory to apply onto the substrate metallic layers which provide a sufficient reservoir of Al or Cr to form Al_2O_3 or Cr_2O_3 diffusion barriers, referred to as thermally grown oxide (TGO) layers. Because the TGO formation improves the adherence of the ceramic top coat (TC), the protective metallic layer in a TBC system is referred to as bond coat (BC). Most BC applications are made either of NiAl, (Ni, Pt)Al, or MCrAlY (M representing Co or Ni). Fig. 10.2 shows a schematic of a thermal TBC system, with typical EB-PVD and APS top coat, respectively. The range of typical values for thermal conductivity, thermal expansion coefficient

(CTE), and Young's modulus are given in Tables 10.1, 10.2, and 10.3, respectively.

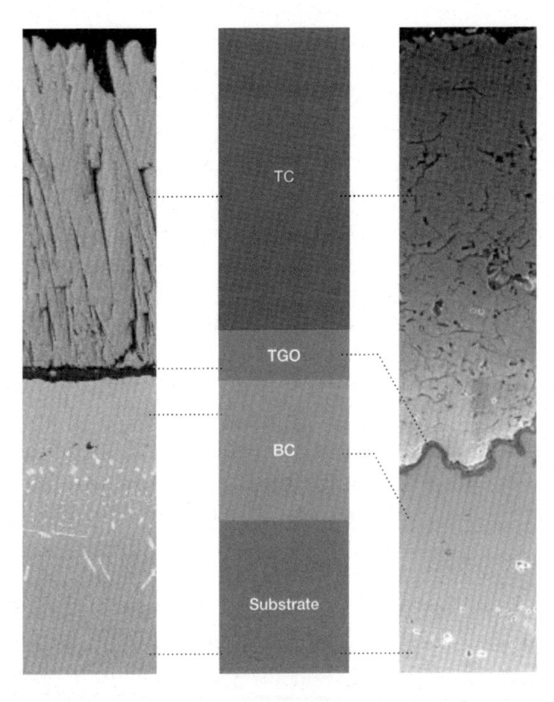

FIGURE 10.2 Typical TBC systems comprising ceramic TC, TGO, and metallic BC applied on a superalloy substrate: (left) EB-PVD processed TC with a (Ni, Pt)Al BC (courtesy of Lara Mahfouz), (right) APS processed TC with an MCrAlY BC; adapted from [1043].

The design of a TBC system, implying the composition and processing of each layer, TC, TGO, BC, and substrate, has a deep impact on the durability of TBC systems. Differences in thermal expansion coefficient (CTE) between the layers and mismatch of their elastic properties result in residual stresses and stress gradients across the layer system during temperature changes, such as cooling after processing or thermal cycling in service. Thermal gradients due to the internal cooling of coated components contribute to high stresses and stress gradients across the TBC system. Phase transformations, such as temperature-dependent ductile-to-brittle transformations, may cause in some BC materials damage and subsequent failure of the TBC system. In service, time-dependent processes, such as the growth of the TGO or sintering of the TC, change the stress field but also material properties, such as strength and fracture toughness. Especially, at the interfaces of different layers diffusion of elements is driven by concentration gradients and may lead to phase reactions or the formation of voids and precipitates. Simultaneously, fatigue loads are effective, such as low cycle fatigue (LCF) associated to an entire engine cycle, e.g., one flight mission of an aeroengine, or high cycle fatigue (HCF), e.g., associated to transient pressure differences.

TBC systems have been intensively investigated because of their importance for the performance of high temperature components made from Ni-base superalloys and because of the implied technological and scientific challenges associated to their processing and use. Tremendous literature has been published on TBC systems for superalloys, including comprehensive overviews, e.g., [329,1043,1190], or focused review articles, e.g., [1260] on processing, [1581] on industrial application, and [411,1574] on lifetime modeling and testing.

However, continuing demands of increasing the turbine inlet temperature for raising the thermal efficiency of the gas turbine process entails new challenges for the coating systems and enhances research and development in processing and characterization of TBC systems. Another strong motivation for innovations is the need for lowering the production costs of high performance components with protective coatings. Main research activities aim at lowering thermal conductivity, improving phase stability at high temperature, limiting sintering of the ceramic TC, controlling TGO formation and behavior during service, and developing measures to avoiding damage by the intake of low fusion point materials, such as volcanic ash or mineral deposits containing CaO–MgO–Al_2O_3–SiO_2. These efforts include innovations in characterization, modeling, and lifetime assessment. Some of the recent developments which aim at allowing higher in-service temperatures and improving resistance to harsh environments are addressed in this chapter. Further, related innovation in characterization, testing, and modeling of protective high temperature coatings is reviewed.

10.2 Innovative materials and coating processes

10.2.1 Corrosion protection layers

As single-crystal superalloys are mainly developed towards high-temperature mechanical properties, their oxidation and corrosion resistance is not sufficient to guarantee long-term operation at elevated temperatures under a corrosive environment. As a result, superalloys are in most cases coated with corrosion-resistant coatings typically consisting of high aluminum and/or chromium containing alloys. The standard deposition methods are chemical vapor deposition (CVD) for aluminide coatings and different thermal spray techniques for NiCoCrAlY coatings. These deposition techniques, as well as alternative routes, are described in the following section. In addition to the corrosion protection, aluminide and MCrAlY coatings often serve as the so-called bond coats for thermal insulation layers, the thermal barrier coatings (TBCs). Due to the rather high temperatures (above 900°C) at the interface bond coat/thermal barrier coating, chromia scale formers are not suitable as the scale growth rates are high and additionally volatile CrO_3 species are formed at temperatures above 1000°C [436].

10.2.1.1 Deposition from the gas phase: chemical vapor deposition (CVD) of aluminide and platinum aluminide coatings

Aluminide coatings are often deposited by CVD methods. A rather simple setup using gas phase deposition is the pack cementation. In this process the parts are embedded into a powder mixture of a pure metal or alloy source (e.g., Al, FeAl, Al-alloys, or also Cr depending on the envisaged composition of the coating), a halide salt activator (e.g., NaCl, NaF, NH_4Cl) and an inert filler material (often Al_2O_3 powder) which guarantees a certain porosity level for the gas phase transport [1599]. Under a controllable atmosphere, usually Ar or H_2/Ar, this powder mixture, the pack, is heated to a temperature of 800°C to 1000°C. At this temperature the master metallic powder reacts with the halide salt activator to form volatile metal halide species providing significant partial pressures. In the case NH_4Cl is used as the activator, the general reaction can be written as

$$M(s) + xNH_4Cl(s, g) = MCl_x(g) + 2xH_2(g) + x/2N_2 \tag{10.1}$$

with M usually being Cr, however, also other metals as Al and Si can be deposited; subscripts (s) and (g) correspond to solid and gaseous state, respectively. On the surface of the component, the metallic chlorides decompose reacting with the nickel of the substrate to form intermetallics (NiAl) or forming metallic chromium layers [173]. The following focuses on aluminide coatings. A better control of the aluminide deposition process compared to the pack cementation is possible in a high temperature furnace in which activator gases such as ammonium halides are passing through a porous bed of the source material (Al) and then flow along the surface of the components where the deposition takes place. Due to the gas flow used for the deposition (non-line-of-sight) also internal borings or complex-shaped parts can be coated.

In the CVD process two extreme cases are established for the aluminide deposition, namely low and high aluminum activity coatings. The first type of coating is deposited at high temperatures (1050–1100°C) which lead to NiAl layers by Ni outward diffusion. The second type of coating is made by deposition at low temperatures (750–900°C) promoting aluminum inward diffusion resulting in Ni_2Al_3 and NiAl layers [1095]. A further improvement of the nickel aluminide can be gained by the addition of platinum. A thin platinum layer is deposited by electroplating up to about 10 μm, see Fig. 10.3. In the subsequent aluminizing process, the aluminum diffusion into the substrate is improved by platinum. Both high and low activity aluminizing processes can be used, leading to (Ni, Pt)Al and/or $PtAl_2$ containing coatings. Platinum significantly improves the oxidation properties of the aluminide coatings, however, too large amounts of $PtAl_2$ might lead to a detrimental embrittlement. Also the formation of brittle interdiffusion zones between aluminide and substrate can be critical [436].

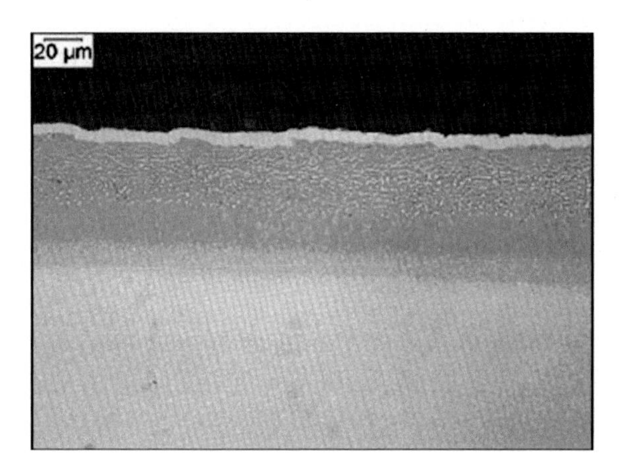

FIGURE 10.3 Optical micrograph of an outwardly grown platinum aluminide diffusion coating. The top thin white layer is electroplated nickel, used for edge retention during metallographic preparation of the sample. Image reproduced from [436].

10.2.1.2 Thermal spray techniques for NiCoCrAlY *coatings*

Different thermal spray techniques are used to deposit NiCoCrAlY coatings on metallic substrates. In these coating processes, a plasma gun which is accelerating and heating the powderous feedstock (particle sizes in the range between 10 and 100 μm) is mounted on a robotic system which moves along the surface of the component and allows the deposition of quite homogeneous coatings in the range of 100 μm and above. Bonding in thermal spray processes is in most cases due to mechanical clamping. Therefore, a rough substrate surface is needed before coating. To achieve that, the substrate is often sand blasted by coarse (0.5 mm) ceramic (e.g., Al_2O_3) grids which gives roughness values of about $R_a \simeq 5$ μm.

The thermal spray technique which gives the lowest impurity content (mostly oxygen, content below 1000 ppm) in combination with low porosity values (below 1%) is vacuum plasma spraying (VPS). A micrograph of a VPS NiCoCrAlY is shown in Fig. 10.4(a). In this process, the plasma gun is operated in a chamber, which is first evacuated to pressures below 10^{-2} mbar to remove mainly oxygen and then refilled by an inert gas up to about 50 mbar for spraying. Before spraying, typically a transverse arc

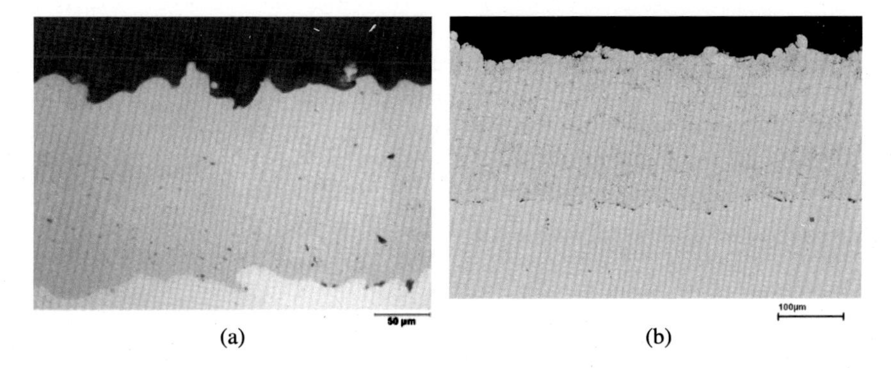

FIGURE 10.4 SEM micrograph of a VPS (a) and a HVOF bond coat (b).

cleaning is frequently applied for an effective removal of contaminations from the surface of the substrate.

Another thermal spray technique which is cheaper than VPS, as it does not need a vacuum chamber, is the high velocity oxygen fuel (HVOF) spraying. In this process, fuel which is kerosene, a hydrocarbon, or hydrogen, is burned at elevated pressures (about 6 bars) in a combustion chamber. The powderous feedstock is added and the particle loaded gas expands through a De Laval nozzle given supersonic conditions and hence very high impact velocities (> 500 m/s). A shroud gas (typically nitrogen) reduces the intermixing with the surrounding air and hence oxygen contamination. In combination with the rather short in flight time the oxygen content is limited (in the order of several thousand ppms) and the high velocity leads to rather dense coatings as seen in Fig. 10.4(b) [1168]. Meanwhile, increasing numbers of especially stationary turbine parts are coated by HVOF spraying instead of VPS.

Another technology more recently used for bond coat application is high velocity air fuel (HVAF) spraying. In this process the pure oxygen of the HVOF process is substituted by air at a higher pressure level (about 10 bars) in special designed nozzles. The additional nitrogen in the combustion process leads to lower temperatures and higher velocities of the coating process (about 1000 m/s). Hence, dense coatings can be produced [1646].

Another process using very fast particles is the cold gas spraying (CGS). Here the pressure level is further increased (in modern facilities about 50 bars). Even though the process is called "cold," the gas is heated up to more than 1000°C to allow the formation of the so-called shear instabilities during impact of the particles which gives dense coatings without melting. The avoidance of high temperatures during spraying reduces the oxygen up-take considerably. As mentioned above, the major bonding mechanism in thermal spray is mechanical clamping. As a result, the bond coats deposited on the superalloys have to show a rather high substrate roughness (well above 5 μm). This is often difficult to obtain with highly kinetic processes as HVAF and, more specifically, CGS. However, adequate selection of deposition parameters can lead to dense coatings with R_a values above 10 μm [298].

The last thermal spray method to discuss here is the atmospheric plasma spraying (APS). It is the cheapest process; however, due to the mixing in of air, a rather high oxygen up-take (in the percentage range) is observed. This leads to reduced oxidation properties and also the density is limited. Despite these disadvantages, APS bond coats are used in commercial applications with limited thermo-mechanical loading. The importance of the oxygen level in the coatings is related to the so-called reactive elements (e.g., Y) in the coatings. These elements improve the bonding of the growing alumina scale to the bond coat and they also reduce the growth rates. If the bond coats take up oxygen during spraying, mainly these reactive elements oxidize and they are no longer available for improving the oxidation behavior. However, also too good spraying conditions might be a problem as typically an excess of reactive elements is in the bond coats, which might lead to increased internal oxidation if not bonded during thermal spraying [1370].

The need for high density and simultaneously high surface roughness is difficult to achieve with only one process and just one particle size. Therefore, the combination of a dense coating (e.g., deposited by VPS using fine particles) with on top a coating with high roughness (deposited, e.g., using large particles with APS) has been developed and is called flash coating. Essential for a good performance of flash coats is the good interconnection of the individual spray splats so that the aluminum can be supplied from the reservoir of the dense MCrAlY layer [1027]. A final aspect which should be addressed here is the necessary compatibility of the bond coat with the substrate. As the chemical composition of bond coat and substrate is rather different, an interdiffusion, e.g., of chromium and aluminum into the substrates and alloying elements of the substrate into the bond coat will take place. This can lead to the formation of brittle phases as topologically close packed (tcp) phases and secondary reaction zones (SRZs) with coarsened γ/γ' microstructure [1384], see more details in Chapter 8. Furthermore, the reduction of the aluminum reservoir in the bond coat might reduce the lifetime of the coating.

10.2.1.3 Other coating techniques

Also, the EB-PVD process can be used to deposit highly dense and homogeneous MCrAlY bond coats with a low amount of roughness. Due to relatively high costs, this coating method is used in research environments [1290]. Also slurries using fine aluminum particles can be used to manufacture aluminide coatings. The slurries are deposited by spraying on nickel-base alloy and performing a heat-treatment for the aluminizing step. If the process is performed in air, the outer aluminum particles can oxidize and form a hollow sphere isolative layer [960].

10.2.1.4 New bond coat approaches

Recently, the rather old idea of oxide dispersion strengthened (ODS) bond coats has been further developed [157]. These materials are adopted in their thermal expansion to the ceramic top coat and often show outstanding oxidation properties. They can be used as flash coats on top of a conventional bond coat. This approach led to excellent performance in TBC systems [1511]. A further interesting new class of materials for the use as bond coats might be MAX phases. It could be demonstrated that these materials form a slow-growing good-adhering scale, e.g., made out of alumina, and can be used as bond coat [530]. In plasma-sprayed systems, the bonding of the ceramic top coat to the bond coat mainly results from an interlocking at the roughness profile of the bond coat. This bonding can be further improved by introducing structures within the bond coat surface, e.g., by laser ablation. Also the structuring of the substrate directly can be used [767].

10.2.2 Thermal insulation

10.2.2.1 EB-PVD and APS as standard deposition methods for yttria stabilized zirconia coatings

For several decades, 6–8 wt.% yttria stabilized zirconia (YSZ) is the material of choice for advanced thermal barrier coatings for a number of reasons. The major reasons are a low thermal conductivity, high thermal expansion coefficient (rather close to the substrate), high sintering resistance, and excellent toughness of YSZ. So the description of the processes will focus on this material, although there are some drawbacks which will be discussed in the sequel.

The EB-PVD process uses an electron beam to heat a porous ingot of the coating material (e.g., YSZ) in a vacuum chamber above the melting temperature up to about 3500°C. The porosity is needed so that the ingot can withstand the fast heating by electron beam. By this procedure, the material is vaporized and condensates in a line-of-sight process on the typically preheated substrates with temperatures between 950°C and 1100°C. During the deposition, a columnar structure with an often <100> texture is formed. The columns show a feather-like outer structure, and their microstructure can be adjusted by the proper choice of process conditions as substrate temperature, evaporation rate, or geometrical conditions (e.g., rotation) during deposition [1290].

Defect-free coatings can be deposited on rather smooth substrates with low roughness values typically below $R_a < 2\,\mu\text{m}$. Here mainly aluminide or platinum aluminide bond coats are used. On the bond coat surface, a thin alumina scale is formed before the EB-PVD deposition during a thermal treatment and further during the EB-PVD coating process, since a substantial partial pressure of oxygen is present in the coating chamber. The thin alumina scale is called thermally grown oxide (TGO) and it ensures a good adhesion of the EB-PVD top coat.

The microstructure of such an EB-PVD system is shown in Fig. 10.5(a). The columnar microstructure reduces tensile stresses in the coating. Such tensile loading is expected in a TBC system at elevated temperatures due to the higher thermal expansion of the substrate compared to the coating.

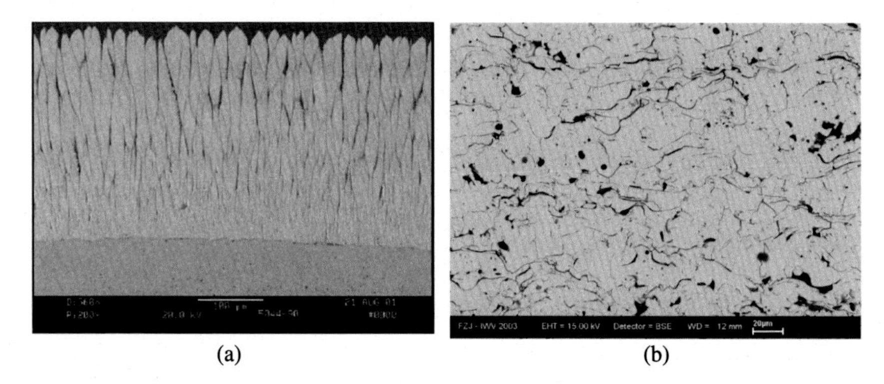

FIGURE 10.5 SEM micrographs of EB-PVD [436] (a) and APS (b) coatings.

In the atmospheric plasma spraying (APS) process, powderous, e.g., fused and crushed or spray dried, feedstocks of YSZ in the size range between 20 and 100 μm are used. The most often produced microstructure contains a lot of microcracks and pores, giving a total porosity level above 10% [1406]. The microcracks are formed due to high tensile stress levels as a result of the fast cooling of the molten splats while the pores are due to the suboptimal filling of gaps between deposited splats by the newly arriving splats. In Fig. 10.5(b) the microstructure of such a TBC is shown. The microcracks are essential for the good performance of APS TBCs as they allow some degree of strain tolerance due to the sliding of splats along each other [9]. In addition, the microcracks are perpendicular to the heat flow direction, leading to rather low thermal conductivities below 1 W/(m K) in the as-sprayed condition. Another microstructure often used for thick coatings in the micrometer range contains segmented or also often called dense vertically cracked (DVC) coatings. In these coatings, high substrate temperatures, in combination with hot plasma conditions and good powder melting, lead to dense coatings under high tensile stress. These stress levels initiate segmentation cracks [563]. The cracks are beneficial under high tensile loading as they can open, leading to reduced stresses. As the columns are rather dense, the thermal conductivity of segmented coatings is higher than that of microcracked TBCs.

10.2.2.2 *Innovative thermal spray techniques*

Instead of using powderous feedstocks, also the injection of liquid suspensions or precursors is employed for the manufacture of thermal barrier coatings. At first, the suspension plasma spraying (SPS) will be described [424]. The suspension consists of fine-scaled particles which allow the deposition of coatings with features in the submicrometer or even nanometer scale range. The suspensions can be introduced as a full stream or as atomized droplets into the plasma plume. In both cases the droplets are further fragmented into finer droplets by the high velocity plasma plume. The degree of atomization depends on many process variables as the viscosity of the suspension, its surface tension, and the plasma conditions. As the droplets are much smaller than typical thermal spray powders, it is often, especially when using a radial injection, more difficult for fine droplets to penetrate into the center part of the plasma plume. This can lead to pore bands in the coatings from improperly molten droplets at the outer fringe of the plasma plume.

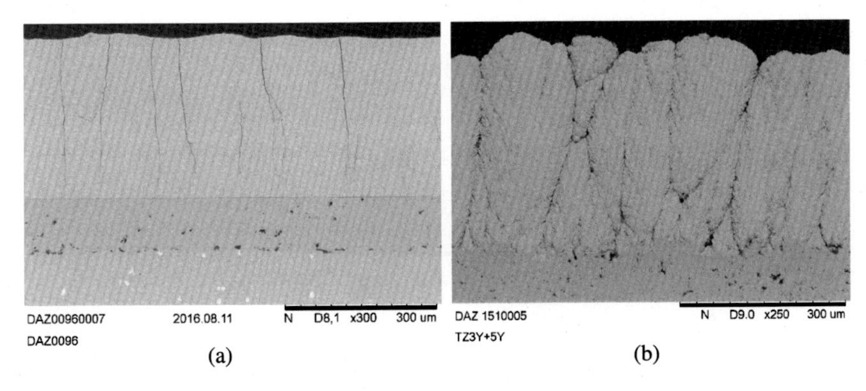

FIGURE 10.6 Suspension plasma sprayed coatings: (a) segmented and (b) columnar structured coatings (courtesy D. Zhou, Forschungszentrum Jülich).

The axial injection is certainly an advantage for the SPS process. After the fragmentation, the evaporation of the liquid carrier of the suspensions leads to a porous agglomeration of particles which sinter and then melt to form a liquid droplet. The SPS process can lead to highly porous coatings with porosity levels even above 40%. With this process, also a highly segmented coating with a certain porosity level can be achieved (see Fig. 10.6). This microstructure shows interesting properties combining low thermal conductivity and Young's modulus with high strain tolerance due to the high number of segmentation cracks [1475]. A further step forward with the technology was possible with the development of columnar structured coatings by SPS. About a decade ago it was found that such coatings can be deposited under certain process conditions. The key for this development was the use of fine droplets with small Stokes numbers as they can be rather simply obtained in SPS. For these conditions, the droplets follow the plasma plume parallel to the substrate instead of just directly impinging on it. They will hit obstacles on the surface of the substrates and these deposited droplets will form columns at these locations [1471]. An example of such a columnar microstructure is shown in Fig. 10.6(b). Meanwhile, the process has been introduced for the coating of gas turbine components.

In the solution precursor plasma spraying (SPPS) process, also liquid feedstocks are employed; however, in this case the final coating material is formed during the deposition process by the decomposition of the precursors. This process offers the possibility to easily change and adjust the coating materials. On the other hand, due to the decomposition and formation step, there are additional process variables which increase the complexity of the process. In the past, it was demonstrated that this process can be used for the manufacture of high-performance TBC systems [668].

Another innovative process, which was especially developed by Oerlikon Metco, Switzerland, is the plasma spray physical vapor deposition process (PS-PVD) [1509]. This process is basically a modified vacuum plasma spraying with powerful pumping units to maintain a low chamber pressure of about 1 mbar even with the inlet of process gas flows of 100 l/min. In addition, very powerful guns, e.g., the O3CP with a power of up to 180 kW, are used. In this combination the plasma can expand to a length of more than 1 m (Fig. 10.7(a)). If specific fine powders are injected in the central part of the plasma, these powders are not only molten but even evaporated to a large degree. So coating deposition takes place from the gas phase similar to PVD, resulting in columnar microstructures (Fig. 10.7(b)). There are certainly differences to a conventional EB-PVD process, especially also due to the process gas flow, leading to a boundary layer which influences the deposition process [592]. Using optimized process conditions, a highly strain tolerant microstructure can be produced which shows extremely good performance in burner rig tests [1216]. Meanwhile, a number of PS-PVD coating facilities are in operation worldwide, especially in China a number of institutions are investigating the technology (e.g., [28]).

One interesting feature of both the PS-PVD and partly also the SPS and SPPS process is their non-line-of-sight characteristic. If particles/droplets to be deposited are small enough (e.g., their Stokes number considerably smaller than 1) they can follow the gas flow along the surface of the substrate before being deposited. In contrast to conventional thermal spray process, this mechanism offers the possibility to coat also complex-shaped parts with hidden surfaces.

An additional cost-efficient technique to apply TBCs is the use of slurry or sol–gel routes. Applications can be made by simple dip coating or spraying with a subsequent heat treatment. It could be demonstrated that coatings survive thermal cycling under rather moderate conditions [1009].

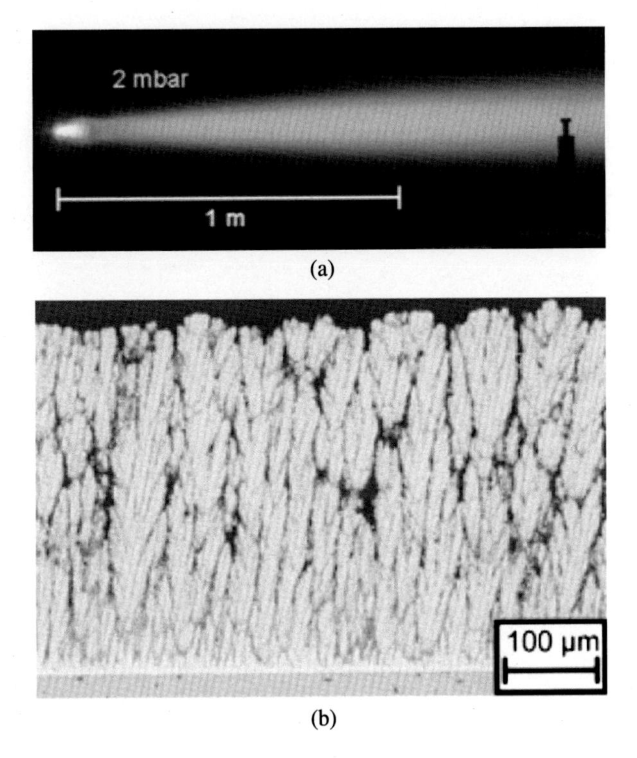

FIGURE 10.7 (a) Photo of the plasma plume in the PS-PVD process, (b) SEM micrograph of the columnar structure of PS-PVD YSZ coating (courtesy H. Moitoux, S. Rezanka, Forschungszentrum Jülich).

10.2.2.3 New materials

For more than two decades, intense investigations have been made to find alternatives to the 4–5 wt.% yttria stabilized zirconia. The reasons are the claimed limited temperature capability of YSZ due to pronounced sintering and phase transition at temperatures above 1200°C. The major identified materials are pyrochlores, perovskites [1474], and zirconia with different dopants as rare earth elements [1670] and five valent cations like tantalum [1112]. A recent overview on the developments is given by Bakan [63]. An issue of many new ceramics especially if not using zirconia as base material is their low toughness which leads to early delamination. As delamination is often located close to the top coat/bond coat interface, a possible solution to overcome this problem is the use of double layer coatings with a tough YSZ layer directly on the bond coat and on top a layer using the new material. In Fig. 10.8 such a double layer coating made by APS using $Gd_2Zr_2O_7$, an often employed new TBC pyrochlore material, as top layer is shown as a double layer system on the left, on the right a double layer using a perovskite sprayed by SPS is shown. Both systems showed an excellent performance in burner rig tests.

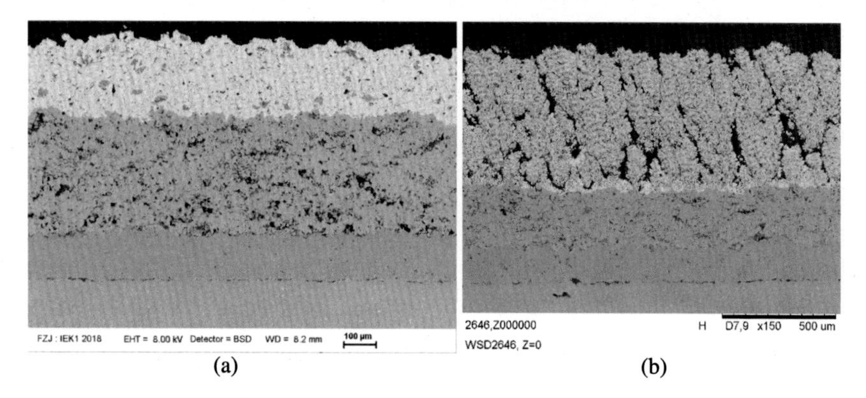

FIGURE 10.8 SEM micrographs of (a) an APS $Gd_2Zr_2O_7$/YSZ and (b) as SPS $La(Al_{0.25}Mg_{0.5}Ta_{0.25})O_3$/YSZ double layer system (courtesy E. Bakan, N. Schlegel, Forschungszentrum Jülich).

10.3 Characterizing properties of protective coating systems

10.3.1 Properties relevant for damage and failure behavior

Protective coatings on superalloy components for high temperature applications are multilayer systems with each layer having its special tasks and different properties. However, all layers have to be compatible during processing and service to maintain structural integrity of the coating system. Depending on the processing conditions transient thermal gradients may occur and lead to high residual stresses. In the case of EB-PVD coatings it is assumed that during the deposition process the layered system is stress free at substrate temperature, which is typically in a range between 950°C and 1100°C. When cooling to ambient temperature, residual stresses build up depending on the CTE, the elastic constants, and yield strength of the individual layer materials. In service residual stresses are present because the temperature distribution in the coated system is always different to the temperature field during coating deposition. Additional stresses in the coating system originate from mechanically induced deformations of coated components, which are generated by, e.g., aerodynamic forces or the rotation of engine parts.

Structural integrity of a TBC is ensured if stress, strain, and fracture mechanical load variables, such as stress intensity or strain energy release rate, are below the respective admissible values. Failure of a TBC system is associated to spallation of the ceramic thermal insulation which may occur due to an initial overload or delayed, as a consequence of stress build up and damage accumulation. For calculating the stress and strain state, knowledge of thermal conductivity, CTE, and the elastic constants is essential. Determination of yield strength is necessary to capture instantaneous stress relaxation and stress redistribution, respectively. Fracture strength and fracture toughness of individual layers and interfaces are key properties for evaluating if a coating may fail under an applied load. These properties are generally observed to decrease with time in service governing duration before failure.

10.3.2 Determination of thermal expansion coefficient and thermal conductivity

For determining the CTE of individual coating layer materials, dilatometry is well established. Since the CTE does not depend on microstructure features such as porosity, it can be measured using bulk material samples having the same crystallographic structure and chemical composition as the coating materials. Another option is X-ray diffractometry which allows for simultaneous determination of the crystallographic structure. For CTE data of substrates, bond coat, and TBC materials, a huge number of data sets exists; however, data can be found in diverse publications but are not yet summarized in central data banks.

Measuring the thermal conductivity of thin layers is still challenging. Microstructure has a crucial impact on the thermal conductivity. Thus, it is not possible to simply use data from reference bulk material, especially not for the porous ceramic topcoat of a TBC system. If gradients in the microstructure occur over the layer thickness, as it is the case for EB-PVD coatings, the thermal conductivity is a function of the layer thickness. Further, the partial transparency of YSZ coatings for infrared radiation has to be considered. Therefore, it is not sufficient to know the thermal conductivity of the solid—the total heat transfer coefficient should be known.

To achieve accurate data for the thermal conductivity of ceramic TBC layers, well-established methods for measuring solid conductivity have been refined. For example, [1185] applied the laser flash method (ASTM, E 1461-01) to determine the thermal diffusivity on free standing EB-PVD processed coating samples. This technique allowed a higher accuracy than using multilayered samples with the coating applied on the superalloy substrate. Coatings of more than 200 μm thickness were separated from the metallic substrate by chemical etching. Very thin layers of 50–200 μm were deposited on a sapphire substrate of about 1–2 mm thickness, which is transparent for the laser light of the laser flash apparatus but ensured that the fragile coating stayed intact. Further, a thin platinum layer was sputtered on both surfaces, which reduced radiation into the sample on the side where the laser hits the sample and improved on the back side the detection of the temperature signal by an infrared detector. The thermal conductivity λ is then calculated via the relationship $\lambda = a \times C_p \times \rho$ with the thermal diffusivity a, specific heat capacity C_p, and density ρ. The specific heat capacity is determined by means of differential scanning calorimetry (DSC) or differential thermal analysis (DTA) and does not depend on the microstructure. The density is typically determined at ambient temperature by established standard methods, and the temperature dependency of the density is derived from temperature-dependent CTE data. The experiments presented in [1185] showed that the thermal conductivity of a 300 μm thick EB-PVD processed YSZ coating was about 40% higher than that of a 50 μm thick coating. This significant difference can be explained by the gradient in the coating microstructure. The much smaller columnar grain size at the interface to the substrate, compared to the grain size in some ten microns distance, results in a higher thermal resistance due to the higher density of grain boundaries.

The effect of heat transfer in ceramic insulation layers by radiation and the factors influencing it have been discussed in a review by [1011]. For reducing radiation, addition of dopant materials and introduction of internal interfaces such as grain boundaries, especially perpendicular to the direction of the heat flux, are proposed. However, quantitative experimental data are rare. One detailed study on a 280 μm thick free standing EB-PVD YSZ coating is presented in [898]. Experimental data of thermal conductivity from laser flash testing and of transmittance and emittance, using the integrated sphere method at ambient temperature and the black boundary condition method for elevated temperatures, were combined to determine the contribution of radiation on the total heat transfer. Depending on thermal history of the coating the contribution of radiation at 1150°C was about 18% (after 100 h at 1100°C) and 24% (after 2 h at 1080°C). The differences between the differently heat treated samples are attributed to microstructural changes, especially of the pore size and shape, due to sintering effects. Comparing EB-PVD and APS coatings, the lamellar microstructure of APS coatings provides much more internal interfaces perpendicular to the heat flux and is expected to be less prone to heat transfer by radiation.

In APS coatings the microstructure is much more homogeneous over the coating thickness than in EB-PVD coatings and does not depend on the coating thickness. Therefore, it may be expected that the experimental procedure for measuring thermal conductivity of APS coatings is quite robust. However, thermal conductivity data reported in the literature show a wide spread, which makes it difficult to decide which values to use for numerical calculations. A database for thermal transport properties of APS coatings is provided in [1531]. Thermal diffusivity data from more than 200 APS coatings with standard composition of ZrO_2 with 7–8 wt.% Y_2O_3 have been collected over 12 years in collaboration of 16 institutes and companies. All data points were included in one diagram (Fig. 10.9), revealing a huge scatter band.

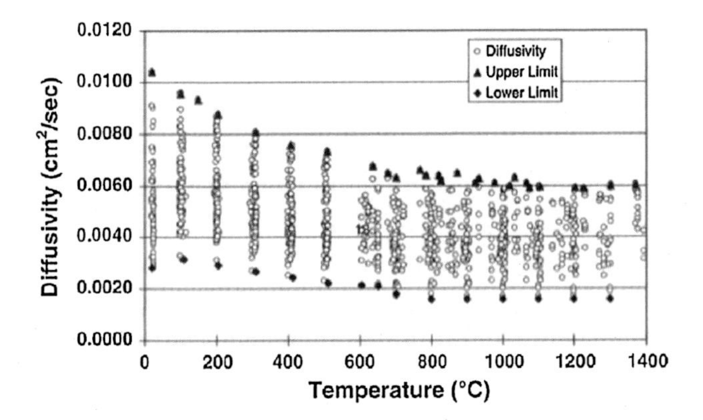

FIGURE 10.9 Spread of thermal diffusivity data of 133 APS 7–8 wt.% YSZ TBC, collected by [1531].

10.3.3 Characterizing mechanical properties of the different coating layers (BC, TGO, TC)

10.3.3.1 Mechanical testing of a TBC system

To assess mechanical properties of a TBC system, testing of a complete TBC system, including the superalloy substrate, enables capturing the effect of constraints and interactions between the layers. For evaluating test results achieved on the complete layered structure, inverse analyses, either by analytical approaches or employing numerical methods such as finite element analyses (FEA), are needed. However, the range of thicknesses in a typical TBC should be kept in mind: 100–200 μm for the EB-PVD TC (up to mm for APS TC), 1–8 μm for the TGO, 50–100 μm for the BC and IDZ, and 0.5–2 mm for the substrate. Thus, mechanical testing of a complete TBC system including the entire substrate would hardly provide relevant information of each layer. To overcome this issue, several authors have developed tests with reduced thickness of the substrate. Taylor et al. have developed a high-temperature creep test based on reduced thickness of the substrate [1404]. Since their work focused on the BC properties, they applied this method on a system without TC and were able to improve the sensitivity of the method [1404]. For mechanical evaluation of a complete TBC system, Hemker and coworkers have developed a microbending facility where the substrate thickness has been further reduced. This approach enables a direct view of the cross-section, and thus digital image correlation (DIC) could be used to measure the strain and strain gradient in each layer, yielding the mechanical behavior of the layers and enabling the assessment of interfacial fracture toughness at room temperature [377]. Using such methodologies at high temperature needs to consider ageing of the layered system due to diffusion and oxidation, respectively. While oxidation can be reduced by testing in vacuum, diffusion processes at the interfaces between the layers cannot be avoided.

Tensile testing of substrates coated with a BC layer only has been conducted in [402] to identify the strain at crack initiation as a function of aging. This is done in using a high-resolution optical microscope, together with a CCD camera, to measure in situ both crack initiation and strain field by DIC. The strain field measurement is a key tool to assess the local critical strain to coating cracking, in particular when strain localization is observed in the single crystal substrate, Fig. 10.10(a). This methodology does not modify the coating and substrate interactions and is very straightforward to handle with the modification of properties with phase transformation due to aging. It has been evidenced that for (Ni,Pt)Al bondcoat the thermal cycling has induced both surface roughness evolution, the so-called rumpling effect, see Fig. 10.10(b), and phase transformation from β to γ' (Fig. 10.10(c)).

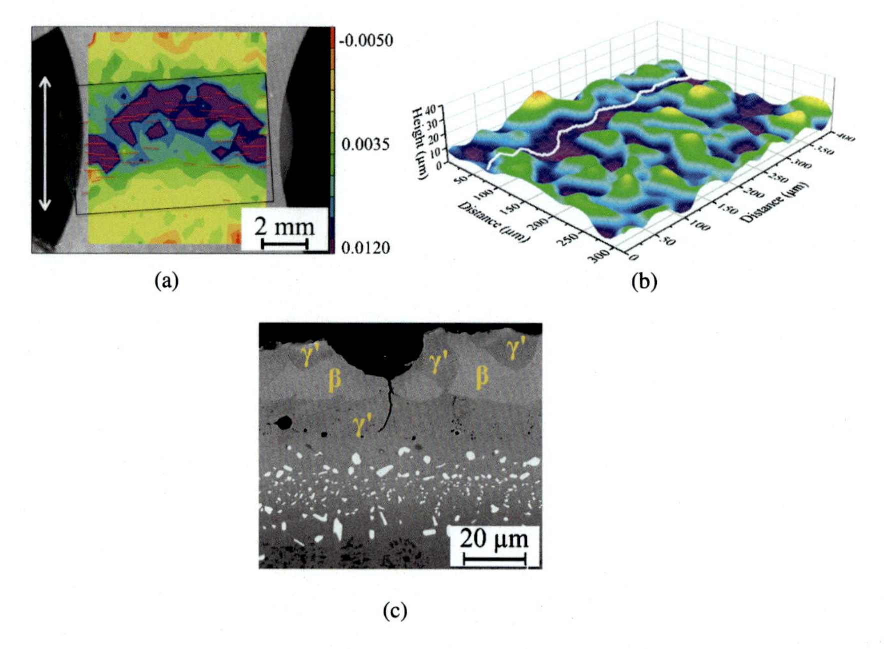

FIGURE 10.10 Tensile room temperature test on a substrate with BC after ageing by thermal cycling: (a) strain field measured by DIC on the surface of the coated specimen with associated BC cracks indicated by short red lines; load direction corresponds to the double arrow, (b) roughness measured on the surface and associated crack path in white, and (c) phase transformation from β to γ' (adapted from [402]).

The use of intermetallic BC, which is brittle in a wide temperature range, can induce crack initiation at a very low strain level [941]. During thermo-mechanical fatigue (TMF), it has been evidenced that the substrate can withstand compressive strain dwell at high temperature, leading to substantial stress relaxation, which results for out-of-phase TMF loading in tensile stress in the low temperature range [1199]. Thus, the substrate can fail due to crack initiation in the metallic bond-coat and further crack propagation into the substrate [407].

Emerging synchrotron facilities providing high energy X-rays enable accessing the local strain in layered systems in situ via X-ray diffraction. At the advanced photon source (APS) of the Argonne National Laboratory, Knipe et al. have assembled a thermal-mechanical fatigue testing equipment and used it to determine the strain evolution in each layer of an EB-PVD coating system during combined thermal and mechanical loading [730]. The strain in BC and TC (see Fig. 10.11) and later in TGO of an aged system with sufficiently thick TGO has been determined based on the θ–2θ method [730,902].

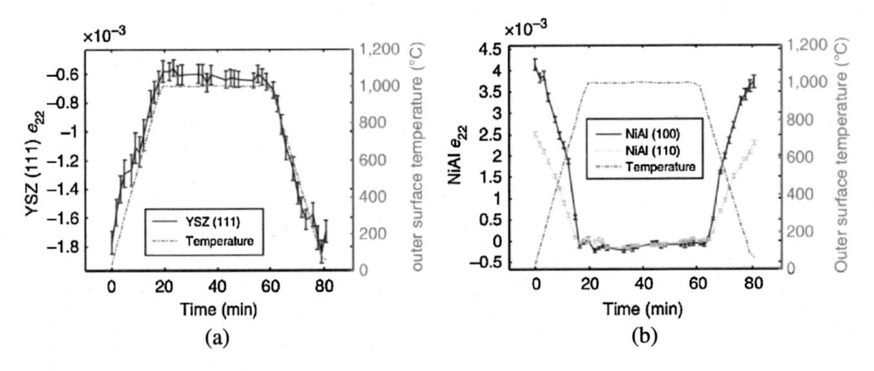

FIGURE 10.11 In situ measurement of temperature and strain in the loading direction in (a) YSZ layer and (b) in NiAl layer [730].

10.3.3.2 Free standing coating

An alternative approach promotes a complete identification of mechanical behavior of the individual coating layers using free-standing coatings [1050,12,1419]. It is possible to obtain small-scale specimen by either manipulating the coating process for inhibiting the adherence of the coating to the substrate by electro-discharge machining of the coating or polishing of a ribbon from the TBC system. This kind of approach enables measuring the stress–strain relationship on free-standing coatings. Due to their ductility, BC samples can be tested under tensile load, see Fig. 10.12(a). Targeted preparation allows also investigating the deformation behavior of the diffusion zone between BC and substrate.

Again, strain measurement is improved by the use of DIC technique for strain field measurement. This strain field measurement enables controlling the homogeneity of the loading within the gage length. This last point is a challenging issue for microtesting. For instance, it has been shown that for (Ni,Pt)Al coating, the aging in isothermal condition leads to a strong modification of the mechanical properties over a wide range of temperature, mainly induced by phase transformation from β to γ' [1050,12]. In contrast, for MCrAlY coating the internal oxidation will drive the mechanical behavior, limiting drastically the ductility increase induced by phase transformation from $\beta+\gamma+\alpha$ to $\beta+\gamma'+\alpha$ [1413]. For a new generation of the so-called γ/γ' coatings, it has been shown that at RT the associated ductility was higher than that observed for (Ni,Pt)Al coatings, but surprisingly at elevated temperature (870°C) some authors have shown that the γ/γ' coatings are brittle and fail by intergranular cracking, which entails cracking of the substrate, while (Ni,Pt)Al coatings are highly ductile at this temperature [12]. In this case, the authors have tested both free-standing coating and coated substrate. Testing of free-standing coating was straightforward to establish the damage mechanism and to confirm for the γ/γ' coating that intergranular failure within the coating was the driving damage mechanism limiting the life of the coated substrate. Comparison of damage behavior of a coating system with thinned substrate and applied (Ni,Pt)Al coating and free-standing BC specimen were consistent, e.g., see Fig. 10.12(b) where cleaved grains are observed in BC and are associated to brittle behavior for tests conducted at low temperature.

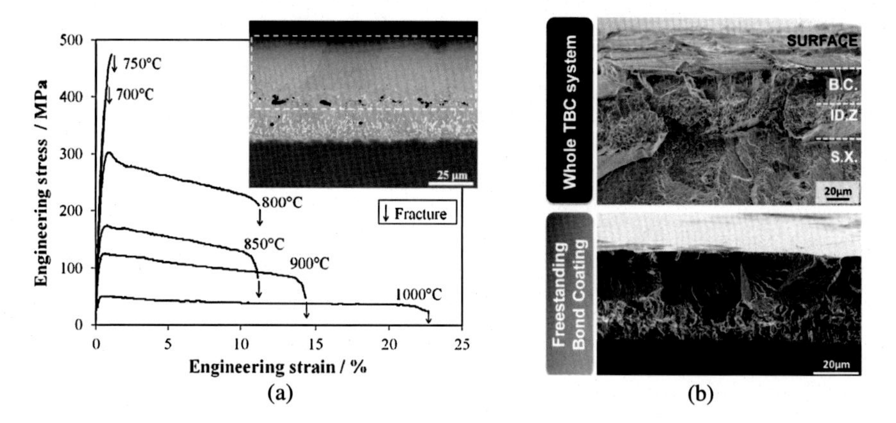

FIGURE 10.12 Tensile test on free standing BC specimens for (Ni,Pt)Al coating: (a) stress–strain curves at different temperatures and (b) fracture surfaces of specimens comprising SX substrate material, BC, and interdiffusion zone between SX and BC (right top) and BC only specimen (right bottom) [1419].

10.3.3.3 *Micro- and nanoindentation*

Micro- and nanoindentation are interesting tools to map room-temperature elastic properties of TC, TGO, BC, interdiffusion zone (IDZ), and secondary reaction zone (SRZ) when developed in the substrate,

Fig. 10.13 [1680,1037]. Besides, a complete analysis of pile-up or sink-in associated to indent yields access to elastoviscoplastic behavior of the studied material [1206]. The major interest of nanoindentation is to achieve local information at a submicron length scale that enables to determine individual phase behavior. Development of such methods with multipoints analysis yields elastic properties mapping in a very short-time [1097]. On the other hand, the characterization of the plastic behavior needs to be coupled with FEA and local measurement by AFM and is subsequently highly time consuming. It is worth noting that the aging of APS YSZ impact elastic modulus significantly [125]. In the field of microtesting, a recent overview addresses strengths and weaknesses of different techniques that can be used [328].

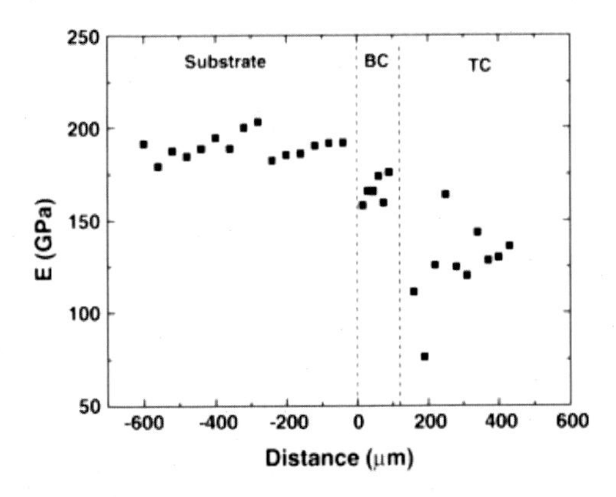

FIGURE 10.13 Young's modulus measurement by nanoindentation across an EB-PVD TBC system, see high dispersion in the TC associated to local intercolumnar voids [1680].

10.3.4 Adherence of the ceramic coating layers – interfacial strength and fracture toughness

One of the major issues in TBC technology is to manage the adherence of the top coat on the substrate in the as-processed and for relevant aging conditions regarding the spallation process. Adherence of the bond coat is seldom an issue since the processing methodology of metallic bond coats has been established for decades. However, the choice of adherence testing methodology is still challenging and the lack of confidence in interfacial adherence measurement is certainly limiting robust modeling of TBC damage until failure by spallation. Tests methods for determining the adherence of a coating are evaluated using criteria of robustness (is the test reproducible?), modification of the TBC system (is there an artifact in testing conditions?), relevance of the loading as compared to in-service loading (how close to reality is the lab testing condition?) and easiness of intrinsic physical/mechanical property measurements (is the result straightforward for further modeling or limited to testing conditions?). To promote spallation of a top coat, the most popular methods are tensile adhesion test (TAT), barb test, shear test, indentation test (including scratch test), bending test, and compression test. The laser adhesion test (LASAT), which is a quite new method, will be described in more detail and evaluated. This method is based on a laser driven shock wave causing debonding of the TC. In the following, each class of adherence tests is briefly described and ranked with respect to the above listed criteria. Note that TBC spallation occurs mostly at low temperature after thermal exposure. Stress relaxation at high temperature entails build up of high compressive stress in the ceramic coating layers (TC and TGO) when cooling. Thus, room-temperature testing for adhesion measurement is straightforward.

10.3.4.1 Tensile adhesion test

For performing the tensile adhesion test (TAT), one cylinder with clamping extension is glued onto the top coat and another cylinder on the opposite side onto the substrate, aligned so that by applying a sufficient force, the coating is separated from the substrate (Fig. 10.14(a)–(b)). The force is controlled by a load cell, and the values measured during the experiment are force and displacement [1130].

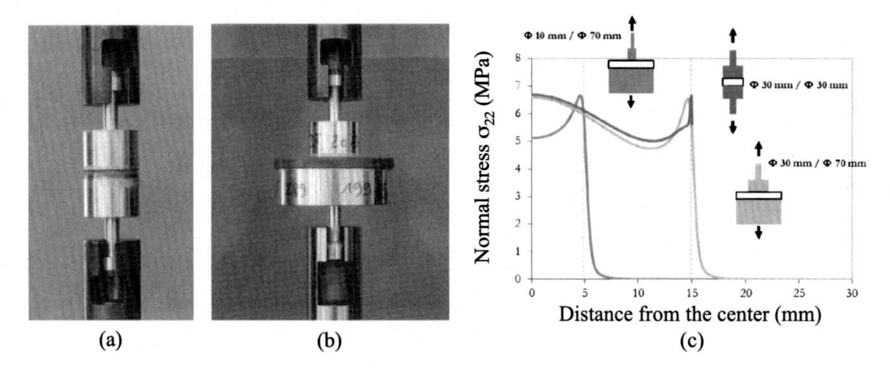

FIGURE 10.14 Tensile adhesion testing: (a) test configuration with both cylinders glued to the coated sample having the same diameter as the circular sample disc, (b) test configuration with a larger cylinder glued to the substrate side and a smaller to the coating side, with the sample diameter similar to that of the larger cylinder, (c) distribution of the stress component in loading direction from the sample center to the edge for different test configurations (adapted from [1130]).

While typically only the force needed for separating the coating from the substrate is exploited, Fig. 10.14(c) reveals that the test configuration, i.e., the diameter ratio between the cylinder glued to the coating and that glued to the substrate, influences the stress distribution over the glued area. Therefore, the separation force is not an intrinsic measure for the

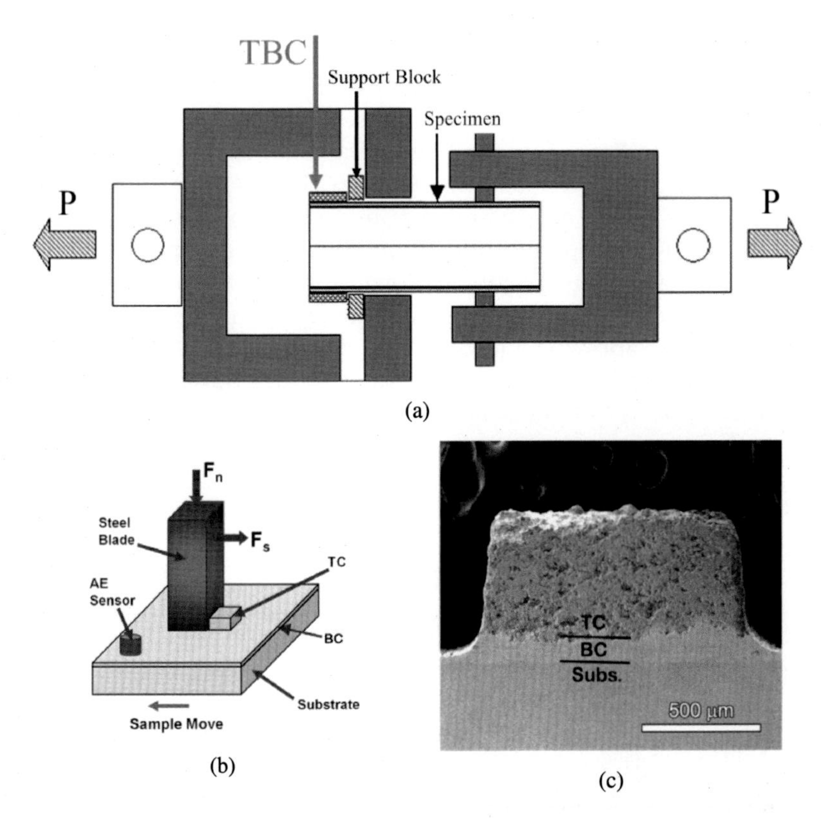

FIGURE 10.15 (a) Barb test from [564] and (b) shear test from [1602], (b) sketch of the setup, and (c) SEM image of the machined specimen, side view.

coating adherence but can rather provide a ranking of coating adherence for one specific test configuration.

10.3.4.2 Shear test

The barb test has been proposed to apply a shear loading on a coated substrate, Fig. 10.15(a) [564]: using a cylindrical substrate, the coating is partially removed by machining. The shear force is controlled by a die moving along the cylinder. The measured values are force and displacement [564]. Alternatively, a pillar can be machined from a TBC system, and shear loading can be obtained by moving a blade applied to the top coat up to delamination of the top coat, Fig. 10.15(b) [1602]. Some authors have also used indentation testing, where indent is achieved on cross-section, or scratch test where indenter is moved along the top coat; recently this kind of test has been modified using laser impulse to promote scratch, e.g., see the review of Chen et al [245].

10.3.4.3 Bending test

Bending and punching tests (where punch is applied at the center of a circular clamped plate) can be either used directly on the TBC system without any modification or used by gluing a stiffening plate on the TC surface and machining a notch, Fig. 10.16 [269,1113,1422]. The stiffener enables accessing energy measurement from simple modeling (analytical or FEA) [1422]. By playing on condition of bending, 3- or 4-points bending, symmetric or asymmetric setup, authors can access to mode I or mode I + II varying the mixity angle, Fig. 10.16(a)–(b) [1481]. Bulge and buckle testing consist in applying a force or a pressure to push the coating from the substrate (using screw, gas or thermal shock), but these techniques are mostly applied to thin films or at least very flexible coating material [245].

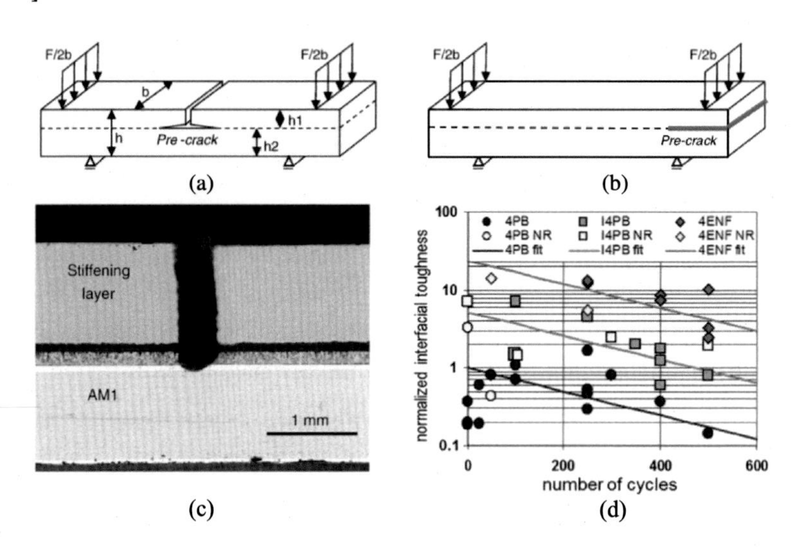

FIGURE 10.16 Four-point bending test with stiffening layer (a) for mode I testing "4PB'," (b) for mode I and mode II "4ENF", (c) detail of notched area for "4PB", test (d) measured interfacial toughness where 4PB corresponds to (a), I4PB same as in (a) except that specimen is upside-down in the setup so as to promote compression in the top coat, and 4ENF corresponds to (b); from [1422,1481].

10.3.4.4 Compressive testing

Compressive testing consists in a strain-controlled test under compression, using cylindrical specimen with a sufficient diameter to length ratio for avoiding macroscopic buckling of the specimen before top coat spallation. This configuration simulates compression induced by CTE mismatch on cooling and promotes mode II decohesion, Fig. 10.17 [295,917,1347, 1674,674].

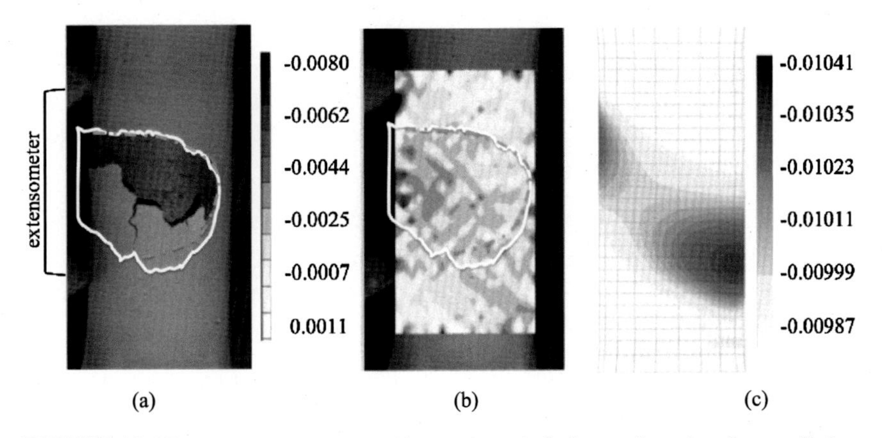

FIGURE 10.17 Compressive testing: (a) in situ optical observation after first spallation event, (b) strain field component in the loading direction measured by DIC: blue line (light grey line in print version) represents failed ceramic observed in (a) and white line corresponds to measurement achieved by SEM, (c) strain field component in the loading direction obtained by FEA considering the misorientation of the single crystalline substrate (adapted from [914]).

10.3.4.5 Laser shock adhesion test (LASAT)

The laser shock adhesion test (LASAT) method consists in a shock wave propagating throughout a TBC system and inducing a tensile stress at the interface. With a nanosecond pulse and a confinement medium (e.g., water or transparent adhesive tape), laser power densities from 0.1 to 8 GW/cm^2 can be introduced at the metallic side of a coated sample with laser spot diameters of a few mm, yielding a strain rate $\sim 10^4$–10^6 s^{-1} in the direction of the shock wave propagation within the substrate. A so-called LASAT threshold can be measured on a coated plate by testing the minimum laser power density needed to debond the TC [99]. LASAT has also been implemented on coated turbine blades and is envisaged as a semidestructive adhesion test [87]. Quantitatively, velocimetry temporal profiles are needed to calibrate the input stress temporal profile with peak stress in GPa and corresponding laser power densities in GW/cm^2. For thick multilayered coatings like TBCs, inverse analysis of shockwave propagation is implemented by 1D finite difference or 2D finite element methods to establish the stress history (mainly in mode I) at the interface for a given input laser energy, especially the critical energy for generating the debonding threshold stress.

Based on 2D FEA, the lateral expansion (orthogonally to the principal direction of propagation) of the shockwave has been evidenced to yield larger delamination than the laser shock diameter for high energy [417]. This effect is detailed as the LASAT 2D effect. These 2D effects can explain the monotonous increasing of the crack size with the laser energy that gave a rise to the so-called "LASAT-2D curve" as experimental output for the LASAT-2D method applied to ceramic coatings [417,87]. It is worth noting that for typical EB-PVD TC, the LASAT induces debonding and consequently blistering of the TC due to the residual stress release. Based on elastic analysis of the blistering condition, it is possible to access the energy release rate at crack arrest making LASAT a very efficient tool to evaluate interfacial toughness [923,561].

10.3.4.6 LASAT to process internal "notch" for further loading

Thanks to the robustness of the LASAT method, it is quite easy to establish a 2D-LASAT curve (in the as-processed or aged condition) yielding the evolution of debonded diameter as a function of the LASER energy, Fig. 10.18(b). It has been proposed to use a decohesion processed by LASAT to obtain a notch-like defect, this defect being circular and "blind" in the sense it is not emerging on any side of the specimen. By this way, NDT techniques can be used to monitor damage evolution from the LASAT defect, Fig. 10.18(c). This methodology has been successfully applied to access damage evolution during either mechanical test [1262], furnace thermal cycling [916], or burner rig test [918]. Associated to cohesive zone model (CZM) or by elastic analysis of the TC blistering, the interfacial toughness can be deduced from these tests as a function of any loading condition (thermal and/or mechanical loading) [1262,561].

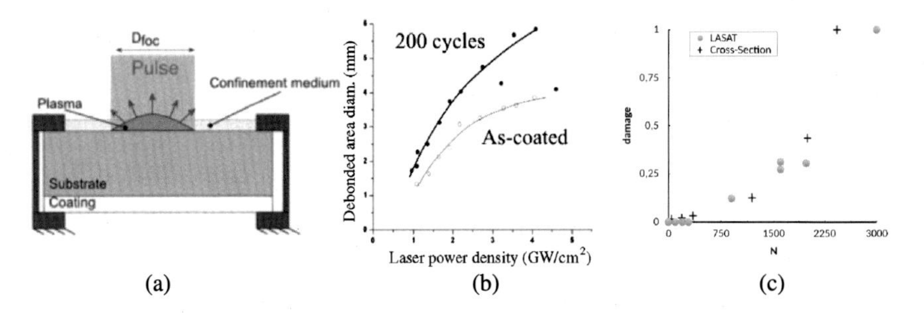

FIGURE 10.18 Laser shock adhesion testing (LASAT) for the same EB-PVD YSZ, (Ni,Pt)Al coating AM1 single crystal system: (a) setup and (b) examples of LASAT-2D curves in as-coated condition and after 200 thermal cycles (adapted from [416,87,561]), (c) comparison of damage size evolution (normalized) from artificial defects processed by LASAT measured by IR top-surface imaging (nondestructive) and on SEM cross-sections (adapted from [1424, 916]).

10.3.4.7 Comparative analysis

Even though tensile adhesion testing is the most popular method to characterize coating adherence, major drawbacks are limiting definitively its robustness: it is rather difficult to reproduce the same condition from test to test because the glueing influences the measurement result. Indeed, glue can penetrate porous coatings, modifying local stiffness, but mainly add a damageable layer modifying both stiffness and damage evolution. Moreover, the quantity to be measured is stress at decohesion which quantity is not seen as an intrinsic characteristic: it has been clearly established that the stress distribution within the specimen is not uniform and depends on the size of the cylinder chosen for both substrate side and top coat side [1130]. The limitation in robustness is also observed for all methods including either gluing of a part (TAT, shear and bending tests when used with stiffener-plate) or machining modifying locally the microstructure of the interface (barb test, bending with notch), Fig. 10.16(c). It is worth noting that the machining will probably induce microcracking at the interface without any possibility to control this microcrack in size and shape, and thus will increase the level of scatter of experiments, Fig. 10.16(d).

However, another strong limitation is due to the relevancy of loading as compared to in-service use. Because damage leading to spallation of a TBC system is mainly induced by CTE mismatch between each layer, in-service loading is associated to mode II cracking in the presence of microcracking mostly located at the TC/TGO interface for EB-PVD coating under thermal cycling [1607]. Therefore, only barb, shear, and compression testing seems to be fully consistent with this point. Barb and shear tests enable accessing directly $K_{II,c}$ but with strong modification of the TBC system that could be a bias to the measured values. Compression testing enables accessing local strain at decohesion and spallation of the top coat, the spallation being driven by local maximum strain arising from localization in the substrate or bond-coat analyzed by DIC, Fig. 10.17 [914]. However, strain at spallation does not take into consideration the actual mechanical state, that is to say, the residual stresses are not considered or should be addressed by modeling. This limitation is also a drawback for other methods. Both indentation and scratch tests have further major drawbacks considering their relevancy compared to in-service loading. For indentation achieved on a cross-section, the indentation location has a major impact on results. This becomes an issue for rough interface, where it is rather difficult to determine precisely the location of measurement. Scratch testing is straightforward to rank different coating solutions, but again it suffers from a very complex loading very far from the in-service one, except in the field of abradable coatings. A comprehensive overview of the role of mode mixing and associated testing methods has been given by Hutchinson and Hutchinson [649].

LASAT method provides a robust methodology to obtain a clear ranking of adhesion of coating systems manufactured by a given processing method. It has been established that the ranking obtained in the as-processed condition for EB-PVD YSZ coating is consistent with both life to spallation under thermal cycling and ranking established after thermal cycling before spallation [561]. The major advantages of this method are to be a noncontact method, limiting the bias induced by specimen preparation, and to be fairly robust as compared to classical TAT [1424]. Whereas the dynamic range of loading prevents from a direct interpretation of the obtained interfacial properties in quasistatic range, the measured dynamic toughness and its evolution with aging is in agreement with any other known mechanical measurement of adherence. Moreover, the notch-like defect processed by LASAT and combined to the nondestructive evaluation of the resulting blister (IRT or optical image analysis and profilometry) has been successfully used to capture interfacial toughness evolution with aging [1424,916].

10.3.5 Advanced 3D characterization of microstructure morphology and cracks

The microstructure of protective coatings plays an important role in initiation and propagation of cracks and final failure by spallation. In a TBC system, cracks may initiate either within the different layers or at the interfaces between the layers. Depending on local stresses and morphology of the microstructure, cracks propagate, leading to coating failure by spallation. Damage and failure mechanisms of TBC systems have been widely investigated mainly based on observation of cross-sections processed by saw or wire-cutting [411]. Such cross-sections provide two-dimensional (2D) data about crack initiation sites and crack paths. However, the precise 3D information of the microstructure and the size and location of cracks especially at the interfacial regions of the layers has to be addressed since the size and morphology of the so-called process zone of a crack modifies the available strain energy density for final buckling and subsequent spallation [1636].

In materials science, emerging 3D techniques are now widely spread for microstructure characterization and available for application on TBC systems. As a nondestructive method, X-ray computed tomography (CT) appears as an attractive method for evaluating TBC systems, but some limitations depending on the X-ray source and the imaging technique have to be considered. In general, the scanned volume and the resolution are reciprocal. Therefore, for achieving high resolution the CT is limited to small volumes. For investigating EB-PVD or APS TBC samples applied on a Ni-base substrate, rods of less than 500 μm in diameter achieved an apparent resolution of about 0.9 μm voxel size [719,8]. Computed laminography (CL), which is similar to CT but with a different X-ray beam angle to the rotational axis of the specimen (90° for CT and typically 60° for CL), enables analyzing a plate with an in-plane extension in the centimeter range [964]. Applied to TBC, CL has provided 3D measurement of progressive damage evolution with thermal cycling [920]. CT and CL have a limited spatial resolution of about 500 nm for CT and a few μm for CL, respectively. As an alternative for achieving high-resolution 3D reconstructions, focused ion beam (FIB) serial sectioning coupled with scanning electron microscopy (SEM) can be used. Although this method is destructive, it is relatively gentle and even brittle, and weakly bonded materials such as the TGO on bare oxidized alloys are preserved [368]. After 3D reconstruction, the spatial resolution is in the range of about 2–20 nm, depending on the distance between the FIB slices. Thus, the FIB-SEM 3D serial sectioning tomography is an attractive method for studying 3D details of the morphology of TBCs.

Recently, results of FIB-SEM 3D serial sectioning tomography on EB-PVD TBC systems have been published. In [556], results from FIB-SEM serial sectioning have been compared to such achieved by CT using different X-ray sources. Besides some pore size distribution analysis of the BC and the diffusion zone between substrate and BC, the 3D representations have been qualitatively compared with 2D SEM images. The work of Dennstedt et al. [333] focuses on the FIB-SEM serial sectioning method and provides quantitative approaches to describe the path of complex delamination cracks. One precondition was a good segmentation between the layers and cracks by using two different detectors for imaging by SEM, the electron backscatter detector and the secondary electron detector, which provide images with complementary contrast information, compare Fig. 10.19(a)–(c). By analyzing further the area fraction of each phase detected in sections parallel to the coating plane, quantitative data to describing the crack path were derived.

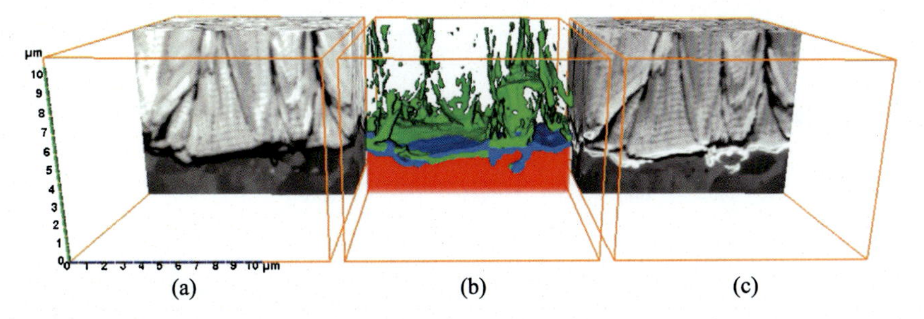

FIGURE 10.19 Reconstructed volume from FIB-SEM serial sectioning (Box size, 11 μm × 14.2 μm × 18.88 μm). Segmentation of the different phases BC, TGO, TC, and cracks or voids was performed by using images from backscatter electron (a) and secondary electron (c) detectors. In (b), the BC is displayed in red, the TGO in blue, cracks (respectively, voids) in green, and the TC transparent to allow for viewing the structure of intercolumnar gaps (courtesy of A. Dennstedt).

10.4 Properties of protective coating systems

Tables 10.1–10.4 present successively the thermal conductivity, coefficient of thermal expansion, Young's modulus, and interfacial toughness for different TBC materials and their respective evolution as functions of temperature. The experimental techniques of characterization are indicated when not usual. Precision is given if microtesting is used, instead of standard specimen. The data are given for only a few temperatures, ignored when not available, and are interpolated when measurements are close to these temperatures.

TABLE 10.1 Thermal conductivity, k (W/(m·K)).

Layer	T (°C)						Ref.
	RT	*600*	*800*	*900*	*1000*	*1100*	
TC							
EB-PVD						*flash* 1316°C/1.8	[682]
EB-PVD (as received heating up)	1.4	1.2	1.2	1.3	1.3	1.3	[1184]
EB-PVD (as received cooling)	1.6	1.4	1.4	1.4	1.4	1.4	[1184]
EB-PVD (200 h 1100°C)	2.1	1.6	1.6	1.6	1.5	1.5	[1184]
EB-PVD	1.6	1.4	1.6	1.3	1.4	1.3	[98]
APS (as received heating up)	0.7	0.6	0.6	0.6	0.6	0.7	[1184]
APS (as received cooling)	0.8	0.7	0.7	0.6	0.7	0.7	[1184]
APS (200 h 1100 °C)	1.4	1.2	1.0	1.0	0.9	0.9	[1184]
SPPS						*flash* 1316°C/0.62	[682]
SPS (columnar)	0.8	1.	0.9	0.9	0.8	0.9	[98]
SPS (compact)	0.7	0.7	0.8	0.7	0.7	0.6	[98]
TGO	33				6.7		[1451]
BC							
NiCoCrAlY	10.8				32.1		[1451]
Substrate							
CMSX4						31.0	[83]

TABLE 10.2 Coefficient of thermal expansion, CTE (10^{-6} K^{-1}).

Layer	T (°C)						Ref.
	RT	*600*	*800*	*900*	*1000*	*1100*	
TC							
EB-PVD	9.0	10.1	10.8		11.7	12.2	[1516,614] cited in [247]
TGO	8.0	8.7	9.0		9.3	9.6	[1516,614] cited in [247]
BC							
(Ni,Pt)Al	13.6	15.2	16.1		17.2	17.6	[1327,1342] cited in [247]
NiCoCrAlY	10.0				17.5		[1451]
NiCoCrAlYTa	15.1	15.1	17.9	22.0			[1416]
Substrate							
AM1	8.0	13.0	14.0		15.4		[576]
CMSX4						16.2	[83]
PWA1480	14.8	16.2	16.9		17.5	18.0	[1015] cited in [247]

In Table 10.3, the average in-plane Young's modulus is given for the coatings. Gradients across the coating thickness, as observed for EB-PVD-TC layers, are not considered. The elastic response of cubic single crystalline materials is described by a stiffness matrix with three independent constants. For comparison reasons, the Young's modulus in the [001] direction is given.

TABLE 10.3 Young's Modulus (GPa); FS indicates free standing coating testing.

Layer	T (°C)						Ref.
	RT	*600*	*800*	*900*	*1000*	*1100*	
TC							
EB-PVD	48	40	34		26	22	[1516,614] cited in [247]
APS	65–110						[492]
TGO	400	370	355		325	320	[1516,614] cited in [247]
BC							
(Ni,Pt)Al	200	160	145		120	110	[1327,1342] cited in [247]
(FS)			55	35	15		[1419]
(FS)	135			110 (*)			[14]
γ–γ' (FS)	130			127 (*)			[14]
				(*) 870°C			
NiCoCrAlY	200				120		[1451]
NiCoCrAlYTa(FS)	182	157	141	128	105		[1415]
Substrate							
AM1 ([001] direction)	137		108		104		[204,576]
CMSX4 ([001] direction)	127	106	95	90	82		[1324]
PWA1480	220	170	155		130	120	[1015] cited in [247]

TABLE 10.4 Interfacial toughness evaluation for different testing and analysis methodology.

System/Testing technique	Analysis	Mixity angle (°)	Aging	Gc (J/m^2)	Ref.
TC					
APS/NiCoCrAlY compression	FEA	-56	no	110-130	[1674]
EB-PVD/CoNiCrAlY push-out/shear	analytical	-90	isothermal 1150 °C	120 (hox=2 μm) 20 (hox=4.5 μm)	[723]
EB-PVD/NiCoCrAlY bending	FEA	20	no	57±21	[377]
4-point bending	analytical	20	no	>81	[62]
		20	1000°C 100 h	63	[62]
		20	1100°C 10 h	37	[62]
EB-PVD/(Ni,Pt)Al 4-point bending	analytical	20	no	110	[1423]
	FEA	20 and 70	FCT 250 cycles (*)	G0/3±3G0 (**)	[1481]
LASAT	analytical	0	FCT 300 cycles (*)	5	[561]
TGO 4-point bending	analytical			110	[263]

(*) *FCT RT* – 1100°C *with 1h per cycle.*
(**) *Only relative values are available.*

10.5 Evolution of TBC microstructure and associated damage under thermal and thermo-mechanical loads

10.5.1 Thermal fatigue/cyclic oxidation testing

The most standard testing configuration for TBC lifetime evaluation consists in a cyclic oxidation test. Standard specimens are button shape coupons (one inch diameter and 1–3 mm thickness) which are chamfered on one edge. The BC is applied all over the specimen, whereas the TC is deposited only on the side with the chamfered edge, to decrease detrimental edge effect. As a standard, electrical furnace is used to prescribe the maximum temperature, the specimens being moved into and out of the furnace for heating and cooling. The applied thermal cycling consists for most of the studies in a direct heating in the furnace, 45–60 minutes dwell at maximum temperature for aeroengine application (could be extended to 24 h and more for reproducing working cycles of stationary gas turbines for energy conversion) and natural or air-assisted cooling. For reducing the time of the cooling sequence, the target minimum temperature could be set for instance to 100–200°C without strong impact in damage evolution [1443].

As a major result of this thermal cycle, the CTE mismatch between each material constitutive of the TBC system (substrate, IDZ, BC, TGO, TC) induces high stresses in each layer which result in crack initiation and subsequent interfacial damage and finally spallation, as described in Section 10.5.3. A thermal cycle can be regarded as representative for a thermo-mechanical fatigue cycle with temperature and mechanical load cycle out-of-phase (OP-cycle). The effect of cyclic thermal loading, which mimics more closely than isothermal loading the in-service condition of an aeroengine, has been systematically investigated in [295]. By varying the dwell time in thermal cycling experiments but keeping the overall time at high temperature the same, it was found that microstructure and damage evolution increase significantly with decreasing the dwell time, or in other words, with increasing the frequency of thermal cycling. Comparison with isothermal oxidation highlights both the detrimental effect of short dwell time in thermal cycling and the modification of damage mechanism [1607,295].

10.5.2 BC and TGO evolution under thermal cycling

Firstly, together with high temperature exposure, TGO growth induces strain growth associated to molar volume difference of the newly formed oxide as compared to the metallic BC [1442]. In addition to TGO growth, diffusion takes place. Mainly, an outward flux of Al, to form the TGO, and an inward flux of Al from the BC to the substrate should be considered. Reciprocally, a Ni flux from the substrate to the BC is observed. These diffusion processes yield phase transformation from β-NiAl to γ'-Ni_3Al in the case of NiAl diffusion coating, Fig. 10.20(a)–(b). For MCrAlY coating, phases evolve from $\beta + \gamma + \alpha$ to $\beta + \gamma' + \alpha + \sigma$ or from $\beta + \gamma$ to $\beta + \gamma'$ for low Cr and Co compositions, Fig. 10.20(e) [1002]. For β-NiAl diffusion coating, the martensitic transformation from B2 to LI_0 structure from high to low temperature has been evidenced [239].

For bare metallic coating, net mass gain (NMG) measurements under thermal cycling shows (i) pure gain of mass due to BC oxidation, (ii) de-

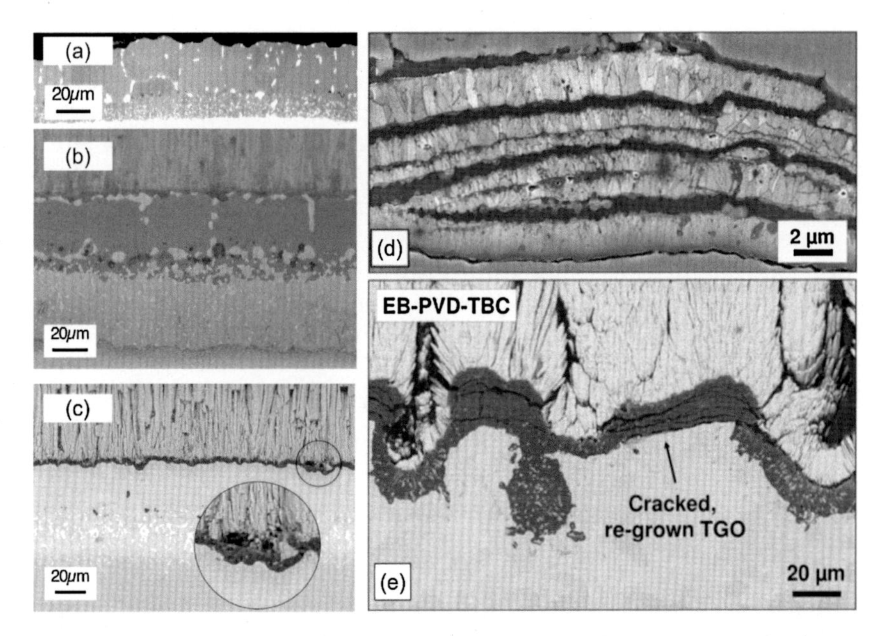

FIGURE 10.20 Evolution of TGO microstructure for thermal cycling (range 100–1100°C) for (Ni,Pt)Al BC: (a) after 1000 cycles without TC observed by optical microscope (OM) (adapted from [1257]), (b) with EB-PVD TC for N = 1200 cycles observed by SEM in BSE mode, (c) SEM image of TBC system from (b) in SE mode (insert corresponds to a zoom of the black circle) (adapted from [920]) and for NiCoCrAlY BC, (d) SEM image in SE mode of oxide fragmentation of a system with APS TC (maximum cycle temperature 1050°C), and (e) SEM image in SE mode of a system with EB-PVD TC (maximum cycle temperature of 1000°C) (adapted from [1002]).

crease of NMG rate due to partial oxide spallation leading sometimes to a plateau (equilibrium of mass gain by oxidation and mass loss by oxide spallation), (iii) decrease of NMG due to an excess of loss as compared to gain, and (iv) catastrophic loss and final breakaway. In this case, the oxide spallation is mostly driven by BC/TGO failure, and the apparent adherence of the oxide decreases with aging under thermal cycling. The testing methodology being detailed in Section 10.7.1.1.

For bare BC (i.e., without ceramic TC), the TGO could spall off from the BC when reaching high thickness and/or more likely reaching a critical interfacial fracture toughness, which leads to spallation during cooling when the stress intensity is the highest. Local discontinuities such as thickness variation of the TGO are increasing local stresses and enhancing the damage process [412].

For improved adhesion of the TGO, as obtained by Pt addition to NiAl, large undulation of the TGO and BC are observed under thermal cycling, with the amplitude of the undulation increasing by the number of cycles. This deformation mechanism is called rumpling. Some authors have claimed that rumpling was induced by either martensitic or β to γ' phase transformation [239]. However, when limiting the partial pressure of O_2 and consequently the TGO growth without substantial modification of phase transformation, the rumpling was drastically limited [1444,915]. Thus, the dominating mechanism of rumpling is the combination of the increase of the TGO volume (and subsequent oxide strain growth) and the low mechanical properties of the BC at the TGO temperature formation. For highly adherent TGO, the oxide growth induces large creep deformation in the BC at high temperature, and for thermal cycling, cumulated plastic straining is likely to amplify the undulation.

10.5.3 TBC evolution under thermal cycling

In the case of a complete TBC system (including ceramic TC), the TGO spallation is confined by the TC and, when further exposed to high temperature strain accumulation, can induce local TGO cracking without spallation, Fig. 10.20(d)–(e). Subsequently, much higher TGO thickness could be observed for adherent TC as compared to bare BC, making the critical thickness of TGO not obvious.

The role of TC in rumpling could be clearly illustrated by introducing an interfacial defect by laser shock before thermal cycling (based on the noncontacting LASAT method described in Section 10.3.4.5). After furnace thermal cycling, three different subsequent morphologies could be observed on a cross-section: (i) along the initial debonding introduced by laser shock the rumpling is very pronounced as usually observed for bare BC, Figs. 10.21(a) and 10.21(c); (ii) "far away" from initial debonded area, no rumpling is observed, but interfacial damage is consistent with damage observed without artificial defect, compare Figs. 10.21(b) and 10.20(c), and (iii) at the transition between large debonding and adherent TC, corresponding to interfacial crack tip, it is observed beginning of large undulation as compared to damage without rumpling, Fig. 10.21(d).

Even though the driving forces for rumpling are present for adherent TC, this example demonstrates that without TC decohesion, no rumpling can take place. Except for TGO growth and rumpling, the microstructure evolution of (Ni,Pt)Al BC is very similar with or without EB-PVD TC layer for both the evolution of the β to γ' phase transformation in the external BC layer and for IDZ evolution, Fig. 10.20(a)–(b).

When exposed to high temperature, sintering is the main process acting on the TC microstructure to induce an evolution of the mechanical properties and in particular an increase of the TC stiffness [871]. The crucial role of CMAS on the change of TC during service will be highlighted in Section 10.6.

For complete TBC including an EB-PVD TC layer, failure yields the final spallation of the TC: spallation is observed on disk shape specimens, see Fig. 10.22(a), and for the same duration and aging condition, both

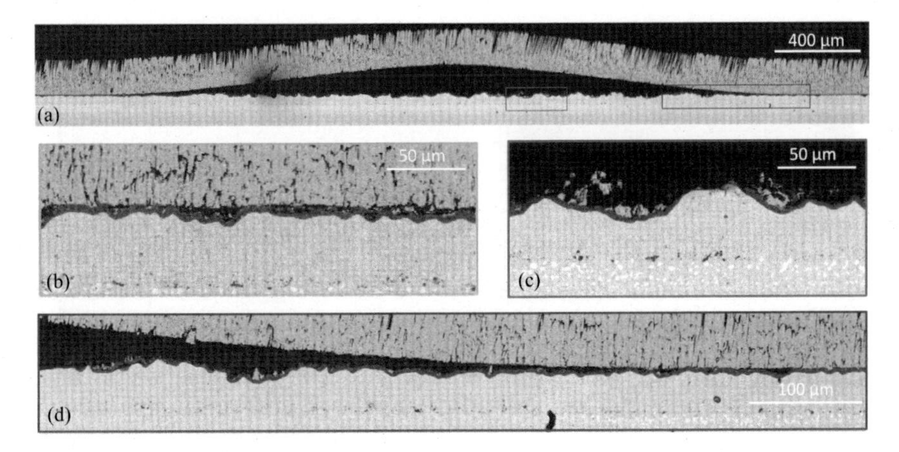

FIGURE 10.21 Interfacial defect introduced by LASAT in the as-processed condition and observed after 2000 thermal cycles at 1100°C: (a) global view of the blister, (b) detail view out of (a): interface evolution without interaction with the initial defect, (c) detail view, see red (mid grey in print version) rectangle in (a): initial area of decohesion, (d) detail view, see blue (mid grey in print version) rectangle in (a): evolution in the vicinity of the interfacial crack tip (courtesy of Lara Mahfouz).

the critical strain to TC failure, Fig. 10.22(b), and the apparent interfacial toughness, Fig. 10.22(c), are observed to decrease. The testing methodology is detailed in section 10.3.4. It has been shown that either the BC/TGO or the TGO/TC interfaces could be the weak interfaces: for isothermal aging and/or low adhesion of the TGO, the BC/TGO damage drives the final failure, whereas for thermal cycling and/or high adhesion of the TGO, the TGO/TC damage dominates the final failure, see Figs. 10.21(b) and 10.21(d), and [1607,295].

Recent investigations of γ–γ' Pt-rich coating during thermal cycling highlights the improvement in lifetime as compared to classical β-(Ni,Pt)Al BC for the same substrate. However, the authors have demonstrated that the critical interface is the BC/TGO interface, where the TGO develops on a γ layer after aging [53].

10.5.4 Damage evolution and lifetime behavior evaluation under mechanical and thermal loading (LCF, TMF, TGMF, and burner rig)

Coated high pressure turbine blades face high mechanical and thermal fatigue loading induced by the expanding hot gas, the transient temperature differences between heated outer surface and cooling channel walls, and additional centrifugal forces in the case of rotating blades. The loading conditions are quite complex and just thermal cycling of coated specimens under laboratory conditions is not sufficient to replicate the damage pro-

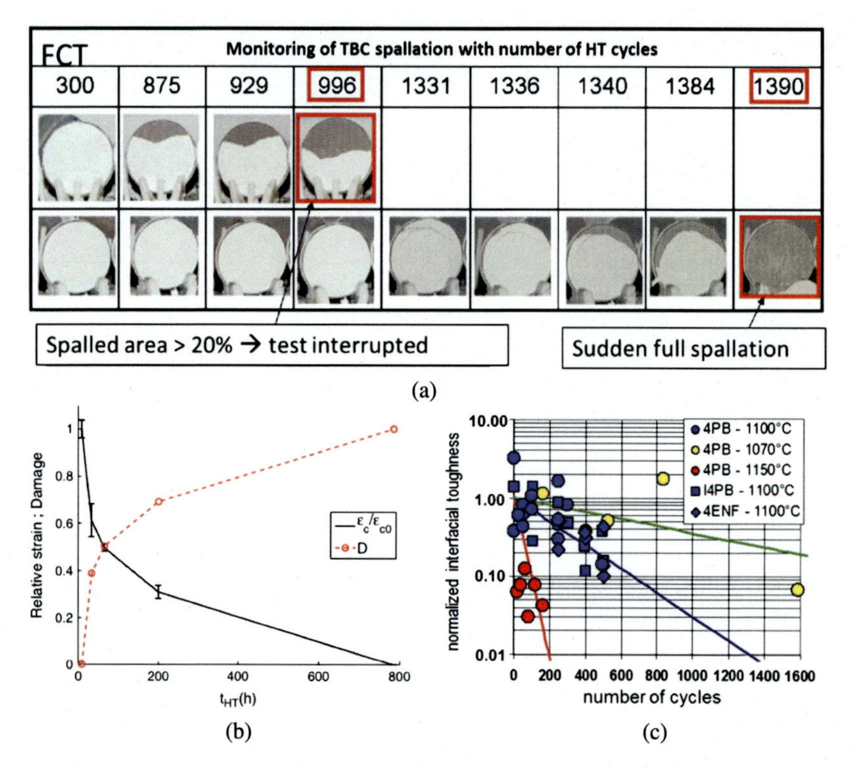

FIGURE 10.22 Effect of thermal furnace cycling for EB-PVD, (Ni,Pt)Al bond-coat and first generation single-crystal superalloy: (a) appearance of TBC spallation observed on button shape specimens as a function of cycle number, [416], (b) evolution of critical strain, ε_c, at spallation and associated damage, D, as a function of the time spent at high temperature (adapted from [295]), and (c) interfacial toughness as a function of the number of furnace cycles and temperature [1481].

cesses occurring in coated turbine blades during gas turbine operating cycles. However, realistic testing in laboratory is difficult to conduct and appropriate test facilities are extremely expensive. Therefore, simplified tests were developed aiming at either ranking of new coatings variants during material development campaigns or at better understanding of damage mechanisms under controlled loading conditions.

Low cycle fatigue (LCF) testing is straightforward to assess the influence of mechanical loads under isothermal conditions on the damage behavior and lifetime of a coated material system. To achieve a stable temperature over the entire test, usually resistance furnaces with a high thermal inertia are used for heating the specimen. It has been observed that under LCF conditions cracks are typically induced at morphological instabilities at the surface, see Fig. 10.23(a), or at the interface between BC-TC [78] or between substrate-BC, see Fig. 10.23(b). The coating could be either beneficial or detrimental depending mainly on substrate composition and type of coating [1167]. This point was seen to be crucial for 3rd and higher generation of single crystal superalloys for which diffusion (Ni,Pt)Al coating was observed to induce secondary reaction zone (SRZ) and drastic decrease of the LCF lifetime [252,211], see also details in chapter 8.

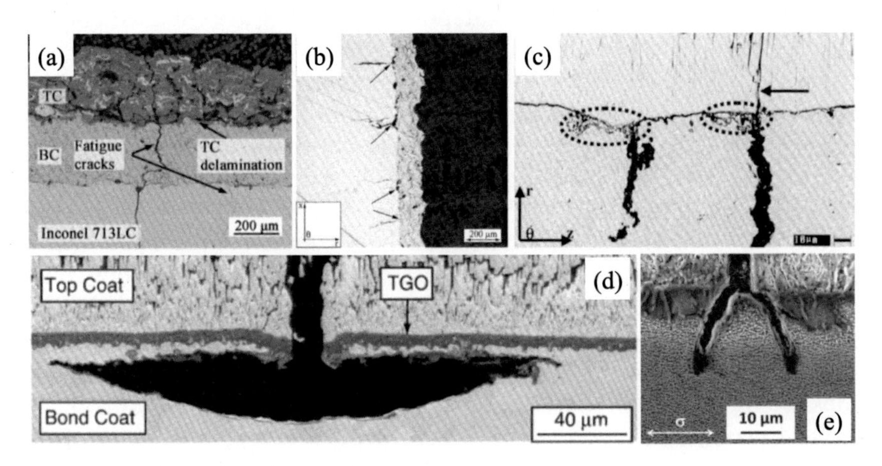

FIGURE 10.23 Exemplary damages observed under different thermal and mechanical testing conditions: (a) APS BC and TC after LCF test with $R_\varepsilon = -1$ [1375], (b) APS BC only, LC after LCF test with $R_\varepsilon = -1$, and $T = 850°C$ [675], (c) EB-PVD TC and APS BC, IP-TMF test: $R_\varepsilon = 0$, $\varepsilon_a = 0.7\%$, and $100 < T < 1000°C$ [1451], (d) IP-TGMF test: $R_\sigma = 0$, $\sigma_a = 100$ MPa and $100 < T < 1000°C$ with $\Delta T = 170°C$ (through thickness) [606], (e) coating system after burner rig test in combustion environment, OP-TGMF test: $R_\sigma = 0$, $\sigma_a = 800$ MPa and $500 < T < 1100°C$ with $\Delta T = 150°C$ (surface) and $\Delta T = 50–70°C$ (through thickness) [911].

The effect of strain dwell under high temperature condition, evaluated in so-called Sustained Peak LCF, was also seen to be crucial in impact on TBC lifetime. This was described to be associated to a mechanism where, in addition to early cracking induced by the TBC, the stress relaxation at high temperature yields tensile stress at zero strain and increases dramatically the fatigue crack growth rate [407]. However, it is worth noting that under LCF condition the driving force for additional damage induced by the TBC is mainly associated to creep of the coating [1669].

To get closer to in-service loading, thermal and mechanical cycling is combined with thermal mechanical fatigue (TMF). Thermal cycles are typically applied by means of induction heating, which allows frequencies in the range of minutes instead of hours as in the case of using resistance furnaces. Usually, in-phase (IP) cycling (maximum temperature and tensile loading are synchronous) and out-of-phase (OP) cycling (maximum temperature and compressive loading are synchronous) are distinguished, but any other phase shift between thermal and mechanical cycle can be applied. For bare BC, it has been observed that phase transformation and rumpling was increased in TMF as compared to pure thermal cycling for both IP and OP conditions [1258]. It has been observed that rumpling under IP TMF results in wrinkles oriented along the mechanical loading direction [70]. In the presence of the ceramic TC layer, it is currently observed that the time to TC spallation is reduced in OP as compared to IP, and that detrimental effects of oxidation and cracking are enhanced under TMF condition, see Fig. 10.23(c)–(d) [1451,435].

More realistic than TMF with homogeneous temperature across the cross-section of the test specimen are experiments replicating the thermal gradient experienced by the turbine blade, the so-called thermal gradient mechanical fatigue (TGMF) tests. Firstly, one should notice that the TC surface is the hottest part of the component, particularly for hollow blades where internal cooling decreases the substrate temperature. Therefore, in TGMF testing of coated specimens, the substrate side has to be cooled and the TC surface has to be heated. No induction heating can be applied, since the electrically conductive substrate is heated, which results in a biased thermal gradient across the coating system, even if the substrate is cooled. In TGMF testing, tubular specimens are used with internal air cooling through a central cooling channel and heating the outer surface by a radiation furnace. For achieving a realistic thermal gradient over the specimen wall, TGMF facilities with a focusing radiation furnace have been developed at the German Aerospace Center [79,77]. The experiments described in [77] were conducted on coated specimens comprising directionally solidified Ni-base superalloy IN 100 as substrate, NiCoCrAlY BC, and YSZ TC, both coatings applied by EB-PVD. The maximum specimen surface temperature in each TGMF cycle was 1000°C, and over the specimen wall a maximum temperature difference of 170°C was achieved. The load cycle was designed to represent the fatigue load at the leading edge of a rotating turbine blade during an entire flight mission of an aeroengine. Compared to thermal cycling experiments with the same maximal temperature a premature spallation of the coating system was observed. The damage feature after TGMF testing differed significantly from that after thermal cycling. After thermal cycling the TC spalled off with adherent TGO (the so-called black failure) without crack formation perpendicular to the coating plane. In contrast, crack initiation in TGMF started in the TGO under tensile stress, creating cracks perpendicular to the coating plane and the applied mechanical load. During propagation into the BC cracks deviated within the BC with the crack path parallel to the interface BC and TGO and changing finally to the interface between BC and TGO. Multiple cracks evolved simultaneously and merged, leading to large scale spallation. At the location of crack initiation, inelastic deformation of the BC occurred, forming a smiley crack feature, Fig. 10.23(e). Further mechanical experiments on the same type of coated specimens were performed with a compact radiation furnace in situ in a synchrotron facility [730,902] in order to observe the deformation behavior of each coating layer during mechanical loading. With these results, it was possible to explain the damage behavior in TGMF testing and to validate predictive simulation models, which captured the observed behavior [607,606].

Another approach to apply thermal gradients is based on burner rig facilities, where a flame (most of facility using natural gas) is heating the coated side, and the substrate side is cooled by pressurized air. These facilities are very convenient to test also the impact of mineral particles which are ingested into the aeroengine from the environment, such as oxides and silicates. With a burner rig facility developed at the Research Centre Jülich, a temperature difference across a coated disc shaped specimen of about 200°C was achieved with a maximum temperature of 1200–1300°C at the TC surface [884], Fig. 10.24(b). However, most burner rig facilities are not capable of applying mechanical loads.

Only a few test rigs are developed, which allow simultaneous mechanical loading combined with a burner rig heating system. One example is the MAATRE burner rig facility [911] in which a specimen can be loaded by means of a servohydraulic fatigue testing machine in a realistic combustion environment using a kerosene burner. The hot gas flow is perpendicular to the length axis of a tubular specimen, which results in different conditions at the leading and trailing edges. Test results on coated specimens revealed severe damage by multiple cracking of the coating with cracks propagating into the substrate, to higher extent at the trailing edge compared to the leading edge, Fig. 10.23(c) [911].

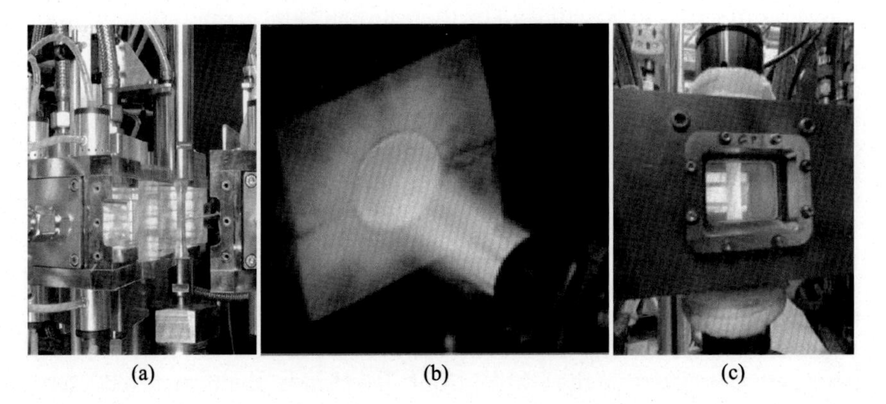

(a) (b) (c)

FIGURE 10.24 TMF testing of TBC in the presence of gradient: (a) TGMF [77], (b) burner rig [884], and (c) combustion environment in TGMF condition [911].

10.6 Challenges due to ingested mineral particles CMAS

10.6.1 Damage mechanisms and TBC degradation due to CMAS attack

Recent enhancement of aircraft performance due to increased turbine inlet temperature has in return generated new and even more critical premature failures of TBC (8YPSZ) that affect its lifetime. In particular, severe degradation is due to mineral particles (sand, dust, volcanic ashes, etc.) ingested with intake air, which subsequently adhere to the hot section components (combustor, high pressure turbine blades and vanes, shrouds) as siliceous deposits [1336,132].

These deposits are mainly constituted of CaO, MgO, Al_2O_3, and SiO_2 (CMAS) but also contain other metallic oxides (Fe_2O_3, NiO, and TiO_2), sometimes in large amounts. CMAS composition strongly depends on flight location [132,1494,1336,149], resulting in various viscosities and melting temperatures [1122]. With the increase in operating temperature, TBC surface temperature during service (1200°C or higher) exceeds most of CMAS deposits melting points. Thus, CMAS melts and infiltrates very rapidly, through capillary action, the TBC open porosity (intercolumnar gaps in the case of EB-PVD coatings), leading to TBC stiffening and loss of its strain tolerance. As a result, delamination cracks can develop in the infiltrated TBC during cooling, leading to progressive spallation with engine thermal cycling [935,760,825]. Chemical degradation is also observed with the dissolution of yttria-stabilized zirconia during interaction with silicate melt, followed by reprecipitation of YSZ with a modified composition and morphology depending on the melt chemistry [762,56,365,926,1494,825,1495]. These two mechanisms are schematically summarized in Fig. 10.25. The thermo-mechanical degradation mode is considered to be predominant compared to the thermo-chemical one. Thus, with the higher operating temperatures, CMAS-induced TBC degradation has become increasingly critical, making it necessary to find alternate TBC compositions, as well as coating morphologies, likely to limit CMAS infiltration.

10.6.2 CMAS resistant compositions for ceramic coatings

The most recent CMAS mitigation strategy relies on the chemical reactivity (dissolution/reprecipitation process) between TBC and CMAS melt. It consists in modifying TBC composition to promote crystallization of phases that will arrests the CMAS front. Two kinds of alternate TBC have been proposed: in the first, rather applicable to APS or SPPS coatings, Al_2O_3 and TiO_2 are incorporated in metastable solid solution in YSZ. During TBC dissolution, alumina combines with CaO and SiO_2 from the melt, resulting in the near-complete CMAS crystallization into anorthite

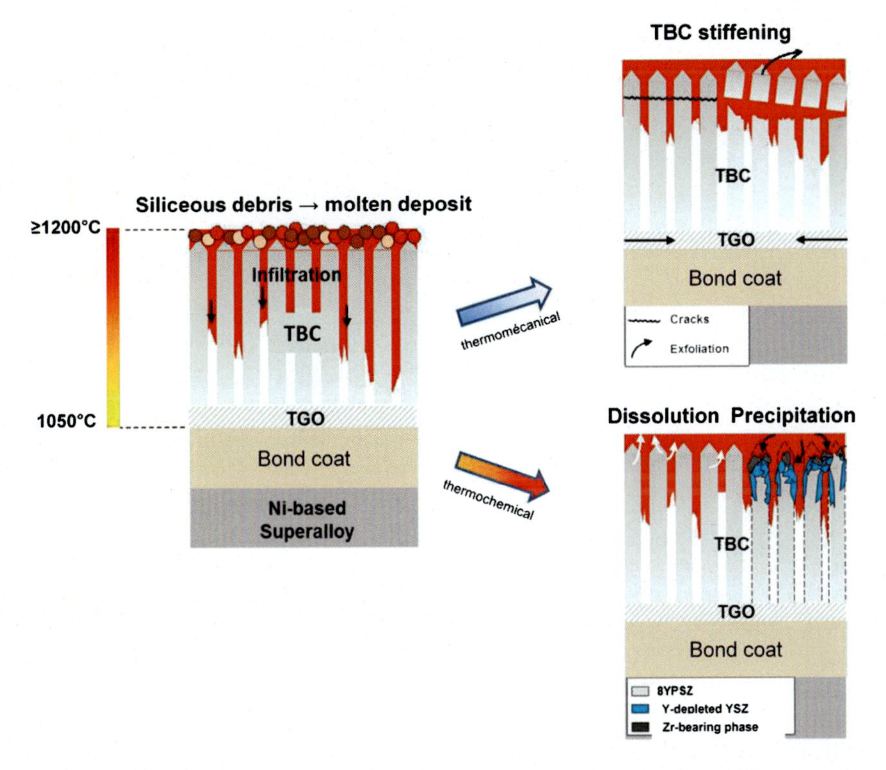

FIGURE 10.25 CMAS-induced degradation mechanisms of EB-PVD thermal barrier coatings.

($CaAl_2Si_2O_8$) [56,365,825]. The second is based on rare-earth zirconates ($RE_2Zr_2O_7$, RE = from La to Yb). Indeed, its dissolution in the CMAS melt leads to the near-simultaneous precipitation of a Zr(RE, Ca)Ox phase and a $Ca_2RE_8(SiO_4)_6O_2$ apatite silicate. In the case of EB-PVD coatings, these two phases fill the TBC intercolumnar gaps, thus blocking further CMAS penetration. The effectiveness of this last strategy has been first demonstrated for $Gd_2Zr_2O_7$ (Fig. 10.26) [761,825].

10.6.3 Characterization of damages by CMAS attack

First investigations have consisted in performing microstructural analysis of CMAS-induced coating degradation on hot section components from the field [1336,132,935,760,149,1494,1566]. Such studies are infrequent as they use destructive methods and because engine parts are not easily available. They have provided very significant source data on CMAS infiltration depth in the TBC, CMAS deposits' composition and distribution at the surface of the components. They have also identified failure mechanisms, as delamination cracks within or below the in-

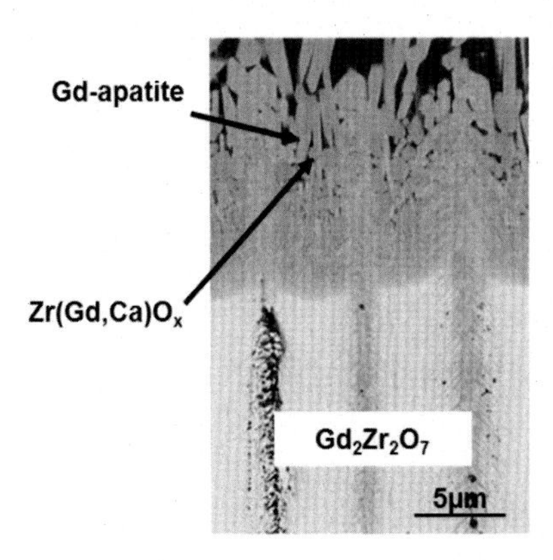

FIGURE 10.26 SEM micrograph showing the sealing of the intercolumnar gaps of an EB-PVD $Gd_2Zr_2O_7$ coating by the combination of Gd-apatite and Zr(Gd,Ca)Ox crystalline phases (based on information from [825]).

filtrated TBC, as well as chemical reaction between 8YPSZ coating and CMAS. Subsequently, simple laboratory furnace tests have been developed, aiming to reproduce the chemical interaction taking place in service [959,365,1547,1580,1147,1495,998,762,1364,56]. They are performed in isothermal controlled conditions at temperatures varying from 1100°C to 1400°C, without thermal gradient in the TBC. In order to eliminate premature TBC spalling during cooling, or degradation of superalloy mechanical properties at temperatures above 1200°C, TBC system samples (TBC coating on bond-coated superalloy substrate) are not used, but rather simplified TBC material, namely ceramic pellets (dense or porous), free standing coatings or deposited on alumina substrate (Fig. 10.27). Due to the numerous oxides contained in CMAS deposit compositions, model CMAS (containing three to five oxides among SiO_2–CaO–Al_2O_3–MgO–Fe_2O_3) are generally used. A classic composition is 33CaO–9MgO–13$AlO_{1.5}$–45SiO_2 [762]. However, tests using sand or volcanic ashes, synthetic or sampled from the field, are also reported [526,926,364,320]. CMAS (powders or plate) is deposited on top of the TBC specimen, and the assembly is heat treated in a furnace for different durations. Though such tests are not representative of the actual engine conditions, the chemical interaction mechanisms are similar to that observed using thermal-gradient tests. The understanding of 8YPSZ/CMAS interaction, as well as the development of alternate coatings, resulted from such tests. Mixtures of TBC and CMAS powders are also used, either with a high ceramic concentration, required for identifying reaction products by X-ray diffraction, or with much lower ceramic concentration, for studies devoted to obtain thermodynamic or kinetic information on these reactions [1121,1079,763].

In the case of TBC coatings, such isothermal tests also provide CMAS penetration depth data and can be used to compare the effectiveness of various alternate TBC compositions or coatings morphology in limiting CMAS infiltration. X-ray elemental mapping (Si or Ca) is usually performed on coating cross-sections to determine CMAS infiltration depth (Fig. 10.27).

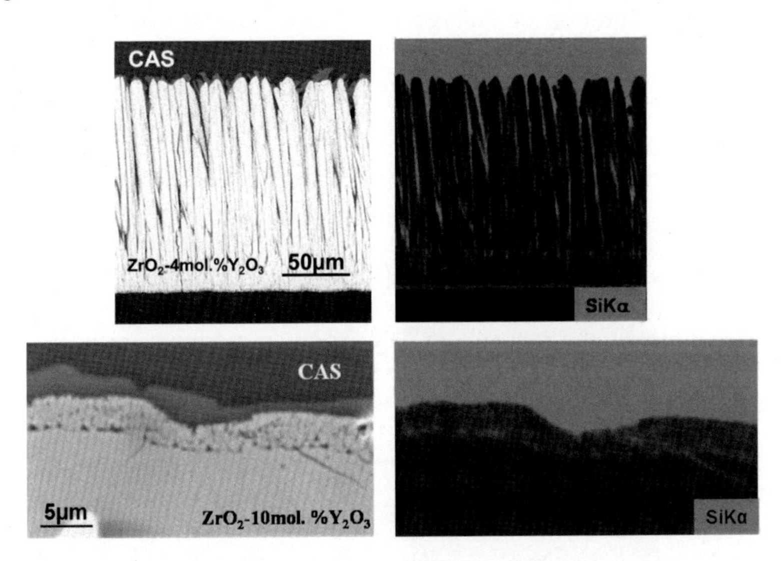

FIGURE 10.27 Cross-sectional SEM micrographs of YSZ EB-PVD coating (top) or dense pellet (bottom) infiltrated by a model CAS melt, and corresponding Si elemental map showing CAS penetration (light contrast).

Concerning experimental characterization of TBC failure mechanisms and evaluation of its lifetime under CMAS attack, isothermal furnace cycle tests have hardly been used as they are too harsh, leading to underestimation of TBC performance, and temperature limited because of the superalloy. Thermal gradient cycling tests with simultaneous CMAS injection, first performed in 2010 in the case of plasma-sprayed coatings [1359,363], are much more relevant but more difficult to develop. The heat source is either a combustion flame (burner rig) [363,1359,884], or a continuous CO_2 laser [825,664,663]. In the case of the burner rig, the CMAS is prepared as an aqueous solution of nitrates or a dispersion of CMAS-frit in water, and is sprayed into the flame. With the laser, CMAS is predeposited on the TBC surface. Such experiments are often analyzed using thermomechanical models adapted or extended from those previously developed for TBC failure mechanisms in the absence of CMAS [408,825,662,664].

10.6.4 Thermal cycling behavior in presence of CMAS

The mechanism generally adopted to describe TBC failure due to CMAS infiltration is a cold shock delamination taking place during engine shutdown and resulting from TBC densification [935]. Indeed, subsurface delamination cracks within or under infiltrated regions of the TBC are observed on ex-service turbine airfoils (Fig. 10.28). They always emanate from surface-connected vertical separations, already existing after elaboration or emerging in service due to sintering or tensile stresses induced upon rapid cooling. The driving force for such TBC delamination and spalling is the release of the elastic strain energy stored in the coating upon cooling, due to the thermal expansion mismatch between the coating and the metal substrate. This elastic energy highly depends on the initial thermal gradient in the TBC coating and is very sensitive to the engine cooling conditions. In particular, slow cooling prevents from the development of vertical separations in the TBC [935,408,825,662,1122]. The CMAS infiltration depth in the coating is also of major importance [825,664]. Indeed, the higher the CMAS penetration depth, the higher the increase in TBC Young's modulus, resulting in a substantial increase in the elastic energy, and thus in the delamination propensity (Fig. 10.29).

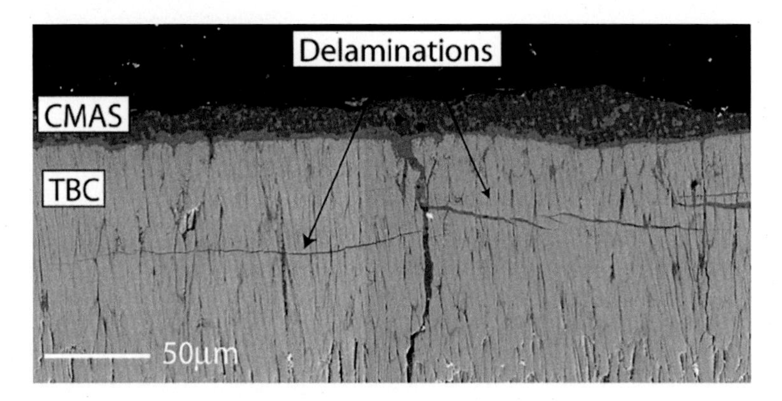

FIGURE 10.28 SEM micrograph on an EB-PVD 8YPSZ cross-section with delamination cracks emanating from surface-connected vertical separations (based on information from [935]).

10.7 Modeling issues

10.7.1 Assessment of microstructure evolution

10.7.1.1 Oxide growth kinetic

First models that can be used to describe microstructure evolution deal with oxide growth. The oxide growth could be cationic, anionic or a mix of both mechanisms [412]. Classical parabolic kinetic of oxide growth has been evidenced for MCrAlY and NiAl bond-coats, following the time evo-

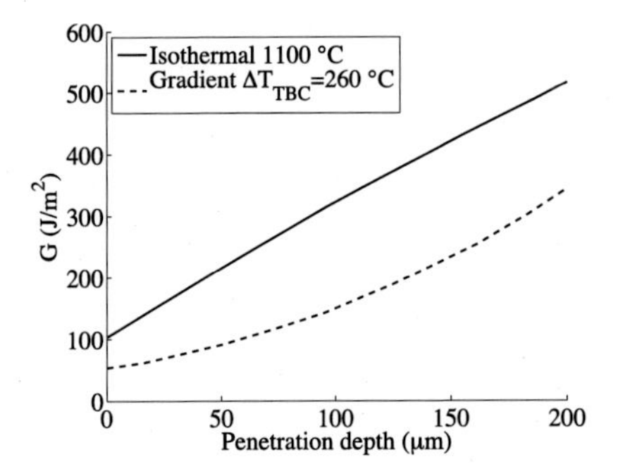

FIGURE 10.29 Energy release rate (G) as a function of CMAS infiltration depth for a delamination crack at the base of a YSZ TBC ($T_{surface}$ = 1300°C, $T_{bond\ coat}$ = 1040°C, and $T_{backside}$ = 800°C); comparison with a TBC isothermally heated at 1100°C (based on information from [1122]).

lution of oxide thickness h_{ox} accordingly to Wagner's theory [1517],

$$h_{\mathrm{ox}}^n = h_{\mathrm{ox},0}^n + k \cdot t, \tag{10.2}$$

where $h_{ox,0}$ is an initial oxide thickness, k and n are material parameters, with n being often close to 2 for isothermal oxidation. In this equation, k is following Arrhenius' law,

$$k = k_0 \cdot \exp\left(- \frac{Q}{RT} \right), \tag{10.3}$$

where k_0 is the kinetic constant of oxide growth, Q is the activation energy of the oxide to be considered and R is the universal gas constant (8.314 kJ·mol^{-1}). For transient temperature, and consequently for thermal cycling, the associated differential equation evaluated at the current time t and temperature T corresponds to

$$dh_{\mathrm{ox}} = \frac{k_0}{n} \exp\left(- \frac{Q}{RT} \right) \frac{dt}{h_{\mathrm{ox}}^{n-1}}, \tag{10.4}$$

where temperature evolution is neglected during the time step dt [295].

The oxide layer is prone to damage (local cracking) and local oxide spallation, yielding a strong modification of subsequent oxidation kinetic [415,412]. The simplest way to model a realistic oxidation kinetic is to associate to the previous parabolic growth a probability of failure with time, the so-called p–kp theory, where p describes this probability of local spallation [1337,1132,962]. The probability of failure at each thermal cycle can be measured experimentally by in situ weighting of specimen, Fig. 10.30(a) [962], and leads to a very good approximation of experimental data, including sensitivity to thermal cycling frequency, Fig. 10.30(b) [1132].

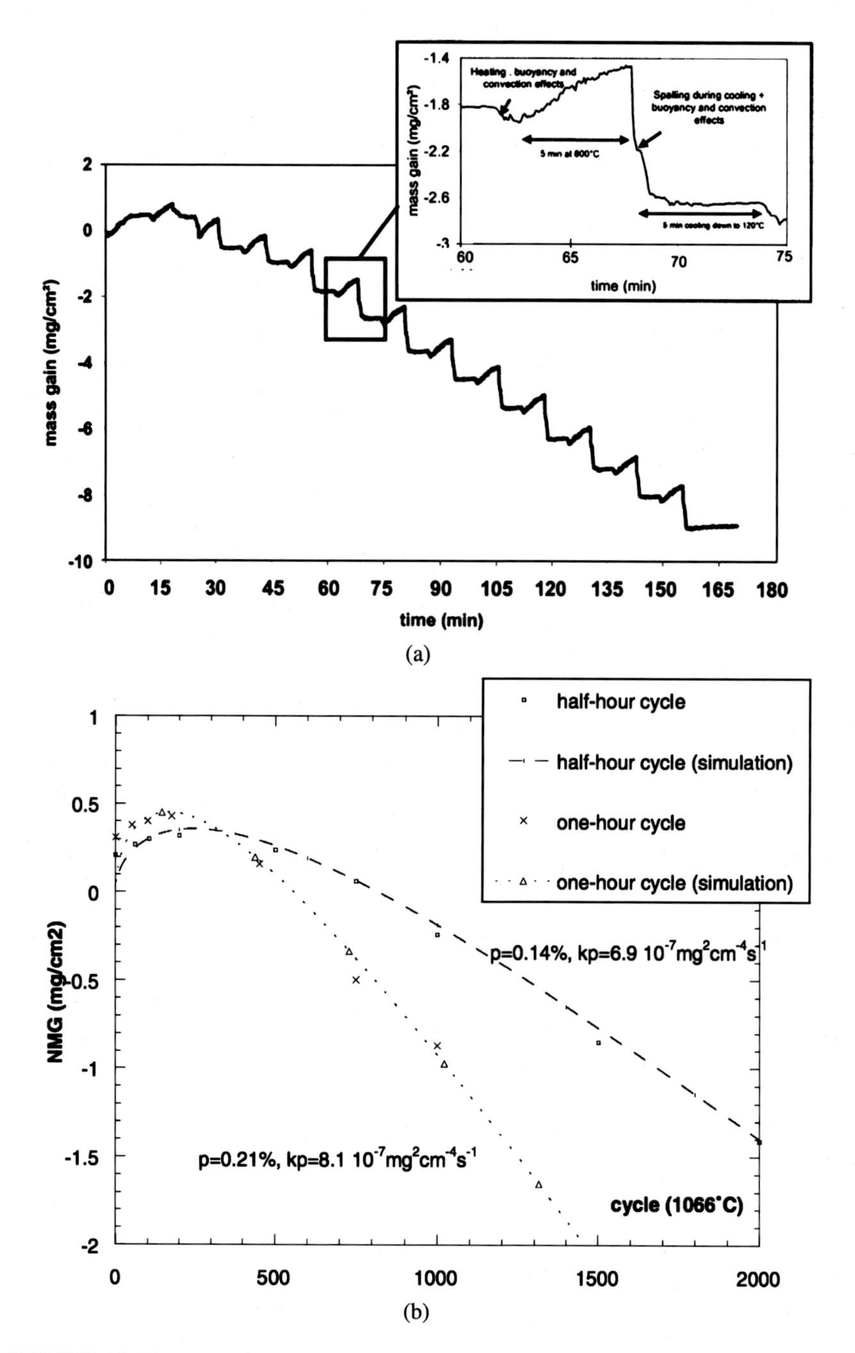

FIGURE 10.30 Net mass gain evolution (a) measured experimentally in situ to derive the oxide growth kinetic k_p and probability of oxide spallation p [962] and (b) measured experimentally ex situ by weighting of specimen after each cycle and modeled by $p - k_p$ [1132].

In the presence of the TC, the limited spallation of oxide strongly modifies a priori critical oxide thickness (see Section 10.5). Thus explicit model of TGO growth accounting for both diffusion and oxide growth kinetics are of interest [175,460,1675]. However, very few studies have been devoted to fully coupled model between mechanical state, oxide failure and oxide kinetic. Strong coupling based on phase field has been for the time being mostly limited to other materials (including NiAl seen as a model material). Modified Fick's law accounting for stress gradient could be coupled to elastoviscoplastic material in the framework of crystal plasticity. For instance, the case of 316L alloy exhibits that oxidation induces strain localization both along grain boundary and in forming intergranular oxide, Fig. 10.31 [318,1410]. This kind of mechanism is likely to induce interfacial damage for most TBC systems, considering intergranular oxidation within the BC.

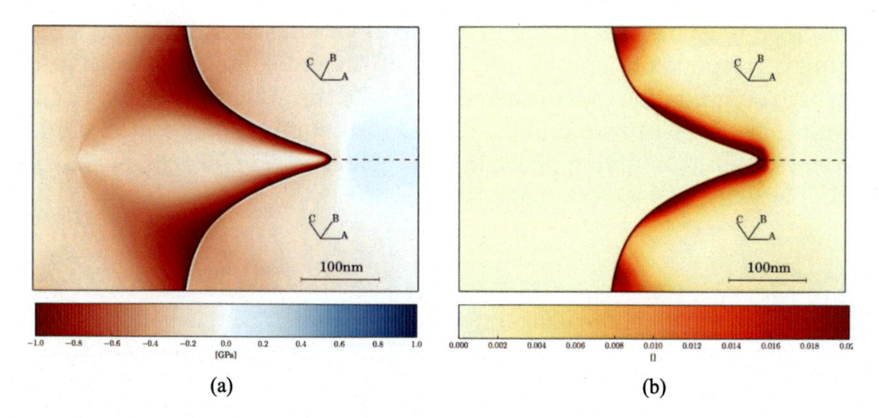

FIGURE 10.31 Phase field model of 316L oxidation accounting for crystal plasticity and diffusion coupled to stress level: (a) hydrostatic pressure and (b) cumulated plastic strain field [318].

10.7.1.2 Phase transformation

Modeling of phase transformation in coated systems is of major interest for optimal design of both the composition of the bond-coat and the target composition of the substrate alloy. Basic composition evaluation could be based on binary phase diagram analysis and diffusion coefficient of Ni and Al [1333,1543,1544]. Moreover, this kind of analytical model could be easily linked to oxide loss analysis to consider evolution of realistic boundary condition for diffusion analysis [1133].

The major concern, similar to TGO growth analysis, is to obtain a realistic value of the diffusion coefficients. Most advanced models for the estimation of diffusion coefficients are based on CALPHAD, considering multicomponent alloys and mobility database to modify the diffusion coefficient as a function of local composition [1632,238,1106] where phase field models are straightforward to model diffusion coefficient [1644].

However, the computation time to obtain converged solution in 2D FEA of diffusion remains an issue to take into consideration complex composition and/or realistic morphology of precipitates. The phase field approach has been intensively used to describe aging leading to rafting in the bare substrate single crystal [503,1645]. Regarding coatings, phase field analysis has been straightforward to consider chemical, interfacial, and elastic energies for simplified materials [1391]. This method provides evolution of microstructure as a function of time, with large modifications of γ' precipitate morphology at and near the interface between substrate and BC, Fig. 10.32, yielding polycrystal formation in the substrate accordingly to the authors analysis [1391]. Moreover, in presence of temperature gradient, the rate of dissolution of γ' in the coating region is increased as compared to isothermal condition. It is worth noting that the height of the figure corresponds to only 50 μm for a 2D model, a limitation owing to the high computation cost of such a model. To the authors knowledge, no results have been provided in the case of full coupling for a coated single crystal substrate implying phase transformation, oxide formation, elasto-viscoplasticity, and local damage.

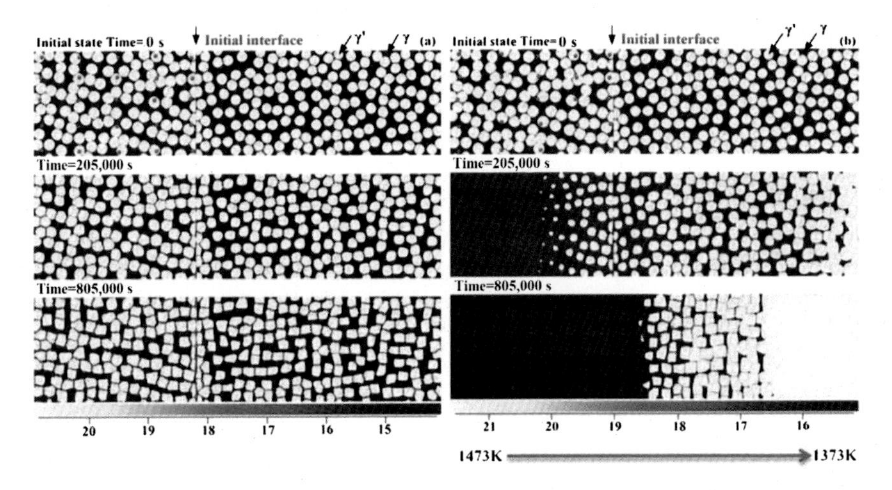

FIGURE 10.32 Phase-field simulated microstructure evolution of microstructure at the interface between γ/γ' bond coat and γ/γ' substrate system (Ni–18.5at.%Al–11.25at.%Cr/Ni–18at.%Al–12at.%Cr) annealed (a) at 1373 K and (b) with an effect of temperature gradient (color bar scale indicates the mole fraction of Al in atom percent, the height of the figure corresponds to 50 μm; vertical arrow indicates initial BC-substrate interface, BC being on left side and substrate on the right side of the pictures, respectively) [1391].

10.7.2 Assessment of mechanical state evolution

10.7.2.1 Flat multilayer model for weak and strong coupling

The simplest way to assess the mechanical state in a multilayered system is to consider a flat multilayer analysis assuming that the total strain is prescribed by the substrate to other layers (TGO/BC/TC). This could be applied to any mechanical behavior for each layer (elasticity, plasticity, viscoplasticity, hardening). For large thickness of the substrate as compared to each layer thickness, this approach is valuable but would be limited for thin substrate (hundreds of microns) cases and would ignore shear stress induced by mechanical properties mismatch at the interface associated to local roughness. This approach is considered as a weak coupling because of the absence of interactions between layers [295].

In the presence of an interfacial crack, elastic mismatch could be easily described by the Dundur's parameters [648]. To account for plasticity in the metal (either BC or substrate) the use of HRR field is prescribed but limited to Ramberg–Osgood nonlinear behavior [1317]. FEA is in this case straightforward by the use of cohesive zone model (CZM): the interfacial behavior is explicitly described by nonlinear laws fitted to describe the interfacial toughness, the behavior being nonlinear after damage initiation for either monotonic or fatigue loading [1224]. In this case, the shear is modeled without coarse assumption and could also integrate description of coupling between local damage and modification of thermal conductivity [1169]. CZM could be applied either to simplified substrate/TC model or to full description of each layer constitutive of the TBC (TGO/BC/TC).

10.7.2.2 Describing interfacial morphology

The assumption of a flat interface between BC and TC is oversimplifying the mechanical state. Especially, APS coatings show a high roughness at the surface and consequently at the interface between BC and TC. EB-PVD coating systems are processed to have a smoother interface. However, in EB-PVD systems localized asperities or undulations occur in industrial coating processes. Particularly, accounting for local roughness could provide a driving force for local damage associated to out-of-plane stress which is ignored in a flat interface assumption. Thus, Chang et al. have proposed to focus the analysis on a simplified sinusoidal 2D pattern representative of the roughness at the interface between top coat and bond coat [231]. In this way, it is possible to address the influence of local roughness and the interaction between each layer constitutive of the TBC system.

Derived from the sinusoidal pattern approach, a class of analytical models has been developed by Balint and Hutchinson [69]. These models are of high interest to account for both shear and out-of-plane interactions between layers, that is straightforward to assess the evolution of rumpling

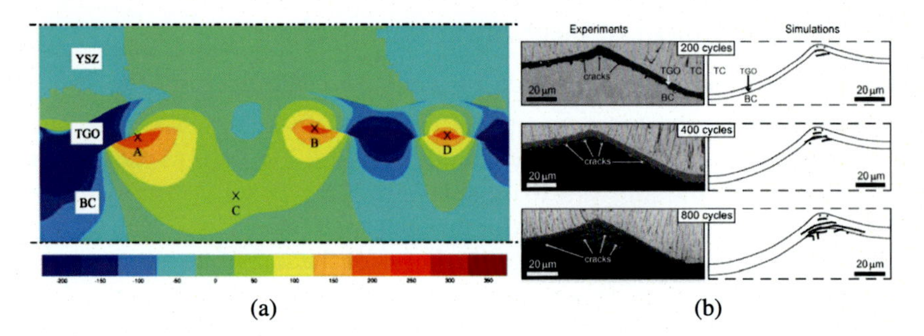

FIGURE 10.33 Modeling of interfacial imperfection by (a) different wave lengths and amplitudes of roughness to highlight local stress concentrations [913] (b) oxide cracking at a single undulation with increasing number of thermal cycles with experimental findings on the left and simulation results on the right [613] (courtesy E. Bakan, N. Schlegel, Forschungszentrum Jülich).

with thermal cycling [71,70,1480] and complex thermal mechanical cycling [1312].

This approach has been very successful and led to most of open literature models with continuous increase in complexity. First, it has been proposed to consider the most advanced mechanical behavior of each layer to obtain relevant stress state in TBC after high temperature stress relaxation [175]. Some authors have also tested the influence of microstructure evolution (oxide growth, martensitic transformation) [179,697], the influence of oxide cracking [613], the influence of 2D and 3D realistic morphologies to access relevant evolution of mechanical state in the TBC [913], Fig. 10.33(a)–(b). While in a coating system with adherent TC the BC is constrained and roughness evolution during thermal and mechanical loading is limited, the situation changes drastically at local separations between TC and BC. In [698] Karlsson et al. developed a model to describe the growth of undulations at local delaminations between BC and an EB-PVD top coat during thermal cycling. They introduced a term for not only thickening but also lengthening of the TGO and were able to capture with their model the experimentally observed amplitude increase of initially existing undulations at local delaminations in a PtAl-BC and EB-PVD TC system. Later, Shi et al. [1312] modeled the amplitude growth of initial undulations for a NiCoCrAlY BC for thermal, thermal gradient and thermal gradient mechanical fatigue loading and could capture well respective experimental results, Fig. 10.34.

10.7.3 Lifetime modeling

Pioneer works have been formulated by Miller [948]: he proposed to view damage growth as a result of the coupling between oxide growth, modeled by the NMG at the current cycle W_N, to mechanical loading, modeled by the thermal expansion mismatch strain ε_r. He therefore ob-

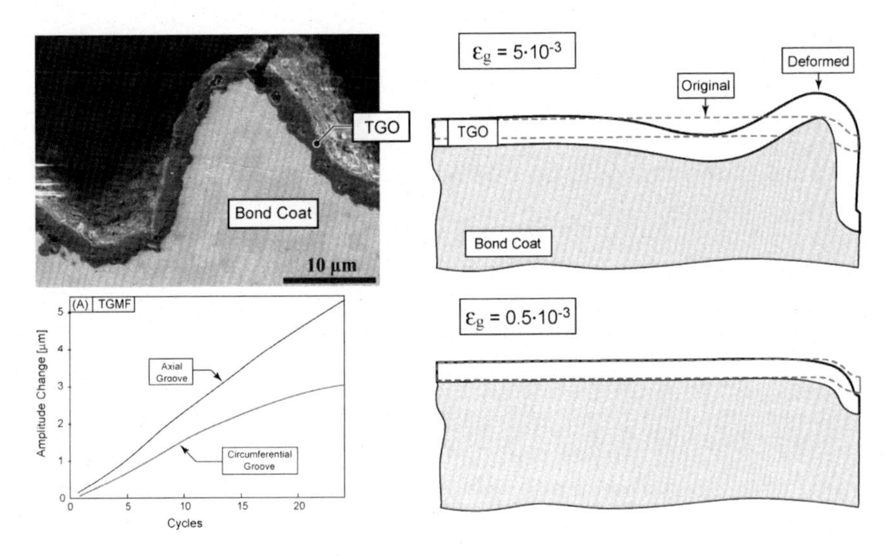

FIGURE 10.34 Rumpling modeling by initial defect meshing and strain increase associated to oxide growth in plane and out of plane (thickening and lengthening strain) and thermal gradient mechanical fatigue (TGMF) loading [1312].

tained a formalism very similar to standard fatigue damage laws,

$$\sum_{N=1}^{N_f} \left\{ \left(1 - \frac{\varepsilon_{f0}}{\varepsilon_r}\right)\left(\frac{W_N}{W_C}\right)^m + \frac{\varepsilon_{f0}}{\varepsilon_r} \right\}^{-b} = 1, \tag{10.5}$$

where W_C is the NMG yielding failure in one cycle, ε_{f0} is the critical strain at failure without aging, m and b being material parameters to be identified.

While Miller in [948] did not consider microstructural features, but used global values for critical strain, Chang et al. [231] calculated local stress concentrations for a simplified 2D model representing undulations as occurring in APS manufactured TBC coatings. However, for this simplified sinusoidal pattern, the link between local failure, assumed to be associated to oxide failure, to the macroscopic TC failure is not obvious. Busso et al. have proposed a model based on mapping of local roughness influence on local stress (out-of-plane component) seen as the driving force for local damage induced by thermal fatigue, to establish a failure criterion [176]. Another approach based on local modeling of oxide cracking by CZM shows spectacular results in capturing experimentally observed microcracking [613]. But for most of these models, the limiting factor is the extremely high sensitivity of prediction to the knowledge of the mechanical behavior of the bond-coat, including the impact of aging and multiaxial and nonproportional loading in a wide range of temperature. As elaborated in Section 10.3.3, limited robust results are available on bond coat materials, especially only few data are reported on properties after aging.

As mentioned before, it is further quite difficult to transpose descriptions of local damage processes to macroscopic spallation of the TBC. For this purpose, models have been developed, which link progressive damage of the interfacial region during cyclic loading to fatigue crack growth at the interface between BC and TC leading to TC spallation.

Miller presented in [948] an approach, which combines the strain induced by oxidation with the mismatch strain between TC and substrate during a thermal cycle to an effective strain ε_e as driving force for the propagation da/dN of a defined crack of length a:

$$\frac{da}{dN} = A\varepsilon_e^b a^d, \tag{10.6}$$

$$\varepsilon_e = (\varepsilon_f - \varepsilon_r)(w/w_c)^m + \varepsilon_r, \tag{10.7}$$

with ε_r being the mismatch strain between TC and substrate, ε_f the failure strain of the TC in the as-processed condition, w the NMG, and w_c a critical mass gain causing TC failure in a single thermal cycle. Other terms are material parameters to be optimized on an experimental database. More recently, it has been established that two damage terms could be used to describe the damage induced by isothermal oxidation D_{ox}, using a direct function of the oxide thickness, and the damage induced by thermal cycling D_r, combining cumulated plasticity in the bond-coat to oxide growth [295],

$$dD_{\text{ox}} = \frac{m}{h_0}\left(\frac{h_{\text{ox}}}{h_0}\right)^{m-1} dh_{\text{ox}}, \tag{10.8}$$

where h_{ox} is the current oxide thickness (accounting for temperature and time increment by incremental formulation of oxidation model seen in Eq. (10.2)), m and h_0 being the material-dependent parameters to be identified. Furthermore,

$$dD_r = N\frac{m}{h_0}\left(\frac{\varepsilon_{\text{cum}}}{\varepsilon_0}\right)^n\left(\frac{h_{\text{ox}}}{h_0}\right)^{m-1} dh_{\text{ox}} + \left(\frac{\varepsilon_{\text{cum}}^0}{\varepsilon_0}\right)^n\left(\frac{h_{\text{ox}}}{h_0}\right)^m dN, \tag{10.9}$$

where ε_{cum} is the cumulated plastic strain evaluated in the BC layer based on a flat multilayer assumption, N is the current cycle, n and ε_0 being the material dependent parameters to be identified. The macroscopic damage parameter D is the combination of integrated values of damage D_{ox} and D_r, with $D = (1 - D_{\text{ox}})(1 - D_r)$.

TC spallation occurs in this case when compressive strain in the TC reaches a critical strain at spallation controlled by interfacial damage, assuming that buckling strain is the critical parameter:

$$\varepsilon_{\text{crit}} = \frac{1.2235}{1+\nu_1}\left(\frac{h}{R_0}\right)^2 (1-D), \tag{10.10}$$

where h is the TC thickness and R_0 is the initial radius of a circular delaminated area with both h and R_0 relevant for buckling [295]. In [914] it was shown that elastic anisotropy and plastic strain localization, as it is the case for single crystalline superalloys, both result in strain variation depending on the orientation and influences the shape of the delamination and the time until buckling occurs.

The models by [948] and [295] explored the interfacial crack growth and TC failure as a stability criterion governed by interfacial toughness evolution as a function of loading history. The reason for this is that it is difficult to observe slow or stable crack growth in EB-PVD systems which have rather smooth interfaces and show more or less spontaneous spallation.

In APS-TBC systems, the coating lifetime is mainly governed by crack growth in the vicinity of the TC/TGO interface and allows for measurement of progressive spallation in the course of thermal cycling [678]. In this study, button shape specimen and edge delamination is considered. The proposed approach yields a fatigue crack growth rate evaluation as

$$\frac{dc}{dN} = A(G)^m, \tag{10.11}$$

where c is a macroscopic crack length, A and m being the classical Paris law parameters, and G the energy release rate. These measurements and model are straightforward for APS YSZ coating because after local spallation at the edge of the specimen, the remaining TC is still adherent to the substrate yielding to a progressive delamination of the TC. Moreover, a linear relationship between interfacial decohesion and spallation has been established, see Fig. 10.35.

Based on four-point bending measurement, Fig. 10.16, Vaunois and coworkers have proposed in [1481] to determine a damage model controlling the decrease of interfacial toughness by mixing out-of-plane stress and plasticity associated to rumpling like damage as developed in Courcier et al. [295].

Inspired by the complex interface analysis detailed above, including a sinusoidal pattern to describe rough interface, it has been proposed to explore the influence of local roughness and defects on crack propagation using CZM [1347]. The main results have shown that the sensitivity to local defects was rather low as compared to the global interfacial toughness. This result fully justifies that, to describe global decohesion of the TC at the length scale of the component, direct CZM models are straightforward [1169,1262]. In this case, the interfacial toughness should be a function of

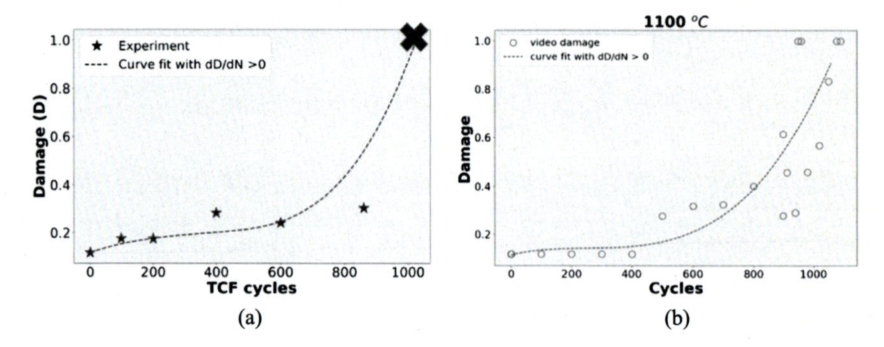

FIGURE 10.35 Fatigue model based on Paris-like crack growth analysis for an APS TBC system under thermal cycling: (a) maximum interfacial crack length and (b) spalled TC area fraction as a function of cycles' number [678].

the aging of the material, described by a lower length scale model such as described in [613] or [1257]. However, a difference should be kept in mind between microstructure evolution and damage evolution, here related to interfacial decohesion: for the same TBC system and similar thermomechanical loading, the respective evolution of phase transformation in the BC, Fig. 10.36(b), and interfacial damage, Fig. 10.36(c), yield different localization paths.

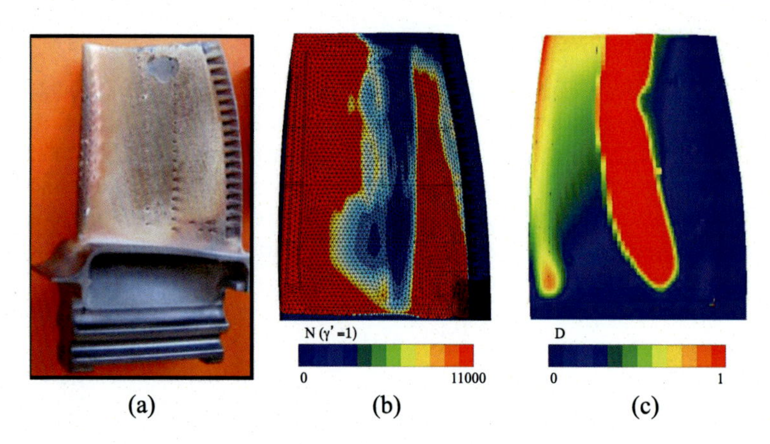

FIGURE 10.36 High pressure turbine blade: (a) photo of a blade after use in-service [1421], (b) phase transformation in bond coat due to thermal exposure, showing the calculated number of cycles to reach a volumic fraction of γ'= 1 (complete transformation from β to γ') [1257], and (c) calculated interfacial damage of the TBC system after thermal exposure [1169].

CZM approach has been successfully applied to describe the influence of biaxial complex loading on a simplified coating system consisting of a ceramic layer directly applied onto a superalloy substrate [1262]. It has been shown that an initial circular defect, here obtained by the LASAT method, turns into an elliptical one for macroscopic combination of tension and compression in the plane of the substrate. This effect was induced by local closure and opening of the delamination crack between ceramic layer and substrate under compression and tension with a more pronounced crack growth in tensile direction than in the compressive direction [1262], see Fig. 10.37. It is worth noting that some similar observation can be made on single crystal substrate when uniaxial loading is applied (isothermal or TMF loading): here the anisotropy of the substrate induces a strong Poisson effect, where for compressive loading the orthogonal direction is bearing tensile strain inducing an elliptical decohesion [1576,914].

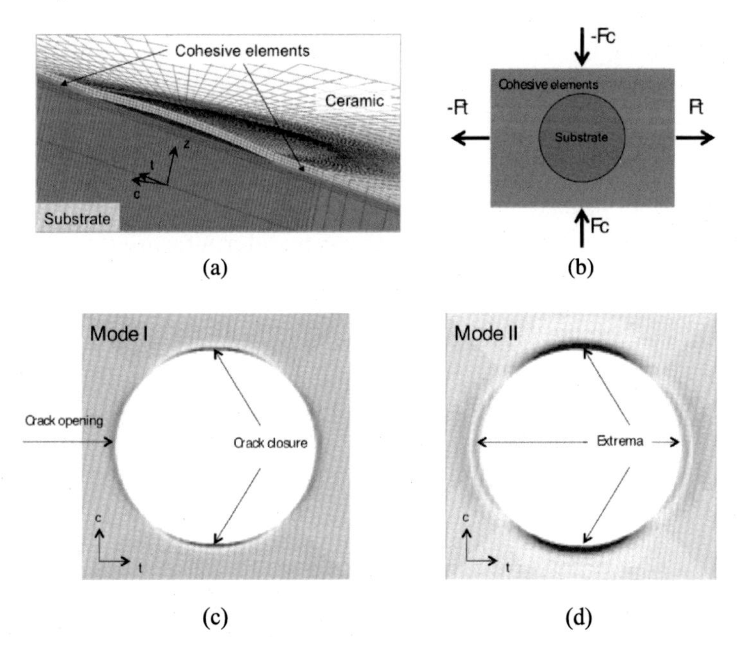

FIGURE 10.37 Modeling of in-plane biaxial testing: (a) perspective view of the half-model highlighting ceramic blister and substrate, (b) schematic top-view without ceramic layer, no cohesive elements are used within the blister making the substrate visible; F_t and F_c indicate in-plane tension and compression, respectively, (c) opening stress component (σ_{zz}), and (d) shear stress component (σ_{tc}) highlighting equivalence in shear mode for c and t directions; adapted from [1262,1261].

10.7.4 Concluding remarks: challenges from theory to application

The fantastic progress in coating technology, physical understanding of microstructure and damage evolution, characterization of TBC and modeling could let us dream about a full model predicting life as a direct function of the chosen manufacturing process and composition of the TBC. However, only partial junctions could be made nowadays including:

- From composition to microstructure and microstructure evolution, diffusion and phase field models are promising to predict aging;
- From microstructure evolution, testing of free standing coatings or complete TBC using synchrotron facilities gives access to layers mechanical properties;
- From morphology and local mechanical properties, FEA yields stress evolution for complex loading conditions;
- From adhesion testing, interfacial toughness assessment enables to predict lifetime of the TBC.

Among missing links, the most critical issue is certainly from microstructure to mechanical properties, including interfacial properties. It is also critical today to give a clear physical basis of the coupling between complex thermo-mechanical loading and observed large impact on phase transformation. However, it is worth noting that a full field model including phase field approach implies a very large CPU time to compute few μm^3. At a larger scale length, for instance, to determine interfacial damage by CZM, implies finite elements sized to few microns to deal with tenths to hundreds of μm^3. As a direct consequence, computing a component could not be achieved with local models but should be based on a homogenized model (analytical or not) to stay within a reasonable computing time for design purposes.

PART III

Appropriate scale modeling, scale bridging methods

CHAPTER

11

Atomic-scale modeling of superalloys

Thomas Hammerschmidt[a], *Jutta Rogal*[b,c], *Erik Bitzek*[d], *and Ralf Drautz*[a]

[a]ICAMS, Ruhr-Universität Bochum, Bochum, Germany, [b]Department of Chemistry, New York University, New York, NY, United States, [c]Fachbereich Physik, Freie Universität Berlin, Berlin, Germany, [d]Department of Materials Science and Engineering, Institute I, Friedrich-Alexander-Universität Erlangen-Nürnberg, Erlangen, Germany

11.1 Introduction

In 1929 Paul Dirac noted that "The underlying physical laws necessary for the mathematical theory of a large part of physics and the whole of chemistry are thus completely known, and the difficulty is only that the exact application of these laws leads to equations much too complicated to be soluble" [350]. He continued to say that "It therefore becomes desirable that approximate practical methods of applying quantum mechanics should be developed, which can lead to an explanation of the main features of complex atomic systems without too much computation.

Three aspects of Dirac's statement largely determine our work today. The first observation is that the fundamental laws, for the modeling of materials with the many-electron Schrödinger equation, are known. As ultimately the behavior of a material is controlled by the bonds between atoms that are mediated by electrons, we are therefore in principle in a position to understand, predict, and design materials from first principles. The caveat, and Dirac was very clear about this in his statement, is that it is strictly impossible to solve the many-electron Schrödinger equation for a superalloy. Therefore today, nearly a century after Dirac's statement, we are still developing approximate models that are rooted in the fundamental laws of nature to support materials design from first principles.

Nickel Base Single Crystals Across Length Scales
https://doi.org/10.1016/B978-0-12-819357-0.00020-2

A breakthrough was made by Kohn and coworkers when they developed density functional theory (DFT) [622,737], today easily the most cited concept in the physical sciences [1470]. The effective one-electron structure of DFT makes it amenable to further approximations that treat the electrons as a mere glue between the atoms, so that effective classical interatomic potentials may be derived.

In this chapter, we summarize the main aspects of the present state-of-the-art of atomistic modeling and simulation for superalloys. We start from small calculations with only a few atoms, for which DFT may be employed, and end with atomistic simulations of the microstructure that require many millions of atoms and that are carried out with classical interatomic potentials, see Fig. 11.1. We list limitations that need to be overcome for the ab initio development of superalloys.

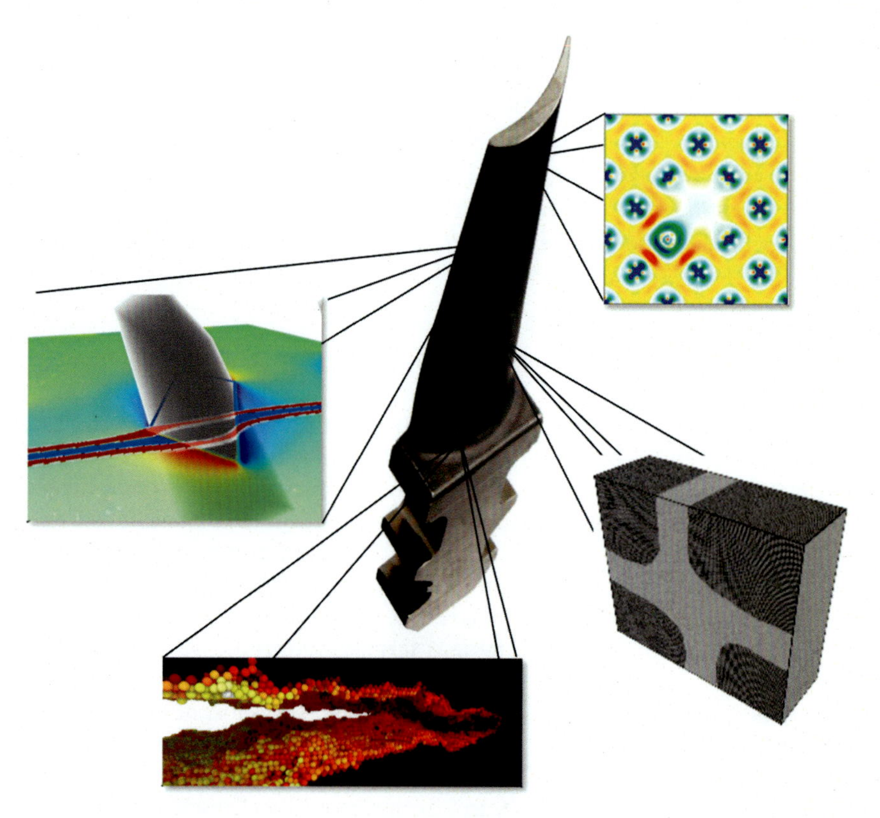

FIGURE 11.1 Aspects of superalloy turbine blades relevant to development, use, and failure that can be addressed by atomistic simulations. From top to bottom: vacancy formation energies as functions of local chemical composition; interaction of a superdislocation in γ' with a γ precipitate and resolved misfit stresses; simulation of γ/γ'-microstructures; fracture.

11.2 Methods

11.2.1 Modeling atomic interactions

The structural stability and mechanical properties of Ni-base superalloys are driven by the distribution of the different chemical elements and the features of the microstructure. Atomistic modeling therefore needs to account for the interaction of the complex alloy chemistry and the diverse geometric features of the atomic structure of imperfections like point defects, interfaces, dislocations, stacking faults, and precipitates. This requires capturing the atomic interactions with appropriate accuracy and tackling simulation cells with a sufficiently large number of atoms. These are antipodal requirements as higher accuracy requires higher complexity and computational effort and therefore leads to smaller simulation cells. The hierarchy of approaches discussed in the following spans the range from highly precise quantum-mechanical calculations at nm length-scales to classical simulations which nowadays can attain µm length-scales.

The most accurate approach discussed here involves DFT calculations that provide a numerical solution to the quantum-mechanical equations for the interaction of electrons and ions. The results for metallic systems like Ni-base compounds are highly reliable but the computational effort sets a practical limit to simulation cells with typically a few hundred atoms. This quantum-mechanical description can be simplified in a second-order expansion of DFT [360] to the tight-binding (TB) bond model [1381]. This increases the tractable system size to several thousand atoms at the expense of reduced accuracy and an additional parameterization step. The required reference data for the latter are typically energies and forces of crystal structures and defects obtained by DFT calculations. The numerical solution of the simplified quantum-mechanical description can be further accelerated by bond-order potentials (BOPs) (see, e.g., [359]). The underlying approximation effectively localizes the range of the quantum-mechanical interaction to a few neighbor shells. With the resulting linear scaling of computational effort with the number of atoms and the possibility for efficient parallelization, the tractable number of atoms is extended to several hundred thousands in a routine simulation [573]. A further level of simplification and millions or even billions of atoms can be reached by giving up even the simplified quantum-mechanical description of TB/BOP for a set of short-ranged, analytical functions that mimic the interatomic interaction. This allows constructing approximate models with simple mathematical forms that are parameterized for specific chemical elements and atomic structures. For metallic systems like Ni-base superalloys, pair potentials like the Finnis–Sinclair potential [443] and embedded-atom method [315] are particularly successful and have been parameterized for studying γ and γ' phases of NiAl [953,1149]. A more recent development are machine-learning potentials for superalloys

[229] that replace the fixed set of short-ranged mathematical functions with artificial intelligence approaches in order to interpolate large sets of DFT data for the atomistic description of the superalloy.

11.2.2 Calculation of structures and energies

The atomistic methods given above start by constructing an atomistic representation of the particular aspect of the superalloy that should be studied. Using the positions and chemical species of the atoms, the different atomistic simulations methods (DFT, TB, BOP, classical potentials) deliver the potential energy of the system, the forces on the individual atoms, and the stresses on the simulation cell.

One of the most basic examples is the computation of the lattice constant of an Ni_3Al γ' phase. The corresponding atomistic representation is a unit cell of an fcc (face-centered cubic) crystal lattice with an $L1_2$-ordered occupation by Ni and Al atoms in a 3:1 ratio subject to periodic boundary conditions. The computation of the total energy with one of the atomistic methods above depends on the input lattice constant that is chosen for the unit cell. The lattice constant where the total energy takes its minimum value is the equilibrium lattice constant and the corresponding energy is the equilibrium energy. The application of strain tensors to the equilibrium unit cell leads to stresses which can be used to determine the elastic constants of the material.

The defect structures discussed in the following sections are represented atomistically by repetitions of the unit cell and additional modifications. For example, a γ' phase of Ni_3Al with 0.2 at.% Ni_{Al} antisite defects is obtained by a five-fold repetition of the $L1_2$-Ni_3Al unit cell with four atoms in [100], [010], and [001] directions followed a replacement of one Al atom by an Ni atom. Similarly, the site preference of alloying elements (see, e.g., [672]) and their influence on structural stability and elastic constants (see, e.g., [721]) can be determined by corresponding total-energy calculations where part of the Ni or Al atoms are replaced by the alloying elements of interest.

In these cases, the simulation cell has more degrees of freedom than only the lattice constant, and a structure relaxation needs to be performed. This minimization of the total energy with respect to all structural degrees of freedom leads to a refinement of unit cell size and shape, and atomic positions. The central results of the relaxation of such structures are the detailed atomic structure including displacements and the energetics in the limit of $T = 0\,K$. This is the starting point for further simulations of finite-temperature effects and defect mobility discussed in the next sections. Examples of large-scale simulation setups for γ/γ' structures, precipitate cutting and fracture are given in Fig. 11.1.

11.2.3 Finite temperature calculations

Total energy calculations as discussed in the previous section provide valuable insight into the structural properties and relative stability of metal alloys. The effect of temperature is, however, not included in these calculations. To be able to compare thermodynamic properties at finite temperatures, it is necessary to determine the different entropy contributions to the free energy. Furthermore, additional simulation tools are required to study the dynamical properties on an atomistic level.

Assuming an adiabatic decoupling of the different degrees of freedom, the free energy can be expressed as

$$F(V,T) = E^{tot}(V) + F^{el}(V,T) + F^{vib}(V,T) + F^{mag}(V,T) + F^{conf}(V,T) + \cdots, \tag{11.1}$$

where E^{tot} is the total energy at $T = 0\,\mathrm{K}$, F^{el} is the electronic, F^{vib} the vibrational, F^{mag} the magnetic, and F^{conf} the configurational contribution to the free energy. The coupling between different degrees of freedom (electronic–vibrational, vibrational–magnetic) would give rise to extra terms in this expansion. Furthermore, any type of defect (like vacancies) would additionally contribute to the free energy. The electronic free energy, F^{el}, for bulk systems can be computed with high accuracy based on ab initio calculations [1653], where methods assuming fixed atomic positions include the self-consistent finite temperature DFT approach, the fixed density of states approximation, and the Sommerfeld approximation. The free energy of atomic vibrations, F^{vib}, usually constitutes the largest contribution at finite temperatures [1049]. Within the harmonic approximation, the vibrational free energy is calculated from the phonon frequencies which are accessible within a first-principles approach either by density functional perturbation theory [76] or by the small displacement approach [765]. To first order, anharmonicity can be included via the quasiharmonic approximation [1049]. Here, the dependence of the phonon frequencies on the volume is explicitly considered. At each temperature, the free energy is then accessible as a function of volume, where the minimum denotes the equilibrium volume at the corresponding temperature. Calculating vibrational free energies including full anharmonicity can be achieved on the basis of molecular dynamics (MD) simulations, discussed below. A widely used approach is thermodynamic integration [469] between a system with known free energy (e.g., within the quasiharmonic approach) and the system of interest. MD simulations based on energies and forces from ab initio calculations (AIMD) can provide very accurate results, but are also computationally very demanding. Including magnetic contributions to the free energy, F^{mag}, from first-principles calculations is rather challenging [748], usually requiring the setup of model Hamiltonians derived from ab initio calculations. These contributions are, however, important in magnetic transition metals such as Ni, Co, and Fe, where they significantly impact thermodynamic quantities such as the free energy and the specific heat [747]. In disordered alloys, the last term, F^{conf}, is often approximated using the entropy of mixing for an ideal solid solution. More sophisticated approaches have been developed based on a cluster expansion approach that aim to sample the space of all possible arrangements of different atom types on a fixed crystal lattice [1463,1654]. Together with Monte Carlo (MC) sampling approaches in different thermodynamic ensembles it becomes possible to investigate phenomena such as order-disorder transitions, precipitate formation, or phase diagrams [534,893].

To investigate the dynamical behavior of materials at finite temperatures, MD simulations [469,18] constitute nowadays the standard workhorse for atomistic simulations. In MD, atoms are treated as classical particles following Newton's equations of motion. Simulations can be performed in various thermodynamic ensembles by using the corresponding thermostats and barostats, including the microcanonical, canonical, and isothermal–isobaric ensembles. Advances in numerical algorithms and increasing computational resources allow performing simulations with up to billions of atoms. The development of elaborate program packages [1116] has contributed to a widespread use of MD simulations to study materials properties. The two main difficulties in MD simulations are an accurate description of the interatomic interactions and the accessible timescales. Ab initio MD [906] is based on highly accurate energies and forces, but system sizes are limited to a few hundred atoms. Empirical potentials, on the other hand, are usually not reliable for complex, multicomponent alloys and often have difficulties to properly represent complex structural environments, such as dislocation cores or during crack propagation. The timescale problem in MD simulations arises when the process of interest involves sizeable energy barriers between two metastable states of the system, e.g., during vacancy mediated diffusion (see also Section 11.4.2). This is due to a separation of timescales between the fast vibrations and the comparably slow changes in structure or exchange of atomic positions. A number of accelerated MD techniques [1515] has been developed to facilitate the escape from metastable states in the course of the simulation while preserving the correct dynamics. Another approach is kinetic Monte Carlo (KMC) [134,437] simulations where a rate constant is determined for each process connecting two metastable states and the dynamics is given by a stochastic state-to-state trajectory that again preserves the correct timescale. If a suitable mapping of the investigated system onto a finite state space can be identified, then the corresponding rate constants for all processes can be determined with high accuracy based on electronic structure calculations. The analysis of dynamical simulations provides insight into the atomistic mechanisms and gives access to macroscopic transport coefficients via the corresponding Green–Kubo relations [469,18], as, e.g., diffusion coefficients, shear viscosity, or thermal conductivity. These quantities can directly be compared to experimental measurements.

11.3 Thermodynamic stability

A central concept for determining the structural stability of crystalline phases by atomistic simulations are differences of total energies from simulations for different systems. As an example, the formation energy of the Ni_3Al γ' phase with respect to the elemental ground states is taken as the energy difference between the $L1_2$ Ni_3Al total energy and the total energy of fcc-Al and fcc-Ni where each total energy is obtained in a separate simulation. With proper accounting of the chemical compositions, one can compare different crystal structures with different chemical composition by convex-hull constructions. The resulting information on the energetically most favorable crystal structure for a given chemical composition at $T = 0\,\mathrm{K}$ and further entropy contributions can be considered in CALPHAD assessments of phase diagrams of compound systems, see Section 2.1.

The concept of energy differences is also used to investigate the stability of defects. The segregation energy of an alloying element to a defect, e.g., is taken as energy difference between the total energy of a supercell with the alloying element positioned at the defect and a second supercell with the alloying element far away from the defect. Similarly, the formation energies of point defects (e.g., vacancies), line defects (e.g., dislocations), planar defects (e.g., stacking faults), and extended defects (e.g., precipitates) are computed from the difference in energy of calculations with and without the corresponding defect.

A certain limitation of the atomistic approaches for the case of typical multicomponent superalloys is a realistic representation of their chemical complexity. Classical interaction models, on the one hand, usually lack reliable multicomponent parameterizations, and DFT calculations, on the other hand, are hardly possible due to (i) the size of the required supercells and (ii) the combinatorially large number of atom distributions within the supercell. An alternative approach are structure maps [1085,1084,572,1297,106] that chart the trends of the structural stability [1086,1298,570] in low-dimensional representations. The application of structure maps to superalloys could, e.g., rationalize the formation of detrimental topologically close-packed (TCP) phases in the CMSX4-like Ni-base superalloy ERBO/1 in the as-cast and heat-treated state [867,571], in low-cycle fatigue-tests [931], and during repair by vacuum plasma spray [689].

11.4 Point defects

In general, there are three types of point defect that we consider in bulk phases: vacancies, where an atom is missing from a particular lattice site; substitutional defects, where a matrix atom is replaced by another element; and interstitial defects, where additional atoms are present in between regular lattice sites. In the following we will focus on the former two, vacancies and substitutional defects.

11.4.1 Thermodynamic properties

One of the important thermodynamic quantities is the defect formation energy, that is, the change in energy due to the creation of a defect. Atomistically, the defect formation energy can be computed within a supercell approach as the difference between the total energy of supercells with and without the corresponding defect, see [473] for a review. The vacancy formation energy (VFE) in pure Ni is, e.g., given by

$$\Delta E^{\mathrm{VFE}} = E_{\mathrm{Ni}_{X-1}\mathrm{Va}} - \frac{X-1}{X} E_{\mathrm{Ni}_X}, \tag{11.2}$$

where X is the number of atoms in the supercell. The size of the supercell determines the minimum defect concentration and needs to be carefully tested to avoid artificial defect–defect interactions [1229]. Similarly, the magnitude of the interaction between point defects can be calculated as the energy difference between an isolated defect, that is a single defect in a large supercell, and two or multiple defects at a certain distance. Attractive interactions indicate binding or clustering, whereas repulsive interactions favor a random distribution of point defects in the matrix. In Fig. 11.2 the interaction energies between a vacancy and solute atoms (Re, W, Mo, Ta) in an Ni matrix are shown. These calculations demonstrate that single Re atoms do not bind vacancies [1291,1651] and that this could be dismissed as a possible hypothesis for the Re effect. Similarly, nonmagnetic electronic structure calculations of interaction energies between Re atoms in Ni revealed that Re does not tend to form clusters [971], eliminating another speculation concerning the cause of the Re effect. For the Ni–Re system the results do, however, strongly depend on the magnetic state of the system [591]. In multicomponent systems, the defect formation energy depends on the local chemical environment as well as on the global composition. In random solid solutions, it is in addition unknown which atom type previously occupied the defect site, which requires considering a properly weighted average over possible configurations [1229]. In chemically ordered phases, the formation energy of antisite defects (an atom of one sublattice occupies a site on another sublattice) determines how the system compensates off-stoichiometric compositions. In the $L1_2$ ordered Ni_3Al γ' phase, the formation energy of antisite defects is much lower than the vacancy formation energy [317,673,539,1650,534] and thus structural vacancies are not observed for off-stoichiometric compositions [61]. Furthermore, a number of first-principle studies have investigated the site preference of ternary alloying elements in the γ' phase [673,1240,1241,720], which was shown to influence the mechanical properties of this phase [1186].

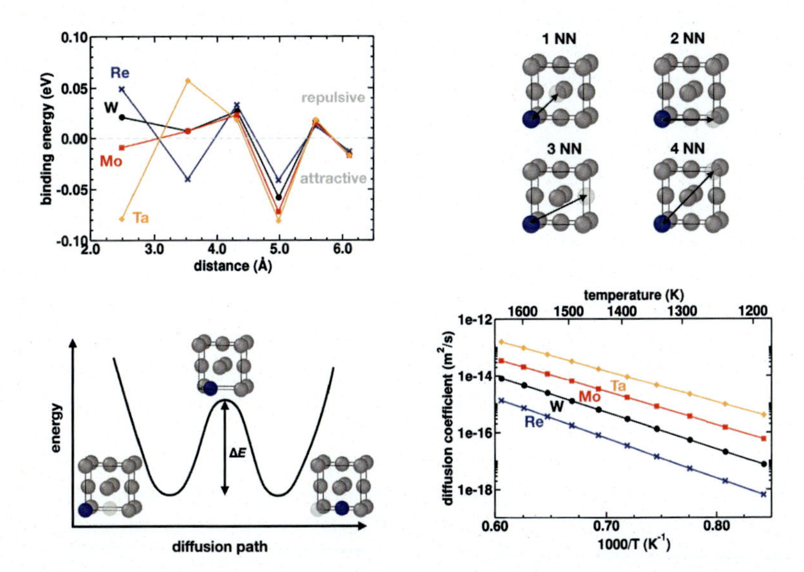

FIGURE 11.2 (Top) Interaction energies between a vacancy and solute atoms (Re, W, Mo, Ta) in an fcc Ni matrix. Overall the interaction energies are small on the order of 50 meV to 100 meV. In the first (1NN) and second (2NN) neighbor shell, the interaction energies are dominated by electronic effects, whereas for larger distances elastic effects due to the difference in size between the solute atoms and Ni become important. In the first neighbor shell, the interaction between a vacancy and Re or W are repulsive, whereas the interaction with Mo or Ta are attractive. The observed trend in the interaction energies can be correlated with the filling of the *d*-band in the electronic structure of the solute atoms. Figure adapted from [1291]. (Bottom left) Schematic representation of the energy along the diffusion path of an atom (blue) exchanging its position with a vacancy (transparent); the energy difference between the initial and transition state corresponds to the energy barrier ΔE of the diffusion process. (Bottom right) Diffusion coefficients of alloying elements (Re, W, Mo, Ta) in fcc Ni as a function of temperature. The diffusion coefficients have been calculated using KMC simulations with barriers from DFT calculations. The diffusion-activation energy Q can be extracted from the slope of the corresponding linear fit to the numerical data in the Arrhenius plot. Figure adapted from [1291].

11.4.2 Mobility

Solid state self-diffusion and the diffusion of substitutional defects are mainly mediated by vacancies. On the atomic scale, a diffusion process involves the exchange of an atom with a neighboring vacancy which is associated with an energy barrier as shown in Fig. 11.2. The minimum energy path along the diffusion process can be determined using the nudged-elastic band (NEB) approach [603] or the string method [1545] with a high degree of accuracy using electronic structure methods. The corresponding microscopic diffusion barriers can either be used in analytical models [813,542] or as input parameters to kinetic Monte Carlo (KMC) [134,1291] simulations to calculate macroscopic diffusion coefficients. In Fig. 11.2 diffusion coefficients of alloying elements in Ni as a function of temperature are shown determined by DFT calculations combined with KMC simulations [1291]. The values for the diffusion activation energies extracted from the Arrhenius plot can directly be compared with experimental measurements of tracer diffusion coefficients. Such simulations also allow directly comparing the mobility of alloying elements in Ni and Co [1007] as a quantity of interest in the investigation of Ni- and Co-base superalloys. Furthermore, the simulations provide insight into the mobility of vacancies and how the presence of alloying elements influences vacancy transport [1291,542,546,543]. In complex alloys, the diffusion properties depend on the composition. An accurate description on an atomistic level requires taking into account interaction energies between solute atoms, as well as their influence on microscopic diffusion barriers [546,543]. This also applies to chemically ordered alloys such as the Ni_3Al γ' phase where diffusion processes might take place on the same sublattice or cause a swap between sublattices [367,539,1650]. In multicomponent alloys, an accurate mapping of all possible diffusion processes becomes rather involved. Here, KMC simulations can be combined with more advanced approaches such as a cluster expansion of the diffusion barriers parameterized by electronic structure calculations [1466].

11.5 Line defects

Line defects, here mainly dislocations, can be studied both by DFT calculations and atomistic simulations with TB/BOP models or classical potentials. The small number of atoms that can be treated with DFT approaches, however, requires advanced methods for the boundary conditions and allows only for static calculations [1572,1398]. The motion of extended dislocations, as well as their interactions with each other and other defects can, so far, only be simulated using interatomic potentials with their known limitations in accuracy and chemical complexity.

11.5.1 Dislocation core structure

Dislocations in pure Ni as model systems for the γ phase have been extensively studied [1390,115,107,1398] also with respect to segregation of solute atoms [857,704]. In atomistic simulations, these dislocations are generated by displacing the atoms according to an analytic solution of the displacement field around the dislocation, followed by a relaxation of the atomic positions. For an edge dislocation this leads to two partial dislocations separated by an intrinsic stacking fault as shown in Fig. 11.3. Dislocations in the γ' phase are much more diverse, including partial

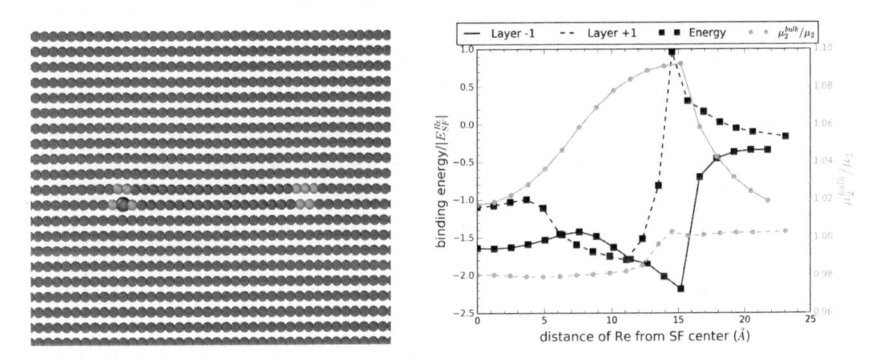

FIGURE 11.3 Atomistic simulation of a Re atom (blue - dark grey in print version) in an edge dislocation complex of two partial dislocations (grey) and a stacking fault (red - mid grey in print version) in fcc Ni (green - lighter mid grey in print version). This structure is obtained by removing a half-plane of atoms from fcc Ni, subsequent relaxation of the atomic positions which leads to the partial dislocations, replacement of one Ni atom by Re, and further atomic relaxation. (Right) The interaction energy (blue - dark grey in print version) of the single Re atom at different positions relative to the dislocation complex shows that the energetically most favorable position is in the tensile layer (layer -1) of the partial dislocation. In this position, the Re atom has the maximum accessible volume factor (orange - light grey in print version) as obtained by TB and BOP calculations. Figure adapted from [704].

dislocations bounding different planar defects (see Section 11.6), different possible glide planes and Burgers vectors [1192]. Consequently, most atomistic studies were directed to reproduce experimentally observed dislocation structures and to determine their properties.

Understanding ⟨110⟩ screw superdislocations in Ni_3Al is of particular importance due to their potential for forming Kear–Wilsdorf (KW) lock configurations [710] and thus leading to the anomalous yield behavior of $L1_2$ alloys. Therefore, much effort has been invested during the 1980s and 1990s to investigate their core structure [1605,1046,1621,1622,423,1623, 1044,1060,1549,1548]. Depending on the simulation details, different core structures are possible, including dissociation of the superdislocation on a {111} plane into a pair of 1/2⟨101⟩ superpartial dislocations separated by an antiphase boundary (APB) and nonplanar configurations. The detailed locking mechanism is commonly believed to involve the thermally-activated formation of Paidar–Pope–Vitek (PPV) locks [1044], which has been recently studied in detail [1060,1008,1177], including the use of NEB calculations to calculate activation energies [1008,1177]. In this context, it is important to note that the activation barriers depend strongly on the studied configuration. Considering, e.g., the interaction with forest dislocations can significantly reduce the activation energy for the critical cross-slip process [1177]. Recent atomistic simulations suggest furthermore, that under applied stress and elevated temperatures also ⟨110⟩ edge superdislocations can show nonplanar dissociation leading to complex Lomer–Cottrell locks [1593,1595,420].

In addition to the usual ⟨110⟩ superdislocations also ⟨100⟩ edge superdislocations were observed to penetrate the γ' particles, in particular during high temperature, low stress creep, where this process is believed to be the rate limiting step [380,1350,847]. The dissociation of a perfect $a\langle 100\rangle\{010\}$ edge superdislocation in Ni_3Al was studied by Kohler et al. [736], who found that it could dissociate in a symmetric dissociation similar to the Hirth lock in fcc structures, an asymmetric dissociation and a dissociation into two interlocked $a/2\langle 110\rangle$ dislocations, with the Hirth lock being the most stable configuration. HRTEM observations and image simulations by Srinivasan et al. [1350], however, showed a somewhat different core structure, which might be related to the coupled glide-climb process necessary for the motion of ⟨100⟩ dislocations.

11.5.2 Dislocation mobility and solid solution strengthening

The Peierls stress τ_P necessary to initiate dislocation motion can be easily determined by quasistatic calculations on infinite straight dislocation lines in a slab geometry with periodic boundary conditions subjected to different shear stresses. The same setup can be used to measure the temperature-dependent drag coefficient B caused by the interaction of phonons with the moving dislocation by performing MD simulations at different temperatures. For pure Ni, these parameters have been determined by Bitzek et al. [114,115]. Alternatively, the Peierls stress can also be determined by DFT calculations of the generalized stacking-fault energy in a Peierls–Nabarro model as, e.g., in [1532] for Ni_3Al. Although an important parameter in discrete dislocation dynamics (DDD) simulations, the drag coefficient B, see Chapter 12, for dislocations or superdislocations in Ni_3Al has not yet been determined by atomistic simulations.

The strengthening effect of different solutes can be predicted, e.g., by combining DFT calculations of solute–dislocation interaction energies with a Labusch-type model [827]. Such an approach has, however, not yet been used to calculate solid solution strengthening in the γ or γ' phases of Ni-base superalloys. Alternatively, direct MD simulations of dislocation motion in solid solutions can be used to determine parameters like static and dynamic threshold stresses and effective "friction coefficients" for dislocations, including superdislocations in γ [1221,905,1649,420]. This approach depends critically on the availability and quality of atomic interaction potentials. Although recently simulations have focused on concentrated solid solutions [1179], no simulations were so far performed on realistic model systems for Ni-base superalloys. Please also refer to Chapter 14 regarding the modeling of solid solution strengthening in crystal plasticity.

It needs to be stressed that the above approaches to study dislocation mobility neglect diffusive processes. The shearing of the γ' phase by partial dislocations, however, requires reordering steps that involve short-range diffusion [756,1665]. Modeling such concerted diffusive–displacive processes at the atomic scale remains one of the fundamental challenges for atomistic simulations [830]. The study of dislocation climb by atomistic simulations is therefore in its infancy [1263,64], and the effects of solute atoms on the climb mobility are yet to be investigated at the atomic scale.

11.6 Interfaces and planar defects

The calculation of the γ/γ' interface of a superalloy with realistic chemical complexity requires large supercells and proper combinatorial sampling of the disordered γ phase. A common approximation is therefore to mimic the γ phase by pure Ni which has been used successfully in calculations with classical EAM (embedded-atom method) potentials of different interfaces [954] and in DFT calculations of the influence of alloying elements on the interface energy [863].

The mechanisms of the plastic deformation of superalloys in the different pressure and temperature regimes are discussed in detail in Chapter 6. The plastic deformation is governed by γ' cutting processes (see, e.g., [7]) and by microtwinning (see, e.g., [756,74]). These involve several steps including dislocation dissociation into partials and the movement of the partials, as well as the segregation to the partial and to stacking faults between partials and behind the trailing partial. The most important planar faults in the context of shearing deformation are superlattice intrinsic stacking faults (SISF), superlattice extrinsic stacking faults (SESF), antiphase boundaries (APB), and pairs of twin boundaries. These phenomena at the atomic level are ideal candidates for atomistic simulations, particularly regarding clarification of (i) the relative formation energy of different planar faults (see, e.g., [1439]), (ii) the influence of alloying elements on the fault energy, and (iii) the segregation of alloying elements to the fault plane (see, e.g., [1499,381]).

The majority of recent calculations for planar faults are, in fact, performed on the basis of DFT calculations due to a lack of reliable alternatives for multicomponent systems. The stacking faults are accessible to atomistic simulations by supercell calculations and, in few cases, by an approximation with an axial Ising model (AIM) [334]. Disorder of the atoms on the crystal lattice (due to, e.g., off-stoichiometry or finite temperature) can be introduced by special quasirandom structures (SQS) [1682], by cluster expansions (CE), or by mean-field approaches like the coherent phase approximation (CPA).

The energy of the APB that is formed during γ' cutting of a dislocation is of central importance for shear resistance, see Chapter 6. Several recent works, including supercell DFT calculations [230,1461,302,775], DFT-based CE [1377], and DFT-based CPA [540], showed consistently that alloying with group IV (e.g., Ti) and group V (e.g., Ta) transition-metal elements can considerably increase the APB energy and hence the shear resistance. Corresponding works for SISF and SESF were carried out with supercell DFT calculations [1550,775] and AIM-based approximations [151]. Considerably fewer works are available that performed corresponding calculations for the microtwinning mechanism during creep. The segregation of selected alloying elements to SESF and SISF was determined by supercell DFT calculations [405] that were recently extended and interpreted in terms of reordering kinetics and intermediate fault structures [1180].

11.7 Microstructure and defect–defect interactions

Simulating the γ/γ' microstructure of Ni-base superalloys, its deformation, and final fracture, as well as the direct interaction between defects, requires atomistic simulations with large numbers of atoms that presently can only be realized with semiempirical potentials. Such simulations are very useful in further developing our mechanistic understanding of deformation processes and can help with the interpretation of experimental observations, which due to their limited time resolution might not capture all relevant details.

11.7.1 Deformation of individual precipitates

Although not directly related to the study of superalloys under typical application conditions, recent experimental [789,882] and MD simulation [21,1320,629] studies on individual γ' nanocubes under compression allowed to investigate the deformation behavior of the pure γ' phase. These defect-free cubes deform by the nucleation of Shockley partial dislocations, leaving behind complex stacking-faults that can at larger strains transform into a pseudotwin structure [21]. The detailed deformation mechanisms, however, depend critically on the used potentials [1320], showing the importance of performing well-controlled experiments to validate interatomic potentials. Combining such well-defined experimental or simulation studies on the individual phases, as well as on the

γ/γ' microstructure can help elucidate the influence of constraints and misfit stresses on the mechanical performance of single-crystalline superalloys [629].

11.7.2 Interfacial dislocation networks

The misfit dislocation network (MFDN) that forms directly upon energy minimization due to the lattice misfit between the γ and the γ' phases has been the subject of many atomistic simulations [1673,1616,1594,1584,1676, 1629,1617,1138,832,343]. These studies have been performed on systems in which the γ phase is represented by pure Ni, whereas the γ' phase is modeled by stoichiometric Ni_3Al. The resulting lattice misfit of $\delta \approx 1.44$–2.8%, depending on the potential, is one order of magnitude larger than in typical superalloys, which furthermore have a negative misfit [1059]. Only recent studies included alloying elements [343] and nonstoichiometric compositions [1138].

The most important critique of these simulations is, however, their artificial construction of the interfacial dislocation network, often using perfectly planar interfaces. In reality, the interfacial dislocation network (IDN) forms by deposition of channel dislocations under creep conditions and their subsequent rearrangement [847]. In this respect, the artificial MFDN can only be seen as an idealized arrangement of interfacial dislocations that would most effectively reduce the misfit stresses. However, in most simulation studies, the IDN dislocations show a compact core, see Fig. 11.4(a). This is due to the fact that the $\{100\}$ interface plane is not a natural glide plane for dislocations in the fcc crystal structure. By using realistic interface morphologies, it was recently shown that dislocations in the MFDN assume configurations in which they spread out on $\{111\}$ planes, and can form stair-rod junctions as also observed in multiple experiments [1138]. In addition, also the Burgers vectors in the MFDN changed due to the interface curvature [1138]. An example for a realistic MFDN can be seen in Fig. 11.4(b).

11.7.3 Dislocation–precipitate interactions

The interaction of γ matrix dislocations with γ' precipitates is the main reason for the high creep strength of Ni-base superalloys. Here, atomistic simulations can in principle be used to identify the conditions under which single dislocations or superdislocations can cut into the γ' phase. The interaction between single, infinitely long edge, or screw dislocations with a regular array of spherical γ' precipitates with diameters up to 6 nm was studied by Proville and Bako [1146]. They showed a transition from dislocation cutting to Orowan bowing with increasing precipitate diameter. Kohler et al. [735] used a similar methodology but focused on even smaller precipitates. The more relevant case of studying the interaction of

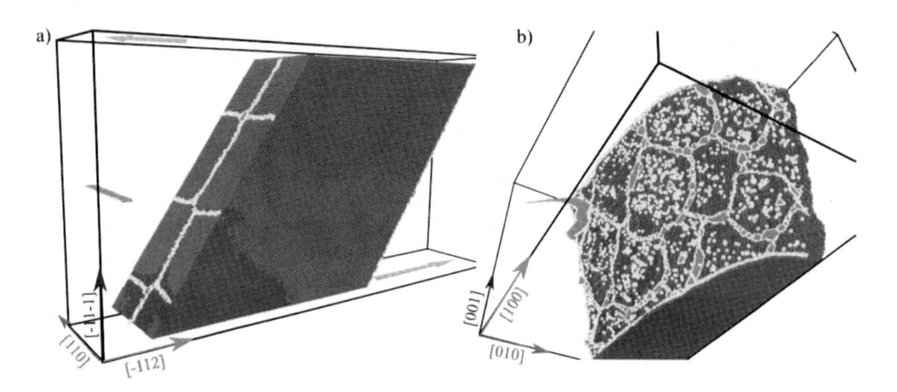

FIGURE 11.4 Simulation setups used to study dislocation–precipitate interactions: (a) typically used setup to determine critical stresses for precipitate cutting, e.g., in [1676], where stresses and/or displacements can be applied on the top and bottom surfaces (symbolized by the grey arrows). These setups with perfectly planar interfaces lead, however, to a square MFDN with unrealistic compact cores; (b) experimentally-informed simulation setup using the precipitate morphology obtained from APT data [1138] with extended MFD cores as observed in experiments (only atoms belonging to defects or the precipitate are shown, atoms in red (light grey in print version) are part of a stacking fault).

super dislocations with spherical precipitates was recently simulated by Hocker et al. [620] and Kirchmayer et al. [725]. The former group used a similar setup as in [1146,735] using regular arrays of spherical precipitates. In that case it was found that, depending on whether the distance between the superpartial dislocations is larger or smaller than the precipitate diameter (corresponding to weak or strong coupling between the superpartial dislocations [650]), different partial dislocations govern the critical resolved shear stress (CRSS) to pass the precipitate. Kirchmayer et al., however, used realistic precipitate morphologies and arrangement obtained from atom probe tomography (APT), and were thus able to show that for a relatively wide distribution of precipitate sizes weak and strong pair coupling can be at play simultaneously [725]. Spherical precipitates were also studied by Takahashi et al. [1393] and Kondo et al. [744], however, in their simulations the precipitates consisted of the γ phase which were cut by superdislocations from the surrounding Ni_3Al γ' phase. Besides providing information about the dislocation–precipitate interaction processes, such simulations can provide – with the usual limitations of the unrealistic compositions – quantitative information on precipitate cutting stresses τ_c and on the relative importance of, e.g., the coherency stresses, the APB energy or the energy to create an interface step.

Single-crystalline superalloys are strengthened by the presence of large, cuboidal γ' precipitates that force the dislocations to glide in narrow channels. Relatively few detailed atomistic studies of dislocation–precipitate interactions in this type of microstructure have been reported so far. Such studies would, however, be important to inform discrete dislocation dynamics simulations, see [1614] and Chapter 12, in particular regarding, e.g., the influence of local interface curvature and chemical composition gradients on dislocation–precipitate interactions. The motion of dislocations into γ channels bounded by two cuboidal γ' precipitates was first studied by Yashiro et al. [1615]. Recently, Xiong et al. [1597] performed a quantitative study to determine the stress required for matrix dislocations to penetrate into the channel in a similar setup, however, with a preexisting MFDN. Overall, they found an inverse proportionality between the critical stress and the channel width. The influence of the MFDN on the interaction of channel dislocations with the γ' precipitate was recently studied using different setups [1676,1138,1597], see also Fig. 11.4. The simulations with a realistic MFDN and dislocation core structures resulting from an APT-informed sample showed that in particular the colinear interaction of the first channel dislocation with a misfit dislocation protects the precipitate from being sheared by a superdislocation as due to the dislocation annihilation no trailing superpartial is available [1138]. Prakash and Bitzek [1137] furthermore studied the interaction of dislocation loops with various shapes and arrangements of γ' precipitates. The different stress fields caused by the misfit stresses lead to significant differences regarding the cutting of precipitates and the penetration of the dislocation into the channels, demonstrating the importance of taking deviations from the idealized morphologies and topologies into account.

11.7.4 Deformation and fracture of superalloys

Interestingly, relatively few large-scale MD studies exist on the plastic deformation and stress–strain response of superalloys with γ/γ' microstructures. Ma et al. [880] performed tensile tests on an Ni/Ni_3Al microstructure, however, made up of an unrealistic checker-board arrangement of cubic γ and γ' crystallites. Using one cubic γ' precipitate surrounded by γ, Li et al. [832] showed a tension compression antisymmetry in the stress-strain response. The result of a recent MD simulation with eight γ' cubes arranged according to an idealized γ/γ' microstructure is shown in Fig. 11.5. All these simulation results are, however, of limited relevance as under these conditions, the deformation is governed by the homogeneous nucleation of dislocations at extremely high strain rates of the order of 10^7 to $10^9\,s^{-1}$. This is in stark contrast to typical experimental conditions and samples that contain preexisting ingrown dislocations. This highlights the need for a multiscale approach in which the information from atomistic simulations are used in, e.g., DDD simulations at lower strain rates, larger sizes, and with more realistic initial dislocation densities. In particular, for the later stages of deformation where creep or fatigue fracture becomes relevant, atomistic simulations are uniquely positioned to provide criteria for crack advance or dislocation nucleation to meso- and

FIGURE 11.5 Dislocation structure after 8.4 % uniaxial tensile strain at T = 1250 K with the EAM potential by Mishin [953] (the structure is periodic in all directions, cube side length 16.4 nm, channel width 4.6 nm, 6.9 million atoms). Only atoms belonging to the dislocation cores and stacking faults as well as the shape of the γ' cubes are shown.

continuum scale models [116,961]. The corresponding atomistic studies regarding crack nucleation and crack propagation in γ/γ' microstructures [864,1628,1306] are, however, still in their infancy.

11.8 Limitations of atomistic simulations for superalloy design

While progress in atomistic theory in the past years means that many properties of superalloys became accessible, other aspects of superalloy performance are not directly accessible by atomistic simulations. Here we summarize the main limitations of present day atomistic simulations for superalloys.

- While DFT is used heavily in atomistic simulations for superalloys, the approximations in current exchange correlation functionals still limit the use of DFT data for alloy design. For example, formation energies and enthalpies computed by DFT can in principle be included in CALPHAD assessments, but accuracy of the DFT data, compatibility to experimental data sets and exhaustive DFT data for multicomponent alloys remain issues that need to be resolved together with an update of the CALPHAD approach to bring it closer to atomistic simulations. The same holds true for elastic constants and interface energies that are required for phase field simulations.

- Superalloys are multicomponent materials. Interatomic potentials for multicomponent materials are not available today. This means that atomistic simulations often cannot provide input for mesoscale or continuum models. The development of quantitatively accurate multielement interatomic potentials remains one of the grand challenges in the field.
- The time scale of atomistic molecular dynamics simulations is limited by the time step for the numerical integration of the atomic trajectories. The time step is on the order of 10^{-15} s. This means that time scales required for studying processes that include diffusion, e.g., dislocation climb-glide creep, are often not accessible. Expanding atomistic simulations to diffusive time scales is one of the major challenges for the field.
- An even greater gap exists between atomistic simulations of microstructural events and the modeling of plasticity. Models of plasticity need to define their parameters explicitly in such a way that atomic-scale simulation procedures may be developed to provide the required parameters.
- Continuum models are sometimes incompatible or at least not directly compatible with atomistic simulations. The transfer of results from atomistic simulations to continuum descriptions remains one of the roadblocks for a first principles guided design of superalloys. Both the atomistic and continuum modeling communities need to make great strides towards a more efficient exchange.

11.9 Summary: modeling Ni-base superalloys from electrons to microstructures

The rapid progress in atomistic modeling and simulation over the past years brings first-principles computational design of superalloys into reach. Today, atomistic modeling enables the prediction of key aspects of superalloy properties, from solute bond chemistry to microstructural properties. We reviewed the different representations of the interatomic interaction that are commonly used for modeling superalloys, briefly introduced atomistic simulation methods in the context of superalloys, and outlined properties of superalloys that can be computed by atomistic simulations.

Limitations of atomistic simulation methods means that many properties of superalloys are not accessible today and the development of "approximate practical methods" that Paul Dirac envisioned nearly a century ago will remain a focus of the atomistic simulations community.

We hope that the present chapter may induce discussions and collaborations that will contribute to further narrow the gap between atomistic simulation and superalloys, and in this way help advance the design of novel superalloys.

Acknowledgments

We are grateful to A.P.A. Subramanyam, H. Lyu, F. Houllé, A. Prakash, and J.J. Möller for their help in preparing some of the figures. We acknowledge financial support from the German Research Foundation (DFG) through projects C1, C2, and C3 of the collaborative research center SFB/TR 103.

CHAPTER

12

Discrete dislocation dynamics

Francesca Boioli[a], *Benoit Devincre*[a], *and Marc Fivel*[b]

[a]LEM, CNRS-ONERA, Chatillon, France, [b]SIMaP, Univ. Grenoble Alpes, CNRS, Grenoble, France

12.1 Introduction

The purpose of this chapter is to present recent progress and open questions on discrete dislocation dynamics (DDD) simulation used to investigate crystal plasticity in the mesoscopic scale (1 μm to 100 μm). Initiated in the late-1980s, DDD simulation are nowadays acknowledged as a key component of multiscale material modeling programs that aim at predicting physically justified constitutive behavior and microstructure based predictions of the mechanical properties of materials. Such a modeling technique is based on a hierarchical multiscale approach bridging the atomistic to the continuum domains. This review is written having in mind specific investigations of the plasticity of multiphase alloys (like Ni-base superalloys) which raise problems different from those related to simulations of single phased materials and nanoobjects.

12.2 Modeling multiphase materials with DDD

12.2.1 Dislocation dynamics simulations

DDD simulations are widely used to investigate many different problems [768]. Their results provide guidelines for modeling material mechanical properties controlled either by the collective behavior of dislocation microstructures or by the interaction between dislocations and other crystal defects such as twins, interphases, grain boundaries, voids, and cracks. In addition, DDD simulations can directly and successfully be compared to experiments either at a macroscale when applied to small crystal

Nickel Base Single Crystals Across Length Scales
https://doi.org/10.1016/B978-0-12-819357-0.00021-4

volume with periodic boundary conditions, or directly to nano- and microscale objects, such as nanoparticles, thin films, and micropillars.

DDD simulations compute plastic strain by integrating the equations of motion for dislocation lines under stress in an elastic continuum. The mutual interactions of dislocations, the formation and destruction of junctions, their line tension, and their interactions with other defects are essentially obtained from the elastic theory. Whereas some differences exist among DDD simulation codes, there are basic features that all of these have in common.

Two-dimensional DDD simulations are not discussed in what follows. These useful simulations allow essentially exploring plasticity issues in a simplified manner before addressing problems in 3D in a more realistic and quantitative way. Crystal plasticity is intrinsically a 3D problem that cannot be reduced to 2D without the omission or oversimplification of major dislocation mechanisms.

DDD simulation codes discretize dislocations into a set of line segments. Such line discretization can be performed essentially in two different manners hence defining two families of simulation codes. In *nodal* simulations (Fig. 12.1(c)), nodes which define the degrees of freedom of the dislocation lines are interconnected by straight segments. The location of these nodes and the orientation of the segments can vary continuously in the simulated volume and thus allows a simple description of dislocations with a continuous curvature. This approach allows for a better resolution of the dislocation stress field close to their singularity line. In *lattice-based* simulations (Fig. 12.1(a)–(b)), dislocations are discretized into a succession of straight segments, which are positioned along preset directions of an underlying simulation lattice. The length of each segment is then a multiple of the lattice translation parallel to the line direction and its direction of motion is by construction normal to it. The simulation lattice parameter is by definition a multiple of the atomic one and constitutes a scaling length of all dimensions and distances. The precision required for a given investigation is then obtained by defining the minimum spacing between adjacent dislocation slip planes on the simulation lattice.

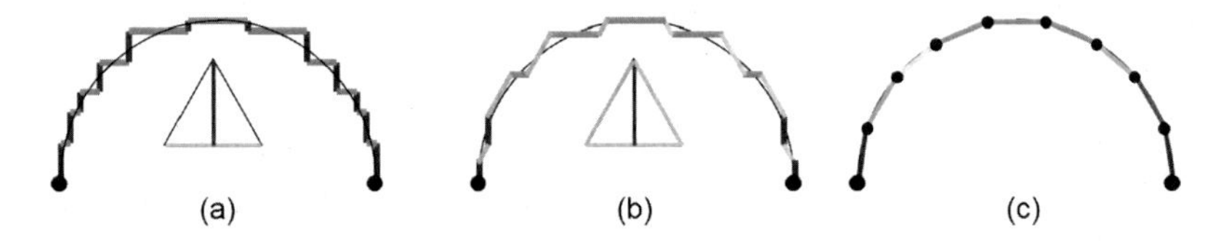

FIGURE 12.1 Discretization of dislocation lines in DDD models. Discrete line orientations are enforced by the edge-screw model (a) [770,1484] and (b) the edge-mixed-screw model [339], whereas the nodal approach (c) allows any possible orientation of the segments connecting two nodes [171,1120,1325,65].

Whatever the discretization rule, all simulation codes share the same strategy going back to the first model [770]. Details on the simulation techniques can be found in textbooks [171,768] and review papers [339, 1120,1325]. First, the forces on dislocations are computed using different solutions to calculate dislocation stress field. Then, the positions of the dislocation segments are updated using explicit time integration algorithms and material-dependent equations of motion. Lastly, contact reaction and local properties controlled by the dislocation core energy are considered with constitutive rules before cycling to the next simulation step increment. During the last 10 years, great progress has been made in each part of the DDD simulation methodology. Within such progress, the most significant advances are described in what follows.

The existence of singularities in the mathematical forms associated with the Volterra dislocation theory used in the early DDD simulations makes some fundamental physical quantities ill-determined (e.g., the dislocations line tension and the energy of intersecting dislocations) and induces numerical problems. Consequently, various methods of regularization of the elastic fields of dislocation segments have been proposed [185,1119]. Those solutions, which are based on the isotropic linear elasticity, are acknowledged today as numerically efficient solutions for most problems. Still, for some materials, the contribution of elastic anisotropy cannot be neglected. In the past few years, semianalytical solutions have been developed for such cases [236,241,1118]. The latter solutions involve extensive additional computations during DDD simulations which explains a rare use. Alternatively, simulations in anisotropic elastic media are possible with the help of coupling methods presented in the next section. The numerical solutions developed many years ago were initially restricted to simulations with few dislocations [551], but the recent development of efficient FFT spectral elastic solvers [101,548] opens the door to massive DDD simulations in the coming years.

The computation of the Peach–Koehler force and the detection of contact reactions are by far the most CPU-consuming part of DDD simulations. The total amount of operations required for both calculations scales as $O(N^2)$ for a system of N segments. Hence, the computational load quickly becomes too expensive in large-scale simulations. The fast multipole algorithm provides an efficient means to account for distant interactions [1319,1538]. In addition, advanced time-integration algorithms have been developed to account for the effect of dislocation patterning and the heterogeneity of dislocation mobility in materials [1326].

Another difficult aspect of DDD simulations is the definition of the constitutive or "local" rules that account for dislocation core properties like dislocation mobility, nucleation, cross-slip, and climb. The last two are discussed in detail in Section 12.3 since they control important features of superalloy plasticity. Special attention must be paid to these local rules, since they are the simulation ingredients controlling the specificities of each material. The validation of these parts of the simulation models can be made either by comparisons with atomistic simulations or from dedicated experiments (see, for instance, [1259,1313]).

12.2.2 Coupling methods for complex boundary problems

Usual formulations of DDD simulations make use of analytical expressions for the computation of the dislocation stress fields. As these expressions are mechanically valid in an infinite media, conventional DDD simulations are theoretically restricted to the modeling of a small volume with periodic boundary conditions (PBC). Hence, one can simulate the deformation of a representative volume (with linear dimensions a few tens of microns) part of a "macroscopic" sample. The use of PBC is then essential, but requires particular caution to avoid artifacts in DDD simulations [331,889].

In order to handle finite boundary conditions and to take into account the effects of free surfaces or internal boundaries, alternative numerical strategies are needed. They are based on a coupling between a DDD code and an elastic solver. In what follows, the two main methodologies used to handle complex boundary value problems are described. The first is the superposition method (SM), the second is the discrete–continuous model (DCM) based on the eigenstrain theory.

The boundary value problem to be solved consists of finding the displacement and stress fields $(\vec{u}, \boldsymbol{\sigma})$ in a finite elastic domain Ω containing displacement jumps $[\![\vec{u}]\!]$ due to the presence of dislocation loops. This boundary value problem $\mathscr{P}$ can be written as follows:

$$\mathscr{P}\left|\begin{array}{ll} \operatorname{Div}\boldsymbol{\sigma} + \vec{f} = \vec{0} & \text{in } \Omega\backslash\{A\}, \\ \boldsymbol{\sigma} = \mathbb{C} : \boldsymbol{\varepsilon} & \text{in } \Omega\backslash\{A\}, \\ [\![\vec{u}]\!] & \text{across } \{A\}, \\ \vec{u} = \vec{u}_0 & \text{on } \partial\Omega_u, \\ \boldsymbol{\sigma}\cdot\vec{n} = \vec{t} & \text{on } \partial\Omega_\sigma. \end{array}\right. \tag{12.1}$$

The boundary $\partial\Omega$ of Ω with outward normal $\vec{n}$ is possibly divided into Ω_u where Dirichlet boundary conditions are applied and Ω_σ (nonoverlapping with Ω_u) where Neumann boundary conditions are applied. At time t, $\{A\}$ represents the area swept by the dislocations since the beginning of the simulation. The displacement jump $[\![\vec{u}]\!]$ is tangent to $\{A\}$, its magnitude and direction are given by the Burgers vector $\vec{b}$. Furthermore, $\mathbb{C}$ is the fourth-order tensor of elasticity, $\boldsymbol{\varepsilon}$ is the infinitesimal elastic strain tensor (the symmetric part of the gradient of the displacement field), $\vec{f}$ represents the body forces, $\vec{t}$ the surface forces applied at Neumann boundaries, and $\vec{u}_0$ is the prescribed displacement at Dirichlet boundaries. Whether one considers the SM or the DCM, both rely on the linearity of the problem and consider a decomposition into subproblems we can solve numerically. In the following, S and NS will refer to the **s**ingular and **n**on**s**ingular dislocation field solutions.

The SM is the most commonly used solution for reason of simplicity. It results from the seminal work of Van der Giessen and Needleman [1465]. This method is routinely used in 3D simulations for different problems using either finite element [331,444,1558] or boundary element [384] elastic solvers. In this method we decompose the problem $\mathscr{P}$ into two subproblems: one in an infinite dislocated medium $\mathscr{P}^\infty$ plus a correction at the boundaries $\mathscr{P}'$. More precisely, the first problem deals with interacting dislocations in a homogeneous, isotropic, infinite solid, and the complementary problem accounts for the initial nonhomogeneous body, but without dislocations and with modified boundary conditions. This decomposition of the problem $\mathscr{P}$ takes the form $\mathscr{P} = \mathscr{P}^\infty + \mathscr{P}'$ with

$$\mathscr{P}^\infty \left| \begin{array}{ll} \mathrm{Div}\,\sigma^\infty = \vec{0} & \text{in } \mathbb{R}^3, \\ \boldsymbol{\sigma}^\infty = \mathbb{C} : \boldsymbol{\varepsilon}^\infty & \text{in } \Omega, \\ [\![\vec{u}^\infty]\!] & \text{across } \{A\}, \end{array} \right. \tag{12.2}$$

and

$$\mathscr{P}' \left| \begin{array}{ll} \mathrm{Div}\,\boldsymbol{\sigma}' + \vec{f} = \vec{0} & \text{in } \Omega, \\ \boldsymbol{\sigma}' = \mathbb{C} : \boldsymbol{\varepsilon}' & \text{in } \Omega, \\ \vec{u}' = \vec{u}_0 - \vec{u}^\infty & \text{at } \partial\Omega_u, \\ \boldsymbol{\sigma}' \cdot \vec{n} = \vec{t} - \boldsymbol{\sigma}^\infty \cdot \vec{n} & \text{at } \partial\Omega_\sigma. \end{array} \right. \tag{12.3}$$

In (12.2), the singular solutions are known analytically for an isotropic homogeneous infinite medium. They contain the displacement jump across $\{A\}$. Here, it must be noted that simulations with the SM dealing with free surfaces or penetrable interfaces involve extra computations to correctly handle the effect of image forces on the dislocations touching boundaries [1156,1546,1558]. Another issue comes from the boundary shape evolution induced by $\vec{u}^\infty$, which is not defined in standard DDD simulations and involves again extra computations [158,445,643].

First developed by Lemarchand et al. [819] and subsequently in several articles [551,670,1476], the discrete–continuous model (DCM) is based on the work of Mura [989], which states that Voltera dislocations can be regarded as inclusions generating uniform plastic shear strain fields. These strain fields belong to a larger group of nonelastic strains named "eigenstrains."[1] Exploiting this idea, the DCM methodology considers the dislocations in Ω when solving its mechanical equilibrium at once.

[1]Other eigenstrains may arise from thermal expansion, phase transformation, or misfit strains.

When a dislocation moves, a corresponding eigenstrain is distributed inside the elastic continuum. This eigenstrain induces eigenstresses which, in rough terms, interact with the boundaries. Consequently, there is no need to apply corrections on the boundary conditions. Most recent versions of the DCM use the nonsingular theory of dislocations and a new formulation of the BVP decomposition. This new formulation essentially avoids using very fine FE meshes to compute the stress field close to the dislocation lines. The latter information is needed for the description of the close-range interactions between dislocations, such as the junction formation. The DCM problem decomposition of $\mathscr{P}$ takes the form $\mathscr{P} = \mathscr{P}^\star + [\mathscr{P}^\infty - \mathscr{P}^{\star\infty}]_{dis}$ with

$$\mathscr{P}^\star \left| \begin{array}{ll} \operatorname{Div}\boldsymbol{\sigma}^\star = \vec{0} & \text{in } \Omega, \\ \boldsymbol{\sigma}^\star = \mathbb{C} : (\boldsymbol{\varepsilon}^\star - \boldsymbol{\varepsilon}^p) & \text{in } \Omega, \\ [\![\vec{u}^\star]\!] & \text{at } \partial\Omega_u, \\ \boldsymbol{\sigma}^\star \cdot \vec{n} = \vec{t} & \text{at } \partial\Omega_\sigma, \end{array} \right. \tag{12.4}$$

and

$$\mathscr{P}^{\star\infty} \left| \begin{array}{ll} \operatorname{Div}\sigma^{\star\infty} = \vec{0} & \text{in } \mathbb{R}^3, \\ \boldsymbol{\sigma}^{\star\infty} = \mathbb{C} : (\boldsymbol{\varepsilon}^{\star\infty} - \boldsymbol{\varepsilon}^p) & \text{in } \Omega. \end{array} \right. \tag{12.5}$$

The DCM can thus be seen as an extension of the SM to which two subproblems are added. Even though at first sight these two subproblems may appear as inducing extra computation, such a solution can actually lead to major computational cost savings. Indeed, the part of the stress field calculation that required analytical forms in the DCM computations is associated with the difference $[\mathscr{P}^\infty - \mathscr{P}^{\star\infty}]_{dis}$. The latter quantity tends towards zero rapidly far away from the dislocation segment, so that it can be truncated at short distances. This mathematical trick is essential as it significantly reduces the number of segment–segment interactions that must be computed in DCM simulations. In addition, the long-range stress field associated with the dislocation microstructure is in the DCM a direct outcome of the elastic solver that can easily take into account-complex boundary conditions. Finally, the coupling of an extra tool like the FEM with DDD simulation codes has a computational cost, but the latter is independent of the number of dislocation segments. This makes a huge difference with standard DDD formulations where the computational complexity increases like $O(n \log(n))$ with a multipole expansion algorithm and where a number of the dislocation microstructure replicas must be accounted for in the stress calculation when periodic boundary conditions are defined. This is why, the DCM is potentially much more numerically efficient at large segment numbers [670]. The initial development of the DCM with elastic solver based on FEM restricted the method to problems with complex boundary shapes and few dislocations, see, for instance, [305,551,866]. In the last years, the development of efficient FFT spectral elastic solvers [101,100,548] opened the door to massive DDD simulations with periodic boundary condition. These new releases of the DCM fit perfectly the need for future investigations of superalloy plasticity at the mesoscopic scale.

12.3 Modeling high temperature dislocation properties with DDD

Dislocations motion and their interactions with other defects are the most important processes governing the mechanical properties of crystalline materials. Several elementary mechanisms, such as dislocation reactions, glide, cross-slip, and climb, take place during material deformation, and they all contribute to the evolution of the microstructure.

At low homologous temperatures ($T < 0.4T_m$, T_m being the melting temperature), dislocation motion is essentially restricted to a few crystallographic planes, and plastic deformation is controlled by dislocations glide. In this regime, DDD simulations have proved to give an accurate description of many plasticity problems in metals [336,338,890], semiconductors [555,1294], as well as in oxides [22]. In Ni-base single-crystal superalloys, the early stages of plastic deformation are controlled by dislocation glide, and several DDD studies were dedicated to simulate this regime, allowing the prediction of plastic strain anisotropy [1478], as well as the critical shear stress for particle shearing as a function of temperature, particle size, volume fraction, etc. [120,638,645,1477,1608,1614].

At high temperature ($T > 0.4T_m$), dislocations can change their glide plane as a result of thermally activated mechanisms such as cross-slip or climb events [188]. Cross-slip can also be active at intermediate temperature if high stresses (either uniform or localized) are applied. Hence, the cross-slip and climb properties are essential for the modeling of superalloys. The initial stages of creep are, for instance, characterized by dislocations gliding in the γ matrix until they reach a γ/γ' interface where they deposit misfit dislocation segments, typically along the $\langle 110 \rangle$ directions. The latter dislocation, pinned in the misfit network, cannot glide anymore and, to promote further plastic deformation, dislocations should move along the interface to reach the nearest γ channel and glide through the channel. However, this type of displacement requires changing dislocations glide plane and can only be achieved by cross-slip or climb mechanisms [494,568,1170].

12.3.1 Cross-slip mechanism

Cross-slip of screw-character dislocation segments is an ubiquitous mechanism in plastic deformation. This process is known to play a role in different aspects of plasticity, such as work hardening during stage I and II of the stress–strain diagram of FCC metals [1247], cell pattern formation [890], and dynamic recovery [771]. Cross-slip has also been recognized as a rate-controlling mechanism in creep deformation at intermediate temperature range, playing an important role in the annihilation, generation, and rearrangement of screw dislocations, as well as in bypassing obstacles.

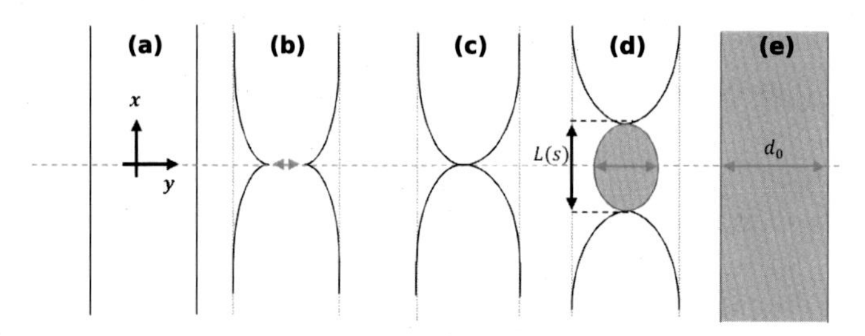

FIGURE 12.2 Sketch of the Friedel–Escaig mechanism for cross-slip in FCC crystals: (a) a screw dislocation is dissociated into two 30° Shockley partials; (b) the two partial recombine into a perfect screw dislocation at a constriction point (c); (d) at the constriction point, the screw dislocation splits into two Shockley partials in the cross-slip plane and expands by moving apart the two newly formed constriction points at the intersection between the glide and the cross-slip plane (e). Figure taken from [895].

Several elementary cross-slip models have been proposed, each being associated with a different minimum free energy path to move from an equilibrium screw segment configuration in the glide plane to one in the cross-slip plane. For a detailed description of such models, we refer to other texts [188,768,1150]. In FCC crystals, both the glide and cross-slip planes are close-packed {111} planes [1283]. Different scenarios of cross-slip have been studied in such crystals (Friedel–Escaig model, Fleisher model, jog-pair mechanisms, etc.), but the most investigated process is the constriction–expansion mechanism first proposed by Friedel–Escaig [131,399,400,1182]. It consists of the formation of a constriction point on a dissociated screw dislocation, followed by the dissociation of the two partials in the cross-slip plane till they reach a critical separation distance between two constriction points (see Fig. 12.2). Even though recent atomistic simulations indicate that the Fleisher mechanism dominates at the low-temperature–high-stress limit [186] and that the jog-pair mechanism is more likely to operate at high-temperature–low-stress limit, DDD studies have been focused mainly on conditions where the Friedel–Escaig mechanism is assumed to dominate.

Cross-slip was first introduced in DDD by Kubin et al. [770] through a semiquantitative rule inspired from Escaig's model. In this approach, the cross-slip probability per time step δt for any screw segment of length L experiencing a local resolved shear stress on its glide plane τ_g is given by

$$P_{cs} = \beta \frac{L}{L_0} \delta t \exp\left(-\frac{\Delta H_0 - (|\tau_g| - \tau_{III})V}{k_B T} \right), \tag{12.6}$$

where T is the temperature, k_B is the Boltzmann constant, ΔH_0 is the activation enthalpy in absence of stresses, V the activation volume, and the effective stress is taken to be $\tau_g^* = |\tau_g| - \tau_{III}$, where the stress τ_{III} is the resolved shear stress at the onset of stage-III for bulk crystals. In the preexponential factor, L_0 is the typical length of a cross-slipping segment [770] and β is a scaling factor. Here, both ΔH_0 and β were adjusted on the basis of experimental observations, since the values of ΔH_0 obtained from atomic-scale calculations for homogeneous cross-slip in FCC metals (i.e., assuming simple straight dislocation geometry) give systematically values too large compared to experimental results [1150,1181]. This model has been the basis for incorporating cross-slip in most DDD simulation methods [42,339,384,1484,1557,1637], where the cross-slip probability P_{cs} is introduced through a classical Monte Carlo scheme. Modified versions of Eq. (12.6) have been suggested thereafter. For example, Déprés et al. [335] considered in Eq. (12.6) the resolved shear stress on the cross-slip plane instead of the stress on the glide plane τ_g. Moreover, Kang et al. [696] suggested that Escaig stresses rather than glide stresses have a dominant effect on cross-slip. Recently, an expression of the cross-slip activation energy and volume that depends on both the glide and Escaig stresses based on a line tension model has been developed in [895]. This model can be implemented in DDD simulations to take into account nonglide stress components.

Most cross-slip processes described in DDD simulations depend only on the local stress field acting on screw dislocation segments and can be referred to as *bulk* cross-slip. Besides bulk cross-slip, molecular static (MS) and nudged elastic band (NEB) calculations showed that the cross-slip activation energy decreases significantly in the presence of specific constriction points along the dislocation line (e.g., in the vicinity of jogs, junction or dislocation cross-states) [1172,1173,1176,1175,1482]. Following these theoretical findings, a configuration-dependent rule for cross-slip has been defined in some DDD simulations [644], taking into account more complex cross-slip mechanisms, such as cross-slip at attractive [1173,1176] and mildly repulsive dislocation intersections [1174] or cross-slip close to free surfaces [1178] (the latter being relevant in the investigation of micropillars or samples characterized by a high surface-to-volume ratio). In such DDD simulations, the energy barrier associated with each cross-slip process is derived from atomistic calculations and is used to calculate the respective frequency of the cross-slip events with a law similar to Eq. (12.6):

$$f_{cs} = \omega_a \frac{L}{L_0} \delta t \exp\left(-\frac{\Delta E_a - (\tau_E^g - \tau_E^{cs}) V_a}{k_B T} \right). \tag{12.7}$$

In Eq. (12.7), E_a and V_a are the activation energy and volume required to form a constriction point on a screw dislocation, which depends on the type of cross-slip under consideration; τ_E^g and τ_E^{cs} are the Escaig stress on the glide plane and on the cross-slip plane, respectively, and the cross-slip frequency ω_a is computed based as $\omega_a = \eta\omega_D$, where ω_D is the Debye frequency of the material and η a scaling factor. This framework was applied by Hussein et al. [645] to simulate the deformation of superalloy microsamples and investigate the role of the distribution of γ' precipitates on the strength and dislocation microstructure evolution. Among other results, they showed that for low volume fraction of the precipitates, no precipitate shearing is observed (in the investigated strain domain) since dislocations can glide around the precipitates or cross-slip in the matrix to avoid the precipitates. On the contrary, a significant decrease of cross-slip frequency is observed for the high volume fraction case, and precipitate shearing becomes the dominant mechanisms. Such results highlight that cross-slip is a key recovery process to reach low-energy dislocation configuration and efficiently relax both the misfit and applied stresses. Lastly, it can be noted that local variation of the solute concentration in metallic alloys can also reduce the barrier to cross-slip [1019], but this effect has not yet been considered in DDD.

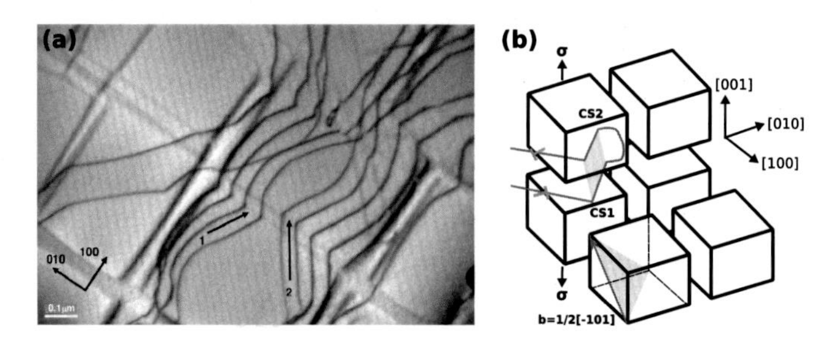

FIGURE 12.3 TEM cross-section showing a typical dislocation configuration in a single-crystal superalloy (TMS 138) after creep at 1100°C, 137 MPa for 2 h. We see interfacial dislocations with Burgers vector $b = a/2[101]$, propagating in the [010] average direction. Cross-slip takes place on the edge of the vertical channels normal to the [100] direction. (b) Sketch of a dislocation loop propagating into a γ channel and cross-slipping at the interface, generating a zigzag-shaped interfacial dislocation. Figure taken from [846].

Modeling of cross-slip in DDD simulations is a key ingredient needed to reproduce many important phenomena involved in the creep behavior of superalloys such as the zigzag shape of the dislocations observed in TEM (see Fig. 12.3). Straightening of the latter interfacial zigzag dislocations into pure edge dislocations along the precipitate [100] directions is expected to occur in the following stages of creep by activation of the climb mechanisms which will be discussed in Section 12.3.2.

12.3.2 Climb and vacancy diffusion

At high temperature, diffusion processes become important and some dislocations may climb out of the slip planes. This climb process is a non-conservative motion (see Fig. 12.4) that requires the absorption or emission of point defects [188,618]. Hence, it involves rare diffusion events, each of them depending on the local atomic configuration, as well as on the long-range strain and vacancy concentration fields. Ideally, the modeling of this process requires characterization at the atomic level by considering jog dynamics on the dislocation lines. However, many studies have proved that this dislocation motion mode can be described on the mesoscopic scale with simpler approaches making use of analytical forms.

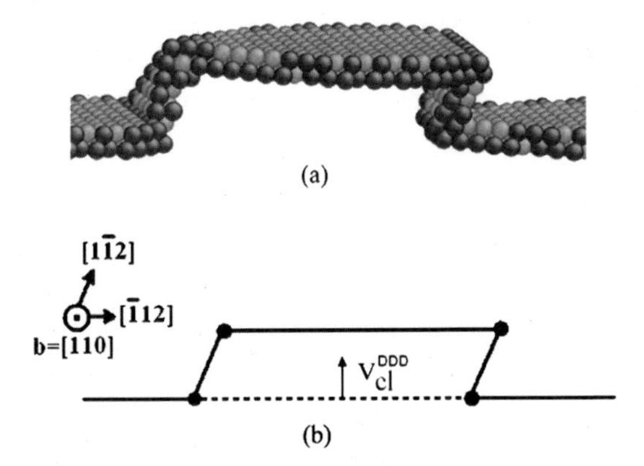

FIGURE 12.4 Sketch of the climb mechanism of edge dislocations in FCC crystallography as observed in MD (a) and implemented in DD (b) by Mordehai et al. [963].

Early attempts to model climb in DDD simulations made use of a phenomenological mobility law where both glide and climb velocities are taken proportional to the Peach–Koehler force and an effective drag tensor is defined for coupled glide-climb mobility (the drag coefficient for climb B_c being defined as a fraction of the drag coefficient for glide B_g) [42,186,515,851,1234,1590,1591].

Still, in order to capture the temperature and strain-rate dependence of creep, more sophisticated models of dislocation climb need to be considered in DDD. Mordehai et al. [963] used an expression for the climb velocity derived from the case of a straight edge dislocation (Eq. (12.8)). This simple solution is related to the diffusion of vacancies from/to the dislocation core, but assuming a constant far field vacancy concentration. Bako et al. [65] extended this solution by considering a local vacancy concentration at the dislocation due to absorption/emission by the dislocation microstructure. For simplicity reason, this approach was extensively used in two-dimensional (2D) DDD simulations to investigate mechanical behavior at elevated temperature in micropillars [713], thin films [314], multilayers structures [636], single crystal [123,308], and polycrystals [637]. More recently, this type of approach could be used in 3D DDD [494,557,856,855]. An alternative approach was also developed by Po et al. [1117] who proposed a method based on Onsager's variational principle and by Gu et al. [557] who derived Green's function formulation. Both solutions provide an expression for the climb velocity of curved dislocations in 3D-DDD.

Nowadays, most existing DDD studies follow the ansatz proposed in Mordehai et al. [963] with additional assumptions, i.e., (1) climb is controlled by vacancy diffusion, with the vacancy concentration being generally larger than the interstitial one, due to their lower formation enthalpy; (2) steady state conditions are met before each climb displacement; (3) dislocation lines are saturated with jogs, meaning that vacancies can be absorbed almost instantaneously once they are in the proximity of the dislocation core region. Following these three assumptions, the net flux of vacancies from and to the dislocation core is known, and the climb velocity v_c takes the form [188,618]:

$$v_c = \eta \frac{D^{sd}}{b} \left(\exp\left(-\frac{\tau_c^* \Omega}{k_B T} \right) - \frac{c(r)}{c_0} \right), \tag{12.8}$$

where D^{sd} is the vacancy self-diffusion coefficient, Ω is the vacancy formation volume, and η is a geometrical factor which depends on the geometry of the flux field [188]. In Eq. (12.8), $c(r)$ is the vacancy concentration at a distance r far from the dislocation, that can be taken as half of the average dislocation distance, while $c_0 = \exp\left(-\Delta H_f/(k_B T)\right)$ is the equilibrium vacancy concentration in bulk condition at a given temperature T. ΔH_f is the vacancy formation enthalpy. The self-diffusion coefficient can also be written as $D^{sd} = c_0 \exp(-\Delta H_m/(k_B T))$, with ΔH_m being the vacancy migration enthalpy.

The two terms in the parentheses of Eq. (12.8) represent the climb driving forces. The first is the Peach–Kohler force, or more exactly the effective stress in the climb direction τ_c^*. This stress is by definition positive when it tends to favor vacancy emission and negative when it tends to favor vacancy absorption. The second term is often referred as a "chemical force" because it arises from any gradient in the vacancy concentration. In supersaturation conditions, i.e., when $c(r) \gg c_0$, this term becomes important and dislocations can climb even in the absence of mechanical forces. A comparison of Eq. (12.8) with atomistic simulations on iron [262,685] and with phase field simulations on aluminum [512] leads to a remarkable agreement.

It must be noted that in Eq. (12.8), $c(r)$ is usually taken as a bulk vacancy concentration. However, this quantity varies in space and depends on the distribution of dislocations that act as sources or sinks of vacancies. Keralavarma et al. (2012) addressed this limitation by connecting the far field vacancy concentration $c(r)$ for each dislocation with an average concentration, $C(R)$, calculated in a finite volume element dV including the dislocations (see Fig. 12.6(a)). Moreover, they proposed to calculate the vacancy concentration field $C(R)$ by solving the diffusion problem with a source/sink term c_{src}. Hence, the governing equations for the vacancy diffusion problem becomes:

$$\dot{C} = \nabla \cdot \mathbf{J} + \dot{C}_{src}, \tag{12.9}$$

$$\mathbf{J} = -\frac{D\Omega}{k_B T}\nabla\mu, \tag{12.10}$$

$$\mu = -\frac{k_B T}{\Omega}\left[\frac{E_f + P\Omega}{k_B T} + \log\frac{C}{(1-C)}\right]. \tag{12.11}$$

Eq. (12.9) follows from mass conservation; $\mathbf{J}$ denotes the vacancy flux per unit volume and C_{src} is the source/sink term, which is given by the sum of the vacancy emitted and absorbed by the dislocations in a given volume element dV. In Eqs. (12.10) and (12.11), μ indicates the chemical potential, E_f the vacancy formation energy, P the hydrostatic pressure. Such coarse-grained solution allowed Keralavarma et al. [713,712], to perform long time-scale 2D simulations and to predict creep rates and stress exponents for aluminum within the experimental ranges. Alternatively, Ayas et al. [55] developed a 2D model solving the fully coupled dislocation motion and vacancy diffusion problem. They took into account the dislocations as discrete sources/sinks of vacancies. Although this solution describes more accurately the cooperative climb motion of dislocations with vacancies, it is computationally more expensive. More recently, Gao et al. [494] and Liu et al. [856] extended the coarse-grained approach developed by [713] to 3D simulations.

Another key issue that needs to be addressed in order to perform DDD simulations that simultaneously treat dislocation glide and climb is the different time scales that characterize both mechanisms. Indeed, glide is several orders of magnitude faster than climb. To solve this problem, iterative schemes are commonly used. First, the creep stress is applied and dislocations are moved by glide only with a small time step dt_g. When the plastic strain saturates, i.e., when the dislocations reach quasiequilibrium configurations, the time step is switched to a larger value Δt_c and dislocations are allowed to climb. When a climb displacement of a predefined amplitude is achieved by at least one dislocation, the time step is switched back to dt_g and dislocations are relaxed again by glide only. This procedure is iterated throughout the entire simulation.

It must be noted that the above climb formulations, all assume that dislocations are perfect sink/sources of vacancies and that vacancy concentration along the dislocation line is at equilibrium. This is because diffusion along the dislocation core (pipe diffusion) is always order of magnitude faster than bulk diffusion [421,1148]. However, to go beyond such approximation, one should consider that dislocations climb by the elementary mechanism of jog migration, resulting from absorption or emission of vacancies at jogs, and that vacancies can migrate by pipe diffusion along the dislocation line. In the last years, only a few attempts have been made to incorporate jog formation and migration or pipe diffusion into climb models. Gao et al. [497,498] developed a 3D dislocation dynamics simulation of pipe diffusion to study the shrinkage and annihilation of prismatic loops, as well as the breaking of parallel edge dislocations into loops. Geslin et al. [513] derived an analytic expression of the climb rate of a jogged dislocation and tested such formulation in a phase field model of pure climb. Additionally, Niu et al. [1018] developed a vacancy-assisted dislocation climb model by modeling the stochastic process of vacancy attachment/detachment at jog. To date, none of these models has been adopted to perform large scale DDD simulations, either because of the excessive computational cost of these approaches or due to the lack of information on the detailed atomic scale processes involved during climb.

12.4 3D-DDD simulations of Ni-base alloys

12.4.1 State-of-the-art

In this section we will exclusively focus on three-dimensional DDD simulations of Ni-base superalloys. Although interesting results could be obtained from 2D simulations [713,712], the limits of the 2D model cannot capture the complex 3D microstructure of superalloys. These works have nevertheless greatly contributed to the development of smart algorithms that were later translated into 3D such as the numerical scheme used to bridge the disparate time scales for dislocation glide and climb we presented in Section 12.3.2.

Using the DCM model, Vattre et al. [1478] incorporated the shearing of precipitates as proposed in [1477] and investigated the orientation dependence of plastic deformation observed at room temperature in the [111] and [001] tensile tests. Typical zigzag dislocation configurations and pseudocubic slip were reproduced in those simulations assuming that dislocation cross-slip is not active at room temperature. Also, the source of hardening anisotropy was explained by differences in the dislocation microstructures deposited at the interfaces when changing the loading conditions. These DD simulation results were later used to define slip system activity in crystal plasticity modeling [1479]. In this modeling, the climb mechanism was neglected.

In 2013, Hafez Haghighat et al. [568] implemented a hybrid effective mobility law coupling glide and climb in a 3D nodal code in order to investigate the effect of the climb mobility on the creep behavior of [001] Ni superalloy of 59% γ' volume fraction. In those simulations, precipitates were taken as square cubes of 420 nm length surrounded by 80 nm γ-channels (see Fig. 12.5). They found that the cubic slip referring to the movement of dislocations in the γ/γ' interfaces could be obtained as a combination of glide and climb. A higher plastic strain rate was obtained when the climb mobility was increased. In this study, the mobility law was purely artificial with no physical consideration on the vacancy concentration and both the misfit stress and precipitate shearing mechanism were neglected.

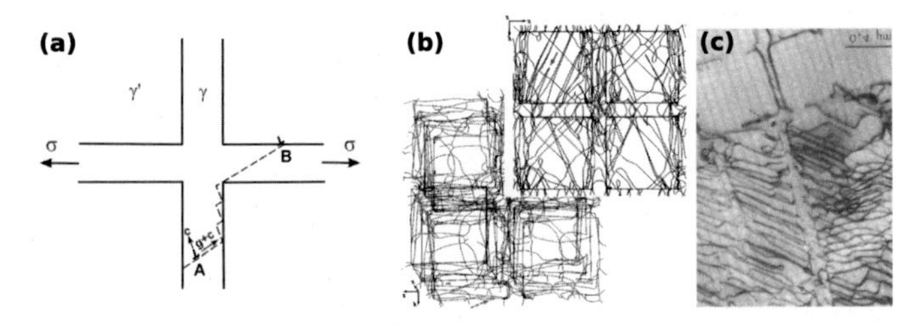

FIGURE 12.5 (a) Sketch of the dynamics of a mixed $1/2\langle 110\rangle\{111\}$ dislocation from position A to position B by glide+climb motion along the γ/γ' interface. (b) Simulated creep microstructure of an Ni-base superalloy (350 MPa along [100]) obtained by using a hybrid glide/climb mobility law, with the ratio between glide and climb drag coefficient being $B_g/B_c = 10$. (c) Experimental creep microstructure of Ni-base superalloy (552 MPa along [001]) taken from [1126]. The solid tail arrows indicate an example of a dislocation squeezing into the γ channel, and the dashed tail arrows indicate a polygonal dislocation loop. Figure taken from [568].

Between 2015 and 2017, Gao et al. developed a dedicated DDD code in order to address the problem of creep in Ni-base superalloys at high temperature [493,495,494]. Using the FFT method, they have computed the misfit stress induced by the lattice mismatch and investigated the effect of the latter on the mechanical response. In a second study [495], the shape of the precipitate was improved by phase field calculations, and they measured the effect of the precipitate distribution on the mechanical response. Finally, in a third paper [494], they have implemented dislocation glide-climb and vacancy diffusion simultaneously in the complex 3D microstructure of superalloys (see Section 12.3.2). As reproduced in Fig. 12.6(c)–(d), both the strain rate and microstructure obtained by DDD creep simulations are significantly different if climb is taken into account or neglected. Indeed, climb allows dislocations to bypass obstacles and to migrate along the interfaces increasing considerably the strain rate, as well as the generated dislocation density, as a function of time. Here, it can be noted that in Gao et al. [493,495,494], the superposition principle to enforce the boundary conditions was made using FFT calculations instead of the usual FEM solution. This new approach reduces the computation cost down by a factor of 100.

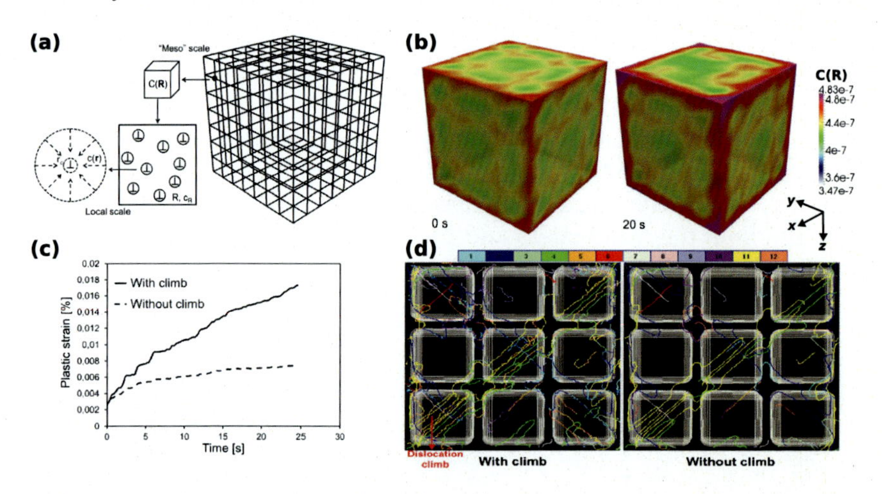

FIGURE 12.6 (a) Sketch of the multiscale model used to solve the vacancy diffusion problem in Gao et al 2017 [494]; $C(R)$ and $c(r)$ represent the mesovacancy concentration field in the whole system and the local vacancy concentration in the vicinity of dislocations, respectively. (b) Evolution of the vacancy concentration distribution $C(R)$ in a DDD creep simulation after 20 s at 1123 K under a tensile stress of 180 MPa along [001]. (c) Global equivalent plastic strains as a function of time for two simulations with and without dislocation climb, respectively. (d) Dislocation configurations obtained at the end of the two creep tests reproduced in (c). The dislocation microstructure is viewed in the [001] direction. Figure taken from [494].

In 2018, Chang et al. [232], using the same DDD creep model as [494], investigated the initial stage of the [001] tension creep. Taking into account the misfit strain induced by the lattice mismatch between the precipitates and the matrix, they showed that the three γ channels of the superalloys are not mechanically equivalent to the dislocation dynamics (see Fig. 12.7(a)). In the case of [001] tensile loading, the dislocations organize mainly at interfaces in the perpendicular channel to form a dislocation network compensating the misfit strain. In doing so, the dislocations induce a 2D plane stress elastic state which leads to a contraction along the loading direction (see Fig. 12.7(b)). The magnitude of this elastic contraction could be higher than the plastic deformation generated by the dislocation slip so that the specimen display a negative creep behavior. The reader should refer to Chapter 7 for experimental evidence of this behavior.

Most recently, Gao et al. [496] investigated the influence of excess volumes induced by Re and W on the creep behavior. The effect of solute solution was taken into account in the DDD model by adding a dragging stress on the edge dislocation segments. The strength of the latter stress was evaluated using DFT calculations. These DD simulations confirm the experimental observations regarding the strengthening effect of the solute atoms. Typically, it was found that a W solute solution leads to a larger strengthening effect than Re solute solution. Because of computational loads, such simulations were limited to low concentrations, and so the results are only qualitative, but they pave the way for more realistic simulations.

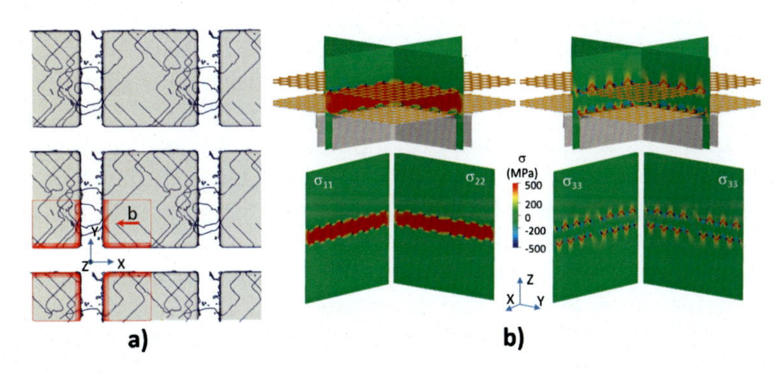

FIGURE 12.7 (a) DDD microstructure developed in the early stage of [001] creep view along the [001] direction. (b) Elastic stress tensor induced by an ideal interface dislocation network. Figures taken from [232].

12.4.2 Limits and perspectives

DDD simulations are limited in terms of the amount of simulated plastic strain. This limitation comes from the fact that plastic deformation is a nonconservative process regarding the dislocation density and the number of dislocation segments used in the simulations is rapidly increasing. Consequently, the time needed to compute each simulation step is increasing throughout the deformation. The recent coupling (see Section 12.2.2) making use of the DCM algorithm and a massively parallel FFT solver allowed bypassing such a problem, and some simulations with more than 10^6 segments are nowadays routinely run. This computational issue is, however, not a major concern for investigating the creep behavior of superalloys. Indeed, the in-service acceptable strain in such materials is limited to a few 10^{-3} in order to avoid failure. This remark emphasizes the usefulness of DDD simulations to investigate this class of materials and the importance to study the early stage of plasticity when the dislocation activity is confined into the γ channels. Surprisingly, many DDD simulations are nevertheless focusing on the precipitate shearing mechanisms which is clearly not the most important phenomenon to investigate for superalloys application domains.

Another limit concerns the assumption that the lattice in which the segments move is supposed to be rigid. In other words, unlike in finite element modeling, most DDD simulations neglect the local crystal distortions and do not take into account large deformations. The reason for this limit arises from the description of the dislocation lines at a mesoscopic scale. The minimum dislocation length is typically a dozen Burgers vectors. If large deformation was to be accounted for, one should first implement jogs' formation that results from dislocation interactions, which size is a few angstroms, and also handle the update of the dislocation network with time. To date, finite strain discrete dislocation plasticity formalism has been proposed [304,658], but such a feature does not yet allow for simulations with a massive number of dislocations in 3D.

One of the most challenging developments for the forthcoming DDD investigations of the superalloys plasticity will certainly deal with the precipitate rafting mechanism. Ideally, such an investigation could be tackled down by DDD provided a strong coupling with a phase field (PF) model is implemented. Toward this direction, a first weak coupling has been realized by Gao et al. [495] where the geometry of the γ' precipitates was computed by PF and then transferred to a DDD simulation. The next step will consist of dynamically coupling DDD and PF so that both the precipitate morphology and the dislocation microstructure are simultaneously computed.

CHAPTER

13

Phase field models for modeling microstructure evolution in single-crystal Ni-base superalloys

Yann Le Bouar[a], *Alphonse Finel*[a], *Benoît Appolaire*[b], *and Maeva Cottura*[b]

[a]Université Paris-Saclay, ONERA, CNRS, LEM, Châtillon, France,
[b]Université de Lorraine, CNRS, IJL, Nancy, France

13.1 Introduction

Mechanical properties of single-crystal Ni-base superalloys are mainly due to their microstructure, i.e., the shape and spatial arrangement of the different phases in the material, and in particular the γ and γ' phases. Phase field methods have proved to be the most efficient methods for studying the evolution of microstructures resulting from phase transformations, in particular when elastic coherency stresses are generated in the material. Therefore, during the past 20 years, these methods have been extensively used to analyze the important mechanisms and driving forces at play during the formation of the initial microstructure in single crystal nickel superalloys and during its evolution in service. The aim of this chapter is to give an overview of the progresses on this topic.

The developments of phase field methods may be traced back to van der Waals theory of liquid gas interfaces [1468], Landau theory of phase transitions [787,788], and Cahn–Hilliard thermodynamics formulation of nonuniform systems [184]. The method consists in introducing a set of continuous fields that represent the local state of the material, heterogeneities such as interfaces being represented by rapid but smooth variations of these fields.

Nickel Base Single Crystals Across Length Scales
https://doi.org/10.1016/B978-0-12-819357-0.00022-6

Phase field models may be classified into two categories, depending on the nature of the phase field used to describe the local state.

In the context of phase transitions, the fields most often refer to a physical property such as local concentrations or the local atomic ordering between the different atomic species that share the same underlying lattice. In this context, the free energy density functional, which is the fundamental ingredient of any phase field model, must be written in such a way that it is invariant under the application of the space group of the phase that, among those involved in the phase transition, displays the highest symmetry. The identification of the necessary fields follow then directly and systematically from symmetry properties (such as group–subgroup and broken symmetry considerations). This is the approach, referred to as "Ginzburg–Landau approach", that has been the most often used in the context of superalloys and it will be presented in the present paper.

In the second phase field approach, the phase fields are simply defined as indicator functions used to identify the phase present at a given point [1559,183,1429,722]. This approach has been initially proposed by Langer [791] in the context of solidification and is nowadays often referred to as the "multiphase-field approach" [1355,1356]. Within these models, the phase fields generically represent the volume fraction of the different phases and an interface is viewed as a mixture of phases, each of which retains its bulk properties even inside the interface. The total free energy is generally written as a sum over convex single-phase free energies with ponderation coefficients simply related (linearly or not) to the phase fields, enabling a conceptually simple way to couple the modeling to thermodynamics databases. This approach has also the advantage that bulk and interface properties may be decoupled. A derivation of this type of models from a grand potential function can also be found in [1114].

The phase field approach as several advantages over other modeling approaches. First, the phase field approach is very efficient when studying microstructures of complex geometry, with topological changes, because there is no need to explicitly track the position of the interfaces. Indeed, interfaces are only implicitly defined as the region of space in which the phase fields vary. In addition, because of the mesoscopic character of the phase fields much larger systems can be addressed than with atomic scale simulations, such as Monte Carlo or molecular dynamics. Another strength of this approach is that it is built within the framework of nonequilibrium thermodynamics, which makes it possible to include, in a coherent framework, a large number of physical phenomena, as well as the couplings between these phenomena. This is particularly suited to the microstructure developments in nickel-base superalloys, which result from the coupling between a phase transition, long-range diffusion, coherency strains, and plasticity.

This chapter presents only one of the phase field models used for studying microstructure evolutions in nickel-base superalloys. The objective is not to describe the numerous developments of the method, or the different subtleties of each model. For a global vision of the development of phase field models in materials science, the reader is referred to the many reviews [1332,958,1537,1355,1145,1356,340].

The chapter is organized as follows. In Section 13.2, the phase field model (of Ginzburg–Landau type) adapted to the study of coherent alloys containing γ and γ' phases is presented, as well as the advantages and limitations of this type of approaches. Then, the extensions of the phase field models to plasticity are presented, first on a continuous scale (Section 13.3), then on the scale of individual dislocations (Section 13.4). In the following sections, we will illustrate how these models have contributed to a better understanding of the consequences of the elastic and plastic driving forces on the microstructure evolution. Finally, the chapter ends with some perspectives for the development of these methods in the context of nickel base superalloys.

13.2 Cuboidal microstructures in Ni-base superalloys

The cuboidal microstructure formation in Ni-base superalloys is a simple example of coherent microstructures. Due to their anisotropic and long range character, the coherency stresses control the microstructure evolution.

The formulation of a phase field model follows three main steps:

(i) The starting point of the modeling is the identification and selection of relevant mesoscopic fields describing, at each position, the crystallographic structure, the concentration, the strains, etc.;

(ii) The second step is the definition of the mesoscopic free energy F, which describes the free energy of an arbitrary microstructure, defined at the mesoscale using the selected fields. This free energy is usually written as a Ginzburg–Landau free energy functional whose shape is dictated by general arguments of symmetry. It contains a so-called homogeneous part that characterizes a system that does not include heterogeneities at the mesoscopic scale and a so-called heterogeneous part that corresponds to the increment of free energy generated by these heterogeneities. It is easy to write this free energy density, for any phase transition, in the form of a polynomial of the relevant fields and their gradients. The approach generally followed in practice is very phenomenological: the polynomial developments are limited to the lowest order compatible with the physical phenomena considered, and the coefficients of the development are chosen so as to reproduce a certain number of macroscopic quantities such as the equilibrium concentrations and interface energies;

(iii) The last step consists in writing the kinetic equations that govern the evolution of the fields, using the thermodynamic driving forces generated by the mesoscopic free energy. These equations must be dissipative, and when the field is conserved, it must fulfill a conservation law. The time derivative of the fields is often assumed linear with respect to the thermodynamics driving forces. Finally, to end up with an efficient algorithm, attention must be paid to the characteristic time scales of the coupled phenomena. When one field relaxes much faster than the others, its evolution can be assumed to follow adiabatically the other fields. For example, it will be the case for the elastic degrees of freedom during a diffusive phase transformation.

13.2.1 Ginzburg–Landau phase field model with elasticity

Ni-base superalloys feature both the disordered γ and ordered γ' phases. In a model binary alloy, such a microstructure can be characterized by a concentration field $c(r,t)$ and order parameters that describe the four types of translation domains that are generated by the $\gamma \to \gamma'$ ordering transition. These order parameters can be defined as the amplitude of the Fourier components of the occupation probabilities of each atomic site [716]. Using this approach, three long-range order (l.r.o.) fields $\phi_i(r,t)$ $(i = 1, 2, 3)$ can be defined. The four translational variants of γ' are described by the following long-range order parameters: $\{\phi_1, \phi_2, \phi_3\} = \{1, 1, 1\}, \{\bar{1}, \bar{1}, 1\}, \{\bar{1}, 1, \bar{1}\}, \{1, \bar{1}, \bar{1}\}$.

As the main ingredient of phase field models, the mesoscopic free energy functional F is taken as the sum of a chemical F_{ch} and an elastic F_{el} contributions. The chemical contribution is expressed as a Ginzburg–Landau free energy functional (see, e.g., [1533,145])

$$F_{ch}(c, \{\phi_i\}) = \int f_{hom}(c, \{\phi_i\}) + \frac{\lambda}{2} |\nabla c|^2 + \frac{\beta}{2} \sum_i |\nabla \phi_i|^2 \; d^3 r, \tag{13.1}$$

where $f_{hom}(c, \{\phi_i\})$ represents the free energy density of a homogeneous system at concentration c and order parameters $\{\phi_i\}$. This quantity can be adjusted on experiments or using a relevant thermodynamic database. However, when studying microstructures with size much larger than interface width (referred to as *well-defined* in the following) the consequence of this energy is mainly to enforce the proper equilibrium concentrations of the coexisting phases. Therefore, the details of its shape are not important and it can be approximated by a simple polynomial expansion, taken due account of the symmetries of the phase transformation. Here, considering the cubic symmetry of the underlying FCC lattice, and limiting the polynomial to the lowest order, we have

$$f_{hom} = \Delta f \left[\frac{1}{2}(c - c^0_\gamma)^2 + \frac{\mathcal{B}}{6}(c_2 - c) \sum_{i=1,3} \phi_i^2 - \frac{\mathcal{C}}{3}\phi_1\phi_2\phi_3 + \frac{\mathcal{D}}{12} \sum_{i=1,3} \phi_i^4 \right], \tag{13.2}$$

where Δf is the free energy density scale, $\mathcal{B}$, $\mathcal{C}$, $\mathcal{D}$, and c_2 are parameters selected to reproduce the equilibrium concentration $c^0_\gamma(T)$ and $c^0_{\gamma'}(T)$. The gradient coefficients in Eq. (13.1) induce a free energy increase in the regions where the fields vary, i.e., in the interfaces. In nickel-base superalloys, because antiphase boundaries in the γ' phase are wetted by the γ phase, only the γ/γ' interfaces are observed in the microstructure. The coefficients λ and β are therefore selected to reproduce the γ/γ' interface free energy, and to ensure a wetting of the antiphase boundaries. In addition, because the small γ' domains are almost spherical, the γ/γ' interface free energy is assumed isotropic. At this point, it is important to stress that the interface free energies are very difficult to measure, in particular at the temperature at which the microstructure evolution occurs. The usual method is to measure the growth law of the average precipitate volume in the coalescence regime and to assume that a simple growth law of the Lifshitz–Slyozov–Wagner type can be used. This first hypothesis is already open to criticism for superalloys (see Section 13.2.2). Moreover, the interface energy enters in the LSW law within the prefactor of the $t^{1/3}$ term together with the interdiffusion coefficient and, more importantly, with the second derivative of the free energy density with respect to the concentration. The latter quantity has an ill-defined status because it depends on the model used to describe the metastable branches of free energy. Adding the fact that the interdiffusion constant itself is tainted with uncertainty, it is clear that the interface energies determined from the prefactor of the growth laws must be taken with caution. As a consequence, a wide range of values of interfacial energies are reported in the literature, ranging from 4 to 80 mJ/m^2 [32,36,143,1552,1592,193]. However, because the evolution of a well-defined coherent microstructure results from the competition between surface energy and volumic elastic energy (presented below), an error in the value of the surface free energy only results in an error on the length scale of the predicted microstructure.

Using the above expression of the chemical free energy (13.1) and of the free energy density (13.2), a simple dimensional analysis reveals that multiplying the gradient coefficients λ and β by a coefficient ζ, and simultaneously dividing the free energy density scale Δf by ζ, does not modify the value of the interface energies, but multiplies the interface width by ζ. This property is used in the numerical implementation of the model to artificially increase the interface width, and thus increase the numerical efficiency of the model. Indeed, depending on the required accuracy, the discretization length must be chosen 5 to 10 times smaller than the interface width. Of course, the numerical interface width must remain much smaller than the smallest length scale of interest in the microstructure. It is nevertheless important to evaluate the sensitivity of the results to the numerical width of the interfaces. Note that a new formulation of phase field models has been very recently proposed, in which the interfaces are resolved with essentially one grid point with no pinning on the grid and an accurate rotational invariance, improving drastically the numerical capabilities of the phase field method [441].

The last contribution is the elastic energy generated by the coherent coexistence of the γ and γ' phases. Due to the very small values of the misfit δ, this energy can be computed in the small strain approximation

$$F_{el}(\boldsymbol{\varepsilon}^e) = -V\boldsymbol{\sigma}^a : \bar{\boldsymbol{\varepsilon}} + \frac{1}{2}\int_V \boldsymbol{\varepsilon}^e : \mathbf{C} : \boldsymbol{\varepsilon}^e \, dV, \tag{13.3}$$

where $\mathbf{C}$ is the elastic tensor field, $\boldsymbol{\sigma}^a$ is the applied stress, and $\bar{\boldsymbol{\varepsilon}}$ is the average deformation of the simulation box of volume V. The elastic strain field $\boldsymbol{\varepsilon}^e$ is given by

$$\boldsymbol{\varepsilon}^e = \boldsymbol{\varepsilon} - \boldsymbol{\varepsilon}^0, \tag{13.4}$$

where $\boldsymbol{\varepsilon}$ is the total strain field and $\boldsymbol{\varepsilon}^0$ is the transformation strain field (or eigenstrain) null in γ and equal to the misfit tensor $\delta\mathbb{I}$ in γ' ($\mathbb{I}$ is the unit second-order tensor). As usual in phase field approaches, the local value of the materials properties (elastic constants, transformation strain, etc.) is interpolated from the values in the phases using the concentration or order parameters. In particular, assuming Vegards law, we write $\boldsymbol{\varepsilon}^0(\mathbf{r}, t) = \delta(c(\mathbf{r}, t) - c_\gamma)/(c_{\gamma'} - c_\gamma)\boldsymbol{I}$. Note that several schemes can be used for the interpolation of the elastic fields between the two phases [20,371,1280].

Finally, the kinetic equations describing the temporal evolution of the microstructure are the Cahn–Hilliard equation for the concentration field

$$\frac{\partial c}{\partial t}(\mathbf{r}, t) = \nabla \cdot \left(M \nabla \frac{\delta F}{\delta c(\mathbf{r}, t)} \right), \tag{13.5}$$

where the mobility M is usually assumed constant. Its value is selected using a relevant diffusion coefficient. The evolution of the order parameter field is given by the Allen–Cahn equation

$$\frac{\partial \phi_i}{\partial t}(\mathbf{r}, t) = -L \frac{\delta F}{\delta \phi_i(\mathbf{r}, t)}, \tag{13.6}$$

where L is a kinetic coefficient whose value is chosen large enough to ensure diffusion limited kinetics.

In addition, because the elastic degrees of freedom equilibrate much faster than the characteristic evolution time of the concentration and order parameter fields, mechanical equilibrium is assumed at each time step and is solved numerically, for example, using efficient fixed point algorithm in Fourier space [973,634].

This simple binary phase field model can be directly applied to the description of the microstructure evolution in binary systems exhibiting a coexistence of γ and γ' phases, such as Ni–Al, Ni–Ge, etc. It can also be applied to the qualitative description of the microstructure of a multi-component alloy, by considering an *effective binary alloy*. The values of the equilibrium phase properties of this alloy (elastic constants, misfit, interface energy) are taken from the multicomponent alloy, but the description of diffusion is limited to a single composition field.

13.2.2 Applications of elastic phase field model

Formation of the cuboidal microstructure

Phase field methods have been extensively used to analyze in detail the consequences of the elastic stresses generated by the coherent coexistence of the γ and γ' phases during an isothermal heat treatment (e.g., [1533,1535,145,777]). As shown in Fig. 13.1, the phase field model presented in the previous section takes into account all the relevant phenomena leading to the formation of a cuboidal microstructure. First, the model describes the existence of four types of γ' domains, called translation variants, which are represented in white, green, red, and blue. Then, the model reproduces the shape transition between small spherical precipitates and large cuboidal precipitates. Finally, we also observe the progressive alignment of precipitates along the cubic directions and the formation of a microstructure in which the γ phase is the matrix, even if the γ' phase fraction exceeds 50%.

One of the interests of modeling is to be able to vary independently all the material parameters in order to highlight their respective roles.

The first important quantity is the anisotropy of the elastic interactions, which originates from the anisotropy of the elastic tensor. Since the γ and γ' phases display cubic symmetry, this anisotropy is characterized by the Zener ratio $A = 2C_{44}/(C_{11} - C_{12})$. For both phases, the ratio is significantly greater than 1 (isotropic case) and increases with temperature [1140,1141,419]. The importance of elastic anisotropy is illustrated in Fig. 13.2(a). In the absence of elastic anisotropy ($A = 1$, Fig. 13.2(a)) the microstructure is isotropic. When an elastic anisotropy is introduced ($A > 1$, Fig. 13.2(b)), we observe the formation of a microstructure containing cuboidal precipitates, which align along the cubic directions. Note, however, that in Fig. 13.2(b) the microstructure displays rather elongated γ' domains, which are not observed in nickel-base superalloys.

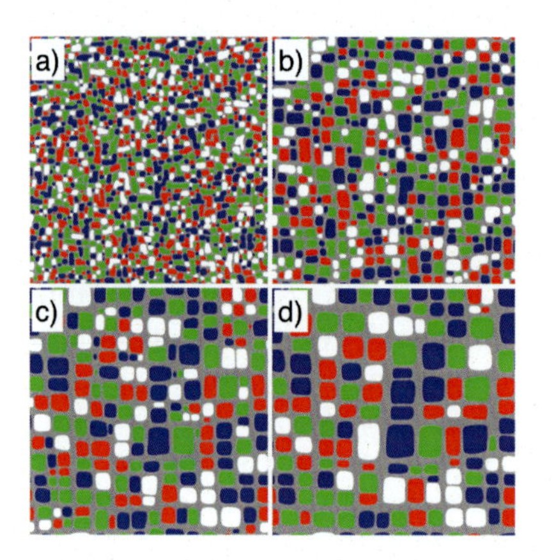

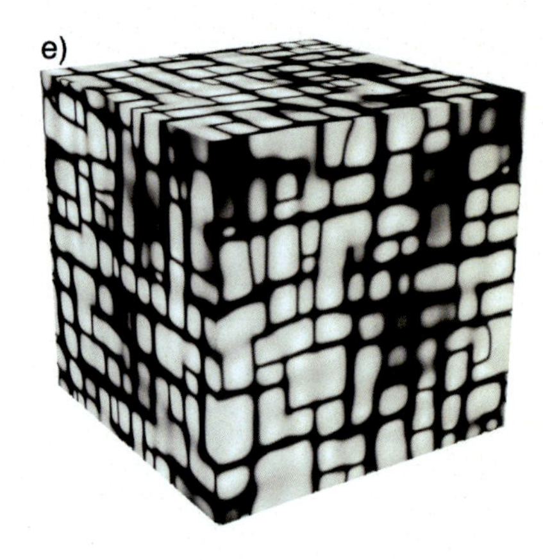

FIGURE 13.1 Phase field simulation of the microstructure evolution in a model nickel-base superalloy in stress-free condition at $T = 950°C$. (a)–(c) 2D simulation in a $(9.2 \times 9.2)\ \mu m^2$ box. The (a)–(d) images correspond to the times $t = 0.06$ h, 0.6 h, 4.5 h and 12.3 h, respectively; (e) 3D microstructure predicted at $t = 9$ h in a $(3.4 \times 3.4 \times 3.4)\ \mu m^3$ box [289,291].

FIGURE 13.2 Microstructure of a coherent mixture of γ and γ' phases: (a) with isotropic elasticity; (b) with anisotropic elasticity ($A = 2$), γ/γ' microstructure obtained in a $(9.2 \times 9.2)\ \mu m^2$ box after 4.8 h for two values of $\Delta C'$ inhomogeneity: 0% (c) and 50% (d).

In addition to the importance of elastic anisotropy, phase field simulations have also demonstrated the sensitivity of the microstructure to the difference in elastic constants between the γ and γ' phases. Although this difference is generally small, it has been shown [289,291] that the coalescence processes between precipitates are sensitive to the difference in elastic constants $C' = (C_{11} - C_{12})/2$ between the γ and γ' phases, called $\Delta C'$. This point is illustrated in Fig. 13.2. With no inhomogeneity, isothermal annealing leads to the formation of a cuboidal microstructure in which some precipitates are significantly elongated in a cubic direction (Fig. 13.2(c)). If one increases the value of the inhomogeneity on C', one observes the formation of a cuboidal microstructure in which the precipitates adopt an almost equiaxed shape (Fig. 13.2(d)). These results show that particular attention must be paid to the determination of the elastic constants of each phase and that the knowledge of the average elastic behavior is not sufficient to correctly predict the microstructure evolution.

Coarsening of the cuboidal microstructure

Numerous experimental and simulation studies have focused on growth kinetics of cuboidal microstructures in superalloys [1535]. This situation differs very significantly from the reference one addressed in the Lifshitz–Slyozov–Wagner (LSW) theory [1510], which concerns very small volume fractions of precipitates and which does not include elasticity. In this theory, the average precipitate size follows a power law with an exponent equal to 1/3. Several theoretical works have proposed extensions of the LSW theory to finite volume fractions, showing that the time invariant precipitate size distribution is slightly broader and more symmetric than the LSW distribution [1510], and to multicomponent alloys [1099]. But, more importantly, it is expected that the long-range anisotropic elastic interactions between precipitates will change the entire late stage phase transformation process. Indeed, strong spatial correlation between precipitates and size-dependent precipitate shapes are induced, so that different coarsening behavior is expected in the regimes where the capillary effects or elastic energy dominates [821,1428].

However, simulations of microstructure evolution representative of nickel-base superalloys show that, in the elastically homogeneous approximation, the average equivalent radius grows according to a power law with an exponent close to 1/3 [464,1458,1428,1535,1672,1252]. This conclusion is consistent with many experimental measurements [34,235,418,33]. Note, however, that in the late stage, in which the dominant driving force is the relaxation of the strain energy, theoretical predictions indicate that the width and length of precipitates follow power laws with different exponents [440]. The exponent is equal to 1/2 for the width and to 1/4 for the length. Therefore, the equivalent radius should follow a growth law with an exponent equal to 3/8 in 2D and 5/12 in 3D. These values are close to 1/3, and it is difficult numerically (and even more experimentally) to measure the growth exponent with sufficient precision to differentiate these values [440,1428].

Finally, note that phase field simulations have also demonstrated that elastic inhomogeneity significantly modifies the coarsening behavior [1036,1253,1252,1671,291]. Indeed, in late stage, softer regions are anisotropically deformed and harder regions tend to be isotropic. The elastic inhomogeneity also significantly change the growth law. In particular, when increasing the shear modulus inhomogeneity, the exponent of the growth law decreases leading to a significant slowdown of the coarsening [1036,1253,1671].

Modeling anisothermal heat treatments

Assuming that temperature is always homogeneous at the scale of the representative volume element, phase field models can be easily extended to study microstructural evolutions resulting from anisothermal heat treatments. It is then sufficient to calibrate the model using mate-

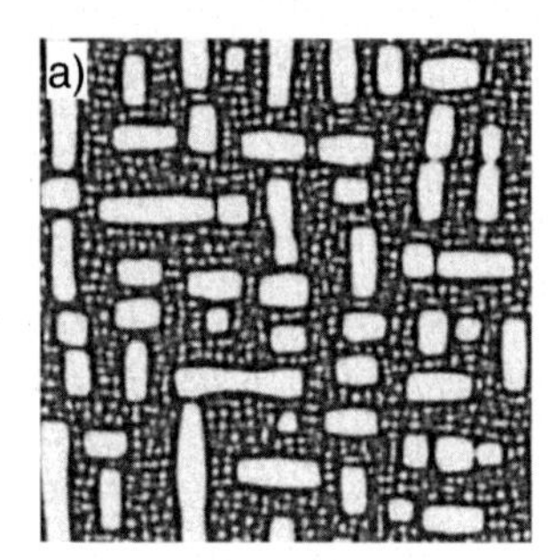

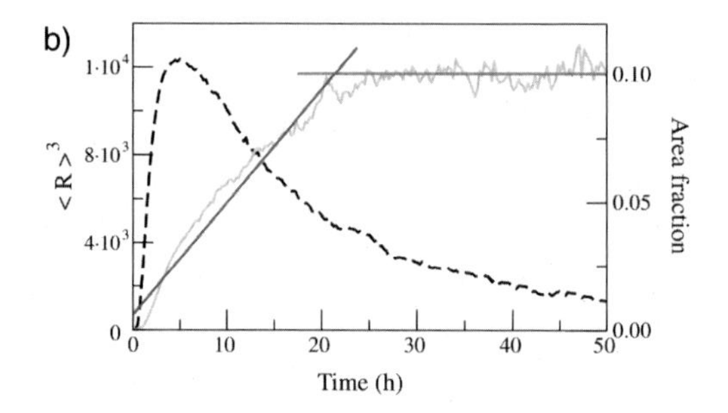

FIGURE 13.3 (a) Bimodal microstructure obtained after a two-step quenching at $T_1 = 1273\,\text{K}$ and $T_2 = 975\,\text{K}$; (b) Mean volume $\langle R(t)\rangle^3$ (red plain line - light grey in print version) and area fraction of the tertiary population (black dashed line) during aging at T_2 [144].

rial parameters whose values depend on temperature. While these data are generally available for misfit [1238,353] and diffusion coefficients, the temperature dependence of the elastic constants of the γ and γ' phases is scarce [605,419,1324]. The γ/γ' interface energy remains the most difficult quantity to estimate, especially if its temperature dependence is required.

Nevertheless, phase field modeling was applied to the study of the formation and evolution of bimodal structures, obtained during a multistep annealing at several temperatures [144]. The model consisted in adding a noise term to the conserved kinetic equation that controls the concentration field. This approach is further commented below. As an example, Fig. 13.3(a) displays a bimodal microstructure, obtained after a two-step heat treatment at $T_1 = 1273\,\text{K}$ and $T_2 = 975\,\text{K}$ in a binary superalloy. The obtained microstructure is composed of large, so-called secondary, cuboidal precipitates formed at T_1 and small, so-called tertiary, precipitates formed at the beginning of annealing at $T_2 = 975\,\text{K}$.

The phase field model was then applied to analyze the unusual coarsening of tertiary precipitates, for which experiments on N18 superalloy revealed that the $t^{1/3}$ growth law does not longer hold after a long aging time [446]. Indeed, the mean size of the tertiary precipitates reaches a plateau and the observed area fraction of the precipitates decreases continuously. As shown in Fig. 13.3(b), the phase field model was able to capture the important characteristics of the coarsening of tertiary precipitates, and has helped clarify the role of diffusion and elastic interactions in this process [144].

Modeling nucleation

As illustrated above, during isothermal or anisothermal heat treatments, new precipitates may appear. Classical (deterministic) phase field models are able to predict the formation of new precipitates only if it results from an instability (spinodal decomposition). The nucleation of precipitates, which requires overcoming energy barriers by thermal fluctuations, cannot be reproduced by a phase field model based on deterministic equations such as the Cahn–Hillard (Eq. (13.5)) and Allen–Cahn (Eq. (13.6)) equations. As shown above, nucleation due to thermal fluctuations may be reproduced by adding a noise term to the equations. The correlation properties of this noise can be chosen to follow the fluctuation–dissipation theorem, which ensures that the thermodynamic equilibrium is characterized by the free energy functional F used to calculate the driving forces. We thus obtain a stochastic phase field model (Langevin equation) which can in principle describe nucleation. The derivation by a coarse graining procedure of a phase field model from a simple atomic kinetic model [159] has shown that the nucleation rate can indeed be correctly reproduced by a mesoscopic model. However, it is also shown in [159] that a consistent derivation of the coarse-grained stochastic dynamics leads to a model in which the amplitude of the noise, the prefactor of the gradient term and the mobility coefficient all depend on the length scale of the coarse-graining and are strongly concentration-dependent. These dependencies, which are generally not taken into account in models, and which are difficult to calibrate without the help of a coarse-graining procedure, seem nevertheless essential to the derivation of a predictive nucleation model.

A very different approach has been developed to allow homogeneous nucleation in a phase field model [1328,1308]. The idea is to periodically add supercritical nuclei in the simulation with a probability given by the classical nucleation theory that depends on the local supersaturation. In order to maintain constant the average concentration during the formation of a nucleus, a depletion layer is introduced around the nucleus. The parameters of the nucleation model are then adjusted to reproduce the nucleation rates observed in some experiments. In the case of superalloys, the order parameter of each nucleus is randomly chosen so as to maintain substantially equal amounts of the four types of antiphase domains of precipitates, which has an effect on the coalescence of particles during their growth. This approach has been applied to superalloys, in particular to study the effect of the cooling rate on the microstructural evolution [1551,1265]. It is demonstrated that bimodal particle size distributions can be achieved at an intermediate cooling rate due to a coupling between diffusion and undercooling. This is caused by soft impingement, followed by a renewal of driving force for nucleation, followed by a subsequent soft impingement.

13.3 Microstructure evolution under stress: extension to continuous plasticity

When exposed to severe conditions, such as high temperatures, loadings, or both in the worst cases, microstructures in Ni-base superalloys are likely to evolve driven by the diffusion of alloying species and by plasticity, most often in a coupled manner. During service, directional coarsening (rafting) of γ' precipitates occurs under creep loading, where plate- or rod-like precipitate shapes develop together with plastic deformation mainly localized within the γ channels. This microstructure evolution is followed by a topological inversion detrimental to the creep properties. Therefore, understanding the strong couplings between diffusion-controlled phase transformations and plasticity in Ni-base superalloys is essential to improve their stability. In the last 20 years, as it became clear that the phase field method is very well suited for predicting morphological evolutions accurately, it appeared natural to extend it to plasticity.

13.3.1 Phase field models coupled with continuum plasticity

As plasticity in crystals is mainly due to the movement of dislocations, several works have aimed at describing explicitly plasticity at the scale of individual dislocations within the phase field approach [1222,1534], exploiting the analogy between a dislocation loop and a thin precipitate. This approach is well adapted to situations where the average distance between individual dislocations is larger than any other internal length scale, such as the distance between interfaces of precipitates. This modeling approach will be described in more details in Section 13.4. We will just mention here that although these models incorporate automatically important physical features, they lack high temperature mechanisms, e.g., climb and cross-slip essential to investigate the influence of plasticity on phase transformations in Ni-base superalloys.

To circumvent these drawbacks, plasticity in evolving microstructures has been addressed by relying on phenomenological models. Works along this route have been proposed by several groups using mesoscale plasticity models differing by their descriptions of hardening, viscosity, and plastic anisotropy. The first attempts to couple a diffuse interface model with an isotropic plasticity model have been proposed in [564] to study stress fields around defects such as holes and cracks, or in [1452] to investigate tin–lead solder joints undergoing thermal cycling. In the context of rafting in Ni-base superalloys, a few works have extended the phase field model to include isotropic plasticity [501] and anisotropic plasticity, either approximately [1668] or relying on a more complete and well-tested crystal plasticity framework [503,502]. It is worth mentioning that in [1668] the yield stress or any hardening effects are not included.

13.3.1.1 *Crystal plasticity*

As described in [503,502], crystal plasticity can be implemented in the phase field model through plastic strain fields defined at mesoscale, supplied by internal variables such as hardening variables. The coupling is then introduced through the decomposition of the total strain field,

$$\underset{\sim}{\varepsilon}(\underline{\mathbf{r}}) = \underset{\sim}{\varepsilon}^{el}(\underline{\mathbf{r}}) + \underset{\sim}{\varepsilon}^{0}(\underline{\mathbf{r}}) + \underset{\sim}{\varepsilon}^{p}(\underline{\mathbf{r}}), \tag{13.7}$$

where $\underset{\sim}{\varepsilon}^{p}(\underline{\mathbf{r}})$ is the plastic strain field. In a crystal plasticity framework, plastic deformation occurs due to slip on crystallographic systems. In the small strain approximation, the plastic deformation tensor reads as the sum of the shear strain γ_s over all slip systems s,

$$\underset{\sim}{\varepsilon}^{p}(\underline{\mathbf{r}}) = \sum_{s} \gamma_s \frac{\mathbf{n}^s \otimes \mathbf{l}^s + \mathbf{l}^s \otimes \mathbf{n}^s}{2}, \tag{13.8}$$

where $\mathbf{n}^s$ denotes the normal of the slip plane and $\mathbf{l}^s$ is the slip direction.

Then, as usual in continuum mechanics, evolution equations in the form of ordinary differential equations are postulated to describe plastic flow and hardening such as those presented in [937,936] (see Chapter 14). This approach has the advantage of phenomenologically including the consequences of all the physical processes at the origin of plasticity.

The viscoplastic parameters of the model are then identified from experimental data. In heterogeneous materials, they are space dependent: in the high temperature and low stress conditions usually considered for Ni-base superalloys, only γ undergoes plastic strain while γ' behaves elastically. To reproduce this behavior, the governing equations of the plasticity model are only activated inside γ.

Despite significant successes achieved by these models, they miss some important features of the plastic behavior in heterogeneous materials such as the plastic size effect. As a region decreases in size, it becomes harder to deform plastically. This effect may be rather important in evolving heterogeneous materials at the mesoscale because the sizes and shapes of the plastic regions may differ and, in addition, evolve during thermomechanical treatments. This size effect can only emerge from a viscoplastic model in which an intrinsic length is included and therefore, the viscoplastic model has to be chosen within the framework of the mechanics of generalized continua as it has been done in [292]. In addition, calibrating the parameters of the phenomenological laws used in these models requires experiments that are unfortunately more complex than usual when considering evolving multiphase materials. Hence, it is highly desirable to resort to plasticity models with firmer physical grounds, i.e., relying on dislocation densities.

13.3.1.2 Dislocation density based crystal plasticity

To go beyond, such dislocation density based plasticity models have been employed in [290]. In this approach, the time derivative $\dot{\gamma}_s$ of the plastic shear strain (see Eq. (13.8)) follows a phenomenological Norton flow rule, in which the critical resolved shear stress τ_f and the backstress τ_b are related to the plastic state, i.e., to the dislocations densities. The evolution of the dislocation densities on each glide system is given by a standard storage-recovery balance equation [927,734] such as

$$\dot{\rho}^{\alpha} = \frac{1}{b}\left(\frac{1}{L^{\alpha}} - 2\, y_c\, \rho^{\alpha}\right) |\dot{\gamma}^{\alpha}|, \tag{13.9}$$

where L^{α} is the dislocation mean free path on slip system α and y_c is a characteristic length associated with the annihilation process.

This model also accounts for previously missing features of the plastic behavior in heterogeneous materials such as the forest hardening resulting from the short-range interactions between dislocations on different glide systems. This phenomenon is included through a hardening matrix in the Taylor relation which describes the critical stress, and whose components are identified on dislocation dynamics calculations taken from the literature. Moreover, the size effect of the plastic behavior is also included through the backstress generated by gradients of geometrically necessary dislocation densities. The model can be used to analyze the microstructure evolution during creep loadings along different axis such as [100] or more complex ones such as [110] or [111] to look into the origin of the sensitivity of the microstructure evolution to the orientation of the loading axis.

13.3.2 Rafting during creep experiments along [100]

Using this coupled model, the microstructure evolution of Ni-base superalloys under service conditions has been investigated. The well-known cuboidal microstructure evolves under a constant uniaxial stress σ^a along the [100] direction while holding the temperature at 950°C (Fig. 13.4). In these conditions, plasticity is only active inside the γ channels, whereas the γ' phase behaves elastically. The precipitates self-organize into rafts that are aligned along the direction perpendicular to the tensile axis, in agreement with previous studies of [100] creep, showing that the elastic and plastic driving forces lead to the same raft orientation [145,1667,292]. The comparison of elastic (Fig. 13.4(a)) and plastic (Fig. 13.4(b)) simulations reveals that viscoplasticity has consequences at different scales. First, kinetics of rafting is strongly modified. Although the rafts appear around the same time with similar morphological features in both simulations, the microstructure continues to evolve towards straight rafts in the elastic case, whereas the wavy rafts are almost frozen soon after being

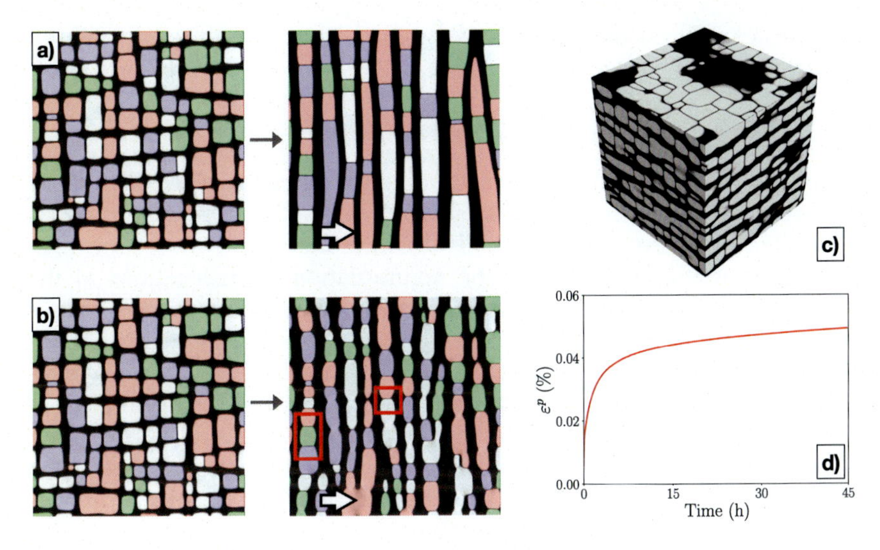

FIGURE 13.4 Model superalloy under 150 MPa at 950°C: 2D evolution after 45 h [290] in a $(2.3 \times 2.3)\,\mu m^2$ periodic box in (a) an elastic and (b) elasto-viscoplastic simulation with its corresponding creep curve shown in (d); (c) 3D $(3.4 \times 3.4 \times 3.4)\,\mu m^3$ snapshot of elasto-viscoplastic simulation after 26 minutes.

formed when viscoplasticity is activated in the γ phase. This last observation is in agreement with the slow evolution of the rafts observed in this alloy during the secondary creep regime [348]. It is also consistent with previous works based on simplified plasticity models [503,292]. Second, viscoplasticity changes the shape of the precipitates which become more curvy (Fig. 13.4(b)). This indicates that the anisotropic coherency stresses at the origin of the cuboidal shapes are partially relaxed. Moreover, it can be observed that plasticity may induce different behaviors depending on the local configuration, where coagulation events are observed to be either accelerated or qualitatively modified with the formation of large domains (white arrows).

The model also has the advantage of allowing a direct comparison with dislocations observed in experiments since it gives access to the distribution of the activated slip systems among the different channels. The creep curve (Fig. 13.4(d)) also allows a comparison between simulation and experiments on a macroscopic scale. The model can also be used to perform 3D simulations (Fig. 13.4(c)), even if the description of dislocation density based crystalline plasticity makes it numerically very expensive.

Beyond the important [100] creep load, phase field models have also been employed to investigate creep along the [110] direction [502,290]. The phase field simulations were in particular able to demonstrate that, for a [110] tensile loading condition, the microstructure evolution is very sensitive to a small misorientations of the loading axis, as it was proposed in [6].

13.3.3 Topological inversion and damage

Beyond the important rafting evolution during creep, phase field models have also been employed to investigate the topological inversion [1668,533,532] where γ' topologically becomes the matrix phase. The main ingredient of this evolution is the accumulation of dislocations at the γ/γ' interfaces during creep, which relax the misfit between the phases and which allow the coalescence of different types of γ' variants. Therefore, the strain energy becomes negligible and the microstructure evolves such that the phase with the higher volume fraction is the matrix. In [532], a phase field approach for the topological inversion is proposed, in which the reduction of the misfit effect and the loss of coherency in the γ/γ' interfaces are modeled by artificially lowering the eigenstrain, as well as the strength of the wetting condition.

With the topological inversion come an increase in the plastic strain rate associated with a massive shearing of the γ' phase and an increase in the volume fraction of voids. It is known as the tertiary creep. In [580,1609], first attempts to include microstructural damage into a phase field model are made using a macroscopic crystal plasticity framework to account for the softening effect of the γ' phase, neglecting porosity and fracture mechanisms.

13.4 Microstructure evolution under stress: plasticity with individual dislocations

Plasticity is an essential driving force for understanding the creep behavior of nickel-base superalloys. In the previous section, we illustrated how phase field methods can incorporate plasticity at a continuous scale. This approach was illustrated by simulations results based on dislocation density models. This approach allows accounting for many aspects of plasticity, especially the difference of plastic behavior of the γ and γ' phases, the high temperature mechanisms (climb, cross-slip), as well as plastic anisotropy. However, this approach is highly phenomenological in such that actual physical mechanisms do not enter explicitly the model but only through the parametrization of the parameters.

One other point is that the use of continuous approaches requires the possibility, at least conceptually, to define continuous dislocation densities. In the case of nickel superalloys under high-temperature creep, the plastic region is limited to the γ channels. In commercial superalloys, the γ phase fraction is high, and the channel width is of the order of 0.1 micron. The use of dislocation densities at such small scales is clearly questionable.

13.4.1 Phase field models with perfect dislocations

Thus, in order to improve the description of plasticity of superalloys, it is desirable to extend the phase field methods to the description of the main mechanism of plasticity, that is to say, the dynamics of dislocations. This extension was realized 20 years ago for the glide of arbitrary dislocations [1222,1534], and requires the introduction of new fields.

On a given glide plane, we define a continuous field $\eta(\mathbf{r}, t)$ which is equal to 1 in the region sheared by the dislocation and to 0 in the unsheared region. The dislocation line is then implicitly localized at the boundary between these two regions, and the dislocation core corresponds to the transition zone between the values 0 and 1 of the field. This approach is easily generalized to several dislocations belonging to the same glide system by allowing the dislocation field to reach any integer value: an integer value n of the field then corresponds to the shear by n successive dislocations. Finally, the complete model containing several glide systems is obtained by introducing as many dislocation fields $\eta_\alpha(\mathbf{r}, t)$ as of glide system.

The next step consist in the introduction of an additional additive term F_{dislo} in the free energy functional,

$$F_{dislo} = \int f_{cryst}(\{\eta_\alpha\}) + f_{grad}(\{\eta_\alpha\})\, dV, \tag{13.10}$$

where f_{cryst} is the crystalline energy density, which must display an infinite sequence of absolute minima for the integer values of the fields. The gradient energy f_{grad} is a gradient term that is chosen to penalize the variations of the dislocation fields *within their glide plane*. The competition of f_{cryst} and f_{grad} controls both the size and energy of the dislocation core. As it is the case for the interfaces in the usual phase field models, the size of the dislocation core is often chosen, for numerical efficiency, much larger than the experimental core size.

The elastic strain is still given by Eq. (13.7) where the plastic strain is the sum of the plastic slip generated by each glide system

$$\varepsilon^p_{ij}(\mathbf{r}, t) = \sum_s \eta_s(\mathbf{r}, t)\, \frac{\mathbf{b}^s \otimes \mathbf{n}^s + \mathbf{n}^s \otimes \mathbf{b}^s}{2 h_s}, \tag{13.11}$$

where, for a given glide system s, $\mathbf{b}^s$ is the Burgers vector, $\mathbf{n}^s$ denotes the normal to the glide plane, and h_s is the interlayer spacing.

Finally, the dislocation dynamics is given by the Allen–Cahn equation (see Eq. (13.6)). This choice corresponds to the assumption that the dislocation velocity is proportional to the resolved shear stress (RSS), and the mobility coefficient entering this equation can be related to an experimentally measured friction term [1242]. We insist that the model described

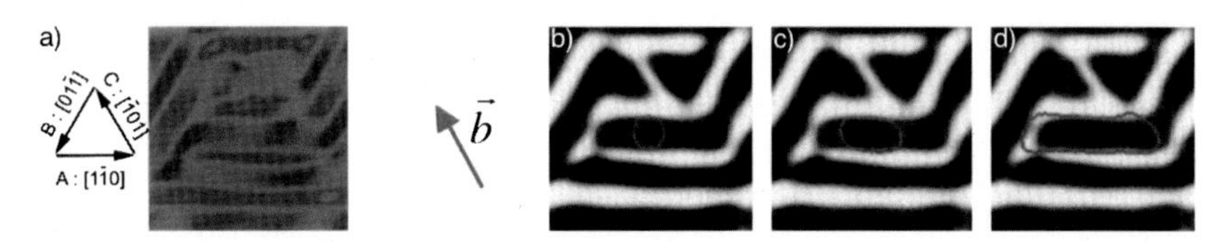

FIGURE 13.5 (a) Resolved shear stress (RSS) map in a γ/γ' microstructure. The RSS varies from $-6.0 \times 10^{-3}C_{44}$ to $-6.0 \times 10^{-3}C_{44}$ when the color varies from blue to red (from dark grey to mid grey in print version); green (light grey in print version) corresponds to zero RSS. (b)–(d) Expansion of the dislocation under an external RSS of $-2.45 \times 10^{-3}C_{44}$ [1223].

above, in which the total free energy is the sum of Eqs. (13.1), (13.3), and (13.10), incorporates the elastic interactions between dislocations, as well as the interactions and the coupling between evolving dislocations, diffusion fields and misfitting precipitates, as first shown in [1222]. It is therefore fully suited to study the glide of dislocations in Ni-base superalloys.

As a first example of application to model superalloys, the expansion of a dislocation loop in a γ channel is shown in Fig. 13.5. A cuboidal microstructure was first generated during an isothermal annealing in stress-free conditions. A slip dislocation loop is then introduced into the (111) plane in the γ phase. Fig. 13.5(a) shows that the RSS stress generated by the microstructure depends on the orientation of the γ channel. In particular, horizontal channels have a negative RSS that tends to extend the dislocation loop. Then, the behavior of the dislocation under an applied RSS of $-2.45 \times 10^{-3}C_{44}$ is shown in Fig. 13.5(b)–(d). This stress is above the activation stress of the loop but below the interface stress. The dislocation therefore expands in the channel, leaving dislocation segments along the interface between channel and inclusion.

A phase field model including dislocation glide was also used to analyze the microstructure evolution during rafting by explicitly introducing dislocations in the γ channels to relax the combination of the misfit and the applied stress [1666]. In this work, the elastic modulus mismatch is ignored, and a rafting similar to the experimental one is reported. This suggests that plasticity plays a dominant role in the rafting process as compared to elastic modulus mismatch.

The phase field model of dislocations has also been recently developed with an improved discretization scheme that explicitly captures the face-centered cubic in a way that allows to consider strongly heterogeneous materials and sharp interfaces (free surfaces, stiffer precipitates, pores, etc.) without generating numerical artifacts [1242]. As shown in Fig. 13.6(a), the model has been applied to the effect of plasticity on the pore closure in nickel-base superalloys during hot isostatic pressing (at $T = 1288°C$) of a model CMSX-4 superalloy. It is shown that special dislocation arrangements are formed at the vicinity of the pore and that the

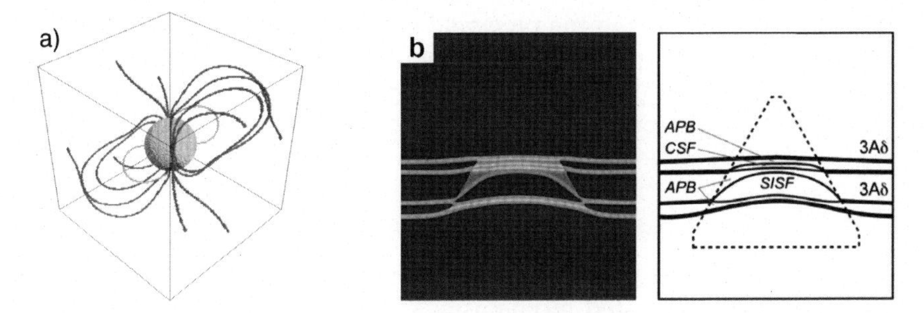

FIGURE 13.6 (a) Dislocation microstructure resulting from the glide of dislocations around a micropore in the γ phase under an isostatic pressure of 100 MPa [1242]; (b) γ' shearing of a pure edge $6A\delta$ dislocation ribbon for 1250 MPa applied stress [1514]. Superlattice intrinsic stacking faults (SISF), complex stacking faults (CSF), and antiphase boundaries (APB) are indicated.

very high elastic anisotropy ($A = 4.6$ at $T = 1288°C$ [390]) strongly modifies the movement of dislocations whose glide plane intersects the pore [1242].

13.4.2 Phase field models with partial dislocations

The phase field dislocation dynamics model described in the previous section is, in fact, close to the Peierls–Nabarro model [1073,997,679]. Both approaches use linear elasticity to describe the strain fields generated outside the dislocation core and a nonlinear part that opposes elastic distortions in the core. The nonlinear part, called crystalline energy, in Eq. (13.10) characterizes the potential energy generated by the shears associated with the dislocation field. This potential energy is therefore directly related to the generalized stacking fault energy (usually referred to as γ-surface), which is a quantity that can be evaluated by *ab initio* calculations [1309]. In FCC phases, the γ-surface exhibits absolute minima for vectors that are multiples of the $1/2\langle 110\rangle$ type Burgers vectors, but also local minima with the $1/6\langle 112\rangle$ vectors corresponding to the stacking faults. Thus, by using this type of γ-surface to construct the crystalline energy f_{cryst}, the phase field model automatically predicts the dissociation of dislocations into Shockley partials [1309]. Of course, such a description of the dislocation core implies the use of a very fine discretization grid, which strongly limits the size of the simulation box.

In addition, if the γ-surface of the γ' phase is also used, the model then reproduces the complex core structure of a dislocation in the γ' phase, under the assumption that the dislocation core remains in the glide plane. This approach was used to study the propagation of global $\langle 112\rangle$ dislocation ribbons in nickel-base superalloys, an important mechanism during primary creep in second generation superalloys in the 750–850°C temperature range [1514]. As shown in Fig. 13.6(b), the shearing of a γ' precipitate by such a ribbon leads in the precipitate to a succession of partial dislocations separated by planar defects: antiphase boundaries and different types of stacking faults. This model allowed evaluating the influence of the character of the dislocations and of the stress applied on the shearing of γ' precipitates [1514].

13.4.3 Perspectives: thermally activated mechanisms

Note that in the works mentioned above, dislocations move only by glide. Thermally activated mechanisms such as cross-slip and climb are not taken into account. However, the phase field method seems particularly suited to the study of dislocation climb because it naturally incorporates diffusion processes. Indeed, phase field models describing the climb of dislocations by vacancy absorption or emission have recently been proposed [512,707]. In particular, the model proposed in [512] incorporates the exact balance between the vacancy flux and the phase field associated with the dislocation evolution, enforced by the conserved character of the total population of vacancies. In addition, the kinetics of the dislocation can be accurately controlled, as shown in [513] where an upscaling procedure is proposed to establish a link between the atomic scale, which controls the mechanisms of absorption/emission of vacancies by the dislocation core, and the mesoscale mobility coefficient that controls the kinematics of the dislocation line within the phase field model.

However, to our knowledge, there is no phase field modeling of microstructure evolution that simultaneously includes glide and climb of individual dislocations. The major difficulty lies in the fact that the fields used for the description of glide are different from those used for the description of climb. New ideas are here needed to propose a fully consistent framework describing simultaneously diffusion-controlled microstructure evolution, dislocation glide, and climb. Such a model would help improve our understanding of the rafting and topological inversion occurring in nickel-base superalloys in creep conditions.

13.5 Perspectives

The present paper is a review of developments over the past 20 years on phase field methods for the evolution of solid state microstructures, focusing on applications to nickel-base superalloys.

The phase field method is now a mature method for modeling microstructure evolution resulting from a phase transformation when elastic effects are taken into account. It allows a deeper understanding of the consequences of the elasticity (anisotropy, inhomogeneity) on the formation and evolution of the cuboidal microstructure. To date, the main results have been obtained on model alloys, most often binary. However, the development of more and more accurate thermodynamic and kinetic databases for multicomponents [631,632] will allow in the near future to calibrate predictive models for commercial alloys. The multiphase field [1358,1356] or the KKS formulation [722] seem particularly adapted to perform a coupling with a multicomponent thermodynamic databases. In fact, such a coupling has already been proposed more than ten years ago [547,1357]. In the context of superalloys, such multicomponent phase field models have only been applied to the description of the solidification processes [1540,778] or to simple configurations at a solid state [729,777,992].

The modeling of the behavior of superalloy under mechanical stress has benefited from the extension of phase field models to plasticity.

On a continuous scale, the use of dislocation density based crystalline plasticity models made it possible to take into account the plastic anisotropy, the difference in visco-plastic behavior between the γ and γ' phases and the size effect of the plastic behavior. However, an important issue still requiring a lot of attention is the inheritance of the plastic deformation: when the interface moves during phase transformations, are plastic strains present in the decreasing phase inherited by the phase front of the increasing phase or not. Although the previous models present certain advantages, these conventional plasticity theories are not well suited to handle such phenomenon. To improve our understanding of the interaction between dislocations and obstacles, where size-dependent effects and, most importantly, transport become fundamental, the development of continuum dislocation density-based models that incorporates dislocation dynamics mechanisms through transport equations are required. A first attempt has been proposed in the context of superalloys [1582,1583], but the plasticity model needs to be improved, for example, by using the dislocation density based plasticity model recently derived using a coarse-graining procedure [1459,552]. A better understanding of the inheritance problem could be reached using such approaches.

At the level of individual dislocations, phase field methods have been extended to slip or climb. The development of a phase field model including simultaneous slip, cross-slip, and climb is, however, still missing. This unified model would make it possible to study more precisely the microstructural evolutions under creep, from rafting to topological inversion. Note also that additional mechanisms such as pipe diffusion along the dislocation core could be easily added to such a model.

Finally, we mention that another interesting perspective is to use phase field models to calibrate large-scale models to predict the evolution of microstructure and mechanical behavior at the scale of the turbine blade.

Acknowledgments

The authors thank Prof. Dr. Ingo Steinbach for valuable discussions and a critical reading of the manuscript.

CHAPTER

14

Crystal plasticity models: dislocation based

Bernard Fedelich

Bundesanstalt für Materialforschung und -prüfung (BAM), Division 5.2: Experimental and Model Based Mechanical Behaviour of Materials, Berlin, Germany

14.1 Introduction

The large number of TEM investigations and the regular microstructure of single-crystal nickel-base superalloys has boosted the development of a number of physically motivated constitutive laws. In contrast to the more phenomenological models discussed in the next chapter, these models use dislocation densities as internal variables. Obvious advantages are that the computed densities can be compared to TEM observations and the deformation mechanisms can be easier translated into mathematical equations. In general, the dislocation densities enter the constitutive equations at (at least) two points:

- The flow rate, which is usually proportional to the density of mobile dislocations. Dislocation multiplication resulting in an apparent softening.
- The work hardening, whereas the increase of the flow stress is usually taken to be proportional to the square-root of an appropriate measure of the dislocation density.

Single-crystal nickel-base superalloys have specific features that need to be accounted for: First, the local stresses are locally strongly fluctuating and differ from the external stress. Second, the activity of the deformation mechanisms, in particular the ways to overcome the precipitates, strongly depend on the temperature and even the stress level, as mentioned in Chapter 6. Note that this chapter does not discuss the modeling of directional coarsening (rafting), which is the focus of Chapter 13. Also the full field models, which apply crystal plasticity and solve the boundary

Nickel Base Single Crystals Across Length Scales
https://doi.org/10.1016/B978-0-12-819357-0.00023-8

value problem at the microstructure level (see, e.g., [259], [178], [1139]) are beyond the scope of this chapter since they cannot be readily applied for the structural analysis of components. The chapter starts with a summary of the most relevant results of dislocation mechanics. The main objective of this section is the identification of the constituents of dislocation-based crystal plasticity models. In the following sections, these constituents are systematically introduced and illustrated by a choice of model equations chosen from the literature.

14.2 From individual dislocations to continuum plasticity

14.2.1 Kinematics of dislocations

An in-depth introduction to dislocations in crystals and their properties is beyond the scope of this chapter. Such an introduction can be found, for example, in [639] or [618]. At the dislocation level, the displacement field $\mathbf{u}^d$ is piecewise smooth and experiences a jump at singular surfaces where shearing (slip) is localized. Let us denote by $\mathcal{S}$ such a singular surface with its normal vector $\mathbf{n}$ directed from the negative side $-$ of the surface to its positive side $+$. The dislocations are the lines that bound these singular surfaces. Since a dislocation cannot end inside the perfect crystal (but only at free surfaces or other defects), the basic object of the theory is a closed single dislocation loop $\mathcal{L}$. The present chapter is restricted to the case of small strains and rotations, i.e., $\|\nabla \mathbf{u}^d\| \ll 1$, which is also the framework in which the elastic theory of discrete dislocations (see, e.g., [618]) is usually formulated.

Let us denote by $\boldsymbol{\xi}$ the unit vector tangent to the dislocation line (positive circuit, see Fig. 14.1). The Burgers vector corresponds to the jump of the displacement across the singular surface $\mathcal{S}$. In accordance with the convention used in [618], it will be defined in the following as an integral along a closed circuit C around the dislocation loop $\mathcal{L}$ taken in the right-handed sense relative to $\boldsymbol{\xi}$, namely

$$\mathbf{b} = \oint_C \frac{\partial \mathbf{u}^d}{\partial l}\, dl = \mathbf{u}^{d-} - \mathbf{u}^{d+}. \tag{14.1}$$

Note that the opposite sign convention is also used in the literature (see, e.g., [1408]).

Introducing $\boldsymbol{\zeta} = \boldsymbol{\xi} \times \mathbf{n}$, the local normal vector to the dislocation line in the slip plane, the dislocation velocity $\mathbf{v}$ can be decomposed in a glide component v_s and a climb component v_c as

$$\mathbf{v} = (\mathbf{v} \cdot \boldsymbol{\zeta})\,\boldsymbol{\zeta} + (\mathbf{v} \cdot \mathbf{n})\,\mathbf{n} = v_s\, \boldsymbol{\zeta} + v_c\, \mathbf{n}. \tag{14.2}$$

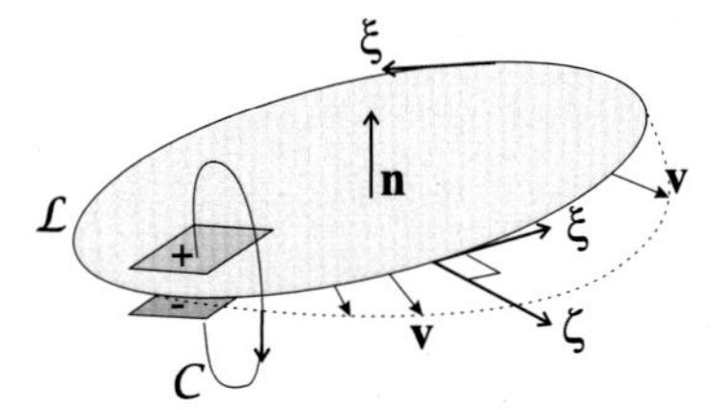

FIGURE 14.1 Expanding dislocation loop.

The transition between the level of discrete dislocations and continuum crystal plasticity requires averaging the fields over space and also over time due to the very fast time fluctuations of the dislocation velocities (coarse graining procedure, see, e.g., [4] or [471]). The resulting problem is complex due to the intricate interactions between dislocations and the derivation of a general solution is still an open issue. Rigorous solutions only exist for simplified situations (2D and straight dislocations for a single slip system, see [1459]). For the present context, we adopt a simple spatial averaging of the microscopic fields over a representative volume element $\mathcal{V}$ (RVE), which is regarded as being representative with respect to the dislocation structure (see Fig. 14.2). At the continuum level, the displacement gradient $\nabla \mathbf{u}$ of the smooth displacement field $\mathbf{u}$ can be written (see, e.g., [1004]) as

$$\nabla \mathbf{u} = \boldsymbol{\beta}^e + \boldsymbol{\beta}^p, \tag{14.3}$$

where

$$\boldsymbol{\beta}^e = \frac{1}{V} \iiint_{\mathcal{V}} \nabla \mathbf{u}^d dV, \qquad \boldsymbol{\beta}^p = -\frac{1}{V} \sum_{I=1}^{N} \iint_{\mathcal{S}^I} \mathbf{b}^I \otimes \mathbf{n}\, dS. \tag{14.4}$$

In the latter term, the sum is taken over all jump surfaces $\mathcal{S}^I$ of the microscopic displacement field $\mathbf{u}^d$. In the case of homogeneous elastic properties at the scale of $\mathcal{V}$, the macroscopic stress is

$$\boldsymbol{\sigma} = \frac{1}{V} \iiint_{\mathcal{V}} \boldsymbol{\sigma}^d dV = \frac{1}{V} \iiint_{\mathcal{V}} \mathbf{C} : \nabla \mathbf{u}^d dV = \frac{1}{V} \mathbf{C} : \iiint_{\mathcal{V}} \nabla \mathbf{u}^d dV, \tag{14.5}$$

so that $\boldsymbol{\beta}^e$ can be indeed identified with the continuum elastic distortion and $\boldsymbol{\beta}^p = -\frac{1}{V} \sum_{I=1}^{N} \iint_{\mathcal{S}^I} \mathbf{b}^I \otimes \mathbf{n} dS$ with the plastic distortion within the additive decomposition of the total displacement gradient. The average plastic distorsion rate in this RVE is obtained by differentiating $\boldsymbol{\beta}^p$ in

Eq. (14.4),

$$\dot{\boldsymbol{\beta}}^p = -\frac{1}{V}\sum_{I=1}^{N}\oint_{\mathcal{L}^I} \mathbf{b}^I \otimes \left(\mathbf{v}^I \times \boldsymbol{\xi}^I\right) dl, \tag{14.6}$$

which with Eq. (14.2) can be decomposed into glide and climb contributions as

$$\begin{aligned}\dot{\boldsymbol{\beta}}^p &= -\frac{1}{V}\sum_{I=1}^{N}\oint_{\mathcal{L}^I} \mathbf{b}^I \otimes \left[v_s^I\, \boldsymbol{\zeta}^I \times \boldsymbol{\xi}^I + v_c^I\, \mathbf{n}^I \times \boldsymbol{\xi}^I\right] dl \\ &= -\frac{1}{V}\sum_{I=1}^{N}\oint_{\mathcal{L}^I} \left[v_s^I \mathbf{b}^I \otimes \mathbf{n}^I - v_c^I \mathbf{b}^I \otimes \boldsymbol{\zeta}^I\right] dl,\end{aligned} \tag{14.7}$$

using the identity $(\mathbf{a} \times \mathbf{b}) \times \mathbf{c} = (\mathbf{a} \cdot \mathbf{c})\mathbf{b} - (\mathbf{b} \cdot \mathbf{c})\mathbf{a}$. In Eq. (14.7), the velocity components v_s^I and v_c^I are expected to be variable along the dislocation loop $\mathcal{L}^I$ and to depend on the local stress $\boldsymbol{\sigma}^d$ at the dislocation level.

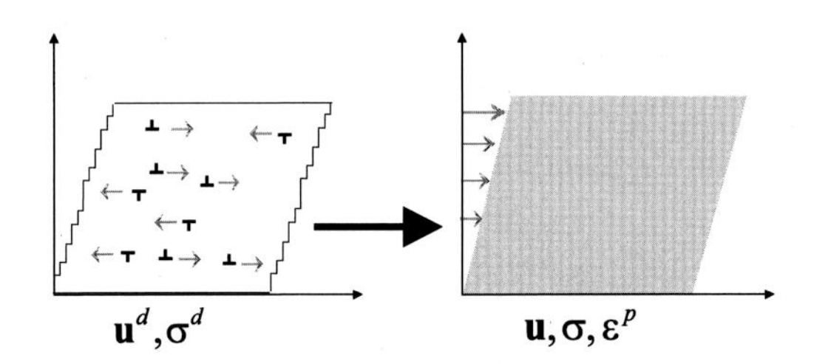

FIGURE 14.2 Averaging of discrete dislocations.

14.2.2 The force on a dislocation

The displacement of an elementary segment $d\mathbf{l} = dl\boldsymbol{\xi}$ of a dislocation by $d\mathbf{r} = dr\boldsymbol{\zeta}$ corresponds to a slipping of the surface element $d\mathbf{a} = d\mathbf{r} \times d\mathbf{l}$ by the Burgers vector $\mathbf{b}$ (see Fig. 14.1). This transformation requires the mechanical work of the local stress $\boldsymbol{\sigma}^d$,

$$dW^d = \mathbf{b} \cdot \boldsymbol{\sigma}^d d\mathbf{a} = \mathbf{b} \cdot \boldsymbol{\sigma}^d \cdot \left(d\mathbf{r} \times d\mathbf{l}\right) = d\mathbf{r} \cdot \left(\mathbf{b} \cdot \boldsymbol{\sigma}^d \times d\mathbf{l}\right). \tag{14.8}$$

From the definition of the mechanical work,

$$dW^d = d\mathbf{f}^d \cdot d\mathbf{r}, \tag{14.9}$$

the force per unit length acting on the dislocation (Peach and Koehler force) can be identified as

$$\mathbf{f}^d = \mathbf{b} \cdot \boldsymbol{\sigma}^d \times \boldsymbol{\xi}. \tag{14.10}$$

It is always directed normally to the tangential vector $\boldsymbol{\xi}$ of the dislocation line. In the case of a segment that is not pure screw, i.e., $\mathbf{b} \times \boldsymbol{\xi} \neq \mathbf{0}$, Peach and Koehler force has a unique decomposition into glide and climb components according to

$$f^{d,g} = \left(\mathbf{b} \cdot \boldsymbol{\sigma}^{d} \times \boldsymbol{\xi}\right) \cdot \boldsymbol{\zeta}, \qquad (14.11)$$
$$f^{d,c} = \left(\mathbf{b} \cdot \boldsymbol{\sigma}^{d} \times \boldsymbol{\xi}\right) \cdot \mathbf{n}.$$

By using the identity $(\mathbf{a} \times \mathbf{b}) \cdot (\mathbf{c} \times \mathbf{d}) = (\mathbf{a} \cdot \mathbf{c})\,(\mathbf{b} \cdot \mathbf{d}) - (\mathbf{b} \cdot \mathbf{c})\,(\mathbf{a} \cdot \mathbf{d})$, the glide component can be further simplified as

$$f^{d,g} = \mathbf{b} \cdot \boldsymbol{\sigma}^{d} \cdot \mathbf{n} = -b\left(\mathbf{m} \cdot \boldsymbol{\sigma}^{d} \cdot \mathbf{n}\right) = -b\tau^{d}, \qquad (14.12)$$

where the unit vector indicating the glide direction is introduced such that $\mathbf{b} = -\mathbf{m}b,\ b > 0$. The glide component of Peach and Koehler force is proportional to the Schmid stress $\tau^{d} = \mathbf{m} \cdot \boldsymbol{\sigma}^{d} \cdot \mathbf{n}$ for the slip system $(\mathbf{n}, \mathbf{m})$. In contrast, the climb component, which can be rewritten as $f^{d,c} = \mathbf{b} \cdot \boldsymbol{\sigma}^{d} \cdot \boldsymbol{\zeta}$, is variable along the dislocation line.

14.2.3 The basic structure of crystal plasticity models: case of pure slip

Due to the dependence of the climb contribution to the plastic distortion and of the climbing force on the vector $\boldsymbol{\zeta}^{I}$, which is variable along the dislocation I, the kinematics if climb is much more complex than that of pure glide. In addition, an inspection of Eq. (14.7) reveals that climb involves volume changes that must be accompanied by vacancies' diffusion. Having this in mind, it is not surprising that most constitutive models neglect the contribution of climb to the total deformation, even during creep at high temperatures. However, there is hardly any experimental evidence that climbing substantially contributes to the overall creep strain. Indeed, when dislocations overcome the γ' precipitates by combined gliding and climbing, it can be expected that the out-of-plane displacements of the dislocations are globally balanced, so that the climbing contributions in Eq. (14.7) vanish after spatial averaging. This is tacitly assumed, for example, in [1677]. Moreover, the investigations of crept specimens of the alloy CMSX-4 at 1288°C reported in [391] suggest that the climbing contribution is limited even at this temperature at which the γ' precipitates are dissolved. In all cases, climb is regarded as a recovery mechanism, which allows for creep continuation. In the following, climb is thus neglected as an active deformation mechanism. Glide takes place on one of the crystallographic slip planes with its normal vector $\mathbf{n}^{s}$ and with the Burgers vector $\mathbf{b}^{s} = -\mathbf{m}^{s} b^{s}, b^{s} > 0$. Both vectors define one of the possible N^{S} slip systems (see Section 14.2.4). The plastic distortion rate (Eq. (14.7)) can be rewritten as

$$\begin{aligned}\dot{\boldsymbol{\beta}}^p &= -\frac{1}{V}\sum_{I=1}^{N}\mathbf{b}^I\otimes\mathbf{n}^I\oint_{\mathcal{L}^I} v_s^I\, dl = -\frac{1}{V}\sum_{I=1}^{N}\mathbf{b}^I\otimes\mathbf{n}^I\,\dot{S}^I \\ &= -\frac{1}{V}\sum_{g=1}^{N^s}\mathbf{b}^s\otimes\mathbf{n}^s\sum_{I\in g}\dot{S}^I = \sum_{g=1}^{N^s}\mathbf{m}^s\otimes\mathbf{n}^s\dot{\gamma}^s,\end{aligned} \tag{14.13}$$

where

$$\dot{\gamma}^s = \frac{b^s}{V}\sum_{I\in g}\dot{S}^I \tag{14.14}$$

can be positive or negative. Note that the apparent simplicity of the result (14.14) is largely deceiving. Indeed, the local velocity of individual dislocation segments can be very different from the average velocity. Parts of the dislocations lines can be arrested at various obstacles (precipitates, forest dislocations, sessile junctions, etc.) while the other parts can move freely. Hence, several models perform a classification in terms of mobile and immobile dislocations. In addition, the dislocation gliding through the matrix and those shearing the precipitates must be distinguished in Ni-base superalloys with a large content of the γ' phase, since the precipitate phase is expected to contribute a significant amount of the total deformation.

Notwithstanding the previous considerations, Eq. (14.14) can be further specialized in the case of straight parallel dislocations. Consider n^s dislocations of length l in the volume $V = l \times A$ with density

$$\rho^s = \frac{n^s}{A}, \tag{14.15}$$

and moving with the average velocity v^s such that $\dot{S}^I = l v^s$. We finally obtain the Orowan equation, which is the starting point of most dislocation-based crystal plasticity models, namely

$$\dot{\gamma}^s = b^s\,\rho^s v^s. \tag{14.16}$$

The plastic strain rate corresponding to the distortion (14.13) is

$$\dot{\boldsymbol{\varepsilon}}^p = \frac{1}{2}\sum_{g=1}^{N^s}\left(\mathbf{m}^s\otimes\mathbf{n}^s + \mathbf{n}^s\otimes\mathbf{m}^s\right)\dot{\gamma}^s, \tag{14.17}$$

while the plastic power for the macroscopic stress $\boldsymbol{\sigma}$ (averaged on an RVE) is

$$\mathfrak{P} = \boldsymbol{\sigma} : \dot{\boldsymbol{\varepsilon}}^p = \sum_{g=1}^{N^s} \mathbf{m}^s \cdot \boldsymbol{\sigma} \cdot \mathbf{n}^s \, \dot{\gamma}^s = \sum_{g=1}^{N^s} \tau^s \, \dot{\gamma}^s, \qquad (14.18)$$

with the macroscopic Schmid stress for the slip system s being

$$\tau^s = \mathbf{m}^s \cdot \boldsymbol{\sigma} \cdot \mathbf{n}^s, \qquad (14.19)$$

which is the macroscopic counterpart of the local shear stress in Eq. (14.12).

Remark 1. So far, nothing has been said about the size of the RVE $\mathcal{V}$. Due to the regular microstructure of Ni-base superalloys, it is appealing to assume an idealized periodic arrangement of precipitates and to regard $\mathcal{V}$ as a periodic cell. This specification corresponds, in fact, to mean field type models, which deal with global quantities for specified regions (for example, the average dislocation density of stress in vertical or horizontal channel, see, e.g., [1388], [425], [1435]). In contrast, the full field models consider dislocation densities and plastic strains at the microstructure scale. This requires averaging over a region that should be much larger than the typical distance between dislocations and much smaller than the characteristic dimensions of the microstructure, for example, the channel width. This condition can hardly be satisfied in the γ channels of width < 100 nm. As a matter of fact, it was shown that conventional crystal plasticity applied at the microstructure level is not able to capture the size effects that appear at this scale. For example, a constant tensile strength instead of a decreasing one is predicted during coarsening [178]. To overcome this drawback, several types of gradient dependent models were proposed ([178], [1435]).

Remark 2. It will be shown that the macroscopic stress $\boldsymbol{\sigma}$ largely differs from $\boldsymbol{\sigma}^d$ at the dislocation level, which is needed in (14.12). A model is needed to estimate $\mathbf{X} = \boldsymbol{\sigma} - \boldsymbol{\sigma}^d$ (backstress).

To conclude this section, dislocation-based crystal plasticity models require the specification of the following components:

- A model for the dislocation structures and the evolution of the corresponding densities ρ^s,
- The definition of the velocity v^s, i.e., a function of the effective shear stress $\tau^{eff,g} = \tau^{d,g} - \tau^{c,g}$, where $\tau^{d,g}$ is the local Schmid stress and $\tau^{c,g}$ represents the resistance against dislocation glide,
- The choice of the deformation mechanisms, i.e., of the active slip systems $(\mathbf{n}^s, \mathbf{m}^s)$,
- An estimate of the local Schmid stress $\tau^{g,d} = \tau^s - x^s$ from the global Schmid stress τ^s, and
- An estimate of the resistance $\tau^{c,g}$ against dislocation glide.

14.2.4 Slip systems in Ni-base superalloys

In fcc crystal structures, the octahedral slip systems (see Fig. 14.3) usually carry most of the deformation. Accordingly, they are also taken into account in most (if not all) crystal plasticity models. Table 14.1 defines the 12 octahedral slip systems according to the denominations of Schmid. Slip traces on 001 planes were also reported on the surface of <111>-oriented specimens [104] after uniaxial straining. Even though the physical origin of these slip bands is somewhat controversial (see, e.g., the discussions in [1502,1478]), many models explicitly consider the six cubic slip systems (see Fig. 14.3). Table 14.2 defines the 6 cubic slip systems. A number of models focusing on creep at intermediate temperatures also consider slip on octahedral planes along <112> directions ([425], [879], [258]).

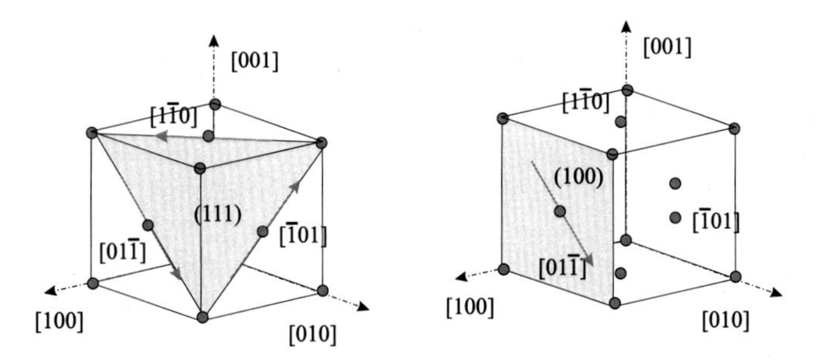

FIGURE 14.3 Octahedral and cubic slip systems.

TABLE 14.1 Definition of the octahedral slip systems.

Slip system	**B4**	**B2**	**B5**	**D4**	**D1**	**D6**
Slip plane	(111)	(111)	(111)	$(\bar{1}1\bar{1})$	$(\bar{1}1\bar{1})$	$(\bar{1}1\bar{1})$
Slip direction	$[\bar{1}01]$	$[0\bar{1}1]$	$[\bar{1}10]$	$[\bar{1}01]$	[011]	[110]
Slip system	**A2**	**A6**	**A3**	**C5**	**C3**	**C1**
Slip plane	$(1\bar{1}\bar{1})$	$(1\bar{1}\bar{1})$	$(1\bar{1}\bar{1})$	$(\bar{1}\bar{1}1)$	$(\bar{1}\bar{1}1)$	$(\bar{1}\bar{1}1)$
Slip direction	$[0\bar{1}1]$	[110]	[101]	$[\bar{1}10]$	[101]	[011]

TABLE 14.2 Definition of the cubic slip systems.

Slip system	α **2**	α **1**	β **4**	β**3**	γ **5**	γ **6**
Slip plane	(100)	(100)	(010)	(010)	(001)	(001)
Slip direction	$[0\bar{1}1]$	[011]	$[\bar{1}01]$	[101]	$[\bar{1}10]$	[110]

In the case of uniaxial tension in the direction of the unit vector $\mathbf{t}$, the stress tensor is $\boldsymbol{\sigma} = \sigma \mathbf{t} \otimes \mathbf{t}$ and the Schmid stress is $\tau^s = \sigma\,(\mathbf{t} \cdot \mathbf{m}^s)\,(\mathbf{t} \cdot \mathbf{n}^s)$. The product $m^s = \cos(\mathbf{t}, \mathbf{n}^s) \cos(\mathbf{t}, \mathbf{m}^s)$ is usually called the Schmid factor for the slip system $(\mathbf{n}^s, \mathbf{m}^s)$. From the knowledge of m^s, the slip systems that are active during uniaxial tension in this direction can be inferred. For example:

- Tension in direction [001]: B4, D4, A3, C3, D1, C1, B2, A2 are activated,
- Tension in direction [$\bar{1}$11]: B4, D4, B5, C5, D1, C1 are activated,
- Tension in direction [011]: B4, B5, A3, A6 are activated.

14.3 Estimation of local stresses

X-ray diffraction measurements, the heterogeneity of the dislocation structures, or structural calculations show that significant internal stresses exist at the scale of the precipitates [772], as well as at the scale of the dendritic structures [392]. In accordance, the local stresses and the macroscopic (external) stress are expected to differ significantly. These internal stresses are mainly due to:

- Differences of the strengths of the γ and the γ' phases,
- Differences between the lattice parameters between both phases in the isolated state (misfit stresses),
- Different superposition of the external and the misfit stress in the three channel types (microstructural effect), and
- Differences between the lattice parameters in the dendritic and the interdendritic zones.

14.3.1 Mean field models. Example of uniform strain approximation

Mean field models estimate average values of the stresses and strains in specific regions of the microstructure, for example, in the precipitates. The simplest estimate, which has been also applied by Dyson [375], assumes that the total strain is uniform over the RVE (Taylor-type interaction between matrix and precipitates). It has in each phase an elastic and a plastic contribution

$$\boldsymbol{\varepsilon} = \boldsymbol{\varepsilon}^e + \boldsymbol{\varepsilon}^p = \boldsymbol{\varepsilon}_\gamma = \boldsymbol{\varepsilon}^e_\gamma + \boldsymbol{\varepsilon}^p_\gamma = \boldsymbol{\varepsilon}_{\gamma'} = \boldsymbol{\varepsilon}^e_{\gamma'} + \boldsymbol{\varepsilon}^p_{\gamma'}. \tag{14.20}$$

Let us denote by f the volume fraction of γ' phase. The macroscopic stress is the volume average of the mean stresses in both phases,

$$\boldsymbol{\sigma} = (1-f)\langle\boldsymbol{\sigma}\rangle_\gamma + f\langle\boldsymbol{\sigma}\rangle_{\gamma'} = (1-f)\boldsymbol{\sigma}_\gamma + f\boldsymbol{\sigma}_{\gamma'}. \tag{14.21}$$

To simplify, a homogeneous elastic stiffness is assumed, which in view of the uncertainties concerning the difference of the elastic constants between both phases (see Chapter 4) is reasonable. Hooke's law reads

$$\boldsymbol{\sigma} = \mathbf{C} : \boldsymbol{\varepsilon}^e. \tag{14.22}$$

The global plastic strain $\boldsymbol{\varepsilon}^p$ is in the case of a homogeneous elasticity also the volume average of the mean plastic stresses in both phases (see Fig. 14.4), that is,

$$\boldsymbol{\varepsilon}^p = (1-f)\boldsymbol{\varepsilon}^p_\gamma + f\boldsymbol{\varepsilon}^p_{\gamma'}. \tag{14.23}$$

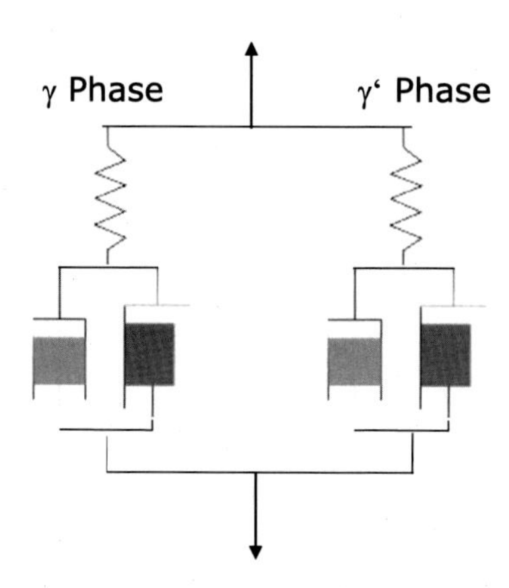

FIGURE 14.4 Rheological model of the γ–γ' microstructure.

From Eqs. (14.20)–(14.23) follow the estimates of the local stresses and the identification of the backstress $\mathbf{X}$ as

$$\begin{aligned} \boldsymbol{\sigma}_\gamma &= \boldsymbol{\sigma} + f\mathbf{C} : \left(\boldsymbol{\varepsilon}^p_{\gamma'} - \boldsymbol{\varepsilon}^p_\gamma\right) = \boldsymbol{\sigma} - \mathbf{X}_\gamma, & \mathbf{X}_\gamma &= f\mathbf{C} : \boldsymbol{\alpha}, \\ \boldsymbol{\sigma}_{\gamma'} &= \boldsymbol{\sigma} + (1-f)\mathbf{C} : (\boldsymbol{\varepsilon}^p_\gamma - \boldsymbol{\varepsilon}^p_{\gamma'}) = \boldsymbol{\sigma} - \mathbf{X}_{\gamma'}, & \mathbf{X}_{\gamma'} &= -(1-f)\mathbf{C} : \boldsymbol{\alpha}, \end{aligned} \tag{14.24}$$

where $\boldsymbol{\alpha} = \boldsymbol{\varepsilon}^p_\gamma - \boldsymbol{\varepsilon}^p_{\gamma'}$ is the difference of the plastic strains in both phases. The misfit δ_u of the lattice parameters between both phases can be taken into account as an additional eigenstrain of the precipitates $\boldsymbol{\varepsilon}^{misfit} = \delta_u \mathbf{1}$ with $\boldsymbol{\alpha} = \boldsymbol{\varepsilon}^p_\gamma - \boldsymbol{\varepsilon}^p_{\gamma'} - \boldsymbol{\varepsilon}^{misfit}$. A drawback of this procedure is that the result is independent of the shape of the precipitates. In addition, the assumption of uniform strains neglects strain accommodation around the particle and the predicted internal stresses are in general largely overestimated. A number of methods that relax the constraint of uniform strains have been proposed, which are listed in the following in the order of increasing computational expenditure:

- Use of a phenomenological equation for the backstress (see the next section);
- Use of a modified Sachs interaction between the precipitates and the matrix, ensuring the compatibility of the deformations at the γ–γ' interfaces [1435];

- Use of the Eshelby fundamental solution for ellipsoidal inclusions in connection with self-consistent approaches ([979], [1183]) or the theory of Mori and Tanaka [965];
- Use of Fourier series for a periodic microstructure [425], further improved in [1479];
- Use of finite element analysis (e.g., [986], [1024], [1139], [178]).

14.3.2 Use of Fourier series

The decomposition of the fields into Fourier series is well suited to account for the cuboidal precipitates and their quasiperiodic arrangement in contrast to the Eshelby solution, which leads to the approximation of spherical precipitates and is therefore better adapted to the case of alloys with a low content of the γ' phase. Within the Fourier series approach, the mesoscopic fields are decomposed into the sum of their average over the periodic cell and the fluctuations around this average. For example, in the case of the mesoscopic eigenstrains,

$$\overline{\boldsymbol{\varepsilon}}^{*}(\mathbf{x}) = \left\langle \overline{\boldsymbol{\varepsilon}}^{*} \right\rangle + \tilde{\overline{\boldsymbol{\varepsilon}}}^{*}(\mathbf{x}) . \tag{14.25}$$

In order to obtain a closed-form solution, a piecewise uniform distribution of the eigenstrains in the microstructure can be assumed. It is most natural to assume a decomposition of the initial microstructure into the precipitates and the three different types of channels, respectively parallel to the (100), (010), and (001) interfaces, i.e.,

$$\overline{\boldsymbol{\varepsilon}}^{p}(\mathbf{x}) = \sum_{K=1}^{4} I_K(\mathbf{x})\, \overline{\boldsymbol{\varepsilon}}^{p}_{K}, \tag{14.26}$$

where $I_K(\mathbf{x})$ is the indicator function of the region V_K. Note that the eigenstrains in the sense of Mura [988] are supplemented here by the plastic strains. After solving the equilibrium equations, the average stresses in each region can be shown to take the form [425] of

$$\overline{\boldsymbol{\sigma}}_K = \boldsymbol{\sigma} - \frac{1}{f_K} \sum_{K=1}^{4} \boldsymbol{\Omega}_{KL}\, \overline{\boldsymbol{\varepsilon}}^{*}_{L}, \tag{14.27}$$

in which the fourth-order influence tensors $\boldsymbol{\Omega}_{KL}$ depend on the geometry of the microstructure and on the elastic constants. As an example, the repartition of the stresses due to the misfit eigenstrain $\overline{\boldsymbol{\varepsilon}}^{*}_{4} = \delta_u \mathbf{1}$ is shown in Fig. 14.5. The predictions of the simple model with Taylor interaction is compared with the Fourier series in the case of an axially plastically-deformed matrix superposed and not superposed with a eigenstrain due to lattice misfit $\delta_u = -0.002$. It can be seen that the Taylor interaction largely overestimates the internal stresses due to the misfit eigenstrain. However, the order of magnitude of the internal stresses resulting from the plastic straining of the matrix alone are reasonably predicted by the simple model.

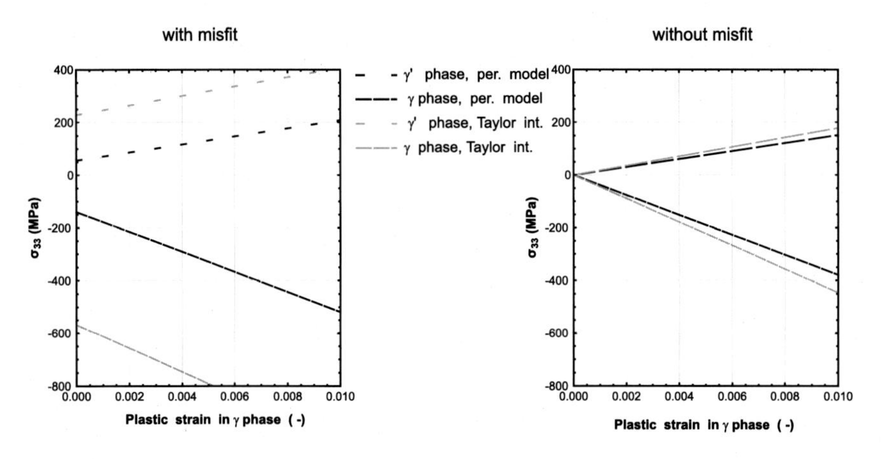

FIGURE 14.5 Average axial stress in the matrix and in the precipitate due to the superposition of an axial plastic strain and the misfit eigenstrain (left side).

An alternative to the direct calculation of the internal stresses is to use of a phenomenological law for the evolution of the backstress as a function of the shear strain. This approach will be described in Chapter 15 (Section 15.3.3).

14.4 Dislocation-based crystal plasticity models

14.4.1 Estimate of the dislocation density

The most essential ingredient of dislocation-based crystal plasticity models is a description of the dislocation densities and their evolution during straining. Most models for single crystal nickel-base superalloys actually resort to the ideas initially developed for pure metals by Kocks and Mecking ([927], [734]) and extend them to the particularities of these alloys.

The dislocation density is at any stage the result of a competition between nucleation and annihilation events. An increment of shear of any slip system $d\gamma$ during dt is associated with an increase of the dislocation density $d\rho^+$ such that $d\gamma = bL_{free}d\rho^+$ (compare with Eq. (14.16)), where L_{free} is the mean free path of the dislocations. Annihilation of mobile dislocations can occur by several mechanisms. For example, spontaneous annihilation can take place if two dislocations with opposite Burgers vectors pass each other with a minimal distance, lower than some critical distance $d_{annihil}$. The probability of this even is $dp = 2v2d_{annihil}dt\frac{\rho}{2}$, where $2d_{annihil}$ is the size of the reaction window, $2v$ the relative velocity of the two dislocations, and $\frac{\rho}{2}$ the density of dislocations with opposite sign. Finally, we obtain for the slip system g that

$$\dot{\rho}^s = \frac{\dot{\gamma}^s}{bL_{free}} - 2v^s d_{annihil}\rho^{g\,2} = \frac{1}{b}\left(\frac{1}{L_{free}} - 2d_{annihil}\rho^s\right)\dot{\gamma}^s. \tag{14.28}$$

If L_{free} is constant, the previous equation can be integrated with respect to the cumulated shear strain (to simplify we assume here monotone loading) by dividing Eq. (14.28) by $d\gamma$. We obtain an evolution of the dislocation density with a saturation

$$\rho^s = \rho^s_\infty + \left(\rho^s_0 - \rho^s_\infty\right)\exp\left(-\frac{2d_{annihil}\gamma^s}{b}\right), \tag{14.29}$$

where ρ^s_0 and $\rho^s_\infty = 1/(2d_{annihil}L_{free})$ is the dislocation density at saturation.

In γ–γ' superalloys, the distinction must be done between the dislocations moving through the channels and those cutting the precipitates. Assuming that $L_{free} \sim 1/\sqrt{\rho}$ scales with the square root of the dislocation density (mean distance between dislocations in a random arrangement), Tinga et al. [1435] and Choi et al. [258] assume an evolution law of the type

$$\dot{\rho}^s_m = \left(k_1\sqrt{\rho^s_m} - k_2\rho^s_m\right)\left|\dot{\gamma}^s\right| \tag{14.30}$$

for the slip system g. The mobile dislocations can be arrested not only by the other dislocations by also at the γ–γ' interfaces. Hence Ma et al. [879] take the free path to be $1/L_{free} = 1/L_\gamma + 1/L_{L_\rho}$, where $1/L_\gamma$ is the average channel width and L_ρ is proportional to the average dislocation spacing.

The previous models do not explicitly describe the dislocation creation mechanism. Rather they implicitly assume the existence of efficient dislocation sources to sustain the deformation. In [426] the following mechanism is described, which is inspired by previous analyses in [216] and [1126]. Starting from a few initially widely-spaced grown-in dislocations, dislocation half-loops start to glide through the narrow γ channels and eventually percolate the matrix. At each crossing, they penetrate new channels, giving rise to new leading half-loops in other channels (see Fig. 14.6). Since the local stresses in the three channel types in general differ, the dislocation densities can also differ. Denoting by L_{cross} the average distance v^s_K, the dislocation velocity for the slip system g in the channel type $K = 1, 2, 3$ and ρ^s_K the corresponding density such that $\dot{\gamma}^s_K = bv^s_K\rho^s_K$,

FIGURE 14.6 Dislocation multiplication in the matrix channels.

we have

$$\dot{\rho}_K^s = \frac{1}{b}\left(\frac{\dot{\gamma}_I^s}{L_{cross}} + \frac{\dot{\gamma}_J^s}{L_{cross}} - 2d_{annihil}\rho_K^s\dot{\gamma}_K^s\right), \quad I \neq J, I \neq K, J \neq K. \tag{14.31}$$

The dislocation production in channel K arises from the dislocation activity in the other channels I and J. In contrast to Eqs. (14.28) and (14.29), Eq. (14.31) predicts no one-to-one correspondence between the resolved shear strain for a slip system and the final dislocation density which is stress dependent. In the extreme case in which only one channel type is plastically yielded (e.g., horizontal channels under low stress creep) there is no dislocation production (incubation). Eventually, the stresses are redistributed in the other channel types and creep can proceed or other mechanisms can operate (climb around the precipitates, shearing of the precipitates). In the other extreme situation (large stresses), the three channels are equally loaded so that $\dot{\gamma}_I^s = \dot{\gamma}_J^s = \dot{\gamma}_K^s$ and Eq. (14.31) becomes practically equivalent to Eq. (14.28).

The gliding dislocations leave behind them segments (60° mixed segment or pure screw segments) that are pressed against the γ–γ' interfaces and immobilized if they cannot penetrate in the precipitates or climb around them. These dislocations can be described by their surface density λ_K^s that can be related to the jump of the plastic shear strains $[\gamma^s]$ at the interface (see, e.g., [988] or [471]) perpendicular to $\mathbf{n}_K$ by

$$\lambda_K^s = \frac{1}{b}\sin\left(\mathbf{n}_K, \mathbf{n}^s\right)\left[\gamma^s\right]. \tag{14.32}$$

The surface density λ_K^s is related to the average distance $d_K^s = 1/\lambda_K^s$ between the dislocation segments deposited in the interfaces (mesh width), which can be measured by TEM. Fig. 14.7 shows the simulated evolution of this distance during creep of a <001> SRR99 specimen. The mesh widths estimated are comparable in their magnitude with measurements in crept specimens [1142]. The dislocation density is higher in the interfaces perpendicular to the load axis than those parallel to it as observed in [434]. This is due to the different superposition of the external and misfit stresses.

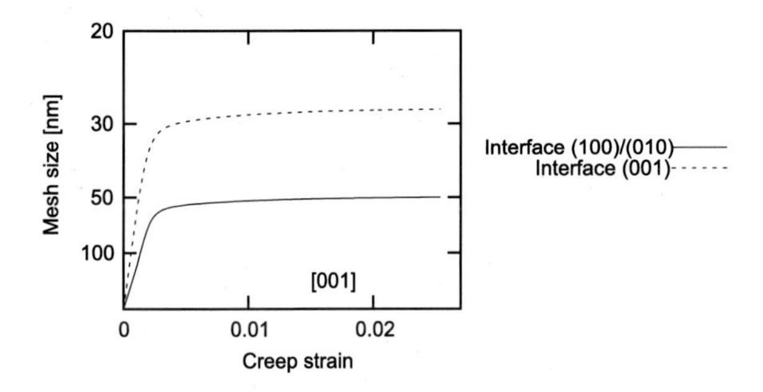

FIGURE 14.7 Evolution of the dislocation mesh width during creep with the alloy SRR-99 loaded under 600 MPa in the [001] direction at 850°C (from [426]).

14.4.2 Modeling the plastic flow in the channels

The resistance experienced by the dislocations against glide through the channels has several contributions:

- Nondirectional resistance τ_c due to solute atoms, interactions with other channel dislocations, etc.;
- Directional resistance $\tau_K^{\text{Orowan},s}$ due to the increase of the dislocation line energy (Orowan stress). It depends on the orientation of the segments deposited in the interface of the channel K behind the leading half-loop of the slip system s, as will be seen below;
- The internal stresses caused by the dislocation structures remaining after straining.

Regarding the nondirectional work hardening, it is customary to assume Taylor-type hardening ([879], [1435], [1354]), which in its basic form reads

$$\tau_c^s = c\mu b\sqrt{\rho^s}, \tag{14.33}$$

where τ_c^s is the increment of the flow stress due to dislocation hardening, c a numerical factor between 0 and 1, and ρ^s an appropriate dislocation density. More refined approaches ([150], [1139], [178]) consider an interaction between the slip systems via an interaction matrix c_{ij}, that is, $\tau_c^i = \mu b\sqrt{\sum_{j=1}^{18} c_{ij}\rho^j}$, thus extending the description of multislip hardening by Franciosi et al. [463] to channel deformation. It is worth mentioning that this description of work hardening was mostly validated by measurements made on pure metals and that its extension to the special geometry of dislocation moving through the narrow channels is questionable. Indeed, within Taylor hardening and its variants, dislocation multiplication necessarily implies cyclic hardening, i.e., an increase of the stress amplitude $\sigma_{max} - \sigma_{min}$ with the number of cycles in the case of cyclic loadings. However, most experimental results only show limited cyclic hardening (see, e.g., Fig. 14.8).

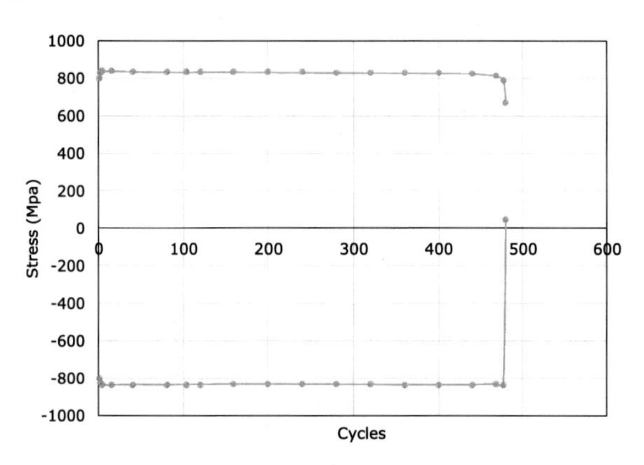

FIGURE 14.8 Cyclic hardening for the alloy CMSX-4 at 850°C, $\dot{\epsilon} = 10^{-3}/s$, $\Delta\epsilon = 0.01$.

14.4.2.1 *Orowan stress*

When a half-loop glides by δl in the matrix channel K of width w in a $< 111 >$ plane (see Fig. 14.9), the work of the local resolved shear stress $(\tau_K^s - \tau_c)\, bw\delta l$ must exceed the increase of the energy of the dislocation line $2T_K^s \delta l$, which leads to the Orowan threshold shear stress for dislocation glide in the channels,

$$\tau_K^s - \tau_c \geqslant \tau_K^{g\,Orowan} = \frac{2T_K^s}{bw}. \tag{14.34}$$

The line energy T_K^s can be written [618] as

$$T_K^s = \frac{K_K^s\, b^2}{4\pi} \ln\left(\frac{R}{r_0}\right) = \frac{1}{2}\alpha\, K_K^s\, b^2, \qquad \alpha = \frac{1}{2\pi} \ln\left(\frac{R}{r_0}\right), \tag{14.35}$$

where R is an outer cut-off radius of magnitude the mean distance between dislocation, and r_0 is an inner cut-off radius of the magnitude of b accounting for the core energy. Taking for R the channel width $R \approx w \approx$ 100 nm and $b = 0.25$ nm, one obtains $\alpha \approx 0.9$. The factor K_K^s depends on the orientation of the trailed segments relative to the Burgers vector and

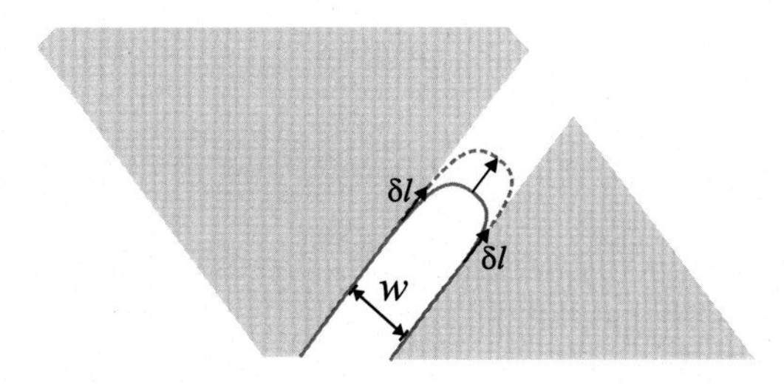

FIGURE 14.9 Increase of the dislocation line length by gliding of half-loops in the matrix channels.

the crystal lattice. For the two cases, we need here the factor which has been calculated by Foreman [453] as

$$K_K^s = K_s = \sqrt{\frac{1}{2}c_{44}(c_{11} - c_{12})}, \qquad \text{for a screw segment,} \tag{14.36}$$

$$K_K^s = \frac{1}{4}(K_x + 2K_y + K_s), \qquad \text{for a } 60^\circ \text{ mixed segment,}$$

where

$$K_x = (\bar{c}'_{11} + c'_{12})\sqrt{\frac{c'_{66}(\bar{c}'_{11} - c'_{12})}{c'_{22}(\bar{c}'_{11} + c'_{12} + 2c'_{66})}}, \quad K_y = \sqrt{\frac{c'_{22}}{c'_{11}}}K_x, \tag{14.37}$$

$$c'_{11} = \frac{1}{2}(c_{11} + c_{12} + 2c_{44}), \; c'_{22} = c_{11}, \; c'_{66} = c_{44},$$

$$c'_{12} = c_{12}, \; \bar{c}'_{11} = \sqrt{c'_{11}c'_{22}}.$$

The Orowan stresses for both types of dislocations are plotted in Fig. 14.10 and compared with the local stresses in the horizontal (001) channels and vertical (010) channels in the case of uniaxial tension. The local stresses are evaluated as in Section 14.3.2. Two conclusions can be drawn:

- Under uniaxial high temperature creep conditions (low stresses), the channels perpendicular to the load axis will be first deformed.
- The slip systems will be first activated in the channels in which screw segments are deposited.

14.4.2.2 *The flow rule*

Following Orowan's equation (Eq. (14.16)), the flow rule requires the knowledge of the dislocation density and the average dislocation velocity. Most models assume the formalism of thermal activation [764] to account for the dependence on the temperature. The dependence on the stress is

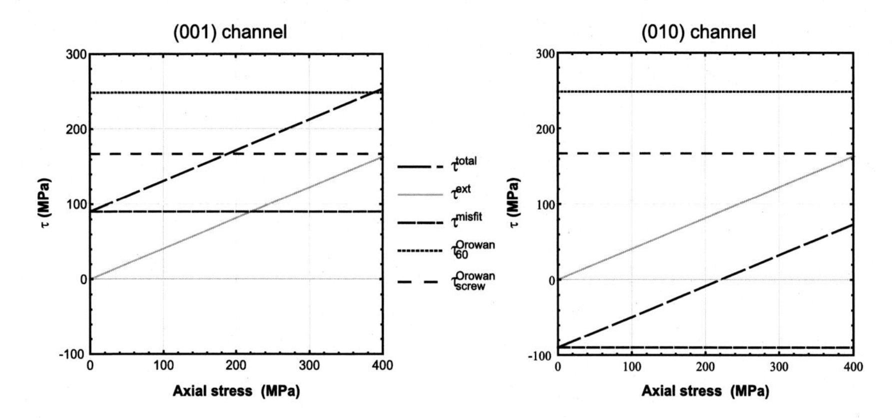

FIGURE 14.10 Local resolved shear stress and Orowan threshold stresses in a (001) channel and in a (010) channel under uniaxial tension along the [001] direction.

usually represented by a power law as in [258], namely

$$v^s = \nu_0' \exp\left(-\frac{Q_m}{kT}\right)\left|\frac{\tau^s}{\tau_c}\right| \operatorname{sign}\left(\tau^s\right), \tag{14.38}$$

while exp or sinh laws have been used, e.g., in [426]

$$v_K^s = v_0 \sinh\left[\frac{\left(\tau_K^s - \tau_K^{gOrowan}\right) V}{kT}\right], \tag{14.39}$$

where V is the activation volume and τ_K^s the local resolved shear stress in the channel type K.

14.4.2.3 The cubic slip phenomenon

Macroscopic evidences of cubic slip, in particular slip traces on cubic planes on the surface of <111>-oriented tension specimens, have been provided by several authors [951,937,104]. In addition, a low tension strength of <111> specimens departing from the Schmid law for the sole octahedral systems was observed and attributed to the activation of cubic slip systems. On the other hand, there is only little evidence of gliding of dislocations on cubic planes through the channels. Instead, a mechanism involving alternating cross-slip of screw dislocations between the γ' on their conjugated octahedral planes (α, β) was postulated to explain zigzag dislocation debris found after creep in <111> CMSX-4 specimens [1502] and tension loading in <111> SC-16 specimens [104]. The macroscopic observations can be explained by the alternating cross-slip as follows:

- The activation of two conjugate octahedral slip systems, with the same glide direction $\mathbf{m}^s$ and their respective slip planes $\mathbf{n}^{s\alpha}$ and $\mathbf{n}^{s\beta}$, and having the same Schmid stress, is after Eq. (14.13) macroscopically equivalent to glide on the plane sign $(\dot{\gamma}^{s\alpha})\,\mathbf{n}^{s\alpha} + \text{sign}\left(\dot{\gamma}^{s\beta}\right)\mathbf{n}^{s\beta}$. In the case of a <111> tension specimen, one can check that the signs of $\dot{\gamma}^{s\alpha}$ and $\dot{\gamma}^{s\beta}$ are such that the later happens to correspond to a cubic plane. Microscopically, this corresponds to the situation in which the screw segment has a positive velocity on its cross-slip plane (see Fig. 14.11).

- The Orowan stress of the dislocations leaving screw dislocations behind them is significantly lower than that of those leaving mixed segments.
- The screw segments that sequentially cross-slip between the phase boundaries to do generate immobilized dislocation segments: yield can proceed without hardening.

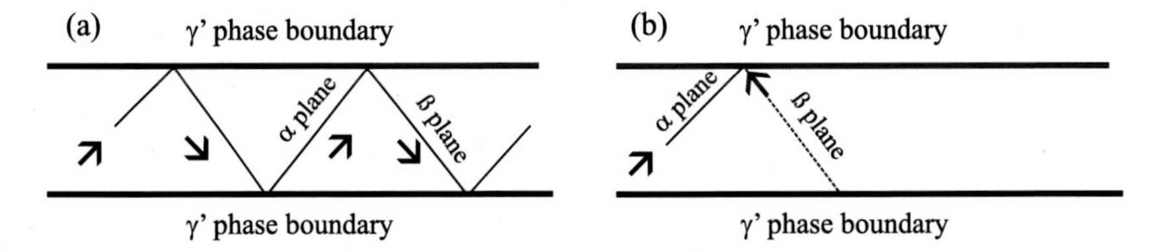

FIGURE 14.11 Alternating cross-slip between two γ–γ' interfaces: (a) case $v^{\alpha}v^{\beta} > 0$, zigzag mechanism is possible; (b) case $v^{\alpha}v^{\beta} < 0$, zigzag mechanism not possible (from [2]).

The deformation rate at the macroscopic level corresponding to this mechanism can be inferred from the general approach presented in Section 14.2.3. It is convenient to introduce the six cubic slip systems defined in Table 14.2. The effective slip rate on these additional systems depend on the frequencies of the cross-slip events $f^{\alpha\to\beta}$ and $f^{\beta\to\alpha}$, and on the glide velocities of the screw segments on the cross-slip planes v^{α} and v^{β}. Unfortunately, cross-slip invokes the local constriction of the partial dislocation segments, which occurs at the atomic scale, and its frequency is not easy to evaluate at the macroscopic scale. In turn, it is assumed in [426] that it only depends on the Schmid stress of the corresponding cubic slip system $\tau^{s,cubic}$, while Tinga et al. [1438] assume that it is function of the sum of the local resolved shear stresses of the two cross-slip systems scaled by the von Mises equivalent stress, i.e., $\left(\tau^{s\alpha} + \tau^{s\beta}\right)/\sigma_{eq}$.

An estimate of the number of dislocations able to contribute to the zigzag mechanism is needed to evaluate Eq. (14.14) or (14.16). It is suggested in [426] that this quantity is proportional to the density of the screw segments of the two cross-slip systems that are deposited in the γ–γ' interfaces, i.e., $\lambda^{s\alpha}$ and $\lambda^{s\beta}$. Finally, the shear rate on the fictive cubic system is obtained as

$$\dot{\gamma}^{s,cubic} = Ab\left(\lambda^{s\alpha} + \lambda^{s\beta}\right) f_0^{cubic} \sinh\left(\frac{\tau^{s,cubic}V^{cubic}}{kT}\right), \tag{14.40}$$

where A is a numerical factor depending on the microstructure geometry and f_0^{cubic} is an attack frequency. Since the cross-slipping segments are deposited by the dislocations gliding through the channels on the octahedral planes, cubic glide and octahedral are closely connected. Note that most models [937,150,178,428,715,1354] assume fully independent and in general similar equations for cubic and octahedral glide.

The DDD simulations performed by Vattré et al. in [1478] showed that repeated cross-slip in the channels is not even necessary to explain the low strength of <111> tension specimens. Instead, it was argued that the apparent departure from the expected Schmid behavior for octahedral glide can be explained by the lack of internal stresses build up by the dislocation segments in the γ–γ' interfaces if two octahedral cross-slip systems are equally activated, which is the case of <111> specimens. Indeed, the internal stresses resulting from the activation of the cross-slip systems compensate exactly in each of the three types of channel. This recognition motivated in [1479] a modification of the computation of the internal stresses by Fourier series as proposed in [425]. Instead of wedge-shaped channels, parallelepipedic channels are assumed in which piecewise uniform plastic strains are taken following (14.26). In accordance, no internal stresses are generated by the simultaneous activation of cross-slip systems as following from the DDD simulations. However, in order to obtain a sufficient flow rate, it is necessary to assume the existence of dislocation sources in all types of channels, in contrast to the dislocation multiplication mechanism expressed by Eq. (14.31), which does not need the existence of dislocation sources but requires that the dislocations are able to move in all channel types. With these additional assumptions, it was possible in [1479] to describe the low strength of <111> tensile specimens without resorting to cubic slip as shown by Fig. 14.12. It can be seen that the high hardening region (2) of the tensile curve, which can be observed for the <001> specimens is absent for the <111> specimens.

14.4.3 Shearing of the precipitates

The precipitates can be sheared by matrix dislocation pairs a<110> or by $a/3$<112> dislocations that leave superlattice intrinsic or extrinsic stacking faults (SISF or SESF) behind them, as described, e.g., in [1160]. Also shearing of the precipitates by a<010> dislocations under high temperature (> 1000°C) creep conditions has been reported by several authors. The relative dominance of each mechanism is strongly temperature and stress dependent. In accordance, the contributions of each mechanism to the overall deformation can hardly be identified unambiguously.

14.4.3.1 Shearing by matrix dislocation pairs

Due to the ordered crystal structure of the γ' phase, a matrix dislocation that enters a precipitate leaves an antiphase boundary (APB) asso-

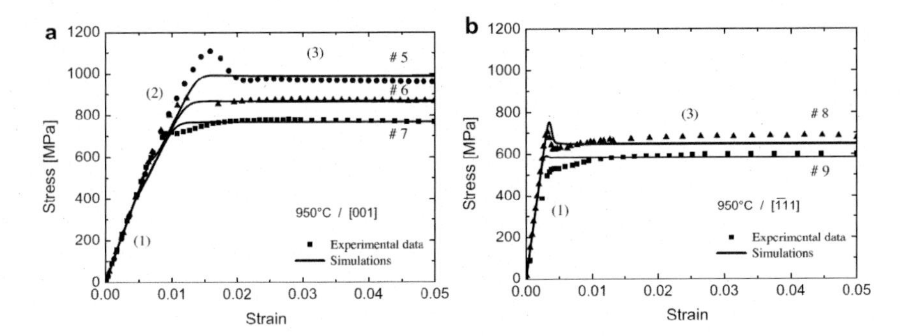

FIGURE 14.12 Simulation of tensile tests with the alloy CMSX-4 at 950°C: (a) <001> specimens with the strain rates 10^{-2}/s (# 5), 10^{-3}/s (# 6), 10^{-4} /s (# 7), (b) <111> specimens with the strain rates 10^{-3}/s (# 8) and 10^{-4}/s (# 9) (from [1479]).

ciated to the surface energy E_{APB}. The value of E_{APB} can be estimated from the distance between two paired dislocations in the γ' phase and is about 0.1 J/m^2. For an infinite straight dislocation to enter a straight and parallel γ–γ' phase boundary, the work of the local stress must exceed the amount of surface energy created, which leads to the threshold value $\tau_{APB} = E_{APB}/b$. With the above value for E_{APB} and $b = 0.25$ nm, one obtains a cutting stress $\tau_{APB} = 400$ MPa. However, the true threshold stress should be considerably lower due the limited dimensions of the precipitates and the effect of the line tension, which tends to reduce the overall dislocation line length. Fig. 14.13 shows possible cutting configurations associated to different amounts of antiphase boundary and dislocation line reduction, thus yielding different critical penetration stresses for the leading matrix dislocation. For the case of the straight dislocation traveling across several precipitates, the amount of APB remains globally almost constant, so that once a sufficiently long dislocation has been formed, it can be expected that it will able to cut many precipitates with little resistance resulting from the APB.

The case of spherical precipitates has received a lot of attention. For large particles, dislocation pairs are expected to travel in the same precipitate ("strong coupling"), separated by the APB. An estimate for the spherical case is presented, for example, by Reed (2006) as

$$\tau_c = \sqrt{\frac{3}{2}} \left(\frac{2\, T_{line}}{b\, r} \right) f_{\gamma'}^{1/2} \frac{\beta}{\pi^{3/2}} \sqrt{\frac{\pi\, E_{APB}\, r}{\beta\, T_{line}} - 1}, \tag{14.41}$$

where r is the particle radius, T_{line} the line tension, and β a numerical factor close to 1. A comparison between the Orowan threshold stress (14.34) and Eq. (14.41) as a function of the particle radius is presented in Fig. 14.14 for $f = 0.7$. The effect of the misfit stress is not taken into account here.

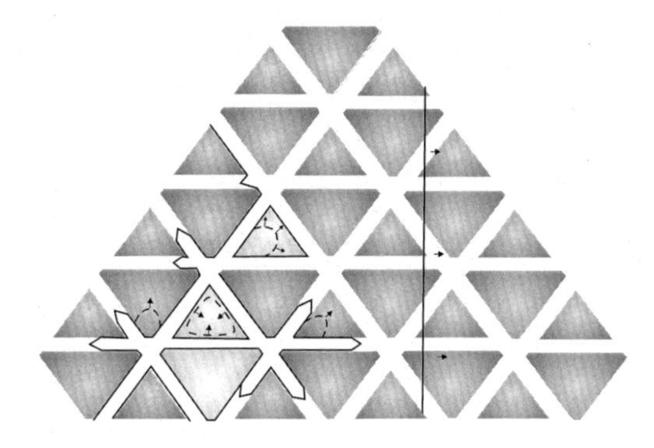

FIGURE 14.13 Possible configurations for precipitate cutting for an alloy with a volume fraction $f_{\gamma'} = 0.6$. Pairs are shown as one dislocation.

In addition, one should keep in mind that the assumption of spherical precipitates represents a very crude simplification, while reliable analytic estimates of the cutting stress for a more realistic geometry are not available.

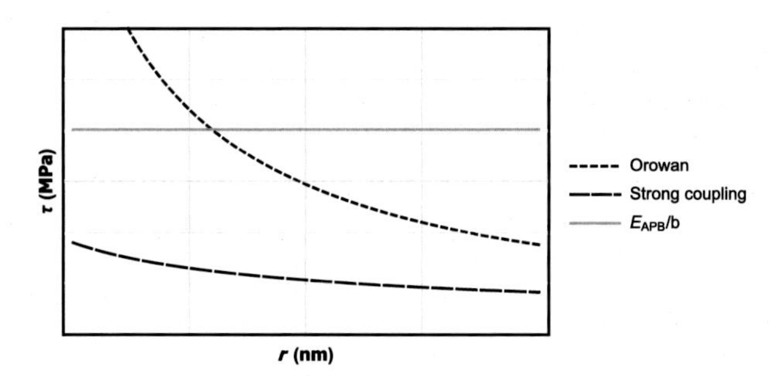

FIGURE 14.14 Comparison of the cutting stress with the Orowan stress, using the line tension of screw segments (see Eq. (14.36)).

An additional strengthening mechanism arises from cross slip on cubic planes within the precipitate phase, yielding to the formation of Kear–Wildorf locks (see, e.g., [1044]).

14.4.3.2 Shearing by <112> dislocations

The dislocations at the γ–γ' interfaces can react, and the reaction products can shear the precipitates, leaving behind them superlattice intrinsic (SISF) or extrinsic stacking faults (SESF) ([709], [1266], [1160]). For example, Sass et al. [1266] and Rae et al. [1160] observed reactions of the type

$$a/2[011] + a/2[\bar{1}01] \rightarrow a/3[\bar{1}12] + a/6[\bar{1}12], \qquad (14.42)$$

where the $a/3[\bar{1}12]$ enters the precipitate, leaving behind it an SISF and the $a/6[\bar{1}12]$ at the matrix precipitate interface. After reaction with additional matrix dislocations, the partial dislocation $a/6[\bar{1}12]$ may also enter the precipitate, leaving behind it a succession of different stacking faults [1160].

The resulting shear rate has been modeled, e.g., by Ma et al. (2008) and Choi et al. (2012, [258]). In the following, the equations of the model by Ma et al. pertaining to shearing of the γ' phase are summarized. For a complete model presentation, the reader may refer to [879]. Denoting by $\dot{\gamma}_{L1_1}$ the shear rate resulting from the $a/3$ <112> dislocations, the dislocation nucleation rate (14.28) is extended to include the reactions at the γ–γ' interfaces, namely

$$\dot{\rho}^{+}_{L1_2} = c_{mult21} \min\left(\rho^{I}_{FCC}, \rho^{II}_{FCC}\right) \Gamma + \frac{c_{mult22}}{b\lambda_{L1_2}} \dot{\gamma}_{L1_2}, \tag{14.43}$$

where the nucleation rate of dislocation partials $a/3\langle\bar{1}12\rangle$ is proportional to the minimal density of the potentially reacting matrix dislocations ρ^{i}_{FCC}. The term Γ represents the stress-dependent reaction frequency, the second term on the right-hand side of Eq. (14.43) is similar to the product term of Eq. (14.28), λ_{L1_2} is the mean free path of the $a/3\langle\bar{1}12\rangle$ partials and c_{mult21}, c_{mult22} are fit parameters. The dislocation annihilation rate has a similar form to Eq. (14.28) with the dislocation density ρ_{L1_2} and the shear rate $\dot{\gamma}_{L1_2}$.

The resistance against glide of the ribbons in the γ' phase is taken as

$$\tau^{c} = c_{pass21} \mu b \sqrt{\rho_{L1_2}} + c_{pass22} \mu \sqrt{\gamma_{L1_2}}, \tag{14.44}$$

as for the matrix (see (14.33)) Taylor hardening is assumed (the first term of the right-hand side of Eq. (14.44)). The second term, which cannot be directly interpreted in terms of dislocation mechanisms, was introduced to better represent the primary creep. Finally, the flow rate in the γ' phase is assumed to be given by (compare with (14.16)) $\dot{\gamma}^{s}_{L1_2} = b^{s} \rho^{s}_{L1_2} v^{s}_{L1_2}$ with an effective velocity

$$v^{s}_{L1_2} = \gamma^{s}_{L1_2} F_{attack} \exp\left(-\frac{Q^{112}_{slip}}{kT} + \frac{|\tau^{s}| - \tau^{c}}{kT} V \right). \tag{14.45}$$

Fig. 14.15 shows that the model is able to predict the large extent of the primary creep at low temperatures.

14.4.4 Climbing over the precipitates

Two different scenarios were considered in the crystal plasticity literature so far:

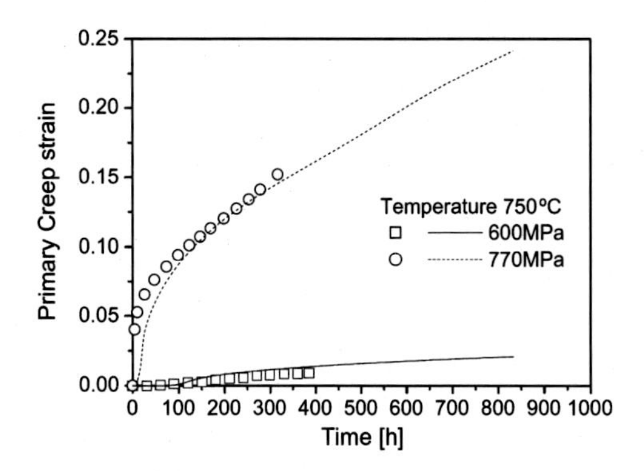

FIGURE 14.15 Creep prediction at 750°C for the alloy CMSX4 for <001> specimens with the model by Ma et al. (2008), from [879].

- The combination of climbing and gliding along the interfaces enables the dislocations to overcome the precipitates. On average, it does not affect the overall dislocation line length ([375], [1677], [1479]). Note that the out-of-plane displacements of the dislocations (pure climbing) are assumed to compensate on average and do not contribute to the deformation.
- Climbing is essentially a recovery mechanism for the dislocation trapped at the γ–γ' interfaces. Dislocation loops are annihilated, thus reducing the internal stresses ([1388], [426]).

14.4.4.1 *The climbing threshold for deformation by climbing*

The combined gliding and climbing of dislocations in precipitates reinforced is usually considered to have a threshold due to the dislocation extension (increase of the total line energy) when the dislocations climb around the obstacles. Estimates of this threshold have been proposed. A review of these models can be found, e.g., in [925] or [1209].

Consider a dislocation of length L climbing over an obstacle, which involves a horizontal displacement δx in the glide plane and a normal displacement δy normal to the glide plane (climb). The total work of the local stress $\tau bL\delta x + \sigma_n bL$, where σ_n is the stress component normal to the climb plane must exceed the increase of the line energy $T_{line}\delta L$. In most models, the work of the climbing force is neglected, so that climbing over the obstacle (precipitate) occurs when

$$\tau > \tau^{cl} = \frac{T_{line}}{bL}\left(\frac{dL}{dx}\right)_{max}, \tag{14.46}$$

where the parameter $R = \left(\frac{dL}{dx}\right)_{max}$, which is often called the climb resistance, describes the rate of increase of the line length L as the dislocation segment climbs over the obstacle. Its value depends on the geometry of the climbing/gliding dislocation segments over the obstacle.

Mukherji and Wahi (1996, [983]) have analyzed the case of cubic precipitates with a high volume fraction of the precipitate phase. They found a minimal climb resistance close to $R \simeq 0.6$ when the climbing dislocation follows the matrix particle interface, while the segments between the precipitates remain in their glide plane (local climb).

14.4.4.2 *The case of creep rate controlled by climbing over the precipitates*

In a series of papers by Dyson and his coworkers [375] [1677], a creep model was developed, which argues that in creep conditions the precipitates are overcome by a combination of glide and climb along the γ–γ' interfaces. The total dislocation density ρ is decomposed into those freely gliding through the channels ρ^g and those that are trapped at the precipitates and climbing along the interface before they can escape ρ^c. Thus at any stage,

$$\rho = \rho^g + \rho^c. \tag{14.47}$$

The rate of change of ρ^g is the difference between the rates of dislocations being released from the interfaces and those becoming arrested at a particle. The rate of released dislocations is the product of the release frequency Γ^e with the fraction trapped dislocations $\frac{2b}{d} f\rho^c$ sufficiently close to the corner of the precipitates of size d to escape. If v^g is the glide velocity in the channels of width w, the rate of dislocations being arrested is $\frac{v^g}{w}\rho^g$. Thus the rate of change of the density of glide dislocations is

$$\dot{\rho}^g = \frac{2b}{d} f\rho^c - \frac{v^g}{w}\rho^g. \tag{14.48}$$

The release rate Γ^e depends on the effective diffusivity for climb D_{eff} via

$$\Gamma^e = \frac{D_{eff}}{b^2} \sinh\left(\frac{\tau b^2 w}{kT}\right). \tag{14.49}$$

It is assumed that the deformation process is controlled by the release rate of the trapped dislocations at the interfaces, and that glide between the particles is sufficiently fast, i.e., $\frac{v^g}{w} \gg \Gamma^e$. In this case, the final shear rate is

$$\dot{\gamma} = 2\rho^m f D_{eff}(1-f)\left(\frac{1}{f^{1/3}} - 1\right) \sinh\left(\frac{\tau b^2 w}{kT}\right). \tag{14.50}$$

The model has been applied to the alloy CMSX-4 in a large range of temperatures. Fig. 14.16 demonstrates the model to describe the time up to 1% creep strain when rafting is and is not taken into account. Note that the model neglects the resistance against climb, that is it assumes no threshold (see Section 14.4.4.1).

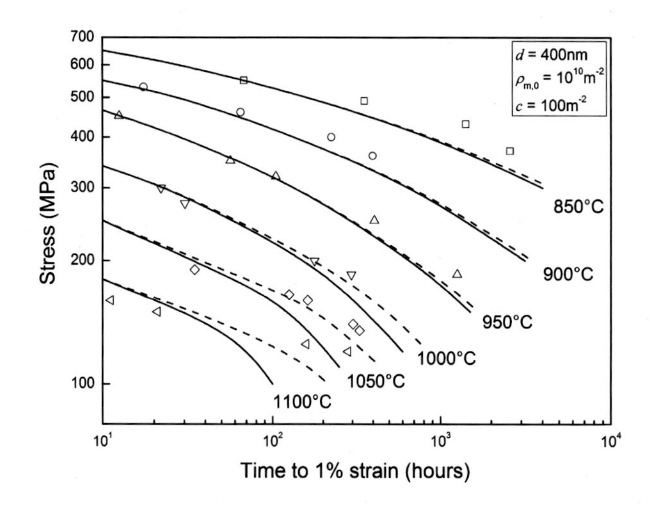

FIGURE 14.16 Comparisons of simulated time to 1% creep strain with experimental data for CMSX-4. Dashed curves demonstrate the predictions with rafting (from [1677]).

14.5 Comparison of some published crystal plasticity models for Ni-base single crystals

Table 14.3 presents a nonexhaustive list of published dislocation-based crystal plasticity models applied to single crystal Ni-base superalloys. The following remarks can be made:

- None of these models covers all relevant known deformation mechanisms. In accordance, none of these models covers the whole range of temperatures and stress levels, even in the restricted range in which the morphological changes can be discarded.
- The assumptions made regarding the dominant deformation mechanisms are largely contradictory.
- The choice of the slip systems (specially cubic slip) is controversial.
- The relevance of Taylor (isotropic) hardening is controversial.
- Some models derive the flow rule from the Orowan equation, which encompasses softening due to dislocation multiplication, while other models assume a flow rate independent of the dislocation density. In

TABLE 14.3 Summary of crystal plasticity models for single crystal nickel-base superalloys.

Authors	Orowan equation	Slip systems	Kinematic hardening	Yield threshold	Isotropic hardening	Loading type
Glatzel et al. [150]	yes	<011> {111}, <011> {001}	no	no	Taylor hardening	creep
Busso et al. [1664]	no	<011> {111}, <011> {001}	Armstrong Frederick	yes	Taylor hardening	creep, tension, fatigue
Choi et al. [258]	yes	<011> {111}, <112> {111}	no	Orowan	Taylor hardening	creep
Dyson [375], Zhu et al. [1677]	yes	<011> {111}	yes	no	no	<001> creep
Fedelich and Vattré [425], [426], [1479]	yes	<011> {111}, <011> {001}	periodic model	Orowan	no	creep, tension, fatigue
Fedelich et al. [430], [428]	yes	<011> {111}, <011> {001}	Armstrong Frederick	Orowan	no	creep, tension, fatigue
Keshavarz [715]	yes	<011> {111}, <011> {001}	no	no	Taylor hardening	tension, fatigue
Ma et al. [879]	yes	<011> {111}, <112> {111}	misfit	Orowan	Taylor hardening	creep
Staroselsky, Cassenti, [1354]	yes	<011> {111}, <011> {001}	Armstrong Frederick	yes	Taylor hardening	creep, tension, fatigue
Tinga et al. [1435], [1438]	no	<011> {111}, <011> {001}	Sachs modified	Orowan	Taylor hardening	creep, tension, fatigue

the latter case, only strain hardening is influenced by the increase of dislocation density.

Even though most of the deformation mechanisms of this class of alloys are now well understood, the quantification of their contribution to the total strain in dependence of the loading conditions is still an open question. This also applies for the formulation of analytical equations that describe dislocation multiplication at the mesoscopic level. Shearing of the precipitates can occur either by matrix dislocation pairs or by complex stacking faults. The simultaneous representation of these competing mechanisms and the identification of the associated model parameters, in particular the corresponding stress thresholds, is also still an open issue. Regarding these uncertainties and the weaknesses of the current dislocation-based constitutive laws, the use of more phenomenological models is still fully justified for engineering applications.

CHAPTER

15

Crystal plasticity models: phenomenological approach

Georges Cailletaud

Mines ParisTech, Centre des Matériaux, Evry, France

15.1 Introduction

Chapter 14 carefully constructed the framework for models based directly on the study of dislocation populations. A different point of view is adopted in the present chapter. The aim is to show how purely macroscopic models, which their authors consider to be "informed" by the microstructure, emerged over the last 60 years. In fact, the models that will be described here have a common basis with those of the previous chapter. They are the founding works of Schmid and Boas (see, for example, the English reprint [1276] of their original 1935 book), then of Taylor, who in 1938 summarized [1403] a series of studies conducted with Elam. It was not until about 30 years later that constitutive equations were clearly formulated in the framework of the mechanics of continuous media. Having in hand the concept of a slip system, and considering that plastic deformation results from the contribution of one or more systems simultaneously, the authors constructed a theory with multiple mechanisms, introducing not only one criterion of plasticity, as in the classical approach of von Mises, but as many criteria as there are slip systems.

By generalizing Koiter's theory [738], Mandel introduced the possibility of multiple interacting criteria [899]. His approach is very generic. It contains, for example, a reference to Tresca's criterion, that can be seen as a multipotential approach. The case of plastic deformation linked to sliding on slip systems is considered in detail, with a study of the existence and uniqueness of the solution as a function of the shape of the hardening matrix that characterizes the interaction between the systems. The work continues with Hill [611] who considers that the resolved shear stress is the driving force for the slip rate on a given system, and leads to a thermodynamic formulation of the models in the 1970s [900,611,612,1218,1219,46]

Nickel Base Single Crystals Across Length Scales
https://doi.org/10.1016/B978-0-12-819357-0.00024-X

then the first applications a few years later [44,43]. This part of the history of crystal plasticity can be reviewed in classical books [584,1409].

A full description of the thermodynamic approach has already been made elsewhere [478], as well as a detailed study of multicriteria models [192]. These theoretical elements will not be repeated here, as the presentation focuses on the phenomenological aspects of the different models available.

The presentation starts with the generic form of the models in time-independent plasticity and viscoplasticity (Section 15.2), and shows the ingredients needed to simulate the mechanical effects observed experimentally. All these terms are then studied in a systematic way in Section 15.3, namely the expressions used for defining the slip resistance, drag stress, backstress, and viscoplastic flow. The section ends with a specific reference to non-Schmid effect. Applications are developed in Section 15.4. They include an illustration of the yield surfaces in two specific cases and the identification of the material parameters for two superalloys, AM1 and CMSX4.

15.2 Generic formulation

The small perturbation formalism is adopted here, so that the strain partition is additive, with an elastic and an inelastic strain

$$\dot{\underset{\sim}{\varepsilon}} = \dot{\underset{\sim}{\varepsilon}}^{\mathrm{e}} + \dot{\underset{\sim}{\varepsilon}}^{\mathrm{p}}. \tag{15.1}$$

Note that the notation $\underset{\sim}{\varepsilon}^{\mathrm{p}}$ will be used for time-independent plastic strain, but also for viscoplastic strain. In order to define the inelastic behavior, the single crystal is seen as a collection of N_S slip systems, defined by their slip planes $\underline{\mathbf{n}}^s$ and slip direction $\underline{\mathbf{l}}^s$. The resolved shear stress τ^{s} on slip system s results from the product of the stress tensor σ and the orientation tensor $\underset{\sim}{\mathbf{m}}^s$:

$$\tau^{\mathrm{s}} = \underset{\sim}{\sigma} : \underset{\sim}{\mathbf{m}}^{\mathrm{s}}, \tag{15.2}$$

$$\text{with} \quad \underset{\sim}{\mathbf{m}}^{\mathrm{s}} = \frac{1}{2}(\underline{\mathbf{n}}^s \otimes \underline{\mathbf{l}}^s + \underline{\mathbf{l}}^s \otimes \underline{\mathbf{n}}^s). \tag{15.3}$$

15.2.1 Time-independent plasticity

The basic time-independent formulation introduces linear isotropic hardening, so that the yield functions are defined for each slip system s as

$$f^s = |\tau^{\mathrm{s}}| - R^{\mathrm{s}} - \tau_y = |\underset{\sim}{\sigma} : \underset{\sim}{\mathbf{m}}^s| - R^{\mathrm{s}} - \tau_y. \tag{15.4}$$

The status of the mechanism representing the slip system s can present (i) an elastic behavior if $f^s < 0$; (ii) elastic unloading if $f^s = 0$ and $\dot{f}^s < 0$;

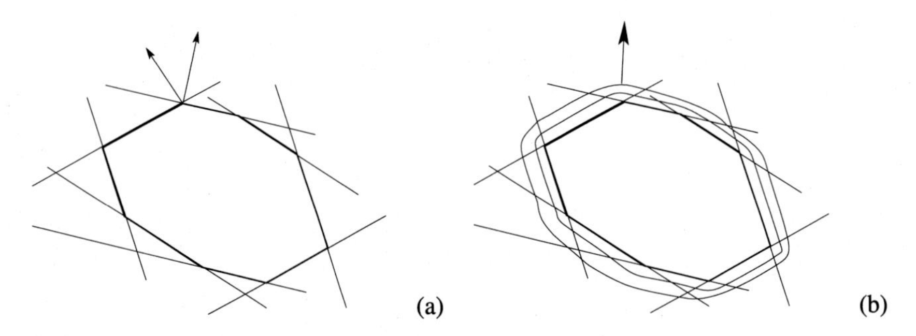

FIGURE 15.1 Illustration of (a) the viscoplastic equipotentials, (b) the elastic domain in time-independent plasticity.

(iii) plastic flow if $f^s = 0$ and $\dot{f}^s = 0$ (consistency condition). The elastic domain is defined by the set of conditions $f^s < 0$ (Fig. 15.1(a)), and the consistency conditions are equivalent to the normality rule, which defines the rate of plastic deformation. Its intensity is given by positive plastic multipliers $\dot{\lambda}^s$, and the flow direction by the orientation tensors that comes from the derivation of the yield function

$$\dot{\underset{\sim}{\varepsilon}}^{\mathrm{p}} = \sum_{s=1}^{N_S} \dot{\lambda}^s \frac{\partial f^s}{\partial \underset{\sim}{\sigma}} = \sum_{s=1}^{N_S} \dot{\lambda}^s \eta^s \underset{\sim}{\mathbf{m}}^{\mathrm{s}} \tag{15.5}$$

with $\eta^s = \mathrm{sign}(\tau^{\mathrm{s}})$. The plastic multiplier on system s is nothing but the rate of accumulated slip p^{s} on this system, which is the absolute value of the plastic slip rate $\dot{\gamma}^{\mathrm{s}}$ itself,

$$\dot{\lambda}^s = \dot{p}^{\mathrm{s}} = |\dot{\gamma}^{\mathrm{s}}|. \tag{15.6}$$

Accumulated slip, which is linked to the dislocation density, is the critical variable that controls hardening, according to Eq. (15.7) for the simple case of linear hardening:

$$R^{\mathrm{s}} = \sum_{r=1}^{N_S} h_{sr}\, p^{\mathrm{r}}, \tag{15.7}$$

where the h_{rs} are the components of the interaction matrix (assumed to be constant for the moment). The diagonal represent self-hardening, meanwhile the off-diagonal terms characterize the so-called latent hardening, produced by each active system on the other systems.

The crucial point of the model is then to determine the set of N active slip systems. The nonzero plastic multipliers are solutions of the linear system formed by the consistency conditions on the active slip systems, $\dot{f}^s = 0$. Several sets of slip systems can produce the same viscoplastic strain rate tensor. An additional condition must be used to select the relevant set of slips. A rather simple computation allows obtaining the relevant system, starting from:

- The strain partition

$$\dot{\underset{\sim}{\sigma}} = \underset{\approx}{\Lambda} : \left(\dot{\underset{\sim}{\varepsilon}} - \sum_r \underset{\sim}{\mathbf{m}}^r \dot{\gamma}^r \right), \tag{15.8}$$

where Λ is the fourth order tensor of the elastic moduli;
- The consistency condition applied to active slip systems,

$$\dot{f}^s = 0 = \underset{\sim}{\mathbf{m}}^s : \dot{\underset{\sim}{\sigma}} - \eta^s \dot{R}^s. \tag{15.9}$$

The following step consists in computing $\underset{\sim}{\mathbf{m}}^s : \dot{\underset{\sim}{\sigma}}$ in Eq. (15.9), introducing h_{sr}, as specified below, and keeping the notation $\dot{p}^r$ for the intensity of plastic shear strain rate on system r,

$$\underset{\sim}{\mathbf{m}}^s : \dot{\underset{\sim}{\sigma}} = \eta^s \dot{R}^s = \eta^s \sum_r h_{sr} \dot{p}^r. \tag{15.10}$$

In the next step, one replaces $\dot{\underset{\sim}{\sigma}}$ by its expression in Eq. (15.8),

$$\underset{\sim}{\mathbf{m}}^s : \underset{\approx}{\Lambda} : \dot{\underset{\sim}{\varepsilon}} - \sum_r \underset{\sim}{\mathbf{m}}^s : \underset{\approx}{\Lambda} : \underset{\sim}{\mathbf{m}}^r \eta^r \dot{p}^r = \sum_r \eta^s h_{sr} \dot{p}^r. \tag{15.11}$$

Under prescribed strain rate, N equations are then defined (N = number of active slip systems) to compute the plastic multipliers,

$$\sum_r \left(\eta^r \underset{\sim}{\mathbf{m}}^s : \underset{\approx}{\Lambda} : \underset{\sim}{\mathbf{m}}^r + \eta^s h_{sr} \right) \dot{p}^r = \underset{\sim}{\mathbf{m}}^s : \underset{\approx}{\Lambda} : \dot{\underset{\sim}{\varepsilon}}. \tag{15.12}$$

The square matrix in the left-hand side of the equation is symmetric, provided that the interaction matrix is also symmetric. This is the case now, but it is no longer true for more complex hardening rules, for instance, with kinematic hardening.

The existence and uniqueness of the solution has been studied by several authors in the past. In fact, the problem can be summarized by

$$\underset{\sim}{\sigma}^* : \dot{\underset{\sim}{\varepsilon}}^p \leqslant \underset{\sim}{\sigma} : \dot{\underset{\sim}{\varepsilon}}^p = \sum_s \tau_y^s \dot{p}^s \leqslant \sum_s \tau_y^s \dot{p}^{s\prime}. \tag{15.13}$$

The solution can be found by minimizing the internal power of the material element [1403]: Among all the sets of active slip systems $\{\dot{p}^{s\prime}\}$, the actual set of active slip systems $\{\dot{p}^s\}$ leads to a minimum of the power computed with the actual resolved shear stresses. Another bound is obtained in terms of stress [113]: Among all the admissible stress tensors $\underset{\sim}{\sigma}$, the real stress tensor provides a maximum of the product $\underset{\sim}{\sigma}^* : \underset{\sim}{\dot{\varepsilon}}^{\mathrm{p}}$. A detailed theoretical discussion can be found in [256]. The consequences in terms of numerical computation and a numerical study of the system selection at the corner of the yield surface can be found elsewhere [174].

15.2.2 Viscoplasticity

As discussed in the previous section for time-independent plasticity, the point representing the admissible stress state must be either within the domain of elasticity or on its boundary. The introduction of viscosity causes equipotentials to be built up outside this domain, along which the equivalent strain rate is identical, as shown in Fig. 15.1(b). This is defined by means of a viscoplastic potential [647,46], the historical formulation of which assumes that the elasticity domain is reduced at the origin of the stress space, so that all the slip systems are active (Eq. (15.14)):

$$\Omega = \sum_s \Omega_s(f^s) = \frac{m}{m+1} \sum_s \left| \frac{\tau^{\mathrm{s}}}{\tau_y^{\mathrm{r}}} \right|^{(m+1)/m}, \tag{15.14}$$

$$\underset{\sim}{\dot{\varepsilon}}^{\mathrm{p}} = \sum_s \frac{\partial \Omega_s}{\partial \underset{\sim}{\sigma}} = \sum_s \frac{\partial \Omega_s}{\partial f^s} \frac{\partial f^s}{\partial \underset{\sim}{\sigma}} = \sum_s \dot{\upsilon}^{\mathrm{s}} \eta^s \underset{\sim}{\mathbf{m}}^{\mathrm{s}}, \tag{15.15}$$

with $\eta^s = \mathrm{sign}(\tau^{\mathrm{s}})$. The slip rate $\dot{\upsilon}^{\mathrm{s}}$ for a given system s is now directly defined by the distance to the boundary of the domain (that is the origin for the historical formulation). It is no longer necessary to solve a system of equations. The slip rate is obtained independently on each system,

$$\dot{\gamma}^{\mathrm{s}} = \dot{\upsilon}^{\mathrm{s}} \eta^s = \left| \frac{\tau^{\mathrm{s}}}{\tau_y^{\mathrm{s}}} \right|^{1/m} \eta^s. \tag{15.16}$$

The model maintains the possibility to introduce a cross-hardening effect, by means of an interaction matrix. A popular expression [1102,45] introduces the total amount of slip V such as $V = \sum_r v^r$ to generate a simple matrix whose diagonal has terms equal to 1 and off-diagonal terms equal to q:

$$\dot{\tau}_y^{\mathrm{s}} = \sum_r h_{sr} \dot{\upsilon}^{\mathrm{r}}, \qquad \text{with} \quad h_{sr} = (q + (1-q)\delta_{sr}) h(V), \tag{15.17}$$

and the hardening function expressed as

$$h(V) = h_0 \mathrm{sech}^{-2}\left(\frac{h_0 V}{\tau^{\mathrm{s}} - \tau_y} \right). \tag{15.18}$$

15.3 Anatomy of the general model formulation

As far as the resolved shear stress is concerned, the elementary models cited in the previous section can be expressed as

$$-\text{ plasticity,} \qquad \tau^{s} = \pm(\tau_y + R^{s}), \tag{15.19}$$

$$-\text{ viscoplasticity,} \quad \tau^{s} = \pm\tau_y^{s}\dot{\gamma}^{s^m}. \tag{15.20}$$

In the first case (Eq. (15.19)), the shear stress is purely defined by the threshold of the elasticity domain, which can be evaluated by the sum of the initial critical resolved stress τ_y and a variable R^s which depends on accumulated slip. One can speak of "additive" isotropic strain-hardening because strengthening occurs through an expansion of the elasticity domain. In the second case (Eq. (15.20)), the hardening that occurs can be described as "multiplicative", since τ_y comes to multiply the term in deformation rate. A constant τ_y would mean that there is no strain-hardening, but simply an influence of the loading rate on the flow, which remains unchanged regardless of the level of deformation.

The use of nickel-base superalloy single crystals is such that they undergo cyclic loading. Isotropic hardening alone is then not appropriate to correctly represent the mechanical response. It is necessary to add the so-called "kinematic" contribution, which allows the description of the translation of the elasticity domain, in order to correctly represent the hysteresis loops. This is why we will discuss in this chapter a general form of the decomposition of the resolved shear stress as shown in Eq. (15.21), where the variable x^s stands for kinematic hardening,

$$\tau^{s} = x^{s} \pm (\tau_y + R^{s}) \pm K^{s}\dot{\gamma}^{s^{1/n}}. \tag{15.21}$$

To have a better compatibility with the equations that will be privileged at the end of the chapter, the notations have been modified here, with the exponent n replacing $1/m$ and K^s replacing τ_y. The term $K^s\dot{\gamma}^{s^{1/n}}$ is the viscous contribution. The use of this expression assumes that a slip system can generate both positive and negative sliding.

The description can be rephrased in the opposite direction, knowing that a general model is defined on each system s by means of:

- An initial elasticity domain (defined by a function f_i^s), which involves a driving term (for the moment, the resolved shear stress) and an initial threshold, τ_y,

$$f_i^{s} = |\tau^{s}| - \tau_y. \tag{15.22}$$

This domain is reduced to the origin in the case of the classical viscoplastic model shown in the preceding section.

- Additive hardening variables, which are introduced to extend f_i^s and to define the evolution of the elasticity domain: the *slip resistance* R^s, figuring isotropic hardening that is either expansion or contraction of the elasticity domain, and the *backstress* x^s, representing kinematic hardening, a shift of the elasticity domain.
- These variables come into f_i^s to generate the yield function f^s,

$$f^s = |\tau^s - x^s| - \tau_y - R^s. \tag{15.23}$$

  The value of x^s defines the center of the actual elasticity domain, the limits of which are then $x^s - R^s - \tau_y$ and $x^s + R^s + \tau_y$. Slip occurs if and only if the value $|\tau^s - x^s| - R^s - \tau_y$ is positive.
- For a time-independent formulation the flow rule is directly defined by the consistency condition which expresses the fact that the active stress must stay at the boundary of the elasticity domain during plastic flow. A viscoplastic model needs an additional rule, which introduces a multiplicative isotropic hardening variable, the *drag stress* K^s. Starting from its initial value, an increase of K^s will reduce the creep rate, producing strain hardening, meanwhile a decrease will generate strain softening, both of them visible on viscoplastic strain rates.

The final general shape, deduced from Eq. (15.21) by inverting the equation, is

$$\dot{\gamma}^s = \left\langle \frac{f^s}{K^s} \right\rangle^n \operatorname{sign}(\tau^s - x^s). \tag{15.24}$$

It will be mentioned later that all the individual elements of the equation might be replaced by alternative formulations to express specific phenomena, namely τ^s to represent the non-Schmid effect in Section 15.3.5, a different kinematic rule, Section 15.3.3, a different viscoplastic flow, Section 15.3.4.

15.3.1 Isotropic hardening: slip resistance

As previously mentioned, isotropic hardening is sensitive to the accumulated slip. Bearing in mind that strain-hardening is produced by the development of dislocation populations, most authors introduce into isotropic hardening the effect of cross-hardening, as already written in Eq. (15.17). Besides the form defined by Eqs. (15.17) and (15.18), authors propose sometimes more sophisticated evolution rules.

The safest way to ensure a reasonable model behavior for anisothermal loadings is to write the evolution of the hardening variables as a function of real state variables, the value of which is not instantaneously influenced by a sudden temperature change [102]. For isotropic hardening, it has been proposed to use an r^s variable that depends on the accumulated slip v^s for

TABLE 15.1 Values of the terms in the first line of the interaction matrix for two geometric assumptions: angles between the vectors normal to the slip planes and angles between the slip direction (line 2), and scalar product of the orientation tensors (line 3). The first line recalls the number of the slip system s. The slip directions are

1:	$\bar{1}01$	2:	$0\bar{1}1$	3:	$\bar{1}10$	4:	$\bar{1}01$	5:	011	6:	110
7:	$0\bar{1}1$	8:	110	9:	101	10:	$\bar{1}10$	11:	101	12:	011

Systems 1–3 are in the plane $\{111\}$, 4–6 in $\{1\bar{1}1\}$, 7–9 in $\{\bar{1}11\}$, and 10–12 in $\{11\bar{1}\}$.

s	**1**	**2**	**3**	**4**	**5**	**6**	**7**	**8**	**9**	**10**	**11**	**12**
$12 \times (\underline{\mathbf{n}}^1 . \underline{\mathbf{n}}^s)(\underline{\mathbf{l}}^1 . \underline{\mathbf{l}}^s)$	12	6	6	4	2	2	2	2	0	2	0	2
$12 \times \underset{\sim}{\mathbf{m}}^1 : \underset{\sim}{\mathbf{m}}^s$	6	3	3	2	1	1	1	3	4	1	4	3

the slip system s [937,478]. It tends to an asymptote for large values of v^s, as prescribed in

$$\dot{r}^s = (1 - Br^s)\dot{v}^s \qquad \text{so that} \qquad r^s = \frac{1}{B}\left(1 - \exp(-Bv^s)\right), \tag{15.25}$$

where B is a material parameter that controls the rate at which equilibrium is reached. Obviously, this variable has to do with dislocation density, without having a direct connection. If needed, a static recovery can be introduced, with two additional material parameters, M_r and m_r, according to

$$\dot{r}^s = (1 - Br^s)\dot{v}^s - \left(\frac{r^s}{M_r}\right)^{m_r}. \tag{15.26}$$

The slip resistance R^s for the system s is then obtained by a linear combination of the r^s variables from all systems,

$$R^s = BQ\sum_r h_{sr} r^r = Q\sum_r h_{sr}\left(1 - \exp(-Bv^r)\right). \tag{15.27}$$

Rather than distinguishing only two types of terms (diagonal and off-diagonal), the h_{rs} terms in the matrix may be chosen according to geometrical arguments. For instance, Weng [1556] introduces a symmetric second-order tensor $\underset{\sim}{\mathbf{M}}$, the components of which characterize the interaction between a system defined by ($\underline{\mathbf{n}}^s$ (normal vector), $\underline{\mathbf{l}}^s$ (slip direction)) and a system defined by ($\underline{\mathbf{n}}^r$ (normal vector), $\underline{\mathbf{l}}^r$ (slip direction)),

$$M_{rs} = (\underline{\mathbf{n}}^r \cdot \underline{\mathbf{n}}^s)(\underline{\mathbf{l}}^r \cdot \underline{\mathbf{l}}^s) \qquad \text{and} \qquad h_{rs} = H(q + (1-q)M_{rs}). \tag{15.28}$$

In fact, this expression leads to predominant diagonal terms. This is summarized in Table 15.1 which shows the values of the interaction terms of all the systems ($s = 1..12$) with the system 1, taken as $\{111\}[\bar{1}01]$. The self-hardening $(1, 1)$ term is equal to 1, coplanar terms $(1, 2)$ and $(1, 3)$ are equal to $1/2$, the collinear term $(1, 4)$ is equal to $1/3$, the other interactions give 1/6, except for the slip direction [101], perpendicular to [$\bar{1}$01], that provides two zero terms in the line. This point was not investigated in details, as the goal of the model was to generate kinematic hardening by considering that a given direction defines two different systems (a system can experience positive slip only; isotropic hardening is obtained for $q = 1$ and purely kinematic hardening for $q = 0$). Note that the third line of the table shows the result of the tensorial product of the orientation tensors, which will be discussed later in Section 15.3.3.

A better solution is to introduce cristallographic properties. One of the first references in this field, after Kocks [732], is Franciosi [462] who distinguishes among off-diagonal terms those designating coplanar systems, collinear systems, deviated slip, Hirth locks, slip junctions and Lomer–Cottrel barriers. The respective values of these different terms have been estimated by many authors with arguments of dislocation physics, or even numerical simulations using the discrete dynamics of dislocations. Of course, the presence of precipitates, and the very specific mechanisms at work in nickel-base superalloys make it important to have a special estimate. But, in fact, the values to be introduced in the interaction matrix is a case that remains much debated. The consequence is that the authors fall back on extreme assumptions in the literature, such as Taylor's [177] (all terms in the matrix are equal to 1, which means that strain-hardening actually depends on the sum of all cumulative slips on all systems), or they consider a purely diagonal matrix [937] (no cross-hardening). From a phenomenological point of view, it is true that this type of hardening is intended to represent cyclic hardening or softening, and the latter is in any case not very pronounced in nickel-base superalloys.

Investigating the problem of cross-softening for CMSX4, Levkovitch et al. [826] mention the fact that a system with a weakly active system can postpone the saturation of an active system for an unacceptable amount of cycles for the case of fatigue loadings. To illustrate the updated model they are using, one can use just two systems. In the classical approach, the interaction is such that the variable on system 2 is R_2, with $R_2/(BQ) = h_{21}r_1 + h_{22}r_2$, and $\dot{r}_i = (1 - Br_i)\dot{v}_i$, for $(i = 1, 2)$. An updated model consists of modifying the interaction term and replacing r_1 by a new variable $r_{1\to 2}$, the influence of which vanishes once system 2 is saturated,

$$R_2/(BQ) = h_{21}r_{1\to 2} + h_{22}r_2 \quad \text{with} \quad \dot{r}_{1\to 2} = (1 - Br_1)(1 - Br_2)\dot{v}_1. \qquad (15.29)$$

In this paper, this expression of the drag stress is combined with the full version of the kinematic hardening shown below in Eq. (15.34).

15.3.2 Isotropic hardening: drag stress

The hardening has to be supported by the drag stress in the case where the viscous part increases of decreases over time. This is the case in the classical approach as shown by Eqs. (15.17) and (15.18). This option has been selected by Jordan and Walker [681], who propose an evolution rule for K^s in Eq. (15.20),

$$\dot{K}^s = \sum_r \left(\beta(q - (1-q)\delta_{sr}) - \eta(K^r - K_0)\right) \dot{v}^r - h(K^r - K_0). \tag{15.30}$$

Such an expression should be manipulated with care at variable temperature. The choice made here allows the authors to represent a variation of K^s from its initial value K_0 to an asymptotic value that will depend on the slip system and the amount of slip. A time recovery term is also present.

A similar expression can be found in [23] or in [924], namely

$$\dot{K}^s = \sum_r \beta(q - (1-q)\delta_{sr}) \left(1 - \frac{K^r}{K^*}\right)^a \dot{v}^r, \tag{15.31}$$

or, more recently [593], where hardening on drag stress is combined with a kinematic hardening, with a zero slip resistance.

15.3.3 Kinematic hardening

As it turns out, the most commonly used rule is nonlinear kinematic hardening, initially proposed in 1966 by Armstrong and Frederick (only much later available in the public literature [465]) and made widely known by Chaboche [226]. It is a purely phenomenological equation which finds for each material its physical meaning. As for isotropic hardening, it is safe to consider state variable evolution. In [937], the hardening variable x^s for the system s is proportional to the state variable α^s, and the evolution of the state variable includes a driving term, a dynamic recovery (corresponding to the shearing of precipitates) and a time controlled static recovery term (e.g., annihilation of dislocations in the γ/γ' interfaces by climb/glide):

$$\dot{\alpha}^s = \dot{\gamma}^s - D\alpha^s \dot{v}^s - \left|\frac{x^s}{M_x}\right|^{m_x} \operatorname{sign}(x^s), \tag{15.32}$$

$$x^s = C\alpha^s. \tag{15.33}$$

Usually authors do not introduce coupling between the systems for the case of kinematic hardening. In the absence of a static recovery term, the asymptotic value of α^s is $1/D$ in the tension going branch ($\dot{\gamma}^s > 0$) and $-1/D$ in the compression going branch ($\dot{\gamma}^s < 0$). Static recovery decreases these values when loading rates are smaller or for long hold times. This expression is found in many papers (see, for instance, [1521,23,1310,1354, 428]).

In fact, a more complex expression of the rate of the kinematic variable comes from [937] and [577]. It includes an evolution of the asymptote with accumulated plastic shear (through a function Φ which initial and final values are respectively 1 and Φ_s and the decay rate ω), and the possibility that it stabilizes around a nonzero value (x_0 in Eq. (15.34)). A full version, expressed in term of rate of x^s, for isothermal conditions, reads

$$\dot{x}^s = C\Phi(v^s)\dot{\gamma}^s - D(x^s - x_0)\dot{v}^s - \left|\frac{x^s}{a}\right|^m \mathrm{sign}(x^s), \tag{15.34}$$

$$\text{with} \quad \Phi(v^s) = \Phi_s + (1 - \Phi_s)\exp(-\omega v^s). \tag{15.35}$$

A different approach is to transfer the kinematic strain-hardening to the level of the single crystal, to express the fact that it results from the effect of the slip on all systems. In this case, the state variable is a tensor that evolves as a function of the plastic deformation instead of being individually dependent on each system, and the kinematic variable $\underset{\sim}{\mathbf{X}}$ is obtained by means of a fourth-order tensor, $\underset{\approx}{\mathbf{C}}$. Eq. (15.36) shows the expression without static recovery for the sake of brevity, which introduces another fourth-order tensor, $\underset{\approx}{\mathbf{D}}$,

$$\dot{\underset{\sim}{\alpha}} = \dot{\underset{\sim}{\varepsilon}}^p - \underset{\approx}{\mathbf{D}} : \underset{\sim}{\alpha} \left(\dot{\underset{\sim}{\varepsilon}}^p : \dot{\underset{\sim}{\varepsilon}}^p\right)^{1/2}, \tag{15.36}$$

$$\underset{\sim}{\mathbf{X}} = \underset{\approx}{\mathbf{C}} : \underset{\sim}{\alpha}. \tag{15.37}$$

Now $\underset{\sim}{\alpha}$ plays the role of an internal strain and $\underset{\sim}{\mathbf{X}}$ that of an internal stress, which has to be projected onto the slip system to form a *resolved internal stress*. The criterion on a system s is $f^s = (\underset{\sim}{\sigma} - \underset{\sim}{\mathbf{X}}) : \underset{\sim}{\mathbf{m}}^s - R^s - \tau_y$, so that the resolved internal stress is

$$x^s = \underset{\sim}{\mathbf{X}} : \underset{\sim}{\mathbf{m}}^s. \tag{15.38}$$

This formulation, which is used by Pilvin [1107], introduces a coupling between all the kinematic variables through the plastic strain tensor. The first line of the interaction matrix is the third line in Table 15.1. As in the case of the scalar products in line 2, it is the self-hardening term that is predominant (it is equal to 1/2). The terms representing coplanar (1/4) and collinear (1/12) interactions are clearly identified. There is no null term, unlike in line 2, and the perpendicular slip directions give a fairly large term (1/3), which is preferable, since according to metallurgical investigations, this term should provide a high degree of interaction.

15.3.4 Flow rule

As nickel-base single crystals are expected to endure high temperatures during operation, viscoplastic flow rules are a critical ingredient of the simulations. Following the historical models, the basic approach introduces a power function to represent the viscoplastic potential, and the intensity of the viscoplastic flow rate $\dot{v}^s$ for a system s is defined by the partial derivative of the potential with respect to the yield function. By extending Eq. (15.15), a large number of models have adopted the general form:

$$\Omega = \sum_s \frac{K}{n+1} \left\langle \frac{f^s}{K} \right\rangle^{n+1}, \qquad f^s = |\tau^s - x^s| - R^s - \tau_y, \tag{15.39}$$

$$\dot{\underset{\sim}{\varepsilon}}^p = \sum_s \frac{\partial \Omega_s}{\partial f^s} \frac{\partial f^s}{\partial \underset{\sim}{\sigma}} = \sum_s \dot{v}^s \eta^s \underset{\sim}{\mathbf{m}}^s, \quad \dot{v}^s = \left\langle \frac{|\tau^s - x^s| - R - \tau_y}{K} \right\rangle^n, \tag{15.40}$$

with $\eta^s = \text{sign}(\tau^s - x^s)$. By construction, such a rule predicts that the ratio between the two viscous stress levels obtained when the equivalent viscoplastic strain rate differs by one order of magnitude is $10^{1/n}$, over the whole range of deformation rates. The ratio is then big for low values of n (e.g., larger than 2 for $n = 3$) and very small for large values of n (1.023 for $n = 100$). The large values of n therefore classically make it possible to find models that are little sensitive to the deformation rate, practically time-independent from a numerical point of view. The lower values are used at high temperature, and in most of the cases, the model is capable of correctly representing databases over a range of strain rates sufficient for use in industrial part simulations.

If there is a need to have a very large domain of identification in terms of strain rates, the constant strain ratio may be a drawback. For this reason, some authors recommend that the equivalent plastic deformation rate be expressed using a saturating function for high values, as follows:

$$\dot{v}^s = \dot{\gamma}_0 \sinh \left\langle \frac{|\tau^s - x^s| - R_0 - \tau_y}{K} \right\rangle^n. \tag{15.41}$$

Following [733], an expression which is based on an estimation of the activation energy reads

$$\dot{v}^s = \dot{\gamma}_0 \exp \left\{ -\frac{F_0}{kT} \left[1 - \left\langle \frac{|\tau^s - x^s| - R_0 - \tau_y}{K} \right\rangle^p \right]^q \right\}. \tag{15.42}$$

It has been used, for instance, in [177] for an NiAl single crystal and in [1664] for CMSX4. This formulation raises a small problem because the slip rate is not naturally zero within the elasticity domain. This condition must therefore be enforced in the numerical process.

More complex expressions have been introduced, with the purpose of explicitly representing the effects of temperature, as follows:

$$\dot{v}^s = \dot{\gamma}_0 \Theta(T) \left\langle \frac{|\tau^s - x^s| - R^s}{K} \right\rangle^n \exp \left\{ B_0 \left\langle \frac{|\tau^s - x^s| - R^s}{K} \right\rangle^{n+1} \right\}. \tag{15.43}$$

This model is proposed by Shenoy et al. [1310] for the directionally solidified (DS) Ni-base superalloy, DS GTD-111, then used for René 88 [776], and for a DS material by [727].

In several successive papers [1438,1437,1436] Tinga et al. investigate the microstructural effect due to rafting, and the so-called non-Schmid effect that will be discussed in the next section. The model is applied to turbine blade calculations. They propose two types of flow, for the matrix (Eq. (15.44)) and for the precipitate (Eq. (15.45)):

$$\dot{v}^s = \dot{\gamma}_0 F(\Omega^s) \left(\frac{|\tau_{eff}|}{K} \right)^m \left\{ 1 - \exp\left(-\frac{|\tau_{eff}|}{K} \right) \right\}^n, \tag{15.44}$$

$$\dot{v}^s = \dot{\gamma}_0 \left\{ 1 - \exp\left(-\frac{|\tau_{eff}|}{K} \right) \right\}^p. \tag{15.45}$$

15.3.5 Non-Schmid effects

In the framework of classical crystal plasticity, the resolved shear stress is the variable which is associated with shear strain rate to estimate internal power. The so-called non-Schmid effect has been considered early in the development of the theory to include, for instance, the effect of cross slip. One of the first papers dealing with the subject [46] defines $\underline{\mathbf{z}}$ as the direction that forms a right-handed triad with the slip direction $\underline{\mathbf{l}}$ and the normal to the slip plane $\underline{\mathbf{n}}$. In these conditions, h being the instantaneous plastic modulus, the plastic strain rate depends classically on the rate of the resolved shear stress τ_{nl}, but also on the linear combination, with weighting factors $\alpha_{\alpha\beta}$, of additional terms $\tau_{\alpha\beta}$, with α and $\beta \in (l, n, z)$,

$$h\dot{\gamma} = \dot{\tau}_{nl} + \alpha_{ll}\dot{\tau}_{ll} + \alpha_{nn}\dot{\tau}_{nn} + \alpha_{zz}\dot{\tau}_{zz} + 2\alpha_{lz}\dot{\tau}_{lz} + 2\alpha_{nz}\dot{\tau}_{nz}. \tag{15.46}$$

The combination includes then two additional shear stresses, and three normal stresses. In the cross-slip process, the shear stress τ_{nz} contributes to the coalescence of the partial dislocation, while τ_{sz} drives the screw dislocation segment along the cross-slip plane. After studying the elastic energy involved, the authors propose values from 0.01 to 0.1 as an order of magnitude of the weighting factors.

Metallurgical arguments have been proposed later by many authors [784,1044], who linked cross-slip with a noticeable different behavior in tension and compression of $L1_2$ materials (specifically [1395] on Ni_3Ga, [784,1045] on Ni_3Al). The mechanism of cross-slip has been described in Chapter 14, so that the detailed analysis is not reproduced here. A classical observation is that, for tests under cyclic loading, the mean stress is positive for near $\langle 100 \rangle$ crystal orientations, but negative near $\langle 111 \rangle$ or $\langle 110 \rangle$, and zero on a circle which goes from $\langle 012 \rangle$ to $\langle \bar{1}23 \rangle$ in the standard triangle. The associated process is as follows: (i) due to the large volume fraction of precipitates, the dislocations are circulating in pairs in the octahedral planes,

(ii) cross-slip is easier if the distance between the dislocations is smaller, (iii) this distance is controlled by the shear normal to the slip system in the octahedral plane. The corresponding model, which is classically expressed in terms of Schmid factors [951] can be rewritten as an equivalent shear stress combining three shear stresses,

$$\tau^* = \alpha_1 \tau_1 + \alpha_2 \tau_2 + \delta \alpha_3 \tau_3, \tag{15.47}$$

where (1) stands for the classical Schmid mechanism (say, for instance, $\{111\}\langle\bar{1}01\rangle$ slip), (2) goes for the same slip system on the corresponding cube plane (in this case $\{010\}\langle\bar{1}01\rangle$), and (3) for the shear that controls the constriction of the dislocations ($\{111\}\langle\bar{1}2\bar{1}\rangle$). The parameter δ is equal to $+1$ in tension and -1 in compression. This model is used by many authors: Osterle et al. [1038] propose a specific investigation of the weighting factor in relation with the specific γ/γ' microstructure of the chromium-rich superalloy SC16, in the framework of an approach which combines and cube slip.

Non-Schmid effect for an $L1_2$-type material has been investigated by Qin and Bassani [1155]. In the model, the classical resolved shear stress on the primary plane in the direction of the Burger vector (index *pb*) is replaced by a linear combination involving also the stress on the cube cross slip plane in the direction of the Burgers vector (index *cb*), the stress on the primary slip plane normal to the Burgers vector, which drives the dislocation splitting (Shockley partial) on the primary slip plane (index *pe*), and the stress on the secondary slip plane normal to the Burgers vector, which drives the dislocation splitting on the secondary slip plane (index *se*). Taking the example of slip in direction $\langle\bar{1}01\rangle$ in plane $\{111\}$, τ_{cb} corresponds to the shear in the same direction on plane $\{010\}$, τ_{pe} to the shear on plane $\{111\}$ in direction $\langle 1\bar{2}1\rangle$, and τ_{se} to the shear in plane $\{\bar{1}1\bar{1}\}$ in direction $\langle\bar{1}\bar{2}\bar{1}\rangle$. This model has been exploited by many authors in different formulations for Ni-base superalloys:

- For CMSX4 at 750°C [1437],

$$\tau^{s*} = \tau^s_{pb} - A\,\mathrm{sign}(\tau^s_{pb})(\tau^s_{pe} - \kappa\tau^s_{se}) - B\tau^s_{cb}; \tag{15.48}$$

- For MD2 at room temperature, the normal stress to the principal slip plane (index *pn*) might be added [816,817],

$$\tau^{s*} = |\tau^s_{pb}| + \kappa_1|\tau^s_{cb}| + \kappa_2|\tau^s_{sb}| + \kappa_3\tau^s_{pe} + \kappa_4\tau^s_{se} + \kappa_5\tau^s_{pn}; \tag{15.49}$$

- A combination of the Schmid resolved shear stress with the normal stress on the principal plane and the so-called coshear (index *pe* in the previous notation) is found in [1360].

15.4 Applications

15.4.1 Yield surfaces

Yield surfaces provide a synthetic information on the multiaxial behavior of the materials. Tension–torsion loadings have already been discussed in this book, for the case of tubes in 5.4.1 of Chapter 5, and cylinder in 17.4.2 of Chapter 17. This is a good illustration, as this loading case underlines the specificities of the single crystal. Many authors in the literature have considered that it is possible to represent the behavior of a single crystal using Hill's criterion (see, for instance, [1663]). This incorrect interpretation of the material's response is not unusual, see [833] or [356]. The main error brought by Hill's criterion on torsional loading is that it predicts an identical plastic flow over the entire circumference of the specimen. This is not the case, as has been shown in elementary finite element calculations [936], and as shown by the experimental result in Fig. 5.6 of Chapter 5. Note that it still remains possible to work with a unique potential, instead of the multipotential approach attached to slip systems. In such a case, the model must include all the invariants that are compatible with the cubic symmetry, see, for instance, [1021,1022,1310].

The present section is intended to give a quick look at the tensile-shear yield surfaces, when the loading frame is positioned in different ways relative to the crystallographic axes. In particular, this will facilitate a good understanding of the locations of the "soft" and "hard" areas in the torsion tests, depending on the orientation of the specimen.

The yield surfaces can be seen as a collection of hyperplanes defined by f_i^s in Eq. (15.22). The components of the orientations tensors $\underset{\sim}{\mathbf{m}}^s$ are given in Table 15.2 for the 12 octahedral slip systems and 6 cube systems. The slip planes and slip directions are given in the caption of Table 15.1 for the octahedral slip, and in the caption of Table 15.2 for the cube slip.

By restricting oneself to a tensile–shear loading, it will be possible to represent this domain in a plane, whose axes are normal stress σ_n and shear σ_t. By calling $\underline{\mathbf{N}}$ the direction of the normal and $\underline{\mathbf{T}}$ the direction of the shear, the stress tensor is written as

$$\underset{\sim}{\sigma} = \sigma_n \, \underline{\mathbf{N}} \otimes \underline{\mathbf{N}} + \sigma_t \, (\underline{\mathbf{N}} \otimes \underline{\mathbf{T}} + \underline{\mathbf{T}} \otimes \underline{\mathbf{N}}). \tag{15.50}$$

All that remains to be done is to define interesting load states in the crystal framework. The following cases are investigated:

1. Normal $\underline{\mathbf{N}}$ in direction $\{001\}$, shear $\underline{\mathbf{T}}$ in direction $\langle 010 \rangle$, which is one of the four "hard" zones for the case of a tube which axis is $\{001\}$;
2. Normal $\underline{\mathbf{N}}$ in direction $\{001\}$, shear $\underline{\mathbf{T}}$ in direction $\langle \bar{1}10 \rangle$, which is one of the four "soft" zones for the case of a tube which axis is $\{001\}$;
3. Normal $\underline{\mathbf{N}}$ in direction $\{111\}$, shear $\underline{\mathbf{T}}$ in direction $\langle \bar{1}01 \rangle$, which is one of the six "soft" zones for the case of a tube which axis is $\{111\}$;

TABLE 15.2 The components of the 12 orientation tensors for octahedral slip systems (columns denoted 1 to 12) and the six components of the cube slip (columns denoted 13 to 18). The octahedral slip systems were defined in the caption of Table 15.1. The cube slip corresponds respectively to $\langle 01\bar{1}\rangle$ and $\langle 011\rangle$ in plane $\{100\}$ (13–14), to $\langle\bar{1}0\bar{1}\rangle$ and $\langle\bar{1}01\rangle$ in plane $\{010\}$ (15–16), and to $\langle\bar{1}10\rangle$ and $\langle 110\rangle$ in plane $\{001\}$ (17–18). To make the writing more compact, the value -1 is denoted by $\bar{1}$. The multiplicative factors come from the normalization of the vectors n and l.

	Octahedral slip													Cube slip						
ij	**Syst. Fact.**	**1**	**2**	**3**	**4**	**5**	**6**	**7**	**8**	**9**	**10**	**11**	**12**	**Syst. Fact.**	**13**	**14**	**15**	**16**	**17**	**18**
11	$\sqrt{6}$	$\bar{1}$	0	$\bar{1}$	$\bar{1}$	0	1	0	$\bar{1}$	$\bar{1}$	$\bar{1}$	1	0	1	0	0	0	0	0	0
22	$\sqrt{6}$	0	$\bar{1}$	1	0	$\bar{1}$	$\bar{1}$	$\bar{1}$	1	0	1	0	1	1	0	0	0	0	0	0
33	$\sqrt{6}$	1	1	0	1	1	0	1	0	1	0	$\bar{1}$	$\bar{1}$	1	0	0	0	0	0	0
12	$2\sqrt{6}$	$\bar{1}$	$\bar{1}$	0	1	1	0	1	0	1	0	1	1	$2\sqrt{2}$	1	1	$\bar{1}$	$\bar{1}$	0	0
23	$2\sqrt{6}$	1	0	1	$\bar{1}$	0	1	0	1	1	$\bar{1}$	1	0	$2\sqrt{2}$	0	0	$\bar{1}$	1	1	1
31	$2\sqrt{6}$	0	1	$\bar{1}$	0	1	1	$\bar{1}$	1	0	1	0	1	$2\sqrt{2}$	$\bar{1}$	1	0	0	$\bar{1}$	1

TABLE 15.3 The components of the normal and shear direction for the loading cases considered, and the components of the resulting stress tensor in the crystallographic axes.

Case	1	2	3	4
N_1	0	0	$1/\sqrt{3}$	$1/\sqrt{3}$
N_2	0	0	$1/\sqrt{3}$	$1/\sqrt{3}$
N_3	1	1	$1/\sqrt{3}$	$1/\sqrt{3}$
T_1	0	$-1/\sqrt{2}$	$-1/\sqrt{2}$	$-1/\sqrt{6}$
T_2	1	$1/\sqrt{2}$	0	$-1/\sqrt{6}$
T_3	0	0	$1/\sqrt{2}$	$2/\sqrt{6}$
σ_{11}	0	0	$\sigma_n/3 - 2\sigma_t/\sqrt{6}$	$\sigma_n/3 - 2\sigma_t/\sqrt{18}$
σ_{22}	0	0	$\sigma_n/3$	$\sigma_n/3 - 2\sigma_t/\sqrt{18}$
σ_{33}	σ_n	σ_n	$\sigma_n/3 + 2\sigma_t/\sqrt{6}$	$\sigma_n/3 + 4\sigma_t/\sqrt{18}$
σ_{12}	0	0	$\sigma_n/3 - \sigma_t/\sqrt{6}$	$\sigma_n/3 - 2\sigma_t/\sqrt{18}$
σ_{23}	σ_t	$\sigma_t/\sqrt{2}$	$\sigma_n/3 + \sigma_t/\sqrt{6}$	$\sigma_n/3 + \sigma_t/\sqrt{18}$
σ_{31}	0	$-\sigma_t/\sqrt{2}$	$\sigma_n/3$	$\sigma_n/3 + \sigma_t/\sqrt{18}$

4. Normal $\underline{\mathbf{N}}$ in direction $\{111\}$, shear $\underline{\mathbf{T}}$ in direction $\langle\bar{1}\bar{1}2\rangle$, which is one of the six "hard" zones for the case of a tube which axis is $\{111\}$;

The value of the components of the vectors $\underline{\mathbf{N}}$ and $\underline{\mathbf{T}}$ are reported in Table 15.3, together with the components of the resulting stress tensor in the crystallographic frame.

The stress states shown in Table 15.3 have to be mapped using the 18 orientation tensors of Table 15.2 to determine the yield domain.

Case 1. The only components of the stress tensor that are nonzero are the terms 33 and 23. The product $\underset{\sim}{\sigma} : \underset{\sim}{\mathbf{m}}^s$ is therefore $\sigma_{33}m_{33} + 2\sigma_{23}m_{23}$, so that, as far as octahedral slip is concerned, the boundary of the domain is defined by:

- $\sigma_n + \sigma_t = \tau_{octa}\sqrt{6}$ (e.g., system 1);
- $\sigma_n - \sigma_t = \tau_{octa}\sqrt{6}$ (e.g., system 4).

Other systems give $\sigma_n = \tau_{octa}\sqrt{6}$ or $\sigma_t = \tau_{octa}\sqrt{6}$. Obviously, the normal stress has no influence on cube slip. The domain for cube slip is then infinite in σ_n direction, and delimited by two horizontal lines:

- $\sigma_t = \pm\tau_{cube}\sqrt{2}$ (systems 15 to 18).

The resulting domain is shown in Fig. 15.2(a). In this plot, it was assumed that $\tau_{octa} = \tau_{cube} = \tau_y$, and the notation $\sigma_y = \tau_y\sqrt{6}$ has been introduced. In the absence of cube slip, the yield domain is a square which corners are at $(\pm\sigma_y, 0)$ and $(0, \pm\sigma_y)$ in the (σ_n, σ_t) plane. Cube slip introduces two horizontal cuts that limit the domain in shear at $\sigma_t = \pm\sigma_y/\sqrt{3}$.
Case 2. The terms coming from σ_n remain unchanged with respect to the first case, but the multiplicative factor of σ_t is not the same. The limits are now:

- $\sigma_n + \sigma_t/\sqrt{2} = \tau_{octa}\sqrt{6}$ (e.g., system 1);
- $\sigma_n - \sigma_t/\sqrt{2} = \tau_{octa}\sqrt{6}$ (e.g., system 4).

An horizontal cut comes for octahedral slip:

- $\sigma_t\sqrt{2} = \tau_{octa}\sqrt{6}$ (e.g., system 3).

The resulting domain for octahedral slip is now hexagonal. As for the first case, it is cut horizontally if the contribution of cube slip is added:

- $\sigma_t = \tau_{cube}$ (system 17 is exactly in the direction of the macroscopic shear).

The resulting domain is plotted in Fig. 15.2(b), using the same assumptions as above regarding the values of critical resolved shear stress. The horizontal cut provided by octahedral slip is at $\sigma_y/\sqrt{2}$ and that of cube slip at $\sigma_y/\sqrt{6}$.
Case 3. The normal stress σ_n is now along $\{111\}$, and the shear stress is in direction $\langle\bar{1}01\rangle$ in the plane normal to this direction. Contrary to the previous case, the geometry of the crystal facilitates sliding on cube systems in tension, and on octahedral systems in shear. As shown in Fig. 15.2(c), the yield domain is defined by an hexagon. The value of the limit in pure tension is $(3/2)\sigma_y$, and in pure shear $\sigma_y/\sqrt{6}$ (as for system 1, $\tau^1 = \sigma_t$). The equations of the inclined boundaries are

- $\pm\sigma_n \pm \sqrt{6}\sigma_t = (3/2)\tau_{octa}\sqrt{6}$.

The hexagon obtained for cube slip is defined by its pure shear values at $\pm\sqrt{3}\tau_{cube}$, its pure tension value at $3\sqrt{2}\tau_{cube}/2$, that is, respectively $\sigma_y/\sqrt{(2)}$ and $\sigma_y\sqrt{3}/2$ with $\sigma_y = \tau_{cube}\sqrt{6}$. The equations of the four inclined lines are

- $\pm\sigma_n \pm (\sqrt{6}/4)\sigma_t = (3\sqrt{3}/2)\tau_{cube}$.

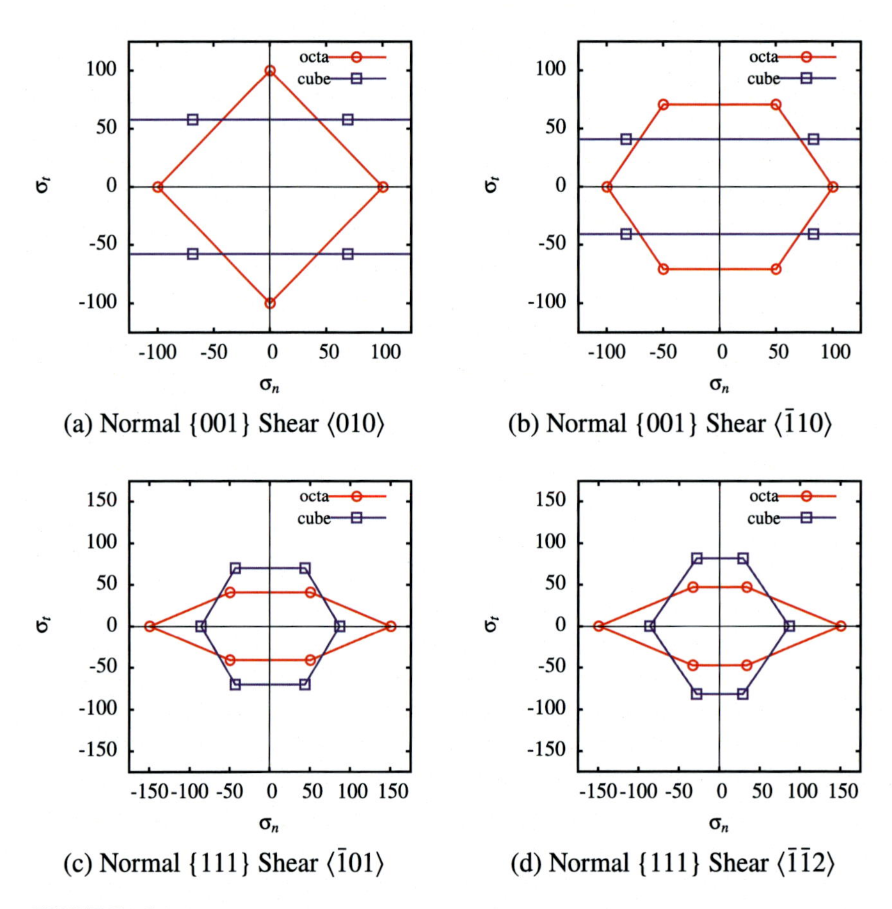

(a) Normal {001} Shear ⟨010⟩

(b) Normal {001} Shear ⟨$\bar{1}$10⟩

(c) Normal {111} Shear ⟨$\bar{1}$01⟩

(d) Normal {111} Shear ⟨$\bar{1}\bar{1}$2⟩

FIGURE 15.2 View of the yield domain for various loading cases in tension–shear. Solid lines are from octahedral systems, dashed lines from cube slip.

Case 4. The general form (Fig. 15.2(d)) is very similar to that of case 3. Obviously, the limit on the tensile axis is the same. The material has a higher shear strength, namely $\sigma_y\sqrt{2}/3$ for octahedral slip and $\sigma_y\sqrt{2/3}$ for cube slip.

15.4.2 Identification strategy

All the models described in this chapter are intended to be implemented in calculation codes in order to carry out numerical simulations by finite elements. Since the loading cases imposed on the structures in which the single crystals are located are anisothermal and multiaxial, and within very large ranges of deformation rates, there are particular problems for identification. It is not essential here to build a very precise model with too many parameters that will be difficult to control over a wide range

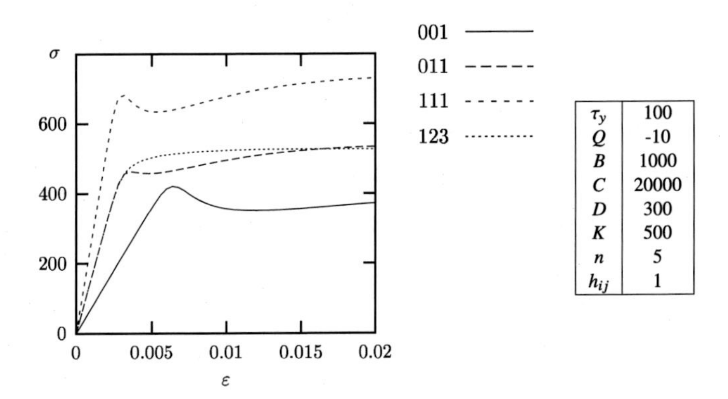

τ_y	100
Q	-10
B	1000
C	20000
D	300
K	500
n	5
h_{ij}	1

FIGURE 15.3 Simulation of the tensile behavior for various crystallographic orientations.

of temperatures and with respect to different types of loading (monotic, cyclic, dwell time, ratcheting, etc.). For reasons of numerical efficiency and reliability with respect to operating conditions, it may be beneficial to be satisfied with a relatively simple framework. However, it is important not to overlook effects that are crucial for controlling behavior. What is important is above all the quality/price ratio from the point of view of technical performance. Concerning the identification of single crystal models under such conditions, three main problems are to be mentioned:

- A certain inconsistency between the first tension, during which the yield strength appears very high, and the cyclic behavior, during which the elasticity domain may appear small. Monotonic tension sometimes shows a rapid softening, which can be modeled by a rapid evolution of the isotropic variable, as shown below in Fig. 15.3. Since industrial loadings are essentially cyclic, it is also possible to disregard this first step and focus only on the steady-state value of the monotonic loading, whose modeling is then consistent with the stress–strain loops obtained under cyclic loading.
- As pointed out in Section 15.3.4, there is sometimes a difficulty to account for viscosity at both very low and high strain rates. This problem might be solved by superimposing the potentials for the same family of slip systems. In this case, if we assume, for example, that $\Omega_1(f_1^s)$ and $\Omega_2(f_2^s)$ are two potentials for octahedral slip, one describing the low strain rates, the other the high strain rates, the viscoplastic flow comes from the extension of Eq. (15.40),

$$\underset{\sim}{\dot{\varepsilon}}^{\mathrm{p}} = \sum_{s} \dot{v}_1^s \eta_1^s \underset{\sim}{\mathbf{m}}_1^s + \sum_{s} \dot{v}_2^s \eta_2^s \underset{\sim}{\mathbf{m}}_2^s. \tag{15.51}$$

- Testing in different crystallographic directions is essential for robust identification. As has been described in several chapters concerning metallurgical aspects, the presence of γ' precipitates disturbs octahedral slip, so that nickel-base superalloys do not react like ordinary fcc materials. As shown in Fig. 15.3, where the model considers octahedral slip only, the strongest direction is $\langle 111 \rangle$ and the softest is $\langle 001 \rangle$. This simulation justifies the fact that it is necessary to introduce into the models an ingredient other than slip on octahedral systems to represent nickel-base single crystals, since in many cases it is crystallographic direction $\langle 001 \rangle$ that presents the highest resistance. As discussed in the previous section, it is possible to replace the critical resolved shear stress on octahedral systems with an expression that takes into account complementary terms. Another consists in introducing into the model a cube slip, see, for instance, [937,680,426,428], (let us speak of "cube" instead of "cubic", following the remark in [1437]), which phenomenologically represents the complex sliding possibilities of dislocations. In such a case, the expression of the flow is formally identical to Eq. (15.51), but with Ω_1 being a potential for octahedral slip and Ω_2 a potential for cube slip. The list of all these slip systems were given in section 14.2.4 of Chapter 14.

In what follows, it is shown that a good quality/price ratio can be obtained with a model which uses on each slip system:

1. An isotropic variable of the "slip resistance" type (Eq. (15.26), without recovery and Eq. (15.27)). The material parameters are then Q, B and h_{rs};
2. A kinematic variable with recovery (Eqs. (15.32) and (15.33)). The material parameters are C, D, M_x, m_x;
3. A viscous flow, represented either by a power function (Eq. (15.40)) or a hyperbolic sinus flow (Eq. (15.41)), with material parameters τ_y, K and n.

It is with such a model that the curves in Fig. 15.3 show the response for tensile loadings in various crystallographic directions, when slip occurs only on octahedral systems. The interaction matrix is full of 1 (Taylor assumption). With the selected parameters, the isotropic variable brings a slight softening just after the onset of plastic flow, which is quite often observed in experiments, especially at high temperature.

In the next section, the model will be used with both cube and octahedral slip system families. A good starting point for the identification consists in considering the two high-symmetry crystallographic directions, for which the response of a specimen in simple tension can be interpreted with an equivalent macroscopic model: in fact, there is no ovalization of the section, and the trace of the plastic deformation tensor for a tension direction 1 is equal to $\varepsilon^p(1; -1/2; -1/2)$, where ε^p is the 11 component. Octahedral slip is the only active family in the case of a tensile loading in $\langle 001 \rangle$ direction (with a Schmid factor $S = 1/\sqrt{6} \approx 0.408$, while it is zero for cube slip). On the contrary, for a tensile test in direction $\langle 111 \rangle$, cube slip is predominant (with a Schmid factor $S = \sqrt{2}/3 \approx 0.471$ for cube slip, and $S = \sqrt{2}/(3\sqrt{3}) \approx 0.272$ for octahedral slip). The number of equivalent systems, N, is 8 octahedral slip systems for a tensile loading in $\langle 001 \rangle$ direction, and 6 cube slip systems for a tensile loading in $\langle 111 \rangle$.

To make a quick comparison with an experiment, it is therefore useful to see in what terms a crystallographic model can be reduced for tensile loading to a classical uniaxial phenomenological response arising from a von Mises-type formulation. A comparison of the two models for a simple tensile case is sufficient to show the equivalence between them. For the crystal plasticity model, the expression of Eq. (15.21) can be simplified, by replacing the sign $\pm$ by $+$, and removing the index s, since all active systems have the same characteristics. The resulting expression is given in Eq. (15.52), together with the formula coming from a von Mises-type viscoplastic model. The latter introduces a kinematic variable X^* and an isotropic variable R^*, the expressions of which are compared with those of the crystallographic model, respectively in Eqs. (15.53) and (15.54). In order to simplify the discussion, and since this is indeed the selected hypothesis for single crystals of superalloys, the interaction matrix is considered as diagonal for the crystallographic model, $h_{rs} = \delta_{rs}$. In the case of off-diagonal terms, the evolution of the isotropic variable would involve the sum of all the terms corresponding to the active systems:

$$\tau = x + \tau_y + R + K\dot{\gamma}^{1/n}, \qquad \sigma = X^* + \sigma_y + R^* + K^*\dot{\varepsilon}^{\mathrm{p}\,1/n}, \tag{15.52}$$

$$x = \frac{C}{D}(1 - \exp(-D\gamma)), \qquad X^* = = \frac{C^*}{D^*}\left(1 - \exp(-D^*\varepsilon^{\mathrm{p}})\right), \tag{15.53}$$

$$R = \; Q(1 - \exp(-B\gamma)), \qquad R^* = \; = Q^*\left(1 - \exp(-B^*\varepsilon^{\mathrm{p}})\right). \tag{15.54}$$

Knowing that $\tau = S\sigma$ and $\varepsilon^{\mathrm{p}} = N_S\gamma$, there is a straightforward equivalence between the formulas, with:

- The same exponents, n and m_x;
- The initial critical shear stress τ_y linked to the initial macroscopic yield stress σ_y by $\tau_y = S\sigma_y$;
- The relations between the other parameters are reported in Table 15.4, with a special reference to both octahedral and cube slip.

15.4.3 Identification on AM1

This section provides some examples of the application of the model to AM1. The first example in Fig. 15.4 shows the stress–strain curves for specimens taken in various crystallographic directions, at a temperature of 950°C [576]. The top right plot is a simple tensile test in the $\langle 011 \rangle$ direction, while the top left, in the $\langle 001 \rangle$ direction, includes a strain rate jump.

TABLE 15.4 Relation between the material parameters of a classical macroscopic model and the parameters of the crystal plasticity model for octahedral slip and cube slip.

Material parameters macro model	Value for multiple slip (N systems, Schmid factor S)	Octa param for ⟨001⟩ tension $N = 8$, $S = 1/\sqrt{6}$	Cube param for ⟨111⟩ tension $N = 6$, $S = \sqrt{2}/3$
K^*	$K/(S(NS)^{1/n})$	$\sqrt{6}K/((8/\sqrt{6})^{1/n})$	$3K/2^{(n+1)/2n}$
σ_y	τ_y/S	$\sqrt{6}\tau_y$	$3\tau_y/\sqrt{2}$
Q^*	Q/S	$\sqrt{6}Q$	$3Q/\sqrt{2}$
B^*	$B/(NS)$	$\sqrt{6}B/8$	$B/\sqrt{2}$
C^*	$C/(NS^2)$	$3C/4$	$3C/2$
D^*	$D/(NS)$	$\sqrt{6}D/8$	$D/\sqrt{2}$
M_x^*	$M_x/(S(NS)^{1/m_x})$	$\sqrt{6}M_x/((8/\sqrt{6})^{1/m_x})$	$3M_x/2^{(m_x+1)/2m_x}$

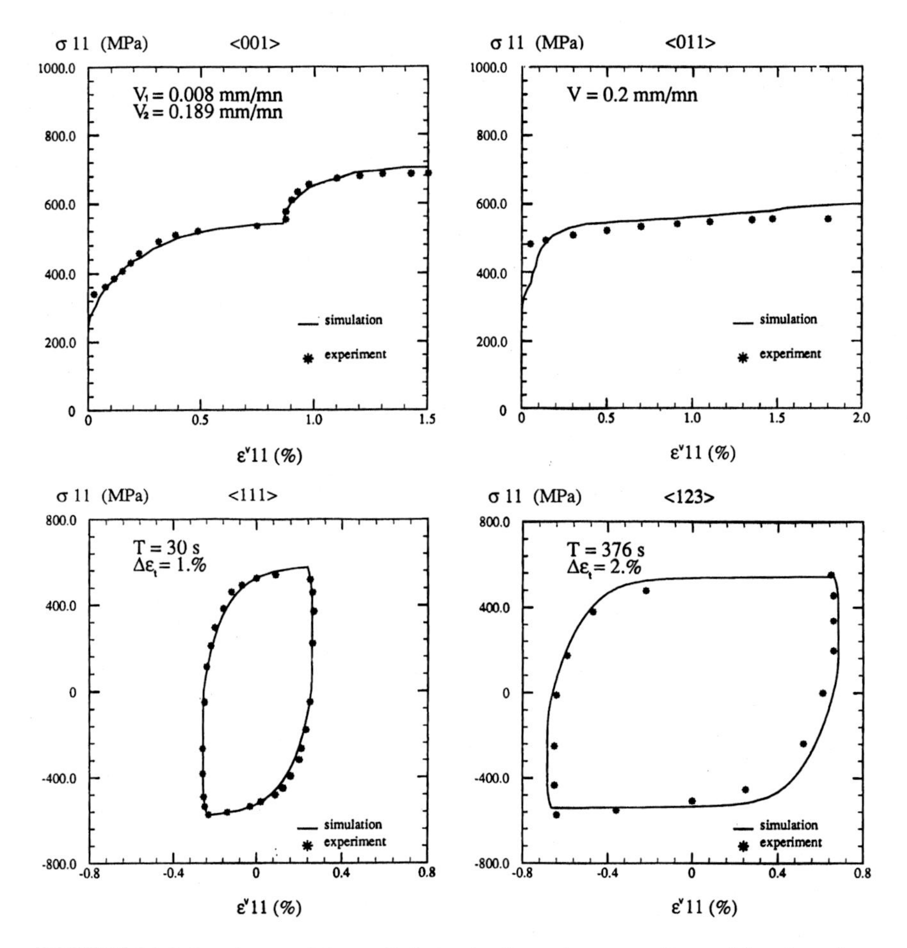

FIGURE 15.4 Simulation of the cyclic behavior for various crystallographic orientations on AM1 specimens [576].

TABLE 15.5 Material parameters from [576] for AM1 superalloy.

	Octahedral slip					Cube slip				
T (°C)	τ_y	K	n	C	D	τ_y	K	n	C	D
20	73	20	10	42000	920	95	18	10	97000	3000
650	72	20	8	95000	2700	85	25	8	127000	3800
950	35	189	5	30000	5800	47	707	5	90500	1100
1100	18	999	2	20000	6300	54	850	2	10000	3300

The two curves at the bottom of the figure represent cycles at the steady state, with moderate loading rates (period of 30 s for the ⟨111⟩ direction at the bottom left, and 30 s for the ⟨123⟩ direction at the bottom right). The material parameters used are shown in Table 15.5. They define the simplest version of the model, which has no isotropic strain-hardening and no recovery on kinematic strain-hardening, limiting the total number of parameters to 10. The viscoplastic flow is defined by a power function, according to Eq. (15.40). A discussion about this identification can be found in [577].

15.4.4 Identification on CMSX4

A few years after the first identifications on AM1 shown in the previous section, it has been shown that the set of parameters fails in correctly representing the viscous effect for a large range of strain rates, and also that the initial yield, which is preferentially calibrated on rapid cyclic tests, is not adequate to predict the behavior for low stresses. The amount of creep is underevaluated, and the relaxation is not large enough. The framework is then slightly modified for an improved identification, including now a hyperbolic sine for the viscous effect (Eq. (15.41)) and some recovery on the kinematic variable (Eq. (15.33)). For the sake of simplicity, and as the cyclic hardening/softening phenomenon is very low in these materials, the isotropic hardening variable is neglected. There are therefore seven parameters to be determined for each family of slip systems.

A very large database on CMSX4 and AM1 was available in the framework of a study between Safran and several academic laboratories [1211]. The data allow obtaining a full characterization, in a 20–1200°C temperature range, of specimens taken in the ⟨001⟩, ⟨011⟩, and ⟨111⟩ directions. The tests include monotonic tensile loadings, creep, cyclic loading without and with dwell, and thermo-mechanical fatigue (TMF).

As for the previous case, the strong behavior in the ⟨001⟩ direction is represented by the octahedral family, while cube slip is the predominant family for simulating the relatively weak behavior in the ⟨111⟩ direction, which allows separate identifications to be made, before carrying both families into the simulations of the tests in direction ⟨011⟩, which exhibit a mixed behavior. In addition, it is possible to coarsely identify the viscosity on the tests under monotonic loadings, and then to refine the value of the initial critical resolved shear stress and the value of the kinematic strain-hardening with cyclic loadings. As far as kinematic strain-hardening is concerned, the static recovery term is only significant for long tests (creep or cyclic relaxation). It is thus this latter type of test which makes it possible to fix the values of the coefficients m_x and M_x at the end of the process.

The parameters are identified separately for each temperature level. Before fixing the final values, it is therefore important to check that the points provided do not lead to abnormal simulations at the intermediate temperatures. A good way to obtain this result is to carry out yield stress simulations. A quick estimation of the value of the engineering yield stress $\sigma_{0.2\%}$ can be easily obtained for a tensile–tensile test in the case of multiple slip. Such an estimation is obtained by replacing ε^{p} by 0.002 and $\dot{\varepsilon}^{\mathrm{p}}$ by $0.001\ \mathrm{s}^{-1}$ (a typical tensile strain rate) in Eqs. (15.55) and (15.56), adapted from (15.52) and (15.53):

$$\sigma = \frac{1}{S}\left\{x + \tau_y + K\left(\operatorname{argsh}\left(\frac{\dot{\varepsilon}^{\mathrm{p}}}{\dot{\gamma}_0 N_S}\right)\right)^{1/n}\right\}, \tag{15.55}$$

$$\text{with} \quad x = \frac{C}{D}\left(1 - \exp\left(-\frac{D\varepsilon^{\mathrm{p}}}{NS}\right)\right). \tag{15.56}$$

Expression (15.56) is valid only if the loading rate is fast enough to neglect the recovery term. If this is not the case, the asymptotic value of x decreases with loading rate, thus for decreasing applied stress. This level can be understood by considering the link between the saturated value of the variable, x_S, and the strain rate in secondary creep $\dot{\varepsilon}^{\mathrm{p}}_S$ under a stress σ_0, which is obtained by annulling the evolution rate in Eq. (15.32):

$$0 = (C - Dx_S)\frac{\dot{\varepsilon}^{\mathrm{p}}_S}{N_S} - \left(\frac{x_S}{M_x}\right)^{m_x}. \tag{15.57}$$

There is a one-to-one relation between σ_0 and the asymptotic value x_S,

$$S\sigma_0 = x_S + \tau_y + K\left(\operatorname{argsh}\left(\frac{\left(\frac{x_S}{M_x}\right)^{m_x}}{\dot{\gamma}_0(C - Dx_S)}\right)\right)^{1/n}. \tag{15.58}$$

There are also a series of rules of thumb that need to be verified:

- The fact that the influence of the hardening variables decreases as the temperature increases, to the point that at very high temperatures (here above 1100°C) the model is almost exclusively viscous.
- The exponent n which characterizes viscosity must decrease smoothly with temperature, until it reaches 1 for the melting temperature,

whereas it takes high values (from 20 to 40) at room temperature to tend towards a behavior practically time independent.
- The kinematic variable defines the shape of the hysteresis loops in the stress–strain plane. In order to improve the representation of the hysteresis loops, it is also possible to "superimpose" several kinematic variables with different values of the decay parameter d. This makes it possible to obtain loops that are more rounded, as shown in the tests when the temperature rises. Note that it is not recommended to significantly modify the d parameter value with temperature [191].
- The value of the parameter K should be small at room temperature, since the viscous effect is almost absent, and also small at very high temperatures, since the stress level itself is low. Depending on the material and the identification strategy, a maximum is classically reached between 900°C and 1000°C, which is the range where the viscous stress takes its highest values.

Three materials were considered, MC2, CMSX4, and AM1. For the sake of brevity, the examples are limited to significant tests performed at ENSMA [278] and Mines ParisTech [751] on CMSX4. The numerical simulations were made by Azzouz [60].

Fig. 15.5 shows an example of monotonic tests on CMSX4. Fig. 15.5(a) shows three tensile tests at 1200°C, with different strain rates (10^{-5}, 10^{-4}, and 10^{-3} s^{-1}). At this temperature, the initial critical resolved shear stress is very low, and the deviation between the curves is largely due to the viscous term. The saturation effect from hyperbolic sine expression is not very significant in the selected strain rate range, since the deviations are about the same for both jumps of an order of magnitude. It is at higher strain rates that this formulation is useful. On the other hand, it is indeed the recovery introduced on the kinematic variable that determines the flow stress at low strain rates. In fact, when the slope of the tensile curves in the strain–stress plane becomes zero, this means that the test becomes a secondary creep test. Under these conditions, the deviation between the curves is directly influenced by the shape of the curve defined by Eq. (15.58). The lower and lower value of x_S when σ_0 is reduced from 515 to 360 MPa allows creep to actually exist at the latter stress level in Fig. 15.5(b), at 950°C.

The low cycle fatigue tests shown in Fig. 15.6 have a trapezoidal imposed strain, with relaxation periods at maximum and minimum strain. Here also the agreement is good, knowing the fact that the material response is much stronger for the $\langle 001 \rangle$ direction in Fig. 15.6(a) than for the $\langle 111 \rangle$ direction in Fig. 15.6(b).

Once the identification has been performed on isothermal tests, the robustness of the parameter set can be assessed on anisothermal loadings. Table 15.6 defines two of them, the first "in-phase" (IP), for which the temperature is maximum when the mechanical strain is maximum, and

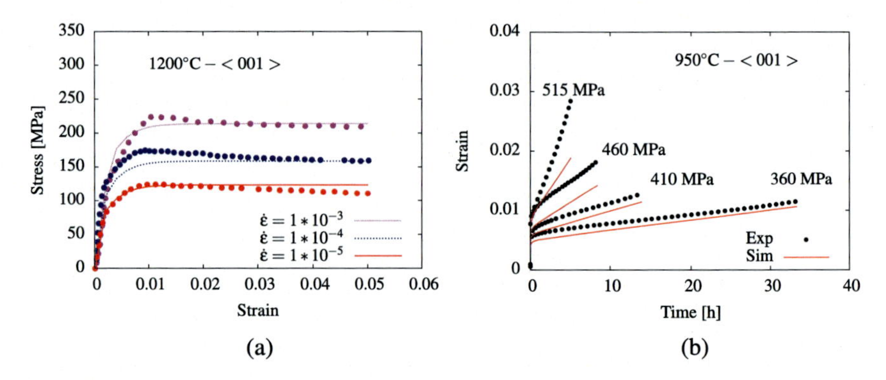

FIGURE 15.5 Numerical simulation of monotonic tests in the ⟨001⟩ direction on a CMSX4 single crystal [60] and comparison with experimental results [278]: (a) tensile tests at 1200°C with various strain rates (10^{-5}, 10^{-4}, 10^{-3} s^{-1}), (b) creep tests at 950°C with various initial stresses (360, 410, 460, 515 MPa) (European project PREMECCY, Predictive methods for combined cycle fatigue in Gas Turbine Blades, 2006–2011).

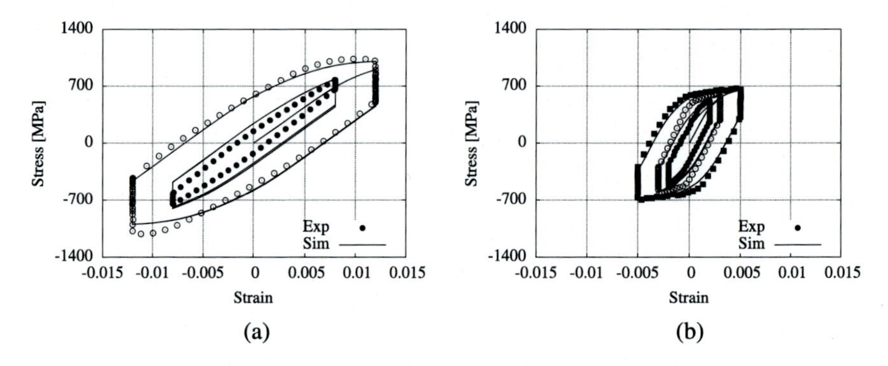

FIGURE 15.6 Numerical simulation of the stress–strain loops at various strain amplitudes for LCF tests at 950°C on a CMSX4 single crystal when steady state is reached [60] and comparison with experimental results [751]. There are dwell time periods of 1800 s at maximum and minimum strain: (a) ⟨001⟩ tensile direction, (b) ⟨111⟩ tensile direction.

TABLE 15.6 Definition of the strain and temperature histories for the in-phase and out-of-phase TMF tests [751].

Time (s)	In-phase test (IP)		Out-of-phase test (OP)	
	ε (mm/mm)	Temperature (°C)	ε (mm/mm)	Temperature (°C)
0	0	700	0	700
50	−0.0045	450	−0.0045	950
150	0.0045	950	−0.0045	950
450	0.0045	950	0.0045	450
500	0	700	0	700

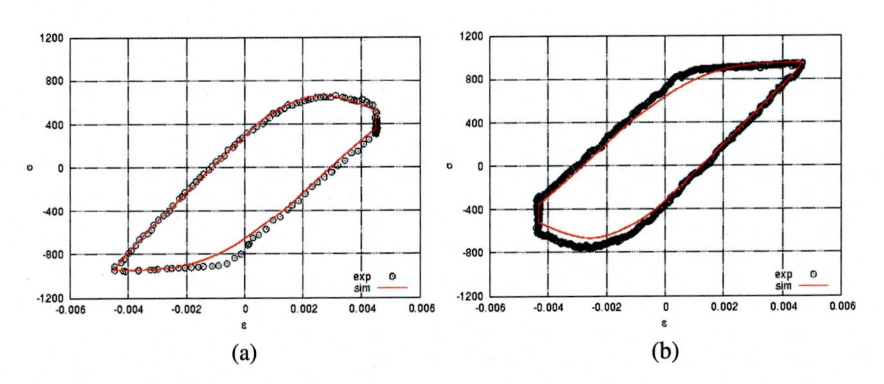

FIGURE 15.7 Numerical simulation of the stress–stress loops for two TMF tests on a CMSX4 single crystal in the ⟨111⟩ direction [60] and comparison with experimental results [751]. The loading histories are in Table 15.6: (a) in-phase test (IP), (b) out-of-phase test (OP).

the second "out-of-phase" (OP), for which the temperature is maximum when the mechanical strain is minimum. There is a relaxation period in both tests, in tension for IP, in compression for OP. The resulting curves for the thermomechanical steady state are shown in Fig. 15.7, where ε on the x-axis denotes the mechanical deformation (thermal expansion is not shown). During cooling from 450°C to 950°C (compression-going branch for IP in Fig. 15.7(a) and tension-going branch for OP in Fig. 15.7(b)), the yield stress is slightly underestimated by the model, but the final stress level at 450°C is correct. During heating, the stress level goes through a maximum at intermediate temperature (around 750°C) for IP, and a minimum for OP. The amount of relaxation is well described in both cases.

Acknowledgments

The collaborative research project "Structures Chaudes" (SAFRAN–ONERA–CNRS) provided an opportunity for productive scientific discussions for academic research fellows and engineers of the industrial group involved. People who have worked in the field of high-temperature single crystals include: at ONERA, Jean-Louis Chaboche, Pascale Kanouté, Franck Gallerneau, Anaïs Gaubert, Florent Fournier; at Safran Helicopter Engine, Zéline Hervier, Elisabeth Ostoja-Kuczynski; at Safran Aircraft Engines, Geoffrey Desmeure; at Mines ParisTech, Farida Azzouz, Alain Koster, Luc Rémy; at ISAE-ENSMA, Jonathan Cormier. The author of this chapter would like to thank SAFRAN for the support provided within the framework of the SAFRAN–Mines ParisTech "CRISTAL" chair on high-temperature materials.

CHAPTER

16

Crystal plasticity and damage at cracks and notches in nickel-base single-crystal superalloys

Samuel Forest

MINES ParisTech, PSL University, Centre des matériaux, CNRS UMR 7633, Evry, France

16.1 Introduction

The complex shape of single crystal turbine blades is a source of strong stress concentrations that may ultimately lead fatigue crack initiation or creep damage. They arise at cooling holes and slits of various sizes and orientations. The amplitude of such stress concentrations must be predicted from the anisotropic constitutive law for single crystals. This is not an easy task due to counterintuitive anisotropy effects and strong dependence on crystal orientation. An example of concentrated load is given by the indentation of a single crystal by a polycrystalline part as it happens in the contact zone between turbine blade and disc in the fir-tree zone of the blade's footing. Crystal plasticity predictions can be validated by comparing the activated slip systems according to the model and the slip lines observed in the experiment, as done in the indentation test of Fig. 16.1. The ideal situation of a rigid cylinder indenting a nickel-base single-crystal superalloy shows the large plastic zone size under the cylinder and the existence of specific zones of slip activation [1250]. Excellent agreement is obtained for the prediction of slip system activation zones using conventional crystal plasticity based on Schmid's law (see Chapters 14 and 15). The slip patterns can be shown to strongly depend on the orientation of the single crystal, see [1250] for an example where the secondary orientation is changed, keeping the [001] indentation axis.

Nickel Base Single Crystals Across Length Scales
https://doi.org/10.1016/B978-0-12-819357-0.00025-1

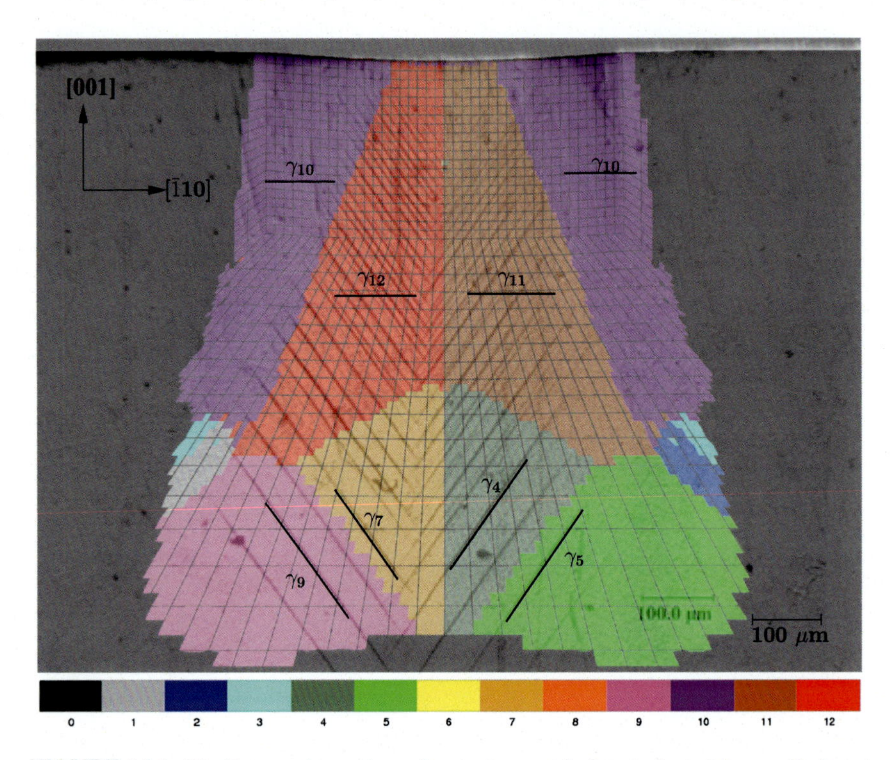

FIGURE 16.1 Slip lines at the surface of a single crystal plate indented by a cylinder at room temperature. The field of activated slip systems as predicted from an FE simulation is superimposed, after [1250].

Little attention has been paid to the observation of the slip line patterns around holes, notches, and cracks in nickel-base single-crystal superalloys. It is a pity since a lot can be learned about the anisotropy of the crystal response. Such observations are reported in this chapter in order to validate the crystal plasticity constitutive framework. Crystallography also plays a fundamental role in the crack propagation paths in nickel-base single-crystal superalloys. Recent continuum damage models incorporate these essential crystallographic features. They are discussed in the second part of the chapter. They account for crack initiation, branching, and bifurcation that are frequently observed in engineering applications.

16.2 Crystal plasticity at a crack tip in single crystals

The problem of the stress field at a crack tip in a single crystal in perfect plasticity under plane strain conditions was solved by Rice in 1987 [1220]. The remarkable solution predicts sectors with uniform stress separated by

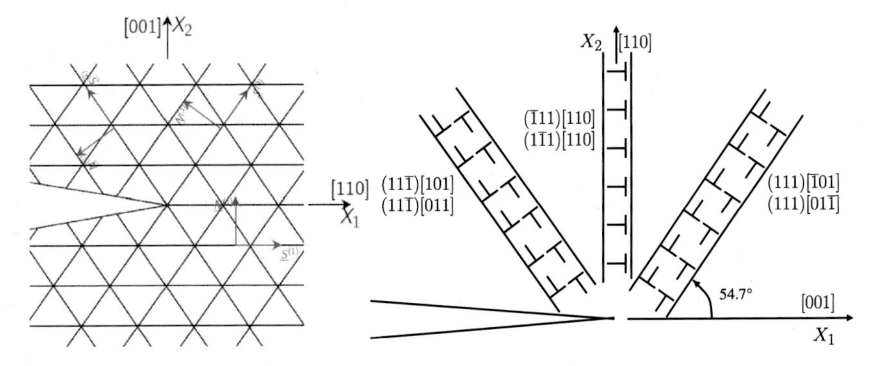

FIGURE 16.2 Theoretical traces of the effective slip planes for plane strain deformation at a crack tip in a single crystal (left). Strain discontinuities accommodated by slip and kink bands (right), as expected from Rice's theoretical analysis of a crack in a FCC single crystal.

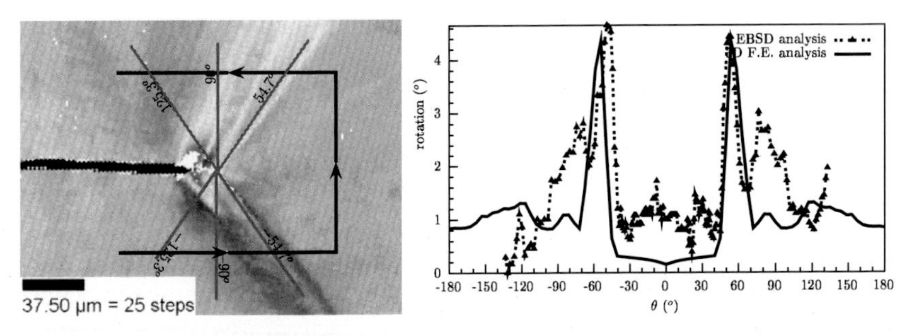

FIGURE 16.3 EBSD mapping of lattice orientation at the crack tip of a nickel-base single-crystal superalloy (left). Comparison between experiment and simulation (right), after [451].

strain discontinuity lines. The existence of such sectors has been confirmed by experiments in CT specimens in [301] for aluminium and copper crystals and for the first time in [451] for nickel-base single-crystal superalloys. A debated question arising from Rice's solution is the existence of kink bands limiting some sectors. Kink bands are strain localization bands that are perpendicular to slip directions (Burgers vectors). This is in contrast to slip bands which are parallel to slip planes as usual. Such slip and kink bands are illustrated in Fig. 16.2 at the crack tip for a specific crystal orientation. Two inclined kink bands and a vertical slip band are expected. They were indeed observed in the experiments reported in [451]. Confirmation of the existence of kink bands was given later on in [1065,1066]. Kink bands can be easily detected by the large amount of lattice rotation that takes place inside such bands, as proved by the EBSD map of Fig. 16.3. Again, the crystal plasticity theory relying on the Schmid law provides excellent predictions of the magnitude of such lattice rotation bands, as reported in Fig. 16.3. Strong lattice rotation gradients develop

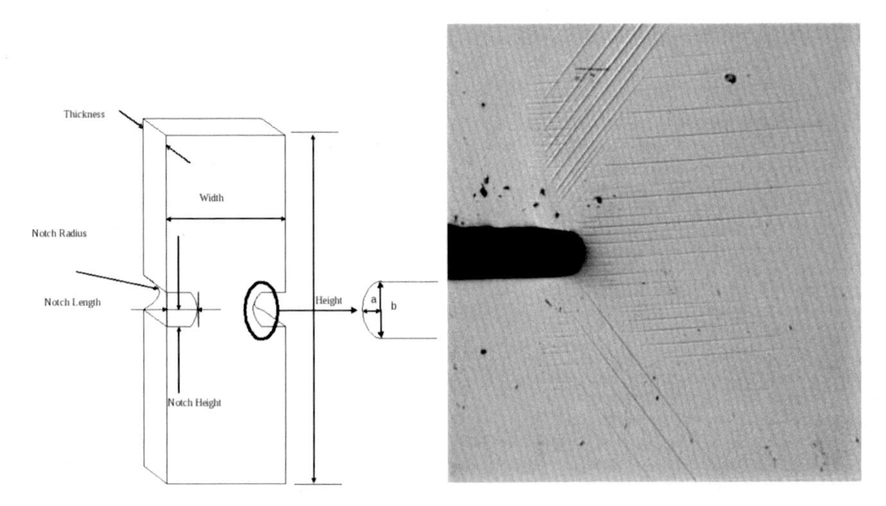

FIGURE 16.4 Slip lines at a notch in a nickel base single crystal superalloy. Specimen orientation (001)[$\bar{1}$10], after [1251].

in kink bands in the form of lattice curvature usually associated with the accumulation of geometrically necessary dislocations (GND) [47]. The magnitude of kink vs. slip bands can be controlled by the amount of GND hardening in strain gradient plasticity models, see [456,1464]. Most observations of crack tip plasticity have been made at low temperatures where slip lines are clearly visible. Results at high temperatures can be found in [903,1554].

16.3 Slip and kink banding at notches in single crystals

Notches in single crystal plates are often studied to mimic the effect of cooling holes. Analytical solutions do not exist in that case. Finite element simulations can be used to interpret the experimental results. The slip patterns at notches share common features with crack tip behavior. The notch region is decomposed into domains with dominant slip systems separated by clear limit lines on the free surface shown in Fig. 16.4. In this picture, two vertical kink bands are visible as predicted by finite element simulations. Such slip patterns were reported in [1064,1251].

16.4 Continuum damage modeling of single-crystal superalloys

A common feature of damage phenomena in nickel-base single-crystal superalloys is the absence of diffuse damage (except perhaps for some very high temperature creep conditions) but rather rapid localization into cracks especially under fatigue loading conditions. Continuum damage models for single crystals were, for instance, proposed in [1154,1449]. However, they do not take this feature into account. Localization of damage in the form of cracks is thought to be a consequence of cumulative slip at some locations of the single crystalline structure. Its prediction requires a strong coupling between crystal plasticity and damage. The structure of such models and their predictions are discussed in this section. Fatigue damage indicators based on crystallographic variables can be used as driving forces for crack initiation and propagation [1565,1564]. A recent account of continuum damage modeling for nickel-base single-crystal superalloys can be found in [693].

16.4.1 Fatigue crack initiation in cylindrical notched specimens

Interesting experimental findings about fatigue crack initiation in notched cylindrical specimens were reported in the PhD work of P. Boubidi based on experiments performed at BAM-Berlin [138]. Testing along the <001> direction reveals that crack initiation does not take place at the notch root but rather slightly above and/or below, see Fig. 16.5. Finite element simulations confirmed that these regions are the location of the highest cumulative slip amount due to the combination of elastic and plastic anisotropy. This is in contrast to the two other tested orientations, as shown again in Fig. 16.5.

16.4.2 Modeling damage mechanisms in fatigue

The complexity of crack paths in nickel-base single-crystal superalloys has been seen very early, for instance, in [300]. The crack path with periodic branching and bifurcation is illustrated in Fig. 16.6 for CMSX4 and two crystal orientations, namely (001)[100] and (001)[011]. The former test was performed at 950°C and the latter at an intermediate temperature, resulting in an even more fragmented crack path. Such complex behavior challenges continuum modeling and fracture mechanics concepts. A common feature, however, is the crystallographic nature of the observed crack planes which are identified as {111} planes. It is apparent in the middle picture of Fig. 16.1 that the fatigue crack follows the path of former intense slip bands and periodically bifurcates along such bands.

Based on these observations, a continuum damage model for nickel-base single-crystal superalloys has been proposed in [50] as follows. Three damage systems are associated to each {111} plane, resulting in 12 damage mechanisms including the 6 following ones as an example:

$$(111)[111],\ (111)[11\bar{2}],\ (111)[1\bar{1}0],\ (11\bar{1})[11\bar{1}],\ (11\bar{1})[112],\ (11\bar{1})[1\bar{1}0]$$

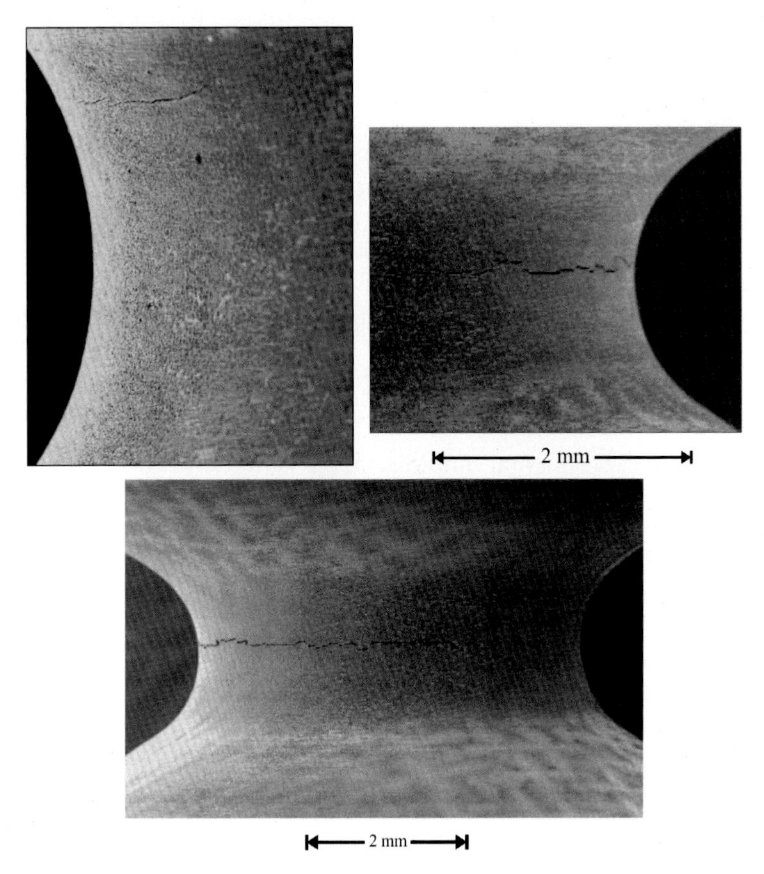

FIGURE 16.5 Fatigue crack initiation and propagation at the surface of notched cylindrical specimens, here SC16 nickel-base single-crystal superalloys, after [138]. Three tensile crystal orientations are considered: <001> (top left), <011> (top right), and <111> bottom.

The total strain rate is decomposed into elastic, plastic, and damage parts (leaving aside the thermal part for simplicity) as

$$\dot{\boldsymbol{\varepsilon}} = \dot{\boldsymbol{\varepsilon}}^e + \dot{\boldsymbol{\varepsilon}}^p + \dot{\boldsymbol{\varepsilon}}^d. \tag{16.1}$$

The plastic strain rate is characterized by the usual kinematics of slip, see Chapter 14. The damage strain results from normal opening of {111} planes and shear accommodation inside each plane, respectively corresponding to mode I, II, and III damage mechanisms,

$$\dot{\boldsymbol{\varepsilon}}^d = \sum_{s=1}^{N^d_{planes}} \dot{\delta}^s_c \boldsymbol{n}^s \otimes \boldsymbol{n}^s + \dot{\delta}^s_1 \boldsymbol{n}^s \otimes \boldsymbol{\ell}^s_1 + \dot{\delta}^s_2 \boldsymbol{n}^s \otimes \boldsymbol{\ell}^s_2, \tag{16.2}$$

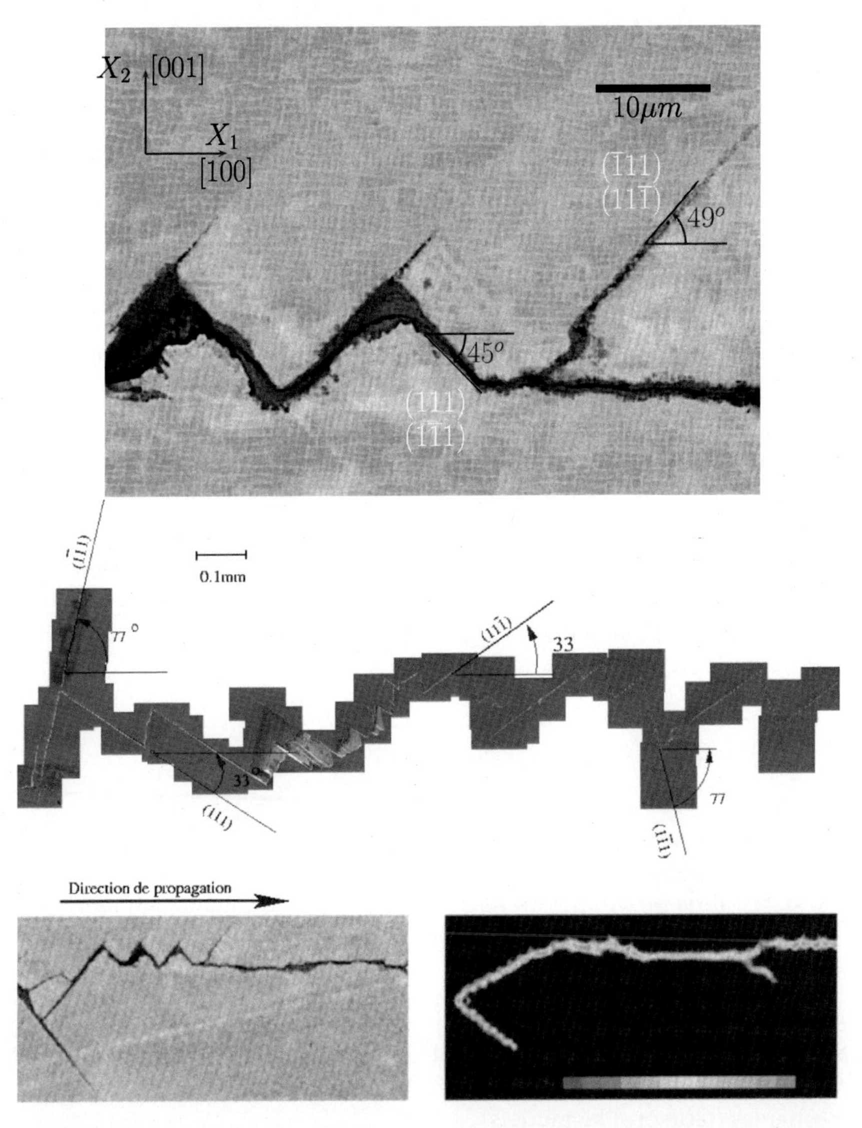

FIGURE 16.6 Fatigue crack paths in nickel-base single-crystal superalloysat intermediate temperatures (source, Centre des Matériaux, Mines ParisTech). In the middle, the width of the picture is 0.5 mm. (Bottom pictures) Qualitative comparison of experimental crack path and a finite element simulation by Ozgur Aslan (2010).

where N^d_{planes} is the number of damage planes, $\boldsymbol{n}^s$ the corresponding normal vectors, and $\boldsymbol{\ell}^s$ the damage shear accommodation directions contained in that plane. In contrast to plastic slip, the damage strain rate is generally not traceless due to the volume change associated with internal crack opening embodied by the $\boldsymbol{n}^s \otimes \boldsymbol{n}^s$ contribution.

The damage criteria are the three following functions introduced for each damage plane:

$$f_c^s = \left| \boldsymbol{n}^s \cdot \boldsymbol{\sigma} \cdot \boldsymbol{n}^s \right| - Y, \quad f_i^s = \left| \boldsymbol{n}^s \cdot \boldsymbol{\sigma} \cdot \boldsymbol{l}_i^s \right| - Y \ (i = 1, 2). \tag{16.3}$$

It means that the driving force for crack opening is the normal stress with respect to the considered cleavage plane. The word *cleavage* is, in fact, reserved to brittle fracture mechanisms which is not the case in general in nickel-base single-crystal superalloys. It is used somewhat abusively in the sequel. The two last functions deal with mode II and III damage mechanisms that are activated once cleavage-like damage has started. The evolution law for crack opening is taken as a usual power law in the form:

$$\dot{\delta}_c^s = \left\langle \frac{f_c^s}{K_d} \right\rangle^{n_d} \operatorname{sign}\left(\boldsymbol{n}^s \cdot \boldsymbol{\sigma} \cdot \boldsymbol{n}^s\right), \quad \dot{\delta}_i^s = \left\langle \frac{f_i^s}{K_d} \right\rangle^{n_d} \operatorname{sign}\left(\boldsymbol{n}^s \cdot \boldsymbol{\sigma} \cdot \boldsymbol{l}_i^s\right), \tag{16.4}$$

thus introducing rate-dependence of damage mechanisms. The essential part of the model is the coupling between plasticity and damage. The cleavage stress Y is a decreasing function of accumulated plastic slip γ_{cum},

$$Y = \sigma_n^c \, e^{-\Theta \gamma_{cum}} + Hd + \sigma_{ult}, \quad \text{with} \quad \dot{\gamma}_{cum} = \sum_{s=1}^{N_{slips}} |\dot{\gamma}^s|. \tag{16.5}$$

The rate of decrease of the cleavage stress with respect to plastic slip is Θ. The modulus H is taken negative and induces softening as soon as damage starts, thus provoking damage localization and microcrack initiation. The cumulative damage variable is d according to

$$\dot{d} = |\dot{\delta}_c^s| + |\dot{\delta}_1^s| + |\dot{\delta}_2^s|. \tag{16.6}$$

Note that it is not a bounded variable like in classical continuum damage mechanics. It is rather a strain-like variable in the spirit of ductile fracture mechanics (Gurson-like models).

An extension of the model at finite deformations within the multiplicative decomposition of the deformation gradient into elastic and inelastic parts, the latter combining the plasticity and damage kinematics, can be found in [48].

16.4.3 Model regularization strategy

Continuum damage models involving softening behavior are known to result in the loss of ellipticity of the boundary value problem, associated with emerging localization modes involving strain rate jumps [102].

A consequence on finite element simulations is that the thickness of simulated cracks is one element. This spurious mesh-dependence of the results precludes any reliable assessment of crack propagation. Regularization procedures are needed to restore the well-posedness of the problem. Gradient approaches introducing penalty terms on the gradient of damage or gradient of strain are good candidates to perform mesh-objective simulations of crack propagation. A general framework has been proposed to implement in an efficient and thermodynamically consistent way such gradient approaches. It is called the *micromorphic approach* which relies on the introduction of additional damage or strain-like degrees of freedom and their gradient in the constitutive framework [454]. This micromorphic regularization technique has been used in several recent continuum damage models [49,1248,152,942], including contributions accounting for finite deformations as it should for crack propagation.

The use of such models requires the choice of an internal length scale. In the case of nickel-base single-crystal superalloys the thickness of cracks is usually very thin reaching the size of precipitates. This is far too small to allow for engineering computations of components and parts. In the spirit of phase field models (cf. Chapter 13), it is recommended to fix a fictitious damaged zone thickness that is larger than the actual one but significantly smaller than the component size. This choice can be made a priori depending on the chosen resolution of the finite element simulations. It must be kept in mind that 2 to 5 elements are necessary to properly resolve the damage zone and obtain converged solutions in the sense of finite elements.

The regularization technique is similar to the so-called phase field damage models [1352] which can be more precisely called gradient damage models.

The ability of such regularized localizing damage models to simulate crack propagation and branching is illustrated by the bottom pictures of Fig. 16.6. The comparison with experiment is qualitative since the simulation was obtained under monotonic loading, whereas the test results are for fatigue cracking, the topic of the next subsection.

16.4.4 Simulation of fatigue crack growth

Continuum damage models applicable to fatigue loading conditions require additional rules allowing for crack closure effects in the form of unilateral damage conditions. This the case in the model presented in the previous section. When the crack path is known, an easier technique is to resort to cohesive zone models. Such models have been developed for nickel-base single-crystal superalloys in [147].

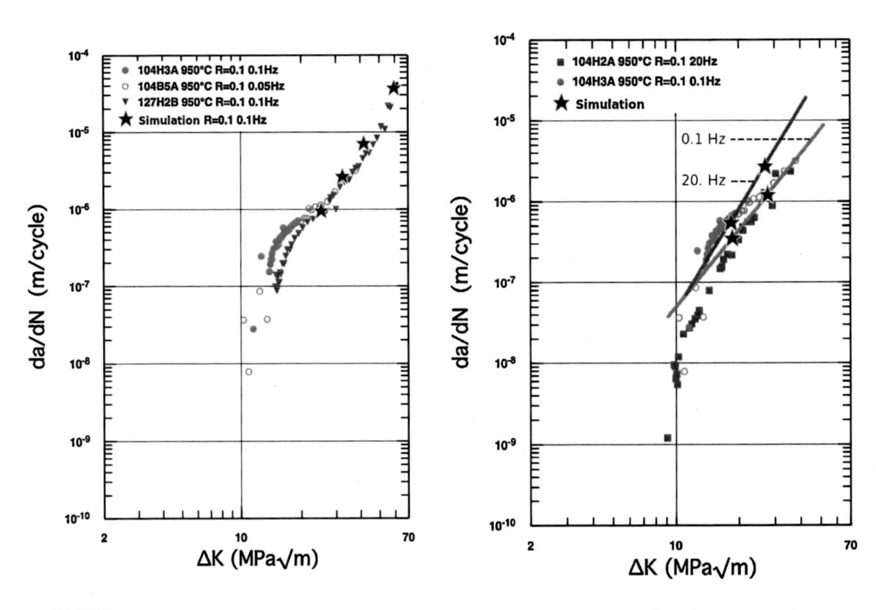

FIGURE 16.7 Paris diagrams for fatigue crack propagation in (001)[100] oriented CT specimens: experimental results and model prediction, after [903].

A typical crack advance mechanism in ductile single crystals like copper or aluminium is the blunting of the crack tip followed by buckling-like phenomenon during reverse loading. This has been exploited in [826] in order to simulate striation fatigue mechanism and estimate crack growth rates. Such blunting is, however, not pronounced enough in nickel-base single-crystal superalloys.

The damage parameters of the continuum model presented in the previous section can be identified from the da/dN curves for single crystal CT specimens, as done on the left of Fig. 16.7. Once calibrated, the model can make predictions regarding orientation dependence and change of loading path, including hold times, for instance. The model predicts a very limited orientation dependence of the Paris diagram in agreement with experimental data for high symmetry crystal orientations. This is due to the wealth of slips systems activated at the crack tip for such orientations. More experiments are needed for more complex and to some extent more realistic cumbersome orientations. The effect of fatigue loading rate predicted by the models contradicts the experimental results [50]: the model predicts faster crack growth rates at higher frequencies, according to Fig. 16.7 on the right. In contrast, the considered experiments are performed under air conditions thus promoting oxidation at slow fatigue rates. Such detrimental effects should be incorporated in the damage model. Orientation effects on crack growth were also tackled in [627] from the experimental perspective.

16.4.5 Creep damage in combined tension and shear

Shear loading conditions are useful to validate modeling approaches and ensure conservative design of components. Early attempts for shear testing of nickel-base single-crystal superalloys were performed on H-shaped samples [922] or four point bending specimens as in Fig. 16.8, within the SOCRAX European project. The latter test was recently simulated based on the previous regularized continuum damage model, see Fig. 16.8 and [1249]. The test was performed at 950°C so that viscoplastic and fatigue effects were combined. A comparison of simulated and experimental results shows good agreement regarding the orientation of the crack path.

The central zone of the specimen with asymmetric notches shown in Fig. 16.9 is subjected to combined shear and tension. Under creep loading, a crack initiates in the notches and then bifurcates to bridge the two notches. This scenario has been properly predicted by the proposed continuum damage model. More quantitative comparisons are necessary to validate this class of models, in the form of strain and lattice orientation field measurements.

Creep in nickel-base single-crystal superalloys sometimes results in ductile fracture associated with the growth and coalescence of small cavities induced by material casting. This can be modeled by extensions of Gurson-like models to single crystals, as done recently in [843]. This promising approach should be combined with a coalescence criterion as recently proposed for single crystals in [642].

Finally, the computational strategy should be applied to arrays of holes (see the experiments in [1553]) as they are encountered in actual single crystal turbine blades.

16.5 Conclusions and prospects

Strain localization phenomena are ubiquitous in single crystals around holes, slits, and cracks in the form of slip or kink bands. They can be predicted with great accuracy based on crystal plasticity constitutive equations. They serve as precursors for crack initiation under fatigue or creep loading conditions. Three-dimensional finite element simulations are required to capture the crystallographic features of plasticity and fracture of nickel-base single-crystal superalloys. Continuum damage models incorporating many crystallographic aspects of material behavior have been reported in this chapter. They are designed to predict crack initiation, propagation, branching, and bifurcation under complex loading paths including rate-dependence effects. They incorporate an intrinsic length scale in order to set a lower limit to the model resolution and produce mesh-independent finite element results. There is still a long way for a sys-

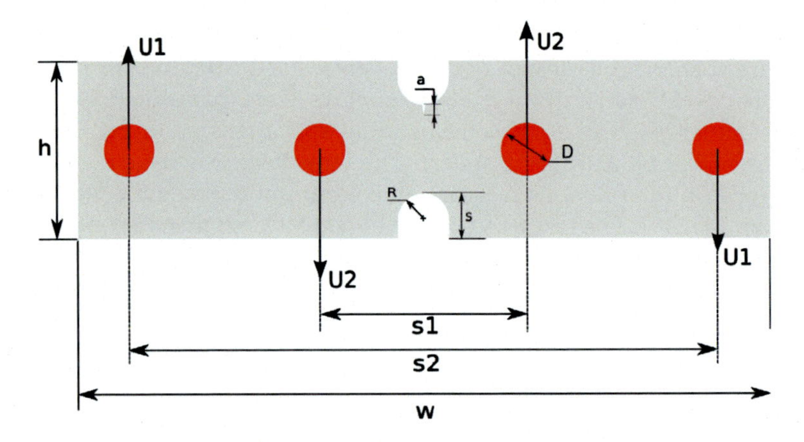

Geometry of the specimen used for testing crack propagation under Mode II loading

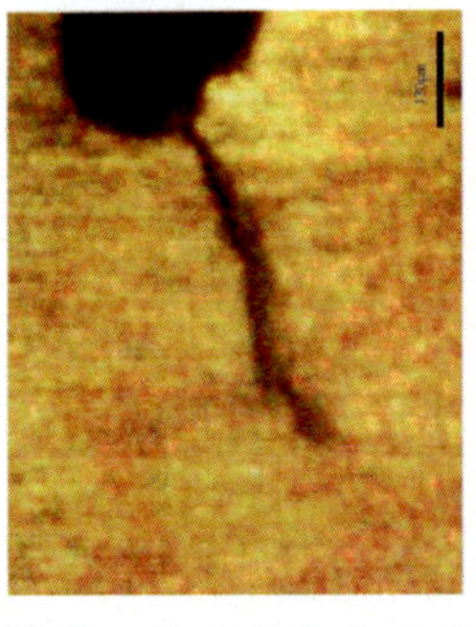

(a) Experimental short crack observation in fatigue; crack length = 500 μm.

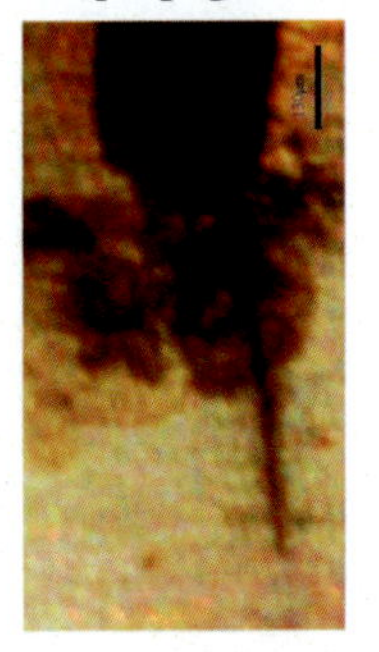

(b) Experimental short crack observation in fatigue; crack length = 650 μm.

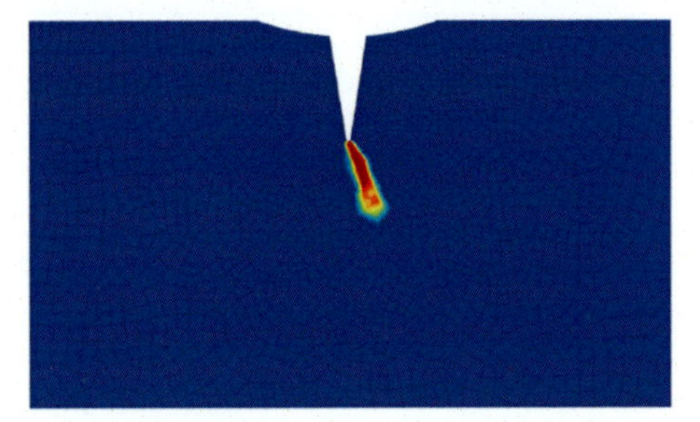

(c) Simulated short crack growth under monotonic loading; crack length = 600 μm.

FIGURE 16.8 Four point shear specimen: geometry, crack propagation at high temperature, and finite element simulation, after [1249].

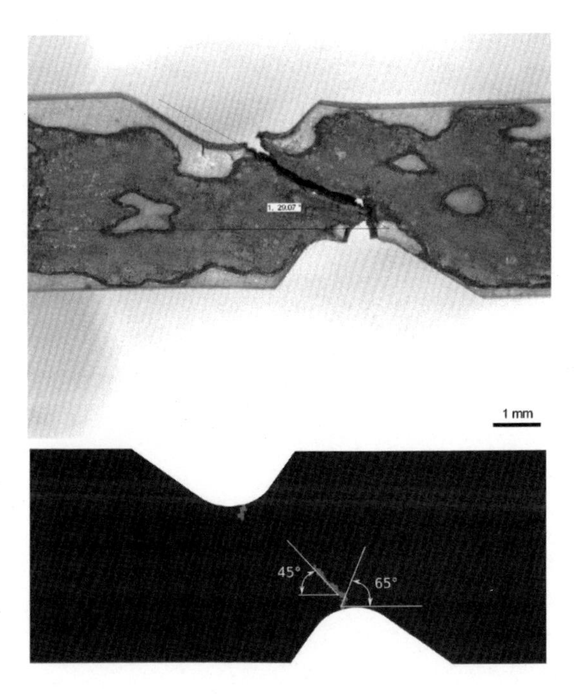

FIGURE 16.9 Combined shear and tension specimen under creep: shear crack formation and propagation. Experiment vs. finite element simulation of the damage zone, after [1249].

tematic use of such models due to their strong nonlinearity, leading to convergence difficulties. Improvements of the constitutive laws are still necessary based on detailed comparison between model prediction and strain, and lattice rotation field measurements. Such models can also be used for explicit computations of damage processes at large strains [1248].

PART IV

Application to engineering cases

CHAPTER

17

Implementation of constitutive equations for single crystals in finite element codes

Jean-Michel Scherer[a,b] *and Jacques Besson*[b]

[a]Université Paris-Saclay, CEA, Service d'Etude des Matériaux Irradiés, Gif-sur-Yvette, France, [b]Centre des Matériaux, Mines ParisTech — PSL Research University, UMR CNRS 7633, Evry, France

Glossary

Orthonormal basis $\mathcal{R} = \{\underline{\mathbf{e}}_1, \underline{\mathbf{e}}_2, \underline{\mathbf{e}}_2\}$

Vector $\underline{\mathbf{a}} = a_i \underline{\mathbf{e}}_i$

2nd-order tensor $\underset{\sim}{\mathbf{A}} = A_{ij} \underline{\mathbf{e}}_i \otimes \underline{\mathbf{e}}_j$

Transpose $\underset{\sim}{\mathbf{A}}^T = A_{ji} \underline{\mathbf{e}}_i \otimes \underline{\mathbf{e}}_j$

Inverse $\underset{\sim}{\mathbf{A}}^{-1} = A_{ij}^{-1} \underline{\mathbf{e}}_i \otimes \underline{\mathbf{e}}_j$

Transpose of the inverse $\underset{\sim}{\mathbf{A}}^{-T} = A_{ji}^{-1} \underline{\mathbf{e}}_i \otimes \underline{\mathbf{e}}_j$

Scalar product $\underline{\mathbf{a}} \cdot \underline{\mathbf{b}} = \sum_{i=1}^{3} a_i b_i$

Double contraction $\underset{\sim}{\mathbf{A}} : \underset{\sim}{\mathbf{B}} = \sum_{i=1}^{3} \sum_{j=1}^{3} A_{ij} B_{ij}$

Tensorial product $\underline{\mathbf{a}} \otimes \underline{\mathbf{b}} = a_i b_j \underline{\mathbf{e}}_i \otimes \underline{\mathbf{e}}_j$

Tensorial product $\underset{\sim}{\mathbf{A}} \otimes \underset{\sim}{\mathbf{B}} = A_{ij} B_{kl} \underline{\mathbf{e}}_i \otimes \underline{\mathbf{e}}_j \otimes \underline{\mathbf{e}}_k \otimes \underline{\mathbf{e}}_l$

Tensorial product $\underset{\sim}{\mathbf{A}} \overline{\otimes} \underset{\sim}{\mathbf{B}} = A_{il} B_{jk} \underline{\mathbf{e}}_i \otimes \underline{\mathbf{e}}_j \otimes \underline{\mathbf{e}}_k \otimes \underline{\mathbf{e}}_l$

Tensorial product $\underset{\sim}{\mathbf{A}} \underline{\otimes} \underset{\sim}{\mathbf{B}} = A_{ik} B_{jl} \underline{\mathbf{e}}_i \otimes \underline{\mathbf{e}}_j \otimes \underline{\mathbf{e}}_k \otimes \underline{\mathbf{e}}_l$

Identity tensor (2nd-order) $\underset{\sim}{\mathbf{1}} = \delta_{ij} \underline{\mathbf{e}}_i \otimes \underline{\mathbf{e}}_j$

Identity tensor (4th-order) $\underset{\approx}{\mathbf{1}} = \frac{1}{2} \left(\delta_{ik} \delta_{jl} + \delta_{il} \delta_{jk} \right) \underline{\mathbf{e}}_i \otimes \underline{\mathbf{e}}_j \otimes \underline{\mathbf{e}}_k \otimes \underline{\mathbf{e}}_l$

Nickel Base Single Crystals Across Length Scales
https://doi.org/10.1016/B978-0-12-819357-0.00027-5

The objective of this chapter is to give a description of the most efficient algorithms for the numerical implementation of crystal plasticity models in finite element codes. Of course, this approach requires a large amount of equations, which are provided in great detail. The specific notations used are grouped in a special section at the end of the chapter.

17.1 General form for the constitutive equations

17.1.1 Variables describing constitutive equations

The eventual programming of material characteristics within a finite element code will ideally respect the physical and mathematical foundations in form. Proper representation naturally leads to a complete interface between the global code design and the local (material). Essentially, this representation entails defining the appropriate variables used to characterize a material behavior [452]. These variables are listed below:

Input variables v_{IN}. They are primary imposed problem variables over which the behavior is integrated. These variables are computed at the element level using the problem degrees of freedom. In standard mechanics, they usually correspond to strains (the measure will depend on the problem formulation).

Output variables v_{OUT}. These variables are the direct result of the time integration of the constitutive equations which are then used to compute the internal forces at the elementary level. In standard mechanics, they usually correspond to stresses (the measure will depend on the problem formulation).

Internal/Integrated variables v_{in}. These variables represent quantities used to describe the material state. They often are expressed as thermodynamic state variables [510,818]. These variables need to be time integrated.

Auxiliary variables v_{aux}. They encode interesting information which does not directly define the material state. They are functions of v_{in} but do not need to be time integrated.

External Parameters EP. These parameters are quantities which are user-defined (prescribed) for the calculation but which influence the behavior. They, e.g., correspond to temperature for a purely mechanical simulation.

Material coefficients CO. These coefficients are used to define the constitutive behavior. Simple examples are elastic moduli, yield stress, hardening modulus, etc. They can depend on other quantities so that

$$\mathrm{CO} = \mathrm{CO}\left(v_{\mathrm{in}}, v_{\mathrm{aux}}, \mathrm{EP}\right). \tag{17.1}$$

The class of behavior is thus characterized by the in–out variable types $v_{\text{IN}}/v_{\text{OUT}}$. In "standard" mechanics they will correspond to small deformation strain tensor / Cauchy stress ($\underset{\sim}{\varepsilon}/\underset{\sim}{\sigma}$), Green–Lagrange strain / second Piola–Kirchhoff stress ($\underset{\sim}{\mathbf{E}}/\underset{\sim}{\mathbf{S}}$), deformation gradient / first Piola–Kirchhoff stress ($\underset{\sim}{\mathbf{F}}/\underset{\sim}{\Pi}$), Hencky's logarithmic strain / work conjugate stress [943] ($\underset{\sim}{\mathbf{H}} = \frac{1}{2}\log\left(\underset{\sim}{\mathbf{F}}^T.\underset{\sim}{\mathbf{F}}\right)/\underset{\sim}{\mathbf{T}}$), etc. One has

$$\rho\dot{\underset{\sim}{\varepsilon}} : \underset{\sim}{\sigma} = \rho_0\dot{\underset{\sim}{\varepsilon}} : \underset{\sim}{\tau} = \rho_0\dot{\underset{\sim}{E}} : \underset{\sim}{\mathbf{S}} = \rho_0\dot{\underset{\sim}{F}} : \underset{\sim}{\Pi} = \rho_0\dot{\underset{\sim}{H}} : \underset{\sim}{\mathbf{T}}, \tag{17.2}$$

where $\underset{\sim}{\tau}$ is the Kirchhoff stress tensor and $\rho/\rho_0 = \det\underset{\sim}{\mathbf{F}}$.

When using implicit simulation codes, it is also required to compute the "consistent" tangent matrix which can be formally expressed as

$$[K] = \frac{d\Delta v_{\text{OUT}}}{d\Delta v_{\text{IN}}}, \tag{17.3}$$

where Δv_{OUT} and Δv_{IN} represent the increments of output/input variables over a finite time increment Δt. In "standard" mechanics, $[K]$ corresponds to a fourth-order tensor.

17.1.2 Integration methods

17.1.2.1 Explicit integration

The evolution of integrated variables v_{in} is usually expressed as a set of differential equations

$$\dot{v}_{\text{in}} = \mathcal{V}(v_{\text{in}}, t, \text{EP}, \text{CO}, v_{\text{aux}}). \tag{17.4}$$

When considering this equation, one must indeed take into account that EP, depends on time, CO on v_{in}, EP, and v_{aux}, and v_{aux} on v_{in}. This can lead to some complex formulation, and it is important to keep track of the dependencies. Eq. (17.4) can therefore be rewritten in a more simple form as

$$\dot{v}_{\text{in}} = \mathcal{V}(v_{\text{in}}, t). \tag{17.5}$$

Eq. (17.5) can be integrated using well-known numerical techniques. Describing these methods is indeed out of the scope of this text. Efficient methods such as the Runge–Kutta methods are, for instance, described in [180], and efficient implementations are now available. The main drawback of this method is that it does not provide a "consistent" tangent matrix $[K]$ so that specific developments are necessary to compute it. Note that a perturbation method can be used, but this tends to be very time-consuming as many integrations must be performed, possibly inaccurate and strongly dependent on the selected perturbation.

17.1.2.2 *Implicit integration*

Eq. (17.5) can be time integrated as

$$\Delta v_{\text{in}} = \mathcal{V}(v_{\text{in}}^{\theta}, t^{\theta})\Delta t, \tag{17.6}$$

where Δt is the time step and θ is a parameter such that $0 \leq \theta \leq 1$. The notation x^{θ} refers to the value of x at time $t^0 + \theta \Delta t$ where t^0 is the time at the beginning of the time step (so that x^0 is the value of x at the beginning of the time step and $x^{\theta} = x^0 + \theta \Delta x$). For $\theta = 0$, Eq. (17.6) corresponds to the simple forward Euler integration scheme which is known to be unstable. Stability is obtained provided $\theta > 1/2$. Eq. (17.6) is then rewritten as

$$\Delta v_{\text{in}} = \mathcal{V}(v_{\text{in}}^{0} + \theta \Delta v_{\text{in}}, t^0 + \theta \Delta t)\Delta t, \tag{17.7}$$

so that it becomes obvious that this equation must be solved for the unknown value Δv_{in}. The equation can be reformulated as

$$R_{\text{in}} = \Delta v_{\text{in}} - \mathcal{V}(v_{\text{in}}^{0} + \theta \Delta v_{\text{in}}, t^0 + \theta \Delta t)\Delta t = 0, \tag{17.8}$$

where R_{in} is the residual vector. Time integrating equation (17.4) becomes equivalent to solving (17.8). This equation is usually highly nonlinear and must be solved using, for instance, an iterative Newton–Raphson scheme. This requires the evaluation of the Jacobian matrix $[J]$ associated to (17.8), which is formally expressed as

$$[J] = \frac{\partial R_{\text{in}}}{\partial \Delta v_{\text{in}}} = [1] - \Delta t \left.\frac{\partial R_{\text{in}}}{\partial v_{\text{in}}}\right|_{\theta} \cdot \frac{\partial v_{\text{in}}}{\partial \Delta v_{\text{in}}} = [1] - \theta \Delta t \left.\frac{\partial R_{\text{in}}}{\partial v_{\text{in}}}\right|_{\theta}. \tag{17.9}$$

The evaluation of $[J]$ can be complex (see examples below). A perturbation technique can be used, but will possibly suffer from the drawbacks that have already been outlined above.

17.1.3 Consistent tangent matrix

Using the evaluation of the Jacobian matrix $[J]$ it becomes possible to compute the consistent tangent matrix at a relatively low cost. Eq. (17.8) was solved for fixed values of the input variables v_{IN}. To evaluate the infinitesimal variation of the integrated variables ($\delta \Delta v_{\text{in}}$) caused by an infinitesimal variation of the input variables ($\delta \Delta v_{\text{IN}}$), one must express that the residual must remain equal to 0. So that

$$\delta R_{\text{in}} = \frac{\partial R_{\text{in}}}{\partial \Delta v_{\text{in}}} \cdot \delta \Delta v_{\text{in}} + \frac{\partial R_{\text{in}}}{\partial \Delta v_{\text{IN}}} \cdot \delta \Delta v_{\text{IN}} = [J]_s \cdot \delta \Delta v_{\text{in}} + [L]_s \cdot \delta \Delta v_{\text{IN}} = 0, \tag{17.10}$$

where the subscript s denotes that the quantities are evaluated for Δv_{in} solution of Eq. (17.8) for the prescribed Δv_{IN}. Note that $[J]$ is a square matrix which can be inverted; $[L]$ is usually not square. For the above equation, the following stiffness matrix can be evaluated:

$$[K]_{\text{in}} = \frac{\partial \Delta v_{\text{in}}}{\partial \Delta v_{\text{IN}}} = -[J]_s^{-1} \cdot [L]_s \,. \tag{17.11}$$

The output variables v_{OUT} are, however, usually not a subset of v_{in}. They are rather expressed explicitly as functions of v_{in} and v_{IN} so that

$$\delta \Delta v_{\text{OUT}} = \frac{\partial \Delta v_{\text{OUT}}}{\partial \Delta v_{\text{IN}}} \cdot \delta \Delta v_{\text{IN}} + \frac{\partial \Delta v_{\text{OUT}}}{\partial \Delta v_{\text{in}}} \cdot \delta \Delta v_{\text{in}} \tag{17.12}$$

$$= \left(\frac{\partial \Delta v_{\text{OUT}}}{\partial \Delta v_{\text{IN}}} + \frac{\partial \Delta v_{\text{OUT}}}{\partial \Delta v_{\text{in}}} \cdot [K]_{\text{in}} \right) \cdot \delta \Delta v_{\text{IN}}, \tag{17.13}$$

which leads to the final expression of the consistent stiffness matrix, namely

$$[K] = \frac{d \Delta v_{\text{OUT}}}{d \Delta v_{\text{IN}}} = \frac{\partial \Delta v_{\text{OUT}}}{\partial \Delta v_{\text{IN}}} + \frac{\partial \Delta v_{\text{OUT}}}{\partial \Delta v_{\text{in}}} \cdot [K]_{\text{in}} \,. \tag{17.14}$$

Examples of these derivations will be given below in the case of the implementation of the constitutive equations for single crystals.

17.2 A small strain constitutive model for single crystals

In the realm of elasto-plasticity of single crystals, it is commonly admitted that elastic deformations are related to the stretching of interatomic bonds, while plastic deformations are due to the gliding of linear defects in the crystal lattice called dislocations. Dislocation glide occurs preferentially on particular planes, which are denoted by unit normal $\underline{\mathbf{n}}^{\alpha}$, and in particular directions, oriented by unit vector $\underline{\mathbf{m}}^{\alpha}$. The type of slip systems (plane/direction) vary upon the kind of crystal lattice (body-centered cubic, face-centered cubic, hexagonal compact, etc.) considered. A convenient way to model plastic deformations in crystals consist in decomposing the overall plastic strain tensor into a sum of shear mechanisms over the slip systems. The motion of a dislocation induces indeed an irreversible shear strain. Locally, for each slip system, the amount of plastic shear strain can be quantified by a scalar value noted γ^{α}. In addition, in order to represent the population of dislocations in the crystal, a scalar dislocation density per slip system ρ^{α} is defined. It corresponds to the length of dislocation line per unit volume for a given systems and thus has a unit of m/m^3, i.e., m^{-2}. Since dislocation densities are for most crystals in the range 10^6–10^{16} m^{-2}, it is numerically more convenient to manipulate dimensionless dislocation densities defined as $r^{\alpha} = \epsilon b^2 \rho^{\alpha}$. The scalar b denotes the norm of the Burgers vector, while ϵ is a nonphysical parameter used in order to scale dimensionless dislocation densities to numerical values that are close to the typical order of magnitudes encountered for strain measures. Typically, $\epsilon = 10^6$ can be used. Within a small strain framework, such a model for the elasto-plasticity of single crystals yields the following sets of input, output, and internal variables:

$$v_{\mathrm{IN}} : \{\underset{\sim}{\varepsilon}\}, \quad v_{\mathrm{OUT}} : \{\underset{\sim}{\sigma}\}, \quad v_{\mathrm{in}} : \{\underset{\sim}{\varepsilon}^e, \gamma^\alpha, r^\alpha\}, \tag{17.15}$$

where $\underset{\sim}{\varepsilon}^e$ denotes the elastic strain tensor. Any increment of the elastic strain tensor is linearly related to an increment of the Cauchy stress tensor $\underset{\sim}{\sigma}$,

$$\Delta\underset{\sim}{\sigma} = \underset{\approx}{\mathbf{C}} : \Delta\underset{\sim}{\varepsilon}^e, \tag{17.16}$$

where $\underset{\approx}{\mathbf{C}}$ denotes the fourth-order elasticity tensor. From a numerical point of view, it means that computing the unique output variable is a mere postprocessing of one of the integration procedure result, namely $\Delta\underset{\sim}{\varepsilon}^e$. As discussed earlier, plastic deformation in crystals is accounted for by a set of shear mechanisms acting on several slip systems. The driving force on a given slip plane in a given slip direction is called the resolved shear stress τ^α and is related to the symmetric Schmid tensor $\underset{\sim}{\mathbf{N}}_s^\alpha$ as follows:

$$\tau^\alpha = \underset{\sim}{\sigma} : \underset{\sim}{\mathbf{N}}_s^\alpha, \qquad \underset{\sim}{\mathbf{N}}_s^\alpha = \frac{1}{2}(\underline{\mathbf{m}}^\alpha \otimes \underline{\mathbf{n}}^\alpha + \underline{\mathbf{n}}^\alpha \otimes \underline{\mathbf{m}}^\alpha). \tag{17.17}$$

Within the small-strain setting considered in this section, an additive split of the strain tensor increment into elastic and plastic parts can be assumed. In light of the plastic deformation mode described above, the plastic increment can in turn be decomposed into a sum of shear modes increments over each slip system. It comes to

$$\Delta\underset{\sim}{\varepsilon} = \Delta\underset{\sim}{\varepsilon}^e + \Delta\underset{\sim}{\varepsilon}^p = \Delta\underset{\sim}{\varepsilon}^e + \sum_{\alpha=1}^{N} \frac{\Delta\gamma^\alpha}{2}(\underline{\mathbf{m}}^\alpha \otimes \underline{\mathbf{n}}^\alpha + \underline{\mathbf{n}}^\alpha \otimes \underline{\mathbf{m}}^\alpha), \tag{17.18}$$

where N denotes the total number of slip systems. Now, the amount of plastic slip increment over each slip system relies on mainly two aspects. First of all, it depends upon the resistance to dislocation glide, i.e., the stress barrier to overcome in order to put dislocations into motion. Such a barrier is called the critical resolved shear stress τ_c^α. Following [463], a form of the critical resolved shear is assumed as

$$\tau_c^\alpha = \tau_0^\alpha + \alpha\mu\sqrt{\sum_{\beta=1}^{N} a^{\alpha\beta} r^\beta}, \tag{17.19}$$

where the first term τ_0^α is a material parameter that varies upon temperature. The second term models the hardening related to forest dislocations.

Dislocation motion in a given slip system can indeed be hindered by dislocations lying in the same or others systems. Such impediments depend upon the nature of the junction or dipoles that two dislocations can form and are characterized by the magnitude of the coefficients involved in the interaction matrix a^{su}. The second aspects on which relies the amount of plastic increment on a given slip system is the resolved shear stress τ^α acting on its slip plane and in its direction of dislocation glide; τ^α needs to be large enough in order to overcome the aforementioned barrier τ_c^α. In other words, the yield locus for a given slip system follows Schmid's law and is given by

$$f^\alpha = |\tau^\alpha| - \tau_c^\alpha. \tag{17.20}$$

Because of the symmetries existing in crystal lattices, slip systems are mostly not unique and hence crystal plasticity inherits a multisurface yield domain by construction. It is well known that such a surface can have sharp edges and even corners. Therefore, if one assumes an associated flow rule, such that the plastic strain rate develops perpendicularly to the yield locus, indeterminacy of the plastic strain rate arises at edges and corners since the normal is not uniquely defined. In rate-independent crystal plasticity, the ill-posed problem of determining the set of active slip systems and associated slip rates was addressed in various manner. Numerical methods involving generalized or pseudoinverses [944,1624], or augmented Lagrangian formulations [1279] were proposed. Alternatively, [458] developed a rate-independent formulation of crystal plasticity based on a smooth elastic-plastic transition which involves a rate-independent overstress. Another possible alternative to overcome slip indeterminacy commonly encountered in the literature consists in smoothing out corners and edges by using viscoplastic flow rules in order to determine active systems and slip rates [174]. The latter option is adopted in what follows. Using a Norton-type viscous flow rule, the plastic slip increment is therefore written as

$$\Delta\gamma^\alpha = \text{sign}\left(\tau^\alpha\right) \Delta t \dot{\gamma}_0 \Phi(f^\alpha), \qquad \Phi(f^\alpha) = \left\langle \frac{f^\alpha}{\tau_0^\alpha} \right\rangle^n, \tag{17.21}$$

where $\dot{\gamma}_0$ and n are viscosity material parameters. The induced viscous overstress is indeed equal to $\tau_0^\alpha(\Delta\gamma^\alpha/(\Delta t\dot{\gamma}_0))^{1/n} = \tau_0^\alpha(\dot{\gamma}^\alpha/\dot{\gamma}_0)^{1/n}$. Using large values for both parameters allows pushing such a rate-dependent model towards a quasi-rate-independent limit. However, from a numerical point of view, increasing these parameters without precaution renders the set of differential equations overly stiff and prevents from the possibility of using large time steps. Finally, in order to close the system of equations, evolution equations for the dislocation densities r^α need to be defined. These evolutionary laws are based on two main phenomena, namely storage and recovery of dislocations [1407,734]. Storage occurs by pinning of dislocations after they have glided over a certain mean free path, controlled by the dimensionless material parameter κ. Recovery takes place by annihilation of screw dislocations thanks to cross-slip. The dimensionless material parameter G_c characterizes the distance at which annihilation can occur. The increment of dislocation density is related to the plastic slip increment on the same slip system and dislocation densities on all slip systems as follows:

$$\Delta r^{\alpha} = |\Delta\gamma^{\alpha}| \left(\frac{1}{\kappa} \sqrt{\sum_{\beta=1}^{N} b^{\alpha\beta} r^{\beta}} - G_c r^{\alpha} \right), \tag{17.22}$$

where b^{sr} denotes another interaction matrix characterizing dislocation storage.

Over a time increment, we recall that the unknowns to be determined are the increments of integrated variables, namely $\Delta\underset{\sim}{\varepsilon}^e$, $\Delta\gamma^{\alpha}$, and Δr^{α}. The set of nonlinear differential equations defined in Eqs. (17.18), (17.21), and (17.22) can be recast in a set of residuals R_{in} as follows:

$$R_{\text{in}} = \begin{cases} R_{\underset{\sim}{\varepsilon}^e} & = \Delta\underset{\sim}{\varepsilon}^e + \Delta\underset{\sim}{\varepsilon}^p - \Delta\underset{\sim}{\varepsilon}, \\ R_{\gamma^{\alpha}} & = \Delta\gamma^{\alpha} - \text{sign}(\tau^{\alpha})\, \Delta t \dot{\gamma}_0 \Phi(f^{\alpha}), \\ R_{r^{\alpha}} & = \Delta r^{\alpha} - |\Delta\gamma^{\alpha}| \left(\frac{1}{\kappa} \sqrt{\sum_{\beta=1}^{N} b^{\alpha\beta} r^{\beta}} - G_c r^{\alpha} \right). \end{cases} \tag{17.23}$$

In order to solve $R_{\text{in}} = 0$ by Newton's method, the Jacobian matrix of Eq. (17.23) can be calculated as follows:

$$[J] = \frac{\partial R_{\text{in}}}{\partial \Delta v_{\text{in}}} = \begin{pmatrix} \frac{\partial R_{\underset{\sim}{\varepsilon}^e}}{\partial \Delta\underset{\sim}{\varepsilon}^e} & \frac{\partial R_{\underset{\sim}{\varepsilon}^e}}{\partial \Delta\gamma^{\beta}} & \frac{\partial R_{\underset{\sim}{\varepsilon}^e}}{\partial \Delta r^{\beta}} \\ \frac{\partial R_{\gamma^{\alpha}}}{\partial \Delta\underset{\sim}{\varepsilon}^e} & \frac{\partial R_{\gamma^{\alpha}}}{\partial \Delta\gamma^{\beta}} & \frac{\partial R_{\gamma^{\alpha}}}{\partial \Delta r^{\beta}} \\ \frac{\partial R_{r^{\alpha}}}{\partial \Delta\underset{\sim}{\varepsilon}^e} & \frac{\partial R_{r^{\alpha}}}{\partial \Delta\gamma^{\beta}} & \frac{\partial R_{r^{\alpha}}}{\partial \Delta r^{\beta}} \end{pmatrix}, \tag{17.24}$$

- $R_{\underset{\sim}{\varepsilon}^e}$

$$R_{\underset{\sim}{\varepsilon}^e} = \Delta\underset{\sim}{\varepsilon}^e + \Delta\underset{\sim}{\varepsilon}^p - \Delta\underset{\sim}{\varepsilon}, \tag{17.25}$$

$$\frac{\partial R_{\underset{\sim}{\varepsilon}^e}}{\partial \Delta\underset{\sim}{\varepsilon}^e} = \underset{\approx}{\mathbf{1}}, \qquad \frac{\partial R_{\underset{\sim}{\varepsilon}^e}}{\partial \Delta\gamma^{\beta}} = \underset{\sim}{\mathbf{N}}_s^{\beta}, \qquad \frac{\partial R_{\underset{\sim}{\varepsilon}^e}}{\partial \Delta r^{\beta}} = 0; \tag{17.26}$$

- $R_{\gamma^{\alpha}}$

$$R_{\gamma^{\alpha}} = \Delta\gamma^{\alpha} - \text{sign}\left(\tau^{\alpha}\right) \Delta t \dot{\gamma}_0 \Phi(f^{\alpha}), \tag{17.27}$$

$$\frac{\partial R_{\gamma^\alpha}}{\partial \Delta \underset{\sim}{\varepsilon}^e} = -\text{sign}\left(\tau^\alpha\right) \Delta t \dot{\gamma}_0 \frac{\partial \Phi^\alpha}{\partial f^\alpha} \frac{\partial f^\alpha}{\partial \tau^\alpha} \frac{\partial \tau^\alpha}{\partial \underset{\sim}{\sigma}} : \frac{\partial \underset{\sim}{\sigma}}{\partial \Delta \underset{\sim}{\varepsilon}^e} \tag{17.28}$$

$$= -\frac{\Delta t \dot{\gamma}_0 n}{(\tau_0^\alpha)^n} \left\langle \frac{f^\alpha}{\tau_0^\alpha} \right\rangle^{n-1} \underset{\sim}{\mathbf{N}}_s^\alpha : \underset{\approx}{\mathbf{C}}, \tag{17.29}$$

$$\frac{\partial R_{\gamma^\alpha}}{\partial \Delta \gamma^\beta} = \delta_{\alpha\beta}, \tag{17.30}$$

$$\frac{\partial R_{\gamma^\alpha}}{\partial \Delta r^\beta} = -\text{sign}\left(\tau^\alpha\right) \Delta t \dot{\gamma}_0 \frac{\partial \Phi^\alpha}{\partial f^\alpha} \frac{\partial f^\alpha}{\partial \tau_c^\alpha} \frac{\partial \tau_c^\alpha}{\partial \Delta r^\beta} \tag{17.31}$$

$$= \text{sign}\left(\tau^\alpha\right) \frac{\Delta t \dot{\gamma}_0 n}{(\tau_0^\alpha)^n} \left\langle \frac{f^\alpha}{\tau_0^\alpha} \right\rangle^{n-1} \frac{\alpha \mu a^{\alpha\beta}}{2} \left(\sum_{u=1}^{N} a^{\alpha u} r^u \right)^{-\frac{1}{2}}; \tag{17.32}$$

- R_{r^α}

$$R_{r^\alpha} = \Delta r^\alpha - |\Delta \gamma^\alpha| \left(\frac{1}{\kappa} \sqrt{\sum_{\beta=1}^{N} b^{\alpha\beta} r^\beta} - G_c r^\alpha \right), \tag{17.33}$$

$$\frac{\partial R_{r^\alpha}}{\partial \Delta \underset{\sim}{\varepsilon}^e} = 0, \tag{17.34}$$

$$\frac{\partial R_{r^\alpha}}{\partial \Delta \gamma^\beta} = -\text{sign}\left(\Delta \gamma^\alpha\right) \delta_{\alpha\beta} \left(\frac{1}{\kappa} \sqrt{\sum_{u=1}^{N} b^{\alpha u} r^u} - G_c r^\alpha \right), \tag{17.35}$$

$$\frac{\partial R_{r^\alpha}}{\partial \Delta r^\beta} = \delta_{\alpha\beta} - |\Delta \gamma^\alpha| \left(\frac{b^{\alpha\beta}}{2\kappa} \left(\sum_{u=1}^{N} b^{\alpha u} r^u \right)^{-\frac{1}{2}} - G_c \delta_{\alpha\beta} \right). \tag{17.36}$$

Following Eq. (17.14), the consistent tangent matrix can finally be computed. As $v_{\text{OUT}} = \underset{\sim}{\sigma}$, $v_{\text{IN}} = \underset{\sim}{\varepsilon}$, and $\underset{\sim}{\sigma} = \underset{\approx}{\mathbf{C}} : \underset{\sim}{\varepsilon}_e$, one has with block-matrix notations:

$$\frac{\partial \Delta v_{\text{OUT}}}{\partial \Delta v_{\text{IN}}} = 0, \qquad \frac{\partial \Delta v_{\text{OUT}}}{\partial \Delta v_{\text{in}}} = \left(\underset{\approx}{\mathbf{C}} \;\middle|\; [0] \right), \qquad [L]_s = \frac{\partial R_{\text{in}}}{\partial \Delta v_{\text{IN}}} = \left(\frac{-\underset{\approx}{\mathbf{1}}}{[0]} \right), \tag{17.37}$$

which leads to

$$[K] = \frac{d \Delta v_{\text{OUT}}}{d \Delta v_{\text{IN}}} = \frac{d \Delta \underset{\sim}{\sigma}}{d \Delta \underset{\sim}{\varepsilon}} = -\left(\underset{\approx}{\mathbf{C}} \;\middle|\; [0] \right).[J]_s^{-1}.\left(\frac{-\underset{\approx}{\mathbf{1}}}{[0]} \right) \tag{17.38}$$

$$= \left(\underset{\approx}{\mathbf{C}} \;\middle|\; [0] \right).[J]_s^{-1}.\left(\frac{\underset{\approx}{\mathbf{1}}}{[0]} \right), \tag{17.39}$$

where $[J]_s^{-1}$ can be expressed as a block matrix

$$[J]_s^{-1} = \left(\begin{array}{c|c} \underset{\approx}{\mathbf{L}}_e & [\ldots] \\ \hline [\ldots] & [\ldots] \end{array} \right) \tag{17.40}$$

so that the consistent tangent matrix can be more simply expressed as a product of fourth-order tensors

$$[K] = \underset{\approx}{\mathbf{C}} : \underset{\approx}{\mathbf{L}}_e. \tag{17.41}$$

17.3 A finite-strain constitutive model for single crystals

The constitutive model of elasto-plasticity in single crystals presented at small strains in the previous section is extended to finite strains in this section. The physical foundations are essentially unchanged and, as a consequence, attention is drawn to the numerical implications of the chosen finite-strain framework. The latter is based on the multiplicative decomposition of the deformation gradient $\underset{\sim}{\mathbf{F}}$ into an elastic part $\underset{\sim}{\mathbf{E}}$ and a plastic part $\underset{\sim}{\mathbf{P}}$ [814,901]. The sets of input, output, and internal variables are respectively chosen as

$$v_{\mathrm{IN}} : \{\underset{\sim}{\mathbf{F}}\}, \quad v_{\mathrm{OUT}} : \{\underset{\sim}{\Pi}\}, \quad v_{\mathrm{int}} : \{\underset{\sim}{\mathbf{E}}, \gamma^\alpha, r^\alpha\}. \tag{17.42}$$

Compared to previous section, $\underset{\sim}{\varepsilon}$ is replaced by $\underset{\sim}{\mathbf{F}}$, and accordingly its work-conjugate $\underset{\sim}{\sigma}$ is replaced by the first Piola–Kirchhoff stress $\underset{\sim}{\Pi}$. It is recalled that $\underset{\sim}{\Pi}$ is connected to the Cauchy stress tensor through

$$\underset{\sim}{\Pi} = \det \underset{\sim}{\mathbf{F}} \underset{\sim}{\sigma} \cdot \underset{\sim}{\mathbf{F}}^{-T}. \tag{17.43}$$

The elastic small strain tensor $\underset{\sim}{\varepsilon}^e$ is replaced by $\underset{\sim}{\mathbf{E}}$. It can already be noted that the stress and strain measures considered in this formulation are no longer necessarily symmetric. According to the multiplicative decomposition of $\underset{\sim}{\mathbf{F}}$, elastic and plastic velocity gradients $\underset{\sim}{\mathbf{L}}^e$ and $\underset{\sim}{\mathbf{L}}^p$ can be introduced as follows:

$$\underset{\sim}{\mathbf{F}} = \underset{\sim}{\mathbf{E}} \cdot \underset{\sim}{\mathbf{P}}, \qquad \underset{\sim}{\mathbf{L}} = \dot{\underset{\sim}{\mathbf{F}}} \cdot \underset{\sim}{\mathbf{F}}^{-1} = \underset{\sim}{\mathbf{L}}^e + \underset{\sim}{\mathbf{E}} \cdot \underset{\sim}{\mathbf{L}}^p \cdot \underset{\sim}{\mathbf{E}}^{-1}, \tag{17.44}$$

$$\underset{\sim}{\mathbf{L}}^e = \dot{\underset{\sim}{\mathbf{E}}} \cdot \underset{\sim}{\mathbf{E}}^{-1}, \qquad \underset{\sim}{\mathbf{L}}^p = \dot{\underset{\sim}{\mathbf{P}}} \cdot \underset{\sim}{\mathbf{P}}^{-1} = \sum_{\alpha=1}^{N} \dot{\gamma}^\alpha \underline{\mathbf{m}}^\alpha \otimes \underline{\mathbf{n}}^\alpha. \tag{17.45}$$

The velocity gradient split is to be put in parallel with its small-strain counterpart in Eq. (17.18). The major difference here is the asymmetric character of the plastic velocity gradient. At this point, the thermodynamical backbone of the present model, that is eluded for conciseness, motivates introducing the elastic Green–Lagrange strain measure $\underset{\sim}{\mathbf{E}}_{GL}^{e}$, the elastic second Piola–Kirchhoff stress measure $\underset{\sim}{\Pi}^{e}$, and the Mandel stress measure $\underset{\sim}{\Pi}^{M}$ as

$$\underset{\sim}{\mathbf{E}}_{GL}^{e} = \frac{1}{2}\left(\underset{\sim}{\mathbf{E}}^{T} \cdot \underset{\sim}{\mathbf{E}} - \underset{\sim}{\mathbf{1}}\right), \tag{17.46}$$

$$J = \det \underset{\sim}{\mathbf{F}} = \det \underset{\sim}{\mathbf{E}} \cdot \underset{\sim}{\mathbf{P}} = \det \underset{\sim}{\mathbf{E}} \det \underset{\sim}{\mathbf{P}} = J_e J_p, \tag{17.47}$$

$$\underset{\sim}{\Pi}^{e} = J_e \underset{\sim}{\mathbf{E}}^{-1} \cdot \underset{\sim}{\sigma} \cdot \underset{\sim}{\mathbf{E}}^{-T} = J_p^{-1} \underset{\sim}{\mathbf{E}}^{-1} \cdot \underset{\sim}{\mathbf{S}} \cdot \underset{\sim}{\mathbf{P}}^{T}, \tag{17.48}$$

$$\underset{\sim}{\Pi}^{M} = \underset{\sim}{\mathbf{E}}^{T} \cdot \underset{\sim}{\mathbf{E}} \cdot \underset{\sim}{\Pi}^{e}. \tag{17.49}$$

The state law equation (17.16) is replaced at finite strains by

$$\Delta \underset{\sim}{\Pi}^{e} = \underset{\approx}{\mathbf{C}} : \Delta \underset{\sim}{\mathbf{E}}_{GL}^{e}, \tag{17.50}$$

while $\underset{\sim}{\Pi}^{e}$ is power-conjugate to the elastic Green–Lagrange strain rate and $\underset{\sim}{\Pi}^{M}$ is power-conjugate to the plastic velocity gradient $\dot{\underset{\sim}{\mathbf{P}}} \cdot \underset{\sim}{\mathbf{P}}^{-1}$. Therefore the driving forces for plastic slip activity, namely the resolved shear stresses τ^{α}, are defined as follows:

$$\tau^{\alpha} = \underset{\sim}{\Pi}^{M} : \underset{\sim}{\mathbf{N}}^{\alpha}, \qquad \underset{\sim}{\mathbf{N}}^{\alpha} = \underline{\mathbf{m}}^{\alpha} \otimes \underline{\mathbf{n}}^{\alpha}, \tag{17.51}$$

where $\underset{\sim}{\mathbf{N}}^{\alpha}$ corresponds to the asymmetric Schmid tensor. It is worth noting that, although the definitions of resolved shear stresses are slightly generalized in Eq. (17.51) compared to (17.17), the formalism setup for crystal plasticity at small strains remains valid at finite strains. In other words, Eqs. (17.19), (17.20), (17.21), and (17.22) stay unchanged. From Eqs. (17.44) and (17.45), and straightforward algebraic manipulations, one obtains the only formally modified residual equation, namely that related to the incremental elastic-plastic split,

$$R_{\underset{\sim}{E}} = \Delta \underset{\sim}{\mathbf{E}} - \Delta \underset{\sim}{\mathbf{F}} \cdot \underset{\sim}{\mathbf{F}}^{-1} \cdot \underset{\sim}{\mathbf{E}} - \underset{\sim}{\mathbf{E}} \cdot \left(\sum_{\alpha=1}^{N} \Delta \gamma^{\alpha} \underset{\sim}{\mathbf{N}}^{\alpha}\right) = 0. \tag{17.52}$$

Since only the first residual equation is formally modified in Eq. (17.23), only the terms in the first row and first column of the Jacobian matrix $[J]$ are affected:

$$[J] = \frac{\partial R_{\text{in}}}{\partial \Delta v_{\text{in}}} = \left(\begin{array}{c|cc} \dfrac{\partial R_{\underset{\sim}{E}}}{\partial \Delta \underset{\sim}{\mathbf{E}}} & \dfrac{\partial R_{\underset{\sim}{E}}}{\partial \Delta \gamma^{\beta}} & \dfrac{\partial R_{\underset{\sim}{E}}}{\partial \Delta r^{\beta}} \\ \hline \dfrac{\partial R_{\gamma^{\alpha}}}{\partial \Delta \underset{\sim}{\mathbf{E}}} & \dfrac{\partial R_{\gamma^{\alpha}}}{\partial \Delta \gamma^{\beta}} & \dfrac{\partial R_{\gamma^{\alpha}}}{\partial \Delta r^{\beta}} \\ \dfrac{\partial R_{r^{\alpha}}}{\partial \Delta \underset{\sim}{\mathbf{E}}} & \dfrac{\partial R_{r^{\alpha}}}{\partial \Delta \gamma^{\beta}} & \dfrac{\partial R_{r^{\alpha}}}{\partial \Delta r^{\beta}} \end{array}\right), \tag{17.53}$$

- $R_{\underset{\sim}{E}}$

$$R_{\underset{\sim}{E}} = \Delta \underset{\sim}{E} - \Delta \underset{\sim}{F} \cdot \underset{\sim}{F}^{-1} \cdot \underset{\sim}{E} + \underset{\sim}{E} \cdot \left(\sum_{\alpha=1}^{N} \Delta \gamma^{\alpha} \underset{\sim}{N}^{\alpha} \right), \tag{17.54}$$

$$\frac{\partial R_{\underset{\sim}{E}}}{\partial \Delta \underset{\sim}{E}} = \underset{\approx}{1} - (\Delta \underset{\sim}{F} \cdot \underset{\sim}{F}^{-1}) \underline{\otimes} \underset{\sim}{1} + \underset{\sim}{1} \underline{\otimes} \left(\sum_{\alpha=1}^{N} \Delta \gamma^{\alpha} \underset{\sim}{N}^{\alpha} \right)^{T}, \tag{17.55}$$

$$\frac{\partial R_{\underset{\sim}{E}}}{\partial \Delta \gamma^{\beta}} = \underset{\sim}{E} \cdot \underset{\sim}{N}^{\beta}, \qquad \frac{\partial R_{\underset{\sim}{E}}}{\partial \Delta r^{\beta}} = 0; \tag{17.56}$$

- $R_{\gamma^{\alpha}}$

$$R_{\gamma^{\alpha}} = \Delta \gamma^{\alpha} - \operatorname{sign}\left(\tau^{\alpha}\right) \Delta t \dot{\gamma}_0 \Phi(f^{\alpha}), \tag{17.57}$$

$$\frac{\partial R_{\gamma^{\alpha}}}{\partial \Delta \underset{\sim}{E}} = -\operatorname{sign}\left(\tau^{\alpha}\right) \Delta t \dot{\gamma}_0 \frac{\partial \Phi^{\alpha}}{\partial f^{\alpha}} \frac{\partial f^{\alpha}}{\partial \tau^{\alpha}} \frac{\partial \tau^{\alpha}}{\partial \underset{\sim}{\Pi}^{M}} : \frac{\partial \underset{\sim}{\Pi}^{M}}{\partial \underset{\sim}{C}^{e}} : \frac{\partial \underset{\sim}{C}^{e}}{\partial \underset{\sim}{E}} : \frac{\partial \underset{\sim}{E}}{\partial \Delta \underset{\sim}{E}}, \tag{17.58}$$

$$\underset{\sim}{C}^{e} = \underset{\sim}{E}^{T} \cdot \underset{\sim}{E} \tag{17.59}$$

$$\frac{\partial \underset{\sim}{\Pi}^{M}}{\partial \underset{\sim}{C}^{e}} = \frac{\partial \left[\underset{\sim}{C}^{e} \cdot \left(\underset{\approx}{C} : \frac{1}{2}(\underset{\sim}{C}^{e} - 1) \right) \right]}{\partial \underset{\sim}{C}^{e}} = (\underset{\sim}{1} \underline{\otimes} \underset{\sim}{\Pi}^{e^{T}}) + \frac{1}{2} (\underset{\sim}{C}^{e} \underline{\otimes} \underset{\sim}{1}) : \underset{\approx}{C}, \tag{17.60}$$

$$\frac{\partial \underset{\sim}{C}^{e}}{\partial \underset{\sim}{E}} = \underset{\sim}{1} \overline{\otimes} \underset{\sim}{E}^{T} + \underset{\sim}{E}^{T} \underline{\otimes} \underset{\sim}{1}, \qquad \frac{\partial \underset{\sim}{E}}{\partial \Delta \underset{\sim}{E}} = \underset{\approx}{1}, \tag{17.61}$$

$$\frac{\partial R_{\gamma^{\alpha}}}{\partial \Delta \underset{\sim}{E}} = -\frac{\Delta t \dot{\gamma}_0 n}{(\tau_0^{\alpha})^{n}} \left\langle \frac{f^{\alpha}}{\tau_0^{\alpha}} \right\rangle^{n-1} \underset{\sim}{N}^{s} : \left(\underset{\sim}{1} \underline{\otimes} \underset{\sim}{\Pi}^{e} + \frac{1}{2} (\underset{\sim}{C}^{e} \underline{\otimes} \underset{\sim}{1}) : \underset{\approx}{C} \right) : (\underset{\sim}{1} \overline{\otimes} \underset{\sim}{E}^{T} + \underset{\sim}{E}^{T} \underline{\otimes} \underset{\sim}{1}); \tag{17.62}$$

- $R_{r^{\alpha}}$

$$R_{r^{\alpha}} = \Delta r^{\alpha} - |\Delta \gamma^{\alpha}| \left(\frac{1}{\kappa} \sqrt{\sum_{\beta=1}^{N} b^{\alpha\beta} r^{\beta}} - G_c r^{\alpha} \right), \tag{17.63}$$

$$\frac{\partial R_{r^{\alpha}}}{\partial \Delta \underset{\sim}{E}} = 0. \tag{17.64}$$

The derivatives which intervene in the consistent tangent matrix are then calculated as follows:

$$\frac{\partial \Delta v_{\text{OUT}}}{\partial \Delta v_{\text{IN}}} = \frac{\partial \Delta \underset{\sim}{\Pi}}{\partial \Delta \underset{\sim}{F}} = \frac{\partial \Delta \left(J \underset{\sim}{\sigma} \cdot \underset{\sim}{F}^{-T} \right)}{\partial \Delta \underset{\sim}{F}} \tag{17.65}$$

$$= J(\underset{\sim}{\sigma} \cdot \underset{\sim}{F}^{-T}) \otimes \underset{\sim}{F}^{-T} + J(\underset{\sim}{\sigma} \underline{\otimes} \underset{\sim}{1}) : (-\underset{\sim}{F}^{-T} \overline{\otimes} \underset{\sim}{F}^{-1}), \tag{17.66}$$

$$\frac{\partial \Delta v_{\mathrm{OUT}}}{\partial \Delta v_{\mathrm{in}}} = \left(\frac{\partial \Delta \underset{\sim}{\Pi}}{\partial \Delta \underset{\sim}{E}} \;\middle|\; [0] \right), \tag{17.67}$$

$$\frac{\partial \Delta \underset{\sim}{\Pi}}{\partial \Delta \underset{\sim}{E}} = \frac{\partial \Delta \left(J \underset{\sim}{\sigma} \cdot \underset{\sim}{F}^{-T} \right)}{\partial \Delta \underset{\sim}{E}} = J(\underset{\sim}{1} \underline{\otimes} \underset{\sim}{F}^{-1}) : \frac{\partial \Delta \underset{\sim}{\sigma}}{\partial \Delta \underset{\sim}{E}}, \tag{17.68}$$

$$\frac{\partial \underset{\sim}{\sigma}}{\partial \underset{\sim}{E}} = -J_e^{-1}(\underset{\sim}{E} \cdot \underset{\sim}{\Pi}^e \cdot \underset{\sim}{E}^T) \otimes \underset{\sim}{E}^{-T} + J_e^{-1} \underset{\sim}{1} \underline{\otimes} (\underset{\sim}{\Pi}^e \cdot \underset{\sim}{E}^T)^T$$
$$+ J_e^{-1}(\underset{\sim}{E} \underline{\otimes} \underset{\sim}{E}) : \frac{\partial \underset{\sim}{\Pi}^e}{\partial \underset{\sim}{E}} + J_e^{-1} \left[(\underset{\sim}{E} \cdot \underset{\sim}{\Pi}^e) \underline{\otimes} \underset{\sim}{1} \right] : (\underset{\sim}{1} \overline{\otimes} \underset{\sim}{1}), \tag{17.69}$$

$$\frac{\partial \underset{\sim}{\Pi}^e}{\partial \underset{\sim}{E}} = \frac{\partial \underset{\sim}{\Pi}^e}{\partial \underset{\sim}{E}^e_{GL}} : \frac{\partial \underset{\sim}{E}^e_{GL}}{\partial \underset{\sim}{E}}, \tag{17.70}$$

$$\frac{\partial \underset{\sim}{\Pi}^e}{\partial \underset{\sim}{E}^e_{GL}} = \underset{\approx}{C}, \tag{17.71}$$

$$\frac{\partial \underset{\sim}{E}^e_{GL}}{\partial \underset{\sim}{E}} = \frac{1}{2}(\underset{\sim}{1} \overline{\otimes} \underset{\sim}{E}^T + \underset{\sim}{E}^T \underline{\otimes} \underset{\sim}{1}), \tag{17.72}$$

$$[L]_s = \frac{\partial R_{\mathrm{in}}}{\partial \Delta v_{\mathrm{IN}}} = \begin{pmatrix} \dfrac{\partial R_{\underset{\sim}{E}}}{\partial \Delta \underset{\sim}{F}} \\ \hline [0] \end{pmatrix}, \tag{17.73}$$

$$\frac{\partial R_{\underset{\sim}{E}}}{\partial \Delta \underset{\sim}{F}} = -\underset{\sim}{1} \underline{\otimes} \left(\underset{\sim}{E}^T \cdot \underset{\sim}{F}^{-T} \right) + (\Delta \underset{\sim}{F} \underline{\otimes} \underset{\sim}{E}^T) : (\underset{\sim}{F}^{-1} \underline{\otimes} \underset{\sim}{F}^{-T}). \tag{17.74}$$

17.4 Applications to a single-crystal turbine blade and a cylinder under torsion

The crystal elasto-plasticity model at finite strain presented in Section 17.3 is implemented in the finite element software Z-set [103,1]. The model is applied to predict the behavior of a single-crystal turbine blade and a single-crystal cylinder under torsion.

17.4.1 Single-crystal turbine blade

An interesting industrial application of the finite element implementation of crystal plasticity constitutive equations is, for instance, the computation of the behavior of a nickel-based superalloy single-crystal turbine blade. In service, such turbine blades are submitted to important centrifugal forces induced by the fast rotation ($\sim$ 20000 RPM) of the turbine disk they are attached to. In addition, during a single flight their operating temperature can vary over three orders of magnitude. For sake of sim-

TABLE 17.1 Numerical values of material parameters used for the simulation of a nickel-based superalloy single-crystal turbine blade.

C_{11} (GPa)	C_{12} (GPa)	C_{44} (GPa)	τ_0 (MPa)	n (—)	$\dot{\gamma}_0$ (s^{-1})	μ (GPa)	G_c (—)	κ (—)
204	125	112	235	15	200	65.6	10.4	42.8

r_0^s	a_1	a_2	a_3	a_4	a_5	a_6	$b^{su}_{s \neq u}$	b^{uu}
5.38e-5	0.124	0.124	0.07	0.625	0.137	0.122	1	0

plicity, we consider here a constant and uniform temperature and focus on the mechanical behavior of the blade as the rotation rate ω of the turbine linearly and indefinitely increases. Of course, such a loading history is unrealistic for nominal in-service conditions and is rather to be seen as a demonstration of the capabilities of the finite element model. Cubic elasticity moduli, critical resolved shear stress, and viscoplastic flow parameters identified at 650°C on the nickel-based superalloy DS200 by [293] are used. Octahedral slip systems families $\{110\}\langle 111\rangle$ are considered. Instead of the phenomenological kinematic hardening law used in [293] based on the work of [936], the dislocation-based isotropic hardening law presented in Section 17.2 is used. The material parameters are presented in Table 17.1. A fictional turbine blade geometry is meshed with 191147 linear tetrahedral elements reduced integrated with one Gauss point. The crystal is oriented such that the crystal directions triplet ([100] – [010] – [001]) coincide with the orthogonal basis vectors triplet ($\underline{\mathbf{X}}_1, \underline{\mathbf{X}}_2, \underline{\mathbf{X}}_3$).

The simulated von Mises equivalent stress field and cumulated plastic slip field are shown in Fig. 17.1 at an angular velocity $\omega = 30000$ RPM. To reach this velocity, a linear ramping was used with $\dot{\omega} = 31.9$ RPM/s. Highly loaded zones are located at notches present in the foot and at the junction between the foot and the body of the turbine blade as expected. Stress concentrations are also visible in the vicinity of cooling holes. Interestingly, the most plastically deformed region lies a little above the junction between the foot and the body of the turbine blade. This region concentrating most of plastic slip spreads across the whole cross-section orthogonal to the radial direction of the turbine $\underline{\mathbf{X}}_3$. In particular, the highest levels of plastic deformations are located in the vicinity of cooling holes nearest to the turbine blade foot.

17.4.2 Single-crystal cylinder under torsion

The authors of [1026,457] investigated experimentally and numerically the behavior of single-crystal wires in torsion. They showed the existence of plastic slip gradients along the radius of wires, as well as along their circumference. The latter gradient is due to the anisotropic activation of slip systems. As already pointed out by [1026], the presence of circumferential strain gradients, visible in experiments, could not be predicted

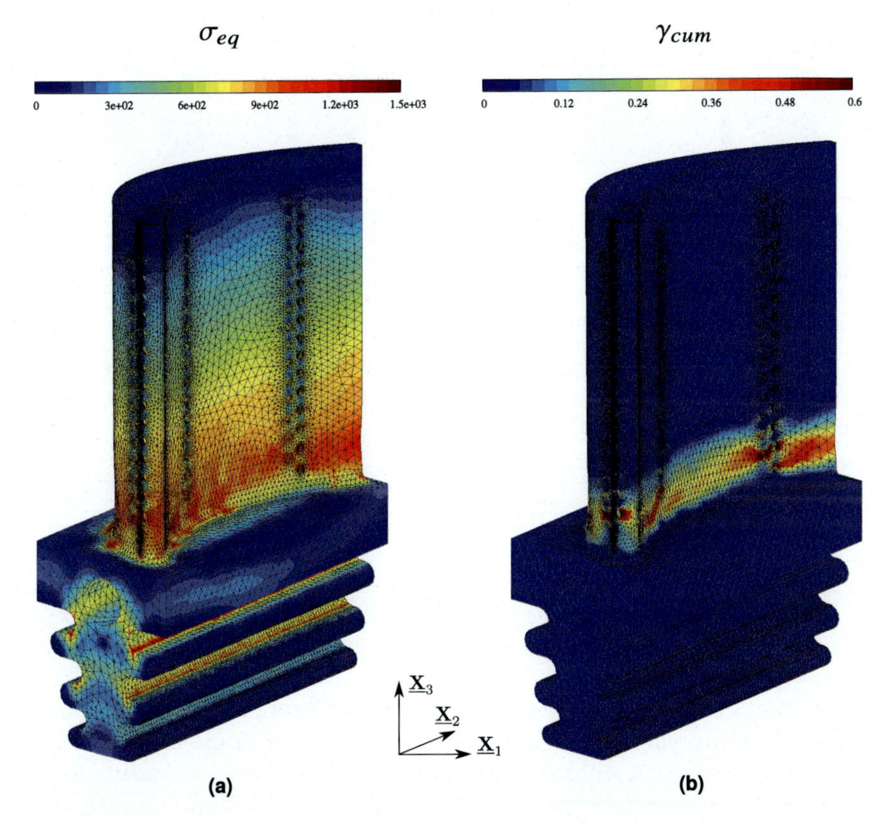

FIGURE 17.1 Simulated (a) von Mises stress and (b) cumulated plastic slip contours in a single crystal turbine blade.

by any quadratic yield criterion, such as, e.g., Hill's criterion. On the contrary, by computing analytically maximum Schmid factor maps, it appears clearly that Schmid's criterion predicts that some regions will yield earlier than others. As an example, octahedral slip systems families $\{110\}\langle 111\rangle$ are considered. Four different orientations of the crystal in the cylinder are chosen such that the crystal directions triplets ([100] – [010] – [001]), ([001] – $[1\bar{1}0]$ – [110]), ($[1\bar{1}0]$ – $[11\bar{2}]$ – [111]), and ($[12\bar{1}]$ – $[\bar{2}10]$ – [125]) respectively coincide with the orthogonal basis vectors triplet $(\underline{\mathbf{X}}_1, \underline{\mathbf{X}}_2, \underline{\mathbf{X}}_3)$. These crystal orientations are later denoted $\langle 100\rangle$, $\langle 110\rangle$, $\langle 111\rangle$, and $\langle 125\rangle$, respectively. The cylinder axis is aligned with $\underline{\mathbf{X}}_3$. Fig. 17.2 displays the maximum Schmid factor maps in the cross-section of these cylinders. Sectors of maximum Schmid factor are indeed visible. Therefore, plastic slip will preferentially be activated in such zones and lead to circumferential plastic strain gradients. It should be noted that in a torsion test, the applied torsion stress $\tau_{\theta z}$ increases linearly with the radial position from the center

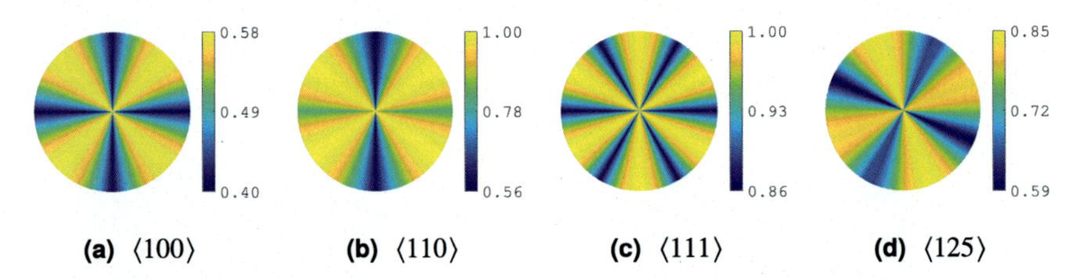

(a) ⟨100⟩ **(b)** ⟨110⟩ **(c)** ⟨111⟩ **(d)** ⟨125⟩

FIGURE 17.2 Maximum Schmid factor maps in the cross-section of single crystal cylinders with the middle axis respectively initially aligned with ⟨100⟩, ⟨110⟩, ⟨111⟩, and ⟨125⟩ crystal directions.

TABLE 17.2 Numerical values of material parameters for the simulation of stainless steel single crystal cylinders in torsion.

C_{11} (GPa)	C_{12} (GPa)	C_{44} (GPa)	τ_0 (MPa)	n (—)	$\dot{\gamma}_0$ (s^{-1})	μ GPa)	G_c (—)	κ (—)
199	136	105	88	20	10^{29}	65.6	10.4	42.8

r_0^s	a_1	a_2	a_3	a_4	a_5	a_6	$b^{su}_{s \neq u}$	b^{uu}
5.38e-5	0.124	0.124	0.07	0.625	0.137	0.122	1	0

of the cylinder. As a consequence, a radial gradient of plastic activity is also to be expected.

The dislocation density based crystal plasticity model presented above is applied to pursue investigation of the behavior of single crystal cylinders under torsion. A single crystal cylinder of length L_0 and radius R_0 is put under torsion by prescribing nil displacements to its bottom face and a rotation rate to its upper face. In cylindrical coordinates, the displacement boundary conditions are written as

$$u_r(r, \theta, z = 0) = 0, \tag{17.75}$$

$$u_\theta(r, \theta, z = 0) = 0, \qquad u_\theta(r, \theta, z = L_0) = r\dot{\theta}, \tag{17.76}$$

$$u_z(r, \theta, z = 0) = 0. \tag{17.77}$$

The rotation rate $\dot{\theta}$ is taken as $2\pi \times 10^{-4}$ s^{-1}. Material parameters relevant for face-centered cubic (FCC) austenitic stainless steel single crystals with octahedral slips systems were used and are listed in Table 17.2. Cylinders are meshed with 12800 quadratic elements reduced integrated with 8 Gauss points.

The fields of cumulated plastic slip obtained with these four orientations are plotted in Fig. 17.3 on the outer surface in (a)–(d) and in the middle cross-section at $z = L_0/2$ in (e)–(h). The fields are plotted on the deformed mesh for a $\theta = 90°$ rotation of the top face with respect to the initial configuration. A characteristic patterning of plastic slip can indeed be observed. First of all, a radial gradient of plastic strain is obtained as expected. Additionally, plastic strains are also heterogeneous on the circumference of the single crystal cylinders because of the heterogeneity of

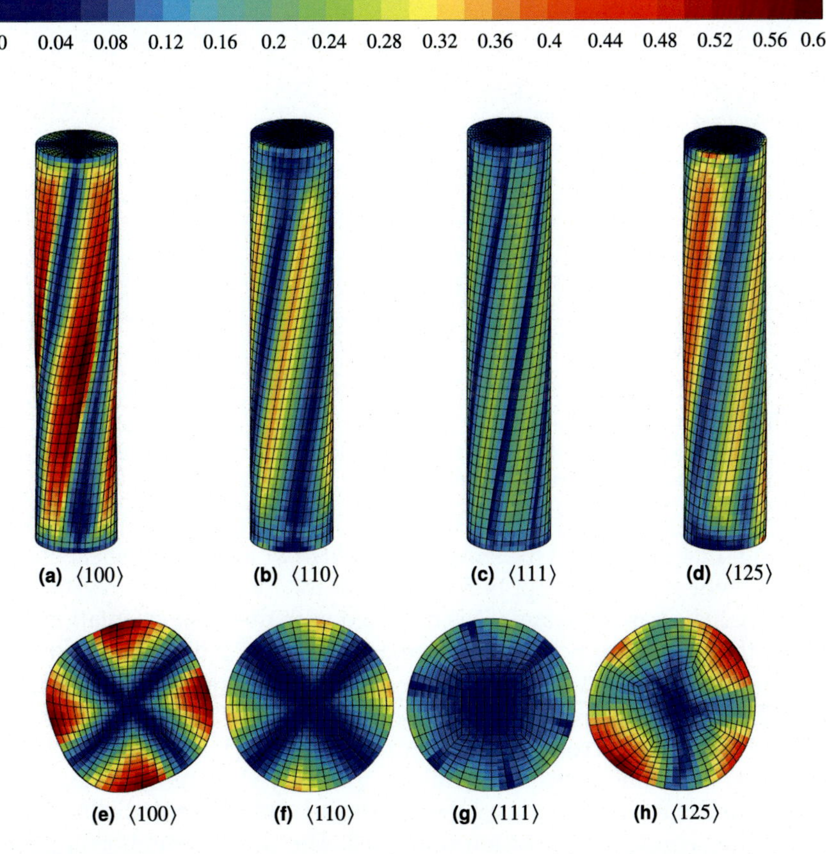

FIGURE 17.3 Simulated cumulated plastic slip contours on the outer surface (a)–(d) and in the middle cross-section (e)–(h) of single crystal cylinders at $\theta = 90°$, with the middle axis respectively initially aligned with ⟨100⟩, ⟨110⟩, ⟨111⟩, and ⟨125⟩ crystal directions and material parameters presented in Table 17.2.

the maximum Schmid factor depicted in Fig. 17.2. "Soft zones" concentrate most of plastic slip, while "hard zones" remain almost completely elastic. The number and intensity of such zones varies with the orientation of the crystal in the cylinder. For orientations ⟨100⟩ and ⟨110⟩, four "soft zones" of equal plastic intensity and four "hard zones" are clearly visible. For the ⟨111⟩ orientation, six zones of each kind and of equal intensities can be observed. The ⟨125⟩ orientation displays a more complex pattern. It is composed of two wide and intensely deformed "soft areas" and two additional "soft zones" which seem to be splitting in two narrower regions. Four "hard zones" are visible for this orientation.

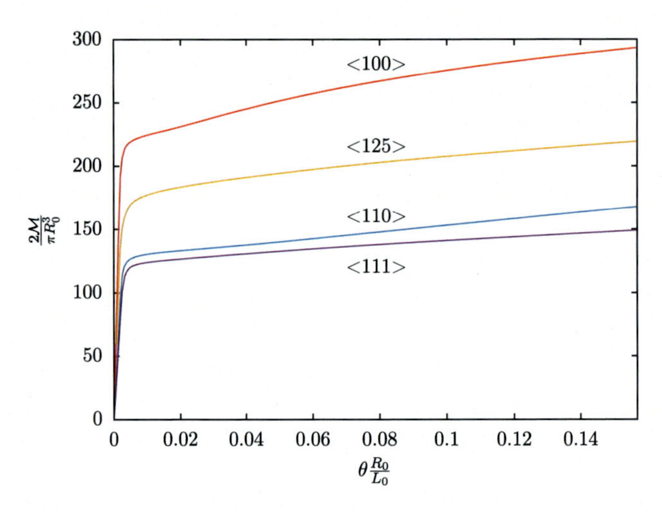

FIGURE 17.4 Simulated torque vs. shear strain curves for single crystal cylinders with the middle axis respectively initially aligned with ⟨100⟩, ⟨110⟩, ⟨111⟩, and ⟨125⟩ crystal directions and material parameters presented in Table 17.2.

In Fig. 17.4, we plotted the shear stress $2\mathcal{M}/\pi R_0^3$ against the shear strain $\theta R_0/L_0$, where $\mathcal{M}$ is the applied torque, for the four different crystal orientations. The hardest response is obtained with the ⟨100⟩ orientation, for which the maximum Schmid factor maps in Fig. 17.2 display the lowest maximum values ($\leq 1/\sqrt{3}$). The second hardest response corresponds to the ⟨125⟩ orientation for which it was shown in Fig. 17.2 that the maximum Schmid factor in the cross section does not exceed 0.85. Furthermore, ⟨110⟩ and ⟨111⟩ orientations display similar apparent yield stresses. These two orientations have indeed a maximum value of the maximum Schmid factors in the cylinder cross-section equal to 1. The ⟨111⟩ orientation appears to be a little softer than the ⟨110⟩ orientation because its average maximum Schmid factor value is the largest. Therefore the plastic slip field is the most heterogeneous for this orientation as seen in Fig. 17.3.

17.5 Extensions of constitutive equations: a reduced micromorphic model for single crystals

Conventional elasto-plasticity is known to be size-independent. Such an assumption is realistic when the size of the medium can be considered large with respect to the characteristic length of deformation mechanisms. However, it is hence unsuited in order to predict size effects arising at small scales. Nonlocal models that naturally introduce material length scales are therefore proficient extensions of conventional theories. They mainly resort on integral or gradient enhancements of conventional constitutive equations. For example, among gradient-type formulations, the micromorphic approach [455] can be followed in order to extend the constitutive model for single crystals presented above. In this section, the finite element implementation of a reduced micromorphic single-crystal model at finite strains based on a scalar nonlocal variable is presented. Following [1588], a possible choice of nonlocal variable in the context of crystal plasticity is the cumulated plastic slip γ_{cum} defined as

$$\gamma_{cum} = \int_0^t \sum_{\alpha=1}^{N} |\dot{\gamma}^\alpha| \, \mathrm{d}t. \tag{17.78}$$

Its micromorphic counterpart, denoted γ_χ, is called microslip. The latter is considered as an additional degree of freedom, on par with displacement degrees of freedom, as well as an input variable of the material behavior. Furthermore, the gradient of microslip $\underline{\mathbf{K}}_\chi = \mathrm{Grad}\,\gamma_\chi$ is equally treated as an input variable. The generalized stresses, which are work conjugates of γ_χ and $\underline{\mathbf{K}}_\chi$, are respectively denoted by S and $\underline{\mathbf{M}}$. Just as the first Piola–Kirchhoff stress, they are output variables of the constitutive model. In addition to the conventional internal variables, the cumulated plastic slip γ_{cum} is treated as an additional variable to be integrated. To summarize, the following sets of input, output, and internal variables are considered:

$$v_{\mathrm{IN}} : \{\underset{\sim}{\mathbf{F}}, \gamma_\chi, \underline{\mathbf{K}}_\chi\}, \quad v_{\mathrm{OUT}} : \{\underset{\sim}{\Pi}, S, \underline{\mathbf{M}}\}, \quad v_{\mathrm{in}} : \{\underset{\sim}{\mathbf{E}}, \gamma^\alpha, r^\alpha, \gamma_{cum}\}. \tag{17.79}$$

The thermodynamical derivation of the present model is detailed in [844,1274]. It is based, first, on enrichment of the principle of virtual power to higher-order contributions [509]. From this generalization, one obtains a supplementary equilibrium equation and, for instance, Neumann boundary condition for any subdomain $\mathcal{D}$ of the material body

$$\begin{aligned} \mathrm{Div}\,\underline{\mathbf{M}} - S &= 0 \qquad \forall \underline{\mathbf{X}} \in \mathcal{D}, \\ M &= \underline{\mathbf{M}} \cdot \underline{\mathbf{n}}_0 \qquad \forall \underline{\mathbf{X}} \in \partial D, \end{aligned} \tag{17.80}$$

where M is a generalized traction scalar, power conjugate to $\dot{\gamma}_\chi$, and $\underline{\mathbf{n}}_0$ the outward normal unit vector in the initial configuration. Solving the weak form of Eq. (17.80), together with the conventional equilibrium equation and boundary condition ($\mathrm{Div}\,(\underset{\sim}{\Pi}) = 0$ and $\underline{\mathbf{T}} = \underset{\sim}{\Pi} \cdot \underline{\mathbf{n}}_0$, in the absence of body forces), by finite elements can be done by following the procedure described in [844]. The second ingredient is a enhanced free energy potential ψ accounting for nonlocal contributions. Assuming a quadratic nonlocal potential, the conventional state law in Eq. (17.50) is complemented by two additional state laws for the generalized stresses S and

$\underline{\mathbf{M}}$,

$$\Delta S = -H_\chi(\Delta\gamma_{cum} - \Delta\gamma_\chi), \tag{17.81}$$

$$\Delta\underline{\mathbf{M}} = A\Delta\underline{\mathbf{K}}_\chi, \tag{17.82}$$

where H_χ and A are higher-order elasticity moduli; H_χ is a penalization parameter which usually takes large values in order to enforce quasiequality between γ_χ and γ_{cum}; A has the units of MPa·mm^2 and therefore bears the material characteristic length. The major outcome of the gradient-enhanced free energy potential is the modification of the residual mechanical dissipation, which now involves higher-order terms. It ensures an extension of the yield criteria in Eq. (17.20) and, equally, of the plastic flow rules in Eq. (17.21) as follows:

$$f_\chi^\alpha = |\tau^\alpha| - (\tau_c^\alpha - S) = |\tau^\alpha| - (\tau_c^\alpha + H_\chi(\gamma_{cum} - \gamma_\chi)), \tag{17.83}$$

$$\Delta\gamma^\alpha = \text{sign}\left(\tau^\alpha\right) \Delta t \dot{\gamma}_0 \Phi(f_\chi^\alpha), \qquad \Phi(f_\chi^\alpha) = \left\langle \frac{f_\chi^\alpha}{\tau_0^\alpha} \right\rangle^n. \tag{17.84}$$

The generalized scalar stress S, depending on its sign, acts locally as an additional hardening or softening contribution. By combining Eqs. (17.80) and (17.82), we obtain $S = \text{Div}\,(A\text{Grad}\,\gamma_\chi)$. Therefore, locally, a positive curvature of γ_χ induces softening; conversely, a locally negative curvature of γ_χ introduces additional hardening. Increments of cumulated plastic slip must satisfy Eq. (17.78), which gives

$$\Delta\gamma_{cum} = \sum_{\alpha=1}^{N} |\Delta\gamma^\alpha|. \tag{17.85}$$

The set of residual equations for the reduced micromorphic single-crystal model at finite strains become

$$R_{\text{in}} = \begin{cases} R_{\underset{\sim}{E}} = \Delta\underset{\sim}{\mathbf{E}} - \Delta\underset{\sim}{\mathbf{F}} \cdot \underset{\sim}{\mathbf{F}}^{-1} \cdot E - \underset{\sim}{\mathbf{E}} \cdot \left(\displaystyle\sum_{\alpha=1}^{N} \Delta\gamma^\alpha \underset{\sim}{\mathbf{N}}^\alpha \right), \\ R_{\gamma^\alpha} = \Delta\gamma^\alpha - \text{sign}\left(\tau^\alpha\right) \Delta t \dot{\gamma}_0 \Phi(f_\chi^\alpha), \\ R_{r^\alpha} = \Delta r^\alpha - |\Delta\gamma^\alpha| \left(\dfrac{1}{\kappa} \sqrt{\displaystyle\sum_{\beta=1}^{N} b^{\alpha\beta} r^\beta} - G_c r^\alpha \right), \\ R_{\gamma_{cum}} = \Delta\gamma_{cum} - \displaystyle\sum_{\alpha=1}^{N} |\Delta\gamma^\alpha|. \end{cases} \tag{17.86}$$

It is remarkable how corresponding residuals in Eq. (17.86) are similar to their conventional counterparts; $R_{\underset{\sim}{E}}$ and R_{r^α} are indeed completely unchanged, while R_{γ^α} is only slightly modified, and the straightforward term $R_{\gamma_{cum}}$ is added. The Jacobian matrix $[J]$ can accordingly be computed without much difficulty, since only the last row and column need to be given, while other terms remain formally unchanged given that f^α is replaced by f_χ^α:

$$[J] = \frac{\partial R_{\text{in}}}{\partial \Delta v_{\text{int}}} = \left(\begin{array}{ccc|c} \dfrac{\partial R_{\underset{\sim}{E}}}{\partial \Delta \underset{\sim}{E}} & \dfrac{\partial R_{\underset{\sim}{E}}}{\partial \Delta \gamma^\beta} & \dfrac{\partial R_{\underset{\sim}{E}}}{\partial \Delta r^\beta} & \dfrac{\partial R_{\underset{\sim}{E}}}{\partial \Delta \gamma_{cum}} \\ \dfrac{\partial R_{\gamma^\alpha}}{\partial \Delta \underset{\sim}{E}} & \dfrac{\partial R_{\gamma^\alpha}}{\partial \Delta \gamma^\beta} & \dfrac{\partial R_{\gamma^\alpha}}{\partial \Delta r^\beta} & \dfrac{\partial R_{\gamma^\alpha}}{\partial \Delta \gamma_{cum}} \\ \dfrac{\partial R_{r^\alpha}}{\partial \Delta \underset{\sim}{E}} & \dfrac{\partial R_{r^\alpha}}{\partial \Delta \gamma^\beta} & \dfrac{\partial R_{r^\alpha}}{\partial \Delta r^\beta} & \dfrac{\partial R_{r^\alpha}}{\partial \Delta \gamma_{cum}} \\ \hline \dfrac{\partial R_{\gamma_{cum}}}{\partial \Delta \underset{\sim}{E}} & \dfrac{\partial R_{\gamma_{cum}}}{\partial \Delta \gamma^\beta} & \dfrac{\partial R_{\gamma_{cum}}}{\partial \Delta r^\beta} & \dfrac{\partial R_{\gamma_{cum}}}{\partial \Delta \gamma_{cum}} \end{array} \right). \tag{17.87}$$

After straightforward derivations, one obtains

$$\frac{\partial R_{\underset{\sim}{E}}}{\partial \Delta \gamma_{cum}} = 0, \tag{17.88}$$

$$\frac{\partial R_{\gamma^\alpha}}{\partial \Delta \gamma_{cum}} = -\text{sign}\left(\tau^\alpha\right) \Delta t \frac{\partial \Phi^\alpha}{\partial f_\chi^\alpha} \frac{\partial f_\chi^\alpha}{\partial \gamma_{cum}} = \text{sign}\left(\tau^\alpha\right) \frac{\Delta t \dot{\gamma}_0 n}{(\tau_0^\alpha)^n} \left\langle \frac{f_\chi^\alpha}{\tau_0^\alpha} \right\rangle^{n-1} H_\chi, \tag{17.89}$$

$$\frac{\partial R_{r^\alpha}}{\partial \Delta \gamma_{cum}} = 0, \tag{17.90}$$

$$\frac{\partial R_{\gamma_{cum}}}{\partial \Delta \underset{\sim}{E}} = 0, \tag{17.91}$$

$$\frac{\partial R_{\gamma_{cum}}}{\partial \Delta \gamma^\beta} = -\text{sign}\left(\Delta \gamma^\beta\right), \tag{17.92}$$

$$\frac{\partial R_{\gamma_{cum}}}{\partial \Delta r^\beta} = 0, \tag{17.93}$$

$$\frac{\partial R_{\gamma_{cum}}}{\partial \Delta \gamma_{cum}} = 1. \tag{17.94}$$

Since additional input and output variables are considered, the consistent tangent operator incorporates the following additional derivatives:

- $\dfrac{\partial \Delta v_{\mathrm{OUT}}}{\partial \Delta v_{\mathrm{IN}}}$

$$
\begin{aligned}
&\frac{\partial \Delta \underset{\sim}{\Pi}}{\partial \Delta \underset{\sim}{\mathbf{F}}} = \text{see Eq. (17.66)}, && \frac{\partial \Delta \underset{\sim}{\Pi}}{\partial \Delta \gamma_{\chi}} = 0, && \frac{\partial \Delta \underset{\sim}{\Pi}}{\partial \Delta \underline{\mathbf{K}}_{\chi}} = 0, \\
&\frac{\partial \Delta S}{\partial \Delta \underset{\sim}{\mathbf{F}}} = 0, && \frac{\partial \Delta S}{\partial \Delta \gamma_{\chi}} = H_{\chi}, && \frac{\partial \Delta S}{\partial \Delta \underline{\mathbf{K}}_{\chi}} = 0, \\
&\frac{\partial \Delta \underline{\underline{\mathbf{M}}}}{\partial \Delta \underset{\sim}{\mathbf{F}}} = 0, && \frac{\partial \Delta \underline{\underline{\mathbf{M}}}}{\partial \Delta \gamma_{\chi}} = 0, && \frac{\partial \Delta \underline{\underline{\mathbf{M}}}}{\partial \Delta \underline{\mathbf{K}}_{\chi}} = A;
\end{aligned}
\tag{17.95}
$$

- $\dfrac{\partial \Delta v_{\mathrm{OUT}}}{\partial \Delta v_{\mathrm{in}}}$

$$
\begin{aligned}
&\frac{\partial \Delta \underset{\sim}{\Pi}}{\partial \Delta \underset{\sim}{\mathbf{E}}} = \text{see Eq. (17.68)}, && \frac{\partial \Delta \underset{\sim}{\Pi}}{\partial \Delta \gamma^{\beta}} = 0, && \frac{\partial \Delta \underset{\sim}{\Pi}}{\partial \Delta r^{\beta}} = 0, && \frac{\partial \Delta \underset{\sim}{\Pi}}{\partial \Delta \gamma_{cum}} = 0, \\
&\frac{\partial \Delta S}{\partial \Delta \underset{\sim}{\mathbf{E}}} = 0, && \frac{\partial \Delta S}{\partial \Delta \gamma^{\beta}} = 0, && \frac{\partial \Delta S}{\partial \Delta r^{\beta}} = 0, && \frac{\partial \Delta S}{\partial \Delta \gamma_{cum}} = -H_{\chi}, \\
&\frac{\partial \Delta \underline{\underline{\mathbf{M}}}}{\partial \Delta \underset{\sim}{\mathbf{E}}} = 0, && \frac{\partial \Delta \underline{\underline{\mathbf{M}}}}{\partial \Delta \gamma^{\beta}} = 0, && \frac{\partial \Delta \underline{\underline{\mathbf{M}}}}{\partial \Delta r^{\beta}} = 0, && \frac{\partial \Delta \underline{\underline{\mathbf{M}}}}{\partial \Delta \gamma_{cum}} = 0;
\end{aligned}
\tag{17.96}
$$

- $[L]_s = \dfrac{\partial R_{\mathrm{in}}}{\partial \Delta v_{\mathrm{IN}}}$

$$
\begin{aligned}
&\frac{\partial R_{\underset{\sim}{E}}}{\partial \Delta \underset{\sim}{\mathbf{F}}} = \text{see Eq. (17.74)}, && \frac{\partial R_{\underset{\sim}{E}}}{\partial \Delta \gamma_{\chi}} = 0, && \frac{\partial R_{\underset{\sim}{E}}}{\partial \Delta \underline{\mathbf{K}}_{\chi}} = 0, \\
&\frac{\partial R_{\gamma^{\alpha}}}{\partial \Delta \underset{\sim}{\mathbf{F}}} = 0, && \frac{\partial R_{\gamma^{\alpha}}}{\partial \Delta \gamma_{\chi}} = -\operatorname{sign}\left(\tau^{\alpha}\right) \frac{\Delta t \dot{\gamma}_0 n}{(\tau_0^{\alpha})^n} H_{\chi}, && \frac{\partial R_{\gamma^{\alpha}}}{\partial \Delta \underline{\mathbf{K}}_{\chi}} = 0, \\
&\frac{\partial R_{r^{\alpha}}}{\partial \Delta \underset{\sim}{\mathbf{F}}} = 0, && \frac{\partial R_{r^{\alpha}}}{\partial \Delta \gamma_{\chi}} = 0, && \frac{\partial R_{r^{\alpha}}}{\partial \Delta \underline{\mathbf{K}}_{\chi}} = 0, \\
&\frac{\partial R_{\gamma_{cum}}}{\partial \Delta \underset{\sim}{\mathbf{F}}} = 0, && \frac{\partial R_{\gamma_{cum}}}{\partial \Delta \gamma_{\chi}} = 0, && \frac{\partial R_{\gamma_{cum}}}{\partial \Delta \underline{\mathbf{K}}_{\chi}} = 0.
\end{aligned}
\tag{17.97}
$$

Bibliography

[1] Zset software, non-linear material & structure analysis suite, http://www.zset-software.com.

[2] W. Österle, D. Bettge, B. Fedelich, H. Klingelhöffer, Modelling the orientation and direction dependence of the critical resolved shear stress of nickel-base superalloy single crystals, Acta Materialia 48 (3) (2000) 689–700.

[3] Abaqus, https://www.3ds.com/products-services/simulia/products/abaqus/, 2018.

[4] A. Acharya, A. Roy, Size effects and idealized dislocation microstructure at small scales: predictions of a phenomenological model of mesoscopic field dislocation mechanics: part I, Journal of the Mechanics and Physics of Solids 54 (8) (2006) 1687–1710.

[5] L. Agudo Jacome, G. Eggeler, A. Dlouhy, Advanced scanning transmission electron microscopy of structural and functional engineering materials, Ultramicroscopy 122 (2012) 48–59.

[6] L. Agudo-Jacome, P. Nörterhäuser, J.-K. Heyer, A. Lahni, J. Frenzel, A. Dlouhy, C. Somsen, G. Eggeler, High-temperature and low-stress creep anisotropy of single-crystal superalloys, Acta Materialia 61 (2013) 2926–2943.

[7] L. Agudo Jacome, P. Nörtershäuser, C. Somsen, A. Dlouhy, G. Eggeler, On the nature of γ' phase cutting and its effect on high temperature and low stress creep anisotropy of Ni-base single crystal superalloys, Acta Materialia 69 (2014) 246–264.

[8] S. Ahmadian, A. Browning, E.H. Jordan, Three-dimensional x-ray micro-computed tomography of cracks in a furnace cycled air plasma sprayed thermal barrier coating, Scripta Materialia 97 (2015) 13–16.

[9] M. Ahrens, R. Vaßen, D. Stöver, S. Lampenscherf, Sintering and creep processes in plasma-sprayed thermal barrier coatings, Journal of Thermal Spray Technology 13 (3) (2004) 432–442.

[10] S. Ai, V. Lupinc, M. Maldini, Creep fracture mechanisms in single crystal superalloys, Scripta Metallurgica 26 (1992) 579–584.

[11] S. Ai, V. Lupinc, G. Onofrio, Influence of precipitate morphology on high temperature fatigue crack growth of a single crystal nickel base superalloy, Scripta Metallurgica 29 (1993) 1385–1390.

[12] M.Z. Alam, S. Kamat, V. Jayaram, D.K. Das, Tensile behavior of a free-standing pt-aluminide (ptal) bond coat, Acta Materialia 61 (4) (2013) 1093–1105.

[13] M.Z. Alam, S. Kamat, V. Jayaram, D.K. Das, Micromechanisms of fracture and strengthening in free-standing Pt-aluminide bond coats under tensile loading, Acta Materialia 67 (2014) 278–296.

[14] M.Z. Alam, C. Parlikar, D. Chatterjee, D.K. Das, Comparative tensile behavior of free-standing γ/γ' and β-(Ni, Pt)Al bond coats and effect on tensile properties of coated superalloy, Materials & Design 114 (2017) 505–514.

[15] M.Z. Alam, S.B. Sarkar, D.K. Das, Tensile behavior of a freestanding γ/γ' bond coat, Surface & Coatings Technology 324 (2017) 486–490.

[16] M. Ali, W. Amin, O. Shchyglo, I. Steinbach, 45-degree rafting in Ni-based superalloys: a combined phase-field and strain gradient crystal plasticity study, International Journal of Plasticity (2020) 102659.

[17] M. Ali, J. Görler, I. Steinbach, Role of coherency loss on rafting behavior of Ni-based superalloys, Computational Materials Science 171 (2020) 102659.

[18] M. Allen, D. Tildesley, Computer Simulation of Liquids, 2nd edition, Oxford University Press, 2017.

[19] R. Amaro, S. Antolovich, R. Neu, P. Fernandez-Zelaia, W. Hardin, Thermomechanical fatigue and bithermal-thermomechanical fatigue of a Ni-base single crystal superalloy, International Journal of Fatigue 42 (2012) 165–171.

[20] K. Ammar, B. Appolaire, G. Cailletaud, S. Forest, Combining phase field approach and homogenization methods for modelling phase transformation in elastoplastic media, European Journal of Computational Mechanics 18 (5–6) (2009).

[21] J. Amodeo, C. Begau, E. Bitzek, Atomistic simulations of compression tests on Ni ₃ Al nanocubes, Materials Research Letters 2 (3) (2014) 140–145.

[22] J. Amodeo, P. Carrez, B. Devincre, P. Cordier, Multiscale modelling of MgO plasticity, Acta Materialia 59 (6) (2011) 2291–2301.

[23] L. Anand, M. Kothari, A computational procedure for rate–independent crystal plasticity, Journal of the Mechanics and Physics of Solids 44 (1996) 525–558.

[24] J. Andersson, J. Ågren, Models for numerical treatment of multicomponent diffusion in simple phases, Journal of Applied Physics 72 (1992) 1350–1355.

[25] E. Andrieu, B. Pieraggi, A. Gourgues, Role of metal-oxide interfacial reactions on the interactions between oxidation and deformation, Acta Materialia 39 (1998) 597–601.

[26] B. Antolovich, A. Saxena, S. Antolovich, Fatigue crack propagation in single-crystal CMSX-2 at elevated temperature, Journal of Materials Engineering and Performance 2 (1993) 489–495.

[27] S. Antonov, Y. Zheng, J.M. Sosa, H.L. Fraser, J. Cormier, P. Kontis, B. Gault, Plasticity assisted redistribution of solutes leading to topological inversion during creep of superalloys, Scripta Materialia 186 (2020) 287–292.

[28] A. Anwaar, L. Wei, Q. Guo, B. Zhang, H. Guo, Novel prospects for plasma spray–physical vapor deposition of columnar thermal barrier coatings, Journal of Thermal Spray Technology 26 (8) (2017) 1810–1822.

[29] D. Apelian, R. DasGupat, M. Gywn, J. Jorstad, R. Monroe, T. Prucha, M. Sahoo, E. Szekeres (Eds.), Casting, ASM Handbook, vol. 15, 2008.

[30] N. Arakere, S. Siddiqui, F. Ebrahimi, Evolution of plasticity in notched ni-base superalloy single crystals, International Journal of Solids and Structures 46 (2009) 3027–3044.

[31] M. Ardakani, M. McLean, B. Shollock, Twin formation during creep in single crystals of nickel-based superalloys, Acta Materialia 47 (1999) 2593–2602.

[32] A.J. Ardell, An application of the theory of particle coarsening: the γ' precipitate in Ni-Al alloys, Acta Metallurgica 16 (1968) 511–516.

[33] A.J. Ardell, Temporal behavior of the number density of particles during Ostwald ripening, Materials Science and Engineering: A 238 (1997) 108–120.

[34] A.J. Ardell, R.B. Nicholson, On the modulated structure of aged Ni-Al alloys, Acta Metallurgica (1966).

[35] A.J. Ardell, R.B. Nicholson, The coarsening of γ' in Ni-Al alloys, Journal of Physics and Chemistry of Solids 27 (1966) 1793–1804.

[36] A.J. Ardell, V. Ozolins, Trans-interface diffusion-controlled coarsening, Nature Materials 4 (2005) 309–316.

[37] D. Argence, C. Vernault, Y. Desvallées, D. Fournier, MC-NG: a 4th generation single crystal superalloy for future aeronautical turbine blades and vanes, in: Proc. 9th Int. Symp. on Superalloys, TMS, 2000.

[38] M. Arnoux, Etude du comportement en fluage à haute température du superalliage monocristallin à base de nickel MCNG: Effet d'une surchauffe, PhD thesis, Université de Poitiers, France, 2009.

[39] K. Arora, K. Kishida, H. Inui, Micro-pillar compression of ni-base superalloy single crystals, in: Symposium JJ - Intermetallic-Based Alloys-Science, Technology and Applications, vol. 1516, Cambridge University Press, 2013, pp. 209–214.

[40] K. Arora, K. Kishida, K. Tanaka, H. Inui, Effects of lattice misfit on plastic deformation behavior of single-crystalline micropillars of Ni-based superalloys, Acta Materialia 138 (2017) 119–130.

[41] D.J. Arrell, J.L. Vallés, Rafting prediction criterion for superalloys under a multiaxial stress, Scripta Materialia 35 (6) (1996) 727–732.

[42] A. Arsenlis, W. Cai, M. Tang, M. Rhee, T. Oppelstrup, G. Hommes, T.G. Pierce, V.V. Bulatov, Enabling strain hardening simulations with dislocation dynamics, Modelling and Simulation in Materials Science and Engineering 15 (6) (2007) 553.

[43] R. Asaro, Crystal plasticity, Journal of Applied Mechanics 50 (1983) 921–934.

[44] R. Asaro, Micromechanics of crystals and polycrystals, Advances in Applied Mechanics 23 (1983) 1–115.

[45] R. Asaro, A. Needleman, Texture development and strain hardening in rate dependent polycrystals, Acta Metallurgica 33 (1985) 923–953.

[46] R. Asaro, J. Rice, Strain localization in ductile single crystals, Journal of the Mechanics and Physics of Solids 25 (1977) 309–338.

[47] M. Ashby, The deformation of plastically non–homogeneous materials, Philosophical Magazine 21 (1970) 399–424.

[48] O. Aslan, N.M. Cordero, A. Gaubert, S. Forest, Micromorphic approach to single crystal plasticity and damage, International Journal of Engineering Science 49 (2011) 1311–1325.

[49] O. Aslan, S. Forest, Crack growth modelling in single crystals based on higher order continua, Computational Materials Science 45 (2009) 756–761.

[50] O. Aslan, S. Quilici, S. Forest, Numerical modeling of fatigue crack growth in single crystals based on microdamage theory, International Journal of Damage Mechanics 20 (2011) 681–705.

[51] P. Audigié, A. Rouaix-Van de Put, A. Malié, P. Bilhé, S. Hamadi, D. Monceau, Observation and modeling of α-NiPtAl and Kirkendall void formations during interdiffusion of a Pt coating with a γ-(Ni-13Al) alloy at high temperature, Surface & Coatings Technology 260 (2014) 9–16.

[52] P. Audigié, A. Rouaix-Van de Put, A. Malié, D. Monceau, High-temperature cyclic oxidation behaviour of Pt-rich γ/γ' coatings. Part I: oxidation kinetics of coated AM1 systems after very long-term exposure at 1100°C, Corrosion Science 144 (2018) 127–135.

[53] P. Audigié, A. Rouaix-Van de Put, A. Malié, C. Thouron, D. Monceau, High-temperature cyclic oxidation behaviour of pt-rich γ-γ' coatings. Part II: effect of pt and al on tbc system lifetime, Corrosion Science 150 (2019) 1–8.

[54] J. Aveson, G. Reinhart, C. Goddard, H. Nguyen-Thi, N. Mangelinck-Noël, A. Tandjaoui, J. Davenport, N. Warnken, F. Di Gioacchino, T. Lafford, On the deformation of dendrites during directional solidification of a nickel-based superalloy, Metallurgical and Materials Transactions A 50 (2019) 5234–5241.

[55] C. Ayas, J. van Dommelen, V. Deshpande, Climb-enabled discrete dislocation plasticity, Journal of the Mechanics and Physics of Solids 62 (2014) 113–136.

[56] A. Aygun, A.L. Vasiliev, N.P. Padture, X. Ma, Novel thermal barrier coatings that are resistant to high-temperature attack by glassy deposits, Acta Materialia 55 (20) (2007) 6734–6745.

[57] D. Ayrault, Fluage à haute température de superalliages base nickel monocristallins, in French, PhD thesis, ENS Mines de Paris, HAL Id: pastel-00002126, 1989.

[58] D. Ayrault, A. Fredholm, J.-L. Strudel, High temperature creep and structural changes in nickel base single crystals, in: T. Khan, A. Lasalmonie (Eds.), Proc. Amer. Soc. Materials Europe Tech. Conf., Ohio, ASM Int., 1987, pp. 71–81.

[59] D. Ayrault, A. Fredholm, J.-L. Strudel, High temperature creep mechanisms in single crystals of some high performance nickel base superalloys, in: J.B. Mariot, H Herz, J. Nihoul, J. Ward (Eds.), High Temperature Alloys, Their Exploitable Potential, Petten Conference 1985, Elsevier Applied Science, 1987, pp. 71–81.

[60] F. Azzouz, Modélisation du comportement mécanique du superalliage monocristallin CMSX4, Technical report, Mines ParisTech, 92372/2661, 2014.

[61] K. Badura-Gergen, H.-E. Schaefer, Thermal formation of atomic vacancies in Ni_3Al, Physical Review B 56 (1997) 3032–3037.

[62] H.-A. Bahr, H. Balke, T. Fett, I. Hofinger, G. Kirchhoff, D. Munz, A. Neubrand, A. Semenov, H.-J. Weiss, Y. Yang, Cracks in functionally graded materials, Materials Science and Engineering: A 362 (1–2) (2003) 2–16.

[63] E. Bakan, R. Vaßen, Ceramic top coats of plasma-sprayed thermal barrier coatings: materials, processes, and properties, Journal of Thermal Spray Technology 26 (6) (2017) 992–1010.

[64] K. Baker, W. Curtin, Multiscale diffusion method for simulations of long-time defect evolution with application to dislocation climb, Journal of the Mechanics and Physics of Solids 92 (2016) 297–312.

[65] B. Bakó, E. Clouet, L.M. Dupuy, M. Blétry, Dislocation dynamics simulations with climb: kinetics of dislocation loop coarsening controlled by bulk diffusion, Philosophical Magazine 91 (23) (2011) 3173–3191.

[66] A. Baldan, Progress in Ostwald ripening theories and their applications to nickel-base superalloys. Part I: Ostwald ripening theories, Journal of Materials Science 37 (11) (2002) 2171–2202.

[67] A. Baldan, Progress in Ostwald ripening theories and their applications to the γ'-precipitates in nickel-base superalloys Part II: nickel-base superalloys, Journal of Materials Science 7 (2002) 2379–2405.

[68] B. Bales, L. Petzold, B.R. Goodlet, W.C. Lenthe, T.M. Pollock, Bayesian inference of elastic properties with resonant ultrasound spectroscopy, The Journal of the Acoustical Society of America 143 (1) (2018) 71–83.

[69] D. Balint, J. Hutchinson, An analytical model of rumpling in thermal barrier coatings, Journal of the Mechanics and Physics of Solids 53 (2005) 949–973.

[70] D. Balint, S.-S. Kim, Y.-F. Liu, R. Kitazawa, Y. Kagawa, A. Evans, Anisotropic tgo rumpling in EB-PVD thermal barrier coatings under in-phase thermomechanical loading, Acta Materialia 59 (6) (2011) 2544–2555.

[71] D.S. Balint, J.W. Hutchinson, Undulation instability of a compressed elastic film on a nonlinear creeping substrate, Acta Materialia 51 (13) (2003) 3965–3983.

[72] N.L. Baluc, Contribution à l'étude des défauts et de la plasticité d'un composé intermétallique ordonné, Technical report, EPFL, 1990.

[73] D. Banerjee, K. Yu, Investment casting, in: K. Yu (Ed.), Modeling for Casting and Solidification Processing, CRC Press, 2002, pp. 333–372.

[74] D. Barba, E. Alabort, S. Pedrazzini, D. Collins, A. Wilkinson, P. Bagot, M. Moody, C. Atkinson, A. Jérusalem, R. Reed, On the microtwinning mechanism in a single crystal superalloy, Acta Materialia 135 (2017) 314–329.

[75] D. Barba, T. Smith, J. Miao, M. Mills, R. Reed, Segregation assisted plasticity in Ni-base superalloys, Materials Transactions A 49:4173–4185 (2018).

[76] S. Baroni, S. de Gironcoli, A. Dal Corso, P. Giannozzi, Phonons and related crystal properties from density-functional perturbation theory, Reviews of Modern Physics 73 (2001) 515–562.

[77] M. Bartsch, B. Baufeld, S. Dalkiliç, L. Chernova, M. Heinzelmann, Fatigue cracks in a thermal barrier coating system on a superalloy in multiaxial thermomechanical testing, International Journal of Fatigue 30 (2) (2008) 211–218.

[78] M. Bartsch, B. Baufeld, S. Dalkiliç, I. Mircea, Testing and characterization of ceramic thermal barrier coatings, in: Materials Science Forum, vol. 492, Trans Tech Publ, 2005, pp. 3–8.

[79] M. Bartsch, G. Marci, K. Mull, C. Sick, Fatigue testing of ceramic thermal barrier coatings for gas turbine blades, Advanced Engineering Materials 2 (1999) 127–129.

[80] M. Bartsch, J. Wischek, C. Meid, K. Knipe, A. Manero, S. Raghavan, A. Karlsson, J. Okasinski, J. Almer, Evaluating deformation behavior of a TBC-system during thermal gradient mechanical fatigue by means of high energy X-ray diffraction, in: Proc. Thermal Barrier Coatings IV, Irsee, Germany, June 2014, pp. 22–24.

[81] H.C. Basoalto, R.N. Ghosh, M.G. Ardakani, B.A. Shollock, M. McLean, Multiaxial creep deformation of single crystal superalloys: modelling and validation, in: Superalloys 2000, Champion, PA (USA), 2000.

[82] C. Batthias, P. Paris, Gigacycle Fatigue in Mechanical Practice, CRC Press, 2004.

[83] B. Baufeld, M. Bartsch, M. Heinzelmann, Advanced thermal gradient mechanical fatigue testing of CMSX-4 with an oxidation protection coating, International Journal of Fatigue 30 (2008) 219–225.

[84] R.H. Baughman, J.M. Shacklette, A.A. Zakhidov, S. Stafström, Negative Poisson's ratios as a common feature of cubic metals, Nature 392 (1998) 362–365.

[85] P. Beardmore, R. Davies, T. Johnston, Temperature dependence of the flow stress of nickel-base alloys, Transactions of the Metallurgical Society of AIME 245 (1969) 1537–1545.

[86] C. Beckermann, J. Gu, W. Boettinger, Development of a freckle predictor via Rayleigh number method for single-crystal nickel-base superalloy castings, Metallurgical and Materials Transactions A 31 (2000) 2545–2557.

[87] G. Bégué, G. Fabre, V. Guipont, M. Jeandin, P. Bilhe, J.Y. Guedou, F. Lepoutre, Laser shock adhesion test (lasat) of EB-PVD tbcs: towards an industrial application, Surface & Coatings Technology 237 (2013) 305–312.

[88] D. Bellet, P. Bastie, A. Royer, J. Lajzerowicz, J. Legrand, R. Bonnet, Small angle neutron scattering (sans) study of γ' precipitates in single crystals of am1 superalloy, Journal de Physique I 2 (6) (1992) 1097–1112.

[89] D. Bellet, A. Royer, P. Bastie, J. Lajzerowicz, J. Legrand, "In situ" small angle neutron scattering study of γ' precipitates in am1 superalloy single crystals, in: Superalloys, The Minerals, Metals & Materials Society, 1992, pp. 547–553.

[90] H. Ben Hamouda, Modélisation et Simulation de la structure de solidification dans les superalliages base-nickel: Application AM1, PhD thesis, MINES ParisTech, 2012.

[91] L.A. Bendersky, F.W. Gayle, Electron diffraction using transmission electron microscopy, Journal of Research of the National Institute of Standards and Technology 106 (6) (2001) 997.

[92] M. Bensch, E. Fleischmann, C. Konrad, M. Fried, C.M.F. Rae, U. Glatzel, Secondary creep of thin-walled specimens affected by oxidation, in: E. Huron, R. Reed, M. Hardy, M. Mills, R. Montero, P. Portella, J. Telesman (Eds.), Superalloys 2012, vol. 6, 2012, pp. 387–394.

[93] M. Bensch, C. Konrad, E. Fleischmann, C. Rae, U. Glatzel, Influence of oxidation on near-surface γ' fraction and resulting creep behaviour of single crystal ni-base superalloy M247LC SX, Materials Science and Engineering: A 577 (2013) 179–188.

[94] M. Bensch, J. Preußner, R. Hüttner, G. Obigodi, S. Virtanen, J. Gabel, U. Glatzel, Modelling and analysis of the oxidation influence on creep behaviour of thin-walled structures of the single-crystal nickel-base superalloy René N5 at 980 °C, Acta Materialia 58 (5) (2010) 1607–1617.

[95] M. Bensch, A. Sato, N. Warnken, E. Affeldt, R.C. Reed, U. Glatzel, Modelling of high temperature oxidation of alumina-forming single-crystal nickel-base superalloys, Acta Materialia 60 (2012) 5468–5480.

[96] M. Benyoucef, A. Coujou, B. Barbker, N. Clement, In situ deformation experiments on a γ/γ' superalloy strengthening mechanisms, Materials Science and Engineering: A 234–236 (1997) 692–694.

[97] M. Benyoucef, M. Legros, A. Coujou, P. Caron, H. Calderon, N. Clément, Micromechanisms involved in rafts of crept MC2 nickel-based single crystal superalloy, in: Materials Sciences Forum, 2003, pp. 779–784.

[98] B. Bernard, A. Quet, L. Bianchi, V. Schick, A. Joulia, A. Malié, B. Rémy, Effect of suspension plasma-sprayed YSZ columnar microstructure and bond coat surface preparation on thermal barrier coating properties, Journal of Thermal Spray Technology 26 (6) (2017) 1025–1037.

[99] L. Berthe, M. Arrigoni, M. Boustie, J.P. Cuq-Lelandais, C. Broussillou, G. Fabre, M. Jeandin, V. Guipont, M. Nivard, State-of-the-art laser adhesion test (LASAT), Nondestructive Testing and Evaluation 26 (2011) 303–317.

[100] N. Bertin, V. Glavas, D. Datta, W. Cai, A spectral approach for discrete dislocation dynamics simulations of nanoindentation, Modelling and Simulation in Materials Science and Engineering 26 (5) (2018) 055004.

[101] N. Bertin, M.V. Upadhyay, C. Pradalier, L. Capolungo, A fft-based formulation for efficient mechanical fields computation in isotropic and anisotropic periodic discrete dislocation dynamics, Modelling and Simulation in Materials Science and Engineering 23 (6) (2015) 065009.

[102] J. Besson, G. Cailletaud, J.-L. Chaboche, S. Forest, M. Blétry, Non–Linear Mechanics of Materials, Solid Mechanics and Its Applications, vol. 167, Springer-Verlag, Berlin Heidelberg, 2009.

[103] J. Besson, R. Foerch, Large scale object–oriented finite element code design, Computer Methods in Applied Mechanics and Engineering 142 (1997) 165–187.

[104] D. Bettge, W. Österle, "Cube slip" in near-[111] oriented specimens of a single-crystal nickel-base superalloy, Scripta Materialia 40 (4) (1999) 389–395.

[105] P. Bhowal, E. Wright, E. Raymond, Effects of cooling rate and γ' morphology on creep and stress-rupture properties of a powder metallurgy superalloy, Metallurgical and Materials Transactions A 21 (1990) 1709–1717.

[106] A. Bialon, T. Hammerschmidt, R. Drautz, Three-parameter crystal-structure prediction for sp-d valent compounds, Chemistry of Materials 28 (2016) 2550–2556.

[107] F. Bianchini, J. Kermode, A. De Vita, Modelling defects in Ni-Al with EAM and DFT calculations, Modelling and Simulation in Materials Science and Engineering 24 (4) (2016).

[108] H. Biermann, M. Strehler, H. Mughrabi, High-temperature measurements of lattice parameters and internal stresses of a creep-deformed monocrystalline nickel-base superalloy, Metallurgical and Materials Transactions A 27 (1996) 1003–1014.

[109] H. Biermann, U. Tetzlaff, B.V. Grossmann, H. Mughrabi, V. Schulze, Rafting in monocrystalline nickel-base superalloys induced by shot peening, Scripta Materialia 43 (2000) 807–812.

[110] H. Biermann, B. Von Grossmann, T. Schneider, H. Feng, H. Mughrabi, Investigation of the γ/γ' morphology and internal stresses in a monocrystalline turbine blade after service: determination of the local thermal and mechanical loads, in: Superalloys 1996: Proc. of the Eighth International Symposium on Superalloys, Champion, USA, 1996.

[111] S. Bigdeli, H. Ehtehsami, Q. Chen, H. Mao, P. Korzhavy, M. Selleby, New description of metastable hcp phase for unaries Fe and Mn: coupling between first-principles calculations and CALPHAD modeling, Physica Status Solidi B 253 (9) (2016) 1830–1836.

[112] S. Bigdeli, H. Mao, M. Selleby, On the third-generation calphad databases: an updated description of Mn, Physica Status Solidi B 252 (10) (2015) 2199–2208.

[113] J. Bishop, R. Hill, A theoretical derivation of the plastic properties of a polycrystalline face–centered metal, Philosophical Magazine 42 (1951) 414–427.

[114] E. Bitzek, P. Gumbsch, Atomistic study of drag, surface and inertial effects on edge dislocations in face-centered cubic metals, Materials Science and Engineering: A 387–389 (1-2 SPEC. ISS.) (2004) 11–15.

[115] E. Bitzek, P. Gumbsch, Dynamic aspects of dislocation motion: atomistic simulations, Materials Science and Engineering: A 400–401 (2005) 40–44.

[116] E. Bitzek, J. Kermode, P. Gumbsch, Atomistic aspects of fractures, International Journal of Fatigue 191 (2015) 13–30.

[117] D. Blavette, A. Bostel, J. Sarrau, Atom-probe microanalysis of a nickel-base superalloy, Metallurgical Transactions A 16 (10) (1985) 1703–1711.
[118] D. Blavette, A. Bostel, J.-M. Sarrau, B. Deconihout, A. Menand, An atom probe for three-dimensional tomography, Nature 363 (6428) (1993) 432–435.
[119] D. Blavette, P. Caron, T. Khan, An atom probe investigation of the role of rhenium additions in improving creep resistance of ni-base superalloys, Scripta Metallurgica 20 (10) (1986) 1395–1400.
[120] P.J. Bocchini, D.C. Dunand, Dislocation dynamics simulations of precipitation-strengthened ni- and co-based superalloys, Materialia 1 (2018) 211–220.
[121] B. Bocklund, R. Otis, A. Egorov, A. Obaied, I. Roslyakova, Z.-K. Liu, Espei for efficient thermodynamic database development, modification, and uncertainty quantification: application to cu-mg, arXiv preprint arXiv:1902.01269, 2019.
[122] J. Böhm, A. Wanner, R. Kampmann, H. Franz, K.-D. Liss, A. Schreyer, H. Clemens, Internal stress measurements by high-energy synchrotron x-ray diffraction at increased specimen-detector distance, Nuclear Instruments & Methods in Physics Research. Section B, Beam Interactions With Materials and Atoms 200 (2003) 315–322.
[123] F. Boioli, P. Carrez, P. Cordier, B. Devincre, M. Marquille, Modeling the creep properties of olivine by 2.5-dimensional dislocation dynamics simulations, Physical Review B 92 (2015) 014115.
[124] B. Bokstein, A. Epishin, T. Link, V. Esin, A. Rodin, I. Svetlov, Model for the porosity growth in single-crystal nickel-base superalloys during homogenization, Scripta Materialia 57 (9) (2007) 801–804.
[125] G. Bolelli, M.G. Righi, M.Z. Mughal, R. Moscatelli, O. Ligabue, N. Antolotti, M. Sebastiani, L. Lusvarghi, E. Bemporad, Damage progression in thermal barrier coating systems during thermal cycling: a nano-mechanical assessment, Materials & Design 166 (2019) 107615.
[126] S.D. Bond, J.W. Martin, Surface recrystallization in a single crystal nickel-based superalloy, Journal of Materials Science 19 (12) (1984) 3867–3872.
[127] B. Bonnand, D. Pacou, Complex thermo-mechanical approaches to study the behavior of high temperature alloys, AerospaceLab 12 (2016).
[128] V. Bonnand, Etude de léndommagement dún superalliage monocristallin en fatigue thermo-mécanique multiaxiale, PhD thesis, ENSMP, 2006.
[129] R. Bonnet, A. Ati, TEM observations of dislocation interactions at γ/γ' interfaces of a two phase superalloy, Acta Metallurgica 37 (1989) 2153–2169.
[130] R. Bonnet, A. Ati, Evidence of microtwinning as a deformation mechanism in CMSX-2 superalloy, Scripta Materialia 25 (1991) 1553–1556.
[131] J. Bonneville, B. Escaig, J. Martin, A study of cross-slip activation parameters in pure copper, Acta Materialia 36 (8) (1988) 1989–2002.
[132] M.P. Borom, C.A. Johnson, L.A. Peluso, Role of environment deposits and operating surface temperature in spallation of air plasma sprayed thermal barrier coatings, Surface & Coatings Technology 86 (1996) 116–126.
[133] L. Bortoluci Ormastroni, L. Mataveli Suave, A. Cervellon, P. Villechaise, J. Cormier, LCF, HCF and VHCF life sensitivity to solution heat treatment of a third-generation Ni-based single crystal superalloy, International Journal of Fracture 130 (2020) 105247.
[134] A. Bortz, M. Kalos, J. Lebowitz, A new algorithm for Monte Carlo simulation of Ising spin systems, Journal of Chemical Physics 17 (1) (1975) 10–18.
[135] S. Bose, High Temperature Coatings, vol. 81, 1st edition, Butterworth-Heinemann, 2007.
[136] S. Bose, High Temperature Corrosion, Chapter 5, Elsevier, 2007, pp. 53–70.
[137] A. Bostel, M. Bouet, J. Sarrau, A software package for atom probe users, Le Journal de Physique Colloques 47 (C7) (1986) C7-521.
[138] P. Boubidi, Experimental characterization and numerical modelling of low cycle fatigue in a nickel base single crystal superalloy under multiaxial loading, Phd thesis, Doctor Communitatis Europae, Ecole des Mines de Paris and BAM Berlin, 2000.

[139] B. Bouchaud, J. Balmain, F. Pedraza, Cyclic and isothermal oxidation at 1100°C of a CVD aluminised directionally solidified Ni superalloy, Oxidation of Metals 69 (2008) 193–210.
[140] B. Bouchaud, J. Creus, C. Rébéré, J. Balmain, F. Pedraza, Controlled stripping of aluminide coatings on nickel superalloys through electrolytic techniques, Journal of Applied Electrochemistry 38 (2008) 817–825.
[141] F. Bourbita, L. Rémy, A combined critical distance and energy density model to predict high temperature fatigue life in notched single crystal superalloy members, International Journal of Fracture 84 (2016).
[142] G. Bouse, J. Mihalisin, Metallurgy of investment cast superalloy components, in: K. Tien, T. Caulfield (Eds.), Superalloys, Supercomposites and Superceramics, Academic Press, 1989, pp. 99–148.
[143] G. Boussinot, Etude du vieillissement des superalliages à base nickel par la méthode des champs de phase, PhD thesis, Université Pierre et Marie Curie, Paris, 2007.
[144] G. Boussinot, A. Finel, Y. Le Bouar, Phase-field modeling of bimodal microstructures in nickel-based superalloys, Acta Materialia 57 (3) (2009) 921–931.
[145] G. Boussinot, Y. Le Bouar, A. Finel, Phase-field simulations with inhomogeneous elasticity: comparison with an atomic-scale method and application to superalloys, Acta Materialia 58 (2010) 4170–4181.
[146] M. Boustie, C. Seymarc, E. Auroux, T. de Rességuier, J.-P. Romain, Coating debonding induced by confined laser shock interpreted in terms of shock wave propagation, in: AIP Conference Proceedings, vol. 429, 1998, p. 985.
[147] J.L. Bouvard, J.L. Chaboche, F. Feyel, F. Gallerneau, A cohesive zone model for fatigue and creep–fatigue crack growth in single crystal superalloys, International Journal of Fatigue 31 (2009) 868–879.
[148] A.D. Brailsford, P. Wynblatt, The dependence of Ostwald ripening kinetics on particle volume fraction, Acta Metallurgica 27 (3) (1979) 489–497.
[149] W. Braue, P. Mechnich, Recession of an EB-PVD YSZ coated turbine blade by CaSO4 and Fe, ti-rich CMAS-type deposits, Journal of the American Ceramic Society 94 (12) (2011) 4483–4489.
[150] H. Brehm, U. Glatzel, Material model describing the orientation dependent creep behavior of single crystals based on dislocation densities of slip systems, International Journal of Plasticity 15 (3) (1999) 285–298.
[151] A. Breidi, J. Allen, A. Mottura, First-principles modeling of superlattice intrinsic stacking fault energies in Ni_3Al based alloys, Acta Materialia 145 (2018) 97–108.
[152] T. Brepols, S. Wulfinghoff, S. Reese, Gradient-extended two-surface damage-plasticity: micromorphic formulation and numerical aspects, International Journal of Plasticity 97 (2017) 64–106.
[153] P. Bridgman, Certain physical properties of single crystals of tungsten, antimony, bismuth, tellurium, cadmium, zinc, and tin, Proceedings of the American Academy of Arts and Sciences 60 (1925) 305–383.
[154] P.W. Bridgman, Effects of high hydrostatic pressure on the plastic properties of metals, Reviews of Modern Physics 17 (Jan 1945) 3–14.
[155] V. Brien, B. Décamps, Low cycle fatigue of a nickel based superalloy at high temperature: deformation microstructures, Materials Science and Engineering: A 316 (1–2) (2001) 18–31.
[156] V. Brien, B. Decamps, A.J. Morton, Microstructural behaviour of a superalloy under repeated or alternate LCF at high temperature, in: Superalloys 1996, number January, Champion, PA (USA), 1996, pp. 313–318.
[157] W.J. Brindley, R.A. Miller, B.J. Aikin, Controlled thermal expansion coat for thermal barrier coatings, Jan. 26, 1999, US Patent 5,863,668.
[158] B. Bromage, E. Tarleton, Calculating dislocation displacements on the surface of a volume, Modelling and Simulation in Materials Science and Engineering 26 (8) (2018) 085007.

[159] Q. Bronchart, Y. Le Bouar, A. Finel, New coarse-grained derivation of a phase field model for precipitation, Physical Review Letters 100 (2008) 015702.
[160] M. Brossard, B. Bouchaud, F. Pedraza, Influence of water vapour on the oxidation behaviour of a conventional aluminide and a new thermal barrier coating system sintered from a slurry, Materials and Corrosion 65 (2014) 161–168.
[161] L. Brown, R. Ham, Dislocation-particle interactions, in: A. Kelly, R. Nicholson (Eds.), Strengthening Methods in Crystals, Elsevier, 1971, p. 11.
[162] U. Brückner, A. Epishin, T. Link, Local x-ray diffraction analysis of the structure of dendrites in single-crystal nickel-base superalloys, Acta Materialia 45 (12) (1997) 5223–5231.
[163] U. Brückner, A. Epishin, T. Link, K. Dressel, The influence of the dendritic structure on the γ/γ'-lattice misfit in the single-crystal nickel-base superalloy CMSX-4, Materials Science and Engineering: A 247 (1998) 23–31.
[164] C. Brundidge, Development of a processing-structure-fatigue property model for single crystal superalloys, PhD thesis, University of Michigan, 2011.
[165] C. Brundidge, D. Drasek, B. Wang, T. Pollock, Structure refinement by a liquid metal cooling solidification process for single-crystal nickel-base superalloys, Metallurgical and Materials Transactions A 43 (2012) 965–976.
[166] C. Brundidge, J. Miller, T. Pollock, Development of dendritic structure in the liquid-metal-cooled, directional-solidification process, Metallurgical and Materials Transactions A 42 (2011) 2723–2732.
[167] M. Brunner, M. Bensch, R. Völkl, E. Affeldt, U. Glatzel, Thickness influence on creep properties for Ni-based superalloy M247LC SX, Materials Science and Engineering: A 550 (2012) 254–262.
[168] I. Buchbender, Laser metal deposition of SX-repair, in: Turbine Forum, Saint-Laurent du Var, 2019.
[169] H. Buck, P. Wollgramm, A.B. Parsa, G. Eggeler, A quantitative metallographic assessment of the evolution of porosity during processing and creep in single crystal Ni-base super alloys, Materialwissenschaft und Werkstofftechnik 46 (2015) 577–590.
[170] J.-Y. Buffiere, M. Ignat, A dislocation based criterion for the raft formation in nickel-based superalloys single crystals, Acta Materialia 43 (5) (1995) 1791–1797.
[171] V. Bulatov, W. Cai, Computer simulations of dislocations, in: A. Sutton, R. Rudd (Eds.), Oxford Series in Materials Modelling, Oxford University Press, 2005.
[172] H.-J. Bunge, Texture Analysis in Materials Science, Butterworth & Co, 1982.
[173] R. Bürgel, H.J. Maier, T. Niendorf Hochtemperaturlegierungen, in: Handbuch Hochtemperatur-Werkstofftechnik, Springer, 2011, pp. 340–484.
[174] E. Busso, G. Cailletaud, On the selection of active slip systems in crystal plasticity, International Journal of Plasticity 21 (2005) 2212–2231.
[175] E. Busso, J. Lin, S. Sakurai, A mechanistic study of oxydation-induced degradation in a plasma-sprayed thermal barrier coating system. Part I: model formulation, Acta Materialia 49 (2001) 1515–1528.
[176] E. Busso, J. Lin, S. Sakurai, A mechanistic study of oxydation-induced degradation in a plasma-sprayed thermal barrier coating system. Part II: life prediction model, Acta Materialia 49 (2001) 1529–1536.
[177] E. Busso, F. McClintock, A dislocation mechanics-based crystallographic model of a B2-type intermetallic alloy, International Journal of Plasticity 12 (1996) 1–28.
[178] E. Busso, F. Meissonnier, N. O'Dowd, Gradient-dependent deformation of two-phase single crystals, Journal of the Mechanics and Physics of Solids 48 (11) (2000) 2333–2361.
[179] E. Busso, L. Wright, H. Evans, L. McCartney, S. Saunders, S. Osgerby, J. Nunn, A physics-based life prediction methodology for thermal barrier coating systems, Acta Materialia 55 (5) (2007) 1491–1503.
[180] J. Butcher, Numerical Methods for Ordinary Differential Equations, Wiley, 1993.

[181] V. Caccuri, J. Cormier, R. Desmorat, γ'-rafting mechanisms under complex mechanical stress state in ni-based single crystalline superalloys, Materials & Design 131 (2017) 487–497.
[182] J. Cadek, Creep in Metallic Materials, Elsevier, 1988.
[183] G. Caginalp, W. Xie, Phase-field and sharp-interface alloy models, Physical Review E 48 (3) (1993) 1897–1909.
[184] J.W. Cahn, J.E. Hilliard, Free energy of a nonuniform system: I. interfacial free energy, Journal of Chemical Physics 28 (2) (1958) 258–267.
[185] W. Cai, A. Arsenlis, C. Weinberger, V. Bulatov, A non-singular continuum theory of dislocations, Journal of the Mechanics and Physics of Solids 54 (2006) 561–587.
[186] W. Cai, V.V. Bulatov, Mobility laws in dislocation dynamics simulations, in: 13th International Conference on the Strength of Materials, Materials Science and Engineering: A 387–389 (2004) 277–281.
[187] D. Caillard, Yield stress anomalies and HT mech props of intermetallics and disordered alloys, Materials Science and Engineering: A 319 (2001) 74–83.
[188] D. Caillard, J. Martin, Thermally Activated Mechanisms in Crystal Plasticity, vol. 8, 1st edition, Elsevier Science, Amsterdam, 2003.
[189] G. Cailletaud, J.-L. Chaboche, Macroscopic description of the microstructural changes induced by varying temperature: example of IN100 cyclic behaviour, in: K. Miller, R. Smith (Eds.), 3rd Int. Conf. on Mechanical Behaviour of Metals, Cambridge, UK, 1979, pp. 23–32, ICM3.
[190] G. Cailletaud, J.-L. Chaboche, S. Forest, L. Rémy, On the design of single crystal turbine blades, Revue de Métallurgie (février 2003) 165–172.
[191] G. Cailletaud, S. Quilici, F. Azzouz, J.L. Chaboche, A dangerous use of the fading memory term for non linear kinematic models at variable temperature, European Journal of Mechanics - A/Solids 54 (2015) 24–29.
[192] G. Cailletaud, K. Sai, L. Taleb, Multi-Mechanism Modeling of Inelastic Material Behavior, ISTE-Wiley, 2018.
[193] H.A. Calderon, P.W. Voorhees, J.L. Murray, G. Kostorz, Ostwald ripening in concentrated alloys, Acta Materialia 42 (3) (1994) 991–1000.
[194] E.C. Caldwell, F.J. Fela, G.E. Fuchs, The segregation of elements in high-refractory-content single-crystal nickel-based superalloys, JOM 56 (9) (Sep 2004) 44–48.
[195] M. Caliez, J.-L. Chaboche, F. Feyel, S. Kruch, Numerical simulation of EBPVD thermal barrier coatings spallation, Acta Materialia 51 (2003) 1133–1141.
[196] C. Campbell, W. Boettinger, U. Kattner, Development of a diffusion mobility database for Ni-base superalloys, Acta Materialia 50 (2002) 775–792.
[197] L. Cao, D. Bürger, P. Wollgramm, K. Neuking, G. Eggeler, Testing of ni-base superalloy single crystals with circular notched miniature tensile creep (CNMTC) specimens, Materials Science and Engineering: A 712 (2018) 223–231.
[198] L. Cao, P. Wollgramm, D. Bürger, A. Kostka, G. Cailletaud, G. Eggeler, How evolving multiaxial stress states affect the kinetics of rafting during creep of single crystal ni-base superalloys, Acta Materialia 158 (2018) 381–392.
[199] J.-M. Cardona, Constitutive behaviour and lifetime of multi-perforated components: application to turbine blades (Comportement et durée de vie des pièces multiperforées: application aux aubes de turbine), PhD thesis, ENSMP, 2000.
[200] L. Caroll, Q. Feng, T. Pollock, Interfacial dislocation networks and creep in directional coarsened Ru-containing nickel-base single-crystal superalloys, Metallurgical and Materials Transactions A 39 (2008) 1290–1307.
[201] P. Caron, Le développement des compositions pour superalliages monocristallins, in: Colloque National Superalliage Monocristallin, Seilh (France), 1995, pp. 65–76.
[202] P. Caron, High γ' solvus new generation nickel-based superalloys for single crystal turbine blade applications, in: T.M. Pollock, R.D. Kissinger, R.R. Bowman, K.A. Green, M. McLean, S.L. Olson, J.J. Schirra (Eds.), Superalloys 2000: Proc. of the 9th International Symposium on Superalloys, Champion, USA, TMS, Minerals, Metals & Materials Society, Sept. 2000, pp. 737–746.

[203] P. Caron, M. Benyoucef, A. Coujou, J. Crestou, N. Clement, Creep behaviour at 1050°C of a Re-containing single crystal superalloy, ONERA, TP no. 2000-230, 2000.

[204] P. Caron, D. Cornu, T. Khan, J. De Monicault, Development of a hydrogen resistant superalloy for single crystal blade application in rocket engine turbopumps, Office national d etudes et de recherches aerospatiales onera-publications-tp, 1996.

[205] P. Caron, P. Henderson, T. Khan, M. McLean, On the effects of heat treatments on the creep behaviour of a single crystal superalloy, Scripta Metallurgica 20 (1986) 875–880.

[206] P. Caron, T. Kahn, P. Vessière, On precipitate shearing by superlattice stacking faults in superalloys, Philosophical Magazine 57:859–875 (1988).

[207] P. Caron, T. Khan, Improvement of creep strength in a nickel-base single-crystal superalloy by heat treatment, Materials Science and Engineering 61 (1983) 173–184.

[208] P. Caron, T. Khan, Tensile behaviour of a nickel-based single crystal superalloy: effects of temperature and orientation, in: Advanced Materials and Processing Techniques for Structural Applications, 1987, pp. 59–70.

[209] P. Caron, T. Khan, Evolution of Ni-based superalloys for single crystal gas turbine blade application, Aerospace Science and Technology 3 (8) (1999) 513–523.

[210] P. Caron, T. Khan, Design of superalloys for single crystal blade applications- a 20-year experience, in: Materials Design Approaches and Experiences (TMS, 2001), ONERA, TP, (2002-3), 2002.

[211] P. Caron, O. Lavigne, Recent studies at Onera on superalloys for single crystal turbine blades, AerospaceLab 3:p (Nov. 2011) 1–14.

[212] P. Caron, Y. Ohta, Y. Nakagawa, T. Khan, Creep deformation anisotropy in single crystal superalloys, in: D. Duhl, G. Maurer, S. Antolovich, C. Lund, S. Reichman (Eds.), Superalloys 1988: Proc. of the Sixth International Symposium on Superalloys, Champion, USA, TMS, Minerals, Metals & Materials Society, Sept. 1988, pp. 215–224.

[213] P. Caron, C. Ramusat, F. Diologent, Influence of the γ' fraction on the γ/γ' topological inversion during high temperature creep of single crystal superalloys, in: Superalloys 2008, Champion, PA (USA), 2008, pp. 159–167.

[214] C. Carry, S. Dermarkar, J.-L. Strudel, B. Wonsiewicz, Internal stresses due to dislocation walls around 2nd phase particles, Materials Science and Engineering: A 10:855–860 (1979).

[215] C. Carry, J.-L. Strudel, Direct observation of $\langle 110 \rangle \{110\}$ slip in fcc single crystals of a nickel base superalloy, Scripta Metallurgica 9 (1975) 731–736.

[216] C. Carry, J.-L. Strudel, Apparent and effective creep parameters in single crystals of a nickel base superalloy – I incubation period, Acta Metallurgica 25 (1977) 767–777.

[217] C.B. Carter, D.B. Williams, Transmission Electron Microscopy: Diffraction, Imaging, and Spectrometry, Springer, 2016.

[218] P. Carter, D. Cox, C.-A. Gandin, R. Reed, Process modelling of grain selection during the solidification of single crystal superalloy castings, Materials Science and Engineering 280 (2000) 233–246.

[219] B. Cassenti, A. Staroselsky, The effect of thickness on the creep response of thin-wall single crystal components, Materials Science and Engineering: A 508 (1–2) (2009) 183–189.

[220] R. Caudron, M. Sarfati, M. Barrachin, A. Finel, F. Solal, In situ diffuse scatterings of neutrons on binary alloys, Physica B: Condensed Matter 180 (1992) 822–824.

[221] A. Cervellon, Propriétés en fatigue à grand et très grand nombre de cycles et à haute température des superalliages base nickel monogranulaires, PhD thesis, ISAE–ENSMA, 2018.

[222] A. Cervellon, J. Cormier, F. Mauget, Z. Hervier, VHCF life evolution after microstructure degradation of a Ni-based single crystal superalloy, International Journal of Fatigue 104 (2017) 251–262.

[223] A. Cervellon, J. Cormier, F. Mauget, Z. Hervier, Y. Nadot, Very high cycle fatigue of Ni-based single crystal superalloys at high temperature, Metallurgical and Materials Transactions A 49 (2018) 3938–3950.

[224] A. Cervellon, S. Hémery, P. Kürnsteiner, B. Gault, P. Kontis, J. Cormier, Crack initiation mechanisms during very high cycle fatigue of Ni-based single crystal superalloys at high temperature, Acta Materialia 188 (2020) 131–144.
[225] A. Cetel, D. Duhl, Cost effective single crystal, in: Proc. 6th Int. Symp. on Superalloys, TMS, 1988.
[226] J.-L. Chaboche, Constitutive equations for cyclic plasticity and cyclic viscoplasticity, International Journal of Plasticity 5 (1989) 247–302.
[227] J.-L. Chaboche, F. Gallerneau, Durability modeling at elevated temperature, Fatigue & Fracture of Engineering Materials & Structures 11 (2000) 1–17.
[228] K. Chan, J. Hack, G. Leverant, Fatigue crack propagation in ni-base superalloy single crystals under multiaxial cyclic loads, Metallurgical and Materials Transactions A 17 (1986) 1739–1750.
[229] M. Chandran, S. Lee, J.-H. Shim, Machine learning assisted first-principles calculation of multicomponent solid solutions: estimation of interface energy in Ni-based superalloys, Modelling and Simulation in Materials Science and Engineering 26 (2018) 025010.
[230] M. Chandran, S. Sondhi, First-principle calculation of APB energy in Ni-based binary and ternary alloys, Modelling and Simulation in Materials Science and Engineering 19 (2011) 025008.
[231] G. Chang, W. Phucharoen, R. Miller, Finite element thermal stress solutions for thermal barrier coatings, Surface & Coatings Technology 32 (1) (1987) 307–325.
[232] H.-J. Chang, M. Fivel, J.-L. Strudel, Micromechanics of primary creep in Ni base superalloys, International Journal of Plasticity 108 (SEP 2018) 21–39.
[233] M. Chase, I. Ansara, A. Dinsdale, G. Eriksson, G. Grimvall, L. Hoglund, H. Yokokawa, Group 1: heat capacity models for crystalline phases from 0 K to 6000 K, Calphad: Computer Coupling of Phase Diagrams and Thermochemistry 19 (4) (1995) 437–447.
[234] E. Chataigner, E. Fleury, L. Rémy, Thermomechanical fatigue of coated and bare nickel base superalloy single crystals, in: J. Bressers, L. Rémy (Eds.), Fatigue Under Thermal and Mechanical Loading, Kluwer Acad. Pub., 1996, pp. 381–392.
[235] D.J. Chellman, A.J. Ardell, The coarsening of γ' precipitates at large volume fraction, Acta Metallurgica (1974).
[236] C. Chen, S. Aubry, T. Oppelstrup, A. Arsenlis, E. Darve, Fast algorithms for evaluating the stress field of dislocation lines in anisotropic elastic media, Modelling and Simulation in Materials Science and Engineering 26 (4) (2018) 045007.
[237] H. Chen, T.H. Hyde, Use of multi-step loading small punch test to investigate the ductile-to-brittle transition behaviour of a thermally sprayed CoNiCrAlY coating, Materials Science and Engineering: A 680 (2017) 203–209.
[238] H. Chen, D. McCartney, Some aspects on modelling of the β-phase depletion behaviour under different oxide growth kinetics in hvof conicraly coatings, Surface & Coatings Technology 313 (2017) 107–114.
[239] M.W. Chen, M.L. Glynn, R.T. Ott, T.C. Hufnagel, K.J. Hemker, Characterization and modeling of a martensitic transformation in a platinum modified diffusion aluminide bond coat for thermal barrier coatings, Acta Materialia 51 (14) (2003) 4279–4294.
[240] M.W. Chen, R.T. Ott, T.C. Hufnagel, P.K. Wright, K.J. Hemker, Microstructural evolution of platinum modified nickel aluminide bond coat during thermal cycling, Surface & Coatings Technology 163 (2003) 25–30.
[241] Q. Chen, S. Biner, Evolution and interaction of dislocations in intermetallics: fully anisotropic discrete dislocation dynamics simulations, in: Symposium II - Advanced Intermetallic-Based Alloys, vol. 980, 2007, pp. 107–112.
[242] Q. Chen, T. Knowles, Superlattice stacking fault formation and twinning during creep in γ/γ' single crystal superalloy CMSX-4, Materials Science and Engineering: A 340 (2003) 88–102.
[243] Q. Chen, B. Sundman, Modeling of thermodynamic properties for Bcc, Fcc, liquid, and amorphous iron, Journal of Phase Equilibria 22 (6) (2001) 631–644.

[244] W. Chen, J.P. Immarigeon, Thickening behavior of γ' precipitates in nickel base superalloys during rafting, Scripta Materialia 39 (2) (1998).

[245] X. Chen, C. Shaw, L. Gelman, K.T. Grattan, Advances in test and measurement of the interface adhesion and bond strengths in coating-substrate systems, emphasising blister and bulk techniques, Measurement (2019).

[246] Y. Chen, X. Fan, Y. Sun, W. Zhang, Effect of tensile load on high temperature oxidation of conicraly coating, Surface & Coatings Technology 352 (2018) 399–405.

[247] J. Cheng, E.H. Jordan, B. Barber, M. Gell, Thermal/residual stress in an electron beam physical vapor deposited thermal barrier coating system, Acta Materialia 46 (16) (1998) 5839–5850.

[248] K.Y. Cheng, C.Y. Jo, T. Jin, Z.Q. Hu, Effect of stress on μ phase formation in single crystal superalloys, Materials Science and Engineering: A 528 (6) (2011) 2704–2710.

[249] K.Y. Cheng, C.Y. Jo, T. Jin, Z.Q. Hu, Precipitation behavior of μ phase and creep rupture in single crystal superalloy CMSX-4, Journal of Alloys and Compounds 509 (25) (2011) 7078–7086.

[250] K.Y. Cheng, C.Y. Jo, T. Jin, Z.Q. Hu, Influence of applied stress on the γ' directional coarsening in a single crystal superalloy, Materials & Design 31 (2) (2010) 968–971.

[251] R. Chieragatti, L. Rémy, Influence of orientation on the low cycle fatigue behaviour of MarM 200 single crystals. Part I: fatigue life behaviour, Materials Science and Engineering: A 141 (1991) 1–9.

[252] M. Chieux, Vieillissement des systèmes Barriere Thermique: transformation de phases, oxydation et effet du soufre sur l'adhérence, in French, PhD thesis, Mines - ParisTech, France, 2010.

[253] M. Chieux, C. Duhamel, R. Molins, F. Jomard, L. Rémy, J.-Y. Guédou, Sulfur localization in NiPtAL superalloy systems after high temperature isothermal oxidation, Oxidation of Metals 81 (2014) 115–125.

[254] M. Chieux, C. Duhamel, R. Molins, L. Rémy, J.-Y. Guédou, Effect of the superalloy composition on the isothermal oxidation behaviour of TBC systems, Oxidation of Metals 81 (1–2) (2014) 57–67.

[255] M. Chieux, R. Molins, L. Rémy, C. Duhamel, Y. Cadoret, Adhesion of thermal barrier coating systems after long term oxidation: influence of preoxidation temperature and surface state of the bond coat, Materials at High Temperatures 26 (2009) 187–194.

[256] G. Chin, W. Mammel, Generalization and equivalence of the minimum work (Taylor) and maximum work (Bishop–Hill) principles for crystal plasticity, Transactions of the Metallurgical Society of AIME 245 (1969) 1211–1214.

[257] M.-S. Chiou, S.-R. Jian, A.-C. Yeh, C.-M. Kuo, J.-Y. Juang, High temperature creep properties of directionally solidified CM-247LC Ni-based superalloy, Materials Science and Engineering: A 655 (2016) 237–243.

[258] Y.S Choi, T.A. Parthasarathy, C. Woodward, D.M. Diminuions, M.D. Uchic, Constitutive model for anisotropic creep behaviors of single-crystal Ni-base superalloys in the low-temperature, high-stress regime, Metallurgical and Materials Transactions A 43 (6) (Jun 2012) 1861–1869.

[259] Y.S Choi, T.A. Parthasarathy, D.M. Diminuions, M.D. Uchic, Microstructural effects in modeling the flow behavior of single-crystal superalloys, Metallurgical and Materials Transactions A 37 (3) (Mar 2006) 545–550.

[260] N. Clément, M. Benyoucef, M. Legros, P. Caron, A. Coujou, In situ deformation at 850°C of standard and rafted microstructures of nickel base superalloys, in: Advanced Structural Materials II, in: Materials Sciences Forum, vol. 509, 2006, pp. 57–62.

[261] N. Clément, D. Caillard, P. Lours, A. Coujou, -TEM in situ straining of a Ni-base γ' single crystal-unlocking of dissociated screw dislocations at low temperature, Scripta Metallurgica 23 (4) (1989) 563–568.

[262] E. Clouet, Predicting dislocation climb: classical modeling versus atomistic simulations, Physical Review B 84 (2011) 092106.

[263] T. Clyne, S. Gill, Residual stresses in thermal spray coatings and their effect on interfacial adhesion: a review of recent work, Journal of Thermal Spray Technology 5 (4) (1996) 401.

[264] J. Coakley, E. Lass, D. Ma, M. Frost, D. Seidman, D. Dunand, H. Stone, Rafting and elastoplastic deformation of superalloys studied by neutron diffraction, Scripta Metallurgica 134 (2017) 110–114.

[265] J. Coakley, E. Lass, D. Ma, M. Frost, H. Stone, D. Seidman, D. Dunand, Lattice parameter misfit evolution during creep of a cobalt-based superalloy single crystal with cuboidal and rafted γ' microstructures, Acta Materialia 136 (2012) 118–125.

[266] J. Coakley, D. Ma, M. Frost, D. Dye, D. Seidman, D. Dunand, H. Stone, Lattice strain evolution and load partitioning during creep of a Ni-based superalloy single crystal with rafted γ' microstructure, Acta Materialia 135 (2017) 77–87.

[267] D. Cockayne, The principles and practice of the weak-beam method of electron microscopy, Journal of Microscopy 98 (2) (1973) 116–134.

[268] D. Cockayne, I. Ray, M. Whelan, Investigations of dislocation strain fields using weak beams, Philosophical Magazine 20 (168) (1969) 1265–1270.

[269] H. Cohrt, F. Thümmler, Degradation mechanisms of thermal barrier coatings in bending tests, Surface & Coatings Technology 32 (1–4) (1987) 339–348.

[270] S. Cole, E. Gray, New NASA Satellite Maps Show Human Fingerprint on Global Air Quality, Release 15-233, updated 2017, 2015.

[271] D. Collins, N. DŚouza, C. Panwisawas, In-situ neutron diffraction during stress relaxation of a single crystal nickel-base superalloy, Scripta Metallurgica (2017) 103–107.

[272] M. Condat, B. Décamps, Shearing of γ'-precipitates by single a/2⟨110⟩ matrix dislocations in a γ/γ' Ni-based superalloy, Scripta Metallurgica 21 (1987) 607–612.

[273] C. Conty, Today's and tomorrow's instruments, Microscopy and Microanalysis 7 (2) (2001) 142–149.

[274] S. Copley, A. Giamei, S. Johnson, M. Hornbecker, The origin of freckles in unidirectionally solidified castings, Metallurgical Transactions 1 (1970) 2193–2204.

[275] F. Cork, Method for removing aluminide coatings from nickel or cobalt base alloys, US patent 42820411, 1981.

[276] F. Cork, Systems and methods for additive manufacturing and repair of metal components, US patent 9522426, 2016.

[277] J. Cormier, Comportement en fluage anisotherme à haute et très haute température du superalliage monocristallin MC2, PhD thesis, Université de Poitiers, France, 2006.

[278] J. Cormier, Technical report, ISAE-ENSMA, 2010.

[279] J. Cormier, Thermal cycling creep resistance of ni-based single crystal superalloys, in: Proc. 13th Int. Symp. on Superalloys, TMS, 2016.

[280] J. Cormier, G. Cailletaud, Constitutive modeling of the creep behavior of single crystal superalloys under non-isothermal conditions inducing phase transformations, Materials Science and Engineering: A 527 (2010) 6300–6312.

[281] J. Cormier, M. Jouiad, F. Hamon, P. Villechaise, X. Milhet, Very high temperature creep behavior of a single crystal Ni-based superalloy under complex thermal cycling conditions, Philosophical Magazine Letters 90 (2010) 611–620.

[282] J. Cormier, X. Milhet, J. Mendez, Effect of very high temperature short exposures on the dissolution of the γ' phase in single crystal MC2 superalloy, Journal of Materials Science 42 (18) (2007) 7780–7786.

[283] J. Cormier, X. Milhet, J. Mendez, Non-isothermal creep at very high temperature of the nickel-based single crystal superalloy MC2, Acta Materialia 55 (18) (2007) 6250–6259.

[284] J. Cormier, X. Milhet, J. Mendez, Anisothermal creep behavior at very high temperature of a Ni-based superalloy single crystal, Materials Science and Engineering: A 483–484 (2008) 594–597.

[285] J. Cormier, X. Milhet, F. Vogel, J. Mendez, Non-isothermal creep behavior of a second generation Ni-based single crystal superalloy: experimental characterization and modeling, in: R.C. Reed, K.A. Green, P. Caron, T.P. Gabb, M.G. Fahrmann, E.S. Huron (Eds.), Superalloys 2008: Proc. of the Eleventh International Symposium on Superalloys, Champion, USA, TMS, Minerals, Metals & Materials Society, Sept. 2008.

[286] J. Cormier, P. Villechaise, X. Milhet, γ'-phase morphology of Ni-based single crystal superalloys as an indicator of the stress concentration in the vicinity of pores, Materials Science and Engineering: A 501 (1–2) (2009) 61–69.

[287] F. Cosentino, N. Warnken, J.-C. Gebelin, R. Reed, Numerical and experimental study of post-heat treatment gas quenching and its impact on microstructure and creep in CMSX-10 superalloy, Journal of Materials Processing Technology 213 (2013) 2350–2360.

[288] F. Cosentino, N. Warnken, J.-C. Gebelin, R. Reed, Numerical modeling of vacuum heat treatment of nickel-based superalloys, Metallurgical and Materials Transactions A 44 (2013) 5154–5164.

[289] M. Cottura, Modélisation champ de phase du couplage entre évolution microstructurale et comportement mécanique, PhD thesis, Université Paris VI, 2013.

[290] M. Cottura, B. Appolaire, A. Finel, Y. Le Bouar, Coupling the phase field method for diffusive transformations with dislocation density-based crystal plasticity: application to Ni-based superalloys, Journal of the Mechanics and Physics of Solids 94 (2016) 473–489.

[291] M. Cottura, Y. Le Bouar, B. Appolaire, A. Finel, Role of elastic inhomogeneity in the development of cuboidal microstructures in Ni-based superalloys, Acta Materialia 94 (2015) 15–25.

[292] M. Cottura, Y. Le Bouar, A. Finel, B. Appolaire, K. Ammar, S. Forest, A phase field model incorporating strain gradient viscoplasticity: application to rafting in Ni-base superalloys, Journal of the Mechanics and Physics of Solids 60 (2012) 1243–1256.

[293] F. Coudon, Comportement mécanique du superalliage base nickel à solidification dirigée DS200+ Hf, PhD thesis, MINES ParisTech, 2017.

[294] C. Coupeau, S. Brochard, F. Pettinari-Sturmel, A. Coujou, J. Grilhe, Short range order heterogeneity on plastic mechanisms in γ-phase nickel-based superalloys, Philosophical Magazine 87 (26) (2007) 3893–3904.

[295] C. Courcier, V. Maurel, L. Remy, S. Quilici, I. Rouzou, A. Phelippeau, Interfacial damage based life model for EB-PVD thermal barrier coating, Surface & Coatings Technology 205 (13–14) (2011) 3763–3773.

[296] D. Cox, B. Roebuck, C. Rae, R. Reed, Recrystallisation of single crystal superalloy CMSX-4, Materials Science and Technology 19 (2003) 440–446.

[297] D.C. Cox, C.M.F. Rae, R.C. Reed, Characterisation of damage accumulation during the creep deformation of CMSX-4 at 1150°C, in: Life Assessment of Hot Section Gas Turbine Components, 1999, pp. 119–133.

[298] V. Crespo, I. Cano, S. Dosta, J. Guilemany, The influence of feedstock powders on the cgs deposition efficiency of bond coats for tbcs, Journal of Alloys and Compounds 622 (2015) 394–401.

[299] J. Crompton, J. Martin, Crack growth in a single crystal superalloy at elevated temperatures, Metallurgical and Materials Transactions A 15 (1984) 1711–1719.

[300] J. Crompton, J. Martin, Crack tip plasticity and crack growth in a single-crystal superalloy at elevated temperatures, Materials Science and Engineering 64 (1984) 37–43.

[301] W. Crone, T. Shield, An experimental study of the effect of hardening on plastic deformation at notch tips in metallic single crystals, Journal of the Mechanics and Physics of Solids 57 (2003) 1623–1647.

[302] D. Crudden, A. Mottura, N. Warnken, B. Raeisinia, R. Reed, Modelling of the influence of alloy composition on flow stress in high-strength nickel-based superalloys, Acta Materialia 75 (2014) 356–370.

[303] A. Cruzado, B. Gan, M. Jimenez, D. Barba, K. Ostolaza, A. Linaza, J. Molina-Aldareguia, J. Llorca, J. Segurado, Multiscale modeling of the mechanical behavior of in718 superalloy based on micropillar compression and computational homogenization, Acta Materialia 98 (2015) 242–253.

[304] Y. Cui, Z. Liu, Z. Zhuang, Quantitative investigations on dislocation based discrete-continuous model of crystal plasticity at submicron scale, International Journal of Plasticity 69 (2015) 54–72.

[305] Y.N. Cui, P. Lin, Z.L. Liu, Z. Zhuang, Theoretical and numerical investigations of single arm dislocation source controlled plastic flow in fcc micropillars, International Journal of Plasticity 55 (2014) 279–292.

[306] J.-P. Culié, G. Cailletaud, A. Lasalmonie, La contrainte interne en viscoplasticité: comparaison des approches mécaniques et microscopiques, La Recherche Aérospatiale 1982 (2) (1982) 51–61.

[307] B. Cullity, S. Stock, Elements of X-ray Diffraction, third edition, Prentice-Hall, 2001.

[308] K. Danas, V.S. Deshpande, Plane-strain discrete dislocation plasticity with climb-assisted glide motion of dislocations, Modelling and Simulation in Materials Science and Engineering 21 (4) (June 2013) 045008.

[309] D.P. Dandekar, A.G. Martin, Single crystal elastic constants of two nickel based superalloys, Journal of Materials Science Letters 8 (10) (Oct 1989) 1172–1173.

[310] J. Dantzig, M. Rappaz, Solidification: Revised & expanded, Technical report, EPFL Press, 2016.

[311] L. Darken, Diffusion, mobility and their interrelation through free energy in binary metallic systems, Transactions of the Metallurgical Society of AIME 175 (1948) 184–201.

[312] R. Darolia, D. Lahrman, R. Field, R. Sisson, Formation of topologically closed packed phases in nickle base single crystal superalloys, in: D. Duhl, G. Maurer, S. Antolovich, C. Lund, S. Reichman (Eds.), Superalloys 1988: Proc. of the Sixth International Symposium on Superalloys, Champion, USA, TMS, Minerals, Metals & Materials Society, Sept. 1988, pp. 255–264.

[313] D.K. Das, B. Gleeson, K.S. Murphy, S. Ma, T.M. Pollock, Formation of secondary reaction zone in ruthenium bearing nickel based single crystal superalloys with diffusion aluminide coatings, Materials Science and Technology 25 (2) (2009) 300–308.

[314] K.M. Davoudi, L. Nicola, J.J. Vlassak, Dislocation climb in two-dimensional discrete dislocation dynamics, Journal of Applied Physics 111 (10) (2012).

[315] M. Daw, S. Foiles, M. Baskes, The embedded-atom method: a review of theory and applications, Materials Science Reports 9 (1993) 251–310.

[316] A. de Bussac, C.-A. Gandin, Prediction of a process window for the investment casting of dendritic single crystals, Materials Science and Engineering 237 (1997) 35–42.

[317] M. de Koning, C. Miranda, A. Antonelli, Atomistic prediction of equilibrium vacancy concentrations in Ni_3Al, Physical Review B 66 (2002) 104110.

[318] V. De Rancourt, K. Ammar, B. Appolaire, E.P. Busso, S. Forest, Modelling stress-diffusion controlled phase transformations: application to stress corrosion cracking, in: CSMA 2013, 11e Colloque National en Calcul des Structures, 2013.

[319] R. De Ridder, S. Amelinckx, Approximate theoretical treatment of weak-beam dislocation images, Physica Status Solidi B 43 (2) (1971) 541–550.

[320] D. De Wet, R. Taylor, F. Stott, Corrosion mechanisms of zro2-y2o3 thermal barrier coatings in the presence of molten middle-east sand, Journal de Physique IV 3 (C9) (1993) C9-655.

[321] B. Décamps, M. Condat, P. Caron, T. Kahn, Dissociated matrix dislocations in a γ/γ' Ni-based single crystal superalloy, Scripta Metallurgica 18 (1984) 1171–1174.

[322] B. Décamps, A. Morton, M. Condat, On the mechanism of shear of γ' precipitates by single $1/2\ \langle 110 \rangle$ dissociated matrix dislocations in Ni-based superalloys, Philosophical Magazine 64 (1991) 641–668.

[323] R. Decker, The evolution of wrought age-hardenable superalloys, JOM 58 (2006) 32–36.

[324] R. Decker, Strengthening mechanisms in nickel-base superalloys, in: Steel Strengthening Mechanism Symposium, Zürich, Switzerland, May 5–6, 1969.

[325] A. Defresne, L. Rémy, Fatigue behaviour of CMSX2 superalloy [001] single crystals at high temperature. I: low cycle fatigue of notched specimens II: fatigue crack growth, Materials Science and Engineering 129 (1990) 45–64.

[326] R. Degeilh, Développement expérimental et modélisation d'un essai de fatigue avec gradient thermique de paroi pour application aube de turbine monocristalline, PhD thesis, ENSC, France, 2013.

[327] M. Degeiter, M. Perrut, B. Appolaire, Y. Le Bouar, A. Finel, A new analysis of the microstructure of ni-based single-crystal superalloys: Relevant topological parameters for efficient microstructural modeling, 2016.

[328] G. Dehm, B.N. Jaya, R. Raghavan, C. Kirchlechner, Overview on micro-and nanomechanical testing: new insights in interface plasticity and fracture at small length scales, Acta Materialia 142 (2018) 248–282.

[329] J. Demasi, K. Sheffler, Mechanisms of thermal barrier coating degradation and failure, NASA (1985).

[330] K. Demtröder, G. Eggeler, J. Schreuer, Influence of microstructure on macroscopic elastic properties and thermal expansion of nickel-base superalloys ERBO/1 and LEK94, Materialwissenschaft und Werkstofftechnik 46 (6) (2015) 563–576.

[331] J. Deng, A. El-Azab, B.C. Larson, On the elastic boundary value problem of dislocations in bounded crystals, Philosophical Magazine 88 (30) (2008) 3527–3548.

[332] J. Dennison, I. Elliott, B. Wilshire, An assessment of hot isostatic pressing and reheat treatment for the regeneration of creep properties of superalloys, in: J.K. Tien, S.T. Wlodek, H. Morrow III, M. Gell, G.E. Maurer (Eds.), Superalloys 1980: Proc. of the Fourth International Symposium on Superalloys, Champion, USA, Amer. Soc. for Metals, Sept. 1980, pp. 671–677.

[333] A. Dennstedt, F. Gaslain, M. Bartsch, V. Guipont, V. Maurel, Three-dimensional characterization of cracks in a columnar thermal barrier coating system for gas turbine applications, Integrating Materials and Manufacturing Innovation (2019) 1–13.

[334] P. Denteneer, W. Van Haeringen, Stacking-fault energies in semiconductors from first-principles calculations, Journal of Physics. C. Solid State Physics 20 (1987) L883–L887.

[335] C. Déprés, C. Robertson, M.C. Fivel, Low-strain fatigue in 316l steel surface grains: a three dimension discrete dislocation dynamics modelling of the early cycles. Part 2: persistent slip markings and micro-crack nucleation, Philosophical Magazine 86 (1) (2006) 79–97.

[336] C. Déprés, C.F. Robertson, M. Fivel, Low-strain fatigue in AISI 316L steel surface grains: a three-dimensional discrete dislocation dynamics modelling of the early cycles I. Dislocation microstructures and mechanical behaviour, Philosophical Magazine 84 (22) (2004) 2257–2275.

[337] R. Desmorat, A. Mattiello, J. Cormier, A tensorial thermodynamic framework to account for the γ' rafting in nickel-based single crystal superalloys, International Journal of Plasticity 95 (2017) 43–81.

[338] B. Devincre, T. Hoc, L. Kubin, Dislocation mean free paths and strain hardening of crystals, Science 320 (2008) 1745–1748.

[339] B. Devincre, R. Madec, G. Monnet, S. Queyreau, R. Gatti, L. Kubin, Modeling crystal plasticity with dislocation dynamics simulations: the 'micromegas' code, in: O. Thomas, A. Ponchet, S. Forest (Eds.), Mechanics of Nano-objects, Presses de l'Ecole des Mines de Paris, 2011.

[340] S. DeWitt, K. Thornton, Phase field modeling of microstructural evolution, in: D. Shin, J. Saal (Eds.), Computational Materials System Design, Springer, 2017, pp. 67–87, Chapter 4.

[341] D.M. Dimiduk, Dislocation structures and anomalous flow in l12 compounds, Journal de Physique III 1 (6) (1991) 1025–1053.

[342] D.M. Dimiduk, M.D. Uchic, T.A. Parthasarathy, Size-affected single-slip behavior of pure nickel microcrystals, Acta Materialia 53 (15) (2005) 4065–4077.

[343] Q. Ding, S. Li, L. Chen, X. Han, Z. Zhang, Q. Yu, J. Li, Re segregation at interfacial dislocation network in a nickel-based superalloy, Acta Materialia 154 (2018) (2018) 137–146.

[344] D. Dingley, A comparison of diffraction techniques for the sem, Scanning Electron Microscopy 4 (1981) 273–286.

[345] D. Dingley, M. Longden, J. Weinbren, J. Alderman, Online analysis of electron back scatter diffraction patterns. 1. Texture analysis of zone refined polysilicon, Scanning Microscopy 1 (2) (1987) 451–456.

[346] A. Dinsdale, Sgte data for pure elements, Calphad: Computer Coupling of Phase Diagrams and Thermochemistry 15 (4) (Oct. 1991) 317–425.

[347] F. Diologent, Comportement en fluage et en traction de superalliages monocristallins a base de nickel, PhD thesis, Université de Paris sud, Centre d'Orsay, France, 2002.

[348] F. Diologent, P. Caron, On the creep behavior at 1033K of new generation single-crystal superalloys, Materials Science and Engineering: A 385 (1) (2004) 245–257.

[349] F. Diologent, P. Caron, T. Almeida, A. Jacques, P. Bastie, The γ/γ' mismatch in Ni-based superalloys: in situ measurements during a creep test, in: Nuclear Instruments and Methods in Physics Research Section B: Beam Interactions with Materials and Atoms, volume Proceedings of the E-MRS 2002 Symposium I on Synchrotron Radiation and Materials Science, 2003, pp. 346–351.

[350] P. Dirac, Quantum mechanics of many-electron systems, Proceedings of the Royal Society of London. Series A 123 (1929) 714.

[351] L. Dirand, A. Jacques, J. Chateau-Cornu, T. Schenk, O. Ferry, P. Bastie, Phase-specific high temperature creep behaviour of a pre-rafted ni-based superalloy studied by x-ray synchrotron diffraction, Philosophical Magazine 93 (2013) 1384–1412.

[352] L. Dirand, Fluage à haute température d'un superalliage monocristallin: expérimentation in situ en rayonnement synchrotron, PhD thesis, Institut National Polytechnique de Lorraine, 2011.

[353] L. Dirand, J. Cormier, A. Jacques, J.-P. Chateau-Cornu, T. Schenk, O. Ferry, P. Bastie, Measurement of the effective γ/γ' lattice mismatch during high temperature creep of Ni-based single crystal superalloy, Materials Characterization 77 (2013) 32–46.

[354] A. Dlouhy, M. Probst-Hein, G. Eggeler, Static dislocation interactions in thin channels between cuboidal particles, Materials Science and Engineering: A 309–310 (2001) 278–282.

[355] M. Donachie, S. Donachie, Superalloys: a technical guide, Technical report, ASM International, 2002.

[356] C. Dong, H. Yu, Y. Li, Fatigue life modeling of a single crystal superalloy and its thin plate with a hole at elevated temperature, Materials & Design 66 (2015) 284–293.

[357] J. Douin, Structure fine des dislocations et plasticité dans Ni (3) Ai, PhD thesis, Poitiers, 1987.

[358] S. Draper, D. Hull, R. Dreshfield, Observations of directional γ' coarsening during engine operation, Metallurgical Transactions A 20 (4) (1989) 683–688.

[359] R. Drautz, T. Hammerschmidt, M. Cak, D. Pettifor, Bond-order potentials: derivation and parameterization for refractory elements, Modelling and Simulation in Materials Science and Engineering 23 (2015) 074004.

[360] R. Drautz, D. Pettifor, Valence-dependent analytic bond-order potential for magnetic transition metals, Physical Review B 84 (2011) 214114.

[361] S. Drawin, J.-F. Justin, Advanced lightweight silicide and nitride based materials for turbo-engine applications, AerospaceLab 3 (2011) 1–13.

[362] G. Drew, R. Reed, K. Kakehi, C. Rae, Single crystal superalloys: the transition from primary to secondary creep, in: K.A. Green, T.M. Pollock, H. Harada, T.E. Howson, R.C. Reed, J.J. Schirra, S. Walston (Eds.), Superalloys 2004: Proc. of the Tenth International Symposium on Superalloys, Champion, USA, TMS, Minerals, Metals & Materials Society, Sept. 2004, pp. 127–136.

[363] J.M. Drexler, A. Aygun, D. Li, R. Vaßen, T. Steinke, N.P. Padture, Thermal-gradient testing of thermal barrier coatings under simultaneous attack by molten glassy deposits and its mitigation, Surface & Coatings Technology 204 (16–17) (2010) 2683–2688.

[364] J.M. Drexler, A.D. Gledhill, K. Shinoda, A.L. Vasiliev, K.M. Reddy, S. Sampath, N.P. Padture, Jet engine coatings for resisting volcanic ash damage, Advanced Materials 23 (21) (2011) 2419–2424.

[365] J.M. Drexler, K. Shinoda, A.L. Ortiz, D. Li, A.L. Vasiliev, A.D. Gledhill, S. Sampath, N.P. Padture, Air-plasma-sprayed thermal barrier coatings that are resistant to high-temperature attack by glassy deposits, Acta Materialia 58 (20) (2010) 6835–6844.

[366] S. Dryepondt, D. Monceau, F. Crabos, E. Andrieu, Static and dynamic aspects of coupling between creep behavior and oxidation on MC2 single crystal superalloy at 1150°C, Acta Metallurgica 53 (15) (sep 2005) 4199–4209.

[367] J. Duan, Atomistic simulations of diffusion mechanisms in off-stoichiometric Al-rich Ni_3Al, Journal of Physics. Condensed Matter 19 (8) (2007) 086217.

[368] C. Duhamel, J. Caballero, T. Couvant, J. Crépin, F. Gaslain, C. Guerre, H.-T. Le, M. Wehbi, Intergranular oxidation of nickel-base alloys: potentialities of focused ion beam tomography, Oxidation of Metals 88 (3–4) (2017) 447–457.

[369] N. Dupin, U.R. Kattner, B. Sundman, M. Palumbo, S.G. Fries, Implementation of an effective bond energy formalism in the multicomponent CALPHAD approach, Journal of Research of NIST 123 (2018).

[370] M. Durand-Charre, The Microstructure of Superalloys, Gordon and Breach Science Publishers, 2013.

[371] A. Durga, P. Wollants, N. Moelans, Evaluation of interfacial excess contributions in different phase-field models for elastically inhomogeneous systems, Modelling and Simulation in Materials Science and Engineering 21 (2013) 055018.

[372] S. Duval, S. Chambreland, P. Caron, D. Blavette, Phase composition and chemical order in the single crystal nickel base superalloy MC2, Acta Materialia 42 (1) (1994) 185–194.

[373] D. Dye, J. Coakley, V. Vorontsov, H. Stone, R. Rogge, Elastic moduli and load partitioning in a single-crystal nickel superalloy, Scripta Metallurgica 61 (2009) 109–112.

[374] D. Dye, A. Ma, R.C. Reed, Numerical modelling of creep deformation in a CMSX-4 single crystal superalloy turbine blade, in: Superalloys 2008, Champion, PA (USA), 2008, pp. 911–919.

[375] B.F. Dyson, Microstructure based creep constitutive model for precipitation strengthened alloys: theory and application, Materials Science and Technology 25 (2) (2009) 213–220.

[376] B.F. Dyson, D. McLean, Creep of Nimonic 80A in torsion and tension, Metal Science 11 (2) (1977) 37–45.

[377] C. Eberl, X. Wang, D.S. Gianola, T.D. Nguyen, M.Y. He, A.G. Evans, K.J. Hemker, In situ measurement of the toughness of the interface between a thermal barrier coating and a ni alloy, Journal of the American Ceramic Society 94 (s1) (2011) s120–s127.

[378] J.W. Edington, Electron diffraction in the electron microscope, in: Electron Diffraction in the Electron Microscope, Springer, 1975, pp. 1–77.

[379] G. Eggeler, A. Dlouhy, γ'-phase cutting during high temperature shear creep deformation of CMSX6 superalloy single crystals, Physica Status Solidi A 149 (1) (1995) 349–353.

[380] G. Eggeler, A. Dlouhy, On the formation of ⟨010⟩-dislocations in the γ'-phase of superalloy single crystals during high temperature low stress creep, Acta Materialia 45 (1997) 4251–4262.

[381] Y. Eggeler, J. Müller, M. Titus, A. Suzuki, T. Pollock, E. Spiecker, Planar defect formation in the γ' phase during high temperature creep in single crystal coni-base superalloys, Acta Materialia 113 (2016) 335–349.

[382] Y. Eggeler, M. Titus, A. Suzuki, T. Pollock, Creep deformation-induced antiphase boundaries in $L1_2$-containing single-crystal cobalt-base superalloys, Acta Materialia 77 (2014) 352–359.

[383] B. Eidel, Crystal plasticity finite-element analysis versus experimental results of pyramidal indentation into (001) fcc single crystal, Acta Materialia 59 (2011) 1761–1771.

[384] J.A. El-Awady, S.B. Biner, N.M. Ghoniem, A self-consistent boundary element, parametric dislocation dynamics formulation of plastic flow in finite volumes, Journal of the Mechanics and Physics of Solids 56 (5) (2008) 2019–2035.

[385] A. Elliott, T. Pollock, S. Tin, W. King, S.-C. Huang, M. Gigliotti, Directional solidification of large superalloy castings with radiation and liquid-metal cooling: a comparative assessment, Metallurgical and Materials Transactions A 35 (2004) 3221–3231.

[386] J.D. Embury, A. Deschamps, Y. Brechet, The interaction of plasticity and diffusion controlled precipitation reactions, Scripta Materialia 49 (10) (2003) 927–932.

[387] C.C. Engler-Pinto Jr, C. Noseda, M.Y. Nazmy, F. Rezai-Aria, Interaction between creep and thermo-mechanical fatigue of CM247LC-DS, in: Superalloys 1996, Champion, PA (USA), 1996, pp. 319–325.

[388] P.J. Ennis, W. Quadakkers, H. Schuster, Effect of selective oxidation of chromium on creep strength of alloy 617, Materials Science and Technology 8 (1992) 78–82.

[389] A. Epishin, U. Brueckner, P. Portella, T. Link, Investigation of porosity in single-crystal nickel-base superalloys, in: Materials for Advanced Power Engineering, 2002.

[390] A. Epishin, B. Fedelich, T. Link, T. Feldmann, I.L. Svetlov, Pore annihilation in a single-crystal nickel-base superalloy during hot isostatic pressing: experiment and modelling, Materials Science and Engineering: A 586 (2013) 342–349.

[391] A. Epishin, B. Fedelich, G. Nolze, S. Schriever, T. Feldmann, M.F. Ijaz, B. Viguier, D. Poquillon, Y. Le Bouar, A. Ruffini, A. Finel, Creep of single crystals of nickel-based superalloys at ultra-high homologous temperature, Metallurgical and Materials Transactions A 49 (9) (Sep 2018) 3973–3987.

[392] A. Epishin, T. Link, Mechanisms of high temperature creep of nickel-base superalloys under low applied stress, in: K.A. Green, T.M. Pollock, H. Harada, T.E. Howson, R.C. Reed, J.J. Schirra, S. Walston (Eds.), Superalloys 2004, Champion, PA (USA), 2004, pp. 137–143.

[393] A. Epishin, T. Link, U. Brückner, P.D. Portella, Kinetics of the topological inversion of the γ/γ'-microstructure during creep of a nickel-based superalloy, Acta Materialia 49 (2001) 4017–4023.

[394] A. Epishin, T. Link, B. Fedelich, I. Svetlov, E. Golubovskiy, Hot isostatic pressing of single-crystal nickel-base superalloys: mechanism of pore closure and effect on mechanical properties, in: Eurosuperalloys 2014, 2nd European Symposium on Superalloys and Their Applications, 2014, p. 08003.

[395] A. Epishin, T. Link, H. Klingelhöffer, B. Fedelich, P. Portella, Creep damage of single-crystal nickel base superalloys: mechanisms and effect on low cycle fatigue, Materials at High Temperatures 27 (1) (2010) 53–59.

[396] A. Epishin, T. Link, M. Nazmy, M. Staubli, H. Klingelhoffer, G. Nolze, Microstructural degradation of cmsx-4: kinetics and effect on mechanical properties, Superalloys 2008 (2008) 725–731.

[397] A. Epishin, T. Link, P.D. Portella, U. Brückner, Evolution of the γ/γ' microstructure during high-temperature creep of a nickel-base superalloy, Acta Materialia 48 (16) (2000) 4169–4177.

[398] G. Erickson, The development and applications of CMSX10, in: Proc. 8th Int. Symp. on Superalloys, TMS, 1996.

[399] B. Escaig, L'activation thermique des déviations sous faibles contraintes dans les structures hc et cc par, Physica Status Solidi B 28 (2) (1968) 463–474.

[400] B. Escaig, Sur le glissement dévié des dislocations dans la structure cubique à faces centrées, Journal de Physique 29 (2–3) (1968) 225–239.

[401] J. Eshelby, The determination of the elastic field of an ellipsoidal inclusion, and related problems, Proceedings of the Royal Society of London 241 (1957) 376–396.

[402] V.A. Esin, V. Maurel, P. Breton, A. Koster, S. Selezneff, Increase in ductility of pt-modified nickel aluminide coating with high temperature ageing, Acta Materialia 105 (2016) 505–518.

[403] M. Eskner, R. Sandstrom, Measurement of the ductile-to-brittle transition temperature in a nickel aluminide coating by a miniaturised disc bending test technique, Surface & Coatings Technology 165 (2003) 71–80.

[404] L. Espié, Experimental study and numerical modeling of the mechanical behavior of superalloys (Étude expérimentale et modélisation numérique du comportement mécanique des superalliages), in French, PhD thesis, ENSMP, 1996.

[405] N. Eurich, P. Bristowe, Segregation of alloying elements to intrinsic and extrinsic stacking faults in γ'-Ni_3Al via first principles calculations, Scripta Metallurgica 102 (2015) 87–90.

[406] A. Evans, M. He, J. Hutchinson, Mechanisms controlling the durability of thermal barrier coatings, Progress in Materials Science 46 (2001) 505–553.

[407] A. Evans, M. He, A. Suzuki, M. Gigliotti, B. Hazel, T. Pollock, A mechanism governing oxidation-assisted low-cycle fatigue of superalloys, Acta Materialia 57 (10) (2009) 2969–2983.

[408] A. Evans, J. Hutchinson, The mechanics of coating delamination in thermal gradients, Surface & Coatings Technology 201 (18) (2007) 7905–7916.

[409] A.G. Evans, G.B. Crumley, R.E. Demaray, On the mechanical behavior of brittle coatings and layers, Oxidation of Metals 20 (5–6) (1983) 193–216.

[410] A.G. Evans, M.Y. He, A. Suzuki, M. Gigliotti, B. Hazel, T.M. Pollock, A mechanism governing oxidation-assisted low-cycle fatigue of superalloys, Acta Materialia 57 (10) (2009) 2969–2983.

[411] A.G. Evans, D.R. Mumm, J.W. Hutchinson, G.H. Meier, F.S. Pettit, Mechanisms controlling the durability of thermal barrier coatings, Progress in Materials Science 46 (5) (2001) 505–553.

[412] H.E. Evans, R.C. Lobb, Conditions for the initiation of oxide-scale cracking and spallation, Corrosion Science 24 (3) (1984) 209–222.

[413] H.E. Evans, M.P. Taylor, Oxidation of high-temperature coatings, Proceedings of the Institution of Mechanical Engineers. Part G, Journal of Aerospace Engineering 220 (1) (2006) 1–10.

[414] R. Evans, B.B. Wilshire, Creep of Metals and Alloys, Institute of Materials, 1985.

[415] U.R. Evans, The mechanism of oxidation and tarnishing, Transactions of the Electrochemical Society 91 (1947) 547.

[416] G. Fabre, Influence des propriétés optiques et de l'endommagement de barrières thermiques EB-PVD pour la mesure d'adhérence par choc laser LASAT-2D (in French), PhD thesis, Mines ParisTech, 2013.

[417] G. Fabre, V. Guipont, M. Jeandin, M. Boustie, J.P. Cuq-Lelandais, L. Berthe, A. Pasquet, J.-Y. Guedou, LAser Shock Adhesion Test (LASAT) of Electron Beam Physical Vapor Deposited Thermal Barrier Coatings (EB-PVD TBCs), in: M. Heilmaier (Ed.), Euro Superalloys 2010, in: Advanced Materials Research, vol. 278, 2011, pp. 509–514.

[418] M. Fahrmann, P. Fratzl, O. Paris, E. Fahrmann, W.C. Johnson, Influence of coherency stress on microstructural evolution in model Ni-Al-Mo alloys, Acta Materialia 43 (1995) 1007.

[419] M. Fahrmann, W. Hermann, E. Fahrmann, A. Boegli, T. Pollock, H. Sockel, Determination of matrix and precipitate elastic constants in γ/γ' Ni-base model alloys, and their relevance to rafting, Materials Science and Engineering: A 260 (1) (1999) 212–221.

[420] Q. Fan, C. Wang, T. Yu, Construction of ternary Ni-Al-Ta potential and its application in the effect of Ta on [110] edge dislocation slipping in γ' (Ni_3Al), Computational Materials Science 118 (2016) (2016) 288–296.

[421] Q.F. Fang, R. Wang, Atomistic simulation of the atomic structure and diffusion within the core region of an edge dislocation in aluminum, Physical Review B 62 (Oct 2000) 9317–9324.

[422] S. Farenc, D. Caillard, A. Couret, An in situ study of prismatic glide in α titanium at low temperatures, Acta Metallurgica Et Materialia 41 (9) (1993) 2701–2709.

[423] D. Farkas, E. Savino, Computer simulation of dislocation core structure in Ni_3Al using local volume dependent potentials, Scripta Metallurgica 22 (1988) 557–560.

[424] P. Fauchais, R. Etchart-Salas, V. Rat, J.-F. Coudert, N. Caron, K. Wittmann-Ténèze, Parameters controlling liquid plasma spraying: solutions, sols, or suspensions, Journal of Thermal Spray Technology 17 (1) (2008) 31–59.

[425] B. Fedelich, A microstructure based constitutive model for the mechanical behavior at high temperatures of nickel-base single crystal superalloys, Computational Materials Science 16 (1) (1999) 248–258.

[426] B. Fedelich, A microstructural model for the monotonic and the cyclic mechanical behavior of single crystals of superalloys at high temperatures, International Journal of Plasticity 18 (1) (2002) 1–49.

[427] B. Fedelich, A. Epishin, T. Link, H. Klingelhöfer, P. Portella, Rafting during high temperature deformation in a single crystal superalloy, in: E.S. Huron, R.C. Reed, M.C. Hardy, M.J. Mills, R.E. Montero, P.D. Portella, J. Telesman (Eds.), Superalloys 2012: Proc. of the Twelfth International Symposium on Superalloys, Champion, USA, TMS, Minerals, Metals & Materials Society, Sept. 2012, pp. 491–500.

[428] B. Fedelich, A. Epishin, T. Link, H. Klingelhöffer, G. Künecke, P.D. Portella, Experimental characterization and mechanical modeling of creep induced rafting in superalloys, Computational Materials Science 64 (2012) 2–6.

[429] B. Fedelich, M. Finn, G. Künecke, Determination of the elastic constants of the alloy CMSX-4 from room temperature to 1300°C by the sonic method, Internal report of the division 5.2 of the BAM, 2010.

[430] B. Fedelich, G. Künecke, A. Epishin, T. Link, P. Portella, Constitutive modelling of creep degradation due to rafting in single-crystalline Ni-base superalloys, Materials Science and Engineering: A 510–511 (2009) 273–277.

[431] J. Feiereisen, O. Ferry, A. Jacques, A. George, Mechanical testing device for in situ experiments on reversibility of dislocation motion in silicon, Nuclear Instruments & Methods in Physics Research. Section B, Beam Interactions With Materials and Atoms 200 (2003) 339–345.

[432] M. Feller-Kniepmeier, U. Hemmersmeier, T. Kuttner, T. Link, Analysis of interfacial dislocations in a single-crystal Ni-base superalloy after [001] creep at 1033 K, Scripta Metallurgica 30 (1994) 1275–1280.

[433] M. Feller-Kniepmeier, T. Link, Correlation of microstructure and creep stages in the $\langle 100 \rangle$ oriented superalloy SRR-99 at 1253 K, Materials Transactions A 20 (1989) 1233–1238.

[434] M. Feller-Kniepmeier, T. Link, Dislocation structures in $\gamma - \gamma'$ interfaces of the single-crystal superalloy srr 99 after annealing and high temperature creep, Materials Science and Engineering: A 113 (1989) 191–195.

[435] H. Feng, N. Ming, L. Jiedong, H. Xu, C. Guofeng, Z. Zhongjiao, Effect of strain ranges and phase angles on the thermomechanical fatigue properties of thermal barrier coating system, Rare Metal Materials and Engineering 46 (12) (2017) 3693–3698.

[436] A. Feuerstein, J. Knapp, T. Taylor, A. Ashary, A. Bolcavage, N. Hitchman, Technical and economical aspects of current thermal barrier coating systems for gas turbine engines by thermal spray and ebpvd: a review, Journal of Thermal Spray Technology 17 (2) (2008) 199–213.

[437] K. Fichthorn, W. Weinberg, Theoretical foundations of dynamical Monte Carlo simulations, Journal of Chemical Physics 95 (2) (1991) 1090–1096.

[438] A. Fick, V. on liquid diffusion, Philosophical Magazine 4 (1855) 30–39.

[439] R. Field, T. Pollock, W. Murphy, The development of γ/γ' interfacial dislocation networks during creep in Ni-base superalloys, in: I. Staff, S. Antolovich (Eds.), Superalloys, 1992: Proc. of the Seventh International Symposium on Superalloys, Champion, USA, TMS, Minerals, Metals & Materials Society, Sept. 1992.

[440] A. Finel, Dynamics in elastically stresses alloys: growth laws in phase separating systems; Monte Carlo simulations, in: P.E.A. Turchi, A. Gonis (Eds.), Phase Transformation and Evolution in Materials, TMS (The Minerals, Metals & Materials Society), 2000, pp. 371–385.

[441] A. Finel, Y. Le Bouar, A. Dabas, B. Appolaire, Y. Yamada, T. Mohri, A sharp-interface phase field method, Physical Review Letters 121 (2018) 025501.

[442] P. Fink, D. Konitzer, J. Miller, Rhenium reduction – alloy design using an economically strategic element, JOM 62 (2010) 55–56.

[443] M. Finnis, J. Sinclair, A simple empirical n-body potential for transition metals, Philosophical Magazine. A 50 (1984) 45–66.

[444] M. Fivel, G.R. Canova, Developing rigourous boundary conditions to simulations of discrete dislocation dynamics, Modelling and Simulation in Materials Science and Engineering 7 (1999) 753–768.

[445] M. Fivel, C. Depres, An easy implementation of displacement calculations in 3d discrete dislocation dynamics codes, Philosophical Magazine 94 (28) (2014) 3206–3214.

[446] B. Flageolet, Effet du vieillissement du superalliage base nickel N18 pour disques de turbines sur sa durabilité en fatigue et en fatigue-fluage à 700°C, PhD thesis, ENSMA, Poitiers, France, 2005.

[447] M. Fleck, F. Schleifer, M. Holzinger, U. Glatzel, Phase-field modeling of precipitation growth and ripening during industrial heat treatments in Ni-base superalloys, Metallurgical and Materials Transactions A 49 (2018) 4146–4157.

[448] E. Fleury, L. Rémy, Low cycle fatigue damage in Ni-base superalloy single crystals at elevated temperature, Materials Science and Engineering: A 167 (1993) 23–30.

[449] E. Fleury, L. Rémy, Behavior of nickel-base superalloy under thermal-mechanical fatigue at elevated temperatures, Metallurgical Transactions A 25 (1994) 99–109.

[450] P.A. Flinn, Theory of deformation in superlattices, Transactions of the American Institute of Mining and Metallurgical Engineers 218 (1) (1960) 145–154.

[451] S. Flouriot, S. Forest, G. Cailletaud, A. Koster, L. Rémy, B. Burgardt, V. Gros, J. Delautre, Strain localization at the crack tip in single crystal CT specimens under monotonous loading: 3D finite element analyses and application to nickel-base superalloys, International Journal of Fracture 124 (2003) 43–77.

[452] R. Foerch, J. Besson, G. Cailletaud, P. Pilvin, Polymorphic constitutive equations in finite element codes, Computer Methods in Applied Mechanics and Engineering 141 (1997) 355–372.

[453] A. Foreman, Dislocation energies in anisotropic crystals, Acta Metallurgica 3 (4) (1955) 322–330.

[454] S. Forest, The micromorphic approach for gradient elasticity, viscoplasticity and damage, Journal of Engineering Mechanics - ASCE 135 (2009) 117–131.

[455] S. Forest, Micromorphic approach for gradient elasticity, viscoplasticity, and damage, Journal of Engineering Mechanics 135 (2009) 117–131.

[456] S. Forest, P. Boubidi, R. Sievert, Strain localization patterns at a crack tip in generalized single crystal plasticity, Scripta Materialia 44 (2001) 953–958.

[457] S. Forest, J. Olschewski, J. Ziebs, H.-J. Kühn, J. Meersmann, H. Frenz, The elastic/plastic deformation behaviour of various oriented SC16 single crystals under combined tension/torsion fatigue loading, in: G. Lütjering, H. Nowack (Eds.), Sixth International Fatigue Congress, Pergamon, 1996, pp. 1087–1092.

[458] S. Forest, M. Rubin, A rate-independent crystal plasticity model with a smooth elastic–plastic transition and no slip indeterminacy, European Journal of Mechanics - A/Solids 55 (2016) 278–288.

[459] F. Förster, Ein neues Meßverfahren zur Bestimmung des Elastizitätsmoduls und der Dämpfung, Zeitschrift für Metallkunde 29 (4) (1937) 109–115.

[460] J. Frachon, Multiscale approach to predict the lifetime of EB-PVD thermal barrier coatings, PhD thesis, Mines ParisTech, 14 Decembre, 2009.

[461] D. François, A. Pineau, A. Zaoui, Mechanical Behaviour of Materials. Volume I: Elasticity and Plasticity, Springer, 1998.

[462] P. Franciosi, The concepts of latent hardening and strain hardening in metallic single crystals, Acta Metallurgica 33 (1985) 1601–1612.

[463] P. Franciosi, M. Berveiller, A. Zaoui, Latent hardening in copper and aluminium single crystals, Acta Metallurgica 28 (3) (1980) 273–283.

[464] P. Fratzl, O. Penrose, Ising model for phase separation in alloys with anisotropic elastic interaction: II a computer experiment, Acta Materialia 44 (8) (1996) 3227.

[465] C. Frederick, P. Armstrong, A mathematical representation of the multiaxial bauschinger effect, Materials at High Temperatures 24 (1) (2007) 1–26.

[466] A. Fredholm, Monocristaux d'alliages base nickel relation entre composition, microstructure et comportement en fluage a haute temperature, Phd dissertation, Ecole Nationale Supérieure des Mines de Paris, France, 1987.

[467] A. Fredholm, J. Strudel, On the creep resistance of some nickel base single crystals, Superalloys 1984 (1984) 211–220.

[468] A. Fredholm, J.-L. Strudel, High temperature creep mechanisms in single crystals of some high performance Ni-base superalloys, in: J.B. Marriott, et al. (Eds.), High Temperature Alloys, Elsevier, 1988, p. 9.

[469] D. Frenkel, B. Smit, Understanding Molecular Simulations: From Algorithms to Applications, 2nd edition, Academic Press, 2001.

[470] H. Frenz, J. Kinder, H. Klingelhoffer, P. Portella, Behaviour of single crystal superalloys under cyclic loading at high temperatures, in: Superalloys, 1996, pp. 305–312.

[471] C. Fressengeas, Mechanics of Dislocations Fields, 1 edition, ISTE Ltd, London, 2017.

[472] L. Freund, O. Messe, J. Barnard, M. Goken, S. Neumeier, C. Rae, Segregation assisted microtwinning during creep of a polycrystalline $L1_2$-hardened Co- base superalloy, Acta Materialia 123 (2017) 295–304.

[473] C. Freysoldt, B. Grabowski, T. Hickel, J. Neugebauer, G. Kresse, A. Janotti, C. Van de Walle, First-principles calculations for point defects in solids, Reviews of Modern Physics 86 (2014) 253–305.

[474] J. Friedel, Dislocations, 1967, 491 pp., New York.

[475] G. Fuchs, B. Boutwell, Calculating solidification and transformation in as-cast CMSX-10, JOM 54 (2002) 45–48.

[476] W. Funk, E. Blank, Shear testing of monocrystalline alloys incorporating the measurement of local and integral strains, Materials Science and Engineering 67 (1984) 1–11.

[477] Y. Furuya, K. Kobayashi, M. Hayakawa, M. Sakamoto, Y. Koizumi, H.H. Harada, High-temperature ultrasonic fatigue testing of single-crystal superalloys, Materials Letters 69 (2012) 1–3.

[478] G. Cailletaud, Basic ingredients, development of phenomenological models and practical use of crystal plasticity, in: R. Pippan, P. Gumbsch (Eds.), Multiscale Modelling of Plasticity and Fracture by Means of Dislocation Mechanics, in: CISM International Centre for Mechanical Sciences, vol. 522, Springer, Vienna, 2010, https://doi.org/10.1007/978-3-7091-0283-1_6.

[479] T. Gabb, J. Gayda, R. Miner, Orientation and temperature dependence of some mechanical properties of the single-crystal superalloy René N4: part II. Low cycle fatigue behavior, Metallurgical Transactions A 17 (1986) 497–505.
[480] D. Gaertner, K. Abrahams, J. Kottke, V. Esin, I. Steinbach, G. Wilde, S. Divinski, Concentration-dependent atomic mobilities in FCC CoCrFeMnNi high-entropy alloys, Acta Materialia 166 (2019) 357–370.
[481] E. Galindo-Nava, L. Connor, C. Rae, On the prediction of the yield stress of unimodal and multimodal γ' nickel-base superalloys, Acta Materialia 98 (2015) 377–390.
[482] T. Galiullin, A. Chyrkin, R. Pillai, R. Vassen, W.J. Quadakkers, Effect of alloying elements in ni-base substrate material on interdiffusion processes in mcraly-coated systems, Surface & Coatings Technology 350 (2018) 359–368.
[483] J. Gallot, et al., Fabrication et mise au point d'une sonde a atomes associee a un microscope ionique a champ, 1975.
[484] B. Gan, H. Murakami, R. Maaß, L. Meza, J.R. Greer, T. Ohmura, S. Tin, Nanoindentation and nano-compresion testing of Ni_3Al precipitates, in: E.S. Huron, R.C. Reed, M.C. Hardy, M.J. Mills, R.E. Montero, P.D. Portella, J. Telesman (Eds.), Superalloys 2012, 2012, https://doi.org/10.1002/9781118516430.ch9.
[485] C.-A. Gandin, From constrained to unconstrained growth during directional solidification, Acta Materialia 48 (2000) 2483–2501.
[486] C.-A. Gandin, J.-L. Desbiolles, M. Rappaz, P. Thevoz, A three-dimensional cellular automation-finite element model for the prediction of solidification grain structures, Metallurgical and Materials Transactions A 30 (1999) 3153–3165.
[487] C.-A. Gandin, M.M. Rappaz, R. Tintillier, Three-dimensional probabilistic simulation of solidification grain structures: application to superalloy precision castings, Materials Transactions A 44 (1993) 467–479.
[488] C.-A. Gandin, M.M. Rappaz, R. Tintillier, 3-dimensional simulation of the grain formation in investment castings, Materials Transactions A 25 (1994) 629–635.
[489] C.-A. Gandin, M. Rappaz, D. West, B. Adams, Grain texture evolution during the columnar growth of dendritic alloys, Metallurgical and Materials Transactions A 26 (1995) 1543–1551.
[490] C.-A. Gandin, R. Schaefer, M.M. Rappaz, Analytical and numerical predictions of dendritic grain envelopes, Acta Materialia 44 (1996) 3339–3347.
[491] J.F. Ganghoffer, A. Hazotte, S. Denis, A. Simon, Finite element calculation of internal mismatch stresses in a single crystal nickel base superalloy, Scripta Materialia 25 (11) (1991) 2491–2496.
[492] L. Gao, H. Guo, L. Wei, C. Li, S. Gong, H. Xu, Microstructure and mechanical properties of yttria stabilized zirconia coatings prepared by plasma spray physical vapor deposition, Ceramics International 41 (7) (2015) 8305–8311.
[493] S. Gao, M. Fivel, A. Ma, A. Hartmaier, Influence of misfit stresses on dislocation glide in single crystal superalloys: a three-dimensional discrete dislocation dynamics study, Journal of the Mechanics and Physics of Solids 76 (2015) 276–290.
[494] S. Gao, M. Fivel, A. Ma, A. Hartmaier, 3d discrete dislocation dynamics study of creep behavior in ni-base single crystal superalloys by a combined dislocation climb and vacancy diffusion model, Journal of the Mechanics and Physics of Solids 102 (2017) 209–223.
[495] S. Gao, M.K. Rajendran, M. Fivel, A. Ma, O. Shchyglo, A. Hartmaier, I. Steinbach, Primary combination of phase-field and discrete dislocation dynamics methods for investigating athermal plastic deformation in various realistic ni-base single crystal superalloy microstructures, Modelling and Simulation in Materials Science and Engineering 23 (7) (sep 2015) 075003.
[496] S. Gao, Y. Zerong, M. Grabowski, J. Rogal, R. Drautz, A. Hartmaier, Influence of excess volumes induced by re and w on dislocation motion and creep in ni-base single crystal superalloys: a 3d discrete dislocation dynamics study, Metals 9 (6) (March 2019) 637.

[497] Y. Gao, Z. Zhuang, Z. Liu, X. You, X. Zhao, Z. Zhang, Investigations of pipe-diffusion-based dislocation climb by discrete dislocation dynamics, International Journal of Plasticity 27 (7) (2011) 1055–1071.

[498] Y. Gao, Z. Zhuang, X. You, A study of dislocation climb model based on coupling the vacancy diffusion theory with 3d discrete dislocation dynamics, International Journal of Multiscale Computational Engineering 11 (1) (2013).

[499] F. Garofalo, Fundamentals of Creep and Creep Rupture, McMillan, 1965.

[500] A. Gaubert, Modélisation des effets de l'evolution microstructurale sur le comportement mécanique du superalliage monocristallin AM1, PhD thesis, Ecole Nationale Supérieure des Mines de Paris, France, 2009.

[501] A. Gaubert, A. Finel, Y. Le Bouar, G. Boussinot, Viscoplastic phase field modelling of rafting in Ni base superalloys, in: D. Jeulin, S. Forest (Eds.), Continuum Models and Discrete Systems CMDS 11, Les Presses de l'École des Mines de, Paris, 2008, pp. 161–166.

[502] A. Gaubert, M. Jouiad, J. Cormier, Y. Le Bouar, J. Ghighi, 3-d imaging and phase field simulations of the microstructure evolution during creep tests of <011> oriented Ni-based superalloys, Acta Materialia 84 (2015) 237–255.

[503] A. Gaubert, Y. Le Bouar, A. Finel, Coupling phase field and viscoplasticity to study rafting in Ni-based superalloys, Philosophical Magazine. B 90 (1–4) (2010) 375–404.

[504] B. Gault, M.P. Moody, J.M. Cairney, S.P. Ringer, Atom Probe Microscopy, vol. 160, Springer Science & Business Media, 2012.

[505] B. Geddes, H. Leon, X. Huang, Superalloys, Alloying and Performances, ASM Int., 2010.

[506] M. Gell, D. Duhl, A. Giamei, The development of single crystal superalloys for turbine blades, in: Proc. 4th Int. Symp. on Superalloys, TMS, 1980.

[507] M. Gell, G. Leverant, C. Wells, The fatigue strength of nickel-base superalloys, in: Achievement of High Fatigue Resistance in Metals and Alloys, vol. STP 467, ASTM, 1970, pp. 113–153.

[508] A. Gemma, J. Phillips, The application of fracture mechanics to life prediction of cooling hole configuration in thermal mechanical fatigue, Engineering Fracture Mechanics 9 (1977) 25–36.

[509] P. Germain, The method of virtual power in continuum mechanics. Part 2: microstructure, SIAM Journal on Applied Mathematics 25 (3) (1973) 556–575.

[510] P. Germain, Q. Nguyen, P. Suquet, Continuum thermodynamics, Journal of Applied Mechanics 5 (1983) 1010–1020.

[511] V. Gerold, H. Karnthaler, On the origin of planar slip in fcc alloys, Acta Metallurgica 37 (8) (1989) 2177–2183.

[512] P.-A. Geslin, B. Appolaire, A. Finel, A phase field model for dislocation climb, Applied Physics Letters 104 (1) (2014) 011903.

[513] P.-A. Geslin, B. Appolaire, A. Finel, Multiscale theory of dislocation climb, Physical Review Letters 115 (2015) 265501.

[514] J. Ghighi, J. Cormier, E. Ostoja-Kuczynski, J. Mendez, G. Cailletaud, F. Azzouz, A microstructure sensitive approach for the prediction of the creep behavior and life under complex loading paths, Technische Mechanik 32 (2012) 205–220.

[515] N.M. Ghoniem, S.-H. Tong, L.Z. Sun, Parametric dislocation dynamics: a thermodynamics-based approach to investigations of mesoscopic plastic deformation, Physical Review B 61 (Jan 2000) 913–927.

[516] R. Ghosh, R. Curtis, M. McLean, Creep deformation of single crystal superalloys-modelling the crystallographic anisotropy, Acta Materialia 38 (10) (1990) 1977–1992.

[517] A. Giamei, Development of single crystal superalloys: a brief history, Advanced Materials & Processes 171 (9) (2013).

[518] A. Giamei, J. Tschinkel, Liquid metal cooling: a new solidification technique, Materials Transactions A 7 (1976) 1427–1434.

[519] C.S. Giggins, F.S. Pettit, Oxidation of Ni-Cr-Al Alloys between 1000 and 1200°C, Journal of the Electrochemical Society 118 (11) (1971) 1782–1790.

[520] R. Giraud, J. Cormier, Z. Hervier, D. Bertheau, K. Harris, J. Wahl, X. Milhet, J. Mendez, A. Organista, Effect of the prior microstructure degradation on the high temperature/low stress non-isothermal creep behavior of CMSX-4® Ni-based single crystal superalloy, in: Superalloys 2012, September 2012, Champion, PA (USA), 2012, pp. 265–274.

[521] R. Giraud, Z. Hervier, J. Cormier, G. Saint-Martin, F. Hamon, X. Milhet, J. Mendez, Strain effect on the γ' dissolution at high temperatures of a nickel-based single crystal superalloy, Metallurgical and Materials Transactions A 44 (1) (2013) 131–146.

[522] R. Glas, M. Jouiad, P. Caron, N. Clément, H. Kirchner, Order and mechanical properties of the γ matrix of superalloys, Acta Materialia 44 (12) (1996) 4917–4926.

[523] M. Glatzel, A. Muller, Neutron scattering experiments with a nickel base superalloy; part 1: material and experiment, Scripta Metallurgica 31 (1994).

[524] U. Glatzel, M. Feller-Kniepmeier, Calculations of internal stresses in the γ/γ' microstructure of a nicle-base superalloy with high volume fraction of γ'-phase, Scripta Metallurgica 23 (6) (1989) 1839–1844.

[525] U. Glatzel, M. Feller-Kniepmeier, Microstructure and dislocation configurations in fatigued [001] specimens of the nickel-based superalloy cmsx-6, Scripta Metallurgica et Materialia 25 (8) (1991) 1845–1850.

[526] A.D. Gledhill, K.M. Reddy, J.M. Drexler, K. Shinoda, S. Sampath, N.P. Padture, Mitigation of damage from molten fly ash to air-plasma-sprayed thermal barrier coatings, Materials Science and Engineering: A 528 (24) (2011) 7214–7221.

[527] B. Gleeson, N. Mu, S. Hayashi, Compositional factors affecting the establishment and maintenance of Al_2O_3 scales on Ni-Al-Pt systems, Journal of Materials Science 44 (7) (2009) 1704–1710.

[528] H. Gleiter, E. Hornbogen, Hardening by coherent precipitates, Acta Metallurgica 13 (1965) 576–578.

[529] H. Gleiter, E. Hornbogen, Precipitation hardening by coherent particles, Materials Science and Engineering: A 2 (1967) 285–302.

[530] T. Go, Y. Sohn, G. Mauer, R. Vaßen, J. Gonzalez-Julian, Cold spray deposition of cr2alc max phase for coatings and bond-coat layers, Journal of the European Ceramic Society 39 (4) (2019) 860–867.

[531] M. Göbel, A. Rahmel, M. Schütze, The cyclic-oxidation behavior of several nickel-base single-crystal superalloys without and with coatings, Oxidation of Metals 41 (3–4) (1994) 271–300.

[532] J. Goerler, I. Lopez-Galilea, L.M. Roncery, O. Shchyglo, W. Theisen, I. Steinbach, Topological phase inversion after long-term thermal exposure of nickel-base superalloys: experiment and phase-field simulation, Acta Materialia 124 (2017) 151–158.

[533] J.V. Goerler, S. Brinckmann, O. Shchyglo, I. Steinbach, γ channel stabilization mechanism in Ni base superalloys, Philosophical Magazine Letters 95 (2015) 519525.

[534] J. Goiri, A. Van der Ven, Phase and structural stability in Ni-Al systems from first principles, Physical Review B 94 (2016) 094111.

[535] J.I. Goldstein, D.E. Newbury, J.R. Michael, N.W. Ritchie, J.H.J. Scott, D.C. Joy, Scanning Electron Microscopy and X-Ray Microanalysis, Springer, 2017.

[536] X. Gong, H. Peng, Y. Ma, H. Guo, S. Gong, Microstructure evolution of an EB-PVD NiAl coating and its underlying single crystal superalloy substrate, Journal of Alloys and Compounds 672 (2016) 36–44.

[537] A.J. Goodfellow, E.I. Galindo-Nava, K.A Christofidou, N.G. Jones, C.D. Boyer, T.L. Martin, P.A.J. Bagot, M.C. Hardy, H.J. Stone, The effect of phase chemistry on the extent of strengthening mechanisms in model Ni-Cr-Al-Ti-Mo based superalloys, Acta Materialia 153 (2018) 290–302.

[538] A.J. Goodfellow, E.I. Galindo-Nava, C. Schwalbe, H.J. Stone, The role of composition on the extent of individual strengthening mechanisms in polycrystalline Ni-based superalloys, Materials & Design 173 (2019) 107760.

[539] P. Gopal, S. Srinivasan, First-principles study of self- and solute diffusion mechanisms in γ'-Ni_3Al, Physical Review B 86 (2012) 014112.

[540] O. Gorbatov, I. Lomaev, Y. Gornostyrev, A. Ruban, D. Furrer, V. Venkatesh, D. Novikov, S. Burlatsky, Effect of composition on antiphase boundary energy in Ni_3Al based alloys: ab initio calculations, Physical Review B 93 (2016) 224106.

[541] J. Görler, I. Lopez-Galilea, L. Mujica, O. Shchyglo, W. Theisen, I. Steinbach, Topological phase inversion after longterm thermal exposure of Ni-base superalloys: experiment and phase field simulation, Acta Materialia 124 (2017).

[542] K. Goswami, A. Mottura, Can slow-diffusing solute atoms reduce vacancy diffusion in advanced high-temperature alloys? Materials Science and Engineering: A 617 (2014) 194–199.

[543] K. Goswami, A. Mottura, A kinetic Monte Carlo study of vacancy diffusion in non-dilute Ni-Re alloys, Materials Science and Engineering: A 743 (2019) 265–273.

[544] M. Goulette, P. Spilling, R. Arthey, Cost effective single crystal, in: Proc. 5th Int. Symp. on Superalloys, TMS, 1984.

[545] G.W. Goward, D.H. Boone, Mechanisms of formation of diffusion aluminide coatings on nickel-base superalloys, Oxidation of Metals 3 (5) (1971) 475–495.

[546] M. Grabowski, J. Rogal, R. Drautz, Kinetic Monte Carlo simulations of vacancy diffusion in nondilute Ni-X (X = Re, W, Ta) alloys, Physical Review Materials 2 (2018) 123403.

[547] U. Grafe, B. Bottger, J. Tiaden, S. Fries, Coupling of multicomponent thermodynamic databases to a phase field model: application to solidification and solid-state transformations of superalloys, Scripta Materialia 42 (2000) 1179–1186.

[548] J.T. Graham, A.D. Rollett, R. LeSar, Fast Fourier transform discrete dislocation dynamics, Modelling and Simulation in Materials Science and Engineering 24 (8) (2016) 085005.

[549] L. Graham, B. Rauguth, Method and device for casting a metal article using a fluidized bed, Technical Report EP 1153681 A1 20011114, European patent, 2000.

[550] B. Grégoire, X. Montero, M. Galetz, G. Bonnet, F. Pedraza, Mechanisms of hot corrosion of pure nickel at 700°C: influence of testing conditions, Corrosion Science 141 (2018) 211–220.

[551] S. Groh, B. Devincre, L. Kubin, A. Roos, F. Feyel, J.-L. Chaboche, Dislocations and elastic anisotropy in heteroepitaxial metallic thin films, Philosophical Magazine Letters 83 (5) (2003) 303–313.

[552] I. Groma, M. Zaiser, P.D. Ispanovity, Dislocation patterning in a two-dimensional continuum theory of dislocations, Physical Review B 93 (214110) (2016).

[553] T. Grosdidier, A. Hazotte, A. Simon, About chemical heterogeneities and γ' precipitate behaviour in single crystal nickel base superalloy, in: High Temperature Materials for Power Engineering 1990, Dordrecht (Netherlands), 1990, pp. 1271–1280.

[554] T. Grosdidier, A. Hazotte, A. Simon, Precipitation and dissolution processes in γ/γ' single crystal nickel-based superalloys, Materials Science and Engineering: A 256 (1998) 183–196.

[555] M. Grydlik, F. Boioli, H. Groiss, R. Gatti, M. Brehm, F. Montalenti, B. Devincre, F. Schäffler, L. Miglio, Misfit dislocation gettering by substrate pit-patterning in SiGe films on Si (001), Applied Physics Letters 101 (1) (2012) 013119.

[556] Y. Gu, B. Ao, G. Wu, W. Wu, Three-dimensional structure analysis of EB-PVD thermal barrier coatings, in: Y. Tian, T. Xiao, P. Liu (Eds.), Second Symposium on Novel Technology of X-Ray Imaging, vol. 11068, International Society for Optics and Photonics, SPIE, 2019, pp. 638–643.

[557] Y. Gu, Y. Xiang, S.S. Quek, D.J. Srolovitz, Three-dimensional formulation of dislocation climb, Journal of the Mechanics and Physics of Solids 83 (2015) 319–337.

[558] Y. Guan, B. Chen, J. Zou, T. Britton, J. Jiang, F. Dunne, Crystal plasticity modelling and HR-DIC measurement of slip activation and strain localization in single and oligo-crystal Ni alloys under fatigue, International Journal of Plasticity 88 (2017) 70–88.

[559] J.-Y. Guédou, Materials evolution in hot parts of aero-turbo engines, in: Proc. of 27th ICAS Conference, ICAS, 2010.

[560] C. Guerre, L. Rémy, R. Molins, Alumina scale growth and degradation modes of a TBC system, Materials at High Temperatures 20 (2003) 481–485.

[561] V. Guipont, G. Bégué, G. Fabre, V. Maurel, Buckling and interface strength analyses of thermal barrier coatings combining Laser Shock Adhesion Test to thermal cycling, Surface and Coatings Technology 378 (2019) 124938.

[562] G. Guliver, The quantitative effect of rapid cooling upon the constitution of binary alloys, Journal of the Institute of Metals 8 (1913) 120–157.

[563] H. Guo, R. Vaßen, D. Stöver, Atmospheric plasma sprayed thick thermal barrier coatings with high segmentation crack density, Surface & Coatings Technology 186 (3) (2004) 353–363.

[564] S. Guo, D. Mumm, A.M. Karlsson, Y. Kagawa, Measurement of interfacial shear mechanical properties in thermal barrier coating systems by a barb pullout method, Scripta Materialia 53 (9) (2005) 1043–1048.

[565] M. Gururajan, T. Abinandanan, Phase field study of precipitate rafting under a uniaxial stress, Acta Materialia 55 (2007) 5015–5026.

[566] P. Haehner, C. Rinaldi, V. Bicego, E. Affeldt, T. Brendel, H. Andersson, T. Beck, H. Klingerhofer, H.-J. Kuhn, A. Koster, M. Loveday, M. Marchionni, C. Rae, Research and development into a European code of practice for strain controlled thermomechanical fatigue testing, International Journal of Fatigue 30 (2008) 372–381.

[567] P. Haener, H. Klingelhoffer, T. Beck, M.S. Loveday, C.C. Rinaldi, High temperature thermo-mechanical fatigue: testing methodology, interpretation of data, and applications, International Journal of Fatigue 30 (2008).

[568] S.H. Haghighat, G. Eggeler, D. Raabe, Effect of climb on dislocation mechanisms and creep rates in -strengthened ni base superalloy single crystals: a discrete dislocation dynamics study, Acta Materialia 61 (10) (2013) 3709–3723.

[569] M. Halbig, M. Jaskowiak, J. Kiser, D. Zhu, Evaluation of ceramic matrix composite technology for aircraft turbine engine applications, Report 010774, NASA, 2013.

[570] T. Hammerschmidt, A. Bialon, D. Pettifor, R. Drautz, Topologically closed-packed phases in binary transition-metal compounds: matching high-throughput ab-initio calculations to an empirical structure-map, New Journal of Physics 15 (2013) 115016.

[571] T. Hammerschmidt, J. Koßmann, C. Zenk, S. Neumeier, M. Göken, I. Lopez-Galilea, L. Mujica Roncery, S. Huth, A. Kostka, W. Theisen, R. Drautz, The role of local chemical composition for TCP phase formation in Ni-base and Co-base superalloys, in: M. Hardy, E. Huron, U. Glatzel, B. Griffin, B. Lewis, C. Rae, V. Seetharaman, S. Tin (Eds.), Superalloys 2016: Proc. of the Thirteenth International Symposium on Superalloys, Champion, USA, Wiley, Minerals, Metals & Materials Society, Sept. 2016, pp. 89–96.

[572] T. Hammerschmidt, B. Seiser, R. Drautz, D. Pettifor, Modelling topologically close-packed phases in superalloys: valence-dependent bond-order potentials based on ab-initio calculations, in: R.C. Reed, K.A. Green, P. Caron, T.P. Gabb, M.G. Fahrmann, E.S. Huron (Eds.), Superalloys 2008: Proc. of the Eleventh International Symposium on Superalloys, Champion, USA, TMS, Minerals, Metals & Materials Society, Sept. 2008, pp. 847–853.

[573] T. Hammerschmidt, B. Seiser, M. Ford, A. Ladines, S. Schreiber, N. Wang, J. Jenke, Y. Lysogorskiy, C. Teijeiro, M. Mrovec, M. Cak, E. Margine, D. Pettifor, R. Drautz, Bopfox program for tight-binding and analytic bond-order potential calculation, Computer Physics Communications 235 (2019) 221–233.

[574] J. Han, Identifikation der elastischen Kennwerte anisotroper Hochtemperaturlegierungen mittels Resonanzmessungen und Finite-Elemente-Simulationen,

PhD thesis, Technical University Berlin, VDI Verlag GmbH, Düsseldorf, 1995, Reihe 5: Grund- und Werkstoffe, Nr. 404.

[575] J. Han, A. Bertram, J. Olschewski, W. Hermann, H.-G. Sockel, Identification of elastic constants of alloys with sheet and fibre textures based on resonance measurements and finite element analysis, Materials Science and Engineering: A 191 (1) (1995) 105–111.

[576] F. Hanriot, Étude du comportement du superalliage AM1 sous sollicitations cycliques, PhD thesis, ENSMP, 1993.

[577] F. Hanriot, G. Cailletaud, L. Rémy, Mechanical behavior of a nickel-base superallot single crystal, in: A. Freed, K. Walker (Eds.), High Temperature Constitutive Modeling – Theory and Application, ASME, 1991.

[578] F. Hanriot, E. Fleury, L. Rémy, Mechanical behaviour of a nickel base superalloy single crystal, in: Proc. High Temperature Materials for Power Engineering, Kluwer, 24–27 Sept. 1990.

[579] M. Hantcherli, M. Pettinari-Sturmel, J. Viguier, B. Douina, A. Coujou, Evolution of interfacial dislocation network during anisothermal high-temperature creep of a nickel-based superalloy, Scripta Metallurgica 66 (2012) 143–146.

[580] R. Harikrishnan, J.-B. le Graverend, A creep-damage phase-field model: predicting topological inversion in ni-based single crystal superalloys, Materials & Design 160 (2018) 405–416.

[581] M. Harvey, C. Courcier, V. Maurel, L. Rémy, Oxide and TBC spallation in β-NiAl coated systems under mechanical loading, Surface & Coatings Technology 203 (5–7) (2008) 432–436.

[582] H. Hasegawa, Y. Tsukamoto, T. Hanada, K. Takita, Research on application of ceramic turbine to jet engine, in: Proc. of 23rd ICAS Conference, ICAS, 2002.

[583] P. Haupt, Continuum Mechanics and Theory of Materials, Springer-Verlag, Berlin, 2000.

[584] K. Havner, Finite Plastic Deformation of Crystalline Solids, Cambridge University Press, 1992.

[585] C. Hawk, Wide gap braze repairs of nickel superalloy gas turbine components, PhD thesis, Colorado School of Mines, USA, 2016.

[586] J.A. Haynes, K.A. Unocic, M.J. Lance, B.A. Pint, Influences of superalloy composition and Pt content on the oxidation behavior of γ/γ' NiPtAl bond coatings, Oxidation of Metals 86 (5–6) (2016) 453–481.

[587] A. Hazotte, Transformations et contraintes de cohérence dans les superalliages et les intermétalliques de base TiAl, Metallurgical Transactions 97 (2009) 23–31.

[588] A. Hazotte, T. Grosdidier, S. Denis, γ' Precipitate splitting in nickel-based superalloys: a 3-D finite element analysis, Scripta Materialia 34 (4) (1996) 601–608.

[589] A. Hazotte, J. Lacaze, Chemically oriented γ'-plate development in a nickel-base superalloy, Scripta Metallurgica 23 (1989) 1877–1882.

[590] A. Hazotte, A. Simon, Interactions contraintes-microstructure dans l'alliage CMSX2 monocristallin, in: Colloque National Superalliages Monocristallin, Villars de Lans, 1986, pp. 210–219.

[591] S. He, P. Peng, O. Gorbatov, A. Ruban, Effective interactions and atomic ordering in Ni-rich Ni-Re alloys, Physical Review B 94 (2016) 024111.

[592] W. He, G. Mauer, M. Gindrat, R. Wäger, R. Vaßen, Investigations on the nature of ceramic deposits in plasma spray–physical vapor deposition, Journal of Thermal Spray Technology 26 (1–2) (2017) 83–92.

[593] Z. He, W. Qiu, Y.-N. Fan, Q.-N. Han, X. Shi, H.-J. nd Ma, Effects of secondary orientation on fatigue crack initiation in a single crystal superalloy, Fatigue & Fracture of Engineering Materials & Structures 41 (2017) 935–948.

[594] R.F.S. Hearmon, The elastic constants of anisotropic materials, Reviews of Modern Physics 18 (1946) 409–440.

[595] M.G. Hebsur, R.V. Miner, High temperature tensile and creep behaviour of low pressure plasma-sprayed NiCoCrAlYcoating alloy, Materials Science and Engineering: A 83 (1986) 239–245.
[596] A. Heckl, R. Rettig, R. Singer, Solidification characteristics and segregation behavior of nickel-base superalloys in dependence on different rhenium and ruthenium contents, Metallurgical and Materials Transactions A 41 (2010) 202.
[597] S. Hegde, R. Kearsey, J. Beddoes, Design of solutionizing heat treatments for an experimental single crystal superalloy, in: R.C. Reed, K.A. Green, P. Caron, T.P. Gabb, M.G. Fahrmann, E.S. Huron (Eds.), Superalloys 2008: Proc. of the Eleventh International Symposium on Superalloys, Champion, USA, TMS, Minerals, Metals & Materials Society, Sept. 2008, pp. 301–310.
[598] T. Helander, J. Ågren, A phenomenological treatment of diffusion in Al-Fe and Al-Ni alloys having B2-B.C.C. ordered structure, Acta Materialia 47 (1999) 1141–1152.
[599] K.J. Hemker, B.G. Mendis, C. Eberl, Characterizing the microstructure and mechanical behavior of a two-phase NiCoCrAlY bond coat for thermal barrier systems, Materials Science and Engineering: A 483–484 (2008) 727–730.
[600] U. Hemmersmeier, M. Feller-Kniepmeier, Element distribution in the macro- and microstructure of nickel base superalloy CMSX-4, Materials Science and Engineering: A 248 (1998) 87–97.
[601] M. Henderson, J. Martin, The influence of crystal orientation on the high temperature fatigue crack growth of a Ni-based single crystal superalloy, Acta Materialia 44 (1996) 111–126.
[602] P. Henderson, L. Berglin, C. Jansson, On rafting in a single crystal nickel-base superalloy after high and low temperature creep, Scripta Materialia 40 (2) (1998).
[603] G. Henkelman, B. Uberuaga, H. Jónsson, A climbing image nudged elastic band method for finding saddle points and minimum energy paths, Journal of Chemical Physics 113 (22) (2000) 9901–9904.
[604] F. Heredia, D. Pope, The tension/compression flow asymmetry in a high γ' volume fraction nickel base alloy, Acta Metallurgica 34 (1986) 279–285.
[605] W. Hermann, H.G. Sockel, J. Han, A. Bertam, Elastic properties and determination of elastic constants of nickel-base superalloys by a free-free beam technique, in: R. Kissinger, D.J. Deye, A. Anton, D.L. Cetal, M. Nathal, T. Pollock, D. Woodford (Eds.), Superalloys 1996: Proc. of the Eighth International Symposium on Superalloys, Champion, USA, TMS, Minerals, Metals & Materials Society, Sept. 1996, pp. 229–238.
[606] M.T. Hernandez, D. Cojocaru, M. Bartsch, A.M. Karlsson, On the opening of a class of fatigue cracks due to thermo-mechanical fatigue testing of thermal barrier coatings, Computational Materials Science 50 (9) (2011) 2561–2572.
[607] M.T. Hernandez, A.M. Karlsson, M. Bartsch, On tgo creep and the initiation of a class of fatigue cracks in thermal barrier coatings, Surface & Coatings Technology 203 (23) (2009) 3549–3558.
[608] T. Hickel, B. Grabowski, F. Koermann, J. Neugebauer, Advancing density functional theory to finite temperatures: methods and applications in steel design, Journal of Physics. Condensed Matter 24 (2011) 053202.
[609] R. Hill, The elastic behaviour of a crystalline aggregate, Proceedings of the Physical Society. Section A 65 (5) (May 1952) 349–354.
[610] R. Hill, Elastic properties of reinforced solids: some theoretical principles, Journal of the Mechanics and Physics of Solids 11 (5) (1963) 357–372.
[611] R. Hill, Generalized constitutive relations for incremental deformation of metal crystals by multislip, Journal of the Mechanics and Physics of Solids 14 (1966) 95–102.
[612] R. Hill, J. Rice, Constitutive analysis of elastic–plastic crystals at arbitrary strains, Journal of the Mechanics and Physics of Solids 20 (1972) 401–413.
[613] T.S. Hille, T.J. Nijdam, A.S. Suiker, S. Turteltaub, W.G. Sloof, Damage growth triggered by interface irregularities in thermal barrier coatings, Acta Materialia 57 (9) (2009) 2624–2630.

[614] R. Hillery, B. Pilsner, R. McKnight, T. Cook, M. Hartle, Thermal barrier coating life prediction model development, 1988.
[615] J. Hillier, On microanalysis by electrons, Physical Review 64 (9–10) (1943) 318.
[616] J. Hillier, R. Baker, Microanalysis by means of electrons, Journal of Applied Physics 15 (9) (1944) 663–675.
[617] M. Hirsch, R. Amaro, S. Antolovich, R. Neu, Coupled thermomechanical high cycle fatigue in a single crystal Ni-base superalloy, International Journal of Fatigue 62 (2014) 53–61.
[618] J. Hirth, J. Lothe, Theory of Dislocations, 2 edition, John Wiley & Sons, New York, 1982.
[619] N. Hiyoshi, M. Sakane, Tension-torsion multiaxial creep-fatigue for CMSX2 nickel base single crystal superalloy, Journal of the Society of Materials Science, Japan 50 (2004) 137–143.
[620] S. Hocker, H. Lipp, S. Schmauder, Precipitation, planar defects and dislocations in alloys: simulations on Ni_3Si and Ni_3Al precipitates, The European Physical Journal Special Topics 227 (14) (2019) 1559–1574.
[621] M. Hofmeister, M. Franke, C. Koerner, R. Singer, Single crystal casting with fluidized carbon bed cooling: a process innovation for quality improvement and cost reduction, Metallurgical and Materials Transactions A 48 (2017) 3132–3142.
[622] P. Hohenberg, W. Kohn, Inhomogeneous electron gas, Physical Review 136 (3B) (1964) B864.
[623] H.U. Hong, J.G. Kang, B.G. Choi, I.S. Kim, Y.S. Yoo, C.Y. Jo, A comparative study on thermomechanical and low cycle fatigue failures of a single crystal nickel-based superalloy, International Journal of Fracture 33 (12) (2011) 1592–1599.
[624] A.A. Hopgood, J.W. Martin, Coarsening of γ'-precipitates in single-crystal superalloy SRR 99, Materials Science and Technology 2 (6) (1986) 543–546.
[625] S. Hopkins, Low-cycle thermal mechanical fatigue testing, in: D. Spera, D. Mowbray (Eds.), Thermal Fatigue of Materials and Components, vol. STP 612, ASTM Int., 1976, pp. 157–169.
[626] O. Horst, B. Ruttert, D. Bürger, L. Heep, H. Wang, A. Dlouhý, W. Theisen, G. Eggeler, On the rejuvenation of crept Ni-base single crystal superalloys (SX) by hot isostatic pressing (HIP), Materials Science and Engineering: A 758 (2019) 202–214.
[627] N. Hou, W. Gou, Z. Wen, Z. Yue, The influence of crystal orientations on fatigue life of single crystal cooled turbine blade, Materials Science and Engineering: A 492 (2008) 413–418.
[628] N. Hou, Q. Yu, Z. Wen, Z. Yue, Low cycle fatigue behavior of single crystal superalloy with temperature gradient, European Journal of Mechanics - A/Solids 29 (2010) 611–618.
[629] F. Houllé, F. Walsh, A. Prakash, E. Bitzek, Atomistic simulations of compression tests on γ-precipitate containing Ni_3Al nanocubes, Metallurgical and Materials Transactions A 49 (9) (2018) 4158–4166.
[630] C. Howland, C. Brown, The effect of orientation on fatigue crack growth in a nickel-based single crystal superalloy, in: C. Beevers (Ed.), Fatigue 84, in: EMAS, vol. 3, 1984.
[631] CALculation of PHAse Diagram.
[632] DICTRA.
[633] http://www.zset-software.com/products/z mat/, 2019, Z-mat user manual. MINES ParisTech – ONERA.
[634] S.Y. Hu, L.Q. Chen, A phase field model for evolving microstructures with strong elastic inhomogeneity, Acta Materialia 49 (2001) 1879–1890.
[635] L. Huang, X. Sun, H. Guan, Z. Hu, Oxidation behavior of the single-crystal Ni-base superalloy DD32 in air at 900, 1000, and 1100°C, Oxidation of Metals 65 (2006) 391–408.

[636] M. Huang, Z. Li, Coupled ddd–fem modeling on the mechanical behavior of microlayered metallic multilayer film at elevated temperature, Journal of the Mechanics and Physics of Solids 85 (2015) 74–97.
[637] M. Huang, Z. Li, J. Tong, The influence of dislocation climb on the mechanical behavior of polycrystals and grain size effect at elevated temperature, International Journal of Plasticity 61 (2014) 112–127.
[638] M. Huang, L. Zhao, J. Tong, Discrete dislocation dynamics modelling of mechanical deformation of nickel-based single crystal superalloys, International Journal of Plasticity 28 (1) (2012) 141–158.
[639] D. Hull, D. Bacon, Introduction to Dislocations, 5 edition, Elsevier, Amsterdam, 2011.
[640] C. Humphreys, Fundamental concepts of stem imaging, Ultramicroscopy 7 (1) (1981) 7–12.
[641] C. Humphreys, J. Spencer, R. Woolf, D. Joy, J. Titchmarsh, G. Booker, Theory and practice of revealing crystallographic defects with the sem by means of diffraction contrast, in: Proceedings of the Fifth Annual Scanning Electron Microscope Symposium, vol. 5, IIT Research Institute Chicago, 1972, pp. 205–214.
[642] J. Hure, A coalescence criterion for porous single crystals, Journal of the Mechanics and Physics of Solids 124 (2019) 505–525.
[643] A.M. Hussein, J.A. El-Awady, Surface roughness evolution during early stages of mechanical cyclic loading, International Journal of Fatigue 87 (2016) 339–350.
[644] A.M. Hussein, S.I. Rao, M.D. Uchic, D.M. Dimiduk, J.A. El-Awady, Microstructurally based cross-slip mechanisms and their effects on dislocation microstructure evolution in fcc crystals, Acta Materialia 85 (2015) 180–190.
[645] A.M. Hussein, S.I. Rao, M.D. Uchic, T.A. Parthasarathy, J.A. El-Awady, The strength and dislocation microstructure evolution in superalloy microcrystals, Journal of the Mechanics and Physics of Solids 99 (2017) 146–162.
[646] N.S. Husseini, D.P. Kumah, J.Z. Yi, C.J. Torbet, D.A. Arms, E.M. Dufresne, T.M. Pollock, J.W. Jones, R. Clarke, Mapping single-crystal dendritic microstructure and defects in nickel-base superalloys with synchrotron radiation, Acta Materialia 56 (17) (2008) 4715–4723.
[647] J. Hutchinson, Bounds and self-consistent estimates for creep of polycrystalline materials, Proceedings of the Royal Society of London. Series A 348 (1966) 101–127.
[648] J. Hutchinson, Z. Suo, Mixed mode cracking in layered materials, in: Advances in Applied Mechanics, vol. 29, Elsevier, 1991, pp. 63–191.
[649] R.G. Hutchinson, J.W. Hutchinson, Lifetime assessment for thermal barrier coatings: tests for measuring mixed mode delamination toughness, Journal of the American Ceramic Society 94 (2011) s85–s95.
[650] W. Hüther, B. Reppich, Interaction of dislocations with coherent, stress-free ordered particles, Zeitschrift für Metallkunde 69 (1978) 628–634.
[651] W. Hüther, B. Reppich, Order hardening of MgO by large precipitated volume fractions of spinel particles, Materials Science and Engineering: A 39 (1979) 247–259.
[652] R. Hüttner, J. Gabel, U. Glatzel, R. Völkl, First creep results on thin-walled single-crystal superalloys, Materials Science and Engineering: A 510–511 (2009) 307–311.
[653] D. Hygate, Engine mro outllook – trends and challenges, in: SR Technics Symposium, Cork, 2011.
[654] T. Ichitsubo, H. Ogi, M. Hirao, K. Tanaka, M. Osawa, T. Yokokawa, T. Kobayashi, H. Harada, Elastic constant measurement of Ni-base superalloy with the RUS and mode selective EMAR methods, Ultrasonics 40 (1) (2002) 211–215.
[655] M. Ignat, J.-Y. Buffiere, J. Chaix, Microstructures induced by a stress gradient in a nickel-based superalloy, Acta Metallurgica Et Materialia 41 (3) (1993) 855–862.
[656] B. Ilschner, Hochtemperatur-Plastizität, Springer, 1973.
[657] ASTM E1875-20a, Standard Test Method for Dynamic Young's Modulus, Shear Modulus, and Poisson's Ratio by Sonic Resonance, ASTM International, West Conshohocken, PA, 2020, www.astm.org, https://doi.org/10.1520/E1875-20A.

[658] N. Irani, Finite strain discrete dislocation plasticity, PhD. thesis, Technische Universiteit Eindhoven, Eindhoven (The Netherlands), 2016.
[659] P. Jablonski, C. Cowen, Homogenizing a nickel-based superalloy: thermodynamic and kinetic simulation and experimental results, Metallurgical and Materials Transactions A 40 (2009) 182–186.
[660] G.A. Jackson, W. Sun, D.G. McCartney, The influence of microstructure on the ductile to brittle transition and fracture behaviour of HVOF NiCoCrAlY coatings determined via small punch tensile testing, Materials Science and Engineering: A 754 (2019) 479–490.
[661] J. Jackson, M. Donachie, M. Gell, R. Henricks, The effects of volume percent of fine γ' on creep in DS Mar-M200+Hf, Metallurgical Transactions 8A (1977) 1615–1620.
[662] R.W. Jackson, M.R. Begley, Critical cooling rates to avoid transient-driven cracking in thermal barrier coating (tbc) systems, International Journal of Solids and Structures 51 (6) (2014) 1364–1374.
[663] R.W. Jackson, E.M. Zaleski, B.T. Hazel, M.R. Begley, C.G. Levi, Response of molten silicate infiltrated gd2zr2o7 thermal barrier coatings to temperature gradients, Acta Materialia 132 (2017) 538–549.
[664] R.W. Jackson, E.M. Zaleski, D.L. Poerschke, B.T. Hazel, M.R. Begley, C.G. Levi, Interaction of molten silicates with thermal barrier coatings under temperature gradients, Acta Materialia 89 (2015) 396–407.
[665] A. Jacques, P. Bastie, The evolution of the lattice parameter mismatch of a nickel-based superalloy during a high-temperature creep test, Philosophical Magazine 83 (2003) 3005–3027.
[666] A. Jacques, F. Diologent, P. Bastie, In situ measurement of the lattice parameter mismatch of a nickel-base single-crystalline superalloy under variable stress, Materials Science and Engineering: A 387–389 (2004) 944–949.
[667] A. Jacques, F. Diologent, P. Caron, P. Bastie, Mechanical behavior of a superalloy with a rafted microstructure: in situ evaluation of the effective stresses and plastic strain rates of each phase, Materials Science and Engineering: A 483–484 (2008) 568–571.
[668] A. Jadhav, N.P. Padture, F. Wu, E.H. Jordan, M. Gell, Thick ceramic thermal barrier coatings with high durability deposited using solution-precursor plasma spray, Materials Science and Engineering: A 405 (1–2) (2005) 313–320.
[669] A. James, Review of rejuvenation process for nickel base superalloys, Materials Science and Technology 17 (2001) 481–486.
[670] O. Jamond, R. Gatti, A. Roos, B. Devincre, Consistent formulation for the discrete-continuous model: improving complex dislocation dynamics simulations, International Journal of Plasticity 80 (2016) 19–37.
[671] J. Jaroszewicz, H. Matysiak, J. Michalski, K. Matuszewski, K. Kubiak, K. Kurzydlowski, Characterization of single-crystal dendrite structure and porosity in nickel-based superalloys using X-ray micro-computed tomography, Advanced Materials Research 278 (2011) 66–71.
[672] C. Jiang, B. Gleeson, Site preference of transition metal elements in Ni_3Al, Scripta Materialia 55 (2006) 433–436.
[673] C. Jiang, D. Sordelet, B. Gleeson, Site preference of ternary alloying elements in Ni_3Al: a first-principles study, Acta Materialia 54 (4) (2006) 1147–1154.
[674] F. Jing, J. Yang, Z. Yang, W. Zeng, Critical compressive strain and interfacial damage evolution of EB-PVD thermal barrier coating, Materials Science and Engineering: A (2020) 139038.
[675] M. Jinnestrand, H. Brodin, Crack initiation and propagation in air plasma sprayed thermal barrier coatings, testing and mathematical modelling of low cycle fatigue behaviour, Materials Science and Engineering: A 379 (1–2) (2004) 45–57.
[676] F. Jioa, D. Bettge, W.W. Osterle, J. Ziebs, Tension-compression asymmetry of the (001) single crystal nickel base superalloy SC16 under cyclic loading at elevated temperatures, Acta Materialia 44 (1996) 3933–3942.

[677] J. Jones, D. McKay, Neural network modelling of the mechanical properties of nickel base superalloys, in: R.D. Kissinger, D.J. Deye, D.L. Anton, A.D. C&l, M.V. Nathal, T.M. Pollock, D.A. Woodford (Eds.), Superalloys 1996: Proc. of the Eighth International Symposium on Superalloys, Champion, USA, TMS, Minerals, Metals & Materials Society, Sept. 1996, pp. 417–424.
[678] K.P. Jonnalagadda, R. Eriksson, X.-H. Li, R.L. Peng, Fatigue life prediction of thermal barrier coatings using a simplified crack growth model, Journal of the European Ceramic Society 39 (5) (2019) 1869–1876.
[679] B. Joós, Q. Ren, M.S. Duesbery, Peierls-Nabarro model of dislocations in silicon with generalized stacking-fault restoring forces, Physical Review B 50 (9) (1994) 5890–5898.
[680] E. Jordan, S. Shi, K. Walker, The viscoplastic behavior of Hastelloy-X single crystal, International Journal of Plasticity 9 (1993) 119–139.
[681] E. Jordan, K. Walker, A viscoplastic model for single crystal, Journal of Engineering Materials and Technology 114 (1992) 19–26.
[682] E.H. Jordan, C. Jiang, J. Roth, M. Gell, Low thermal conductivity yttria-stabilized zirconia thermal barrier coatings using the solution precursor plasma spray process, Journal of Thermal Spray Technology 23 (5) (2014) 849–859.
[683] B. Jouffrey, R.A. Portier, Diffraction dans les métaux et alliages: conditions de diffraction, 2007.
[684] M. Jouiad, J. Ghighi, J. Cormier, E. Ostoja-Kuczynski, G. Lubineau, J. Mendez, 3D imaging using x-ray tomography and SEM combined FIB to study non isothermal creep damage of (111) oriented samples of γ/γ' nickel base single crystal superalloy MC2, Materials Sciences Forum 2400 (2012) 706–709.
[685] M. Kabir, T.T. Lau, D. Rodney, S. Yip, K.J. Van Vliet, Predicting dislocation climb and creep from explicit atomistic details, Physical Review Letters 105 (Aug 2010) 095501.
[686] H. Kagawa, Y. Mukai, The effect of crystal orientation and temperature on fatigue crack growth of ni-based single crystal superalloy, in: E.S. Huron, R.C. Reed, M.C. Hardy, M.J. Mills, R.E. Montero, P.D. Portella, J. Telesman (Eds.), Superalloys 2012: Proc. of the Twelfth International Symposium on Superalloys, Champion, USA, TMS, Minerals, Metals & Materials Society, Sept. 2012.
[687] K. Kakehi, Influence of secondary precipitates and crystallographic orientation on the strength of single crystals of a Ni-based superalloy, Metallurgical and Materials Transactions A 30 (5) (1999) 1249–1259.
[688] K. Kakehi, Effect of primary and secondary precipitates on creep strength of Ni-base superalloy single crystals, Materials Science and Engineering: A 278 (1–2) (2000) 135–141.
[689] T. Kalfhaus, M. Schneider, B. Ruttert, D. Sebold, T. Hammerschmidt, J. Frenzel, R. Drautz, W. Theisen, G. Eggeler, O. Guillon, R. Vassen, Repair of Ni-based single-crystal superalloys using vacuum plasma spray, Materials & Design 168 (2019) 107656.
[690] M. Kamaraj, Rafting in single crystal nickel-base superalloys – an overview, Sadhana-Academy Proceedings in Engineering Sciences 28 (2003) 115–128.
[691] M. Kamaraj, C. Mayr, M. Kolbe, G. Eggeler, On the influence of stress state on rafting in the single crystal superalloy CMSX-6 under conditions of high temperature and low stress creep, Scripta Metallurgica 38 (1998) 589–594.
[692] M. Kamaraj, K. Serin, M. Kolbe, G. Eggeler, Shear creep deformation of the superalloy single crystal CMSX-4, Materialwissenschaft und Werkstofftechnik 48 (2003) 469–477.
[693] M. Kaminski, P. Kanouté, S. Kruch, E. Busso, J.-L. Chaboche, A high temperature fatigue damage model for single crystal superalloys, Materials at High Temperatures 33 (2016) 412–424.
[694] R. Kampmann, T. Lippmann, J. Burmester, J. dos Santos, H. Franz, M. Haese-Seiller, M. Marmotti, Upgrading of the petra-2 beamline at hasylab for materials science analyses, Nuclear Instruments & Methods in Physics Research. Section A, Accelerators, Spectrometers, Detectors and Associated Equipment 467 (2001) 1261–1264.

[695] M. Kanda, M. Sakane, T. Ohnami, M. Hasebe, High temperature multiaxial low cycle fatigue of CMSX-2 Ni-base single crystal superalloy, Journal of Engineering Materials and Technology 119 (1997) 153–160.

[696] K. Kang, J. Yin, W. Cai, Stress dependence of cross slip energy barrier for face-centered cubic nickel, Journal of the Mechanics and Physics of Solids 62 (2014) 181–193, Sixtieth anniversary issue in honor of Professor Rodney Hill.

[697] A.M. Karlsson, A. Evans, A numerical model for the cyclic instability of thermally grown oxides in thermal barrier systems, Acta Materialia 49 (10) (2001) 1793–1804.

[698] A.M. Karlsson, J. Hutchinson, A. Evans, A fundamental model of cyclic instabilities in thermal barrier systems, Journal of the Mechanics and Physics of Solids 50 (8) (2002) 1565–1589.

[699] S. Karthikeyan, R. Unocic, P. Sarosi, G. Viswanathan, D. Whitis, M. Mills, Modeling microtwinning during creep in Ni-based superalloys, Scripta Metallurgica 54 (2006) 1157–1162.

[700] M. Karunaratne, P. Carter, R. Reed, Interdiffusion in the faced-centered cubic phase of the Ni-Re, Ni-Ta and Ni-W systems between 900 and 1300°C, Materials Science and Engineering: A 281 (2000) 229–233.

[701] M. Karunaratne, D. Cox, P. Carter, R. Reed, Modelling of the microsegregation in CMSX-4 superalloy and its homogenisation during heat treatment, in: T.M. Pollock, R.D. Kissinger, R.R. Bowman, K.A. Green, M. McLean, S.L. Olson, J.J. Schirra (Eds.), Superalloys 2000: Proc. of the 9th International Symposium on Superalloys, Champion, USA, TMS, Minerals, Metals & Materials Society, Sept. 1992, pp. 263–272.

[702] M. Karunaratne, R. Reed, Interdiffusion in the FCC-A1 phase of the Ni-Re, Ni-Ta and Ni-W systems between 900 and 1300°C, Materials Science and Engineering: A 281 (2000) 229–233.

[703] M. Karunaratne, R. Reed, Interdiffusion of the platinum group metals in nickel at elevated temperatures, Acta Materialia 51 (2003) 905–919.

[704] S. Katnagallu, L. Stephenson, I. Mouton, C. Freysoldt, A. Subramanyam, J. Jenke, A. Ladines, S. Neumeier, T. Hammerschmidt, R. Drautz, J. Neugebauer, F. Vurpillot, D. Raabe, B. Gault, Imaging individual solute atoms at crystalline imperfections in metals, New Journal of Physics 21 (2019) 123020.

[705] L. Kaufman, H. Bernstein, Computer Calculations of Phase Diagrams, Academic Press, New York, 1970.

[706] K. Kawagishi, A.-C. Ye, T. Yokokawa, Y. Koizumi, H. Hrada, Development of an oxidation resistant high strength sixth generation superalloys TMS238, in: E. Huron, R. Reed, M. Hardy, M. Mills, R. Montero, P. Portella, J. Telesman (Eds.), Proc. 12th Int. Symp. on Superalloys, TMS, John Wiley& Sons, 2012.

[707] J.H. Ke, A. Boyne, Y. Wang, C.R. Kao, Phase field microelasticity model of dislocation climb: methodology and applications, Acta Materialia 79 (2014) 396–410.

[708] B. Kear, G. Leverant, J. Oblak, An analysis of creep-induced intrinsic/extrinsic fault pairs in a precipitation hardened nickel-base alloy, Transactions of American Society for Metals 62 (1969) 639–650.

[709] B. Kear, J. Oblak, A. Giamei, Stacking faults in γ' $Ni_3(Al, Ti)$ precipitation hardened Ni-base alloys, Metallurgical Transactions 1 (1970) 2477–2486.

[710] B. Kear, H. Wilsdorf, Dislocation configurations in plastically deformed Cu_3Au alloys, Transactions of American Society for Metals 224 (1962) 382–386.

[711] R. Keller, H.H.J. Maier, H. Mughrabi, Characterization of interfacial dislocation networks in a creep deformed Ni-base superalloy, Scripta Metallurgica 28 (1993) 23–28.

[712] S. Keralavarma, A. Benzerga, High-temperature discrete dislocation plasticity, Journal of the Mechanics and Physics of Solids 82 (2015) 1–22.

[713] S.M. Keralavarma, T. Cagin, A. Arsenlis, A.A. Benzerga, Power-law creep from discrete dislocation dynamics, Physical Review Letters 109 (2012) 265504.

[714] R. Kersey, A. Staroselsky, D. Dudzinski, M. Genest, Thermomechanical fatigue crack growth from laser drilled holes in single crystal nickel based superalloy, International Journal of Fatigue 55 (2013) 183–193.
[715] S. Keshavarz, S. Ghosh, A.C. Reid, S.A. Langer, A non-Schmid crystal plasticity finite element approach to multi-scale modeling of nickel-based superalloys, Acta Materialia 114 (Supplement C) (2016) 106–115.
[716] A.G. Khachaturyan, The Theory of Structural Transformations in Solids, Wiley, New York, 1983.
[717] T. Khan, P. Caron, Effect of processing conditions and heat treatments on mechanical properties of single-crystal superalloy cmsx-2, Materials Science and Technology 2 (5) (1986) 486–492.
[718] T. Khan, P. Caron, Single crystal superalloys for turbine blades in advanced aircraft engines, in: Proc. of 15th ICAS Conference, CAS, 1986.
[719] D. Khoshkhou, M. Mostafavi, C. Reinhard, M. Taylor, D. Rickerby, I. Edmonds, H. Evans, T. Marrow, B. Connolly, Three-dimensional displacement mapping of diffused pt thermal barrier coatings via synchrotron x-ray computed tomography and digital volume correlation, Scripta Materialia 115 (2016) 100–103.
[720] D. Kim, S.-L. Shang, Z.-K. Liu, Effects of alloying elements on thermal expansions of γ-Ni and γ'-Ni_3Al by first-principles calculations, Acta Materialia 60 (4) (2012) 1846–1856.
[721] D. Kim, S. Zhang, Z. Liu, Effects of alloying elements on elastic properties of Ni_3Al by first-principles calculations, Intermetallics 18 (2010) 1163–1171.
[722] S.G. Kim, W.T. Kim, T. Suzuki, Phase-field model for binary alloys, Physical Review E 60 (6) (1999) 7186–7197.
[723] S.-S. Kim, Y.-F. Liu, Y. Kagawa, Evaluation of interfacial mechanical properties under shear loading in EB-PVD tbcs by the pushout method, Acta Materialia 55 (11) (2007) 3771–3781.
[724] J. Kimmel, Z. Mutasim, W. Brentnall, Effects of alloy composition on the performance of yttria stabilized zirconia—thermal barrier coatings, Journal of Engineering for Gas Turbines and Power 122 (3) (2000) 393–400.
[725] A. Kirchmayer, H. Lyu, M. Pröbstle, F. Houllé, A. Förner, D. Hünert, M. Göken, P. Felfer, E. Bitzek, S. Neumeier, Combining experiments and atom probe tomography-informed simulations on γ' precipitation strengthening in the polycrystalline Ni-base superalloy A718Plus, Advanced Engineering Materials (2020), in press.
[726] M. Kirka, K. Brindley, R. Neu, S. Antolovich, S. Shinde, P. Gravett, Influence of coarsened and rafted microstructures on the thermomechanical fatigue of a ni-base superalloy, International Journal of Fracture 81 (2015) 191–201.
[727] M. Kirka, D. Smith, R. Neu, Efficient methodologies for determining temperature-dependent parameters of a Ni-base superalloy crystal viscoplasticity model for cyclic loadings, Journal of Engineering Materials and Technology 136 (2014) 041001.
[728] J. Kirkaldy, D. Young, Diffusion in the Condensed State, CRC Press, 1987.
[729] T. Kitashima, H. Harada, A new phase-field method for simulating γ' precipitation in multicomponent nickel-base superalloys, Acta Materialia 57 (2009) 2020–2028.
[730] K. Knipe, A. Manero, S.F. Siddiqui, C. Meid, J. Wischek, J. Okasinski, J. Almer, A.M. Karlsson, M. Bartsch, S. Raghavan, Strain response of thermal barrier coatings captured under extreme engine environments through synchrotron x-ray diffraction, Nature Communications 5 (1) (2014) 1–7.
[731] M. Knoll, Aufladepotentiel und sekundäremission elektronenbestrahlter körper, Zeitschrift für technische Physik 16 (1935) 467–475.
[732] U. Kocks, The relation between polycrystal deformation and single-crystal deformation, Metallurgical Transactions 1 (1970) 1121–1143.
[733] U. Kocks, A. Argon, M. Ashby, in: B. Chalmers, J. Christian, T. Massalski (Eds.), Thermodynamics and Kinetics of Slip, in: Progress in Materials Science, vol. 19, Pergamon Press, 1975.

[734] U. Kocks, H. Mecking, Physics and phenomenology of strain hardening: the fcc case, Progress in Materials Science 48 (3) (2003) 171–273.
[735] C. Kohler, P. Kizler, S. Schmauder, Atomistic simulation of the pinning of edge dislocations in Ni by Ni3Al precipitates, Materials Science and Engineering: A 400 (4001) (2005) 481–484.
[736] C. Kohler, T. Link, A. Epishin, Dissociation of a ⟨100⟩ edge superdislocations in the γ'-phase of nickel-base superalloys, Philosophical Magazine 86 (32) (2006) 5103–5121.
[737] W. Kohn, L. Sham, Self-consistent equations including exchange and correlation effects, Physical Review 140 (4A) (1965) A1133.
[738] W. Koiter, General theorems for elastic–plastic solids, in: Progress in Solid Mechanics, vol. 6, North–Holland Publishing Company, 1960, pp. 167–221.
[739] Y. Koizumi, T. Kobayashi, T. Yokokawa, J. Zhang, M. Osawa, H. Harada, Y. Aoki, M. Arai, Development of next generation Ni-base single crystal superalloys, in: K.A. Green, T.M. Pollock, H. Harada, T.E. Howson, R.C. Reed, J.J. Schirra, S. Walston (Eds.), Superalloys 2004: Proc. of the Tenth International Symposium on Superalloys, Champion, USA, TMS, Minerals, Metals & Materials Society, Sept. 2004, pp. 35–43.
[740] M. Kolbe, The high temperature decrease of the critical resolved shear stress in nickel-base superalloys, Materials Science and Engineering: A 319–321 (2001) 383–387.
[741] M. Kolbe, A. Dlouhy, G. Eggeler, Dislocation reactions at γ/γ'-interfaces during shear creep deformation in the macroscopic crystallographic shear system (001)[110] of CMSX6 superalloy single crystals at 1025°C, Materials Science and Engineering: A 246 (1–2) (1998) 133–142.
[742] M. Kolbe, K. Neuking, G. Eggeler, Dislocation reactions and microstructural instability during 1025°C shear creep testing of superalloy single crystals, Materials Science and Engineering: A 234–236 (1997) 877–879.
[743] J. Komenda, P. Henderson, Growth of pores during the creep of a single crystal nickel-base superalloy, Scripta Metallurgica 37 (1997) 1821–1826.
[744] H. Kondo, M. Wakeda, I. Watanabe, Atomic study on the interaction between superlattice screw dislocation and γ-Ni precipitate in γ'-Ni_3Al intermetallics, Intermetallics 102 (July 2018) 1–5.
[745] P. Kontis, Interactions of solutes with crystal defects: a new dynamic design parameter for advanced alloys, Scripta Materialia 194 (2021) 113626.
[746] P. Kontis, Z. Li, D.M. Collins, J. Cormier, D. Raabe, B. Gault, The effect of chromium and cobalt segregation at dislocations on nickel-based superalloys, Scripta Materialia 145 (2018) 76–80.
[747] F. Körmann, A. Dick, T. Hickel, J. Neugebauer, Role of spin quantization in determining the thermodynamic properties of magnetic transition metals, Physical Review B 83 (2011) 165114.
[748] F. Körmann, T. Hickel, J. Neugebauer, Influence of magnetic excitations on the phase stability of metals and steels, Current Opinion in Solid State & Materials Science 20 (2) (2016) 77–84.
[749] C. Körner, M. Ramsperger, C. Mied, D. Bürger, P. Wollgramm, M. Bartsch, G. Eggeler, Microstructure and mechanical properties of CMSX-4 single crystals prepared by additive manufacturing, Metallurgical and Materials Transactions A 49 (9) (2018) 3781–3792.
[750] R. Kornfeld, Fluoride ion cleaning as a prebraze process, Heat Treating Progress 38–39 (2006).
[751] A. Koster, Micromod-sx project, predictive microstructural assessment and micromechanical modelling of deformation and damage accumulation in single crystal gas turbine blading, be 96-3911, impcol, armines, immg, kwu, egt, cnr-itm, ean, 1997-2001, Technical report, Mines ParisTech, 2001.
[752] A. Koster, E. Fleury, E. Vasseur, L. Rémy, Automation in fatigue and fracture: testing and analysis, in: C. Amzallag (Ed.), Thermal-Mechanical Fatigue Testing, vol. STP 1231, ASTM, 1994, pp. 559–576.

[753] G. Kostorz, 13 - x-ray and neutron scattering, in: D.E. Laughlin, K. Hono (Eds.), Physical Metallurgy, fifth edition, Elsevier, Oxford, 2014, pp. 1227–1316.

[754] A. Kounitzky, J. Wortmann, P. Agarwal, A single crystal casting process for high-temperature components, Materials & Design 12 (1991) 323–330.

[755] L. Kovarik, R. Unocic, J. Li, M. Mills, The intermediate temperature deformation of Ni- based superalloys: importance of reordering, JOM 61 (2009) 42–48.

[756] L. Kovarik, R. Unocic, J. Li, P. Sarosi, C. Shen, Y. Wanga, M. Mills, Microtwinning and other shearing mechanisms at intermediate temperatures in Ni-based superalloys, Progress in Materials Science 54 (2009) 839–873.

[757] R. Kozar, A. Suzuki, W. Milligan, J. Shirra, M. Savage, T. Pollock, Strengthening mechanisms in polycrystalline multimodal nickel-base superalloys, Metallurgical and Materials Transactions A 40 (2009) 1588–1603.

[758] A. Kracke, Superalloys, the most successful alloy system of modern times-past, present and future, in: Proc. 7th Int. Symp. on Superalloys 718 and Derivatives, TMS, 2010.

[759] S. Kraft, I. Altenberger, H. Mughrabi, Directional γ/γ' coarsening in a monocrystalline nickel-based superalloy during low-cycle thermomechanical fatigue, Scripta Materialia 32 (3) (1995) 411–416.

[760] S. Krämer, S. Faulhaber, M. Chambers, D. Clarke, C. Levi, J. Hutchinson, A. Evans, Mechanisms of cracking and delamination within thick thermal barrier systems in aero-engines subject to calcium-magnesium-alumino-silicate (CMAS) penetration, Materials Science and Engineering: A 490 (1–2) (2008) 26–35.

[761] S. Krämer, J. Yang, C.G. Levi, Infiltration-inhibiting reaction of gadolinium zirconate thermal barrier coatings with CMAS melts, Journal of the American Ceramic Society 91 (2) (2008) 576–583.

[762] S. Krämer, J. Yang, C.G. Levi, C.A. Johnson, Thermochemical interaction of thermal barrier coatings with molten CaO–MgO–Al2O3–SiO2 (CMAS) deposits, Journal of the American Ceramic Society 89 (10) (2006) 3167–3175.

[763] A.R. Krause, X. Li, N.P. Padture, Interaction between ceramic powder and molten calcia-magnesia-alumino-silicate (CMAS) glass, and its implication on CMAS-resistant thermal barrier coatings, Scripta Materialia 112 (2016) 118–122.

[764] A.S. Krausz, H. Eyring, Deformation Kinetics, 1 edition, John Wiley and Sons, New York, 1975.

[765] G. Kresse, J. Furthmüller, J. Hafner, Ab initio force constant approach to phonon dispersion relations of diamond and graphite, Europhysics Letters 32 (9) (1995) 729–734.

[766] N. Krieger Lassen, The relative precision of crystal orientations measured from electron backscattering patterns, Journal of Microscopy 181 (1) (1996) 72–81.

[767] R. Kromer, J. Cormier, S. Costil, D. Courapied, L. Berthe, P. Peyre, High temperature durability of a bond-coatless plasma-sprayed thermal barrier coating system with laser textured ni-based single crystal substrate, Surface & Coatings Technology 337 (2018) 168–176.

[768] L. Kubin, Dislocations, mesoscale simulations and plastic flow, in: A. Sutton, R. Rudd (Eds.), Oxford Series on Materials Modelling, Oxford University Press, 2013.

[769] L. Kubin, B. Lisiecki, P. Caron, Octahedral slip instabilities in γ/γ'superalloy single crystals cmsx-2 and am3, Philosophical Magazine. A 71 (5) (1995) 991–1009.

[770] L.P. Kubin, G. Canova, M. Condat, B. Devincre, V. Pontikis, Y. Bréchet, Dislocation microstructures and plastic flow: a 3d simulation, in: Non Linear Phenomena in Materials Science II, in: Solid State Phenomena, vol. 23, Trans Tech Publications, 1992, pp. 455–472.

[771] L.P. Kubin, B. Devincre, T. Hoc, Inhibited dynamic recovery and screw dislocation annihilation in multiple slip of fcc single crystals, Philosophical Magazine 86 (25–26) (2006) 4023–4036.

[772] H.-A. Kuhn, H. Biermann, T. Ungar, H. Mughrabi, An x-ray study of creep-deformation induced changes of the lattice mismatch in the γ' -hardened monocrystalline nickel-base superalloy srr 99, Acta Metallurgica 39 (11) (1991) 2783–2794.
[773] H.-A. Kuhn, H.G. Sockel, Contributions of the different phases of two nickel-base superalloys to the elastic behaviour in a wide temperature range, Physica Status Solidi A 119 (1) (1990) 93–105.
[774] U. Kulkarni, S. Muralidhar, S. Banerjee, Computer simulation of the early stages of ordering in ni mo alloys, Physica Status Solidi A 110 (2) (1988) 331–345.
[775] K. Kumar, R. Sankarasubramanian, U. Waghmareb, Tuning planar fault energies of Ni_3Al with substitutional alloying: first-principles description for guiding rational alloy design, Scripta Materialia 142 (2018) 74–78.
[776] R. Kumar, A.-J. Wang, D. McDowell, Effects of microstructure variability on intrinsic fatigue resistance of nickel-base superalloys – a computational micromechanics approach, International Journal of Fatigue 137 (2006) 173–210.
[777] J. Kundin, L. Mushongera, T. Goehler, H. Emmerich, Phase-field modeling of the γ'-coarsening behavior in Ni-based superalloys, Acta Materialia 60 (2012) 3758–3772.
[778] J. Kundin, R. Siquieri, Phase-field model for multiphase systems with different thermodynamic factors, Physica D 240 (2011) 459–469.
[779] K. Kunze, S. Wright, B.L. Adams, D.J. Dingley, Advances in automatic ebsp single orientation measurements, in: Textures and Microstructures, vol. 20, 1970.
[780] W. Kurz, D. Fisher, Fundamentals of Solidification, Switzerland Trans Tech Publications Ltd, Aedermannsdorf, 1989.
[781] R. Labusch, A statistical theory of solid solution hardening, Physica Status Solidi 41 (1970) 659–669.
[782] G. Lai, Hot corrosion in gas turbines, Chapter 9, ASM International, 2007, pp. 249–258.
[783] B. Lakshminarayana, Fluid Dynamics and Heat Transfer of Turbomachinery, Wiley&Sons, 1996.
[784] C. Lall, S. Chin, D. Pope, The orientation and temperature dependence of the yield stress of $Ni_3(AlNb)$ single crystals, Metallurgical Transactions A 10 (1979) 1323–1332.
[785] M. Lamm, R. Singer, The effect of casting conditions on the high-cycle fatigue properties of the single-crystal nickel-base superalloy PWA 1483, Metallurgical and Materials Transactions A 38 (2007) 1177–1183.
[786] L. Landau, E. Lifshitz, Statistical Physics Part 1, third revised edition, Pergamon, Oxford, 1980, 1959.
[787] L.D. Landau, On the theory of phase transitions I, Physikalische Zeitschrift der Sowjetunion 11 (1937) 26.
[788] L.D. Landau, Zur theorie der phasenumwandlungen II, Physikalische Zeitschrift der Sowjetunion 11 (1937) 545.
[789] A. Landefeld, W. Mook, J. Rösler, J. Michler, Compression experiments on γ'-nanoparticles, ISRN Nanomaterials 2012 (2012) 1–4.
[790] C. Lane, The development of a 2D ultrasonic array inspection for single crystal superalloy, PhD thesis, University of Bristol, England, 2013.
[791] J. Langer, Models of pattern formation in first-order phase transitions, in: G.M.G. Grinstein (Ed.), Directions in Condensed Matter Physics, World Scientific, Singapore, 1986, pp. 165–185.
[792] L. Langston, Each blade a single crystal, Proceedings of the American Academy of Arts and Sciences 103 (2015) 30–37.
[793] G. Laplanche, N. Wieczorek, F. Fox, S. Berglund, J. Pfetzing-Micklich, K. Kishida, H. Inuic, G. Eggeler, On the influence of crystallography and dendritic microstructure on micro shear behavior of single crystal ni-based superalloys, Acta Materialia 160 (2018) 173–184.
[794] A. Lasalmonie, J.-L. Strudel, Interfacial dislocation networks around γ' precipitates in nickel-base alloys, Philosophical Magazine 32 (1975) 937–949.

[795] N.K. Lassen, Automatic crystal orientation determination from ebsps, Micron and Microscopica Acta 23 (1992) 191–192.
[796] A. Laurence, J. Cormier, P. Villechaise, T. Billot, J.-M. Franchet, F. Pettinari-Sturmel, M. Hantcherli, F. Mompiou, A. Wessman, Impact of the solution cooling rate and of thermal aging on the creep properties of the new cast & wrought René 65 Ni-based superalloys, in: 8th Int. Symp. on Superalloy 718 and Derivatives, Pittsburg, USA, TMS, 2014, pp. 333–348.
[797] J.-C. Lautridou, J.-Y.D.J. Guedou, Comparison of single crystal superalloys for turbine blades through TMF tests, in: Proc. Fatigue under Thermal and Mechanical Loading: Mechanisms, Mechanics and Modelling, May 22–24, 1995.
[798] O. Lavigne, C. Ramusat, S. Drawin, P. Caron, D. Boivin, J. Pouchou, Relationships between microstructural instabilities and mechanical behaviour in new generation nickel-based single crystal superalloys, in: Superalloys 2004, 2004, p. 667.
[799] J.-B. le Graverend, Experimental analysis and modeling of the effects of high temperature incursions on the mechanical behavior of a single crystal superalloy for turbine blades, PhD thesis, ISAE-ENSMA, France, 2013.
[800] J.-B. le Graverend, J. Adrien, J. Cormier, Ex-situ x-ray tomography characterization of porosity during high-temperature creep in a ni-based single-crystal superalloy: toward understanding what is damage, Materials Science and Engineering: A 695 (2017) 367–378.
[801] J.-B. le Graverend, J. Cormier, P. Caron, S. Kruch, F. Gallerneau, J. Mendez, Numerical simulation of γ/γ' microstructural evolutions induced by TCP-phase in the MC2 nickel base single crystal superalloy, Materials Science and Engineering: A 528 (6) (2011) 2620–2634.
[802] J.-B. le Graverend, J. Cormier, F. Gallerneau, S. Kruch, J. Mendez, Highly non-linear creep life induced by a short close γ'-solvus overheating and a prior microstructure degradation on a nickel-based single crystal superalloy, Materials & Design 56 (2014) 990–997.
[803] J.-B. le Graverend, J. Cormier, F. Gallerneau, S. Kruch, J. Mendez, Strengthening behavior in non-isothermal monotonic and cyclic loading in a ni-based single crystal superalloy, International Journal of Fracture 91 (2016) 257–263.
[804] J.B. le Graverend, J. Cormier, F. Gallerneau, P. Paulmier, Dissolution of fine γ' precipitates of mc2 ni-based single-crystal superalloy in creep-fatigue regime, in: Advanced Materials Research, vol. 278, Trans Tech Publ, 2011, pp. 31–36.
[805] J.-B. le Graverend, J. Cormier, F. Gallerneau, P. Villechaise, S. Kruch, J. Mendez, A microstructure-sensitive constitutive modeling of the inelastic behavior of single crystal nickel-based superalloys at very high temperature, International Journal of Plasticity 59 (2014) 55–83.
[806] J.-B. le Graverend, J. Cormier, M. Jouiad, F. Gallerneau, P. Paulmier, F. Hamon, Effect of fine γ' precipitation on non-isothermal creep and creep-fatigue behaviour of nickel base superalloy MC2, Materials Science and Engineering: A 527 (20) (2010) 5295–5302.
[807] J.-B. le Graverend, J. Cormier, S. Kruch, F. Gallerneau, J. Mendez, Microstructural parameters controlling high-temperature creep life of the nickel-base single-crystal superalloy MC2, Metallurgical and Materials Transactions A 43 (11) (2012) 3988–3997.
[808] J.-B. le Graverend, L. Dirand, A. Jacques, J. Cormier, O. Ferry, T. Schenk, F. Gallerneau, S. Kruch, J. Mendez, In situ measurement of the γ' lattice mismatch evolution of a Ni-based single-crystal superalloy during non-isothermal very high-temperature creep experiments, Metallurgical and Materials Transactions A 43 (2012) 3946–3951.
[809] J.-B. le Graverend, A. Jacques, J. Cormier, O. Ferry, T. Schenk, J. Mendez, Creep of a nickel-based single-crystal superalloy during very high temperature jumps followed by synchrotron x-ray diffraction, Acta Materialia 84 (2015) 65–79.
[810] J.-B. le Graverend, S. Lee, Phenomenological modeling of the effect of oxidation on the creep response of Ni-based single-crystal superalloys, in: Superalloys 2020, Springer, Cham, 2020, pp. 282–291.

[811] J.-B. le Graverend, F. Pettinari-Sturmel, J. Cormier, M. Hantcherli, P. Villechaise, J. Douin, Mechanical twinning in Ni-based single crystal superalloys during multiaxial creep at 1050°C, Materials Science and Engineering: A 722 (February) (2018) 76–87.

[812] Y. Le Guevel, B. Grégoire, B. Bouchaud, P. Bilhé, A. Pasquet, M. Thiercelin, F. Pedraza, Influence of the oxide scale features on the electrochemical descaling and stripping of aluminide coatings, Surface & Coatings Technology 292 (2016) 1–10.

[813] A. LeClaire, A. Lidiard, LIII. Correlation effects in diffusion in crystals, Philosophical Magazine 1 (6) (1956) 518–527.

[814] E.H. Lee, Elastic-plastic deformation at finite strains, Journal of Applied Mechanics 36 (1) (1969) 1–6.

[815] M. Legros, N. Clement, P. Caron, A. Coujou, In-situ observation of deformation micromechanisms in a rafted γ/γ' superalloy at 850°C, Materials Science and Engineering: A 337 (2002) 160–169.

[816] D. Leidermark, J. Moverare, S. Johansson, K. Simonsson, S. Sjöström, Tension/compression asymmetry of a single-crystal superalloy in virgin and degraded condition, Acta Materialia 58 (2010) 4986–4997.

[817] D. Leidermark, M. Segersäll, Modelling of thermomechanical fatigue stress relaxation in a single-crystal nickel-base superalloy, Computational Materials Science 90 (2014) 61–70.

[818] J. Lemaitre, J. Chaboche, Mécanique des Matériaux Solides, Dunod, 1985.

[819] C. Lemarchand, B. Devincre, L. Kubin, Homogenization method for a discrete-continuum simulation of dislocation dynamics, Journal of the Mechanics and Physics of Solids 49 (2001) 1969.

[820] D. Lempidaki, E. Busso, N. O'Dowd, Application of non-contact strain measurement techniques to a single crystal alloy at elevated temperature, in: Proc. 11th International Conference on Fracture–ICF 11, Torino, Italy, 2005.

[821] P.H. Leo, W.W. Mullins, R.F. Sekerka, J. Vinals, Effect of elasticity on the late stage coarsening, Acta Materialia 38 (8) (1990) 1573–1580.

[822] B. Lerch, S. Antolovich, Fatigue crack propagation behavior of a single crystalline superalloy, Metallurgical and Materials Transactions A 21 (1990) 2169–2177.

[823] G. Leverant, B. Kear, The mechanism of creep in γ' precipitation-hardened nickel-base alloys at intermediate temperature, Metallurgical Transactions 1 (1970) 491–498.

[824] G. Leverant, B. Kear, J. Oblak, Creep of precipitation hardened Ni-base alloy single crystals at high temperatures, Metallurgical Transactions 4 (1973) 355–362.

[825] C.G. Levi, J.W. Hutchinson, M.-H. Vidal-Sétif, C.A. Johnson, Environmental degradation of thermal-barrier coatings by molten deposits, MRS Bulletin 37 (10) (2012) 932–941.

[826] V. Levkovitch, R. Sievert, B. Svendsen, Simulation of deformation and lifetime behavior of a fcc single crystal superalloy at high temperature under low-cycle fatigue loading, International Journal of Fatigue 28 (2006) 1791–1801.

[827] G. Leyson, W. Curtin, L. Hector, C. Woodward, Quantitative prediction of solute strengthening in aluminium alloys, Nature Materials 9 (9) (oct 2010) 750–755.

[828] J. Li, Impression creep and other localized tests, Materials Science and Engineering: A 322 (2010) 23–42.

[829] J. Li, H. Proudhon, A. Roos, V. Chiaruttini, S. Forest, Crystal plasticity finite element simulation of crack growth in single crystals, Computational Materials Science 94 (2014) 191–197.

[830] J. Li, S. Sarkar, W. Cox, T. Lenosky, E. Bitzek, Y. Wang, Diffusive molecular dynamics and its application to nanoindentation and sintering, Physical Review B 84 (5) (2011) 1.

[831] M. Li, X. Sun, Z. Li, Z. Zhang, T. Jin, H. Guan, Z. Hu, Oxidation behavior of a single-crystal Ni-base superalloy in air. I: at 800 and 900°C, Oxidation of Metals 59 (2003) 591–605.

[832] N. Li, W. Wu, K. Nie, Molecular dynamics study on the evolution of interfacial dislocation network and mechanical properties of Ni-based single crystal superalloys, Physics Letters A 382 (20) (2018) 1361–1367.

[833] S. Li, D. Smith, Development of an anisotropic constitutive model for single-crystal superalloy for combined fatigue and creep loading, International Journal of Mechanical Sciences 40 (1998) 937–948.

[834] X. Li, B. Sun, H. You, L. Wang, Evolution of Rolls Royce air-cooled turbine blades and feature analysis, Procedia Engineering 99 (2015) 1482–1491.

[835] X. Li, L. Wang, J. Dong, L. Lou, J. Zhang, Evolution of micro-pores in a single-crystal nickel-based superalloy during solution heat treatment, Metallurgical and Materials Transactions 48 (2017) 2682–2686.

[836] X. Li, X. Zhang, C. Liu, C. Wang, T. Yu, Z. Zhang, Regular γ/γ' phase interface instability in a binary model nickel-based single-crystal alloy, Journal of Alloys and Compounds 633 (2015) 366–369.

[837] Z. Li, S. Bigdeli, H. Mao, Q. Chen, M. Selleby, Thermodynamic evaluation of pure Co for the third generation of thermodynamic databases, Physica Status Solidi B (2016).

[838] Z. Li, X. Fan, B. Liu, Influence of deformation temperature on recrystallization in a Ni-based single crystal superalloy, Materials Letters 160 (2015) 318–322.

[839] J. Liburdi, 40 years of reliable life extension of industrial and aero turbine components, in: Turbine Forum 2017, Nice, 2017.

[840] J. Liburdi, J. Wilson, Guidelines for reliable extension of turbine blade life, in: Proc. of the Twelfth Turbomachinery Symposium, College Station, Texas, 1984, pp. 21–30.

[841] I.M. Lifshitz, V.V. Slyozov, The kinetics of precipitation from supersaturated solid solutions, Journal of Physics and Chemistry of Solids 19 (1–2) (1961) 35–50.

[842] K. Lillerud, P. Kofstad, Sulfate-induced hot corrosion of nickel, Oxidation of Metals 21 (1984) 233–270.

[843] C. Ling, J. Besson, S. Forest, B. Tanguy, F. Latourte, E. Bosso, An elastoviscoplastic model for porous single crystals at finite strains and its assessment based on unit cell simulations, International Journal of Plasticity 84 (2016) 58–87.

[844] C. Ling, S. Forest, J. Besson, B. Tanguy, F. Latourte, A reduced micromorphic single crystal plasticity model at finite deformations. Application to strain localization and void growth in ductile metals, International Journal of Solids and Structures 134 (2018) 43–69.

[845] T. Link, A. Epishin, U. Brückner, P. Portella, Increase of misfit creep of superalloys and its correlation with deformation, Acta Materialia 48 (8) (2000) 1981–1994.

[846] T. Link, A. Epishin, B. Fedelich, Inhomogeneity of misfit stresses in nickel-base superalloys: effect on propagation of matrix dislocation loops, Philosophical Magazine 89 (13) (2009) 1141–1159.

[847] T. Link, A. Epishin, M. Klaus, U. Brückner, A. Reznicek, ⟨100⟩ Dislocations in nickel-base superalloys: formation and role in creep deformation, Materials Science and Engineering: A 405 (1–2) (2005) 254–265.

[848] T. Link, S. Zabler, A. Epishin, A. Haibel, M. Bansal, X. Thibault, Synchrotron tomography of porosity in single-crystal nickel-base superalloys, Materials Science and Engineering: A 425 (1) (2006) 47–54.

[849] K.-D. Liss, A. Bartels, A. Schreyer, H. Clemens, High-energy x-rays: a tool for advanced bulk investigations in materials science and physics, Textures and Microstructures 35 (3–4) (2003) 219–252.

[850] K.-D. Liss, A. Royer, T. Tschentscher, P. Suortti, A. Williams, On high-resolution reciprocal-space mapping with a triple-crystal diffractometer for high-energy x-rays, Journal of Synchrotron Radiation 5 (2) (1998) 82–89.

[851] B. Liu, D. Raabe, F. Roters, A. Arsenlis, Interfacial dislocation motion and interactions in single-crystal superalloys, Acta Materialia 79 (2014) 216–233.

[852] C. Liu, X. Zhang, L. Ge, S. Liu, C. Wang, T. Yu, Y. Zhang, Z. Zhang, Effect of rhenium and ruthenium on the deformation and fracture mechanism in nickel-based model single crystal superalloys during the in-situ tensile at room temperature, Materials Science and Engineering: A 682 (2017) 90–97.

[853] D. Liu, Z. Wen, Z. Yue, Creep damage mechanism and phase morphology of a V-notched round bar in Ni-based single crystal superalloys, Materials Science and Engineering: A 605 (2014) 215–221.

[854] D. Liu, D. Zhang, L. Liang, Z. Wen, Z. Yue, Prediction of creep rupture life of a V-notched bar in dd6 Ni-based single crystal superalloy, Materials Science and Engineering: A 615 (2014) 14–21.

[855] F. Liu, Z. Liu, X. Pei, J. Hu, Z. Zhuang, Modeling high temperature anneal hardening in au submicron pillar by developing coupled dislocation glide-climb model, International Journal of Plasticity 99 (2017) 102–119.

[856] F. Liu, Z.-l. Liu, P. Lin, Z. Zhuang, Numerical investigations of helical dislocations based on coupled glide-climb model, International Journal of Plasticity 92 (2017) 2–18.

[857] F. Liu, C. Wang, Electronic structure and multi-scale behaviour for the dislocation-doping complex in the gamma phase of nickel-base superalloys, RCS Advances 7 (2017) 19124–19135.

[858] F. Liu, Z. Wang, S. Ai, Y.C. Wang, X. Sun, T. Jin, H. Guan, Thermo-mechanical fatigue of single crystal nickel-based superalloy DD8, Scripta Metallurgica 48 (2003) 1265–1270.

[859] J.L. Liu, T. Jin, X.F. Sun, J.H. Zhang, H.R. Guan, Z.Q. Hu, Anisotropy of stress rupture properties of a Ni base single crystal superalloy at two temperatures, Materials Science and Engineering: A 479 (1–2) (2008) 277–284.

[860] J.L. Liu, T. Jin, J.J. Yu, X.F. Sun, H.R. Guan, Z.Q. Hu, Effect of thermal exposure on stress rupture properties of a Re bearing Ni base single crystal superalloy, Materials Science and Engineering: A 527 (4–5) (2010) 890–897.

[861] L. Liu, N. Husseini, C. Torbet, D. Kumah, R. Clarke, T. Pollock, J. Jones, In situ imaging of high cycle fatigue crack growth in single crystal nickel-base superalloys by synchrotron X-radiation, Journal of Engineering Materials and Technology 130 (2008) 021008.

[862] L. Liu, T. Jin, N. Zhao, X. Sun, H. Guan, Z. Hu, Formation of carbides and their effects on stress rupture of a Ni-base single crystal superalloy, Materials Science and Engineering: A 361 (2003) 191–197.

[863] X. Liu, S.-L. Shang, Y.-J. Hu, Y. Wang, Y. Du, Z.-K. Liu, Insight into γ-Ni/γ'-Ni_3Al interfacial energy affected by alloying elements, Materials & Design 133 (2017) 39–46.

[864] Z. Liu, C. Wang, T. Yu, Influence of Re on the propagation of a Ni/Ni_3Al interface crack by molecular dynamics simulation, Modelling and Simulation in Materials Science and Engineering 21 (4) (2013).

[865] Z.-K. Liu, Y. Wang, Computational Thermodynamics of Materials, Cambridge University Press, 2016.

[866] Z.L. Liu, X.M. Liu, Z. Zhuang, X.C. You, A multi-scale computational model of crystal plasticity at submicron-to-nanometer scales, International Journal of Plasticity 25 (8) (2009) 1436–1455.

[867] I. Lopez-Galilea, J. Koßmann, A. Kostka, R. Drautz, L. Mujica Roncery, T. Hammerschmidt, S. Huth, W. Theisen, The thermal stability of topologically close-packed phases in the single crystal Ni-base superalloy ERBO/1, Journal of Materials Science 51 (2016) 2653–2664.

[868] F. Louchet, M. Ignat, TEM analysis of square-shape dislocation configurations in the γ' phase of a Ni-based superalloy, Acta Metallurgica 34 (1986) 1681–1686.

[869] P. Lours, A. Coujou, B.d. Mauduit, On the high-temperature stress-induced spreading of superlattice intrinsic stacking faults in γ'nickel-based single crystals, Philosophical Magazine. A 62 (2) (1990) 253–266.

[870] T. Lu, R. Unocic, H. Deutchman, M. Mills, Creep deformation mapping in nickel base disk superalloys, Materials at High Temperatures 33 (2016) 372–383.
[871] V. Lughi, V.K. Tolpygo, D.R. Clarke, Microstructural aspects of the sintering of thermal barrier coatings, Materials Science and Engineering: A 368 (1–2) (2004) 212–221.
[872] H. Lukas, S. Fries, B. Sundman, Computational Thermodynamics: The Calphad Method, Cambridge University Press, 2007.
[873] P. Lukas, L. Kunz, M. Svoboda, High cycle fatigue of superalloy single crystals at high mean stresses, Materials Science and Engineering 387–389 (2004) 505–510.
[874] P. Lukas, L. Kunz, M. Svoboda, High-temperature ultra-high cycle fatigue damage of notched single crystal superalloys at high mean stresses, International Journal of Fatigue 27 (2005) 1535–1540.
[875] P. Lukas, P. Preclik, A. Cadek, Notch effects on creep behaviour of CMSX4 superalloy single crystals, Materials Science and Engineering: A 298 (2001) 84–89.
[876] A. Lund, P.W. Voorhees, The effects of elastic stress on microstructural development: the three-dimensional microstructure of a γ-γ' alloy, Acta Materialia 50 (10) (2002) 2585–2598.
[877] A. Lund, P.W. Voorhees, A quantitative assessment of the three-dimensional microstructure of a γ-γ' alloy, Philosophical Magazine 83 (14) (2003) 1719–1733.
[878] G. Lvov, V. Levit, M. Kaufman, Mechanism of primary MC carbide decomposition in Ni-base superalloys, Metallurgical and Materials Transactions A 35 (2004) 1669–1679.
[879] A. Ma, D. Dye, R. Reed, A model for the creep deformation behaviour of single-crystal superalloy CMSX-4, Acta Materialia 56 (8) (2008) 1657–1670.
[880] L. Ma, S. Xiao, H. Deng, W. Hu, Tensile mechanical properties of Ni-based superalloy of nanophases using molecular dynamics simulation, Physica Status Solidi B 253 (4) (2016) 726–732.
[881] X. Ma, H. Shi, J. Gu, G. Chen, O. Luesebrink, H. Harders, In-situ observations of the effects of orientation and carbide on low cycle fatigue crack propagation in a single crystal superalloy, Procedia Engineering 2 (2010) 2287–2295.
[882] R. Maaß, L. Meza, B. Gan, S. Tin, J. Greer, Ultrahigh strength of dislocation-free Ni_3Al nanocubes, Small 8 (2012) 1869–1875.
[883] L. Machon, G. Sauthoff, Deformation behaviour of Al-containing C14 Laves phase alloys, Intermetallics 4 (6) (1996) 469–481.
[884] D.E. Mack, T. Wobst, M.O.D. Jarligo, D. Sebold, R. Vaßen, Lifetime and failure modes of plasma sprayed thermal barrier coatings in thermal gradient rig tests with simultaneous CMAS injection, Surface & Coatings Technology 324 (2017) 36–47.
[885] R.A. MacKay, L.J. Ebert, Factors which influence directional coarsening of γ during creep in nickel-base superalloy single crystals, in: Superalloys 1984, Champion, PA (USA), 1984, pp. 135–144.
[886] R.A. MacKay, L.J. Ebert, The development of γ/γ' lamellar structures in a nickel-base superalloy during elevated temperature mechanical testing, Metallurgical Transactions A 16 (11) (1985) 1969–1982.
[887] D. MacLachlan, D. Knowles, Fatigue behaviour and lifing of two single crystal superalloys, Fatigue & Fracture of Engineering Materials & Structures 24 (2004) 503–521.
[888] D.W. MacLachlan, L.W. Wright, S.S.K. Gunturi, D.M. Knowles, Modelling the anisotropic and biaxial creep behaviour of Ni-base single crystal superalloys CMSX-4 and SRR99 at 1223K, in: Superalloys 2000, Champion USA, 2000, pp. 357–366.
[889] R. Madec, B. Devincre, L. Kubin, On the use of periodic boundary conditions in dislocation dynamics simulations, in: H. Kitagawa, Y. Shibutani (Eds.), IUTAM Symposium on Mesoscopic Dynamics of Fracture Process and Materials Strength, in: Solid Mechanics and Its Applications, vol. 115, Kluwer Academic Publishers, NL-Dordrecht, 2004, pp. 35–44.
[890] R. Madec, B. Devincre, L.P. Kubin, From dislocation junctions to forest hardening, Physical Review Letters 89 (Dec 2002) 255508.

[891] J. Madison, J. Spowart, D. Rowenhorst, L.K. Aagesen, K. Thornton, T.M. Pollock, Modeling fluid flow in three-dimensional single crystal dendritic structures, Acta Materialia 58 (8) (2010) 2864–2875.

[892] R. Mahnken, Anisotropic creep modeling based on elastic projection operators with applications to CMSX-4 superalloy, Computer Methods in Applied Mechanics and Engineering 191 (15) (2002) 1611–1637.

[893] S. Maisel, N. Schindzielorz, A. Mottura, R. Reed, S. Müller, Nickel-rhenium compound sheds light on the potency of rhenium as a strengthener in high-temperature nickel alloys, Physical Review B 90 (2014) 094110.

[894] S. Makineni, A. Kumar, M. Lenz, P. Kontis, T. Meiners, C. Zenk, S. Zaefferer, G. Eggeler, S. Neumeier, E. Spiecker, D. Raabe, B. Gault, On the diffusive phase transformation mechanism assisted by extended dislocations during creep of a single crystal CoNi-based superalloy, Acta Materialia 155 (2018) 362–371.

[895] A. Malka-Markovitz, D. Mordehai, Cross-slip in face-centered cubic metals: a general escaig stress-dependent activation energy line tension model, Philosophical Magazine 98 (5) (2018) 347–370.

[896] J.-L. Malpertu, L. Rémy, Influence of tests parameters on the thermal-mechanical fatigue behavior of a superalloy, Metallurgical Transactions A 21 (1990) 389–399.

[897] G. Mälzer, R. Hayes, T. Mack, G. Eggeler, Miniature specimen assessment of creep of the single-crystal superalloy LEK 94 in the 1000°C temperature range, Metallurgical and Materials Transactions A 38 (2007) 314–327.

[898] J. Manara, M. Arduini-Schuster, H.-J. Rätzer-Scheibe, U. Schulz, Infrared-optical properties and heat transfer coefficients of semitransparent thermal barrier coatings, Surface & Coatings Technology 203 (8) (2009) 1059–1068.

[899] J. Mandel, Une généralisation de la théorie de la plasticité de W.T. Koiter, International Journal of Solids and Structures 1 (1965) 273–295.

[900] J. Mandel, Plasticité classique et viscoplasticité, Cours du CISM, vol. 97, 1971.

[901] J. Mandel, Equations constitutives et directeurs dans les milieux plastiques et viscoplastiques, International Journal of Solids and Structures 9 (6) (1973) 725–740.

[902] A. Manero, S. Sofronsky, K. Knipe, C. Meid, J. Wischek, J. Okasinski, J. Almer, A.M. Karlsson, S. Raghavan, M. Bartsch, Monitoring local strain in a thermal barrier coating system under thermal mechanical gas turbine operating conditions, JOM 67 (7) (2015) 1528–1539.

[903] N. Marchal, S. Flouriot, S. Forest, L. Remy, Crack–tip stress–strain fields in single crystal nickel–base superalloys at high temperature under cyclic loading, Computational Materials Science 37 (2006) 42–50.

[904] G. Marci, K. Mull, C. Sick, M. Bartsch, New testing facility and concept for life prediction of TBC turbine engine components, in: H. Sehitoglu, H. Maier (Eds.), Proc. Third Symposium on Thermo-Mechanical Fatigue Behaviour of Materials, West Conshohocken (PA), 1998, pp. 296–303.

[905] J. Marian, A. Caro, Moving dislocations in disordered alloys: connecting continuum and discrete models with atomistic simulations, Physical Review B 74 (2) (2006) 1.

[906] D. Marx, J. Hutter, Ab initio Molecular Dynamics: Basic Theory and Advanced Methods, 1st edition, Cambridge University Press, 2009.

[907] N. Matan, Rationalisation of the Creep Performance of the CMSX-4 Single Crystal Superalloy, PhD thesis, University of Cambridge, 1999, unpublished.

[908] N. Matan, D. Cox, P. Carter, M. Rist, C. Rae, R. Reed, Creep of CMSX-4 superalloy single crystals: effects of misorientation and temperature, Acta Materialia 47 (1999) 1549–1563.

[909] N. Matan, D.C. Cox, C.M.F. Rae, R.C. Reed, On the kinetics of rafting in CMSX-4 superalloy single crystals, Acta Materialia 47 (7) (1999) 2031–2045.

[910] L. Mataveli-Suave, High Temperature Durability of DS200+Hf Alloy, PhD thesis, ISAE-ENSMA, 2017.

[911] F. Mauget, F. Hamon, M. Morisset, J. Cormier, F. Riallant, J. Mendez, Damage mechanisms in an EB-PVD thermal barrier coating system during TMF and TGMF testing conditions under combustion environment, International Journal of Fatigue 99 (2017) 225–234.

[912] F. Mauget, D. Marchand, G. Benoit, M. Morisset, D. Bertheau, J. Cormier, J. Mendez, Z. Hervier, E. Ostoja-Kuczynski, C. Moriconi, Development and use of a new burner rig facility to mimic service loading conditions of Ni-based single crystal superalloys, in: 2nd European Symposium on Superalloys and Their Applications, vol. 14, 2014.

[913] V. Maurel, E. Busso, J. Frachon, J. Besson, F. N'Guyen, A methodology to model the complex morphology of rough interfaces, International Journal of Solids and Structures 51 (19) (2014) 3293–3302.

[914] V. Maurel, P. de Bodman, L. Remy, Influence of substrate strain anisotropy in tbc system failure, Surface & Coatings Technology 206 (7) (2011) 1634–1639.

[915] V. Maurel, V.A. Esin, P. Sallot, F. Gaslain, S. Gailliegue, L. Rémy, Rumpling of nickel aluminide coatings: a reassessment of respective influence of thermal grown oxide and phase transformations, Materials at High Temperatures 33 (4–5) (2016) 318–324.

[916] V. Maurel, V. Guipont, M. Theveneau, B. Marchand, F. Coudon, Thermal cycling damage monitoring of thermal barrier coating assisted with lasat (laser shock adhesion test), Surface & Coatings Technology 380 (2019) 125048.

[917] V. Maurel, M. Harvey, L. Rémy, Aluminium oxide spallation on NiAl coating induced by compression surface, Surface & Coatings Technology 205 (2011) 3158–3166.

[918] V. Maurel, L. Mahfouz, V. Guipont, B. Marchand, F. Gaslain, A. Koster, A. Dennstedt, M. Bartsch, F. Coudon, Recent progress in local characterization of damage evolution in thermal barrier coating under thermal cycling, Superalloys (2020).

[919] V. Maurel, L. Rémy, M. Harvey, H. Tezenas du Montcel, A. Koster, The respective roles of thermally grown oxide roughness and NiAl coating anisotropy in oxide spallation, Surface & Coatings Technology 215 (2013) 52–61.

[920] V. Maurel, R. Soulignac, L. Helfen, T. Morgeneyer, A. Koster, L. Remy, Three-dimensional damage evolution measurement in ebpvd tbc using synchrotron laminography, Oxidation of Metals 79 (3-4, SI) (2013) 313–323.

[921] C. Mayr, G. Eggeler, A. Dlouhy, Analysis of dislocation structures after double shear creep deformation of CMSX6-superalloy single crystals at temperatures above 1000°C, Materials Science and Engineering: A 207 (1) (1996) 51–63.

[922] C. Mayr, G. Eggeler, G.A. Webster, G. Peter, Double shear creep testing of superalloy single crystals at temperatures above 1000°C, Materials Science and Engineering: A 199 (2) (1995) 121–130.

[923] J.P. McDonald, M. Thouless, S.M. Yalisove, Mechanics analysis of femtosecond laser-induced blisters produced in thermally grown oxide on si(100), Journal of Materials Research 25 (6) (2010) 1087–1095.

[924] R. McGinty, D. McDowell, A semi-implicit integration scheme for rate independent finite crystal plasticity, International Journal of Plasticity 22 (2006) 996–1025.

[925] M. McLean, On the threshold stress for dislocation creep in particle strengthened alloys, Acta Metallurgica 33 (4) (1985) 545–556.

[926] P. Mechnich, W. Braue, U. Schulz, High-temperature corrosion of EB-PVD yttria partially stabilized zirconia thermal barrier coatings with an artificial volcanic ash overlay, Journal of the American Ceramic Society 94 (3) (2011) 925–931.

[927] H. Mecking, U. Kocks, Kinetics of flow and strain-hardening, Acta Metallurgica 29 (11) (1981) 1865–1875.

[928] M.M. Mehrabadi, S.C. Cowin, Eigentensors of linear anisotropic elastic materials, Quarterly Journal of Mechanics and Applied Mathematics 43 (1) (1990) 15–41.

[929] H. Mehrer, Diffusion in Solids: Fundamentals, Methods, Materials, Diffusion-Controlled Processes, Springer, 2007.

[930] C. Meid, Einfluss der Mikrostruktur auf das Ermüdungsverhalten einkristalliner Nickelbasis-Superlegierungen bei Hochtemperatur, PhD thesis, Ruhr-Universität, Bochum (Germany), 2018.
[931] C. Meid, M. Eggeler, P. Watermeyer, A. Kostka, T. Hammerschmidt, R. Drautz, G. Eggeler, M. Bartsch, Stress-induced formation of tcp phases during high temperature low cycle fatigue loading of the single-crystal Ni-base superalloy ERBO/1, Acta Materialia 168 (2019) 343–352.
[932] C. Meid, U. Waedt, A. Subramaniam, J. Wischek, M. Bartsch, P. Terberger, R. Vaßen, Miniaturization of lcf-testing of single crystal superalloys at high temperature for uncoated and coated specimens, Material Science& Engineering Technology 50 (2019) 777–787.
[933] J. Meng, T. Jin, X. Sun, Z. Hu, Effect of surface recrystallization on the creep rupture properties of a nickel-base single crystal superalloy, Materials Science and Engineering: A i527 (2010) 6119–6122.
[934] J. Meng, T. Jin, X. Sun, Z. Hu, Surface recrystallization of a single crystal nickel-base superalloy, International Journal of Minerals, Metallurgy, and Materials 18 (2011) 197–202.
[935] C. Mercer, S. Faulhaber, A. Evans, R. Darolia, A delamination mechanism for thermal barrier coatings subject to calcium–magnesium–alumino-silicate (CMAS) infiltration, Acta Materialia 53 (4) (2005) 1029–1039.
[936] L. Méric, G. Cailletaud, Single crystal modeling for structural calculations: part 2 – finite element implementation, Journal of Engineering Materials and Technology - ASME 113 (1991) 171–182.
[937] L. Méric, P. Poubanne, G. Cailletaud, Single crystal modeling for structural calculations. Part 1: model presentation, Journal of Engineering Materials and Technology 113 (1991) 162–170.
[938] U. Messerschmidt, D. Baither, M. Bartsch, B. Baufeld, B. Geyer, S. Guder, A. Wasilkowska, A. Czyrska-Filemonowicz, M. Yamaguchi, M. Feuerbacher, et al., High-temperature in situ straining experiments in the high-voltage electron microscope, Microscopy and Microanalysis 4 (1998) 226–234.
[939] U. Messerschmidt, M. Bartsch, High-temperature straining stage for in situ experiments in the high-voltage electron microscope, Ultramicroscopy 56 (1–3) (1994) 163–171.
[940] R. Mevrel, State of the art on high-temperature corrosion-resistant coatings, Materials Science and Engineering: A 120 (1989) 13–24.
[941] R. Mevrel, R. Pichoir, Les revêtements par diffusion, Materials Science and Engineering: A 88 (1987) 1–9.
[942] C. Miehe, F. Aldakheel, S. Teichtmeister, Phase-field modeling of ductile fracture at finite strains: a robust variational-based numerical implementation of a gradient-extended theory by micromorphic regularization, International Journal for Numerical Methods in Engineering 111 (9) (2017) 816–863.
[943] C. Miehe, N. Apel, M. Lambrecht, Anisotropic additive plasticity in the logarithmic strain space: modular kinematic formulation and implementation based on incremental minimization principles for standard materials, Computer Methods in Applied Mechanics and Engineering 191 (2002) 5383–5425.
[944] C. Miehe, J. Schröder, A comparative study of stress update algorithms for rate-independent and rate-dependent crystal plasticity, International Journal for Numerical Methods in Engineering 50 (2) (2001) 273–298.
[945] A. Migliori, Implementation of a modern resonant ultrasound spectroscopy system for the measurement of the elastic moduli of small solid specimens, Review of Scientific Instruments 76 (12) (2005) 121301.
[946] X. Milhet, M. Arnoux, J. Cormier, J. Mendez, C. Tromas, On the influence of the dendritic structure on the creep behavior of a Re-containing superalloy at high temperature/low stress, Materials Science and Engineering: A 546 (2012) 139–145.

[947] X. Milhet, J. Cormier, M. Arnoux, C. Tromas, On the influence of the dendritic structure on the creep behavior of a Re containing superalloy at high temperature/low stress, in: 12th International Conference on Creep and Fracture of Engineering Materials and Structures, The Japan Institute of Metals, Kyoto (Japan), 2012, pp. 1–4.
[948] R. Miller, Oxidation-based model for thermal barrier coating life, Journal of the American Ceramic Society 67 (8) (1984) 517–521.
[949] S. Miller, J. Rosier, C. Sommer, W. Hartnagel, The influence of load ratio, temperature, orientation, and hold time on fatigue crack growth of CMSX-4, in: T.M. Pollock, R.D. Kissinger, R.R. Bowman, K.A. Green, M. McLean, S.L. Olson, J.J. Schirra (Eds.), Superalloys 2000: Proc. of the 9th International Symposium on Superalloys, Champion, USA, TMS, Minerals, Metals & Materials Society, Sept. 1992.
[950] W.W. Milligan, S.D. Antolovich, Yielding and deformation behavior of the single crystal superalloy pwa 1480, Metallurgical Transactions A 18 (1) (1987) 85–95.
[951] R. Miner, T. Gabb, J. Gayda, K. Hemker, Orientation dependence of some mechanical properties of the single crystal nickel base superalloy René N4 – part III: tension-compression anisotropy, Metallurgical Transactions A 17 (1986) 507–512.
[952] R. Miner, J. Gayda, M. Hebsur, Creep-fatigue behavior of Ni-Co-Cr-Al-Y coated PWA 1480 superalloy single crystals, in: H.D. Solomon, G.R. Halford, L. Kaisand, B. Leis (Eds.), Low-Cycle Fatigue, vol. STP 942, ASTM, 1988, pp. 371–384.
[953] Y. Mishin, Atomistic modelling of the γ and γ' phases of the Ni-Al system, Acta Materialia 52 (2004) 1451–1467.
[954] Y. Mishin, Calculation of the γ/γ' interface free energy in the Ni-Al system by the capillary fluctuation method, Modelling and Simulation in Materials Science and Engineering 22 (4) (jun 2014) 045001.
[955] N. Mishra, S. Ranganathan, Electron microscopy and diffraction of ordering in a ni-25wt.% mo alloy, Materials Science and Engineering: A 150 (1) (1992) 75–85.
[956] A. Misra, D. Whittle, W. Worrell, Thermodynamics of molten sulfate mixtures, Journal of the Electrochemical Society 129 (1982) 1840–1845.
[957] N. Miura, K. Nakata, M. Miyazaki, Y. Hayashi, Y. Kondo, Morphology of γ' precipitates in second stage high pressure turbine blade of single crystal nickel-based superalloy after serviced, Materials Sciences Forum 638–642 (2010) 2291–2296.
[958] N. Moelans, B. Blanpain, P. Wollants, An introduction to phase-field modeling of microstructure evolution, Calphad: Computer Coupling of Phase Diagrams and Thermochemistry 32 (2008) 268.
[959] P. Mohan, B. Yuan, T. Patterson, V. Desai, Y.H. Sohn, Degradation of yttria stabilized zirconia thermal barrier coatings by molten CMAS (CaO-MgO-Al2O3-SiO2) deposits, in: Materials Science Forum, vol. 595, Trans Tech Publ, 2008, pp. 207–212.
[960] M. Mollard, F. Pedraza, B. Bouchaud, X. Montero, M. Galetz, M. Schütze, Influence of the superalloy substrate in the synthesis of the pt-modified aluminide bond coat made by slurry, Surface & Coatings Technology 270 (2015) 102–108.
[961] J. Möller, E. Bitzek, R. Janisch, H. Hassan, A. Hartmaier, Fracture ab initio: a force-based scaling law for atomistically informed continuum models, Journal of Materials Research 33 (22) (2018) 3750–3761.
[962] D. Monceau, D. Poquillon, Continuous thermogravimetry under cyclic conditions, Oxidation of Metals 61 (2004) 143–163.
[963] D. Mordehai, C. Emmanuel, M. Fivel, M. Verdier, Introducing dislocation climb by bulk diffusion in discrete dislocation dynamics, Philosophical Magazine 88 (2008) 899–925.
[964] T. Morgeneyer, J. Besson, Flat to slant ductile fracture transition: tomography examination and simulations using shear-controlled void nucleation, Scripta Materialia 65 (11) (2011) 1002–1005.
[965] T. Mori, K. Tanaka, Average stress in matrix and average elastic energy of materials with misfitting inclusions, Acta Metallurgica 21 (5) (1973) 571–574.

[966] M. Morinaga, N. Yukawa, H. Adachi, H. Ezaki, New Phacomp and its applications to alloy design, in: Superalloys 1984, Champion, PA (USA), 1984, pp. 523–532.

[967] J. Morniroli, Cbed and lacbed characterization of crystal defects, Journal of Microscopy 223 (3) (2006) 240–245.

[968] J.-P. Morniroli, Large-angle convergent-beam electron diffraction (LACBED), Technical report, The French Society of Microscopy, Paris, 2002.

[969] D. Morris, M. Munoz-Morris, Thermal and mechanical disordering of ordered alloys, in: Y. Bréchet, et al. (Eds.), Solid Phase Transformations in Inorganic Materials, vol. 1, Warrendale, USA, TMS, Minerals, Metals & Materials Society, 2005.

[970] S. Moss, G. Webster, E. Fleury, Creep deformation and crack growth behavior of a single-crystal nickel-base superalloy, Metallurgical and Materials Transactions A 27 (1996) 829–837.

[971] A. Mottura, M. Finnis, R. Reed, On the possibility of rhenium clustering in nickel-based superalloys, Acta Materialia 60 (6) (2012) 2866–2872.

[972] A. Mottura, R.C. Reed, What is the role of rhenium in single crystal superalloys? in: MATEC Web of Conferences, vol. 14, EDP Sciences, 2014, p. 01001.

[973] H. Moulinec, P. Suquet, A fast numerical method for computing the linear and nonlinear properties of composites, Comptes rendus de l'Académie des Sciences Paris, Série II 318 (1994) 1417–1423.

[974] H. Moulinec, P. Suquet, A numerical method for computing the overall response of nonlinear composites with complex microstructure, Computer Methods in Applied Mechanics and Engineering 157 (1) (1998) 69–94.

[975] J. Moverare, S. Johansson, R. Reed, Deformation and damage mechanisms during thermal-mechanical fatigue of a single-crystal superalloy, Acta Materialia 57 (2009) 2266–2276.

[976] J. Moverare, C. Roger, R. Reed, Thermomechanical fatigue in single crystal superalloys, in: Proc. MATEC Web of Conferences, 14, 2014, p. 0600.

[977] J. Moverare, M. Segersäll, A. Sato, S. Johansson, R. Reed, Thermomechanical fatigue of single crystal superalloys: influence of composition and microstructure, in: E. Huron, R. Reed, M. Hardy, M. Mills, R. Montero, P. Portella, J. Telesman (Eds.), Superalloys 2012: Proc. of the 12th International Symposium on Superalloys, Champion, USA, TMS, Minerals, Metals & Materials Society, Sept. 2012.

[978] B. Mueller, R. Spicer, Land-based turbine casting initiative, in: Advanced Turbine Systems Annual Program Review Meeting, Oak Ridge National Lab, USA, 1995, pp. 161–170.

[979] R. Mueller, D. Gross, A time-dependent constitutive law for materials with microstructural evolution, Mechanics of Materials 33 (2) (2001) 63–76.

[980] H. Mughrabi, Microstructural aspects of high temperature deformation of monocrystalline nickel base superalloys: some open problems, Materials Science and Technology 25 (2) (2009) 191–204.

[981] H. Mughrabi, The importance of sign and magnitude of γ/γ' prime misfit in superalloys with special reference to the new γ' hardened Co-based superalloys, Acta Materialia 81 (2014) 21–29.

[982] L. Mujica-Roncery, I. Lopez-Galilea, B. Ruttert, S. Huth, W. Theisen, Influence of temperature, pressure, and cooling rate during hot isostatic pressing on the microstructure of a SX Ni-base superalloy, Materials & Design 97 (2016) 544–552.

[983] D. Mukherji, R. Wahi, Some implications of the particle and climb geometry on the climb resistance in nickel-base superalloys, Acta Materialia 44 (4) (1996) 1529–1539.

[984] E.W. Müller, Field ion microscopy, Science 149 (3684) (1965) 591–601.

[985] L. Müller, U. Glatzel, M. Feller-Kniepmeier, Modelling thermal misfit stresses in nickel-base superalloys containing high volume fraction of γ' phase, Acta Materialia 40 (6) (1992) 1321–1327.

[986] L. Müller, U. Glatzel, M. Feller-Kniepmeier, Calculation of the internal stresses and strains in the microstructure of a single crystal nickel-base superalloy during creep, Acta Materialia 41 (12) (1993) 3401–3411.
[987] L. Müller, T. Link, M. Feller-Kniepmeier, Temperature dependence of the thermal lattice mismatch in single crystal nickel-base superalloy measured by neutron diffraction, Scripta Metallurgica 26 (1992).
[988] T. Mura, Micromechanics of Defects in Solids, 2 edition, Springer-Verlag, Berlin Heidelberg, 1987.
[989] T. Mura, Micromechanics of Defects in Solids, Kluwer Academic Publishers, Dordrecht (The Netherlands), 1991.
[990] T. Murakumo, T. Kobayashi, Y. Koizumi, H. Harada, Creep behaviour of Ni-base single-crystal superalloys with various γ' volume fraction, Acta Materialia 52 (12) (2004) 3737–3744.
[991] H.J. Murphy, C.T. Sims, A.M. Beltran, Phacomp revisited, JOM 20 (11) (1968) 46–53.
[992] L. Mushongera, M. Fleck, J. Kundin, Y. Wang, H. Emmerich, Effect of Re on directional γ'-coarsening in commercial single crystal Ni-base superalloys: a phase field study, Acta Materialia 93 (2015) 60–72.
[993] R. Muñoz, Cleaning of turbine engines: a simple and trivial process, in: Turbine Forum, Saint-Laurent du Var, 2019.
[994] F. Nabarro, Rafting in superalloys, Metallurgical and Materials Transactions A 27 (1996) 513–530.
[995] F. Nabarro, C. Cress, P. Kotschy, Thermodynamic driving force for rafting in superalloys, Acta Materialia 44 (1996) 3189–3198.
[996] F. Nabarro, F. De Villiers, The Physics of Creep, Taylor & Francis, 1995.
[997] F.R.N. Nabarro, Dislocations in a simple cubic lattice, Proceedings of the Royal Society of London 59 (1947) 256–272.
[998] R. Naraparaju, M. Hüttermann, U. Schulz, P. Mechnich, Tailoring the EB-PVD columnar microstructure to mitigate the infiltration of CMAS in 7YSZ thermal barrier coatings, Journal of the European Ceramic Society 37 (1) (2017) 261–270.
[999] M. Nathal, R. Mackay, R. Garlick, Temperature dependence of γ/γ' lattice mismatch in nickel-base superalloys, Materials Science and Engineering: A 75 (1985) 195–205.
[1000] M.V. Nathal, R.A. Mackay, The stability of lamellar γ/γ' structures, Materials Science and Engineering: C 85 (C) (1987) 127–138.
[1001] D. Naumenko, B.A. Pint, W.J. Quadakkers, Current thoughts on reactive element effects in alumina-forming systems: in memory of John Stringer, Oxidation of Metals 86 (1–2) (2016) 1–43.
[1002] D. Naumenko, V. Shemet, L. Singheiser, W.J. Quadakkers, Failure mechanisms of thermal barrier coatings on mcraly-type bondcoats associated with the formation of the thermally grown oxide, Journal of Materials Science 44 (7) (Apr 2009) 1687–1703.
[1003] P. Nellist, S. Pennycook, Incoherent imaging using dynamically scattered coherent electrons, Ultramicroscopy 78 (1–4) (1999) 111–124.
[1004] S. Nemat-Nasser, M. Hori, Micromechanics: Overall Properties of Heterogeneous Materials, second edition, Elsevier, 1998.
[1005] E. Nembach, Particle Strengthening of Metals and Alloys, John Riley and Sons, 1997.
[1006] R. Neu, Crack paths in single-crystal ni-base superalloys under isothermal and thermomechanical fatigue, International Journal of Fatigue 123 (2019) 268–278.
[1007] S. Neumeier, F. Pyczak, M. Göken, The influence of ruthenium and rhenium on the local properties of the γ-and γ'-phase in nickel-base superalloys and their consequences for alloy behavior, Superalloys 2008 (2008) 109–119.
[1008] A. Ngan, M. Wen, C. Woo, Atomistic simulations of Paidar-Pope-Vitek lock formation in Ni_3Al, Computational Materials Science 29 (2004) (2004) 259–269.
[1009] P. Nguyen, A.G. Kotousov, S.Y. Ho, S. Wildy, Investigation of thermo-mechanical properties of slurry based thermal barrier coatings under repeated thermal shock, in: Key Engineering Materials, vol. 417, Trans Tech Publ, 2010, pp. 197–200.

[1010] J.R. Nicholls, Designing oxidation-resistant coatings, JOM 52 (1) (2000) 28–35.
[1011] J.R. Nicholls, K. Lawson, A. Johnstone, D. Rickerby, Methods to reduce the thermal conductivity of EB-PVD tbcs, Surface & Coatings Technology 151 (2002) 383–391.
[1012] T.J. Nijdam, W.G. Sloof, Modelling of composition and phase changes in multiphase alloys due to growth of an oxide layer, Acta Materialia 56 (18) (2008) 4972–4983.
[1013] P. Niranatlumpong, C.B. Ponton, H.E. Evans, Failure of protective oxides on plasma-sprayed NiCrAlY overlay coatings, Oxidation of Metals 53 (3) (2000) 241–258.
[1014] K. Nishimoto, K. Saida, Y. Fujita, Crystal growth in laser surface melting and cladding of Ni-base single crystal superalloy, Welding in the World 52 (2008) 64–78.
[1015] D. Nissley, T. Meyer, K.P. Walker, Life prediction and constitutive models for engine hot section anisotropic materials program, 1992.
[1016] B. Nithin, A. Samanta, S.K. Makineni, T. Alam, P. Pandey, A.K. Singh, R. Banerjee, K. Chattopadhyay, Effect of Cr addition on γ/γ' cobalt-based Co-Mo-Al-Ta class of superalloys: a combined experimental and computational study, Journal of Materials Science 52 (SEP 2017) 11036–11047.
[1017] A. Nitz, U. Lagerpusch, E. Nembach, CRSS anisotropy and tension/compression asymmetry of a commercial superalloy, Acta Materialia 46 (1998) 4769–4779.
[1018] X. Niu, T. Luo, J. Lu, Y. Xiang, Dislocation climb models from atomistic scheme to dislocation dynamics, Journal of the Mechanics and Physics of Solids 99 (2017) 242–258.
[1019] W.G. Nöhring, W.A. Curtin, Thermodynamic properties of average-atom interatomic potentials for alloys, Modelling and Simulation in Materials Science and Engineering 24 (4) (2016) 045017.
[1020] P. Nörtershäuser, J. Frenzel, A. Ludwig, K. Neuking, G. Eggeler, The effect of cast microstructure and crystallography on rafting, dislocation plasticity and creep anisotropy of single crystal Ni-base superalloys, Materials Science and Engineering: A 626 (2015) 305–312.
[1021] D. Nouailhas, Un modèle de viscoplasticité cyclique pour matériaux anisotropes à symétrie cubique, Comptes rendus de l'Académie des Sciences Paris, Série II 310 (1990) 887–890.
[1022] D. Nouailhas, G. Cailletaud, Comparaison de divers critères anisotropes pour monocristaux cubiques à faces centrées (CFC), Comptes rendus de l'Académie des Sciences Paris, Série II 315 (1992) 1573–1579.
[1023] D. Nouailhas, G. Cailletaud, Tension-torsion behavior of single-crystal superalloys: experiment and finite element analysis, International Journal of Plasticity 11 (4) (1995) 451–470.
[1024] D. Nouailhas, S. Lhuillier, On the micro-macro modelling of $\gamma - \gamma'$ single crystal behavior, Computational Materials Science 9 (1) (1997) 177–187.
[1025] D. Nouailhas, D. Pacou, G. Cailletaud, F. Hanriot, L. Rémy, Experimental study of the anisotropic behavior of the CMSX2 single-crystal superalloy under tension-torsion loadings, in: D. McDowell, J. Ellis (Eds.), Advances in Multiaxial Fatigue, ASTM International, West Conshohocken, PA, 1993, pp. 244–258, https://doi.org/10.1520/STP24805S.
[1026] D. Nouailhas, J.-P. Pierre Culié, G. Cailletaud, L. Méric, F.E. analysis of the stress–strain behaviour of single–crystal tubes, European Journal of Mechanics - A/Solids 14A (1) (1995) 137–154.
[1027] W. Nowak, D. Naumenko, G. Mor, F. Mor, D.E. Mack, R. Vassen, L. Singheiser, W. Quadakkers, Effect of processing parameters on mcraly bondcoat roughness and lifetime of aps–tbc systems, Surface & Coatings Technology 260 (2014) 82–89.
[1028] N. Ohno, J. Wang, Kinematic hardening rules with critical state for the activation of dynamic recovery. Part I: formulation and basic features for ratchetting behaviour, International Journal of Plasticity 9 (1993) 375–390.
[1029] M. Okazaki, Creep-fatigue small crack propagation in a single crystal Ni-base superalloy, CMSX-2: microstructural influences and environmental effects, International Journal of Fatigue 21 (1999) 79–86.

[1030] M. Okazaki, M. Sakaguchi, Thermo-mechanical fatigue failure of a single crystal Ni-based superalloy, International Journal of Fatigue 30 (2008) 318–323.

[1031] J. Olfe, H. Neuhäuser, Dislocation groups, multipoles, and friction stresses in α-cuzn alloys, Physica Status Solidi A 109 (1) (1988) 149–160.

[1032] J. Olschewski, Das elastische Materialverhalten kubisch einkristalliner Körper, Technical report, Federal Institute for Materials Research and Testing, BAM, Berlin, Unter den Eichen, 87, 12205 Berlin, 1992.

[1033] J. Olschewski, Der kubisch-flächenzentrierte Einkristall in Theorie und Experiment, Technical report, Federal Institute for Materials Research and Testing, BAM, Berlin, Unter den Eichen, 87, 12205 Berlin, 1997.

[1034] L. Onsager, Reciprocal relations in irreversible processes. I, Physical Review 37 (1931) 405–426.

[1035] L. Onsager, Reciprocal relations in irreversible processes. II, Physical Review 38 (1931) 2265–2279.

[1036] A. Onuki, H. Nishimori, Anomalously slow domain growth due to a modulus inhomogeneity in phase separating alloys, Physical Review B 43 (16) (1991) 13649.

[1037] C. Oskay, M. Rudolphi, E. Affeldt, M. Schütze, M. Galetz, Evolution of microstructure and mechanical properties of nial-diffusion coatings after thermocyclic exposure, Intermetallics 89 (2017) 22–31.

[1038] W. Österle, D. Bettge, B. Fedelich, H. Klingelhöffer, Modelling the orientation and direction dependence of the critical resolved shear stress of nickel-base superalloy single crystals, Acta Materialia 48 (2000) 689–700.

[1039] R. Otis, Z.-K. Liu, Pycalphad: Calphad-based computational thermodynamics in python, Journal of Open Research Software 5 (1) (2017).

[1040] M. Ott, H. Mughrabi, Dependence of the high-temperature low-cycle fatigue behaviour of the monocrystalline nickel-base superalloys CMSX-4 and CMSX-6 on the γ/γ'-morphology, Materials Science and Engineering: A 272 (1999) 24–30.

[1041] M. Ott, U. Tetzlaff, H. Mughrabi, Influence of directional coarsening on the isothermal high-temperature fatigue behaviour of the monocrystalline nickel-base superalloys CMSX-6 and CMSX-4, in: Microstructure and Mechanical Properties of Metallic High-Temperature Materials, 1999, Research report.

[1042] G. Ouyang, High temperature structure materials beyond nickel base superalloys, Journal of Materials Science and Nanomaterials 1 (2) (2017) e107.

[1043] N.P. Padture, M. Gell, E.H. Jordan, Thermal barrier coatings for gas-turbine engine applications, Science 296 (5566) (2002) 280–284.

[1044] V. Paidar, D. Pope, V. Vitek, A theory of the anomalous yield behavior in $L1_2$ ordered alloys, Acta Materialia 32 (3) (1984) 435–448.

[1045] V. Paidar, D. Pope, V. Vitek, A theory of the anomalous yield behavior in ł12 ordered alloys, Acta Metallurgica 32 (1984) 435–448.

[1046] V. Paidar, M. Yamaguchi, D. Pope, V. Vitek, Dissociation and core structure of $\langle 110 \rangle$ screw dislocations in $L1_2$ ordered alloys II. Effects of an applied shear stress, Philosophical Magazine. A 45 (5) (1982) 883–894.

[1047] J. Pal, D. Srinivasan, E. Cheng, Effect of rejuvenation heat treatment and aging on the microstructural evolution in Rene N5 single crystal Ni base superalloy blades, in: M. Hardy, et al. (Eds.), Superalloys 2016: Proc. of the Sixteenth International Symposium on Superalloys, Champion, USA, TMS, Minerals, Metals & Materials Society, Sept. 2016.

[1048] F. Palmert, J. Johan Moverare, D. Gustafsson, Thermomechanical fatigue crack growth in a single crystal nickel base superalloy, International Journal of Fatigue 122 (2019) 79–86.

[1049] M. Palumbo, B. Burton, A. Costa e Silva, B. Fultz, B. Grabowski, G. Grimvall, B. Hallstedt, O. Hellman, B. Lindahl, A. Schneider, P. Turchi, W. Xiong, Thermodynamic modelling of crystalline unary phases, Physica Status Solidi B 251 (1) (2014) 14–32.

[1050] D. Pan, M.W. Chen, P.K. Wright, K.J. Hemker, Evolution of a diffusion aluminide bond coat for thermal barrier coatings during thermal cycling, Acta Materialia 51 (8) (2003) 2205–2217.

[1051] H. Pang, H. Dong, R. Beanland, H. Stone, C. Rae, P. Midgley, G. Brewster, N. D'Souza, Microstructure and solidification sequence of the interdendritic region in a third generation single-crystal nickel-base superalloy, Metallurgical and Materials Transactions A 30 (2009) 1660–1669.

[1052] H. Pang, N. D'Souza, H. Dong, H. Stone, C. Rae, Detailed analysis of the solution heat treatment of a third-generation single-crystal nickel-based superalloy cmsx-10k®, Metallurgical and Materials Transactions A 47 (2016) 889–906.

[1053] H. Pang, L. Zhang, R. Hobbs, H. Stone, C. Rae, Solution heat treatment optimization of fourth-generation single-crystal nickel-base superalloys, Metallurgical and Materials Transactions A 43 (2012) 3264–3282.

[1054] C. Panwisawas, H. Mathur, J. Gebelin, D. Putman, P. Withey, N. Warnken, C. Rae, R. Reed, Prediction of plastic strain for recrystallisation during investment casting of single crystal superalloys, in: E.S. Huron, R.C. Reed, M.C. Hardy, M.J. Mills, R.E. Montero, P.D. Portella, J. Telesman (Eds.), Superalloys 2012: Proc. of the Twelfth International Symposium on Superalloys, Champion, USA, TMS, Minerals, Metals & Materials Society, Sept. 2012, pp. 547–556.

[1055] O. Paris, M. Fa, E. Fa, T. Pollock, P. Fratzl, et al., Early stages of precipitate rafting in a single crystal nialmo model alloy investigated by small-angle x-ray scattering and tem, Acta Materialia 45 (3) (1997) 1085–1097.

[1056] M. Parlier, R. Valle, L. Perrière, S. Lartigue-Korinek, L. Mazerolles, Potential of directionally solidified eutectic ceramics for high temperature applications, AerospaceLab 3 (2011) 1–13.

[1057] A. Parsa, M. Ramsperger, A. Kostka, C. Somsen, C. Körner, G. Eggeler, Transmission electron microscopy of a CMSX-4 Ni-base superalloy produced by selective electron beam melting, Metals 6 (2016) 258.

[1058] A. Parsa, P. Wollgramm, H. Buck, A. Kostka, C. Somsen, A. Dlouhy, G. Eggeler, Ledges and grooves at γ/γ' interfaces of single crystal superalloys, Acta Materialia 90 (2015) 105–117.

[1059] A.B. Parsa, P. Wollgramm, H. Buck, C. Somsen, A. Kostka, I. Povstugar, P. Choi, D. Raabe, A. Dlouhy, J. Mueller, E. Spiecker, K. Demtröder, J. Schreuer, K. Neuking, G. Eggeler, Advanced scale bridging microstructure analysis of single crystal Ni-base superalloys, Advanced Engineering Materials 17 (2015) 216–230.

[1060] T. Parthasarathy, D. Dimiduk, Atomistic simulations of the structure and stability of "PPV" locks in an L1_2 compound, Acta Materialia 44 (1996) 2237–2247.

[1061] T. Parthasarathy, S. Rao, D. Dimiduk, Xx, in: K.A. Green, T.M. Pollock, H. Harada, T.E. Howson, R.C. Reed, J.J. Schirra, S. Walston (Eds.), Superalloys 2004: Proc. of the Tenth International Symposium on Superalloys, Champion, USA, TMS, Minerals, Metals & Materials Society, Sept. 2004, pp. 887–896.

[1062] B. Passilly, P. Kanoute, F.-H. Leroy, R. Mévrel, High temperature instrumented microindentation: applications to thermal barrier coating constituent materials, Philosophical Magazine 86 (2006) 5739–5752.

[1063] S. Patel, A century of discoveries, inventors and new nickel alloys, JOM 58 (2006) 18–20.

[1064] S. Patil, R. Narasimhan, P. Biswas, R. Mishra, Crack tip fields in a single crystal edge notched aluminum single crystal specimen, Journal of Engineering Materials and Technology - ASME 130 (2008) 021013.

[1065] S. Patil, R. Narasimhan, R. Mishra, A numerical study of crack tip constraint in ductile single crystals, Journal of the Mechanics and Physics of Solids 56 (2008) 2265–2286.

[1066] S. Patil, R. Narasimhan, R. Mishra, Observation of kink shear bands in an aluminium single crystal fracture specimen, Scripta Materialia 61 (2009) 465–468.

[1067] A. Paul, T. Laurila, V. Vuorinen, S. Divinski, Thermodynamics, Diffusion and the Kirkendall Effect in Solids, Springer, 2014.
[1068] U. Paul, P. Sahm, D. Goldschmidt, Inhomogeneities in single-crystal components, Materials Science and Engineering 173 (1993) 49–54.
[1069] E. Payton, P. Phillips, M. Mills, Semi-automated characterization of the γ' phase in ni-based superalloys via high-resolution backscatter imaging, Materials Science and Engineering: A 527 (10–11) (2010) 2684–2692.
[1070] P. Pedrak, A. Nowotnik, M. Góral, K. Kubiak, M. Drajewicz, X. Sieniawski, The technology of TBC deposition by EB-PVD method, Solid State Phenomena 227 (2015) 377–380.
[1071] F. Pedraza, Protection against high temperature environments, Master's course "Materials Science and Engineering", University of La Rochelle, France, 2019.
[1072] F. Pedraza, C. Tuohy, L. Whelan, A. Kennedy, High quality aluminide and thermal barrier coatings deposition for new and service exposed parts by CVD techniques, Materials Sciences Forum 461–464 (2004) 305–312.
[1073] R. Peierls, The size of a dislocation, Proceedings of the Royal Society of London 52 (1940) 34–37.
[1074] A. Pelton, Phase Diagrams and Thermodynamic Modeling of Solutions, Elsevier, 2018.
[1075] Z.F. Peng, Y.Y. Ren, Q.S. Mei, B.Z. Fan, P. Yan, J.C. Zhao, Y.Q. Wang, J.H. Sun, Directional coarsening of γ' precipitates in typical regions of original dendritic structure of CMSX-2, Scripta Materialia 42 (11) (2000) 1059–1064.
[1076] R.C. Pennefather, D.H. Boone, Mechanical degradation of coating systems in high-temperature cyclic oxidation, Surface & Coatings Technology 76–77 (1995) 47–52.
[1077] R. Pepinsky, in: H. Lipson, W. Cochran (Eds.), The Determination of Crystal Structures, in: Lawrence Bragg (Ed.), The Crystalline State, vol. iii, G. Bell/Macmillan, London/New York, 1954.
[1078] P. Perruchaud, Influence of a coating and oxidation on the fatigue-creep damage at high temperature of a single crystal, PhD thesis, Univ. Poitiers, France, 1997.
[1079] F. Perrudin, C. Rio, M. Vidal-Sétif, C. Petitjean, P.-J. Panteix, M. Vilasi, Gadolinium oxide solubility in molten silicate: dissolution mechanism and stability of Ca_2Gd_8 (SiO_4) $6O_2$ and Ca_3Gd_2 $(Si_3O_9)_2$ silicate phases, Journal of the European Ceramic Society 37 (7) (2017) 2657–2665.
[1080] M. Pessah-Simonetti, P. Caron, T. Khan, Effect of mu phase on the mechanical properties of a nickel-base single crystal superalloy, in: Superalloys 1992, Champion, PA (USA), 1992, pp. 567–576.
[1081] M. Pessah-Simonetti, P. Caron, T. Khan, Effect of a long-term prior aging on the tensile behaviour of a high-performance single crystal superalloy, Journal de Physique IV 3 (C7) (1993) C7-347.
[1082] M. Pessah-Simonetti, P. Donnadieu, P. Caron, T.C.P. phase particles embedded in a superalloy matrix: interpretation and prediction of the orientation relationships, Scripta Materialia 30 (12) (1994) 1553–1558.
[1083] D. Peter, F. Otto, T. Depka, P. Nörtershäuser, G. Eggeler, High temperature test rig for inert atmosphere miniature specimen creep testing, Materialwissenschaft und Werkstofftechnik 42 (2011) 493–499.
[1084] D. Pettifor, The structures of binary compounds: I. Phenomenological structure maps, Journal of Physics. C. Solid State Physics 19 (1986) 285–313.
[1085] D. Pettifor, R. Podloucky, Microscopic theory of the structural stability of pd-bonded AB compounds, Physical Review Letters 53 (1984) 1080–1083.
[1086] D. Pettifor, R. Podloucky, The structures of binary compounds: II. Theory of the pd-bonded AB compounds, Journal of Physics. C. Solid State Physics 19 (1986) 315–331.
[1087] F. Pettinari, A. Couret, D. Caillard, G. Molénat, N. Clément, A. Coujou, Quantitative measurements in in situ straining experiments in transmission electron microscopy, Journal of Microscopy 203 (1) (2001) 47–56.

[1088] F. Pettinari, J. Douin, G. Saada, P. Caron, A. Coujou, N. Clement, Stacking fault energy in short-range ordered γ-phases of ni-based superalloys, Materials Science and Engineering: A 325 (1–2) (2002) 511–519.

[1089] F. Pettinari, M. Prem, G. Krexner, P. Caron, A. Coujou, H. Kirchner, N. Clément, Local order in industrial and model γ phases of superalloys, Acta Materialia 49 (13) (2001) 2549–2556.

[1090] F. Pettinari-Sturmel, M. Benyoucef, J. Douin, P. Caron, D. Locq, N. Clément, A. Coujou, Some deformation micromechanisms in ni-based superalloys evidenced using tem in situ experiments, in: Advanced Materials Research, vol. 278, Trans Tech Publ, 2011, pp. 90–95.

[1091] F. Pettinari-Sturmel, J. Douin, A. Coujou, N. Clément, Characterisation of short-range order using dislocations, Zeitschrift für Metallkunde 97 (3) (2006) 200–204.

[1092] F. Pettinari-Sturmel, J. Douin, F. Krieg, E. Fleischmann, U. Glatzel, Evidence of short-range order (sro) by dislocation analysis in single-crystal ni-based matrix alloys with varying re content after creep, in: Superalloys 2020, Springer, 2020, pp. 253–259.

[1093] F. Pettinari-Sturmel, J. Douin, D. Locq, P. Caron, A. Coujou, Decorrelated dislocation movement in the γ-matrix channels of a Ni-based superalloy: experiment and dislocation dynamics simulation, Applied Mechanics Reviews 278 (2011) 13–18.

[1094] F. Pettinari-Sturmel, G. Saada, J. Douin, A. Coujou, N. Clément, Quantitative analysis of dislocation pile-ups in thin foils compared to bulk, Materials Science and Engineering: A 387 (2004) 109–114.

[1095] F. Pettit, G. Goward, Oxidation–corrosion–erosion mechanisms of environmental degradation of high-temperature materials, in: Coatings for High-Temperature Applications, 1983, pp. 1–32.

[1096] F. Pettit, G. Meier, Oxidation and hot corrosion of superalloys, in: M. Gell, S. Kortovich, H. Bricknell, B. Kent, J. Radavich (Eds.), Superalloys 1984: Proc. of the Fifth International Symposium on Superalloys, Champion, USA, Metall. Soc. of AIME, Oct. 1984.

[1097] P.S. Phani, W. Oliver, A critical assessment of the effect of indentation spacing on the measurement of hardness and modulus using instrumented indentation testing, Materials & Design 164 (2019) 107563.

[1098] J. Philibert, Atom Movements: Diffusion and Mass Transport in Solids, Editions de Physique, 1991.

[1099] T. Philippe, P.W. Voorhees, Ostwald ripening in multicomponent alloys, Acta Materialia 61 (2013) 4237–4244.

[1100] P. Phillips, M. Brandes, M. Mills, M. De Graef, Diffraction contrast stem of dislocations: imaging and simulations, Ultramicroscopy 111 (9–10) (2011) 1483–1487.

[1101] P.J. Phillips, M. De Graef, L. Kovarik, A. Agrawal, W. Windl, M. Mills, Atomic-resolution defect contrast in low angle annular dark-field stem, Ultramicroscopy 116 (2012) 47–55.

[1102] D. Pierce, R. Asaro, A. Needleman, Material rate dependence and localized deformation in crystalline solids, Acta Metallurgica 31 (1985) 1951.

[1103] S. Pierret, Microstructure of single crystal Ni-based superalloys during blade manufacturing process and creep deformation: a study using X-rays and neutrons, PhD thesis, Ecole Polytechnique Fédérale de Lausanne (EPFL), 2012.

[1104] S. Pierret, T. Etter, A. Evans, H. Van Swygenhoven, Origin of localized rafting in Ni-based single crystal turbine blades before service and its influence on the mechanical properties, Acta Materialia 61 (2013) 1478–1488.

[1105] G.D. Pigrova, TCP-phases in nickel-base alloys with elevated chromium content, Metal Science and Heat Treatment 47 (11–12) (2005) 544–551.

[1106] R. Pillai, M. Taylor, T. Galiullin, A. Chyrkin, E. Wessel, H. Evans, W. Quadakkers, Predicting the microstructural evolution in a multi-layered corrosion resistant coating on a ni-base superalloy, Materials at High Temperatures 35 (1–3) (2018) 78–88.

[1107] P. Pilvin, Une approche simplifiée pour schématiser l'effet de surface sur le comportement mécanique d'un polycristal, Journal de Physique IV 8 (1998) Pr4-33-42.

[1108] A. Pineau, Influence of uniaxial stress on the morphology of coherent precipitates during coarsening - elastic energy considerations, Acta Metallurgica 24 (1976) 559–564.

[1109] A. Pineau, G. Guillemot, D. Tourret, A. Karma, C.-A. Gandin, Growth competition between columnar dendritic grains-cellular automaton versus phase field modeling, Acta Materialia 155 (2018) 286–301.

[1110] B.A. Pint, Optimization of reactive-element additions to improve oxidation performance of alumina-forming alloys, Journal of the American Ceramic Society 86 (4) (2003) 686–695.

[1111] B.A. Pint, J.A. Haynes, Y. Zhang, Effect of superalloy substrate and bond coating on TBC lifetime, Surface & Coatings Technology 205 (5) (2010) 1236–1240.

[1112] F.M. Pitek, C.G. Levi, Opportunities for tbcs in the zro2–yo1. 5–tao2. 5 system, Surface & Coatings Technology 201 (12) (2007) 6044–6050.

[1113] P. Planques, V. Vidal, P. Lours, V. Proton, F. Crabos, J. Huez, B. Viguier, Characterization of the mechanical properties of thermal barrier coatings by 3 points bending tests and modified small punch tests, Surface & Coatings Technology 332 (2017) 40–46.

[1114] M. Plapp, Unified derivation of phase-field models for alloy solidification from a grand-potential functional, Physical Review E 84 (2011) 031601.

[1115] J. Plessing, C. Achmus, H. Neuhäuser, B. Schönfeld, G. Kostorz, Short-range order and the mode of slip in concentrated cu-based alloys, Zeitschrift für Metallkunde 88 (8) (1997) 630–635.

[1116] S. Plimpton, Fast parallel algorithms for short-range molecular dynamics, Journal of Chemical Physics 117 (1) (1995) 1–19, http://lammps.sandia.gov.

[1117] G. Po, N. Ghoniem, A variational formulation of constrained dislocation dynamics coupled with heat and vacancy diffusion, Journal of the Mechanics and Physics of Solids 66 (2014) 103–116.

[1118] G. Po, M. Lazar, N.C. Admal, N. Ghoniem, A non-singular theory of dislocations in anisotropic crystals, International Journal of Plasticity 103 (2018) 1–22.

[1119] G. Po, M. Lazar, D. Seif, N. Ghoniem, Singularity-free dislocation dynamics with strain gradient elasticity, Journal of the Mechanics and Physics of Solids 68 (2014) 161–178.

[1120] G. Po, M. Mohamed, T. Crosby, C. Erel, A. El-Azab, N. Ghoniem, Recent progress in discrete dislocation dynamics and its applications to micro plasticity, JOM 66 (10) (2014) 2108–2120.

[1121] D.L. Poerschke, T.L. Barth, C.G. Levi, Equilibrium relationships between thermal barrier oxides and silicate melts, Acta Materialia 120 (2016) 302–314.

[1122] D.L. Poerschke, R.W. Jackson, C.G. Levi, Silicate deposit degradation of engineered coatings in gas turbines: progress toward models and materials solutions, Annual Review of Materials Research 47 (2017) 297–330.

[1123] J.-P. Poirier, High-Temperature Deformation Processes in Metals, Ceramics and Minerals, Cambridge University Press, 1985.

[1124] T. Pollock, A. Argon, Directional coarsening in Ni-base single-crystals with high volume fractions of coherent precipitates, Acta Materialia 42 (1994) 1859–1874.

[1125] T. Pollock, R. Field, Dislocations and high-temperature plastic deformation of superalloy single crystals, in: F. Nabarro, M. Duesbery (Eds.), Dislocations in Solids, Elsevier, 2002, pp. 566–568.

[1126] T.M. Pollock, A.S. Argon, Creep resistance of CMSX-3 nickel base superalloy single crystals, Acta Materialia 40 (1) (1992) 1–30.

[1127] T.M. Pollock, A.S. Argon, Directional coarsening in nickel-base single crystals with high volume fractions of coherent precipitates, Acta Materialia 42 (6) (1994) 1859–1874.

[1128] T.M. Pollock, B. Laux, C.L. Brundidge, A. Suzuki, M.Y. He, Oxide-assisted degradation of Ni-base single crystals during cyclic loading: the role of coatings, Journal of the American Ceramic Society 94 (S1) (jun 2011) s136–s145.
[1129] T.M. Pollock, S. Tin, Nickel-based superalloys for advanced turbine engines: chemistry, microstructure, and properties, Journal of Propulsion and Power 22 (2) (2006) 361–374.
[1130] E. Pons, Propriétés d'adhérence de revêtements projetés plasma sur substrats fragiles: caractérisation et identification de lois d'interface par Modèles de Zones Cohésives, PhD thesis, Université Grenoble Alpes, 2016.
[1131] D. Poquillon, D. Monceau, Application of a simple statistical spalling model for the analysis of high-temperature, cyclic-oxidation kinetics data, Oxidation of Metals 59 (3–4) (2003) 409–431.
[1132] D. Poquillon, D. Monceau, Initiation of geometric roughening in polycrystalline metal films, Oxidation of Metals 59 (3–4) (2003) 409–431.
[1133] D. Poquillon, N. Vialas, D. Monceau, Numerical modelling of diffusion coupled with cyclic oxidation. application to alumina-forming coatings used for industrial gas turbine blades, in: Materials Science Forum, vol. 595, Trans Tech Publ, 2008, pp. 159–168.
[1134] D. Porter, K. Easterling, M. Sherif, Phase Transformation in Metals and Alloys, 3rd edition, CRC Press, 2009.
[1135] S. Poupard, J.-F. Martinez, F. Pedraza, Soft chemical stripping of aluminide coatings and oxide products on Ni superalloys, Surface & Coatings Technology 202 (2008) 3100–3108.
[1136] B. Power, F. Pedraza, B. Bouchaud, Electrolytic stripping, European patent 2679705B1, 2015.
[1137] A. Prakash, E. Bitzek, Idealized vs. realistic microstructures: an atomistic simulation case study on γ/γ' microstructures, Materials 10 (1) (2017).
[1138] A. Prakash, J. Guénolé, J. Wang, J. Müller, E. Spiecker, M. Mills, I. Povstugar, P. Choi, D. Raabe, E. Bitzek, Atom probe informed simulations of dislocation-precipitate interactions reveal the importance of local interface curvature, Acta Materialia 92 (2015) 33–45.
[1139] J. Preußner, Y. Rudnik, H. Brehm, R. Völkl, U. Glatzel, A dislocation density based material model to simulate the anisotropic creep behavior of single-phase and two-phase single crystals, International Journal of Plasticity 25 (5) (2009) 973–994.
[1140] S.V. Prikhodko, J.D. Carnes, D.G. Isaak, A.J. Ardell, Elastic constants of a Ni-12.69 at. % Al alloy from 295 to 1300K, Scripta Materialia 38 (1) (1998) 67–72.
[1141] S.V. Prikhodko, J.D. Carnes, D.G. Isaak, H. Yang, A.J. Ardell, Temperature and composition dependence of the elastic constants of Ni_3Al, Metallurgical and Materials Transactions A 30A (September 1999) 2403–2408.
[1142] M. Probst-Hein, A. Dlouhy, G. Eggeler, Interface dislocations in superalloy single crystals, Acta Materialia 47 (8) (1999) 2497–2510.
[1143] M. Probst-Hein, A. Dlouhy, G. Eggeler, Dislocation interactions in γ-channels between γ'-particles of superalloy single crystals, Materials Science and Engineering: A 319–321 (2001) 379–382.
[1144] N.A. Protasova, I.L. Svetlov, M.B. Bronfin, N.V. Petrushin, Lattice-parameter misfits between the γ and γ' phases in single crystals of nickel superalloys, The Physics of Metals and Metallography 106 (5) (2008) 495–502.
[1145] N. Provatas, K. Elder, Phase Field Methods in Materials Science and Engineering, Wiley-VCH, Germany, 2010.
[1146] L. Proville, B. Bakó, Dislocation depinning from ordered nanophases in a model fcc crystal: from cutting mechanism to Orowan looping, Acta Materialia 58 (2010) 5565–5571.
[1147] G. Pujol, F. Ansart, J.-P. Bonino, A. Malié, S. Hamadi, Step-by-step investigation of degradation mechanisms induced by CMAS attack on YSZ materials for tbc applications, Surface & Coatings Technology 237 (2013) 71–78.

[1148] G.P. Pun, Y. Mishin, A molecular dynamics study of self-diffusion in the cores of screw and edge dislocations in aluminum, Acta Materialia 57 (18) (2009) 5531–5542.

[1149] G. Purja Pun, Y. Mishin, Development of an interatomic potential for the Ni-Al system, Philosophical Magazine 89 (2009) 3245–3267.

[1150] W. Püschl, Models for dislocation cross-slip in close-packed crystal structures: a critical review, Progress in Materials Science 47 (4) (2002) 415–461.

[1151] F. Pyczak, A. Bauer, M. Goeken, S. Neumeier, U. Lorenz, M. Oehring, N. Schell, A. Schreyer, A. Stark, F. Symanzik, Plastic deformation mechanisms in a crept $l1_2$ hardened Co-base superalloy, Materials Science and Engineering: A 571 (2013) 13–18.

[1152] F. Pyczak, H. Mughrabi, CBED-Measurement of Residual Internal Strains in the Neighbourhood of TCP-Phases in Ni-Base Superalloys, John Wiley & Sons, Ltd, 2000, pp. 91–95.

[1153] F. Pyczak, S. Neumeier, M. Goeken, Influence of lattice misfit on the internal stress and strain states before and after creep investigated in nickel-base superalloys containing rhenium and ruthenium, Materials Science and Engineering: A 510–511 (2009) 295–300.

[1154] W. Qi, A. Bertram, Anisotropic continuum damage modeling for single crystals at high temperatures, International Journal of Plasticity 15 (1999) 1197–1215.

[1155] Q. Qin, J. Bassani, Non-Schmid yield behavior in single crystals, Journal of the Mechanics and Physics of Solids 40 (1992) 813–833.

[1156] S. Queyreau, J. Marian, B.D. Wirth, A. Arsenlis, Analytical integration of the forces induced by dislocations on a surface element, Modelling and Simulation in Materials Science and Engineering 22 (3) (2014) 035004.

[1157] S. Rabbolini, P. Luccarelli, S. Beretta, S. Folettia, H. Sehitoglu, Near-tip closure and cyclic plasticity in Ni-based single crystals, International Journal of Fatigue 89 (2016) 53–65.

[1158] C. Rae, N. Matan, D. Cox, M. Rist, R. Reed, On the primary creep of CMSX-4 superalloy single crystals, Materials Science and Engineering: A 31 (2000) 2219–2228.

[1159] C. Rae, N. Matan, R. Reed, The role of stacking fault shear in the primary creep of [001]-oriented single crystal superalloys at 750°c and 750 MPa, Materials Science and Engineering: A 300 (2001) 125–134.

[1160] C. Rae, R. Reed, Primary creep in single crystal superalloys: origins, mechanisms and effects, Acta Materialia 55 (2007) 1067–1081.

[1161] C.M.F. Rae, M.S. Hook, R.C. Reed, The effect of TCP morphology on the development of aluminide coated superalloys, Materials Science and Engineering: A 396 (1–2) (2005) 231–239.

[1162] C.M.F. Rae, M.S.A. Karunaratne, C.J. Small, R.W. Broomfield, C.N. Jones, R.C. Reed, Topologically close packed phases in an experimental rhenium-containing single crystal superalloy, in: Superalloys 2000, Champion, PA (USA), 2000, pp. 767–776.

[1163] C.M.F. Rae, R.C. Reed, The precipitation of topologically close-packed phases in rhenium-containing superalloys, Acta Materialia 49 (19) (2001) 4113–4125.

[1164] A. Raffaitin, F. Crabos, E. Andrieu, D. Monceau, Advanced burner-rig test for oxidation-corrosion resistance evaluation of MCrAlY/superalloys systems, Surface & Coatings Technology 201 (2006) 3829–3835.

[1165] A. Raffaitin, F. Crabos, E. Andrieu, D. Monceau, The effect of thermal cycling on the high-temperature creep behaviour of a single crystal nickel-based superalloy, Scripta Metallurgica 56 (2007) 277–280.

[1166] A. Raffaitin, D. Monceau, E. Andrieu, F. Crabos, Cyclic oxidation of coated and uncoated single-crystal nickel-based superalloy MC2 analyzed by continuous thermogravimetry analysis, Acta Materialia 54 (17) (2006) 4473–4487.

[1167] K. Rahmani, S. Nategh, Influence of aluminide diffusion coating on low cycle fatigue properties of rené 80, Materials Science and Engineering: A 486 (1–2) (2008) 686–695.

[1168] B. Rajasekaran, G. Mauer, R. Vaßen, Enhanced characteristics of hvof-sprayed mcraly bond coats for tbc applications, Journal of Thermal Spray Technology 20 (6) (2011) 1209–1216.

[1169] N. Rakotomalala, Coupled thermomechanical simulation of the failure of thermal barrier coatings of turbine blades (in French), PhD thesis, Mines ParisTech, 2014.

[1170] F. Ram, Z. Li, S. Zaefferer, S.M.H. Haghighat, Z. Zhu, D. Raabe, R.C. Reed, On the origin of creep dislocations in a ni-base, single-crystal superalloy: an ecci, ebsd, and dislocation dynamics-based study, Acta Materialia 109 (2016) 151–161.

[1171] J. Ramirez, C. Beckermann, Evaluation of a Rayleigh-number-based freckle criterion for Pb-Sn alloys and Ni-base superalloys, Metallurgical and Materials Transactions A 34 (2003) 1525–1536.

[1172] S. Rao, D. Dimiduk, J. El-Awady, T. Parthasarathy, M. Uchic, C. Woodward, Atomistic simulations of cross-slip nucleation at screw dislocation intersections in face-centered cubic nickel, Philosophical Magazine 89 (34–36) (2009) 3351–3369.

[1173] S. Rao, D. Dimiduk, J. El-Awady, T. Parthasarathy, M. Uchic, C. Woodward, Activated states for cross-slip at screw dislocation intersections in face-centered cubic nickel and copper via atomistic simulation, Acta Materialia 58 (17) (2010) 5547–5557.

[1174] S. Rao, D. Dimiduk, J. El-Awady, T. Parthasarathy, M. Uchic, C. Woodward, Spontaneous athermal cross-slip nucleation at screw dislocation intersections in FCC metals and L1 2 intermetallics investigated via atomistic simulations, Philosophical Magazine 93 (22) (2013) 3012–3028.

[1175] S. Rao, D. Dimiduk, J. El-Awady, T. Parthasarathy, M. Uchic, C. Woodward, Screw dislocation cross slip at cross-slip plane jogs and screw dipole annihilation in fcc cu and ni investigated via atomistic simulations, Acta Materialia 101 (2015) 10–15.

[1176] S. Rao, D. Dimiduk, T. Parthasarathy, J. El-Awady, C. Woodward, M. Uchic, Calculations of intersection cross-slip activation energies in fcc metals using nudged elastic band method, Acta Materialia 59 (19) (2011) 7135–7144.

[1177] S. Rao, D. Dimiduk, T. Parthasarathy, M. Uchic, C. Woodward, Atomistic simulations of intersection cross-slip nucleation in $L1_2$ Ni_3Al, Scripta Materialia 66 (6) (2012) 410–413.

[1178] S. Rao, D. Dimiduk, T. Parthasarathy, M. Uchic, C. Woodward, Atomistic simulations of surface cross-slip nucleation in face-centered cubic nickel and copper, Acta Materialia 61 (7) (2013) 2500–2508.

[1179] S. Rao, C. Woodward, T. Parthasarathy, O. Senkov, Atomistic simulations of dislocation behavior in a model FCC multicomponent concentrated solid solution alloy, Acta Materialia 134 (aug 2017) 188–194.

[1180] Y. Rao, T. Smith, M. Mills, M. Ghazisaeidi, Segregation of alloying elements to planar faults in γ'-Ni_3Al, Acta Materialia 148 (2018) 173–184.

[1181] T. Rasmussen, K.W. Jacobsen, T. Leffers, O.B. Pedersen, Simulations of the atomic structure, energetics, and cross slip of screw dislocations in copper, Physical Review B 56 (6) (1997) 2977.

[1182] G.M. Rassweiler, W.L. Grube, Internal Stresses and Fatigue in Metals: Proceedings, Elsevier Pub. Co., 1959.

[1183] N. Ratel, P. Bastie, T. Mori, P. Withers, Application of anisotropic inclusion theory to the energy evaluation for the matrix channel deformation and rafting geometry of $\gamma - \gamma'$ ni superalloys, Materials Science and Engineering: A 505 (1) (2009) 41–47.

[1184] H.-J. Rätzer-Scheibe, U. Schulz, The effects of heat treatment and gas atmosphere on the thermal conductivity of aps and EB-PVD PYSZ thermal barrier coatings, Surface & Coatings Technology 201 (18) (2007) 7880–7888.

[1185] H.-J. Rätzer-Scheibe, U. Schulz, T. Krell, The effect of coating thickness on the thermal conductivity of EB-PVD PYSZ thermal barrier coatings, Surface & Coatings Technology 200 (18–19) (2006) 5636–5644.

[1186] R. Rawlings, A. Staton-Bevan, The alloying behaviour and mechanical properties of polycrystalline Ni_3Al (γ' phase) with ternary additions, Journal of Materials Science 10 (3) (1975) 505–514.

[1187] I. Ray, D. Cockayne, The observation of dissociated dislocations in silicon, Philosophical Magazine 22 (178) (1970) 853–856.

[1188] D. Raynor, J. Silcock, Strengthening mechanisms in γ' precipitating alloys, Metal Science Journal 4 (1970) 121–130.

[1189] P. Reed, X. Wu, I. Sinclair, Fatigue crack path prediction in UDIMET 720 nickel based alloy single crystals, Metallurgical and Materials Transactions A 31 (2000) 109–123.

[1190] R. Reed, The Superalloys – Fundamentals and Applications, Cambridge University Press, 2006.

[1191] R. Reed, D. Cox, C. Rae, Kinetics of rafting in a single crystal superalloy: effects of residual microsegregation, Materials Science and Technology 23 (2007) 893–902.

[1192] R. Reed, C. Rae, Physical metallurgy of Ni-based superalloys, in: D. Laughlin, K. Hono (Eds.), Physical Metallurgy, 5th edition, Springer, 2014, pp. 2215–2290.

[1193] R. Reed, T. Tao, N. Warnken, Alloys-by-design: application to nickel-based single crystal superalloys, Acta Materialia 57 (2009) 5898–5913.

[1194] R.C. Reed, D.C. Cox, C.M.F. Rae, Damage accumulation during creep deformation of a single crystal superalloy at 1150°C, Materials Science and Engineering: A 448 (1–2) (2007) 88–96.

[1195] M. Reger, L. Rémy, High temperature low cycle fatigue behaviour of IN100 superalloy. I. Influence of temperature on the low cycle fatigue behaviour, Materials Science and Engineering: A 101 (1988) 47–57;
M. Reger, L. Rémy, High temperature low cycle fatigue behaviour of IN100 superalloy. II. Influence of frequency and environment at high temperature, Materials Science and Engineering: A 101 (1988) 55–63.

[1196] M. Reid, M.J. Pomeroy, J.S. Robinson, Microstructural instability in coated single crystal superalloys, Journal of Materials Processing Technology 153 (2004) 660–665.

[1197] L. Reinhard, B. Schönfeld, G. Kostorz, W. Bührer, Short-range order in α-brass, Physical Review B 41 (4) (1990) 1727.

[1198] G. Reinhart, D. Grange, L. Abou-Khalil, N. Mangelinck-Noël, N. Niane, V. Maguin, G. Guillemot, C.-A. Gandin, H. Nguyen-Thi, Impact of solute flow during directional solidification of a Ni-based alloy: in situ and real-time X-radiography, Acta Materialia 194 (2020) 68–79, https://doi.org/10.1016/j.actamat.2020.04.003.

[1199] L. Remy, Oxidation effects in high temperature creep and fatigue of engineering alloys, in: Corrosion–Deformation Interactions, Fontainebleau, France, 5–7 Oct. 1992, Ecole Nationale Superieure des Mines de Paris, Les Editions de Physique, 1993, Avenue du Hoggar, Zone Industrielle de Courtaboeuf, B.P. 112, F-91944 Les Ulis Cedex A, France, 5–7 Oct. 1992.

[1200] L. Rémy, Thermal mechanical fatigue (including thermal shock), in: R.R.I. Milne, R. Karihakoo (Eds.), Comprehensive Structural Integrity, vol. 5, Elsevier, 2003, pp. 113–200.

[1201] L. Rémy, A. Alam, A. Bickard, Thermo-mechanical creep-fatigue of coated systems, in: J. Bressers, M.A. McGaw, S. Kalluri, S. Peteves (Eds.), Thermomechanical Fatigue Behavior of Materials, vol. STP 1428, ASTM, 2003, pp. 98–111.

[1202] L. Rémy, M. Geuffrard, A. Alam, A. Koster, E. Fleury, Effect of microstructure in high temperature fatigue: lifetime to crack initiation of a single crystal superalloy in high temperature low cycle fatigue, International Journal of Fracture 57 (2013).

[1203] L. Rémy, C. Guerre, I. Rouzou, R. Molins, Assessment of TBC oxidation induced degradation using compression tests, Oxidation of Metals 81 (2014) 3–15.

[1204] L. Rémy, N. Haddar, A. Alam, A. Koster, N. Marchal, Growth of small cracks and prediction of lifetime in high temperature alloys, Materials Science and Engineering: A 468 (2007) 40–50.

[1205] L. Rémy, J. Petit, Temperature-Fatigue Interaction, vol. 29, Esis Publication, 2002.
[1206] E. Renner, Y. Gaillard, F. Richard, F. Amiot, P. Delobelle, Sensitivity of the residual topography to single crystal plasticity parameters in Berkovich nanoindentation on fcc nickel, International Journal of Plasticity 77 (2016) 118–140.
[1207] B. Reppich, Some new aspects concerning particle hardening mechanisms in γ' precipitating Ni-base alloys – I. Theoretical concept, Acta Materialia 30 (1982) 87–94.
[1208] B. Reppich, Negative creep, Zeitschrift für Metallkunde 75 (1984) 193–202.
[1209] B. Reppich, Particle strengthening, in: H. Mughrabi (Ed.), Plastic Deformation and Fracture, in: R.W. Cahn, P. Haasen, E.J. Kramer (Eds.), Materials Science and Technology, vol. 6, Wiley-VCH, 1993, p. 312.
[1210] B. Reppich, P. Schepp, G. Wehner, Some new aspects concerning particle hardening mechanisms in γ' precipitating Ni-base alloys–II. Experiments, Acta Materialia 30 (1982) 95–104.
[1211] Research project, Structures Chaudes, SAFRAN–ONERA–CNRS, 2005–2010.
[1212] L. Rettberg, B. Goodlet, T. Pollock, Detecting recrystallization in a single crystal Ni-base alloy using resonant ultrasound spectroscopy, NDTE International 83 (2016) 68–77.
[1213] L. Rettberg, M. Tsunekane, T. Pollock, Rejuvenation of nickel-based superalloys GTD444(DS) and René N5(SX), in: E.S. Huron, R.C. Reed, M.C. Hardy, M.J. Mills, R.E. Montero, P.D. Portella, J. Telesman (Eds.), Superalloys 2012: Proc. of the Twelfth International Symposium on Superalloys, Champion, USA, TMS, Minerals, Metals & Materials Society, Sept. 2012.
[1214] R. Rettig, N. Ritter, F. Müller, M. Franke, R. Singer, Optimization of the homogenization heat treatment of nickel-based superalloys based on phase-field simulations: numerical methods and experimental validation, Materials & Design 46 (2015) 5842–5855.
[1215] M. Rezaei, A. Kermanpur, F. Sadeghi, A three-dimensional cellular automation-finite element model for the prediction of solidification grain structures, Journal of Crystal Growth 485 (2018) 19–27.
[1216] S. Rezanka, G. Mauer, R. Vaßen, Improved thermal cycling durability of thermal barrier coatings manufactured by ps-pvd, Journal of Thermal Spray Technology 23 (1–2) (2014) 182–189.
[1217] T.N. Rhys-Jones, Coatings for blade and vane applications in gas turbines, 1989.
[1218] J. Rice, On the structure of stress-strain relations for time–dependent plastic deformation in metals, Journal of Applied Mechanics 37 (1970) 728.
[1219] J. Rice, Inelastic constitutive relations for solids: an internal variable theory and its application to metal plasticity, Journal of the Mechanics and Physics of Solids 19 (1971) 433–455.
[1220] J. Rice, Tensile crack tip fields in elastic-ideally plastic crystals, Mechanics of Materials 6 (1987) 317–335.
[1221] E. Rodary, D. Rodney, L. Proville, Y. Bréchet, G. Martin, Dislocation glide in model Ni(Al) solid solutions by molecular dynamics, Physical Review B 70 (5) (aug 2004) 1–11.
[1222] D. Rodney, A. Finel, Phase field methods and dislocations, in: M. Aindow, M. Asta, M. Glazov, D. Medlin, A. Rollet, M. Zaiser (Eds.), Influences of Interface and Dislocation Behavior on Microstructure Evolution, in: MRS Proceedings, vol. 652, 2000.
[1223] D. Rodney, Y. Le Bouar, A. Finel, Phase field methods and dislocations, Acta Materialia 51 (2003) 17–30.
[1224] K. Roe, T. Siegmund, An irreversible cohesive zone model for interface fatigue crack growth simulation, Engineering Fracture Mechanics 70 (2) (2003) 209–232.
[1225] B. Roebuck, D. Cox, R. Reed, The temperature dependence of γ' volume fraction in a ni base single crystal superalloy from resistivity measurements, Scripta Metallurgica 44 (2001) 917–921.

[1226] B. Roebuck, D. Cox, R. Reed, An innovative device for the mechanical testing of miniature specimens of superalloys, in: K.A. Green, T.M. Pollock, H. Harada, T.E. Howson, R.C. Reed, J.J. Schirra, S. Walston (Eds.), Superalloys 2004: Proc. of the Tenth International Symposium on Superalloys, Champion, USA, TMS, Minerals, Metals & Materials Society, Sept. 2004.

[1227] B. Roebuck, D. Cox, R.C. Reed, The temperature dependence of γ' volume fraction in a Ni-based single crystal superalloy from resistivity measurements, Scripta Materialia 44 (6) (2001) 917–921.

[1228] H. Roelofs, B. Schönfeld, G. Kostorz, W. Bührer, J. Robertson, P. Zschack, G. Ice, Atomic and magnetic short-range order in cu-17 at.% mn, Scripta Materialia 34 (9) (1996).

[1229] J. Rogal, S. Divinski, M. Finnis, A. Glensk, J. Neugebauer, J. Perepezko, S. Schuwalow, M. Sluiter, B. Sundman, Perspectives on point defect thermodynamics, Physica Status Solidi B 251 (1) (2014) 97–129.

[1230] I. Roslyakova, B. Sundman, H. Dette, L. Zhang, I. Steinbach, Modeling of Gibbs energies of pure elements down to 0 k using segmented regression, CALPHAD Journal 55 (2016).

[1231] I. Roslyakova, S. Zomorodpoosh, Y. Jiang, A. Obaied, B. Bocklund, R. Otis, L. Zhang, Z. Liu, Third generation calphad databases: new unary database and its application for re-assessment of binary systems (oral presentation), in: CALPHAD XLVII Conference, Juriquilla, Queretaro, Mexico, 2018.

[1232] E. Ross, K. O'Hara, René N4: a first generation single crystal turbine airfoil with improved oxidation resistance, low angle boundary strength and superior long time rupture strength, in: Proc. 8th Int. Symp. on Superalloys, TMS, 1996.

[1233] L. Rossman, B. Sarley, J. Hernandez, P. Kenesei, A. Köster, J. Wischek, J. Almer, V. Maurel, M. Bartsch, S. Raghavan, Method for conducting in situ high-temperature digital image correlation with simultaneous synchrotron measurements under thermomechanical conditions, Review of Scientific Instruments 91 (2020) 033705.

[1234] F. Roters, D. Raabe, G. Gottstein, Calculation of stress—strain curves by using 2 dimensional dislocation dynamics, Composite Materials Science 7 (1) (1996) 56–62, Zeolites and related materials: trends, targets and challenges.

[1235] A. Rowe, J. Wells, G. West, R. Thompson, Microstructural evolution of single crystal and directionally solidified rejuvenated nickel superalloys, in: E.S. Huron, R.C. Reed, M.C. Hardy, M.J. Mills, R.E. Montero, P.D. Portella, J. Telesman (Eds.), Superalloys 2012: Proc. of the Twelfth International Symposium on Superalloys, Champion, USA, TMS, Minerals, Metals & Materials Society, Sept. 2012.

[1236] N. Roy, R.N. Ghosh, Modelling effects of specimen size and shape on creep of metals and alloys, Scripta Materialia 36 (12) (1997) 1367–1372.

[1237] A. Royer, P. Bastie, D. Bellet, J.-L. Strudel, Temperature dependence of the lattice mismatch of the am1 superalloy influence of the γ' precipitates morphology, Philosophical Magazine 72 (1995) 669–689.

[1238] A. Royer, P. Bastie, M. Véron, In situ determination of γ' phase volume fraction and of relations between lattice parameters and precipitate morphology in Ni-based single crystal superalloy, Acta Materialia 46 (15) (1998) 5357–5368.

[1239] A. Royer, A. Jacques, P. Bastie, M. Veron, The evolution of the lattice parameter mismatch of nickel based superalloy: an in situ study during creep deformation, Materials Science and Engineering: A 319–321 (2001) 800–804.

[1240] A. Ruban, V. Popov, V. Portnoi, V. Bogdanov, First-principles study of point defects in Ni_3Al, Philosophical Magazine 94 (1) (2014) 20–34.

[1241] A. Ruban, H. Skriver, Calculated site substitution in ternary γ'-Ni_3Al: temperature and composition effects, Physical Review B 55 (1997) 856–874.

[1242] A. Ruffini, Y. Le Bouar, A. Finel, Three-dimensional phase-field model of dislocations for a heterogeneous face-centered cubic crystals, Journal of the Mechanics and Physics of Solids 105 (2017) 95–115.

[1243] B. Ruttert, D. Bürger, L. Roncery, A. Parsa, P. Wollgramm, G. Eggeler, W. Theisen, Rejuvenation of creep resistance of a Ni-base single-crystal superalloy by hot isostatic pressing, Materials & Design 134 (2017) 418–425.

[1244] B. Ruttert, O. Horst, I. Lopez-Galilea, D. Langenkämper, A. Kostka, C. Somsen, J.V. Goerler, M. Ali, O. Shchyglo, I. Steinbach, G. Eggeler, W. Theisen, Rejuvenation of single-crystal Ni-base superalloy turbine blades: unlimited service life? Metallurgical and Materials Transactions A 49 (2018) 4262–4273.

[1245] B. Ruttert, M. Ramsperger, L. Roncery, I. Lopez-Galilea, C. Körner, W. Theisen, Impact of hot isostatic pressing on microstructures of CMSX-4 Ni-base superalloy fabricated by selective electron beam melting, Materials & Design 110 (2016) 720–727.

[1246] A. Saad, C.-A. Gandin, M. Bellet, N. Shevchenko, S. Eckert, Simulation of channel segregation during directional solidification of In-75 wt% Ga. Qualitative comparison with insitu observations, Metallurgical and Materials Transactions A 46 (2015) 4886–4897.

[1247] G. Saada, P. Veyssière, Chapter 61 work hardening of face centred cubic crystals. dislocations intersection and cross slip, in: F. Nabarro, M. Duesbery (Eds.), Dislocations in Solids, in: Dislocations in Solids, vol. 11, Elsevier, 2002, pp. 413–458.

[1248] K. Saanouni, M. Hamed, Micromorphic approach for finite gradient-elastoplasticity fully coupled with ductile damage: formulation and computational aspects, International Journal of Solids and Structures 50 (2013) 2289–2309.

[1249] P. Sabnis, S. Forest, J. Cormier, Microdamage modelling of crack initiation and propagation in fcc single crystals under complex loading conditions, Computer Methods in Applied Mechanics and Engineering 312 (2016) 468–491.

[1250] P.A. Sabnis, S. Forest, N.K. Arakere, V. Yastrebov, Crystal plasticity analysis of cylindrical indentation on a Ni-base single crystal superalloy, International Journal of Plasticity 51 (2013) 200–213.

[1251] P.A. Sabnis, M. Mazière, S. Forest, N.K. Arakere, F. Ebrahimi, Effect of secondary orientation on notch–tip plasticity in superalloy single crystals, International Journal of Plasticity 28 (2012) 102–123.

[1252] C. Sagui, D. Orlikowski, A.M. Somoza, C. Roland, Three-dimensional simulations of Ostwald ripening with elastic effects, Physical Review E 58 (4) (1998) 4092–4095.

[1253] C. Sagui, A.M. Somoza, R.C. Desai, Spinodal decomposition in an order-disorder phase transition with elastic effects, Physical Review E 50 (6) (1994) 4865–4879.

[1254] F. Saint-Antonin, J.-L. Strudel, Stress relaxation in a Ni-base superalloy after low initial straining, in: B. Whilshire, R. Evans (Eds.), Proc. of the Fourth Int. Conf. on Creep and Fracture of Engineering Materials and Structures, Swansea, 1990, pp. 303–312.

[1255] M. Sakaguchi, T. Tsuru, M. Okazaki, Fatigue crack propagation in thin-wall superalloys components: experimental investigation via miniature CT specimen, in: E.S. Huron, R.C. Reed, M.C. Hardy, M.J. Mills, R.E. Montero, P.D. Portella, J. Telesman (Eds.), Superalloys 2012: Proc. of the Twelfth International Symposium on Superalloys, Champion, USA, TMS, Minerals, Metals & Materials Society, Sept. 2012.

[1256] M. Sakane, S. Zhang, A. Yoshinari, N. Matsuda, N. Isobe, Multiaxial low cycle fatigue for ni-base single crystal super alloy at high temperature, Advanced Materials Modelling for Structures 19 (2013) 297–305.

[1257] P. Sallot, Modélisation de la durée de vie d'un revêtement aluminoformeur en conditions de sollicitations thermo-mécaniques (in French), PhD thesis, Mines ParisTech, 2012.

[1258] P. Sallot, V. Maurel, L. Rémy, F. N'Guyen, A. Longuet, Microstructure evolution of a platinum-modified nickel-aluminide coating during thermal and thermo-mechanical fatigue, Metallurgical and Materials Transactions A 46 (10) (2015) 4589–4600.

[1259] V. Samaee, R. Gatti, B. Devincre, T. Pardoen, D. Schryvers, H. Idrissi, Dislocation driven nanosample plasticity: new insights from quantitative in-situ tem tensile testing, Scientific Reports 8 (1) (2018) 12012.

[1260] S. Sampath, U. Schulz, M.O. Jarligo, S. Kuroda, Processing science of advanced thermal-barrier systems, MRS Bulletin 37 (10) (2012) 903–910.

[1261] H. Sapardanis, Fissuration à l'interface d'un revêtement plasma céramique et d'un substrat métallique sous sollicitations dynamique et quasi-statique multiaxiales (in French), PhD thesis, Paris Sciences et Lettres, 2016.

[1262] H. Sapardanis, V. Maurel, A. Köster, S. Duvinage, F. Borit, V. Guipont, Influence of macroscopic shear loading on the growth of an interfacial crack initiated from a ceramic blister processed by laser shock, Surface & Coatings Technology 291 (2016) 430–443.

[1263] S. Sarkar, J. Li, W. Cox, E. Bitzek, T. Lenosky, Y. Wang, Finding activation pathway of coupled displacive-diffusional defect processes in atomistics: dislocation climb in fcc copper, Physical Review B 86 (1) (2012) 1.

[1264] P. Sarosi, R. Srinivasan, G. Eggeler, M. Nathal, M. Mills, Observations of a< 0 1 0> dislocations during the high-temperature creep of ni-based superalloy single crystals deformed along the [0 0 1] orientation, Acta Materialia 55 (7) (2007) 2509–2518.

[1265] P.M. Sarosi, B. Wang, J.P. Simmons, Y. Wang, M.J. Mills, Formation of multimodal size distributions of γ' in a nickel-base superalloy during interrupted continuous cooling, Scripta Materialia 57 (2007) 767–770.

[1266] V. Sass, U. Glatzel, M. Feller-Kniepmeier, Anisotropic creep properties of the Ni-base superalloy CMSX-4, Acta Materialia 44 (1996) 1967–1977.

[1267] V. Sass, W. Schneider, H. Mughrabi, On the orientation dependence of the intermediate-temperature creep behaviour of a monocrystalline nickel-base superalloy, Scripta Metallurgica 31 (1994) 885–890.

[1268] A. Sato, H. Harada, A.-C. Yeh, K. Kawagishi, T. Kobayashi, Y. Koizumi, T. Yokokawa, J.-X. Zhang, A 5th generation SC superalloy with balanced high temperature properties and processability, in: Proc. 12th Int. Symp. on Superalloys, TMS, 2012.

[1269] N. Saunders, A.e. Miodownik, CALPHAD: Calculation of Phase Diagrams, A Comprehensive Guide, Pergamon, 1998.

[1270] A. Sawant, S. Tin, J. Zhao, High temperature nanoindentation of Ni-base superalloys, in: R.C. Reed, K.A. Green, P. Caron, T.P. Gabb, M.G. Fahrmann, E.S. Huron (Eds.), Superalloys 2008: Proc. of the Eleventh International Symposium on Superalloys, Champion, USA, TMS, Minerals, Metals & Materials Society, Sept. 2008.

[1271] R. Schaefer, M. Vaudin, B. Mueller, C. Choi, J. Szekely, Generation of defects in single crystal components by dendrite remelting, in: 7th Conference on Modeling of Casting, Welding and Advanced Solidification Processes, Warrendale, PA, USA, 1995, pp. 593–600.

[1272] E. Scheil, Bemerkungen zur schichtkristallbildung, Zeitschrift für Metallkunde 34 (1942) 69.

[1273] T. Schenk, R. Trehorel, L. Dirand, A. Jacques, Dislocation densities and velocities within the γ channels of an sx superalloy during in situ high-temperature creep tests, Materials 11 (9) (2018) 1527.

[1274] J.-M. Scherer, J. Besson, S. Forest, J. Hure, B. Tanguy, Strain gradient crystal plasticity with evolving length scale: application to voided irradiated materials, European Journal of Mechanics - A/Solids 77 (2019) 103768.

[1275] G. Scheunemann-Frerker, H. Gabrisch, M. Feller-Kniepmeier, Dislocation microstructure in a single-crystal nickel-based superalloy after tensile testing at 823 k in the [001] direction, Philosophical Magazine. A 65 (6) (1992) 1353–1368.

[1276] E. Schmid, W. Boas, Plasticity of Crystals with Special Reference to Metals, Chapman&Hall, 1968. Initial title, Kristallplastizität: mit der besonderer Berücksichtigung Metalle, 1935.

[1277] N. Schmidt, J. Bilde-Sørensen, D.J. Jensen, D. Joy, J. Hjelen, Band positions used for on-line crystallographic orientation determination from electron back scattering patterns. Discussion, Scanning Microscopy 5 (3) (1991) 637–643.

[1278] R. Schmidt, M. Feller-Kniepmeier, Phase chemistry of a nickel-base superalloy after creep experiments, Scripta Metallurgica 29 (1993).

[1279] M. Schmidt-Baldassari, Numerical concepts for rate-independent single crystal plasticity, Computer Methods in Applied Mechanics and Engineering 192 (11–12) (2003) 1261–1280.

[1280] D. Schneider, O. Tschukin, T. Bohlke, B. Nestler, Phase field elasticity model based on mechanical jump conditions, Computational Mechanics (2015).

[1281] K. Schneider, H.W. Grünling, Mechanical aspects of high temperature coatings, Thin Solid Films 107 (4) (1983) 395–416.

[1282] W. Schneider, H. Mughrabi, J. Hammer, Creep deformation and rupture behaviour of the mono crystalline superalloy CMSX-4 – a comparison with the alloy SRR 99, in: I. Staff, S. Antolovich (Eds.), Superalloys, 1992: Proc. of the Seventh International Symposium on Superalloys, Champion, USA, TMS, Minerals, Metals & Materials Society, Sept. 1992, pp. 589–598.

[1283] G. Schoeck, A. Seeger, Defects in Crystalline Solids, Physical Society, London, 1955.

[1284] A. Scholz, A. Schmidt, H. Walther, M. Schein, M. Schwienheer, Experiences in the determination of TMF, LCF, and creep life of CMSX-4 in four-point-bending experiments, International Journal of Fatigue 30 (2008) 357–362.

[1285] B. Schönfeld, Local atomic arrangements in binary alloys, Progress in Materials Science 44 (5) (1999) 435–543.

[1286] B. Schönfeld, L. Reinhard, G. Kostorz, W. Bührer, Short-range order and atomic displacements in ni–20 at% cr single crystals, Physica Status Solidi B 148 (2) (1988) 457–471.

[1287] H.L. Schreyer, Q.H. Zuo, Anisotropic yield surfaces based on elastic projection operators, Journal of Applied Mechanics 62 (3) (Sep 1995) 780–785.

[1288] F. Schubert, T. Rieck, P. Ennis, The growth of small cracks in the single crystal superalloy CMSX-4 at 750 and 1000°C, in: T.M. Pollock, R.D. Kissinger, R.R. Bowman, K.A. Green, M. McLean, S.L. Olson, J.J. Schirra (Eds.), Superalloys 2000, Proc. of the 9th International Symposium on Superalloys, Champion, USA, TMS, Minerals, Metals & Materials Society, Sept. 1992.

[1289] U. Schulz, C. Leyens, K. Fritscher, M. Peters, B. Saruhan-Brings, O. Lavigne, J.-M. Dorvaux, M. Poulain, R. Mévrel, M. Caliez, Some recent trends in research and technology of advanced thermal barrier coatings, Aerospace Science and Technology 7 (2003) 73–80.

[1290] U. Schulz, M. Menzebach, C. Leyens, Y. Yang, Influence of substrate material on oxidation behavior and cyclic lifetime of EB-PVD tbc systems, Surface & Coatings Technology 146 (2001) 117–123.

[1291] S. Schuwalow, J. Rogal, R. Drautz, Vacancy mobility and interaction with transition metal solutes in Ni, Journal of Physics. Condensed Matter 26 (48) (2014).

[1292] P. Schwander, B. Schönfeld, G. Kostorz, Configurational energy change caused by slip in short-range ordered ni–mo, Physica Status Solidi B 172 (1) (1992) 73–85.

[1293] L.H. Schwartz, J.B. Cohen, The Nature of Diffraction, Springer Berlin Heidelberg, Berlin, Heidelberg, 1987, pp. 46–76.

[1294] K.W. Schwarz, Discrete dislocation dynamics study of strained-layer relaxation, Physical Review Letters 91 (Oct 2003) 145503.

[1295] R. Schwarz, R. Labusch, Dynamic simulation of solution hardening, Journal of Applied Physics 49 (1978) 5174–5187.

[1296] W. Schweika, H.-G. Haubold, Short-range order and atomic interaction in nicr x, in: Atomic Transport and Defects in Metals by Neutron Scattering, Springer, 1986, pp. 22–27.

[1297] B. Seiser, R. Drautz, D.G. Pettifor, TCP phase predictions in Ni-based superalloys: structure maps revisited, Acta Materialia 59 (2) (2011) 749–763.

[1298] B. Seiser, T. Hammerschmidt, A. Kolmogorov, R. Drautz, D. Pettifor, Theory of structural trends within 4d and 5d transition metals topologically close-packed phases, Physical Review B 83 (2011) 224116.

[1299] C. Seitz, M. Weisser, M. Gomm, R. Hock, A. Magerl, A high-energy triple-axis x-ray diffractometer for the study of the structure of bulk crystals, Journal of Applied Crystallography 37 (6) (2004) 901–910.

[1300] S.L. Semiatin, R.C. Kramb, R.E. Turner, F. Zhang, M.M. Antony, Analysis of the homogenization of a nickel-base superalloy, Scripta Materialia 51 (6) (2004) 491–495.

[1301] A. Sengupta, S. Putatunda, L. Bartosiewicz, J. Hangas, P. Nailos, M. Peputapeck, F. Alberts, Tensile behavior of a new single crystal nickel-based superalloy (CMSX-4) at room and elevated temperatures, Journal of Materials Engineering and Performance 3 (1994) 73–81.

[1302] K. Serin, G. Göbenli, G. Eggeler, On the influence of stress state, stress level and temperature on γ-channel widening in the single crystal superalloy cmsx-4, Materials Science and Engineering: A 387 (2004) 133–137.

[1303] P.A. Shade, Small scale mechanical testing techniques and application to evaluate a single crystal nickel superalloy, PhD thesis, The Ohio State University, USA, 2008.

[1304] D. Shah, D. Duhl, The effect of orientation, temperature and γ' size on the yield strength of Ni_base superalloy, in: M. Gell, S. Kortovich, H. Bricknell, B. Kent, J. Radavich (Eds.), Superalloys 1984: Proc. of the Fifth International Symposium on Superalloys, Metall. Soc. of AIME, Oct. 1984, pp. 105–114.

[1305] D. Shah, S. Vega, S. Woodward, A. Cetel, Primary creep in nickel-base superalloys, in: K.A. Green, T.M. Pollock, H. Harada, T.E. Howson, R.C. Reed, J.J. Schirra, S. Walston (Eds.), Superalloys 2004: Proc. of the Tenth International Symposium on Superalloys, Champion, USA, TMS, Minerals, Metals & Materials Society, Sept. 2004, pp. 197–206.

[1306] J. Shang, F. Yang, C. Li, N. Wei, X. Tan, Size effect on the plastic deformation of pre-void Ni/Ni_3Al interface under uniaxial tension: a molecular dynamics simulation, Composite Materials Science 148 (2018) (2018) 200–206.

[1307] S.-L. Shang, Y. Wang, Z.-K. Liu, ESPEI: extensible, self-optimizing phase equilibrium infrastructure for magnesium alloys, 2010.

[1308] C. Shen, J.P. Simmons, Y. Wang, Effect of elastic interaction on nucleation: II implementation of strain energy of nucleus formation in the phase field method, Acta Materialia 55 (2007) 1457–1466.

[1309] C. Shen, Y. Wang, Incorporation of γ-surface to phase field model of dislocations: simulating dislocation dissociation in fcc crystal, Acta Materialia 52 (2004) 683–691.

[1310] M. Shenoy, D. McDowell, R. Neu, Transversely isotropic viscoplasticity model for a directionally solidified Ni-base superalloy, International Journal of Plasticity 22 (2005) 2301–2326.

[1311] D. Shi, C. Wang, X. Yang, S. Li, Failure assessment of the first stage high-pressure turbine blades in an aero-engine turbine, Fatigue & Fracture of Engineering Materials & Structures 40 (2017) 2092–2106.

[1312] J. Shi, A.M. Karlsson, B. Baufeld, M. Bartsch, Evolution of surface morphology of thermo-mechanically cycled nicocraly bond coats, Materials Science and Engineering: A 434 (1–2) (2006) 39–52.

[1313] X.J. Shi, L. Dupuy, B. Devincre, D. Terentyev, L. Vincent, Interaction of $< 100 >$ dislocation loops with dislocations studied by dislocation dynamics in α-iron, Journal of Nuclear Materials 460 (2015) 37–43.

[1314] Z. Shi, J. Li, S. Liu, M. Han, High cycle fatigue behavior of the second generation single crystal superalloy DD6, Transactions of Nonferrous Metals Society of China 21 (2011) 998–1003.

[1315] Z. Shi, S. Liu, J. Li, Rejuvenation heat treatment of the second-generation single-crystal superalloy DD6, Acta Metallurgica Sinica (English Letters) 28 (2015) 1278–1285.

[1316] Z. Shi, S. Liu, X. Wang, J. Li, Effects of heat treatment on surface recrystallization and stress rupture properties of a fourth-generation single-crystal superalloy after grit blasting, Acta Metallurgica Sinica (English Letters) 30 (2017) 614–620.

[1317] C. Shih, Cracks on bimaterial interfaces: elasticity and plasticity aspects, Materials Science and Engineering: A 143 (1–2) (1991) 77–90.

[1318] R. Shimizu, Development of electron probe instrumentation during those early days when professor castaing visited Japan, Microscopy and Microanalysis 7 (2) (2001) 119–123.

[1319] C.S. Shin, M.C. Fivel, M. Verdier, S.C. Kwon, Numerical methods to improve the computing efficiency of discrete dislocation dynamics simulations, Journal of Computational Physics 215 (2) (2006) 417–429.

[1320] K. Shreiber, D. Mordehai, Dislocation-nucleation-controlled deformation of Ni_3Al nanocubes in molecular dynamics simulations, Modelling and Simulation in Materials Science and Engineering 23 (8) (2015) 85004.

[1321] L. Shui, T. Jin, S. Tian, Z. Hu, Influence of precipitate morphology on tensile creep of a single crystal nickel-base superalloy, Materials Science and Engineering: A 454–455 (72) (2007) 461–466.

[1322] L. Shui, S. Tian, T. Jin, Z. Hu, Influence of pre-compression on microstructure and creep characteristic of a single crystal nickel-base superalloy, Materials Science and Engineering: A 418 (1–2) (2006) 229–235.

[1323] N. Siderey, M. Boufoussi, S. Denis, J. Lacaze, Dendritic growth and crystalline quality of nickel-base single grains, Journal of Crystal Growth 130 (1993) 132–146.

[1324] D. Siebörger, H. Knake, U. Glatzel, Temperature dependence of the elastic moduli of the nickel-base superalloy CMSX-4 and its isolated phases, Materials Science and Engineering: A 298 (2001) 26–33.

[1325] R. Sills, W. Kuykendall, A. Aghaei, W. Cai, Fundamentals of dislocation dynamics simulations, in: C. Weinberger (Ed.), Multiscale Materials Modelling for Nanomechanics, in: Springer Series in Materials Science, 2016.

[1326] R.B. Sills, A. Aghaei, W. Cai, Advanced time integration algorithms for dislocation dynamics simulations of work hardening, Modelling and Simulation in Materials Science and Engineering 24 (4) (2016) 045019.

[1327] G. Simmons, H. Wang, Single Crystal Elastic Constants and Calculated Aggregated Properties: A Handbook, The MIT Press, Cambridge, MA, 1971.

[1328] J.P. Simmons, C. Shen, Y. Wang, Phase field modeling of simultaneous nucleation and growth by explicitly incorporating nucleation events, Scripta Materialia 43 (2000) 935–942.

[1329] M. Simonetti, P. Caron, Role and behaviour of μ phase during deformation of a nickel-based single crystal superalloy, Materials Science and Engineering: A 254 (1–2) (1998) 1–12.

[1330] C. Sims, A history of superalloy metallurgy for superalloys metallurgists, in: M. Gell, S. Kortovich, H. Bricknell, B. Kent, J. Radavich (Eds.), Superalloys 1984: Proc. of the Fifth International Symposium on Superalloys, Champion, USA, Metall. Soc. of AIME, Oct. 1984.

[1331] C.T. Sims, N.S. Stoloff, W.C. Hagel, Superalloys II: High-Temperature Materials for Aerospace and Industrial Power, 2nd edition, Wiley, 1987.

[1332] I. Singer-Loginova, H.M. Singer, The phase field technique for modeling multiphase materials, Reports on Progress in Physics 71 (2008) 106501.

[1333] M. Singleton, J. Murray, P. Nash, Aluminum-nickel, Binary Alloy Phase Diagrams 1 (1986) 140–143.

[1334] W.G. Sloof, T.J. Nijdam, On the high-temperature oxidation of MCrAlY coatings, International Journal of Materials Research 100 (10) (2009) 1318–1330.

[1335] J. Smialek, Moisture-induced delayed spallation and interfacial hydrogen embrittlement of alumina scales, JOM 58 (2006) 29–35.

[1336] J.L. Smialek, The chemistry of Saudi Arabian sand: a deposition problem on helicopter turbine airfoils, in: 3rd International SAMPE Metals Conference, 1992, M63.

[1337] J.L. Smialek, Universal characteristics of an interfacial spalling cyclic oxidation model, Acta Materialia 52 (8) (2004) 2111–2121.

[1338] J.L. Smialek, N.S. Jacobson, Oxidation of high-temperature aerospace materials, in: Y. Bar-Cohen (Ed.), High Temperature Materials and Mechanisms, 2014, pp. 95–162, number February, chapter Oxidation.

[1339] T. Smith, B. Esser, N. Antolin, A. Carlsson, R. Williams, A. Wessman, T. Hanlon, H. Fraser, W. Windl, D. McComb, M. Mills, Phase transformation strengthening of high-temperature superalloys, Nature Communications 7 (2016) 1–7.

[1340] T. Smith, B. Esser, N. Antolin, G. Viswanathan, T. Hanlon, A. Wessman, D. Mourer, W. Windl, D. McComb, M. Mills, Segregation and η-phase formation along stacking faults during creep at intermediate temperatures in a Ni-based superalloy, Acta Materialia 100 (2015) 19–31.

[1341] T. Smith, B. Esser, B. Good, M. Hooshmand, G. Viswanathan, C. Rae, M. Ghazisaeidi, D. McComb, M. Mills, Segregation and phase transformations along superlattice intrinsic stacking faults in Ni-based superalloys, Metallurgical and Materials Transactions A 49 (2018) 4186–4198.

[1342] C. Smithells, in: E.A. Brandes (Ed.), Metals Reference Book, 1983.

[1343] S. Socrate, D. Parks, Numerical determination of the elastic driving force for directional coarsening in Ni-superalloys, Acta Materialia 41 (1993) 2185–2209.

[1344] J.M. Sosa, D.E. Huber, B. Welk, H.L. Fraser, Development and application of mipar™: a novel software package for two-and three-dimensional microstructural characterization, Integrating Materials and Manufacturing Innovation 3 (1) (2014) 10.

[1345] M. Soucail, Y. Bienvenu, Dissolution of the γ' phase in a nickel base superalloy at equilibrium and under rapid heating, Materials Science and Engineering: A 220 (1–2) (1996) 215–222.

[1346] R. Soulignac, Prévision de la durée de vie à lécaillage des barrières thermiques, in French, PhD thesis, Paris, ENMP, 2014.

[1347] R. Soulignac, V. Maurel, L. Remy, A. Koster, Cohesive zone modelling of thermal barrier coatings interfacial properties based on three-dimensional observations and mechanical testing, Surface & Coatings Technology 237 (2013) 95–104.

[1348] J. Spencer, A brief history of CALPHAD, Calphad: Computer Coupling of Phase Diagrams and Thermochemistry 32 (2008) 1–8.

[1349] L. Spitz McDonald, D.D. Sangeeta, M. Rosenzweig, Method for removing an aluminide coating from a substrate, US patent 6494960B1, 2012.

[1350] R. Srinivasan, G. Eggeler, M. Mills, Gamma prime cutting as rate-controlling recovery process during high-temperature and low-stress creep of superalloy single crystals, Acta Materialia 48 (2000) 4867–4878.

[1351] A. Srivastava, S. Gopagoni, A. Needleman, V. Seetharaman, A. Staroselsky, R. Banerjee, Effect of specimen thickness on the creep response of a Ni-based single-crystal superalloy, Acta Materialia 60 (16) (2012) 5697–5711.

[1352] A. Staroselsky, R. Acharya, B. Cassenti, Phase field modeling of fracture and crack growth, Engineering Fracture Mechanics 205 (2019) 268–284.

[1353] A. Staroselsky, B. Cassenti, Mechanisms for tertiary creep of single crystal superalloy, Mechanics of Time-Dependent Materials 12 (4) (2008) 275–289.

[1354] A. Staroselsky, B. Cassenti, Creep, plasticity, and fatigue of single crystal superalloy, International Journal of Solids and Structures 48 (13) (2011) 2060–2075.

[1355] I. Steinbach, Phase-field models in materials science, Modelling and Simulation in Materials Science and Engineering 17 (2009) 073001.

[1356] I. Steinbach, Phase-field model for microstructure evolution at the mesoscopic scale, Annual Review of Materials Research 43 (2013) 89–107.

[1357] I. Steinbach, B. Bottger, J. Eiken, N. Warnken, S.G. Freies, Calphad and phase-field modeling: a successful liaison, Journal of Phase Equilibria and Diffusion 28 (1) (2007) 101–106.

[1358] I. Steinbach, F. Pezzolla, B. Nestler, M. Seeselberg, R. Prieler, G.J. Schmitz, J.L.L. Rezende, A phase field concept for multiphase systems, Physica D 94 (3) (1996) 135–147.

[1359] T. Steinke, D. Sebold, D.E. Mack, R. Vaßen, D. Stöver, A novel test approach for plasma-sprayed coatings tested simultaneously under CMAS and thermal gradient cycling conditions, Surface & Coatings Technology 205 (7) (2010) 2287–2295.

[1360] P. Steinmann, E. Kuhl, E. Stein, Aspects of non-associated single crystal plasticity: influence of non-Schmid effects and localization analysis, International Journal of Solids and Structures 35 (1998) 4437–4456.

[1361] S. Steuer, Z. Hervier, S. Thabart, C. Castaing, T. Pollock, J. Cormier, Creep behavior under isothermal and non-isothermal conditions of am3 single crystal superalloy for different solutioning cooling rates, Materials Science and Engineering: A 601 (2014) 145–152.

[1362] S. Steuer, P. Villechaise, T. Pollock, J. Cormier, Benefits of high gradient solidification for creep and low cycle fatigue of AM1 single crystal superalloy, Materials Science and Engineering 645 (2015) 109–115.

[1363] W. Stobbs, C. Sworn, The weak beam technique as applied to the determination of the stacking-fault energy of copper, Philosophical Magazine 24 (192) (1971) 1365–1381.

[1364] F. Stott, D. De Wet, R. Taylor, Degradation of thermal-barrier coatings at very high temperatures, MRS Bulletin 19 (10) (1994) 46–49.

[1365] A. Strang, E. Lang, Effect of coatings on the mechanical properties of superalloys, in: R. Brunetaud, D. Coutsouradis, T. Gibbons, Y. Lindblom, D. Meadowcroft, R. Stickler (Eds.), High Temperature Alloys for Gas Turbines 1982, Springer, 1982, pp. 469–506.

[1366] J. Stringer, High-temperature corrosion of superalloys, Materials Science and Technology 3 (7) (1987) 482–493.

[1367] J.-L. Strudel, Mechanical properties of multiphase alloys, in: R. Khan, P. Haasen (Eds.), Physical Metallurgy, 3rd edition, Springer, 1996, pp. 2105–2206.

[1368] J.-L. Strudel, Nickel-base superalloys – an engineering and scientific challenge, in: B. Raj, S. Ranganathan, S. Mannan, K. Bhanu Sankara Rao, M. Matthew, P. Shankar (Eds.), Frontiers in the Design of Materials, CRC Press, 2007, pp. 195–209.

[1369] Y. Su, S. Tian, H. Yu, L. Yu, Effect of pre-compressive treatment on creep behavior of a <011> - oriented single-crystal ni-based superalloy, Scripta Materialia 93 (2014) 24–27.

[1370] M. Subanovic, D. Sebold, R. Vassen, E. Wessel, D. Naumenko, L. Singheiser, W. Quadakkers, Effect of manufacturing related parameters on oxidation properties of mcraly-bondcoats, Materials and Corrosion 59 (6) (2008) 463–470.

[1371] R. Subramanian, Y. Mori, S. Yamagishi, M. Okazaki, Thermo-mechanical fatigue failure of thermal barrier coated superalloy specimen, Metallurgical and Materials Transactions A 46 (9) (2015) 3999–4012.

[1372] T. Sugui, Z. Huihua, Z. Jinghua, Y. Hongcai, X. Yongbo, H. Zhuangqi, Formation and role of dislocation networks during high temperature creep of a single crystal nickel-base superalloy, Materials Science and Engineering: A 279 (1–2) (2000) 160–165.

[1373] T. Sugui, W. Minggang, L. Tang, Q. Benjiang, X. Jun, Influence of TCP phase and its morphology on creep properties of single crystal nickel-based superalloys, Materials Science and Engineering: A 527 (21–22) (2010) 5444–5451.

[1374] M. Sujata, M. Madan, K. Raghavendra, M.A. Venkataswamy, S.K. Bhaumik, Identification of failure mechanisms in nickel base superalloy turbine blades through microstructural study, Engineering Failure Analysis 17 (6) (2010) 1436–1446.

[1375] I. Šulák, K. Obrtlík, L. Čelko, T. Chráska, D. Jech, P. Gejdoš, Low cycle fatigue performance of ni-based superalloy coated with complex thermal barrier coating, Materials Characterization 139 (2018) 347–354.

[1376] D. Sun, L. Liu, T. Huang, W. Yang, C. He, X. Li, J. Zhang, H. Fu, Formation of lateral sliver defects in the platform region of single-crystal superalloy turbine blades, Metallurgical and Materials Transactions A 50 (2019) 1119–1124.

[1377] R. Sun, C. Woodward, A. van de Walle, First-principles study on Ni_3Al (111) antiphase boundary with Ti and Hf impurities, Physical Review B 95 (2017) 214121.

[1378] B. Sundman, U. Kattner, M. Palumbo, S. Fries, OpenCalphad-a free thermodynamic software, Integrating Materials and Manufacturing Innovation 4 (1) (2015) 1.

[1379] B. Sundman, U. Kattner, C. Sigli, M. Stratmann, R. Le Tellier, M. Palumbo, S. Fries, The OpenCalphad thermodynamic software interface, Computational Materials Science 125 (2016) 188–196.

[1380] S. Sutcliffe, Spectral decomposition of the elasticity tensor, Journal of Applied Mechanics 59 (4) (Dec 1992) 762–773.

[1381] A. Sutton, M. Finnis, D. Pettifor, Y. Ohta, The tight-binding bond model, Journal of Physics. C. Solid State Physics 21 (1988) 35–66.

[1382] A. Suzuki, G. DeNolf, T. Pollock, Flow stress anomalies in γ/γ' two-phase Co-Al-W-base alloys, Scripta Metallurgica 56 (2007) 385–388.

[1383] A. Suzuki, T. Pollock, High-temperature strength and deformation of γ/γ' two-phase Co-Al-W alloys, Acta Materialia 56 (2008) 1288–1297.

[1384] A. Suzuki, C. Rae, Secondary reaction zone formations in coated ni-base single crystal superalloys, Journal of Physics. Conference Series 165 (2009) 012002, IOP Publishing.

[1385] S. Suzuki, H. Inui, T. Pollock, $L1_2$-strengthened Co-base superalloys, Annual Review of Materials Research 45 (2015) 345–368.

[1386] S. Suzuki, M. Sakaguchi, H. Inoue, Temperature dependent fatigue crack propagation in a single crystal Ni-base superalloy affected by primary and secondary orientations, Materials Science and Engineering: A 724 (2018) 559–565.

[1387] I.L. Svetlov, B.A. Golovko, A.I. Epishin, N.P. Abalakin, Diffusional mechanism of γ'-phase particles coalescence in single crystals of nickel-base superalloys, Scripta Materialia 26 (9) (1992) 1353–1358.

[1388] J. Svoboda, P. Lukas, Modelling of recovery controlled creep in nickel-base superalloy single crystals, Acta Materialia 45 (1) (1997) 125–135.

[1389] A. Szczotok, H. Reichel, Methodology for revealing the phases and microstructural constituents of the cmsx-4 nickel-based superalloy implicating their computer-aided detection for image analysis, Materials 13 (2) (2020) 341.

[1390] P. Szelestey, M. Patriarca, K. Kaski, Computational study of core structure and Peierls stress of dissociated dislocations in nickel, Modelling and Simulation in Materials Science and Engineering 11 (6) (2003) 883–895.

[1391] N. Ta, L. Zhang, Y. Du, A trial to design γ/γ' bond coat in ni–al–cr mode tbcs aided by phase-field simulation, Coatings 8 (12) (2018) 421.

[1392] H. Takagi, M. Fujiwara, K. Kakehi, Measuring Young's modulus of Ni-based superalloy single crystals at elevated temperatures through microindentation, Materials Science and Engineering: A 33–35 (2004) 387–389.

[1393] A. Takahashi, Y. Terada, Molecular dynamics simulation of dislocation — γ-precipitate interactions in γ'-precipitates, Key Engineering Materials 462–463 (2011) (2011) 425–430.

[1394] T. Takaki, S. Sakane, M. Ohno, Y. Shibuta, T. Aoki, C.-A. Gandin, Competitive grain growth during directional solidification of a polycrystalline binary alloy: three-dimensional large-scale phase-field study, Materialia 1 (2018) 104–113.

[1395] S. Takeuchi, E. Kuramoto, Temperature and orientation dependence of the yield stress in ni_3ga single crystals, Acta Metallurgica 21 (1973) 415–425.

[1396] H. Tamaki, K. Fujita, A. Okayama, N. Matsuda, Y. Yoshinari, K. Kakehi, A study of bending deformation behavior of Ni-based DS and SC superalloys, in: K.A. Green, T.M. Pollock, H. Harada, T.E. Howson, R.C. Reed, J.J. Schirra, S. Walston (Eds.), Superalloys 2004: Proc. of the Tenth International Symposium on Superalloys, Champion, USA, TMS, Minerals, Metals & Materials Society, Sept. 2004, pp. 145–154.

[1397] Y. Tamarin, Protective Coatings for Turbine Blades, ASM International, 2002.

[1398] A.-M. Tan, C. Woodward, D. Trinkle, Dislocation core structures in Ni-based superalloys computed using a density functional theory based flexible boundary condition approach, Physical Review Materials 3 (3) (2019) 1–8.

[1399] M. Tanaka, Convergent-beam electron diffraction, Acta Crystallographica. Section A, Foundations of Crystallography 50 (3) (1994) 261–286.

[1400] M. Tanaka, K. Tsuda, Convergent-beam electron diffraction, Journal of Electron Microscopy 60 (suppl_1) (2011) S245–S267.

[1401] M.N. Task, B. Gleeson, F. Pettit, G. Meier, The effect of microstructure on the type II hot corrosion of Ni-base MCrAlY alloys, Oxidation of Metals 80 (2013) 125–146.

[1402] A. Taylor, R. Floyd, The constitution of nickel-rich alloys of the nickel chromium-aluminum system, Journal of the Institute of Metals 81 (1952) 451–464.

[1403] G. Taylor, Plastic strain in metals, Journal of the Institute of Metals 62 (1938) 307–324.

[1404] M. Taylor, H. Evans, C. Ponton, J. Nicholls, A method for evaluating the creep properties of overlay coatings, Surface & Coatings Technology 124 (2000) 13–18.

[1405] M.P. Taylor, H.E. Evans, E.P. Busso, Z.Q. Qian, Creep properties of a Pt–aluminide coating, Acta Materialia 54 (12) (2006) 3241–3252.

[1406] T. Taylor, D. Appleby, A. Weatherill, J. Griffiths, Plasma-sprayed yttria-stabilized zirconia coatings: structure-property relationships, Surface & Coatings Technology 43 (1990) 470–480.

[1407] C. Teodosiu, A physical theory of the finite elastic-viscoplastic behaviour of single crystals, Engineering Transactions 23 (1975) 151–184.

[1408] C. Teodosiu, Elastic Models of Crystal Defects, 1 edition, Springer-Verlag, Berlin Heidelberg, 1982.

[1409] C. Teodosiu (Ed.), Constitutive Modeling of Polycrystalline Metals at Large Strains, CISM Courses and Lectures, vol. 376, Springer Verlag, 1997.

[1410] B. Ter-Ovanessian, A. Villani, É. Andrieu, S. Forest, Oxidation-assisted cracking, in: Mechanics-Microstructure-Corrosion Coupling, Elsevier, 2019, pp. 339–358.

[1411] U. Tetzlaff, H. Mughrabi, Enhancement of the high-temperature tensile creep strength of monocrystalline nickel-base superalloys by pre-rafting in compression, in: Superalloys 2000, vol. 1, Champion, PA (USA), 2000, pp. 273–282.

[1412] D. Texier, Mesure et évolution des gradients de propriétés mécaniques dans le système superalliages à base de nickel MC2 revêtu MCrAlY, PhD thesis, Université de Toulouse, France, 2013.

[1413] D. Texier, C. Cadet, T. Straub, C. Eberl, V. Maurel, Tensile behavior of air plasma spray mcraly coatings: role of high temperature agings and process defects, Metallurgical and Materials Transactions A 51 (2020) 2766–2777.

[1414] D. Texier, D. Monceau, E. Andrieu, Z. Hervier, Creep behaviour of a nickel-based single crystal superalloy at the dendritic scale using micro-tensile specimen, in: Creep 2012/JIMIS 11: 12th International Conference on Creep and Fracture of Engineering Materials and Structures, Kyoto (Japan), The Japan Institute of Metals, 2012, pp. 1–4.

[1415] D. Texier, D. Monceau, F. Crabos, E. Andrieu, Tensile properties of a non-line-of-sight processed beta-gamma-gamma prime MCrAlY coating at high temperature, Surface & Coatings Technology 326 (Oct. 2017) 28–36.

[1416] D. Texier, D. Monceau, Z. Hervier, E. Andrieu, Effect of interdiffusion on mechanical and thermal expansion properties at high temperature of a MCrAlY coated Ni-based superalloy, Surface & Coatings Technology 307 (Dec. 2016) 81–90.

[1417] D. Texier, D. Monceau, R. Mainguy, E. Andrieu, Evidence of high-temperature strain heterogeneities in a nickel-based single-crystal superalloy, Advanced Engineering Materials 16 (1) (2014) 60–64.

[1418] D. Texier, D. Monceau, J.-C. Salabura, R. Mainguy, E. Andrieu, Micromechanical testing of ultrathin layered material specimens at elevated temperature, Materials at High Temperatures 33 (2016) 325–337.

[1419] D. Texier, D. Monceau, S. Selezneff, A. Longuet, E. Andrieu, High temperature micromechanical behavior of a pt-modified nickel aluminide bond-coating and of its interdiffusion zone with the superalloy substrate, Materials Science and Engineering: A (2020) 1–6.

[1420] L. Thébaud, P. Villechaise, C. Crozet, A. Devaux, D. Béchet, J.-M. Franchet, A.-L. Rouffié, M. Mills, J. Cormier, Is there an optimal grain size for creep resistance in ni-based disk superalloys? Materials Science and Engineering: A 716 (2018) 274–283.

[1421] P.-Y. Théry, Adhérence de barrières thermiques pour aube de turbine avec couche de liaison β-(Ni, Pt) Al ou β-NiAl (Zr) (in French), PhD thesis, 2007.

[1422] P.Y. Thery, M. Poulain, M. Dupeux, M. Braccini, Adhesion energy of a YPSZ EB-PVD layer in two thermal barrier coating systems, Surface & Coatings Technology 202 (4–7) (2007) 648–652.

[1423] P.Y. Thery, M. Poulain, M. Dupeux, M. Braccini, Spallation of two thermal barrier coating systems: experimental study of adhesion and energetic approach to lifetime during cyclic oxidation, Journal of Materials Science 44 (7) (2009) 1726–1733.

[1424] M. Theveneau, V. Guipont, B. Marchand, F. Coudon, V. Maurel, Damage monitoring and thermal cycling life of thermal barrier coating involving lasat (laser shock adhesion test), in: 46th ICMCTF, San Diego, CA, 2019.

[1425] D.J. Thoma, Intermetallics: Laves phases, in: K.H.J. Buschow, R.W. Cahn, M.C. Flemings, B. Ilschner, E.J. Kramer, S. Mahajan, P. Veyssière (Eds.), Encyclopedia of Materials: Science and Technology, Elsevier, Oxford, 2001, pp. 4205–4213.

[1426] M. Thompson, C. Su, P.W. Voorhees, The equilibrium shape of a misfitting precipitate, Acta Metallurgica Et Materialia 42 (6) (1994) 2107–2122.

[1427] W. Thomson (Lord Kelvin), Elements of a mathematical theory of elasticity, Philosophical Transactions of the Royal Society of London 146 (1856) 481–498.

[1428] K. Thornton, N. Aikawa, P.W. Voorhees, Large-scale simulations of Ostwald ripening in elastically stressed solids: II. Coarsening kinetics and particle size distribution, Acta Materialia 52 (2004) 1365–1378.

[1429] J. Tiaden, B. Nestler, H.J. Diepers, I. Steinbach, The multiphase-field model with an integrated concept for modelling solute diffusion, Physica D 115 (1998) 73–86.

[1430] S.G. Tian, Z.G. Guo, H.C. Yu, Y.C. Xue, S. Zhang, Y. Su, F.L. Meng, Microstructure evolution and FEM analysis of a [011] oriented single crystal nickel-based superalloy during compressive creep, Journal of Alloys and Compounds 563 (2013) 135–142.

[1431] J. Tien, R. Gamble, The room temperature nickel-base superalloy fatigue behavior of crystals at ultrasonic frequency, Metallurgical Transactions 2 (1971) 1933–1938.

[1432] J.K. Tien, S.M. Copley, The effect of uniaxial stress on the periodic morphology of coherent γ' precipitates in nickel-base superalloy crystals, Metallurgical Transactions 2 (1) (1971) 215–219.

[1433] S. Tin, T. Pollock, Phase instabilities and carbon additions in single-crystal nickel-base superalloys, Materials Science and Engineering: A 348 (2003) 111–121.

[1434] S. Tin, T. Pollock, Predicting freckle formation in single crystal Ni-base superalloys, Journal of Materials Science 39 (2004) 7199–7205.

[1435] T. Tinga, W. Brekelmans, M. Geers, Incorporating strain gradient effects in a multiscale constitutive framework for nickel-base superalloys, Philosophical Magazine 88 (30–32) (2008) 3793–3825.

[1436] T. Tinga, W. Brekelmans, M. Geers, Application of a multiscale constitutive framework to real gas turbine components, Applied Mechanics Reviews 278 (2010) 253–258.

[1437] T. Tinga, W. Brekelmans, M. Geers, Cube slip and non-Schmid effects in single crystal Ni-base superalloys, Modelling and Simulation in Materials Science and Engineering 18 (2010) 015005.

[1438] T. Tinga, W.A.M. Brekelmans, M.G.D. Geers, Directional coarsening in nickel-base superalloys and its effect on the mechanical properties, Computational Materials Science 47 (2) (2009) 471–481.

[1439] M. Titus, Y. Eggeler, A. Suzuki, T. Pollock, Creep-induced planar defects in $L1_2$-containing Co- and CoNi-base single-crystal superalloys, Acta Materialia 82 (2015) 530–539.

[1440] M. Titus, A. Mottura, G. Visvanathan, A. Suzuki, M. Mills, T. Pollock, High resolution energy dispersive spectroscopy mapping of planar defects in $L1_2$-containing Co-base superalloys, Acta Materialia 89 (2015) 423–437.

[1441] M. Titus, S. Suzuki, T. Pollock, Creep and directional coarsening in single crystals of new γ/γ' Co-base alloys, Scripta Metallurgica 66 (2012) 574–577.

[1442] V. Tolpygo, D. Clarke, Wrinkling of α-alumina films grown by oxidation. II. Oxide separation and failure, Acta Materialia 46 (14) (1998) 5167–5174.

[1443] V.K. Tolpygo, D. Clarke, K. Murphy, Evaluation of interface degradation during cyclic oxidation of EB-PVD thermal barrier coatings and correlation with tgo luminescence, Surface & Coatings Technology 188 (2004) 62–70.

[1444] V.K. Tolpygo, D.R. Clarke, On the rumpling mechanism in nickel-aluminide coatings: part I: an experimental assessment, Acta Materialia 52 (17) (2004) 5115–5127.

[1445] H. Tomimitsu, K. Lijima, K. Aizawa, A. Yoshinari, Neutron diffraction topography observation of a Ni-based superalloy single crystal, Physical Review B 213–214 (1995) 818–820.

[1446] F. Touratier, Etude des mécanismes de déformation et d'endommagement du superalliage base nickel MC2 en fluage aux très hautes températures, PhD thesis, INP, Toulouse, 2008.

[1447] R. Trehorel, Comportement mécanique haute température du superalliage monocristallin AM1: étude in situ par une nouvelle technique de diffraction en rayonnement synchrotron, PhD thesis, 2018.

[1448] R. Tréhorel, G. Ribarik, T. Schenk, A. Jacques, Real-time study of transients during high-temperature creep of an ni-base superalloy by far-field high-energy synchrotron x-ray diffraction, Journal of Applied Crystallography 51 (5) (2018) 1274–1282.

[1449] B. Trinh, K. Hackl, A model for high temperature creep of single crystal superalloys based on nonlocal damage and viscoplastic material behavior, Continuum Mechanics and Thermodynamics 26 (2014) 551–562.

[1450] M.A. Tschopp, M.A. Groeber, J.P. Simmons, A.H. Rosenberger, C. Woodward, Automated extraction of symmetric microstructure features in serial sectioning images, Materials Characterization 61 (12) (2010) 1406–1417.

[1451] E. Tzimas, H. Mullejans, S.D. Peteves, J. Bressers, W. Stamm, Failure of thermal barrier coating systems under cyclic thermomechanical loading, Acta Materialia 48 (18–19) (2000) 4699–4707.

[1452] R.L.J.M. Ubachs, P.J.G. Schreurs, M.G.D. Geers, Phase field dependent viscoplastic behaviour of solder alloys, International Journal of Solids and Structures 42 (2005) 2533–2558.

[1453] M. Uchic, M. De Graef, R. Wheeler, D. Dimiduk, Microstructural tomography of a ni70cr20al10 superalloy using focused ion beam microscopy, Ultramicroscopy 109 (10) (2009) 1229–1235.

[1454] M. Uchic, D. Dimiduk, R. Wheeler, P. Shade, H. Fraser, Application of micro-sample testing to study fundamental aspects of plastic flow, Scripta Metallurgica 54 (2006) 759–764.

[1455] M. Uchic, P. Shade, D. Dimiduk, Plasticity of micrometer-scale single crystals in compression, Annual Review of Materials Research 39 (2009) 361–386.

[1456] S. Utada, J. Rame, S. Hamadi, J. Delautre, P. Villechaise, J. Cormier, Kinetics of creep damage accumulation induced by a room-temperature plastic deformation introduced during processing of am1 ni-based single crystal superalloy, Materials Science and Engineering: A (2020) 139571.

[1457] S. Utada, J. Rame, S. Hamadi, J. Delautre, L. Mataveli Suave, P. Villechaise, J. Cormier, High-temperature pre-deformation and rejuvenation treatment on the microstructure and creep properties of Ni-based single-crystal superalloys, in: S. Tin, et al. (Eds.), Superalloys 2020, in: The Minerals, Metals & Materials Series, Springer, Cham, 2020, pp. 240–252, https://doi.org/10.1007/978-3-030-51834-9_23.

[1458] V. Vaithyanathan, L.Q. Chen, Coarsening of ordered intermetallic precipitates with coherency stress, Acta Materialia 50 (2002) 4061–4073.

[1459] P.-L. Valdenaire, Y. Le Bouar, B. Appolaire, A. Finel, Density-based crystal plasticity: from the discrete to the continuum, Physical Review B 93 (2016) 214111.

[1460] J.L. Vallés, D.J. Arrell, Monte Carlo simulation of anisotropic coarsening in nickel-base superalloys, Acta Materialia 42 (9) (1994) 2999–3008.

[1461] K. Vamsi, S. Karthikeyan, Modelling ternary effects on antiphase boundary energy of Ni_3Al, MATEC Web of Conferences 14 (2014) 11005.

[1462] K. Vamsi, S. Karthikeyan, High-throughput estimation of planar fault energies in a3b compounds with l12 structure, Acta Materialia 145 (2018) 532–542.

[1463] A. van de Walle, G. Ghosh, M. Asta, Ab initio modeling of alloy phase equilibria, in: G. Bozzolo, R. Noebe, P. Abel, D. Vij (Eds.), Applied Computational Materials Modeling: Theory, Simulation and Experiment, Springer US, Boston, MA, 2007, pp. 1–34.

[1464] E. Van der Giessen, V. Deshpande, H. Cleveringa, A. Needleman, Discrete dislocation plasticity and crack tip fields in single crystals, Journal of the Mechanics and Physics of Solids 49 (2001) 2133–2153.

[1465] E. van der Giessen, A. Needleman, Discrete dislocation plasticity: a simple planar model, Modelling and Simulation in Materials Science and Engineering 3 (1995) 689–735.

[1466] A. Van der Ven, G. Ceder, M. Asta, P. Tepesch, First-principles theory of ionic diffusion with nondilute carriers, Physical Review B 64 (2001) 184307.

[1467] G.F. der Van Voort, Metallography, Principles and Practice, ASM International, 1999.

[1468] J.D. van der Waals, The thermodynamic theory of capillarity under the hypothesis of a continuous variation of density, Koninklijke Nederlandse Akademie van Wetenschappen. Verhandelingen, Afd. Natuurkunde. Eerste Reeks 1 (8) (1893) 56.

[1469] G. Van Drunen, J. Liburdi, Rejuvenation of used turbine blades by hot isostatic processing, in: Proc. of the Sixth Turbomachinery Symposium, College Station, Texas, 1977, pp. 55–60.

[1470] R. Van Noorden, B. Maher, R. Nuzzo, The top 100 papers, Nature 514 (2014) 550.

[1471] K. VanEvery, M.J. Krane, R.W. Trice, H. Wang, W. Porter, M. Besser, D. Sordelet, J. Ilavsky, J. Almer, Column formation in suspension plasma-sprayed coatings and resultant thermal properties, Journal of Thermal Spray Technology 20 (4) (2011) 817–828.

[1472] C. Varvenne, Etude des effets de taille atomiques sur les diagrammes de phases et les microstructures, PhD thesis, Paris 6, 2010.

[1473] C. Varvenne, A. Finel, Y. Le Bouar, M. Fevre, Alloy microstructures with atomic size effects: a Monte Carlo study under the lattice statics formalism, Physical Review B 86 (18) (2012) 184203.

[1474] R. Vassen, X. Cao, F. Tietz, D. Basu, D. Stöver, Zirconates as new materials for thermal barrier coatings, Journal of the American Ceramic Society 83 (8) (2000) 2023–2028.

[1475] R. Vaßen, H. Kaßner, G. Mauer, D. Stöver, Suspension plasma spraying: process characteristics and applications, Journal of Thermal Spray Technology 19 (1–2) (2010) 219–225.

[1476] A. Vattré, B. Devincre, F. Feyel, R. Gatti, S. Groh, O. Jamond, A. Roos, Modelling crystal plasticity by 3d dislocation dynamics and the finite element method: the discrete-continuous model revisited, Journal of the Mechanics and Physics of Solids 63 (2014) 491–505.

[1477] A. Vattré, B. Devincre, A. Roos, Dislocation dynamics simulations of precipitation hardening in ni-based superalloys with high volume fraction, Intermetallics 17 (12) (2009) 988–994.

[1478] A. Vattré, B. Devincre, A. Roos, Orientation dependence of plastic deformation in nickel-based single crystal superalloys: discrete–continuous model simulations, Acta Materialia 58 (6) (2010) 1938–1951.

[1479] A. Vattré, B. Fedelich, On the relationship between anisotropic yield strength and internal stresses in single crystal superalloys, Mechanics of Materials 43 (12) (2011) 930–951.

[1480] J.-R. Vaunois, J.-M. Dorvaux, P. Kanouté, J.-L. Chaboche, A new version of a rumpling predictive model in thermal barrier coatings, European Journal of Mechanics - A/Solids 42 (2013) 402–421.

[1481] J.-R. Vaunois, M. Poulain, P. Kanouté, J.-L. Chaboche, Development of bending tests for near shear mode interfacial toughness measurement of EB-PVD thermal barrier coatings, Engineering Fracture Mechanics 171 (2017) 110–134.

[1482] T. Vegge, T. Rasmussen, T. Leffers, O. Pedersen, K.W. Jacobsen, Atomistic simulations of cross-slip of jogged screw dislocations in copper, Philosophical Magazine Letters 81 (3) (2001) 137–144.

[1483] J. Venables, C. Harland, Electron back-scattering patterns—a new technique for obtaining crystallographic information in the scanning electron microscope, Philosophical Magazine 27 (5) (1973) 1193–1200.

[1484] M. Verdier, M. Fivel, I. Groma, Mesoscopic scale simulation of dislocation dynamics in fcc metals: principles and applications, Modelling and Simulation in Materials Science and Engineering 6 (6) (1998) 755.

[1485] M. Véron, Etude et modélisation de la coalescence orientée dans les superalliages à base de nickel, PhD thesis, Grenoble INPG, 1995.

[1486] M. Véron, P. Bastie, Strain induced directional coarsening in nickel based superalloys: investigation on kinetics using the small angle neutron scattering (SANS) technique, Acta Materialia 45 (1997) 3277–3282.

[1487] M. Véron, Y. Bréchet, F. Louchet, Directional coarsening of Ni-based superalloys – computer simulation at the mesoscopic level, Acta Materialia 44 (1996) 3633–3641.

[1488] M. Véron, Y. Bréchet, F. Louchet, Strain induced directional coarsening in Ni-based superalloys, Scripta Metallurgica 34 (1996) 1883–1886.

[1489] F. Versnyder, R. Guard, Directional solidification of columnar grains in Ni-Cr-Al alloy, Transactions of American Society for Metals 52 (1960) 485–493.

[1490] F. VerSnyder, M.M. Shank, The development of columnar grain and single crystal high temperature materials through directional solidification, Materials Science and Engineering 6 (1970) 213–247.

[1491] J.M. Veys, R. Mévrel, Influence of protective coatings on the mechanical properties of CMSX-2 and Cotac 784, Materials Science and Engineering: A 88 (1987) 253–260.

[1492] P. Veyssiere, J. Douin, P. Beauchamp, On the presence of super lattice intrinsic stacking faults in plastically deformed ni3al, Philosophical Magazine. A 51 (3) (1985) 469–483.

[1493] N. Vialas, D. Monceau, Effect of Pt and Al content on the long-term, high temperature oxidation behavior and interdiffusion of a Pt-modified aluminide coating deposited on Ni-base superalloys, Surface & Coatings Technology 201 (7 SPEC. ISS.) (2006) 3846–3851.

[1494] M. Vidal-Setif, N. Chellah, C. Rio, C. Sanchez, O. Lavigne, Calcium–magnesium–alumino-silicate (CMAS) degradation of EB-PVD thermal barrier coatings: characterization of CMAS damage on ex-service high pressure blade tbcs, Surface & Coatings Technology 208 (2012) 39–45.

[1495] M. Vidal-Setif, C. Rio, D. Boivin, O. Lavigne, Microstructural characterization of the interaction between 8ypsz (EB-PVD) thermal barrier coatings and a synthetic cas, Surface & Coatings Technology 239 (2014) 41–48.

[1496] B. Viguier, F. Touratier, E. Andrieu, High-temperature creep of single-crystal nickel-based superalloy: microstructural changes and effects of thermal cycling, Philosophical Magazine 01 (2011) 4427–4446.

[1497] A. Villemiane, Analyse du comportement mécanique d'alliages pour couches de liaison de barrière thermique par microindentation instrumentée à haute température, PhD thesis, INP Lorraine, 2008.

[1498] G. Viswanathan, S. Karthikeyan, P. Sarosi, R. Unocic, M. Mills, Microtwinning during intermediate temperature creep of polycrystalline ni-based superalloys: mechanisms and modelling, Philosophical Magazine 86 (2006) 4823–4840.

[1499] G. Viswanathan, R. Shi, A. Genc, V. Vorontsov, L. Kovarik, C. Rae, M. Mills, Segregation at stacking faults within the γ'-phase of two Ni-base superalloys following intermediate temperature creep, Scripta Materialia 94 (2015) 5–8.

[1500] A. Volek, R.F. Singer, Influence of solidification conditions on TCP phase formation, casting porosity and high temperature mechanical properties in a Re-containing nickel-base superalloy with columnar grain structure, in: Superalloys 2004, Champion, PA (USA), 2004, pp. 713–718.

[1501] A. Volek, R.F. Singer, R. Buergel, J. Grossmann, Y. Wang, Influence of topologically closed packed phase formation on creep rupture life of directionally solidified nickel-base superalloys, Metallurgical and Materials Transactions A 37 (2) (2006) 405–410.

[1502] R. Völkl, U. Glatzel, M. Feller-Kniepmeier, Analysis of matrix and interfacial dislocations in the nickel base superalloy cmsx-4 after creep in [1–11] direction, Scripta Materialia 31 (11) (1994) 1481–1486.

[1503] R. Völkl, U. Glatzel, M. Feller-Kniepmeier, Measurement of the lattice misfit in the single crystal nickel based superalloys CMSX-4, SRR99 and SC16 by convergent beam electron diffraction, Acta Materialia 46 (12) (1998) 4395–4404.

[1504] R. Völkl, U. Glatzel, M. Feller-Kniepmeier, Measurement of the unconstrained misfit in the nickel-base superalloy CMSX-4 with CBED, Scripta Materialia 38 (6) (1998) 893–900.

[1505] M. Von Ardenne, Das elektronen-rastermikroskop. Praktische ausführung, Zeitschrift für technische Physik 19 (11) (1938) 407–416.

[1506] M. von Ardenne, Das rasterelektronenmikroskop theoretische grundlagen, Zeitschrift für Physik 109 (1938) 553–572.

[1507] B. Von Grossmann, H. Biermann, H. Mughrabi, Measurement of service-induced internal elastic strains in a single-crystal nickel-based turbine blade with convergent-beam electron diffraction, Philosophical Magazine 80 (2000) 1743–1757.

[1508] A. Von Keitz, Laves phases for high temperatures – part II: stability and mechanical properties, Intermetallics 10 (5) (2002) 497–510.

[1509] K. Von Niessen, M. Gindrat, A. Refke, Vapor phase deposition using plasma spray-pvd, Journal of Thermal Spray Technology 19 (1–2) (2010) 502–509.

[1510] P.W. Voorhees, The theory of Ostwald ripening, Journal of Statistical Physics (1985).

[1511] C. Vorkötter, D. Mack, O. Guillon, R. Vaßen, Superior cyclic life of thermal barrier coatings with advanced bond coats on single-crystal superalloys, Surface & Coatings Technology 361 (2019) 150–158.

[1512] V. Vorontsov, L. Kovarik, M. Mills, C. Rae, High-resolution electron microscopy of dislocation ribbons in a CMSX-4 superalloy single crystal, Acta Materialia 60 (2012) 4866–4878.

[1513] V. Vorontsov, R. Voskoboinikov, C. Rae, Shearing of γ' precipitates in Ni-base superalloys: a phase field study incorporating the effective g-surface, Philosophical Magazine 92 (2012) 608–634.

[1514] V.A. Vorontsov, C. Shen, Y. Wang, D. Dye, C.M.F. Rae, Shearing of γ' precipitates by a<112> dislocation ribbons in Ni-base superalloys: a phase field approach, Acta Materialia 58 (2010) 4110–4119.

[1515] A. Voter, F. Montalenti, T. Germann, Extending the time scale in atomistic simulation of materials, Annual Review of Materials Research 32 (1) (2002) 321–346.

[1516] J. Wachtman, Mechanical properties of Ceramics, 1996.

[1517] C. Wagner, Theorical analysis of the diffusion processes determining the oxidation rate of alloys, Journal of the Electrochemical Society 99 (10) (1952) 369–380.

[1518] C. Wagner, Theorie der Alterung von Niederschlägen durch Umlösen (Ostwald-Reifung), Zeitschrift für Elektrochemie, Berichte der Bunsengesellschaft für physikalische Chemie 65 (7–8) (1961) 581–591.

[1519] J. Wahl, K. Harris, New single crystal superalloys CMSX7 and CMSX8, in: Proc. 12th Int. Symp. on Superalloys, TMS, 2012.

[1520] J. Wahl, K. Harris, CMSX4PLUS single crystal alloy development, characterization and applications development, in: Proc. 13th Int. Symp. on Superalloys, TMS, 2016.

[1521] K.P. Walker, E.M. Jordan, Biaxial constitutive modelling and testing of a single crystal superalloy at elevated temperatures, in: M.W. Brown, K. Miller (Eds.), Biaxial and Multiaxial Fatigue, EGF 3, London, Mechanical Engineering Publications, Ltd., London, 1989, pp. 145–170.

[1522] L. Walpole, Fourth-rank tensors of the thirty-two crystal classes: multiplication tables, Proceedings of the Royal Society of London. Series A: Mathematical, Physical and Engineering Sciences 391 (1800) (1984) 149–179.

[1523] L. Walpole, Evaluation of the elastic moduli of a transversely isotropic aggregate of cubic crystals, Journal of the Mechanics and Physics of Solids 33 (6) (1985) 623–636.

[1524] B. Walser, F. Staub, R. Wege, J. Wortmann, An alternative single crystal manufacturing process for gas turbine blades, Report 1, MTU focus, 1989.

[1525] W. Walsh, K. Thole, C. Joe, Effects of sand ingestion on the blockage of cooling holes, in: Proceedings of the ASME Turbo Expo: Power for Land, Sea and Air, vol. 4238, 2006, pp. 81–90.

[1526] S. Walston, A. Cetel, R. MacKay, K. O'Hara, D. Duhl, R. Dreshfield, Joint development of a fourth generation single crystal superalloy, in: Proc. 10th Int. Symp. on Superalloys, TMS, 2004.

[1527] S. Walston, K. O'Hara, E. Ross, T. Pollock, W. Murphy, René N6: third generation single crystal superalloy, in: Proc. 8th Int. Symp. on Superalloys, TMS, 1996.

[1528] W. Walston, J. Schaeffer, W. Murphy, A new type of microstructural instability in superalloys-SRZ, in: Superalloys 1996: Proc. of the Eighth International Symposium on Superalloys, Champion, USA, Sept. 1996, pp. 9–18.

[1529] J. Wan, Z. Yue, A low-cycle fatigue life model of nickel-based single crystal superalloys under multiaxial stress state, Materials Science and Engineering: A 392 (2005) 145–149.

[1530] D. Wang, J. Zhang, L.H. Lou, On the role of μ phase during high temperature creep of a second generation directionally solidified superalloy, Materials Science and Engineering: A 527 (20) (2010) 5161–5166.

[1531] H. Wang, R.B. Dinwiddie, W.D. Porter, Development of a thermal transport database for air plasma sprayed zro 2-y 2 o 3 thermal barrier coatings, Journal of Thermal Spray Technology 19 (5) (2010) 879–883.

[1532] J. Wang, H. Sehitoglu, Dislocation slip and twinning in Ni-based $L1_2$ type alloys, Intermetallics 52 (2014) 20–31.

[1533] J.S. Wang, A.G. Evans, Measurement and analysis of buckling and buckle propagation in compressed oxide layers on superalloy substrates, Acta Materialia 46 (14) (1998) 4993–5005.

[1534] Q. Wang, D. Apelian, D. Lados, Fatigue behavior of a356-t6 aluminum cast alloys. Part I. Effect of casting defects, Journal of Light Metals 1 (1) (2001) 73–84.

[1535] T. Wang, G. Sheng, Z.K. Liu, L.-Q. Chen, Coarsening kinetics of γ' precipitates in the Ni-Al-Mo system, Acta Materialia 56 (2008) 5544–5551.

[1536] W. Wang, M. Sakane, T. Itoh, A. Yoshinari, N. Isobe, N. Matsuda, High temperature multiaxial creep-fatigue life prediction for YH61 nickel-base single crystal superalloy, Applied Mechanics Reviews (2014) 1027–1032.

[1537] Y. Wang, J. Li, Phase field modeling of defects and deformation, Acta Materialia 58 (4) (2010) 1212–1235.

[1538] Z. Wang, N.M. Ghoniem, R. Lesar, Multipole representation of the elastic field of dislocation ensembles, Physical Review B 69 (2004) 174102.

[1539] Y. Wang-Koh, in preparation, adv. C. Rae, PhD thesis, University of Cambridge, 2021.

[1540] N. Warnken, D. Ma, A. Drevermann, R. Reed, S. Fries, I. Steinbach, Phase-field modelling of as-cast microstructure evolution in nickel-based superalloys, Acta Materialia 57 (2009) 5862–5875.

[1541] B. Warren, X-Ray Diffraction, Addison–Wesley Series in Metallurgy and Materials, 1969.

[1542] P. Warren, A. Cerezo, G. Smith, An atom probe study of the distribution of rhenium in a nickel-based superalloy, Materials Science and Engineering: A 250 (1998) 88–92.

[1543] H. Wei, X. Sun, Q. Zheng, H. Guan, Z. Hu, Estimation of interdiffusivity of the nial phase in ni-al binary system, Acta Materialia 52 (9) (2004) 2645–2651.

[1544] H. Wei, X. Sun, Q. Zheng, G. Hou, H. Guan, Z. Hu, An inverse method for determination of the interdiffusivity in aluminide coatings formed on superalloy, Surface & Coatings Technology 182 (1) (2004) 112–116.

[1545] E. Weinan, W. Ren, E. Vanden-Eijnden, Simplified and improved string method for computing the minimum energy paths in barrier-crossing events, Journal of Chemical Physics 126 (16) (2007) 164103.

[1546] C.R. Weinberger, S. Aubry, S.-W. Lee, W.D. Nix, W. Cai, Modelling dislocations in a free-standing thin film, Modelling and Simulation in Materials Science and Engineering 17 (2009) 075007.

[1547] R. Wellman, G. Whitman, J. Nicholls, CMAS corrosion of EB PVD TBCs: identifying the minimum level to initiate damage, International Journal of Refractory Metals & Hard Materials 28 (2010) 124–132.

[1548] M. Wen, S. Li, Computer simulation of superdislocation dissociation in Ni_3Al, Acta Materialia 46 (12) (1998) 4351–4355.

[1549] M. Wen, D. Lin, Effect of elastic center on dislocation core structure in Ni_3Al, Acta Materialia 45 (3) (1997) 1005–1008.

[1550] Y.-F. Wen, J. Sun, J. Huang, First-principles study of stacking fault energies in Ni_3Al intermetallic alloys, Transactions of Nonferrous Metals Society of China 22 (2012) 661–664.

[1551] Y.H. Wen, J.P. Simmons, C. Shen, C. Woodward, Y. Wang, Phase-field modeling of bimodal particle size distributions during continuous cooling, Acta Materialia 51 (2003) 1123–1132.

[1552] Y.H. Wen, B. Wang, J.P. Simmons, Y. Wang, A phase-field model for heat treatment applications in Ni-based alloys, Acta Materialia 54 (2006) 2087–2099.

[1553] Z. Wen, H. Pei, H. Yang, Y. Wu, Z. Yue, A combined cp theory and tcd for predicting fatigue lifetime in single-crystal superalloy plates with film cooling holes, International Journal of Fatigue 111 (2018) 243–255.

[1554] Z. Wen, Z. Yue, Fracture behaviour of the compact tension specimens of nickel–based single crystal superalloys at high temperature, Materials Science and Engineering: A 456 (2007) 189–201.

[1555] Z. Wen, D. Zhang, S. Li, Z. Yue, J. Gao, Anisotropic creep damage and fracture mechanism of nickel-base single crystal superalloy under multiaxial stress, Journal of Alloys and Compounds 692 (2017) 301–312.

[1556] G. Weng, Constitutive equations of single crystals and polycrystalline aggregates under cyclic loadings, International Journal of Engineering Science 18 (1980) 1385–1397.

[1557] D. Weygand, L.H. Friedman, E.V. der Giessen, A. Needleman, Aspects of boundary-value problem solutions with three-dimensional dislocation dynamics, Modelling and Simulation in Materials Science and Engineering 10 (4) (2002) 437.

[1558] D. Weygand, E. Van der Giessen, A. Needleman, Discrete dislocation modeling in 3D confined volumes, Materials Science and Engineering: A 309–310 (2001) 420.

[1559] A.A. Wheeler, W.J. Boettinger, G.B. McFadden, Phase field model for isothermal phase transitions in binary alloys, Physical Review A 45 (10) (1992) 7424–7439.

[1560] H. Whitesell, R. Overfelt, Influence of solidification variables on the microstructure, macrosegregation, and porosity of directionally solidified mar-m247, Materials Science and Engineering: A 318 (1) (2001) 264–276.

[1561] D.B. Williams, C.B. Carter, Transmission electron microscopy: a textbook for materials science, 2009.

[1562] B. Wilson, E. Cutler, G. Fuchs, Effect of solidification parameters on the microstructures and properties of CMSX-10, Materials Science and Engineering: A 479 (2008) 356–364.

[1563] B.C. Wilson, J.A. Hickman, G.E. Fuchs, The effect of solution heat treatment on a single-crystal Ni-based superalloy, JOM 55 (3) (Mar 2003) 35–40.

[1564] D. Wilson, F. Dunne, A mechanistic modelling methodology for microstructure-sensitive fatigue crack growth, Journal of the Mechanics and Physics of Solids 124 (2019) 827–848.

[1565] D. Wilson, Z. Zheng, F.P. Dunne, A microstructure-sensitive driving force for crack growth, Journal of the Mechanics and Physics of Solids 121 (2018) 147–174.

[1566] G. Witz, V. Shklover, W. Steurer, S. Bachegowda, H.-P. Bossmann, High-temperature interaction of yttria stabilized zirconia coatings with CaO–MgO–Al2O3–SiO2 (CMAS) deposits, Surface & Coatings Technology 265 (2015) 244–249.

[1567] P. Wollgramm, H. Buck, K. Neuking, A. Parsa, S. Shuwalow, J. Rogal, R. Drautz, G. Eggeler, On the role of re in the stress and temperature dependence of creep of Ni-base single crystal superalloys, Materials Science and Engineering: A 628 (2015) 382–395.

[1568] P. Wollgramm, H. Buck, A. Parsa, S. Schuwalow, R. Drautz, G. Eggeler, The effect of stress, temperature and loading direction on the creep behaviour of ni-base single crystal superalloy miniature tensile specimens, Materials Science and Engineering: A 628 (2015) 382–395.

[1569] P. Wollgramm, D. Bürger, A. Parsa, K. Neuking, G. Eggeler, The effect of stress, temperature and loading condition on the creep behavior of Ni-base single crystal superalloy miniature tensile specimens, Materials at High Temperatures 33 (2016) 346–360.

[1570] P. Wollgramm, P. Nörtershäuser, G. Eggeler, On the coarsening of the γ/γ'-microstructure in crack tip stress fields of differently oriented single crystalline miniature CT-specimens at 1020°c, in: Proc. 12th Int. Conf. Creep and Fracture of Engineering Materials and Structures, Kyoto, Japan, Sept. 2012.

[1571] M.I. Wood, The mechanical properties of coatings and coated systems, Materials Science and Engineering: A 121 (1989) 633–643.

[1572] C. Woodward, D. Trinkle, L. Hector, D. Olmsted, Prediction of dislocation cores in aluminum from density functional theory, Physical Review Letters 100 (4) (2008) 1.

[1573] P. Wright, Oxidation-fatigue interactions in a single-crystal superalloy, in: L. Kaisand, D. Solomon, G.R. Halford, B. Leis (Eds.), Low-Cycle Fatigue, vol. STP 942, ASTM, 1988, pp. 558–575.

[1574] P. Wright, A. Evans, Mechanisms governing the performance of thermal barrier coatings, Current Opinion in Solid State & Materials Science 4 (3) (1999) 255–265.

[1575] P. Wright, M. Jain, D. Cameron, High cycle fatigue in a single crystal superalloy: time dependence at elevated temperature, in: K.A. Green, T.M. Pollock, H. Harada, T.E. Howson, R.C. Reed, J.J. Schirra, S. Walston (Eds.), Superalloys 2004: Proc. of the Tenth International Symposium on Superalloys, Champion, USA, TMS, Minerals, Metals & Materials Society, Sept. 2004.

[1576] P.K. Wright, Influence of cyclic strain on life of a pvd tbc, Materials Science and Engineering: A 245 (2) (1998) 191–200.

[1577] S.I. Wright, B.L. Adams, Automatic analysis of electron backscatter diffraction patterns, Metallurgical Transactions A 23 (3) (1992) 759–767.

[1578] E. Wu, G. Sun, B. Chen, T. Pirling, D. Hughes, S. Wang, J. Zhang, A neutron diffraction study of lattice distortion, mismatch and misorientation in a single-crystal superalloy after different heat treatments, Acta Materialia 61 (2013) 2308–2319.

[1579] E. Wu, G. Sun, B. Chen, J. Zhang, V. Ji, V. Klosek, M. Mathon, Neutron diffraction study of strain/stress states and subgrain defects in a creep-deformed, single-crystal superalloy, Metallurgical and Materials Transactions A 45 (2014) 139–146.

[1580] J. Wu, H.-B. Guo, Y.-Z. Gao, S.-K. Gong, Microstructure and thermo-physical properties of yttria stabilized zirconia coatings with CMAS deposits, Journal of the European Ceramic Society 31 (10) (2011) 1881–1888.

[1581] J. Wu, X. Wei, N.P. Padture, P.G. Klemens, M. Gell, E. García, P. Miranzo, M.I. Osendi, Low-thermal-conductivity rare-earth zirconates for potential thermal-barrier-coating applications, Journal of the American Ceramic Society 85 (12) (2002) 3031–3035.

[1582] R. Wu, S. Sandfeld, Insights from a minimal model of dislocation-assisted rafting in single crystal nickel-based superalloys, Scripta Materialia 123 (2016) 42–45.

[1583] R. Wu, Z. Yue, M. Wang, Effect of initial γ/γ' microstructure on creep of single crystal nickel-based superalloys: a phase-field simulation incorporating dislocation dynamics, Journal of Alloys and Compounds 779 (2019) 326–334.

[1584] W. Wu, Y. Guo, Y. Wang, R. Mueller, D. Gross, Molecular dynamics simulation of the structural evolution of misfit dislocation networks at γ/γ' phase interfaces in Ni-based superalloys, Philosophical Magazine 91 (3) (2011) 357–372.

[1585] X. Wu, A. Dlouhy, Y. Eggeler, E. Spiecker, A. Kostka, C. Somsen, G. Eggeler, On the nucleation of planar faults during low temperature and high stress creep of single crystal Ni-base superalloys, Acta Materialia 144 (2018) 642–655.

[1586] X. Wu, S. Makineni, P. Kontis, G. Dehm, D. Raabe, B. Gault, G. Eggeler, On the segregation of Re at dislocations in the γ' phase, Materialia 4 (2018) 109–114.

[1587] X. Wu, P. Wollgramm, C. Somsen, A. Dlouhy, A. Kostka, G. Eggeler, Double minimum creep of single crystal Ni-base superalloys, Acta Materialia 112 (2016) 242–260.

[1588] S. Wulfinghoff, T. Böhlke, Equivalent plastic strain gradient enhancement of single crystal plasticity: theory and numerics, Proceedings of the Royal Society of London. Series A 468 (2145) (2012) 2682–2703.

[1589] A.M. Wusatowska-Sarnek, M.J. Blackburn, M. Aindow, Techniques for microstructural characterization of powder-processed nickel-based superalloys, Materials Science and Engineering: A 360 (1–2) (2003) 390–395.

[1590] Y. Xiang, L.-T. Cheng, D.J. Srolovitz, W. E, A level set method for dislocation dynamics, Acta Materialia 51 (18) (2003) 5499–5518.

[1591] Y. Xiang, D.J. Srolovitz, Dislocation climb effects on particle bypass mechanisms, Philosophical Magazine 86 (25–26) (2006) 3937–3957.

[1592] S.Q. Xiao, P. Haasen, HREM investigation of homogeneous decomposition in a Ni-12 at. % alloy, Acta Materialia 39 (4) (1991) 651–659.

[1593] H. Xie, L. Bo, T. Yu, Molecular dynamics simulation of an edge dislocation slipping on a cubic plane of Ni_3Al, Modelling and Simulation in Materials Science and Engineering 19 (6) (2011).

[1594] H. Xie, C. Wang, T. Yu, Motion of misfit dislocation in an Ni/Ni_3Al interface: a molecular dynamics simulations study, Modelling and Simulation in Materials Science and Engineering 17 (5) (2009).

[1595] H. Xie, T. Yu, C. Tang, Motion mechanism of the edge dislocation slipping in the cubic plane of Ni_3Al single crystals, Modelling and Simulation in Materials Science and Engineering 21 (5) (2013).

[1596] Y. Xingfu, T. Sugui, D. Hongqiang, Y. Huichen, W. Minggang, S. Lijuan, C. Shusen, Microstructure evolution of a pre-compression nickel-base single crystal superalloy during tensile creep, Materials Science and Engineering: A 506 (1–2) (2009) 80–86.

[1597] J. Xiong, Y. Zhu, Z. Li, M. Huang, Quantitative study on interactions between interfacial misfit dislocation networks and matrix dislocations in Ni-based single crystal superalloys, Acta Mechanica Solida Sinica 30 (4) (2017) 345–353.

[1598] B. Xu, X. Wang, B. Zhao, Z. Yue, Study of crystallographic creep parameters of nickel-based single crystal superalloys by indentation method, Materials Science and Engineering: A 478 (2008) 187–194.

[1599] H. Xu, H. Guo, Thermal Barrier Coatings, Elsevier, 2011.

[1600] J. Xu, S. Reuter, W. Rothkegel, Tensile and bending thermo-mechanical fatigue testing on cylindrical and flat specimens of CMSX4 for design of turbine blades, International Journal of Fatigue 30 (2008) 363–371.

[1601] Q. Xu, C. Yang, H. Zhang, X. Yan, N. Tang, B. Liu, Multiscale modeling and simulation of directional solidification process of ni-based superalloy turbine blade casting, Metals 8 (2018) 632.

[1602] Z.-H. Xu, Y. Yang, P. Huang, X. Li, Determination of interfacial properties of thermal barrier coatings by shear test and inverse finite element method, Acta Materialia 58 (18) (2010) 5972–5979.

[1603] W. Xuan, J. Lan, H. Liu, C. Li, J. Wang, W. Ren, Y. Zhong, X. Li, Z.Z. Ren, Effects of a high magnetic field on the microstructure of Ni-based single-crystal superalloys during directional solidification, Metallurgical and Materials Transactions A 48 (2017) 3804–3813.

[1604] F. Xue, C. Zenk, L. Freund, M. Hoelzel, S. Neumeier, M. Goeken, Double minimum creep in the rafting regime of a single-crystal Co-base superalloy, Scripta Metallurgica 142 (2018) 129–132.

[1605] M. Yamaguchi, V. Paidar, D. Pope, V. Vitek, Dissociation and core structure of $\langle 110 \rangle$ screw dislocations in $L1_2$ ordered alloys I. Core structure in an unstressed crystal, Philosophical Magazine. A 45 (5) (1982) 867–882.

[1606] M. Yamashita, K. Kakehi, Tension-compression asymmetry in yield and creep strengths of ni-based superalloy with a high amount of tantalum, Scripta Metallurgica 55 (2006) 139–142.

[1607] N. Yanar, F. Pettit, G. Meier, Failure characteristics during cyclic oxidation of yttria stabilized zirconia thermal barrier coatings deposited via electron beam physical vapor deposition on platinum aluminide and on nicocraly bond coats with processing modifications for improved performances, Metallurgical and Materials Transactions A 37 (5) (2006) 1563–1580.

[1608] H. Yang, Z. Li, M. Huang, Modeling dislocation cutting the precipitate in nickel-based single crystal superalloy via the discrete dislocation dynamics with SISF dissociation scheme, Composite Materials Science 75 (2013) 52–59.

[1609] M. Yang, J. Zhang, H. Wei, W. Gui, H. Su, T. Jin, L. Liu, A phase-field model for creep behavior in nickel-base single-crystal superalloy: coupled with creep damage, Scripta Materialia 147 (2018) 16–20.

[1610] Z. Yao, C. Degnan, M. Jepson, R. Thomson, Effect of rejuvenation heat treatments on γ' distributions in a Ni-based superalloy for power plant applications, Materials Science and Technology 29 (2013) 775–780.

[1611] Z. Yao, C.C. Degnan, M.A. Jepson, R.C. Thomson, Microstructural and chemical rejuvenation of a ni-based superalloy, Metallurgical and Materials Transactions A 47 (12) (2016) 6330–6338.

[1612] Z. Yao, M. Jepson, R. Thomson, C. Degnan, Microstructural evolution in a nickel based superalloy for power plant applications as a consequence of high temperature degradation and rejuvenation heat treatments, in: Advances in Materials Technology for Fossil Power Plants: Proc. from the Seventh Int. Conf. (EPRI 2013), Waikoloa, Hawaii, 2013, pp. 424–435.

[1613] V. Yardley, I. Povstugar, P.-P. Choi, D. Raabe, A. Parsa, A. Kostka, C. Somsen, A. Dlouhy, K. Neuking, E. George, G. Eggeler, On local phase equilibria and the appearance of nanoparticles in the microstructure of single-crystal Ni-base superalloys, Advanced Engineering Materials 18 (2016) 1556–1567.

[1614] K. Yashiro, F. Kurose, Y. Nakashima, K. Kubo, Y. Tomita, H. Zbib, Discrete dislocation dynamics simulation of cutting of γ' precipitate and interfacial dislocation network in Ni-based superalloys, International Journal of Plasticity 22 (4) (apr 2006) 713–723.

[1615] K. Yashiro, M. Naito, Y. Tomita, Molecular dynamics simulation of dislocation nucleation and motion at γ/γ' interface in Ni-based superalloy, International Journal of Mechanical Sciences 44 (9) (2002) 1845–1860.

[1616] K. Yashiro, Y. Suzuki, J. Pangestu, Y. Tomita, Molecular dynamics study on characteristics of misfit dislocations in Ni-based superalloys, Key Engineering Materials 346 (2007) 951–954.

[1617] X. Ye, C. Liu, W. Zhong, Y. Du, Precipitate size dependence of Ni/Ni_3Al interface energy, Physics Letters A 379 (1–2) (2015) 37–40.

[1618] A. Yeh, C. Rae, S. Tin, High temperature creep of Ru-bearing Ni-base single crystal superalloys, in: K.A. Green, T.M. Pollock, H. Harada, T.E. Howson, R.C. Reed, J.J. Schirra, S. Walston (Eds.), Superalloys 2004: Proc. of the Tenth International Symposium on Superalloys, Champion, USA, TMS, Minerals, Metals & Materials Society, Sept. 2004, pp. 677–685.

[1619] J. Yi, C. Torbet, Q. Feng, T. Pollock, J. Jones, Ultrasonic fatigue of a single crystal superalloy at 1000°C, Materials Science and Engineering: A 443 (2007) 142–149.

[1620] M. Yoo, Dislocation configurations and work-hardening in Cu_3Al crystals, Acta Materialia 12 (1964) 555–569.

[1621] M. Yoo, On the theory of anomalous yield behavior of Ni_3Al - effect of elastic anisotropy, Scripta Metallurgica 20 (6) (1986) 915–920.

[1622] M. Yoo, Stability of superdislocations and shear faults in $L1_2$ ordered alloys, Acta Materialia 35 (7) (1987) 1559–1569.

[1623] M. Yoo, M. Daw, M. Baskes, Atomistic simulation of superdislocation dissociation in Ni_3Al, in: D. Srolovitz, V. Vitek (Eds.), Atomistic Simulation of Materials – Beyond Pair Potentials, Springer, 1989, pp. 401–410.

[1624] K. Yoshida, A plastic flow rule representing corner effects predicted by rate-independent crystal plasticity, International Journal of Solids and Structures 120 (2017) 213–225.

[1625] D.J. Young, High Temperature Oxidation and Corrosion of Metals, 2nd edition, Elsevier Science, 2016.

[1626] H. Yu, Y. Su, N. Tian, S. Tian, Y. Li, X. Yu, L. Yu, Microstructure evolution and creep behavior of a [111] oriented single crystal nickel-based superalloy during tensile creep, Materials Science and Engineering: A 565 (2013) 292–300.

[1627] J. Yu, Y. Sun, X. Sun, T. Jin, Z. Hu, Anisotropy of high cycle fatigue behavior of a Ni-base single crystal superalloy, Materials Science and Engineering: A 56 (2013) 90–95.

[1628] J. Yu, Q. Zhang, R. Liu, Z. Yue, M. Tang, X. Li, Molecular dynamics simulation of crack propagation behaviors at the Ni/Ni_3Al grain boundary, RCS Advances 4 (62) (2014) 32749–32754.

[1629] J. Yu, Q. Zhang, Z. Yue, R. Liu, M. Tang, X. Li, Microstructure evolution and mechanical behavior of the Ni/Ni_3Al interface under thermal-mechanical coupling, Materials Express 5 (4) (2015) 343–350.

[1630] Q. Yu, Y. Wang, Z. Wen, Z. Yue, Z. Wen, D. Zhang, S. Li, Z. Yue, J.J. Gao, Anisotropic creep damage and fracture mechanism of nickel base single crystal superalloy under multiaxial stress, Journal of Alloys and Compounds 692 (2017) 301–312.

[1631] Z.Y. Yu, X.M. Wang, H. Liang, Z.X. Li, L. Li, Z.F. Yue, Thickness debit effect in Ni-based single crystal superalloys at different stress levels, International Journal of Mechanical Sciences 170 (2020) 105357.

[1632] K. Yuan, R. Eriksson, R.L. Peng, X.-H. Li, S. Johansson, Y.-D. Wang, Modeling of microstructural evolution and lifetime prediction of mcraly coatings on nickel based superalloys during high temperature oxidation, Surface & Coatings Technology 232 (2013) 204–215.

[1633] Z.F. Yue, Z.Z. Lu, Rafting prediction criterion for nickel-base single crystals under multiaxial stresses and crystallographic orientation dependence of creep behavior, Acta Metallurgica Sinica 12 (2) (1999) 149–154.

[1634] S. Zaefferer, N.-N. Elhami, Theory and application of electron channelling contrast imaging under controlled diffraction conditions, Acta Materialia 75 (2014) 20–50.
[1635] C. Zambaldi, F. Roters, D. Raabe, U. Glatzel, Modeling and experiments on the indentation deformation and recrystallization of a single-crystal nickel-base superalloy, Materials Science and Engineering: A 454–455 (2007) 433–440.
[1636] P.D. Zavattieri, L.G. Hector Jr, A.F. Bower, Cohesive zone simulations of crack growth along a rough interface between two elastic–plastic solids, Engineering Fracture Mechanics 75 (15) (2008) 4309–4332.
[1637] H.M. Zbib, M. Rhee, J.P. Hirth, On plastic deformation and the dynamics of 3d dislocations, International Journal of Mechanical Sciences 40 (2) (1998) 113–127.
[1638] C. Zenk, S. Neumeier, N. Engl, S. Fries, O. Dolotko, M. Weiser, S. Virtanen, M. Goeken, Intermediate Co/Ni-base model superalloys – thermophysical properties, creep and oxidation, Scripta Metallurgica 112 (2016) 83–86.
[1639] J. Zhang, Y. Koizumi, T. Kobayashi, T. Murakamo, H. Harada, Strengthening by γ/γ' interfacial dislocation networks in TMS-162 – toward a fifth-generation single-crystal superalloy, Metallurgical and Materials Transactions A 35 (2004) 1911–1915.
[1640] J.X. Zhang, T. Murakumo, H. Harada, Y. Koizumi, Dependence of creep strength on the interfacial dislocations in a fourth generation SC superalloy TMS-138, Scripta Materialia 48 (3) (2003) 287–293.
[1641] J.X. Zhang, T. Murakumo, Y. Koizumi, H. Harada, The influence of interfacial dislocation arrangements in a fourth generation single crystal TMS-138 superalloy on creep properties, Journal of Materials Science 38 (24) (2003) 4883–4888.
[1642] J.X. Zhang, T. Murakumo, Y. Koizumi, T. Kobayashi, H. Harada, S. Masaki, Interfacial dislocation networks strengthening a fourth-generation single-crystal TMS-138 superalloy, Metallurgical and Materials Transactions A 33 (12) (2002) 3741–3746.
[1643] J.X. Zhang, J. Wang, H. Harada, Y. Koizumi, The effect of lattice misfit on the dislocation motion in superalloys during high-temperature low-stress creep, Acta Materialia 53 (2005) 4623–4633.
[1644] L. Zhang, Y. Du, Q. Chen, I. Steinbach, B. Huang, Atomic mobilities and diffusivities in the fcc, l12 and b2 phases of the ni-al system, International Journal of Materials Research 101 (12) (2010) 1461–1475.
[1645] L. Zhang, I. Steinbach, Y. Du, Phase-field simulation of diffusion couples in the ni–al system, International Journal of Materials Research 102 (4) (2011) 371–380.
[1646] P. Zhang, E. Sadeghimeresht, S. Chen, X.-H. Li, N. Markocsan, S. Joshi, W. Chen, I.A. Buyanova, R.L. Peng, Effects of surface finish on the initial oxidation of hvaf-sprayed nicocraly coatings, Surface & Coatings Technology 364 (2019) 43–56.
[1647] Q. Zhang, Y. Chang, L. Gu, Y. Luo, B. Ge, Study of microstructure of nickel based superalloys at high temperatures, Scripta Materialia 126 (2017) 55–57.
[1648] W.-J. Zhang, Thermal mechanical fatigue of single crystal superalloys: achievements and challenges, Materials Science and Engineering: A 650 (2016) 389–395.
[1649] X. Zhang, H. Deng, S. Xiao, X. Li, W. Hu, Atomistic simulations of solid solution strengthening in Ni-based superalloy, Composite Materials Science 68 (2013) (2013) 132–137.
[1650] X. Zhang, H. Deng, S. Xiao, J. Tang, L. Deng, W. Hu, First-principles calculation of self-diffusion coefficients in Ni_3Al, Journal of Alloys and Compounds 612 (2014) 361–364.
[1651] X. Zhang, H. Deng, S. Xiao, Z. Zhang, J. Tang, L. Deng, W. Hu, Diffusion of Co, Ru and Re in Ni-based superalloys: a first-principles study, Journal of Alloys and Compounds 588 (2014) 163–169.
[1652] X. Zhang, B. Grabowski, F. Koermann, A. Ruban, Y. Gong, R. Reed, T. Hickel, J. Neugebauer, Temperature dependence of the stacking-fault Gibbs energy for Al, Cu, and Ni, Physical Review B 98 (2018) 224106.
[1653] X. Zhang, B. Grabowski, F. Körmann, C. Freysoldt, J. Neugebauer, Accurate electronic free energies of the $3d$, $4d$ and $5d$ transition metals at high temperatures, Physical Review B 95 (2017) 165126.

[1654] X. Zhang, M. Sluiter, Cluster expansions for thermodynamics and kinetics of multicomponent alloys, Journal of Phase Equilibria and Diffusion 37 (1) (2016) 44–52.
[1655] X. Zhang, P.R. Stoddart, J.D. Comins, A.G. Every, High-temperature elastic properties of a nickel-based superalloy studied by surface Brillouin scattering, Journal of Physics. Condensed Matter 13 (10) (2001) 2281.
[1656] X. Zhang, C. Wang, W. Chen, L. Ye, Y. Mai, Investigation of short fatigue cracks in nickel-based single crystal superalloy sc16 by in-situ sem fatigue testing, Scripta Materialia 44 (2001) 2443–2448.
[1657] Y. Zhang, J.A. Haynes, B.A. Pint, I.G. Wright, W.Y. Lee, Martensitic transformation in CVD NiAl and (Ni, Pt) Al bond coatings, Surface & Coatings Technology 163 (2003) 19–24.
[1658] Y. Zhang, L. Liu, T. Huang, Y. Li, Z. Jie, J. Zhang, W. Yang, H. Fu, Investigation on remelting solution heat treatment for nickel-based single crystal superalloys, Scripta Metallurgica 136 (2017) 74–77.
[1659] Y. Zhang, L. Liu, T. Huang, Q. Yue, D. Sun, J. Zhang, W. Yang, H. Su, H. Fu, Investigation on a ramp solution heat treatment for a third generation nickel-based single crystal superalloy, Journal of Alloys and Compounds 723 (2017) 922–929.
[1660] Y. Zhang, W. Qiu, H. Shi, C. Li, K. Kadau, O. Luesebrink, Effects of secondary orientations on long fatigue crack growth in a single crystal superalloy, Engineering Fracture Mechanics 136 (2015) 172–184.
[1661] Y. Zhang, Z. Wen, H. Pei, Y. Zhao, Z. Li, Z. Yue, Microstructure evolution mechanisms in nickel-based single crystal superalloys under multiaxial stress state, Journal of Alloys and Compounds 797 (2019) 1059–1077.
[1662] Y.H. Zhang, D.M. Knowles, P.J. Withers, Microstructural development in Pt-aluminide coating on CMSX-4 superalloy during TMF, Surface & Coatings Technology 107 (1) (1998) 76–83.
[1663] Z. Zhang, I. Cinoglu, S. Charbal, N. Vermaak, L. Lou, J. Zhang, Cyclic inelastic behavior and shakedown response of a 2nd generation nickel-base single crystal superalloy under tension-torsion loadings: experiments and simulations, European Journal of Mechanics - A/Solids 80 (2020) 103895.
[1664] L. Zhao, N. O'Dowd, E. Busso, A coupled kinetic-constitutive approach to the study of high temperature crack initiation in single crystal nickel-base superalloys, Journal of the Mechanics and Physics of Solids 54 (2006) 288–309.
[1665] N. Zhou, C. Shen, M. Mills, J. Li, Y. Wang, Modeling displacive-diffusional coupled dislocation shearing of γ' precipitates in Ni-base superalloys, Acta Materialia 59 (9) (2011) 3484–3497.
[1666] N. Zhou, C. Shen, M. Mills, Y. Wang, Phase field modeling of channel dislocation activity and γ' rafting in single crystal Ni-Al, Acta Materialia 55 (16) (2007) 5369–5381.
[1667] N. Zhou, C. Shen, M. Mills, Y. Wang, Contributions from elastic inhomogeneity and from plasticity to γ' rafting in single-crystal Ni-Al, Acta Materialia 56 (2008) 6156–6173.
[1668] N. Zhou, C. Shen, M. Mills, Y. Wang, Large-scale three-dimensional phase field simulation of γ' rafting and creep deformation, Philosophical Magazine. B 90 (1–4) (2010) 405–436.
[1669] D. Zhu, R.A. Miller, Investigation of thermal fatigue behavior of thermal barrier coating systems, Surface & Coatings Technology 94 (1997) 94–101.
[1670] D. Zhu, R.A. Miller, Development of advanced low conductivity thermal barrier coatings, International Journal of Applied Ceramic Technology 1 (1) (2004) 86–94.
[1671] J. Zhu, L.Q. Chen, J. Shen, Morphological evolution during phase separation and coarsening with strong inhomogeneous elasticity, Modelling and Simulation in Materials Science and Engineering 9 (2001) 499–511.
[1672] J.Z. Zhu, T. Wang, A.J. Ardell, S.H. Zhou, Z.K. Liu, L.Q. Chen, Three-dimensional phase-field simulations of coarsening kinetics of γ' particles in binary Ni-Al alloys, Acta Materialia 52 (2004) 2837–2845.

[1673] T. Zhu, C. Wang, Misfit dislocation networks in the γ/γ' phase interface of a Ni-based single-crystal superalloy: molecular dynamics simulations, Physical Review B 72 (1) (2005) 1.

[1674] W. Zhu, L. Yang, J. Guo, Y. Zhou, C. Lu, Determination of interfacial adhesion energies of thermal barrier coatings by compression test combined with a cohesive zone finite element model, International Journal of Plasticity 64 (2015) 76–87.

[1675] W. Zhu, Z. Zhang, L. Yang, Y. Zhou, Y. Wei, Spallation of thermal barrier coatings with real thermally grown oxide morphology under thermal stress, Materials & Design 146 (2018) 180–193.

[1676] Y. Zhu, Z. Li, M. Huang, Atomistic modeling of the interaction between matrix dislocation and interfacial misfit dislocation networks in Ni-based single crystal superalloy, Composite Materials Science 70 (2013) 178–186.

[1677] Z. Zhu, H. Basoalto, N. Warnken, R. Reed, A model for the creep deformation behaviour of nickel-based single crystal superalloys, Acta Materialia 60 (12) (2012) 4888–4900.

[1678] J. Ziebs, J. Meersmann, H. Kühn, H. Klinglehöffer, Multiaxial thermo-mechanical deformation behavior of IN738 LC and SC16, in: Thermo-mechanical Fatigue Behavior of Materials: Third Volume, vol. STP 1371, ASTM, 2000, pp. 257–278.

[1679] M. Zietara, S. Neumeier, M. Göken, A. Czyrska-Filemonowicz, Characterization of γ and γ' phases in 2nd and 4th generation single crystal nickel-base superalloys, Metals and Materials International 23 (2017) 126–131.

[1680] N. Zotov, M. Bartsch, G. Eggeler, Thermal barrier coating systems analysis of nanoindentation curves, Surface & Coatings Technology 203 (14) (2009) 2064–2072.

[1681] J. Zrnik, P. Strunz, M. Maldini, V. Davydov, SANS investigation of γ' precipitate morphology evolution in creep exposed single crystal ni base superalloy, Materials Sciences Forum 636–637 (2010) 1475–1482.

[1682] A. Zunger, S.-H. Wei, L. Ferreira, J. Bernard, Special quasirandom structures, Physical Review Letters 65 (1990) 353–356.

Subject index

E

F

G

H

I

K

L

M

N

O

P

T

U

Printed in the United States
by Baker & Taylor Publisher Services